Taylor expansions

$$f(x) = \sum_{n=0}^{\infty} \frac{1}{n!} \left(\frac{d^n f}{dx^n} \right)_a (x-a)^n$$

$$e^x = 1 + x + \tfrac{1}{2}x^2 + \cdots \qquad \ln x = (x-1) - \tfrac{1}{2}(x-1)^2 + \tfrac{1}{3}(x-1)^3 - \tfrac{1}{4}(x-1)^4 + \cdots$$

$$\ln(1+x) = x - \tfrac{1}{2}x^2 + \tfrac{1}{3}x^3 - \cdots \qquad \frac{1}{1+x} = 1 - x + x^2 + \cdots$$

Derivatives

$$d(f+g) = df + dg \qquad d(fg) = f\,dg + g\,df$$

$$d\frac{f}{g} = \frac{1}{g}df - \frac{f}{g^2}dg \qquad \frac{df}{dt} = \frac{df}{dg}\frac{dg}{dt}$$

$$\frac{dx^n}{dx} = nx^{n-1} \qquad \frac{d}{dx}e^{ax} = ae^{ax} \qquad \frac{d}{dx}\ln x = \frac{1}{x}$$

Integrals

$$\int x^n\,dx = \frac{x^{n+1}}{n+1} + \text{constant} \qquad \int \frac{1}{x}\,dx = \ln x + \text{constant}$$

$$\int_0^\infty x^n e^{-ax}\,dx = \frac{n!}{a^{n+1}} \qquad \int \sin^2 ax\,dx = \tfrac{1}{2}x - (1/4a)\sin 2ax + \text{constant}$$

$$\int \sin ax \sin bx\,dx = \frac{\sin(a-b)x}{2(a-b)} - \frac{\sin(a+b)x}{2(a+b)} + \text{constant, for } a^2 \neq b^2$$

Prefixes

z	zepto	10^{-21}
a	atto	10^{-18}
f	femto	10^{-15}
p	pico	10^{-12}
n	nano	10^{-9}
μ	micro	10^{-6}
m	milli	10^{-3}
c	centi	10^{-2}
d	deci	10^{-1}
da	deca	10^{1}
k	kilo	10^{3}
M	mega	10^{6}
G	giga	10^{9}
T	tera	10^{12}
P	peta	10^{15}

Greek alphabet

A, α	alpha
B, β	beta
Γ, γ	gamma
Δ, δ	delta
E, ε	epsilon
Z, ζ	zeta
H, η	eta
Θ, θ	theta
I, ι	iota
K, κ	kappa
Λ, λ	lambda
M, μ	mu
N, ν	nu
Ξ, ξ	xi
O, o	omicron
Π, π	pi
P, ρ	rho
Σ, σ	sigma
T, τ	tau
Y, υ	upsilon
Φ, φ	phi
X, χ	chi
Ψ, ψ	psi
Ω, ω	omega

Physical Chemistry for the Life Sciences

Second edition

Peter Atkins
Professor of Chemistry, Oxford University

Julio de Paula
Professor of Chemistry, Lewis & Clark College

OXFORD
UNIVERSITY PRESS

OXFORD
UNIVERSITY PRESS

Great Clarendon Street, Oxford OX2 6DP

Oxford University Press is a department of the University of Oxford.
It furthers the University's objective of excellence in research, scholarship,
and education by publishing worldwide in

Oxford New York

Auckland Cape Town Dar es Salaam Hong Kong Karachi
Kuala Lumpur Madrid Melbourne Mexico City Nairobi
New Delhi Shanghai Taipei Toronto

With offices in

Argentina Austria Brazil Chile Czech Republic France Greece
Guatemala Hungary Italy Japan Poland Portugal Singapore
South Korea Switzerland Thailand Turkey Ukraine Vietnam

Oxford is a registered trade mark of Oxford University Press
in the UK and in certain other countries

Published in the United States
by W.H. Freeman and Company, New York

British Library Cataloguing in Publication Data
Data available

Library of Congress Cataloging in Publication Data
Data available

Typeset by Graphicraft Limited, Hong Kong
Printed in Italy by L.E.G.O. S.p.A

ISBN 978-0-19-956428-6

10 9 8 7 6 5 4 3 2 1

Contents in brief

Full contents

Preface

The second edition of this text—like the first edition—seeks to present all the material required for a course in physical chemistry for students of the life sciences, including biology and biochemistry. To that end we have provided the foundations and biological applications of thermodynamics, kinetics, quantum theory, and molecular spectroscopy.

The text is characterized by a variety of pedagogical devices, most of them directed toward helping with the mathematics that must remain an intrinsic part of physical chemistry. One such new device is the *Mathematical toolkit*, a boxed section that—as we explain in more detail in the 'About the book' section below—reviews concepts of mathematics just where they are needed in the text.

Another device that we continue to invoke is *A note on good practice*. We consider that physical chemistry is kept as simple as possible when people use terms accurately and consistently. Our *Notes* emphasize how a particular term should and should not be used (by and large, according to IUPAC conventions). Finally, new to this edition, each chapter ends with a *Checklist of key concepts* and a *Checklist of key equations*, which together summarize the material just presented. The latter is annotated in many places with short comments on the applicability of each equation.

Elements of biology and biochemistry continue to be incorporated in the text's narrative in a number of ways. First, each numbered section begins with a statement that places the concepts of physical chemistry about to be explored in the context of their importance to biology. Second, the narrative itself shows students how physical chemistry gives quantitative insight into biology and biochemistry. To achieve this goal, we make generous use of *A brief illustration* sections (by which we mean quick numerical exercises) and *Worked examples*, which feature more complex calculations than do the illustrations. Third, a unique feature of the text is the use of *Case studies* to develop more fully the application of physical chemistry to a specific biological or biomedical problem, such as the action of ATP, pharmacokinetics, the unique role of carbon in biochemistry, and the biochemistry of nitric oxide. Finally, the new *In the laboratory* sections highlight selected experimental techniques in modern biochemistry and biomedicine, such as differential scanning calorimetry, gel electrophoresis, electron microscopy, and magnetic resonance imaging.

All the illustrations (nearly 500 of them) have been redrawn and are now in full color. Another innovation in this edition is the *Atlas of structures*, in the *Resource section* at the end of the book. Many biochemically important structures are referred to a number of times in the text, and we judged it appropriate and convenient to collect them all in one place. The *Resource section* also includes data used in a variety of places in the text.

A text cannot be written by authors in a vacuum. To merge the languages of physical chemistry and biochemistry we relied on a great deal of extraordinarily useful and insightful advice from a wide range of people. We would particularly like to acknowledge the following people, who reviewed draft chapters of the text:

Professor Björn Åkerman, Chalmers University of Technology

Dr Perdita Barran, University of Edinburgh

Professor Bo Carlsson, University of Kalmar

Dr Monique Cosman, California State University, East Bay

Dr Erin E. Dahlke, Loras College

Prof Roger DeKock, Calvin College

Professor Steve Desjardins, Washington and Lee University

Dr Bridgette Duncombe, University of Edinburgh

Dr Niels Engholm Henriksen, Technical University of Denmark

Professor Andrew Fisher, University of California, Davis

Dr Peter Gardner, Royal Holloway University of London

Dr Anton Guliaev, San Francisco State University

Dr Magnus Gustafsson, University of Gothenburg

Dr Hal Harris, University of Missouri- St. Louis

Dr Lars Hemmingsen, Copenhagen University

Dr Hans A. Heus, Radboud University Nijmegen

Dr Martina Huber, Leiden University

Dr Eihab Jaber, Worcester State College

Dr Ryan R. Julian, University of California, Riverside

Professor Tim Keiderling, University of Illinois at Chicago

Dr Paul King, Birkbeck College

Professor Krzysztof Kuczera, University of Kansas

Professor H.E. Lundager Madsen, University of Copenhagen

Dr Jeffrey Mack, California State University, Sacramento

Dr Jeffry Madura, Duquesne University

Dr John Marvin, Brescia University

Dr Stephen Mezyk, California State University, Long Beach

Dr Yorgo Modis, Yale University

Dr Lee Reilly, University of Warwick

Dr Brent Ridley, Biola University

Dr Jens Risbo, University of Copenhagen

Dr Martha Sarasua, University of West Florida

Prof Steve Scheiner, Utah State University

Dr Andrew Shaw, University of Exeter

Dr Suzana K. Straus, University of British Columbia

Dr Cindy Tidwell, University of Montevallo

Professor Geoff Thornton, University College London

Dr Andreas Toupadakis, University of California, Davis

Dr Jeffrey Watson, Gonzaga University

Dr Andrew Wilson, University of Leeds

We have been particularly well served by our publishers, and would wish to acknowledge our gratitude to our editors Jonathan Crowe of Oxford University Press and Jessica Fiorillo of W.H. Freeman and Company, who helped us achieve our goal. We also thank Valerie Walters for proofreading the text so carefully and Charles Trapp and Marshall Cady for compiling the solutions manual and making very helpful comments in the course of its development.

PWA, Oxford   JdeP, Portland

About the book

Numerous features in this text are designed to help you learn physical chemistry and its applications to biology, biochemistry, and medicine. One of the problems that makes the subject so daunting is the sheer amount of information. To help with that problem, we have introduced several devices for organizing the material in your mind: see *Organizing the information*. We appreciate that mathematics is often troublesome, and therefore have included several devices for helping you with this enormously important aspect of physical chemistry: see *Mathematics support*. Problem solving, especially, 'where do I start?', is often a problem, and we have done our best to help you find your way over the first hurdle: see *Problem solving*. Finally, the web is an extraordinary resource, but you need to know where to go for a particular piece of information; we have tried to point you in the right direction: see *Using the Web*. The following paragraphs explain the features in more detail.

Organizing the information

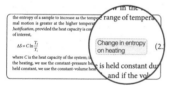

Equation and concept tags The most significant equations and concepts—and which we urge you to make a particular effort to remember—are flagged with an annotation, as shown here.

Checklist of key concepts Here we collect together the major concepts that we have introduced in the chapter. You might like to check off the box that precedes each entry when you feel that you are confident about the topic.

Checklist of key equations This is a collection of the most important equations introduced in the chapter.

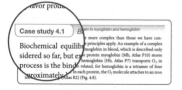

Case studies We incorporate general concepts of biology and biochemistry throughout the text, but in some cases it is useful to focus on a specific problem in some detail. A Case study contains some background information about a biological process, such as the action of adenosine triphosphate or the metabolism of drugs, and may be followed by a series of calculations that give quantitative insight into the phenomena.

In the laboratory Here we describe some of the modern techniques of biology, biochemistry, and medicine. In many cases, you will use these techniques in laboratory courses, so we focus not on the operation of instruments but on the physical principles that make the instruments perform a specific task.

Notes on good practice Science is a precise activity, and using its language accurately can help you to understand the concepts. We have used this feature to help you to use the language and procedures of science in conformity to international practice and to avoid common mistakes.

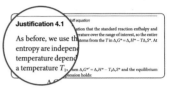

Justifications On first reading you might need the 'bottom line' rather than a detailed development of a mathematical expression. However, once you have collected your thoughts, you might want to go back to see how a particular expression was obtained. The Justifications let you adjust the level of detail that you require to your current needs. However, don't forget that the development of results is an essential part of physical chemistry, and should not be ignored.

Further information In some cases, we have judged that a derivation is too long, too detailed, or too different in level for it to be included in the text. In these cases, you will find the derivation at the end of the chapter.

Mathematics support

A brief comment A topic often needs to draw on a mathematical procedure or a concept of physics; a brief comment is a quick reminder of the procedure or concept.

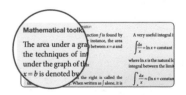

Mathematical toolkit It is often the case that you need a more full-bodied account of a mathematical concept, either because it is important to understand the procedure more fully or because you need to use a series of tools to develop an equation. The Mathematical toolkit sections are located in the chapters, primarily where they are first needed.

Problem solving

Brief illustrations A Brief illustration (don't confuse this with a diagram!) is a short example of how to use an equation that has just been introduced in the text. In particular, we show how to use data and how to manipulate units correctly.

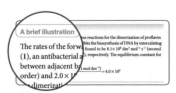

Examples An Example is a much more structured form of Brief illustration, often involving a more elaborate procedure. Every Example has a Strategy section to suggest how you might set up the problem (you might prefer another way: setting up problems is a highly personal business). Then we provide the worked-out Answer.

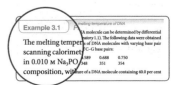

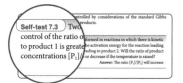

Self-tests Every Example has a Self-test, with the answer provided, so that you can check whether you have understood the procedure. There are also free-standing Self-tests where we thought it a good idea to provide a question for you to check your understanding. Think of Self-tests as in-chapter Exercises designed to help you to monitor your progress.

Discussion questions The end-of-chapter material starts with a short set of questions that are intended to encourage you to think about the material you have encountered and to view it in a broader context than is obtained by solving numerical problems.

Exercises The real core of testing your progress is the collection of end-of-chapter Exercises. We have provided a wide variety at a range of levels.

Projects Longer and more involved exercises are presented as Projects at the end of each chapter. In many cases, the projects encourage you to make connections between concepts discussed in more than one chapter, either by performing calculations or by pointing you to the original literature.

Media and supplements

Oxford University Press has developed an extensive package of electronic resources and printed supplements to accompany the second edition of *Physical Chemistry for the Life Sciences*.

The Online Resource Centre

The Online Resource Centre provides teaching and learning resources to augment the printed book. It is free of charge, and contains additional material for download, much of which can be incorporated into a virtual learning environment. The Online Resource Centre can be accessed by visiting

 www.oxfordtextbooks.co.uk/orc/lchem2e/

Note that lecturer resources are available only to registered adopters of the textbook. To register simply visit www.oxfordtextbooks.co.uk/orc/lchem2e/ and follow the appropriate links. You will be given the opportunity to select your own username and password, which will be activated once your adoption has been verified.

For Students

Living Graphs A living graph can be used to explore how a property changes as a variety of parameters are changed. To encourage the use of this resources (and the more extensive *Explorations in Physical Chemistry 2.0*; below), we have included a suggested interactivity to many of the illustrations in the text, iconed in the book.

Animated Molecules A visual representation of each molecule found throughout the text is also available in the Online Resource Centre, courtesy of ChemSpider, the popular online search engine that aggregates chemical structures and their associated information from all over the web into a single searchable repository. You'll also find 2D and 3D representations, as well as information on each structures' inherent properties, identifiers, and references. For more information on ChemSpider, visit **www.chemspider.com**.

For Lecturers

Textbook Images Almost all of the figures, tables, and images from the text are available for download in both .JPEG and PowerPoint® format. These can be use for lectures without charge, but not for commercial purposes without specific permission.

Other supplements

Explorations in Physical Chemistry 2.0

Valerie Walters, Julio de Paula, and Peter Atkins

Explorations in Physical Chemistry 2.0 consists of interactive Mathcad® worksheets, interactive Excel® workbooks, and stimulating exercises, designed to motivate students to simulate physical, chemical, and biochemical phenomena with their personal computers. Students can manipulate over 75 graphics, alter simulation parameters, and solve equations, to gain deeper insight into physical chemistry. It covers:

- Thermodynamics, including applications to biological processes.
- Quantum chemistry, including interactive three-dimensional renderings of atomic and molecular orbitals.
- Atomic and molecular spectroscopy, including tutorials on Fourier-transform techniques in modern spectroscopy.
- Properties of materials, including metals, polymers, and biological macromolecules.
- Chemical kinetics and dynamics, including enzyme catalysis, oscillating reactions, and polymerization reactions.

Explorations of Physical Chemistry 2.0 is available as a CD-ROM.
ISBN 978-0-19-928894-6

Solutions Manual for Physical Chemistry for the Life Sciences, Second Edition

Charles Trapp, University of Louisville, and Marshall Cady, Indiana University Southeast. ISBN: 978-0-19-960032-8

The Solutions Manual contains complete solutions to the end-of-chapter exercises, discussion questions, and projects from each chapter in the textbook. These worked-out-solutions will guide you through each step and help you refine your problem-solving skills.

Prolog

Chemistry is the science of matter and the changes it can undergo. **Physical chemistry** is the branch of chemistry that establishes and develops the principles of the subject in terms of the underlying concepts of physics and the language of mathematics. Its concepts are used to explain and interpret observations on the physical and chemical properties of matter.

This text develops the principles of physical chemistry and their applications to the study of the life sciences, particularly biochemistry and medicine. The resulting combination of the concepts of physics, chemistry, and biology into an intricate mosaic leads to a unique and exciting understanding of the processes responsible for life.

The structure of physical chemistry

Like all scientists, physical chemists build descriptions of nature on a foundation of careful and systematic inquiry.

(a) The organization of science

The observations that physical chemistry organizes and explains are summarized by scientific laws. A **law** is a summary of experience. Thus, we encounter the *laws of thermodynamics*, which are summaries of observations on the transformations of energy. Laws are often expressed mathematically, as in the *perfect gas law* (or *ideal gas law*; see Section F.2), $pV = nRT$. This law is an approximate description of the physical properties of gases (with p the pressure, V the volume, n the amount, R a universal constant, and T the temperature). We also encounter the *laws of quantum mechanics*, which summarize observations on the behavior of individual particles, such as molecules, atoms, and subatomic particles.

The first step in accounting for a law is to propose a **hypothesis**, which is essentially a guess at an explanation of the law in terms of more fundamental concepts. Dalton's *atomic hypothesis*, which was proposed to account for the laws of chemical composition and changes accompanying reactions, is an example. When a hypothesis has become established, perhaps as a result of the success of further experiments it has inspired or by a more elaborate formulation (often in terms of mathematics) that puts it into the context of broader aspects of science, it is promoted to the status of a **theory**. Among the theories we encounter are the theories of *chemical equilibrium*, *atomic structure*, and the *rates of reactions*.

A characteristic of physical chemistry, like other branches of science, is that to develop theories, it adopts models of the system it is seeking to describe. A **model** is a simplified version of the system that focuses on the essentials of the problem. Once a successful model has been constructed and tested against known observations and any experiments the model inspires, it can be made more sophisticated and incorporate some of the complications that the original model ignored.

Thus, models provide the initial framework for discussions, and reality is progressively captured rather like a building is completed, decorated, and furnished. One example is the *nuclear model* of an atom, and in particular a hydrogen atom, which is used as a basis for the discussion of the structures of all atoms. In the initial model, the interactions between electrons are ignored; to elaborate the model, repulsions between the electrons are taken into account progressively more accurately.

(b) The organization of our presentation

The text begins with an investigation of **thermodynamics**, the study of the transformations of energy, and the relations between the bulk properties of matter. Thermodynamics is summarized by a number of laws that allow us to account for the natural direction of physical and chemical change. Its principal relevance to biology is its application to the study of the deployment of energy by organisms.

We then turn to **chemical kinetics**, the study of the rates of chemical reactions. We shall establish how the rates of reactions can be determined and how experimental data give insight into the molecular processes by which chemical reactions occur. To understand the molecular mechanism of change, we also explore how molecules move, either in free flight in gases or by diffusion through liquids. Chemical kinetics is a crucial aspect of the study of organisms because the array of reactions that contribute to life form an intricate network of processes occurring at different rates under the control of enzymes.

Next, we develop the principles of **quantum theory** and use them to describe the structures of atoms and molecules, including the macromolecules found in biological cells. Quantum theory is important to the life sciences because the structures of its complex molecules and the migration of electrons cannot be understood except in its terms. We extend these theories of structure to solids, principally because that most revealing of all structural techniques, X-ray diffraction, depends on the availability and features of crystalline samples.

Finally, we explore the information about biological structure and function that can be obtained from **spectroscopy**, the study of interactions between molecules and electromagnetic radiation. The spectroscopic techniques available for the investigation of structure, which includes shape, size, and the distribution of electrons in ground and excited states, make use of most of the electromagnetic spectrum. We conclude with an account of perhaps the most important of all spectroscopies, nuclear magnetic resonance (NMR).

Applications of physical chemistry to biology and medicine

Here we discuss some of the important problems in biology and medicine being tackled with the tools of physical chemistry. We shall see that physical chemists contribute importantly not only to fundamental questions, such as the unravelling of intricate relationships between the structure of a biological molecule and its function, but also to the application of biochemistry to new technologies.

(a) Techniques for the study of biological systems

Many of the techniques now employed by biochemists were first conceived by physicists and then developed by physical chemists for studies of small molecules and chemical reactions before they were applied to the investigation of complex biological systems. Here we mention a few examples of physical techniques that are used routinely for the analysis of the structure and function of biological molecules.

X-ray diffraction and **nuclear magnetic resonance (NMR) spectroscopy** are two very important tools commonly used for the determination of the three-dimensional arrangement of atoms in biological assemblies. An example of the power of the X-ray diffraction technique is the recent determination of the three-dimensional structure of the ribosome, a complex of protein and ribonucleic acid with a molar mass exceeding 2×10^6 g mol^{-1} that is responsible for the synthesis of proteins from individual amino acids in the cell. This work led to the 2009 Nobel Prize in Chemistry, awarded to Venkatraman Ramakrishnan, Thomas Steitz, and Ada Yonath. Nuclear magnetic resonance spectroscopy has also advanced steadily through the years and now entire organisms may be studied through **magnetic resonance imaging** (MRI), a technique used widely in the diagnosis of disease. Throughout the text we shall describe many tools for the structural characterization of biological molecules.

Advances in biotechnology are also linked strongly to the development of physical techniques. The ongoing effort to characterize the entire genetic material, or **genome**, of organisms as simple as bacteria and as complex as *Homo sapiens* will lead to important new insights into the molecular mechanisms of disease, primarily through the discovery of previously unknown proteins encoded by the deoxyribonucleic acid (DNA) in genes. However, decoding genomic DNA will not always lead to accurate predictions of the amino acids present in biologically active proteins. Many proteins undergo chemical modification, such as cleavage into smaller proteins, after being synthesized in the ribosome. Moreover, it is known that one piece of DNA may encode more than one active protein. It follows that it is also important to describe the **proteome**, the full complement of functional proteins of an organism, by characterizing the proteins directly after they have been synthesized and processed in the cell.

The procedures of **genomics** and **proteomics**, the analysis of the genome and proteome, of complex organisms are time-consuming because of the very large number of molecules that must be characterized. For example, the human genome contains about 20 000 to 25 000 protein-encoding genes and the number of active proteins is likely to be much larger. Success in the characterization of the genome and proteome of any organism will depend on the deployment of very rapid techniques for the determination of the order in which molecular building blocks are linked covalently in DNA and proteins. An important tool is **gel electrophoresis**, in which molecules are separated on a gel slab in the presence of an applied electrical field. It is believed that **mass spectrometry**, a technique for the accurate determination of molecular masses, will be of great significance in proteomic analysis. We discuss the principles and applications of gel electrophoresis and mass spectrometry in Chapters 8 and 11, respectively.

(b) Protein folding

Proteins consist of flexible chains of amino acids. However, for a protein to function correctly, it must have a well-defined conformation. Although the amino acid sequence of a protein contains the necessary information to create the active conformation of the protein from a newly synthesized chain, the prediction of the conformation from the sequence, the so-called **protein folding problem**, is extraordinarily difficult and is still the focus of much research. Solving the problem of how a protein finds its functional conformation will also help us to understand why some proteins fold improperly under certain circumstances. Misfolded proteins are thought to be involved in a number of diseases, such as cystic fibrosis, Alzheimer's disease, and 'mad cow' disease (variant Creutzfeldt–Jakob disease, v-CJD).

To appreciate the complexity of the mechanism of protein folding, consider a small protein consisting of a single chain of 100 amino acids in a well-defined sequence. Statistical arguments lead to the conclusion that the polymer can exist in about 10^{49} distinct conformations, with the correct conformation corresponding to a minimum in the energy of interaction between different parts of the chain and the energy of interaction between the chain and surrounding solvent molecules. In the absence of a mechanism that streamlines the search for the interactions in a properly folded chain, the correct conformation can be attained only by sampling every one of the possibilities. If we allow each conformation to be sampled for 10^{-20} s, a duration far shorter than that observed for the completion of even the fastest of chemical reactions, it could take more than 10^{21} years, which is much longer than the age of the Universe, for the proper fold to be found. However, it is known that proteins can fold into functional conformations in less than 1 s.

The preceding arguments form the basis for *Levinthal's paradox* and lead to a view of protein folding as a complex problem in thermodynamics and chemical kinetics: how does a protein minimize the energies of all possible molecular interactions with itself and its environment in such a relatively short period of time? It is no surprise that physical chemists are important contributors to the solution of the protein-folding problem.

We discuss the details of protein folding in Chapters 8 and 11. For now, it is sufficient to outline the ways in which the tools of physical chemistry can be applied to the problem. Computational techniques that employ both classical and quantum theories of matter provide important insights into molecular interactions and can lead to reasonable predictions of the functional conformation of a protein. For example, in a **molecular mechanics** simulation, mathematical expressions from classical physics are used to determine the structure corresponding to the minimum in the energy of molecular interactions within the chain at the absolute zero of temperature. Such calculations are usually followed by **molecular dynamics** simulations, in which the molecule is set in motion by heating it to a specified temperature. The possible trajectories of all atoms under the influence of intermolecular interactions are then calculated by consideration of Newton's equations of motion. These trajectories correspond to the conformations that the molecule can sample at the temperature of the simulation. Calculations based on quantum theory are more difficult and time-consuming, but theoretical chemists are making progress toward merging classical and quantum views of protein folding.

As is usually the case in physical chemistry, theoretical studies inform experimental studies and vice versa. Many of the sophisticated experimental techniques in chemical kinetics to be discussed in Chapter 6 continue to yield details of the mechanism of protein folding. For example, the available data indicate that, in a number of proteins, a significant portion of the folding process occurs in less than 1 ms (10^{-3} s). Among the fastest events is the formation of helical and sheet-like structures from a fully unfolded chain. Slower events include the formation of contacts between helical segments in a large protein.

(c) Rational drug design

The search for molecules with unique biological activity represents a significant portion of the overall effort expended by pharmaceutical and academic laboratories to synthesize new drugs for the treatment of disease. One approach consists of extracting naturally occurring compounds from a large number of organisms and testing their medicinal properties. For example, the drug paclitaxel (sold under the tradename Taxol), a compound found in the bark of the Pacific yew tree, has been found to be effective in the treatment of ovarian cancer. An alternative approach to the discovery of drugs is **rational drug design**, which begins with the identification of molecular characteristics of a disease-causing agent—a microbe, a virus, or a tumor—and proceeds with the synthesis and testing of new compounds to react specifically with it. Scores of scientists are involved in rational drug design, as the successful identification of a powerful drug requires the combined efforts of microbiologists, biochemists, computational chemists, synthetic chemists, pharmacologists, and physicians.

Many of the targets of rational drug design are **enzymes**, proteins, or nucleic acids that act as biological catalysts. The ideal target is either an enzyme of the host organism that is working abnormally as a result of the disease or an enzyme unique to the disease-causing agent and foreign to the host organism. Because enzyme-catalyzed reactions are prone to inhibition by molecules that interfere with the formation of product, the usual strategy is to design drugs that are specific inhibitors of specific target enzymes. For example, an important part of the treatment of acquired immune deficiency syndrome (AIDS) involves the steady administration of a specially designed protease inhibitor. The drug inhibits an enzyme that is key to the formation of the protein envelope surrounding the genetic material of the human immunodeficiency virus (HIV). Without a properly formed envelope, HIV cannot replicate in the host organism.

The concepts of physical chemistry play important roles in rational drug design. First, the techniques for structure determination described throughout the text are essential for the identification of structural features of drug candidates that will interact specifically with a chosen molecular target. Second, the principles of chemical kinetics discussed in Chapters 6 and 7 govern several key phenomena that must be optimized, such as the efficiency of enzyme inhibition and the rates of drug uptake by, distribution in, and release from the host organism. Finally, and perhaps most importantly, the computational techniques discussed in Chapters 10 and 11 are used extensively in the prediction of the structure and reactivity of drug molecules. In rational drug design, computational chemists are often asked to predict the structural features that lead to an efficient drug by considering the nature of a receptor site in the target. Then synthetic chemists make the proposed molecules, which are in turn tested by biochemists and pharmacologists for efficiency. The process is often iterative, with experimental results feeding back into additional calculations, which in turn generate new proposals for efficient drugs, and so on. Computational chemists continue to work very closely with experimental chemists to develop better theoretical tools with improved predictive power.

(d) Biological energy conversion

The unraveling of the mechanisms by which energy flows through biological cells has occupied the minds of biologists, chemists, and physicists for many decades. As a result, we now have a very good molecular picture of the physical and chemical events of such complex processes as oxygenic photosynthesis and carbohydrate metabolism:

$$6\,CO_2(g) + 6\,H_2O(l) \underset{\substack{\text{carbohydrate}\\\text{metabolism}}}{\overset{\substack{\text{oxygenic}\\\text{photosynthesis}}}{\rightleftharpoons}} C_6H_{12}O_6(s) + 6\,O_2(g)$$

where $C_6H_{12}O_6$ denotes the carbohydrate glucose. In general terms, oxygenic photosynthesis uses solar energy to transfer electrons from water to carbon dioxide. In the process, high-energy molecules (carbohydrates, such as glucose) are synthesized in the cell. Animals feed on the carbohydrates derived from photosynthesis. During carbohydrate metabolism, the O_2 released by photosynthesis as a waste product is used to oxidize carbohydrates to CO_2. This oxidation drives biological processes, such as biosynthesis, muscle contraction, cell division, and nerve conduction. Hence, the sustenance of much of life on Earth depends on a tightly regulated carbon–oxygen cycle that is driven by solar energy.

We shall encounter photosynthesis and carbohydrate metabolism throughout the text. As we shall see in Chapter 12, the harvesting of solar energy during photosynthesis occurs very rapidly and efficiently. Within about 100–200 ps (1 ps = 10^{-12} s) of the initial light absorption event, more than 90 per cent of the energy is trapped within the cell and is available to drive the electron transfer reactions that lead to the formation of carbohydrates and O_2. Sophisticated spectroscopic techniques pioneered by physical chemists for the study of chemical reactions are being used to track the fast events that follow the absorption of solar energy.

The electron transfer processes of photosynthesis and carbohydrate metabolism drive the flow of protons across the membranes of specialized cellular compartments. The *chemiosmotic theory*, discussed in Chapter 5, describes how the energy stored in a proton gradient across a membrane can be used to synthesize adenosine triphosphate (ATP), a mobile energy carrier. Intimate knowledge of thermodynamics and chemical kinetics is required to understand the details of the theory and the experiments that eventually verified it.

The structures of nearly all the proteins associated with photosynthesis and carbohydrate metabolism have been characterized by X-ray diffraction or NMR techniques. Together, the structural data and the mechanistic models afford a nearly complete description of the relations between structure and function in biological energy conversion systems. This knowledge is now being used to design and synthesize molecular assemblies that can mimic oxygenic photosynthesis. The goal is to construct devices that trap solar energy in products of light-induced electron transfer reactions. One example is light-induced water splitting:

$$H_2O(l) \xrightarrow{\text{light}} \tfrac{1}{2}O_2(g) + H_2(g)$$

The hydrogen gas produced in this manner can be used as a fuel in a variety of other devices. The preceding is an example of how a careful study of the physical chemistry of biological systems can yield not only surprising insights but also new technologies.

Fundamentals

We begin by reviewing material fundamental to the whole of physical chemistry and its application to biology, but which should be familiar from introductory courses. Matter and energy are the principal focus of our discussion.

F.1 Atoms, ions, and molecules

Atoms, ions, and molecules are the currency of discourse in the whole of chemistry and of biochemistry in particular. These concepts will be familiar from introductory chemistry and need little review here. However, it is important to keep in mind the following points.

Atoms are characterized by their **atomic number**, Z, the number of protons in the nucleus. According to the nuclear model of an atom, a nucleus of charge Ze and containing most of the mass of the atom is surrounded by Z electrons, each of charge $-e$. **Isotopes** are atoms of the same atomic number but different **mass number** (or **nucleon number**), A, the total number of protons and neutrons in the nucleus. The loss of electrons results in **cations** (such as Na^+ and Ca^{2+}) and the gain of electrons results in **anions** (such as Cl^- and O^{2-}). When atoms are arranged in the order of increasing atomic number their properties show periodicities that are summarized by the periodic table with its familiar groups and periods (see inside the back cover).

(a) Bonding and nonbonding interactions

There are three types of interaction that result in atoms bonding together into more elaborate structures. **Ionic bonds** arise from the electrostatic attraction between cations and anions and give rise to typically hard, brittle arrays known as 'ionic solids'. **Covalent bonds** are due to the sharing of electrons and are responsible for the existence of discrete molecules, such as H_2O and elaborate proteins. **Metallic bonds** arise when atoms are able to pool one or more of their electrons into a common sea and give rise to metals with their characteristic lustre and electrical conductivity.

Covalent bonding is of the greatest importance in biology as it is responsible for the stabilities of the frameworks of organic molecules, such as DNA and proteins. However, there are interactions between regions of molecules that although much weaker than covalent bonding play a very important role in determining their shapes, and in biology molecular shape is closely allied with function. One such interaction is the **hydrogen bond**, A–H⋯B, where A and B are one of the atoms N, O, or F. Although only about 10 per cent as strong as a covalent bond, hydrogen bonding plays a major role in determining the shape of a biological macromolecule. Moreover, because it is quite weak, it permits the changes of

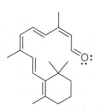

1 Water, H_2O

2 Acetic acid, CH_3COOH

3 Ethene, C_2H_4

4 Retinal (one form)

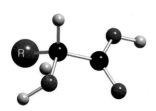

5 Amino acid, $NH_2CHRCOOH$

6 The peptide link, –CONH–

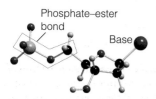

7 β-D-2-Deoxyribose phosphate

shape that allow an enzyme or nucleic acid to function. Weaker still are non-bonding interactions, commonly called **van der Waals interactions**, which are attractions between groups of atoms in different regions of a macromolecule or between different molecules. These forces also contribute to the shapes of molecules and the interactions between them, as we shall see.

The **connectivity** of a molecule, the pattern of covalent bonds it forms, is commonly represented by a Lewis structure, in which bonds are shown by lines, with two lines for double bonds (two shared electron pairs) and three lines for triple bonds (three shared pairs). **Lone pairs**, electron pairs not involved directly in bonding are also shown in Lewis structures, such as that for water (**1**) and acetic acid (**2**). Structural formulas of organic molecules are essentially Lewis structures without the explicit display of lone pairs. The rules for writing Lewis structures (such as the 'octet rule' relating to the number of electrons around each atom) should be familiar from introductory chemistry courses. A crucially important aspect of a double bond between two atoms, such as that in ethene (**3**) and on a more extensive scale in the visual pigment retinal (**4**), is that it confers torsional rigidity (resistance to twisting) in the region of the bond.

Lewis structures of all but the simplest molecules do not show the shape of the molecule. A collection of rules known as **valence-shell electron repulsion theory** (VSEPR theory), in which regions of electron density (attached atoms and lone pairs) are supposed to adopt positions that minimize their repulsions, is often a helpful guide to the local shape at an atom, such as the tetrahedral arrangement of single bonds around a carbon atom. This theory should also be familiar from introductory chemistry courses.

(b) Structural and functional units

Biochemistry effectively elaborates the concept of atoms by recognizing that characteristic groups of molecules can be regarded as building blocks from which the elaborate structures characteristic of organisms are constructed. These building blocks include the amino acids from which proteins are built as polypeptides, the bases that decorate the DNA double helix and constitute the genetic code, and carbohydrate molecules, such as glucose, that link together to form polysaccharides.

It will already be familiar from introductory courses that proteins, which are either structural or biochemically active molecules, are **polypeptides** formed from different α-amino acids of general form $NH_2CHRCOOH$ (**5**) strung together by the **peptide link**, –CONH– (**6**). Each monomer unit in the chain is referred to as a peptide **residue**. About 20 amino acids occur naturally and differ in the nature of the group R. These fundamental building blocks are illustrated in the *Atlas of structures*, Section A, in the *Resource section* at the end of the text.

Nucleic acids, which primarily store and transmit genetic information, are **polynucleotides** in which base–sugar–phosphate units are connected by **phosphodiester bonds** built from phosphate–ester links like that shown in (**7**). In DNA the sugar is β-D-2-deoxyribose (as shown in **8**) and the bases are adenine (A), cytosine (C), guanine (G), and thymine (T); see the *Atlas of structures*, Section B. In RNA the sugar is β-D-ribose and uracil (U) replaces thymine.

Polysaccharides are polymers of simple carbohydrates, such as glucose (**9**), linked together by C–O–C groups. They perform a variety of structural and functional roles in the cell, including energy storage and the mediation of interactions between cells (including those involved in immunological response). See the *Atlas of structures*, Section S.

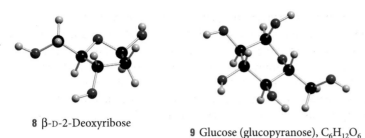

8 β-D-2-Deoxyribose

9 Glucose (glucopyranose), $C_6H_{12}O_6$

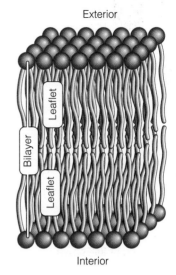

Third among the major structural units are the **lipids**, which are long-chain hydrocarbons, typically in the range C_{14}–C_{24}, with a variety of polar head groups at one end of the chain, such as $-CH_2CH_2N(CH_3)_3^+$ and $-COOH$. The basic structural element of a cell membrane is a **phospholipid**, in which one or more hydrocarbon chains are attached to a phosphate group (see the *Atlas of structures*, Section L). Phospholipids form a membrane by stacking together to form a **lipid bilayer**, about 5 nm across (Fig. F.1), leaving the polar groups exposed to the aqueous environment on either side of the membrane.

Fig. F.1 The long hydrocarbon chains of a phospholipid can stack together to form a bilayer structure with the polar groups (represented by the spheres) exposed to the aqueous environment.

(c) Levels of structure

The concept of the 'structure' of a biological macromolecule takes on different meanings for the different levels at which we think about the spatial arrangement of the polypeptide chain:

- The **primary structure** of a macromolecule is the sequence in which the units are linked in the polymer (Fig. F.2a).

- The **secondary structure** of a macromolecule is the (often local) spatial arrangement of the chain.

Examples of secondary structure motifs are random coils and ordered structures, such as helices and sheets, held together primarily by hydrogen bonds (Fig. F.2b). The secondary structure of DNA arises primarily from the winding of two poly-nucleotide chains around each other to form a double helix (Fig. F.3) held

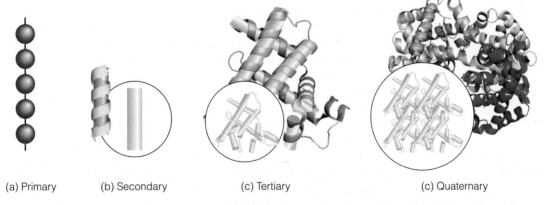

(a) Primary   (b) Secondary   (c) Tertiary   (c) Quaternary

Fig. F.2 The structural hierarchy of a biological macromolecule, in this case a protein, and a simplified representation in terms of cylinders. (a) The primary structure, the sequence of amino acid residues; (b) the local secondary structure (in this case a helix); (c) the tertiary structure: several helical segments connected by short random coils pack together; (d) the quaternary structure: several subunits with specific structures pack together.

Fig. F.3 The DNA double helix, in which two polynucleotide chains are linked together by hydrogen bonds between adenine (A) and thymine (T), and between cytosine (C) and guanine (G).

together by hydrogen bonds involving A–T and C–G base pairs that lie parallel to each other and perpendicular to the major axis of the helix.

- The **tertiary structure** is the overall three-dimensional structure of a macromolecule.

The hypothetical protein shown in Fig. F.2c has helical regions connected by short random-coil sections. The helices interact to form a compact tertiary structure.

- The **quaternary structure** of a macromolecule is the manner in which large molecules are formed by the aggregation of others.

Figure F.2d shows how several molecular subunits, each with a specific tertiary structure, aggregate together.

F.2 Bulk matter

Atoms, ions, and molecules cohere to form bulk matter. The broadest classification of the resulting materials is as gas, liquid, or solid. The term 'state' has many different meanings in chemistry, and it is important to keep them all in mind. Here we review the terms 'state of matter' and 'physical state'.

(a) States of matter

At a 'macroscopic' (observational) level, we distinguish the three **states of matter** by noting the behavior of a substance enclosed in a rigid container:

A **gas** is a fluid form of matter that fills the container it occupies.

A **liquid** is a fluid form of matter that possesses a well-defined surface and (in a gravitational field) fills the lower part of the container it occupies.

A **solid** retains its shape regardless of the shape of the container it occupies.

One of the roles of physical chemistry is to establish the link between the properties of bulk matter and the behavior of the particles of which it is composed. As we work through this text, we shall gradually establish and elaborate the following models for the states of matter at a 'microscopic' (atomic) level:

Mathematical toolkit F.1   *Quantities and units*

The result of a measurement is a **physical quantity** that is reported as a numerical multiple of a unit:

physical quantity = numerical value × unit

It follows that units are treated like algebraic quantities and may be multiplied, divided, and canceled. Thus, the expression (physical quantity)/unit is the numerical value (a dimensionless quantity) of the measurement in the specified units. For instance, the mass m of an object could be reported as $m = 2.5$ kg or $m/\text{kg} = 2.5$. See *Resource section* 2 for a list of units.

Units may be modified by a prefix that denotes a factor of a power of 10. Among the most common prefixes

are those listed in Table 3 of *Resource section* 2. Examples of the use of these prefixes are:

$$1 \text{ nm} = 10^{-9} \text{ m}$$
$$1 \text{ ps} = 10^{-12} \text{ s}$$
$$1 \text{ } \mu\text{mol} = 10^{-6} \text{ mol}$$

Powers of units apply to the prefix as well as the unit they modify. For example, $1 \text{ cm}^3 = 1 \text{ (cm)}^3$ and $(10^{-2} \text{ m})^3 = 10^{-6} \text{ m}^3$. But note that 1 cm^3 does not mean $1 \text{ c(m}^3)$. When carrying out numerical calculations, it is usually safest to write out the numerical value of an observable as powers of 10.

A gas is composed of widely separated particles in continuous rapid, disordered motion. A particle travels several (often many) diameters before colliding with another particle. For most of the time the particles are so far apart that they interact with each other only very weakly.

A liquid consists of particles that are in contact but are able to move past one another in a restricted manner. The particles are in a continuous state of motion but travel only a fraction of a diameter before bumping into a neighbor. The overriding image is one of movement but with molecules jostling one another.

A solid consists of particles that are in contact and unable to move past one another. Although the particles oscillate around an average location, they are essentially trapped in their initial positions and typically lie in ordered arrays.

The main difference between the three states of matter is the freedom of the particles to move past one another. If the average separation of the particles is large, there is hardly any restriction on their motion, and the substance is a gas. If the particles interact so strongly with one another that they are locked together rigidly, then the substance is a solid. If the particles have an intermediate mobility between these extremes, then the substance is a liquid. We can understand the melting of a solid and the vaporization of a liquid in terms of the progressive increase in the liberty of the particles as a sample is heated and the particles become able to move more freely.

(b) Physical state

By **physical state** (or just 'state') is meant a specific condition of a sample of matter that is described in terms of its physical form (gas, liquid, or solid) and the volume, pressure, temperature, and amount of substance present. (The precise meanings of these terms are described below.) So, 1 kg of hydrogen gas in a container of volume 10 dm^3 at a specified pressure and temperature is in a particular state. The same mass of gas in a container of volume 5 dm^3 is in a different state. Two samples of a given substance are in the same state if they are the same state of matter (that is, are both present as gas, liquid, or solid) *and* if they have the same mass, volume, pressure, and temperature.

To report the physical state of a sample we need to specify a number of properties in terms of their appropriate units. The manipulation of units, which almost always will be from the **International System** of units (SI, from the French Système International d'Unités) described in the *Resource section*, is explained in *Mathematical toolkit* F.1. These properties and their units include the following:

- **Mass**, m, is a measure of the quantity of matter a sample contains. Unit: 1 kg.

Thus, 2 kg of lead contains twice as much matter as 1 kg of lead and indeed twice as much matter as 1 kg of anything. For typical laboratory-sized samples it is usually more convenient to use a smaller unit and to express mass in grams (g), where 1 kg = 10^3 g.

- **Volume**, V, is a measure of the space a sample occupies. Unit: 1 m^3.

For volume we write $V = 100$ cm^3 if the sample occupies 100 cm^3 of space. Units used to express volume include cubic meters (m^3), cubic decimeters (dm^3), liters (L), and milliliters (mL). The liter is not an SI unit, but is exactly equal to 1 dm^3.

- **Amount of substance**, n, is a measure of the number of specified entities a sample contains. Unit: 1 mol.

A note on good practice
Physical quantities are denoted by italic, and sometimes Greek, letters (as in m for mass or ρ for mass density). Units are denoted by Roman letters (as in m for meter).

Table F.1 Pressure units and conversion factors*

pascal, Pa	$1\ Pa = 1\ N\ m^{-2}$
bar	$1\ bar = 10^5\ Pa$
atmosphere, atm	$1\ atm = \mathbf{101.325}\ kPa$
	$= \mathbf{1.013\ 25}\ bar$
torr, Torr[†]	$\mathbf{760}\ Torr = 1\ atm$
	$1\ Torr = 133.32\ Pa$

*Values in bold are exact.
†The name of the unit is torr; its symbol is Torr.

A brief comment
We shall see later (in Section F.3b) that temperature determines how molecules populate the energy levels available to them. Related to this interpretation is the fact that for molecules in a gas, the temperature determines their mean or average speed ($c_{average} \propto (T/M)^{1/2}$).

The amount is expressed in moles (mol), where 1 mole is defined as the same number of specified entities as there are atoms in exactly 12 g of carbon-12. In practice, the amount of substance is related to the number of entities, N, by $n = N/N_A$, where N_A is **Avogadro's constant** ($N_A = 6.022 \times 10^{23}\ mol^{-1}$). Note that N_A is a constant with units, not a pure number.

To convert from an amount to an actual number, N, of entities we write

$$N = nN_A \qquad \text{Relation between amount and number} \qquad \text{(F.1)}$$

To express a known mass of matter as an amount we use the molar mass, M, of the entities:

$$n = \frac{m}{M} \qquad \text{Relation between mass and amount} \qquad \text{(F.2)}$$

The molar mass, M, is the mass of a sample of an element or compound divided by the amount of atoms, molecules, or formula units it contains:

$$M = \frac{m}{n} \qquad \text{Definition of molar mass} \qquad \text{(F.3)}$$

The *atomic weight* of an element is the numerical value of the molar mass of the atoms it contains, the *molecular weight* of a molecular compound is the numerical value of the molar mass of its molecules, and the *formula weight* of an ionic compound is the molar mass of a specified formula unit of the compound. In each case 'numerical value' means $M/(g\ mol^{-1})$.

- **Pressure**, p, is the force a sample is subjected to divided by the area to which that force is applied. Unit: 1 Pa.

Because force (see later) is measured in newtons (1 N = 1 kg m s^{-2}), pressure is reported in newtons per square meter, or pascals (1 Pa = 1 N m^{-2}). The atmosphere (atm) is commonly used as a unit of pressure, but is not an SI unit. To convert between atmospheres and pascals use 1 atm = 101.325 kPa exactly. See Table F.1.

If an object is immersed in a gas, it experiences a pressure over its entire surface because molecules collide with it from all directions and exert a force during every collision. We are incessantly battered by molecules of gas in the atmosphere and experience this battering as 'atmospheric pressure'. The pressure is greatest at sea level because the density of air, and hence the number of colliding molecules, is greatest there. The pressure of the atmosphere at sea level is about 100 kPa. When a gas is confined to a cylinder fitted with a movable piston, the position of the piston adjusts until the pressure of the gas inside the cylinder is equal to that exerted by the atmosphere. When the pressures on either side of the piston are the same, we say that the two regions on either side are in **mechanical equilibrium** (Fig. F.4).

- **Temperature**, T, is the property of an object that determines in which direction energy will flow when it is in contact with another object: energy flows from higher temperature to lower temperature. Unit: 1 K.

When the two bodies have the same temperature, there is no net flow of energy between them. In that case we say that the bodies are in **thermal equilibrium** (Fig. F.5). The symbol T is used to denote the **thermodynamic temperature**, which is an absolute scale with $T = 0$ as the lowest point. Temperatures above

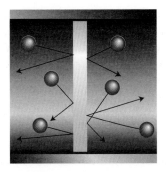

Fig. F.4 A system is in mechanical equilibrium with its surroundings if it is separated from them by a movable wall and the external pressure is equal to the pressure of the gas in the system.

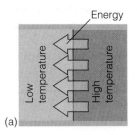
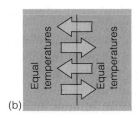

Fig. F.5 The temperatures of two objects act as a signpost showing the direction in which energy will flow as heat through a thermally conducting wall: (a) heat always flows from high temperature to low temperature. (b) When the two objects have the same temperature, although there is still energy transfer in both directions, there is no net flow of energy.

$T = 0$ are then most commonly expressed by using the **Kelvin scale**, in which the gradations of temperature are called kelvin (K). The Kelvin scale is defined by setting the triple point of water (the temperature at which ice, liquid water, and water vapour are in mutual equilibrium) at exactly 273.16 K. The freezing point of water (the melting point of ice) at 1 atm is then found experimentally to lie 0.01 K below the triple point, so the freezing point of water is approximately 273.15 K. The Kelvin scale is unsuitable for everyday measurements of temperature, and it is common to use the **Celsius scale**, which is defined in terms of the Kelvin scale as

$$\theta/°C = T/K - 273.15 \qquad \text{Relation between Kelvin and Celsius scales} \qquad (F.4)$$

(The 273.15 is exact in this definition.) Thus, the freezing point of water is 0°C and its boiling point (at 1 atm) is found to be 100°C. Note that in this text T invariably denotes the thermodynamic (absolute) temperature and that temperatures on the Celsius scale are denoted θ (theta).

Self-test F.1 Use eqn F.4 to express body temperature, 37°C, in kelvins.

Answer: 310 K

Temperature is an example of an **intensive property**, a property that is independent of the size of the sample. A property that does depend on the size ('extent') of the sample is called an **extensive property**. More formally, if we think of a sample as being divided into portions ('subsystems'), then the value of an extensive property is the sum of the contribution from each of the subsystems. For instance, the mass of a 10 mg sample of a protein is the sum of the masses of the 10 portions, each of 1 mg, into which it can be imagined as being divided. The value of an intensive property is the same for each of the subsystems and of the overall system itself. For instance, the temperature of a uniform 100 cm³ flask of water is the same as that of each of the 10 regions, each of volume 10 cm³, into which it can be regarded as being divided. Mass, volume, and amount of substance are all extensive properties. Temperature and pressure are intensive properties. Molar mass is intensive because the size-dependence of m and n cancel in the ratio m/n. All molar properties, $X_m = X/n$, where X is an extensive property, are intensive for the same reason. Mass density, $\rho = m/V$, is also intensive.

(c) Equations of state

Although the state of any sample of substance can be specified by giving the values of its volume, the pressure, the temperature, and the amount of substance, a remarkable experimental fact is that *these four quantities are not independent of one another*. For instance, we cannot arbitrarily choose to have a sample of 5.5 mmol H_2O in a volume of $100\ cm^3$ at 100 kPa and 500 K: it is found *experimentally* that that state simply does not exist. If we select the amount, the volume, and the temperature, then we find that we have to accept a particular pressure (in this case, close to 230 kPa). The same is true of all substances, but the pressure in general will be different for each one. This experimental generalization is summarized by saying the substance obeys an **equation of state**, an equation of the form

$$p = f(n, V, T) \qquad \text{A general equation of state} \qquad \text{(F.5)}$$

This expression tells us that the pressure is some function of amount, volume, and temperature, and that if we know those three variables, then the pressure can have only one value.

The equations of state of most substances are not known, so in general we cannot write down an explicit expression for the pressure in terms of the other variables. However, certain equations of state are known. In particular, the equation of state of a low-pressure gas is known and proves to be very simple and very useful:

$$p = \frac{nRT}{V} \qquad \text{Perfect gas equation of state} \qquad \text{(F.6)}$$

where R is the gas constant $R = 8.314\ J\ K^{-1}\ mol^{-1}$ (for values of R in other and sometimes more convenient units see Table F.2). Although the properties of gases might seem to be of little direct relevance to biochemistry, this equation is used to describe the behavior of gases taking part in a variety of biologically important processes (such as respiration), the properties of the gaseous environment we inhabit (the atmosphere), and as a starting point for the discussion of the properties of species in aqueous environments (such as the cell).

The perfect gas equation of state—more briefly, the 'perfect gas law'—is so-called because it is an idealization of the equations of state that gases actually obey. Specifically, it is found that all gases obey the equation ever more closely as the pressure is reduced toward zero. That is, eqn F.6 is an example of a **limiting law**, a law that becomes increasingly valid as the pressure is reduced and is obeyed exactly at the limit of zero pressure.

A hypothetical substance that obeys eqn F.6 at *all* pressures is called a **perfect gas**.[1] From what has just been said, an actual gas, which is termed a **real gas**, behaves more and more like a perfect gas as its pressure is reduced toward zero. In practice, normal atmospheric pressure at sea level ($p \approx 100$ kPa) is already low enough for most real gases to behave almost perfectly and, unless stated

Table F.2 The gas constant in various units

$R =$	
8.314 47	$J\ K^{-1}\ mol^{-1}$
8.314 47	$kPa\ dm^3\ K^{-1}\ mol^{-1}$
$8.205\ 74 \times 10^{-2}$	$atm\ dm^3\ K^{-1}\ mol^{-1}$
62.364	$Torr\ dm^3\ K^{-1}\ mol^{-1}$
1.987 21	$cal\ K^{-1}\ mol^{-1}$

---

[1] The term 'ideal gas' is also widely used.

otherwise, we shall always assume in this text that the gases we encounter behave like a perfect gas. The reason why a real gas behaves differently from a perfect gas can be traced to the attractions and repulsions that exist between actual molecules and that are absent in a perfect gas (Chapter 11).

A brief illustration

Consider the calculation of the pressure in kilopascals exerted by 1.25 g of nitrogen gas in a flask of volume 250 mL (0.250 dm³) at 20°C. The amount of N_2 molecules (of molar mass $M = 28.02$ g mol^{-1}) present is

$$n = \frac{m}{M} = \frac{1.25 \text{ g}}{28.02 \text{ g mol}^{-1}} = \frac{1.25}{28.02} \text{ mol}$$

The temperature of the sample is $T/K = 20 + 273.15$. Therefore, from $p = nRT/V$,

$$p = \frac{\overbrace{(1.25/28.02) \text{ mol}}^{n} \times \overbrace{(8.3145 \text{ kPa dm}^3 \text{ K}^{-1} \text{ mol}^{-1})}^{R} \times \overbrace{(20 + 273.15 \text{ K})}^{T = 293 \text{ K}}}{\underbrace{0.250 \text{ dm}^3}_{V = 250 \text{ mL}}}$$

$$= 435 \text{ kPa}$$

where we have used more convenient units for the constant R. Note how all units (except kPa in this instance) cancel like ordinary numbers (see *Mathematical toolkit* F.1).

A note on good practice
It is best to postpone the actual numerical calculation to the last possible stage and carry it out in a single step. This procedure avoids rounding errors.

Self-test F.2 Calculate the pressure exerted by 1.22 g of carbon dioxide confined to a flask of volume 500 mL at 37°C.

Answer: 143 kPa

The **molar volume**, V_m, is the volume a substance (not just a gas) occupies per mole of molecules. It is calculated by dividing the volume of the sample by the amount of molecules it contains:

$$V_m = \frac{V}{n} \qquad \boxed{\text{Definition of molar volume}} \quad (F.7)$$

The perfect gas law can be used to calculate the molar volume of a perfect gas at any temperature and pressure. When we combine eqns F.6 and F.7, we get

$$V_m = \frac{V}{n} = \frac{nRT}{np} = \frac{RT}{p} \qquad \boxed{\text{Molar volume of a perfect gas}} \quad (F.8)$$

This expression lets us calculate the molar volume of any gas (provided it is behaving perfectly) from its pressure and its temperature. It also shows that, for a given temperature and pressure, provided they are behaving perfectly, all gases have the same molar volume.

Chemists have found it convenient to report much of their data at a particular set of *standard* conditions, as summarized in Table F.3. The 'standard state' of a substance (at a specified temperature, not necessarily 298 K) is discussed further in Section 1.7. The condition SATP for the discussion of gases is now favored over

Table F.3 A summary of standard conditions

Name	Conditions	Comment
Standard pressure, $p^{\oplus}$	$p^{\oplus} = 1$ bar	1 bar is exact
Standard ambient temperature and pressure (SATP)	25°C (more precisely, 298.15 K) and 1 bar	At SATP, $V_m = 24.79$ dm^3 mol^{-1} for a perfect gas
Standard temperature and pressure (STP)	0°C and 1 atm	At STP, $V_m = 22.41$ dm^3 mol^{-1} for a perfect gas
Standard state	Pure substance at 1 bar	Temperature to be specified. See Section 1.7.

the earlier STP on account on the shift of emphasis from 1 atm to 1 bar in the specification of standard states.

A mixture of perfect gases, such as to a good approximation the atmosphere, behaves like a single perfect gas. According to **Dalton's law**, the total pressure of such a mixture is the sum of the **partial pressures** of the constituents, the pressure to which each gas would give rise if it occupied the container alone:

$$p = p_A + p_B + \cdots \qquad \boxed{\text{Dalton's law}} \quad \text{(F.9)}$$

Each partial pressure, p_J, can be calculated from the perfect gas law in the form $p_J = n_J RT/V$.

F.3 Energy

A property that will continue to occur in just about every chapter of the following text is 'energy'. Indeed, we begin the text with a discussion of the deployment of energy in living organisms. **Energy**, E, is the capacity to do work. **Work** is the process of moving against an opposing force. A fully wound spring can do more work than a half-wound spring (that is, it can raise a weight through a greater height or move a greater weight through a given height). A hot object has the potential for doing more work than the same object when it is cool and therefore has a higher energy.

In his formulation of classical mechanics Isaac Newton focused on the role of **force**, F, an agent that changes the state of motion of a body. His mechanics was built on three laws, the second of which relates the **acceleration**, a, the rate of change of velocity, of a body of mass m to the strength of the force it experiences:

$$F = ma \qquad \boxed{\text{Newton's second law}} \quad \text{(F.10)}$$

A brief illustration

A stationary ball of mass 150 g is hit by a bat, and in 0.20 s reaches a speed of 80 km h^{-1} (8.0×10^4 m/3600 s = 22 m s^{-1}) before being slowed down by air resistance. The initial acceleration of the ball is (22 m s^{-1})/(0.20 s) = 110 m s^{-2}. The force exerted by the bat on the ball is therefore

$$F = (0.150 \text{ kg}) \times (110 \text{ m s}^{-2}) = 16.5 \text{ kg m s}^{-2} = 16.5 \text{ N}$$

We have expressed the result in newtons, with 1 N = 1 kg m s^{-2}.

Force, like acceleration, is actually a 'vector' quantity, a quantity with direction as well as magnitude, but in most instances in this text we need consider only its magnitude.

The magnitude of the work done in moving against a constant opposing force, w, is the product of the distance moved, d, and the strength of the force:

$$w = Fd \qquad \boxed{\text{Definition of work}} \quad \text{(F.11)}$$

A brief illustration

A bird of mass 50 g flies from the ground to a branch 10 m above. The force of gravity on an object of mass m close to the surface of the Earth is mg, where g is the 'acceleration of free fall': $g = 9.81$ m s^{-2}. Therefore, the work it has to do against gravity is

$$w = mgd = (0.050\ \text{kg}) \times (9.81\ \text{m s}^{-2}) \times (10\ \text{m}) = 4.9\ \text{kg m}^2\ \text{s}^{-2}$$

We would report this value as 4.9 J, where J = 1 kg m^2 s^{-2}.

As implied in the *brief illustration*, the SI unit of energy is the joule (J), named after the nineteenth-century scientist James Joule, who helped to establish the concept of energy (see Chapter 1). It is defined as 1 J = 1 N m = 1 kg m^2 s^{-2}. A joule is quite a small unit, and in chemistry we often deal with energies of the order of kilojoules (1 kJ = 10^3 J).

(a) Varieties of energy

We need to distinguish the energies possessed by matter and due to radiation. The **kinetic energy**, E_k, is the energy of a body due to its motion. For a body of mass m moving at a speed v,

$$E_k = \tfrac{1}{2}mv^2 \qquad \boxed{\begin{array}{c}\text{Definition of} \\ \text{kinetic energy}\end{array}} \quad \text{(F.12)}$$

That is, a heavy object moving at the same speed as a light object has a higher kinetic energy, and doubling the speed of any object increases its kinetic energy by a factor of 4. A ball of mass 1 kg traveling at 1 m s^{-1} has a kinetic energy of 0.5 J.

The **potential energy**, E_p (and commonly V), of a body is the energy it possesses due to its position. The precise dependence on position depends on the type of force acting on the body. An important type of potential energy is the Coulombic potential energy of interaction between two electric charges Q_1 and Q_2 separated by a distance r:

$$E_p = \frac{Q_1 Q_2}{4\pi\varepsilon_0 r} \qquad \boxed{\begin{array}{c}\text{Coulombic potential} \\ \text{energy}\end{array}} \quad \text{(F.13)}$$

The fundamental constant ε_0 is called the **vacuum permittivity**; its value (and those of other fundamental constants) is given inside the front cover. With the charges in coulombs (C) and the distance in meters, the energy is obtained in joules. Equation F.13 is based on the convention of taking the potential energy to be zero when the charges are infinitely apart. The Coulombic potential energy will inform our discussion of a range of topics, from atomic structure to the nature of interactions that give rise to levels of structure in biological assemblies.

A mass m close to the surface of the Earth has a potential energy that is proportional to its height above the ground, h:

$$E_p = mgh$$

Gravitational potential energy   (F.14)

The constant $g = 9.81$ m s^{-2} is called the **acceleration of free fall**. It depends on the location on the Earth's surface, but the variation is quite small. In this case, the arbitrary zero of potential energy is taken as being at the surface of the Earth (at $h = 0$).

A brief illustration

The potential energy of the 50-g bird mentioned in the preceding *brief illustration* is higher by 4.9 J when it is on the branch than when it is on the ground. The potential energy of this book (of mass about 1 kg) is higher by about 10 J when it is on a table 1 m above the floor than when it is on the floor.

The **total energy**, E, of a material body is the sum of its kinetic and potential energies:

$$E = E_k + E_p$$

Total energy   (F.15)

Provided no external forces are acting on the body, its total energy is constant. This remark is elevated to a central statement of classical physics known as the **law of the conservation of energy**. Potential and kinetic energy may be freely interchanged, for instance a falling ball loses potential energy but gains kinetic energy as it accelerates, but its total energy remains constant provided the body is isolated from external influences, such as air resistance.

Energy may also be present even in the absence of matter in the form of **electromagnetic radiation**, a wave of electric and magnetic fields traveling through a vacuum at the 'speed of light', $c = 2.998 \times 10^8$ m s^{-1}. The wave is characterized by its amplitude, frequency, and wavelength. The **amplitude** of the wave is the maximum displacement, and the perceived intensity of the wave is proportional to the square of the amplitude. The **frequency**, v (nu), is a measure of the rate at which the field goes through a complete cycle of orientations. The SI unit of frequency is 1 hertz (1 Hz), which corresponds to one cycle per second: $1 \text{ Hz} = 1 \text{ s}^{-1}$. The **wavelength**, λ (lambda), is the distance between neighboring peaks of the wave (Fig. F.6). The frequency and wavelength are related by

$$\lambda v = c$$

Relation between frequency and wavelength   (F.16)

That is, high frequencies correspond to short wavelengths, and vice versa. This expression also applies to sound waves, with c interpreted as the speed of sound.

The **electromagnetic spectrum** runs—as far as we know—over all frequencies. Each range of frequencies is classified as shown in Fig. F.7. The boundaries between each region are only approximate. The **visible region** of the spectrum, the region to which our eyes are sensitive, occupies a very narrow band between 400 and 700 nm. As we shall see in later chapters, each region of the spectrum excites, or is excited by, different types of nuclear, atomic, or molecular transition. For instance, electronic excitations, where electrons are redistributed into

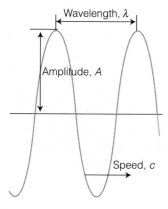

Fig. F.6 An electromagnetic wave is characterized by its amplitude, A, wavelength, λ, and frequency, v; the frequency is related to the wavelength by $v = c/\lambda$.

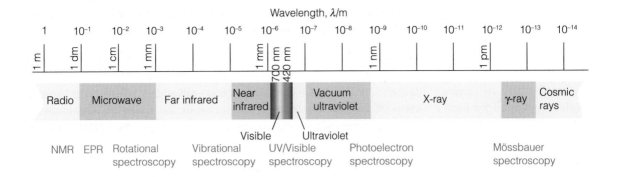

Fig. F.7 The regions of the electromagnetic spectrum and some of the spectroscopic techniques that make use of them.

different regions of the molecule, are stimulated by or give rise to visible and ultraviolet radiation. Because the separation between energy levels is dictated by the arrangement of atoms in a molecule, measuring the frequencies of transitions facilitates the study of molecular structure and reactivity. Ultraviolet radiation can also cause such extreme electron redistributions that bonds are broken.

We need to be aware that electromagnetic energy is delivered in packets known as **photons**. The energy of a photon of electromagnetic radiation is related to the frequency of the radiation by

$$E = h\nu \qquad \boxed{\text{Energy of a photon}} \qquad (F.17)$$

where h is a fundamental constant known as **Planck's constant** ($h = 6.626 \times 10^{-34}$ J s). In terms of photons, an intense ray of light consists of numerous photons, each of the same energy and each moving at the speed c. The higher the frequency of the radiation, the greater is the energy carried by each photon. Photons of visible light are sufficiently energetic to stimulate the processes of vision; photons of ultraviolet radiation are so energetic that they can destroy tissue.

$\boxed{\text{A brief illustration}}$

The energy of a photon of 350 nm ultraviolet radiation is

$$E = \frac{hc}{\lambda} = \frac{(6.626 \times 10^{-34} \text{ J s}) \times (2.998 \times 10^{8} \text{ m s}^{-1})}{3.50 \times 10^{-7} \text{ m}} = 5.68 \times 10^{-19} \text{ J}$$

corresponding to 0.568 aJ. To know the energy per mole of photons, which helps us to assess the chemical potency of the radiation, we multiply by Avogadro's constant:

$$E = \frac{hcN_A}{\lambda} = \frac{(6.626 \times 10^{-34} \text{ J s}) \times (2.998 \times 10^{8} \text{ m s}^{-1}) \times (6.022 \times 10^{23} \text{ mol}^{-1})}{3.50 \times 10^{-7} \text{ m}}$$

$$= 342 \text{ kJ mol}^{-1}$$

A note on good practice It is best to carry out a numerical calculation in one step or at least to avoid rounding at an intermediate stage.

(b) The Boltzmann distribution

One of the most important expressions in science, the 'Boltzmann distribution', helps to elucidate the concept of temperature as well as underlying virtually all the bulk properties and reactions of matter and their variation with temperature.

Mathematical toolkit F.2   *Exponential functions*

In preparation for the large number of occurrences of exponential functions throughout the text, it will be useful to know the shape of exponential functions. Here we deal with two types, e^{-ax} and e^{-ax^2}. An **exponential function** of the form e^{-ax} starts off at 1 when $x = 0$ and decays toward zero, which it reaches as x approaches infinity (see the illustration). This function approaches zero more rapidly as a increases. The Boltzmann distribution is an example of an exponential function. The function e^{-ax^2} is called a **Gaussian**

function. It also starts off at 1 when $x = 0$ and decays to zero as x increases, however, its decay is initially slower but then plunges down more rapidly than e^{-ax}. Gaussian functions will appear several times through the text. The illustration also shows the behavior of the two functions for negative values of x. The exponential function e^{-ax} rises rapidly to infinity, but the Gaussian function falls back to zero and traces out a bell-shaped curve.

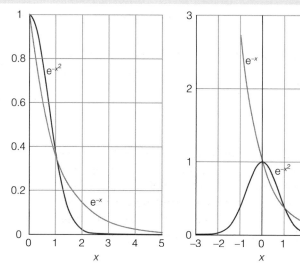

The exponential function, e^{-x}, and the bell-shaped Gaussian function, e^{-x^2}. Note that both are equal to 1 at $x = 0$, but the exponential function rises to infinity as $x \to -\infty$.

It should be familiar from introductory courses, and will be explained in detail later in the text, that atoms and molecules can possess only discrete amounts of energy. For instance, an electron in a hydrogen atom can possess only the energies 2.17 aJ, 0.54 aJ, 0.24 aJ, . . . (where 1 aJ, 1 attojoule = 10^{-18} J) below that of a widely separated proton and electron, and a C–H bond in a molecule can vibrate only with the energies 0.029 aJ, 0.086 aJ, 0.144 aJ, . . . Intermediate values of the energy are simply not allowed. The precise values of the allowed energies depend on the details of molecular structure, but it is generally the case that electronic energy levels are most widely spaced, then the energies of molecular vibration, and then the energies with which molecules rotate (Fig. F.8). The energies of translational motion are so close together even on an atomic scale (for instance, of the order of 10^{-44} J for a CO_2 molecule in a region 10 cm wide) that they may be treated as continuous.

The apparently random motion that molecules undergo at $T > 0$ is called **thermal motion**. The energy associated with this motion is the energy of thermal motion, but is commonly called simply **thermal energy**. A useful rule of thumb is that the order of magnitude of the energy that a molecule possesses as a result of its thermal motion is kT, where $k = 1.381 \times 10^{-23}$ J K^{-1} is a fundamental constant called **Boltzmann's constant**. The gas constant R is simply the 'molar' form of Boltzmann's constant:

$$R = N_A k$$

Relation between the gas constant and Boltzmann's constant   (F.18)

Thermal motion ensures that molecules will be found spread over the energy levels available to them such that their mean energy is of order kT. The population of each energy level depends on the temperature, and a very important result is that in a system at a temperature T, the ratio of populations N_2 and N_1 in states with energies E_1 and E_2 is given by the **Boltzmann distribution**, one form of which is

$$\frac{N_2}{N_1} = e^{-(E_2 - E_1)/kT}$$

The Boltzmann distribution   (F.19a)

This form of the distribution applies when the E_i are actual energies (in joules, for instance); when the E_i are molar quantities (in joules or kilojoules per mole, for instance), we use

$$\frac{N_2}{N_1} = e^{-(E_2 - E_1)/RT}$$

(F.19b)

with R in place of k. We see that the greater the energy separation $E_2 - E_1$, the smaller the ratio of populations. Alternatively, for a given separation, the ratio becomes smaller as the temperature is lowered. In other words, as the temperature is lowered, more and more molecules are found in their lowest energy levels and fewer are found in high energy levels. The temperature, we see, is the single parameter we need in order to state the relative populations of energy levels.

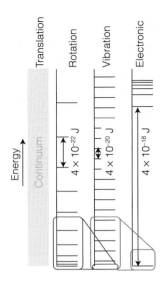

Fig. F.8 The energy level separations (in joules) typical of four types of motion.

A brief illustration

Suppose that two conformations of neighboring peptide groups in a polypeptide differ in energy by 7.5 kJ mol^{-1}, with conformation A higher in energy than conformation B. At body temperature (37°C, corresponding to 310 K) the ratio of populations of the two conformations is

$$\frac{N_A}{N_B} = e^{-(7500\,\text{J mol}^{-1})/(8.3145\,\text{J K}^{-1}\,\text{mol}^{-1} \times 310\,\text{K})} = 0.054$$

That is, conformation B is about 18 times more abundant than conformation A.

The importance of the Boltzmann distribution will become apparent as the following chapters unfold. We shall see that it accounts for the stability of matter, for very few molecules are found in highly excited states at ordinary temperatures, but it allows for the possibility of reaction, as *some* molecules will be found with sufficient energy to react, and the proportion that can react increases as the temperature is raised. Already we are beginning to see why chemical reactions proceed more quickly as the temperature is raised.

We can obtain insight into the molecular origins of temperature by using the simple but powerful **kinetic model of gases** (also called the 'kinetic molecular theory', KMT, of gases), which is based on a model of a gas that we mentioned earlier, in which the molecules are in ceaseless random motion, do not interact with one another except during collisions, and are much smaller than the average distance traveled between collisions (Fig. F.9). Different speeds correspond to different energies, so the Boltzmann formula can be used to predict the proportions of molecules having a specific speed at a particular temperature. The expression giving the fraction of molecules that have a particular speed is called the **Maxwell distribution** (sometimes the *Maxwell–Boltzmann distribution*) and has the features summarized in Figs F.10 and F.11. The Maxwell distribution, which is discussed more fully in *Further information 7.1*, can be used to show that the mean speed, $\bar{c}$, of the molecules depends on the temperature T and their molar mass M as

$$\bar{c} = \left(\frac{8RT}{\pi M} \right)^{1/2}$$

$\boxed{\text{Mean speed according to the Maxwell distribution}}$ (F.20)

Thus, the mean or average speed is high for light molecules at high temperatures. The distribution itself gives more information. For instance, the tail towards high speeds is longer at high temperatures than at low, which indicates that at high temperatures more molecules in a sample have speeds much higher than average.

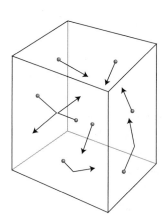

Fig. F.9 The model used for discussing the molecular basis of the physical properties of a perfect gas. The pointlike molecules move randomly with a wide range of speeds and in random directions, both of which change when they collide with the walls or with other molecules.

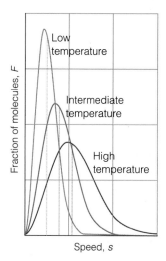

Fig. F.10 The Maxwell distribution of speeds and its variation with the temperature. Note the broadening of the distribution and the shift of the mean speed (denoted by the locations of the vertical dotted lines) to higher values as the temperature is increased.

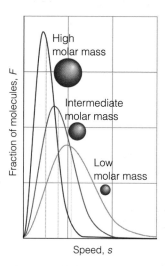

Fig. F.11 The Maxwell distribution of speeds also depends on the molar mass of the molecules. Molecules of low molar mass have a broad spread of speeds, and a significant fraction may be found traveling much faster than the mean speed. The distribution is much narrower for heavy molecules, and most of them travel with speeds close to the mean value (denoted by the locations of the vertical dotted lines).

Checklist of key concepts

☐ 1. In the nuclear model, an atom of atomic number Z consists of a nucleus of charge $+Ze$ surrounded by Z electrons each of charge $-e$.

☐ 2. Proteins, nucleic acids, and polysaccharides are long molecular chains with different levels of three-dimensional structure.

☐ 3. Cell membranes are formed by the stacking of lipid molecules into a bilayer structure.

☐ 4. The states of matter are gas, liquid, and solid.

☐ 5. An equation of state is an equation relating pressure, volume, temperature, and amount of a substance.

☐ 6. The perfect gas equation of state is a limiting law applicable as $p \to 0$.

☐ 7. Energy is the capacity to do work.

☐ 8. Work is done when a body is moved against an opposing force.

☐ 9. The contributions to the energy of matter are the kinetic energy (the energy due to motion) and the potential energy (the energy due to position).

☐ 10. The total energy of an isolated system is conserved, but kinetic and potential energy may be interchanged.

☐ 11. Electromagnetic radiation is characterized by its amplitude, frequency, and wavelength.

☐ 12. Electromagnetic radiation consists of photons, packets of energy of magnitude $h\nu$ and traveling at the speed of light.

☐ 13. The Boltzmann distribution gives the relative numbers of molecules in the energy levels available to them.

☐ 14. The mean speed of molecules is proportional to the square root of the (absolute) temperature and inversely proportional to the square root of the molar mass.

☐ 15. The properties of the Maxwell distribution of speeds are summarized in Figs F.10 and F.11.

Checklist of key equations

Property or process	Equation	Comment
Relation between number and amount	$N = nN_A$	N_A is Avogadro's constant
Molar quantity	$X_m = X/n$	Molar mass is denoted M
Temperature conversion	$\theta/°C = T/K - 273.15$	273.15 is exact
Equation of state	$p = nRT/V$	Perfect (ideal) gas
Molar volume	$V_m = RT/p$	Perfect (ideal) gas
Dalton's law	$p = p_A + p_B + \cdots$	Perfect (ideal) gas
Newton's second law	$F = ma$	
Work	$w = Fd$	F is the opposing force
Kinetic energy	$E_k = \frac{1}{2}mv^2$	
Coulomb potential energy	$E_p = Q_1Q_2/4\pi\varepsilon_0 r$	Charges in a vacuum
Gravitational potential energy	$E_p = mgh$	Close to surface of the Earth
Relation between wavelength and frequency	$\lambda\nu = c$	c is speed of propagation (e.g. speed of light)
Energy of a photon	$E = h\nu$	h is Planck's constant
Boltzmann distribution	$N_2/N_1 = e^{-(E_2-E_1)/kT}$	k is Boltzmann's constant, $R = N_A k$
Mean speed of molecules	$\bar{c} = (8RT/\pi M)^{1/2}$	Perfect (ideal) gas

Discussion questions

F.1 Distinguish between ionic bonds, covalent bonds, hydrogen bonds, and van der Waals interactions.

F.2 Distinguish between polypeptides, polynucleotides, and polysaccharides.

F.3 Describe the main structural features of a lipid bilayer.

F.4 Distinguish between primary, secondary, tertiary, and quaternary levels of structure in biological macromolecules.

F.5 Explain the differences between gases, liquids, and solids at macroscopic and microscopic levels.

F.6 Define the terms force, work, energy, kinetic energy, potential energy, and the energy of thermal motion.

F.7 Distinguish between mechanical and thermal equilibrium.

F.8 Describe the main features of electromagnetic radiation and the electromagnetic spectrum.

F.9 Use the Boltzmann distribution to provide a molecular interpretation of temperature.

Exercises

Treat all gases as perfect unless instructed otherwise.

F.10 You will see Lewis structures throughout the text. Using your knowledge of introductory chemistry, draw the Lewis structures of (a) SO_3^{2-}, (b) XeF_4, (c) P_4, (d) O_3, (e) ClF_3^+, and (f) N_3^-.

F.11 Using your knowledge of VSEPR theory from introductory chemistry, predict the shapes of (a) PCl_3, (b) PCl_5, (c) XeF_2, (d) XeF_4, (e) H_2O_2, (f) FSO_3^-, (g) KrF_2, and (h) PCl_4^+.

F.12 Express (a) 110 kPa in torr, (b) 0.997 bar in atmospheres, (c) 2.15×10^4 Pa in atmospheres, and (d) 723 Torr in pascals.

F.13 Given that the Celsius and Fahrenheit temperature scales are related by $\theta_{Celsius}/°C = \frac{5}{9}(\theta_{Fahrenheit}/°F - 32)$, what is the temperature of absolute zero ($T = 0$) on the Fahrenheit scale?

F.14 Imagine that Pluto is inhabited and that its scientists use a temperature scale in which the freezing point of liquid nitrogen is 0°P (degrees Plutonium) and its boiling point is 100°P. The inhabitants of Earth report these temperatures as −209.9°C and −195.8°C, respectively. What is the relation between temperatures on (a) the Plutonium and Kelvin scales, and (b) the Plutonium and Fahrenheit scales?

F.15 Much to everyone's surprise, nitrogen monoxide (nitric oxide, NO) has been found to act as a neurotransmitter. To prepare to study its effect, a sample was collected in a container of volume 250.0 cm³. At 19.5°C its pressure is found to be 24.5 kPa. What amount (in moles) of NO has been collected?

F.16 The effect of high pressure on organisms, including humans, is studied to gain information about deep-sea diving and anesthesia. A sample of air occupies 1.00 dm³ at 25°C and 1.00 atm. What pressure is needed to compress it to 100 cm³ at this temperature?

F.17 You are warned not to dispose of pressurized cans by throwing them onto a fire. The gas in an aerosol container exerts a pressure of 125 kPa at 18°C. The container is thrown on a fire, and its temperature rises to 700°C. What is the pressure at this temperature?

F.18 Until we find an economical way of extracting oxygen from seawater or lunar rocks, we have to carry it with us to inhospitable places and do so in compressed form in tanks. A sample of oxygen

at 101 kPa is compressed at constant temperature from 7.20 dm³ to 4.21 dm³. Calculate the final pressure of the gas.

F.19 Hot-air balloons gain their lift from the lowering of density of air that occurs when the air in the envelope is heated. To what temperature should you heat a sample of air, initially at 340 K, to increase its volume by 14 per cent?

F.20 At sea level, where the pressure was 104 kPa and the temperature 21.1°C, a certain mass of air occupied 2.0 m³. To what volume will the region expand when it has risen to an altitude where the pressure and temperature are (a) 52 kPa, −5.0°C and (b) 880 Pa, −52.0°C?

F.21 A diving bell has an air space of 3.0 m³ when on the deck of a boat. What is the volume of the air space when the bell has been lowered to a depth of 50 m? Take the mean density of seawater to be 1.025 g cm⁻³ and assume that the temperature is the same as on the surface.

F.22 Calculate the work that a person of mass 65 kg must do to climb between two floors of a building separated by 3.5 m.

F.23 What is the kinetic energy of a tennis ball of mass 58 g served at 30 m s⁻¹?

F.24 A car of mass 1.5 t (1 t = 10^3 kg) traveling at 50 km h⁻¹ must be brought to a stop. How much kinetic energy must be dissipated?

F.25 Consider a region of the atmosphere of volume 25 dm³, which at 20°C contains about 1.0 mol of molecules. Take the average molar mass of the molecules as 29 g mol⁻¹ and their average speed as about 400 m s⁻¹. Estimate the energy stored as molecular kinetic energy in this volume of air.

F.26 Calculate the minimum energy that a bird of mass 25 g must expend in order to reach a height of 50 m.

F.27 The potential energy of a charge Q_1 in the presence of another charge Q_2 can be expressed in terms of the *Coulomb potential*, ϕ (phi):

$$V = Q_1\phi \qquad \phi = \frac{Q_2}{4\pi\varepsilon_0 r}$$

The units of potential are joules per coulomb, J C⁻¹, so when ϕ is multiplied by a charge in coulombs, the result is in joules. The combination joules per coulomb occurs widely and is called a volt (V), with $1\ V = 1\ J\ C^{-1}$. Calculate the Coulomb potential due to the nuclei at a point in a LiH molecule located 200 pm from the Li nucleus and 150 pm from the H nucleus. *Hint:* Use $Q = +Ze$, where Z is the atomic number and e is the elementary charge.

F.28 Plot the Coulomb potential (see *Exercise* F.27) due to the nuclei at a point in a Na^+Cl^- ion pair located on a line half-way between the nuclei (the internuclear separation is 283 pm) as the point approaches from infinity and ends at the mid point between the nuclei.

F.29 What is the wavelength of the radiation used by an FM radio transmitter broadcasting at 92.0 MHz?

F.30 What is the energy of (a) a single photon and (b) 1.00 mol of photons of wavelength 670 nm?

F.31 Suppose that a macromolecule can exist either as a random coil or fully stretched out, with the latter conformation $2.4\ kJ\ mol^{-1}$ higher in energy. What is the ratio of the two conformations at 20°C?

F.32 An electron spin can adopt either of two orientations in a magnetic field, and its energies are $\pm\mu_B\mathcal{B}$, where $\mu_B = 9.274 \times 10^{-24}\ J\ T^{-1}$ is the Bohr magneton and $\mathcal{B}$ is the intensity of the magnetic field, often reported in teslas ($1\ T = 1\ kg\ s^{-2}\ A^{-1}$). Calculate the relative populations of the spin states at (a) 4.0 K and (b) 298 K, when $\mathcal{B} = 1.0\ T$.

F.33 The composition of planetary atmospheres is determined in part by the speeds of the molecules of the constituent gases because the faster-moving molecules can reach escape velocity and leave the planet. Calculate the mean speed of (a) He atoms and (b) CH_4 molecules at (i) 77 K, (ii) 298 K, and (iii) 1000 K.

Project

F.34 You will now explore the gravitational potential energy in some detail, with an eye toward discovering the origin of the value of the constant g, the acceleration of free fall, and the magnitude of the gravitational force experienced by all organisms on the Earth.

(a) The gravitational potential energy of a body of mass m at a distance r from the center of the Earth is $-Gmm_E/r$, where m_E is the mass of the Earth and G is the gravitational constant (see inside front cover). Consider the difference in potential energy of the body when it is moved from the surface of the Earth (radius r_E) to a height h above the surface, with $h \ll r_E$, and find an expression for the acceleration of free fall, g, in terms of the mass and radius of the Earth. *Hint:* Use the approximation $(1 + h/r_E)^{-1} \approx 1 - h/r_E$. (See *Mathematical toolkit 3.2* for more information on series expansions and the approximations that can be made by using expansions.)

(b) You need to assess the fuel needed to send the robot explorer *Spirit*, which has a mass of 185 kg, to Mars. What was the energy needed to raise the vehicle itself from the surface of the Earth to a distant point where the Earth's gravitational field was effectively zero? The mean radius of the Earth is 6371 km and its average mass density is $5.5170\ g\ cm^{-3}$. *Hint:* Use the full expression for the gravitational potential energy in part (a).

(c) Given the expression for the gravitational potential energy in part (a), (i) what is the gravitational force on an object of mass m at a distance r from the center of the Earth? (ii) What is the gravitational force that you are currently experiencing? For data on the Earth, see part (b).

PART 1 Biochemical Thermodynamics

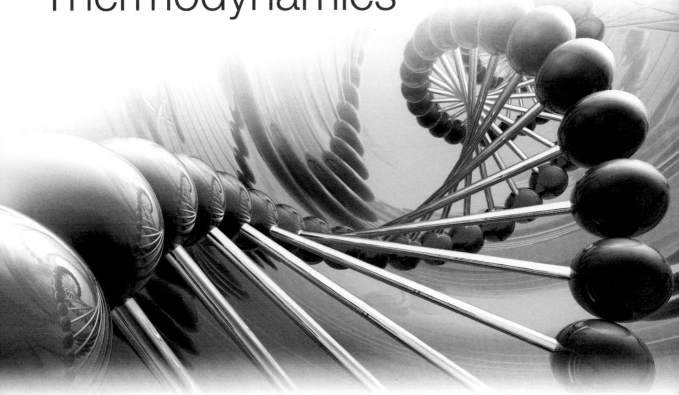

The branch of physical chemistry known as **thermodynamics** is concerned with the study of the transformations of energy. That concern might seem remote from chemistry, let alone biology. Indeed, thermodynamics was originally formulated by physicists and engineers interested in the efficiency of steam engines. However, thermodynamics has proved to be of immense importance in both chemistry and biology. Not only does it deal with the energy output of chemical reactions but it also helps to answer questions that lie right at the heart of biochemistry, such as how energy flows in biological cells and how large molecules assemble into complex structures like the cell.

The First Law

Classical thermodynamics, the thermodynamics developed during the nineteenth century, stands aloof from any models of the internal constitution of matter: we could develop and use thermodynamics without ever mentioning atoms and molecules. However, the subject is greatly enriched by acknowledging that atoms and molecules do exist and interpreting thermodynamic properties and relations in terms of them. Wherever it is appropriate, we shall cross back and forth between thermodynamics, which provides useful relations between observable properties of bulk matter, and the properties of atoms and molecules, which are ultimately responsible for these bulk properties.

Throughout the text we shall pay special attention to **bioenergetics**, the deployment of energy in living organisms. We shall initiate discussions of thermodynamics with the perfect gas as a model system. Although a perfect gas may seem far removed from biology, its properties are crucial to the formulation of thermodynamics of systems in aqueous environments, such as biological cells. First, it is quite simple to formulate the thermodynamic properties of a perfect gas. Then—and this is the crucially important point—because a perfect gas is a good approximation to a vapor and a vapor may be in equilibrium with a liquid, the thermodynamic properties of a perfect gas are mirrored (in a manner we shall describe) in the thermodynamic properties of the liquid. In other words, we shall see that a description of the gases (or 'vapors') that hover above a solution opens a window onto the description of physical and chemical transformations occurring in the solution itself.

Once we become equipped with the formalism to describe chemical reactions in solution, it will be easy to apply the concepts of thermodynamics to the complex environment of a biological cell. That is, we need to make a modest investment in the study of systems that may seem removed from our concerns so that, in the end, we can collect sizable dividends that will enrich our understanding of energy trapping and utilization in biological cells.

The conservation of energy

Almost every argument and explanation in chemistry boils down to a consideration of some aspect of a single property: the *energy*. Energy determines what molecules can form, what reactions can occur, how fast they can occur, and (with a refinement in our conception of energy) in which direction a reaction has a tendency to occur.

As we saw in *Fundamentals*:

energy is the capacity to do work
work is the process of moving against an opposing force

These definitions imply that a raised weight has more energy than one of the same mass resting on the ground because the former has a greater capacity to do work: it can do work as it falls to the level of the lower weight. The definition also implies that a gas at a high temperature has more energy than the same gas at a low temperature: the hot gas has a higher pressure and can do more work in driving out a piston. In biology, we encounter many examples of the relation between energy and work. As a muscle contracts and relaxes, energy stored in its protein fibers is released as the work of walking, lifting a weight, and so on. In biological cells, nutrients, ions, and electrons are constantly moving across membranes and from one cellular compartment to another. The synthesis of biological molecules and cell division are also manifestations of work at the molecular level. The energy that produces all this work in our bodies comes from food.

People struggled for centuries to create energy from nothing, for they believed that if they could create energy, then they could produce work (and wealth) endlessly. However, without exception, despite strenuous efforts, many of which degenerated into deceit, they failed. As a result of their failed efforts, we have come to recognize that energy can be neither created nor destroyed but merely converted from one form into another or moved from place to place. This 'law of the conservation of energy' is of great importance in chemistry. Most chemical reactions—including the majority of those taking place in biological cells—release energy or absorb it as they occur; so according to the law of the conservation of energy, we can be confident that all such changes—including the vast collection of physical and chemical changes we call life—must result only in the *conversion* of energy from one form to another or its transfer from place to place, not its creation or annihilation.

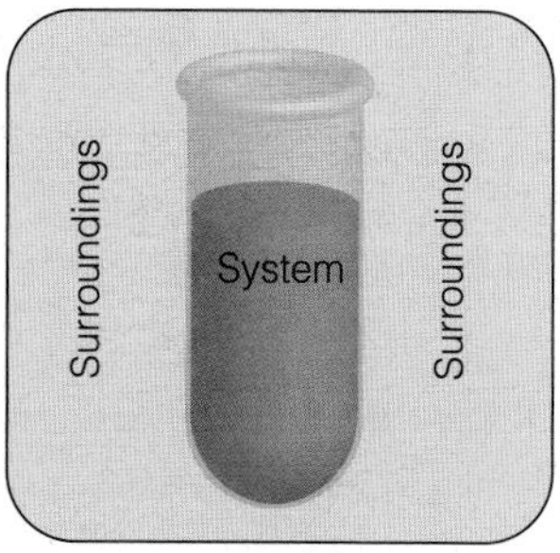

1.1 Systems and surroundings

We need to understand the unique and precise vocabulary of thermodynamics before applying it to the study of bioenergetics.

In thermodynamics, a **system** is the part of the world in which we have a special interest. The **surroundings** are where we make our observations (Fig. 1.1). The surroundings, which can be modeled as a large water bath, remain at constant temperature regardless of how much energy flows into or out of them. They are so huge that they also have either constant volume or constant pressure regardless of any changes that take place to the system. Thus, even though the system might expand, the surroundings remain effectively the same size.

We need to distinguish three types of system (Fig. 1.2):

An **open system** can exchange both energy and matter with its surroundings.

A **closed system** is a system that can exchange energy but not matter with its surroundings.

An **isolated system** is a system that can exchange neither matter nor energy with its surroundings.

An example of an open system is a flask that is not stoppered and to which various substances can be added. A biological cell is an open system because nutrients and waste can pass through the cell wall. You and I are open systems: we ingest, respire, perspire, and excrete. An example of a closed system is a stoppered flask: energy can be exchanged with the contents of the flask because the walls may be able to conduct heat. An example of an isolated system is a sealed flask that is thermally, mechanically, and electrically insulated from its surroundings.

Fig. 1.1 The sample is the system of interest; the rest of the world is its surroundings. The surroundings are where observations are made on the system. They can often be modeled by a large water bath. The universe consists of the system and its surroundings.

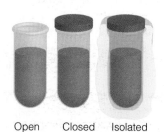

Open   Closed   Isolated

Fig. 1.2 A system is *open* if it can exchange energy and matter with its surroundings, *closed* if it can exchange energy but not matter, and *isolated* if it can exchange neither energy nor matter.

1.2 Work and heat

Organisms can be regarded as systems that exchange energy with their surroundings, and we need to understand how those transfers take place.

Energy can be exchanged between a closed system and its surroundings by doing work or by the process called 'heating'. A system does work when it causes motion against an opposing force. We can identify when a system does work by noting whether the process can be used to change the height of a weight somewhere in the surroundings. **Heating** is the process of transferring energy as a result of a temperature difference between the systems and its surroundings. To avoid a lot of awkward language, it is common to say that 'energy is transferred as work' when the system does work and 'energy is transferred as heat' when the system heats its surroundings (or vice versa). However, we should always remember that 'work' and 'heat' are *modes of transfer* of energy, not *forms* of energy.

(a) Exothermic and endothermic processes

Walls that permit heating as a mode of transfer of energy are called **diathermic** (Fig. 1.3). A metal container is diathermic and so is our skin or any biological membrane. Walls that do not permit heating even though there is a difference in temperature are called **adiabatic**.[1] The double walls of a vacuum flask are adiabatic to a good approximation.

A process in a system that transfers energy as heat to the surroundings (we commonly say 'releases heat into the surroundings') is called **exothermic**. A process in a system that absorbs energy as heat from the surroundings (we commonly say 'absorbs heat from the surroundings') is called **endothermic**. All combustions are exothermic. The reactions leading to the oxidative breakdown of nutrients in organisms are also exothermic. These reactions include oxidation of the carbohydrate glucose ($C_6H_{12}O_6$, Atlas S4) and of the fat tristearin ($C_{57}H_{110}O_6$):

$$C_6H_{12}O_6(s) + 6\,O_2(g) \rightarrow 6\,CO_2(g) + 6\,H_2O(l)$$
$$2\,C_{57}H_{110}O_6(s) + 163\,O_2(g) \rightarrow 114\,CO_2(g) + 110\,H_2O(l)$$

Endothermic reactions are much less common. The endothermic dissolution of ammonium nitrate in water is the basis of the instant cold packs that are included in some first-aid kits. They consist of a plastic envelope containing water dyed blue (for psychological reasons) and a small tube of ammonium nitrate, which is broken when the pack is to be used.

As an example of these terms, consider a chemical reaction that is a net producer of gas, such as the combustion of urea, $(NH_2)_2CO$, to yield carbon dioxide, water, and nitrogen:

$$(NH_2)_2CO(s) + \tfrac{3}{2}\,O_2(g) \rightarrow CO_2(g) + 2\,H_2O(l) + N_2(g)$$

Suppose first that the reaction takes place inside a cylinder with diathermic walls and fitted with a movable piston, then the gas produced drives out the piston and raises a weight in the surroundings (Fig. 1.4). In this case, energy has migrated to the surroundings as a result of the system doing work because a weight has been raised in the surroundings: that weight can now do more work, so it possesses more energy. Because the reaction is exothermic and walls are diathermic, some energy also migrates into the surroundings as heat. We can detect that transfer of energy by immersing the reaction vessel in an ice bath and noting how

---

[1] The word is derived from the Greek words for 'not passing through'.

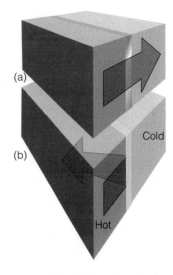

Fig. 1.3 (a) A diathermic wall permits the passage of energy as heat; (b) an adiabatic wall does not, even if there is a temperature difference across the wall.

Fig. 1.4 When urea reacts with oxygen, the gases produced (carbon dioxide and nitrogen) must push back the surrounding atmosphere (represented by the weight resting on the piston) and hence must do work on its surroundings. This is an example of energy leaving a system as work.

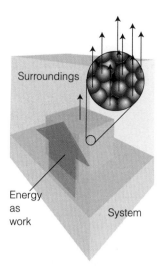

Fig. 1.5 Work is the transfer of energy that causes or utilizes uniform motion of atoms in the surroundings. For example, when a weight is raised, all the atoms of the weight (shown magnified) move in unison in the same direction.

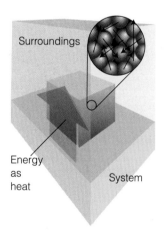

Fig. 1.6 Heat is the transfer of energy that causes or utilizes random motion in the surroundings. When energy leaves the system (the green region), it generates random motion in the surroundings (shown magnified).

much ice melts. Alternatively, we could let the same reaction take place in a diathermic vessel with a piston locked in position. No work is done because no weight is raised. However, because it is found that more ice melts than in the first experiment, we can conclude that more energy has migrated to the surroundings as heat.

(b) The molecular interpretation of work and heat

The clue to the molecular nature of work comes from thinking about the motion of a weight in terms of its component atoms. When a weight is raised, all its atoms move in the same direction. This observation suggests that *work is the transfer of energy that achieves or utilizes uniform motion in the surroundings* (Fig. 1.5). Whenever we think of work, we can always think of it in terms of uniform motion of some kind. Electrical work, for instance, corresponds to electrons being pushed in the same direction through a circuit. Mechanical work corresponds to atoms being pushed in the same direction against an opposing force.

Now consider the molecular nature of heating. When energy is transferred as heat to the surroundings, the atoms and molecules oscillate more rapidly around their positions or move from place to place more vigorously. The key point is that the motion stimulated by the arrival of energy from the system as heat is random, not uniform as in the case of doing work. This observation suggests that *heat is the mode of transfer of energy that achieves or utilizes random motion in the surroundings* (Fig. 1.6). A fuel burning, for example, generates random molecular motion in its vicinity.

An interesting historical point is that the molecular difference between work and heat correlates with the chronological order of their application. The release of energy when a fire burns is a relatively unsophisticated procedure because the energy emerges in a disordered fashion from the burning fuel. It was developed— stumbled upon—early in the history of civilization. The generation of work by a burning fuel, in contrast, relies on a carefully controlled transfer of energy so that vast numbers of molecules move in unison. Apart from Nature's achievement of work through the evolution of muscles, the large-scale transfer of energy by doing work was achieved thousands of years later than the liberation of energy by heating, for it had to await the development of the steam engine.

(c) The molecular interpretation of temperature

We are now also in a position to understand the molecular basis of temperature (a concept first introduced in *Fundamentals* F.3). To do so, we consider an isolated system composed of N molecules. Although the total energy is constant at E, it is not possible to be definite about how that energy is shared between the molecules. Collisions result in the ceaseless redistribution of energy not only between the molecules but also among their different modes of motion (translation, rotation, and vibration). The closest we can come to a description of the distribution of energy is to report the **population** of a state, the average number of molecules that occupy it, and to say that on average there are N_i molecules in a state of energy ε_i. The populations of the states remain almost constant, but the precise identities of the molecules in each state may change at every collision.

Any individual molecule may exist in states with energies $\varepsilon_0, \varepsilon_1, \ldots$. At any instant there are N_0 molecules in the state with energy ε_0 (the 'ground state'), N_1 with ε_1 (the 'first excited state'), and so on. The specification of the set of populations $N_0, N_1, \ldots$ in the form $\{N_0, N_1, \ldots\}$ is a statement of the 'instantaneous configuration' of the system. The instantaneous configuration fluctuates with time because the populations change. We can picture a large number of different instantaneous configurations of 100 molecules. One configuration, for example, might be {98,0,2,...}, corresponding to every molecule except two being in the ground state. Another of the same total energy might be {96,1,1,1,1,...}, in which four molecules occupy the first four excited states. The latter configuration is intrinsically more likely to be found than the former because it can be achieved in more ways: {98,0,2,...} can be achieved in 4950 different ways but {96,1,1,1,1,...} can be achieved in 94 109 400 different ways, which is over 19 000 times more ways. (These numbers are obtained by counting how many ways there are of selecting molecules at random from 100.) If, as a result of molecular jostling, the system were to fluctuate between the configurations {98,0,2,...} and {96,1,1,1,1,...}, it would almost always be found in the second, more likely, configuration. In other words, a system free to switch between the two configurations would show properties characteristic almost exclusively of the second configuration. It should be easy to believe that there may be other configurations that have a much greater likelihood of occurring than both.

When the statistics of the distributions are analyzed, with energy distributed purely at random subject to its total being fixed at a certain value E, one configuration can be obtained in so many ways that it overwhelms all the rest in importance to such an extent that the system will almost always be found in it. The properties of the system will therefore be characteristic of that particular dominating configuration. The ratio of populations that correspond to this dominating configuration turns out to be given by the Boltzmann distribution, which we introduced and illustrated in *Fundamentals* F.3:

$$\frac{N_2}{N_1} = e^{-(E_2-E_1)/kT} = e^{-\Delta E/kT}$$
    The Boltzmann distribution   (1.1)

where k is Boltzmann's constant ($k = 1.381 \times 10^{-23}$ J K^{-1}). We now see that the Boltzmann distribution, which is one of the most important concepts in the whole of physical chemistry, specifies the most probable distribution of molecules over their available energy levels subject only to the requirement that the total energy has a certain value. We also see that *the temperature is a parameter that characterizes that distribution*. A low temperature implies that only low-energy states are occupied; a high temperature indicates that high-energy states are also occupied. Zero temperature ($T = 0$) indicates that only the ground state is occupied. Infinite temperature ($T = \infty$) indicates that all available states are equally occupied. We can now begin to see that molecules are stable at low temperature because they occupy only low energy states; as the temperature is increased, they occupy more states of high energy and as a result can undergo reaction or, in the case of macromolecules, lose their secondary and higher levels of structure.

Case study 1.1   *Energy conversion in organisms*

Figure 1.7 outlines the main processes of **metabolism**, the collection of chemical reactions that trap, store, and utilize energy in biological cells. Most chemical reactions taking place in biological cells are either endothermic or exothermic, and cellular processes can continue only as long as there is a steady supply of energy to the cell. Furthermore, as we shall see in Section 1.6, only the *conversion* of the supplied energy from one form to another or its transfer from place to place is possible.

Fig. 1.7 Diagram demonstrating the flow of energy in living organisms. Arrows point in the direction in which energy flows. We focus only on the most common processes and do not include less ubiquitous ones, such as bioluminescence. (Adapted from D.A. Harris, *Bioenergetics at a glance*, Blackwell Science, Oxford (1995).)

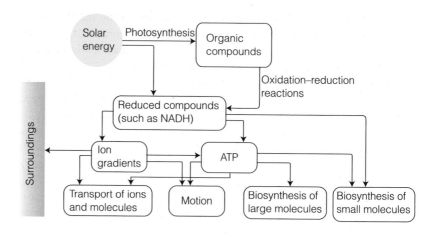

The primary source of energy that sustains the bulk of plant and animal life on Earth is the Sun.[2] We saw in the *Prologue* that energy from solar radiation is ultimately stored during photosynthesis in the form of organic molecules, such as carbohydrates, fats, and proteins, that are subsequently oxidized to meet the energy demands of organisms. **Catabolism** is the collection of reactions associated with the oxidation of nutrients in the cell and may be regarded as highly controlled reactions, with the energy liberated as work rather than as heat. Thus, even though the free combustion of a carbohydrate or fat to carbon dioxide and water is highly exothermic, in cells the equivalent oxidation is highly controlled and much of the energy is expended as useful work. The net outcome is the conversion of energy from controlled oxidation of nutrients to energy for doing work in the cell, including the transport of ions and neutral molecules (such as nutrients) across cell membranes, the physical motion of the organism (for example through the contraction of muscles), and **anabolism**, the biosynthesis of small and large molecules. The biosynthesis of DNA may be regarded as an anabolic process in which energy is converted ultimately to useful information, the genome of the organism.

Figure 1.7 also shows how organisms distribute the energy stored in a variety of ways. Because energy is extracted from organic compounds by oxidation, the initial energy carriers are reduced species, species that have gained electrons, such as NADH, the reduced form of nicotinamide adenine dinucleotide (NAD[+], Atlas N4). Light-induced electron transfer in photosynthesis also leads to the formation of reduced species, such as NADPH from NADP[+] (Atlas N5), the phosphorylated derivative of NAD[+]. The details of the reactions leading to the production of NADH and NADPH are discussed in Chapter 5. Oxidation–reduction reactions ('redox reactions') transfer energy out of NADH and other reduced species, storing it in the mobile carrier adenosine triphosphate, ATP (Atlas N3), and in ion gradients across membranes. As we shall see in Chapter 4, the essence of ATP's action is the loss of its terminal phosphate group in an energy-releasing reaction. Ion gradients arise from the movement of charged species across a membrane and we shall see in Chapter 5 how they store energy that can be used to drive biochemical processes and the synthesis of ATP.

---

[2] Some ecosystems near volcanic vents in the dark depths of the oceans do not use sunlight as their primary source of energy.

Living organisms are not perfectly efficient machines, for not all the energy available from the Sun and oxidation of organic compounds is used to perform work as some is lost as heat. The dissipation of energy as heat is advantageous because it can be used to control the organism's temperature. However, energy is eventually transferred as heat to the surroundings. In Chapter 2 we shall explore the origin of the incomplete conversion of energy supplied by heating into energy that can be used to do work, a feature that turns out to be common to all energy conversion processes.

1.3 The measurement of work

In bioenergetics, the most useful outcome of the breakdown of nutrients during metabolism is work, so we need to know how work is measured.

We saw in Section F.3 that if the force is the gravitational attraction of the Earth on a mass m, then the force opposing raising the mass vertically is mg, where g is the acceleration of free fall (9.81 m s^{-2}). Therefore, the work needed to raise the mass through a height h on the surface of the Earth is

$$\text{work} = mgh \qquad \boxed{\text{Work of raising a weight}} \quad (1.2)$$

It follows that we have a simple way of measuring the work done by or on a system: we measure the height through which a weight is raised or lowered in the surroundings and then use eqn 1.2.

A brief illustration

Nutrients in the soil are absorbed by the root system of a tree and then rise to reach the leaves through a complex vascular system in its trunk and branches. From eqn 1.2, the work required to raise 10 g of liquid water (corresponding to a volume of about 10 mL) through the trunk of a 20-m tree from its roots to its topmost leaves is

$$\text{work} = (1.0 \times 10^{-2}\,\text{kg}) \times (9.81\,\text{m s}^{-2}) \times (20\,\text{m}) = 2.0\,\text{kg m}^2\,\text{s}^{-2} = 2.0\,\text{J}$$

It should be easy for you to show that this quantity of work is equivalent to the work of raising a book like this one (of mass about 1.0 kg) through a vertical distance of 20 cm (0.20 m).

A note on good practice
Whenever possible, find a relevant derived unit that corresponds to the collection of base units in a result. We used 1 kg m^2 s^{-2} = 1 J, hence verifying that the answer has units of energy.

(a) Sign conventions

So far, we have referred only to the magnitude of the work done; now we need to consider its sign. When a system does work, such as by raising a weight in the surroundings or forcing the movement of an ion across a biological membrane, the energy transferred as work, w, is reported as a negative quantity. For instance, if a system raises a weight in the surroundings and in the process does 100 J of work (that is, 100 J of energy leaves the system by doing work), then we write $w = -100$ J. When work is done on the system—for example, when we stretch a muscle from its relaxed position—w is reported as a positive quantity. We write $w = +100$ J to signify that 100 J of work has been done on the system (that is, 100 J of energy has been transferred to the system by doing work). The sign convention is easy to follow if we think of changes to the energy of the system: its energy

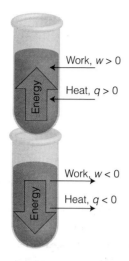

Work, $w > 0$

Heat, $q > 0$

Energy

Work, $w < 0$

Heat, $q < 0$

Energy

Fig. 1.8 The sign convention in thermodynamics: w and q are positive if energy enters the system (as work and heat, respectively) but negative if energy leaves the system.

decreases (w is negative) if energy leaves it as work and its energy increases (w is positive) if energy enters it as work (Fig. 1.8).

(b) Expansion work

To see how energy flow as work can be determined experimentally, we deal first with **expansion work**, the work done when a system expands against an opposing pressure. In bioenergetics we are not generally concerned with expansion work, which can occur as a result of gas-producing or gas-consuming chemical reactions, but rather with the work of making and moving molecules in the cell, muscle contraction, or cell division. However, even though we might not be explicitly interested in it, expansion work is done in any chemical reaction that involves gases, such as the oxidation of fuels and photosynthesis, and for a proper analysis of energy resources it must be taken into account. We shall see that that can be done automatically in the following section, which will build on the material developed here.

Consider the combustion of urea illustrated in Fig. 1.4 as an example of a reaction in which expansion work is done in the process of making room for the gaseous products, carbon dioxide and nitrogen in this case. We show in the following *Justification* that when a system expands through a volume ΔV against a constant external pressure p_{ex}, the work done is

$$w = -p_{ex}\Delta V$$

Work of expansion against a constant pressure (1.3)

Justification 1.1 *Expansion work*

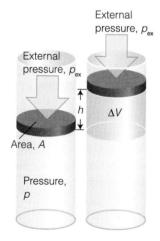

External pressure, p_{ex}

External pressure, p_{ex}

h

ΔV

Area, A

Pressure, p

Fig. 1.9 When a piston of area A moves out through a distance h, it sweeps out a volume $\Delta V = Ah$. The external pressure p_{ex} opposes the expansion with a force $p_{ex}A$.

To calculate the work done when a system expands from an initial volume V_i to a final volume V_f, a change $\Delta V = V_f - V_i$, we consider a piston of area A moving out through a distance h (Fig. 1.9). There need not be an actual piston: we can think of the piston as representing the boundary between the expanding gas and the surrounding atmosphere. However, there may be an actual piston, such as when the expansion takes place inside an internal combustion engine.

The force opposing the expansion is the constant external pressure p_{ex} multiplied by the area of the piston (because force is pressure times area; *Fundamentals* F.2). The work done is therefore

work done by the system = distance (h) × opposing force ($p_{ex}A$)

$$= h \times p_{ex}A = p_{ex} \times (hA) = p_{ex} \times \Delta V$$

The last equality follows from the fact that hA is the volume of the cylinder swept out by the piston as the gas expands, so we can write $hA = \Delta V$. That is, for expansion work,

work done by the system = $p_{ex}\Delta V$

Now consider the sign. A system does work and thereby loses energy (that is, w is negative) when it expands (when ΔV is positive). Therefore, we need a negative sign in the equation to ensure that w is negative when ΔV is positive, so we obtain eqn 1.3.

According to eqn 1.3, the *external* pressure determines how much work a system does when it expands through a given volume: the greater the external pressure, the greater the opposing force and the greater the work that a system does.

When the external pressure is zero, $w = 0$. In this case, the system does no work as it expands because it has nothing to push against. Expansion against zero external pressure is called **free expansion**.

A brief illustration

Exhalation of air during breathing requires work because air must be pushed out from the lungs against atmospheric pressure. Consider the work of exhaling 0.50 dm³ (5.0×10^{-4} m³) of air, a typical value for a healthy adult, through a tube into the bottom of the apparatus shown in Fig. 1.9 and against an atmospheric pressure of 1.00 atm (101 kPa). The exhaled air lifts the piston so the change in volume is $\Delta V = 5.0 \times 10^{-4}$ m³ and the external pressure is $p_{ex} = 101$ kPa. From eqn 1.3 the work of exhaling is

$$w = -p_{ex}\Delta V = -(1.01 \times 10^5 \text{ Pa}) \times (5.0 \times 10^{-4} \text{ m}^3) = -51 \text{ Pa m}^3 = -51 \text{ J}$$

where we have used the relation 1 Pa m³ = 1 J. That value (51 J) might not seem much, but you should use eqn 1.2 to show that −51 J is approximately the same as the work of lifting seven books like this one (a total of 7.0 kg) from the ground to the top of a standard desk (a vertical distance of 0.75 m).

A note on good practice
Always keep track of signs by considering whether stored energy has left the system as work (w is then negative) or has entered it (w is then positive).

Self-test 1.1   Calculate the work done by a system in which a reaction results in the formation of 1.0 mol CO_2(g) at 25°C and 100 kPa. (*Hint*: The increase in volume will be 25 dm³ under these conditions if the gas is treated as perfect; use the relation 1 Pa m³ = 1 J.)

Answer: $w = -2.5$ kJ

(c) Maximum work

Equation 1.3 can be used to show us how to get the *least* expansion work from a system: we just reduce the external pressure—which provides the opposing force—to zero. But how can we achieve the *greatest* work for a given change in volume? According to eqn 1.3, the system does maximum work when the external pressure has its maximum value. The force opposing the expansion is then the greatest and the system must exert most effort to push the piston out. However, that external pressure cannot be greater than the pressure, p, of the gas inside the system, for otherwise the external pressure would compress the gas instead of allowing it to expand. Therefore, *maximum work is obtained when the external pressure is only infinitesimally less than the pressure of the gas in the system*. In effect, the two pressures must be adjusted to be the same at all stages of the expansion: the external pressure must be progressively reduced so that it remains only infinitesimally lower than the pressure of the gas at each stage. As we remarked in *Fundamentals* F.2, this balance of pressures corresponds to a state of mechanical equilibrium. Therefore, we can conclude that *a system that remains in mechanical equilibrium with its surroundings at all stages of the expansion does maximum expansion work*.

There is another way of expressing this condition. Because the external pressure is infinitesimally less than the pressure of the gas at some stage of the expansion, the piston moves out. However, suppose we increased the external pressure so that at that stage of the expansion it became infinitesimally greater than the pressure of the gas; now the piston moves in. That is, *when a system is in a state of mechanical equilibrium, an infinitesimal change in the pressure results in opposite directions of change.* A change that can be reversed by an *infinitesimal* change in a variable—in this case, the pressure—is said to be **reversible**. In everyday life 'reversible' means a process that can be reversed; in thermodynamics it has a stronger meaning—it means that a process can be reversed by an *infinitesimal* modification in some variable (such as the pressure).

We can summarize this discussion in the following remarks:

1) A system does maximum expansion work when the external pressure is equal to that of the system at every stage of the expansion ($p_{ex} = p$).

2) A system does maximum expansion work when it is in mechanical equilibrium with its surroundings at every stage of the expansion.

3) Maximum expansion work is achieved in a reversible change.

All three statements are equivalent, but they reflect different degrees of sophistication in the way the point is expressed. The last statement is particularly important in our discussion of bioenergetics, especially when we consider how the reactions of catabolism drive anabolic processes. The arguments we have developed lead to the conclusion that maximum work (whether it is expansion work or some other type of work) is done if the cellular process is reversible.

1.4 The measurement of heat

A thermodynamic assessment of energy output during metabolic processes requires knowledge of ways to measure the energy transferred as heat.

We use the same sign convention for energy transferred by heating, q, as we do for work. Thus, we write $q = -100$ J if 100 J of energy leaves the system by heating its surroundings, so reducing the energy of the system, and $q = +100$ J if 100 J of energy enters the system when it is heated by the surroundings.

In certain cases, we can relate the value of q to the change in volume of a system and so can calculate, for instance, the flow of energy as heat into the system when a gas expands. The simplest case is that of a perfect gas undergoing isothermal expansion. Because the expansion is isothermal, the temperature of the gas is the same at the end of the expansion as it was initially, therefore the mean speed of the molecules of the gas is the same before and after the expansion. That implies in turn that the total kinetic energy of the molecules is the same. But for a perfect gas, the *only* contribution to the energy is the kinetic energy of the molecules, so we have to conclude that the *total* energy of the gas is the same before and after the expansion. Energy has left the system as work, therefore a compensating amount of energy must have entered the system as heat. We can therefore write

$$q = -w \qquad \boxed{\text{Isothermal expansion of a perfect gas}} \qquad (1.4)$$

A brief illustration

If we find that $w = -100$ J for a particular expansion (meaning that 100 J has left the system as a result of the system doing work), then we can conclude that $q = +100$ J (that is, 100 J must enter as heat). For free expansion, $w = 0$, so we conclude that $q = 0$ too: there is no influx of energy as heat when a perfect gas expands against zero pressure.

(a) Heat capacity

When a substance is heated, its temperature typically rises.[3] However, the change in temperature, ΔT, depends on the 'heat capacity' of the substance. The **heat capacity**, C, is defined as

$$C = \frac{q}{\Delta T}$$ Definition of the heat capacity   (1.5a)

where the temperature change may be expressed in kelvins (ΔT) or degrees Celsius ($\Delta \theta$); because the size of a kelvin is the same as that of a degree Celsius, the same numerical value is obtained but with the units joules per kelvin (J K^{-1}) and joules per degree Celsius (J °C^{-1}), respectively. It follows that we have a simple way of measuring the energy absorbed or released by a system as heat: we measure a temperature change and then use the appropriate value of the heat capacity and eqn 1.5a rearranged into

$$q = C\Delta T$$ (1.5b)

> **A brief illustration**
>
> If the heat capacity of a beaker of water is 0.50 kJ K^{-1} and we observe a temperature rise of 4.0 K, then we can infer that the heat transferred to the water is
>
> $$q = (0.50\ \text{kJ K}^{-1}) \times (4.0\ \text{K}) = +2.0\ \text{kJ}$$

Heat capacities occur in many places in the following sections and chapters, and we need to be aware of their properties and how their values are reported. First, we note that the heat capacity is an extensive property: 2 kg of iron has twice the heat capacity of 1 kg of iron, so twice as much heat is required to change its temperature to the same extent. It is more convenient to report the heat capacity of a substance as an intensive property. We therefore use either the **specific heat capacity**, C_s, the heat capacity divided by the mass of the sample ($C_s = C/m$, in joules per kelvin per gram, J K^{-1} g^{-1}), or the **molar heat capacity**, C_m, the heat capacity divided by the amount of substance ($C_m = C/n$, in joules per kelvin per mole, J K^{-1} mol^{-1}). In common usage, the specific heat capacity is often called simply the *specific heat*.

For reasons that will be explained shortly, the heat capacity of a substance depends on whether the sample is maintained at constant volume (like a gas in a sealed vessel) as it is heated or whether the sample is maintained at constant pressure (like water in an open container) and free to change its volume. The latter is a more common arrangement, and the values given in Table 1.1 are for the **heat capacity at constant pressure**, C_p. The **heat capacity at constant volume** is denoted C_V.

> **A brief illustration**
>
> The high heat capacity of water is ecologically advantageous because it stabilizes the temperatures of lakes and oceans: a large quantity of energy must be

Table 1.1 Heat capacities of selected substances*

Substance	Molar heat capacity, $C_{p,m}$/ (J K^{-1} mol^{-1})
Air	29
Benzene, C_6H_6(l)	136.1
Ethanol, C_2H_5OH(l)	111.46
Glycine, $CH_2(NH_2)COOH$(s)	99.2
Oxalic acid, $(COOH)_2$(s)	117
Urea, $CO(NH_2)_2$(s)	93.14
Water, H_2O(s)	37
H_2O(l)	75.29
H_2O(g)	33.58

*For additional values, see the *Resource section*.

---

[3] We say 'typically' because the temperature does not always rise. The temperature of boiling water, for instance, remains unchanged as it is heated (see Chapter 3).

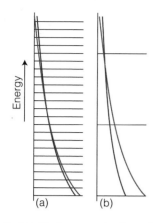

Fig. 1.10 The heat capacity depends on the availability of levels as explained in the text. In each case the blue line is the distribution at low temperature and the red line that at higher temperature.

lost or gained before there is a significant change in temperature. The molar heat capacity of liquid water at constant pressure, $C_{p,m}$, is 75 J K^{-1} mol^{-1}. It follows that the increase in temperature of 100 g of water (5.55 mol H$_2$O) when 1.0 kJ of energy is supplied by heating a sample free to expand is approximately

$$\Delta T = \frac{q}{C_p} = \frac{q}{nC_{p,m}} = \frac{1.0 \times 10^3 \, \text{J}}{(5.55 \, \text{mol}) \times (75 \, \text{J K}^{-1} \, \text{mol}^{-1})} = +2.4 \, \text{K}$$

(b) The molecular interpretation of heat capacity

The molecular reason why different substances have different molar heat capacities can be traced to differences in the separations of their energy levels. As remarked in *Fundamentals* F.3, molecules can exist with only certain energies. When the available energy levels are close together, a given quantity of energy arriving as heat can be accommodated with little adjustment of the populations and hence with little modification of the temperature that occurs in the Boltzmann distribution and specifies the distribution of populations. The relative insensitivity of temperature to the arrival of energy corresponds to a high heat capacity (Fig. 1.10). When the energy levels are widely separated, the arriving energy must be accommodated by making use of the high energy levels with a consequent greater 'reach' of the Boltzmann distribution and hence a greater modification of the temperature. That is, widely spaced energy levels correlate with a low heat capacity. The translational energy levels of molecules in a gas are very close together, and all monatomic gases have similar molar heat capacities. The separation of the vibrational energies of atoms bound together in solids depends on the stiffness of the bonds between them and on the masses of the atoms. As we shall see in Sections 9.6 and 12.3, the stronger the bond and the lighter the atoms in a bond, the larger is the separation between vibrational energy levels. As a result, solids show a wide range of molar heat capacities. Biological macromolecules have large numbers of atoms and can vibrate in many different ways. Many of these ways correspond to the collective motion of many atoms, so the vibrational energies are spaced closely. Hence, heat capacities of biological macromolecules may be large.

Water, as so often, is anomalous. It is a small, rigid molecule but has a high heat capacity. Once again, the anomaly can be traced to hydrogen bonds in the liquid. These bonds link many molecules together into clusters that vibrate in numerous ways. Consequently, the vibrational energies are close together, and the heat capacity of water is larger than expected for a substance consisting of small molecules interacting weakly.

Internal energy and enthalpy

Heat and work are *equivalent* ways of transferring energy into or out of a system in the sense that once the energy is inside, it is stored simply as 'energy', regardless of how the energy was supplied, as work or as heat, it can be released in either form. The experimental evidence for this **equivalence of heat and work** goes all the way back to the experiments done by James Joule in the nineteenth century, who in effect showed that the same rise in temperature of a sample of water is brought about by transferring a given quantity of energy either as heat or as work.

1.5 The internal energy

To understand how biological processes can store and release energy, we need to be familiar with a very important law that relates work and heat to changes in the energy of all the constituents of a system.

We need some way of keeping track of the energy changes in a system. This is the job of the property called the **internal energy**, U, of the system, the sum of all the kinetic and potential contributions to the energy of all the atoms, ions, and molecules in the system. The internal energy is the grand total energy of the system with a value that depends on the temperature and, in general, the pressure. It is an extensive property because 2 kg of iron at a given temperature and pressure, for instance, has twice the internal energy of 1 kg of iron under the same conditions. The **molar internal energy**, $U_m = U/n$, the internal energy per mole of atoms or molecules, is an intensive property.

(a) Changes in internal energy

In practice, we do not know and cannot measure the absolute value of the total energy of a sample because it includes the kinetic and potential energies of all the electrons and all the components of the atomic nuclei. Nevertheless, there is no problem with dealing with the *changes* in internal energy, ΔU, because we can determine those changes by monitoring the energy supplied or lost as heat or as work. All practical applications of thermodynamics deal with ΔU, not with U itself. A change in internal energy is written

$$\Delta U = w + q \qquad \boxed{\text{Change in internal energy in terms of heat and work}} \qquad (1.6)$$

where w is the energy transferred to the system by doing work and q is the energy transferred to it by heating. The internal energy is an accounting device, like a country's gold reserves, which are used for monitoring transactions with the outside world (the surroundings) involving either currency (heat or work).

We have seen that a feature of a perfect gas is that for any *isothermal* expansion the total energy of the sample remains the same and that $q = -w$. That is, any energy lost as work is restored by an influx of energy as heat. We can express this property in terms of the internal energy, for it implies that the internal energy remains constant when a perfect gas expands isothermally: from eqn 1.6 we can write

$$\Delta U = 0 \qquad \boxed{\text{Change of internal energy during isothermal expansion of a perfect gas}} \qquad (1.7)$$

In other words, *the internal energy of a sample of perfect gas at a given temperature is independent of the volume it occupies.* We can understand this independence by realizing that when a perfect gas expands isothermally, the only feature that changes is the average distance between the molecules; their average speed and therefore total kinetic energy remains the same. However, as there are no intermolecular interactions, the total energy is independent of the average separation, so the internal energy is unchanged by expansion.

The definition of ΔU in terms of w and q points to a very simple method for measuring the change in internal energy of a system when a reaction takes place. We have seen already that the work done by a system when it pushes against a fixed external pressure is proportional to the change in volume. Therefore, if we carry out a reaction in a container of constant volume, the system can do no expansion work and provided it can do no other kind of work (so-called non-expansion work, such as electrical work), we can set $w = 0$. Then eqn 1.6 simplifies to

$$\Delta U = q_V \qquad \boxed{\text{Constant volume, no non-expansion work}} \quad (1.8)$$

The subscript V signifies that the volume of the system is constant. An example of a system that can be approximated as a constant-volume container is an individual biological cell.

Example 1.1 *Calculating the change in internal energy*

Nutritionists are interested in the use of energy by the human body, and we can consider our own body as a thermodynamic 'system'. Suppose in the course of an experiment you do 622 kJ of work on an exercise bicycle and lose 82 kJ of energy as heat. What is the change in your internal energy? Disregard any matter loss by perspiration.

Strategy This example is an exercise in keeping track of signs correctly. When energy is lost from the system, w or q is negative. When energy is gained by the system, w or q is positive.

Solution To take note of the signs, we write $w = -622$ kJ (622 kJ is lost by doing work) and $q = -82$ kJ (82 kJ is lost by heating the surroundings). Then eqn 1.6 gives us

$$\Delta U = w + q = (-622 \text{ kJ}) + (-82 \text{ kJ}) = -704 \text{ kJ}$$

We see that your internal energy falls by 704 kJ. Later, that energy will be restored by eating.

Self-test 1.2 An electric battery is charged by supplying 250 kJ of energy to it as electrical work (by driving an electric current through it), but in the process it loses 25 kJ of energy as heat to the surroundings. What is the change in internal energy of the battery?

Answer: +225 kJ

A note on good practice
Always attach the correct signs: use a positive sign when there is a flow of energy into the system and a negative sign when there is a flow of energy out of the system. Also, the quantity ΔU always carries a sign explicitly, even if it is positive: we never write $\Delta U = 20$ kJ, for instance, but always +20 kJ.

We can use eqn 1.8 to obtain more insight into the heat capacity of a substance. The definition of heat capacity is given in eqn 1.5 ($C = q/\Delta T$). At constant volume, q may be replaced by the change in internal energy of the substance, so

$$C_V = \frac{\Delta U}{\Delta T} \qquad \boxed{\text{Definition of the constant-volume heat capacity}} \quad (1.9a)$$

The expression on the right is the slope of the graph of internal energy plotted against temperature, with the volume of the system held constant, so C_V tells us how the internal energy of a constant-volume system varies with temperature. If, as is generally the case, the graph of internal energy against temperature is not a straight line, we interpret C_V as the slope of the tangent to the curve at the temperature of interest (Fig. 1.11). That is, the constant-volume heat capacity is the derivative of the function U with respect to the variable T at a specified volume (see *Mathematical toolkit* 1.1):

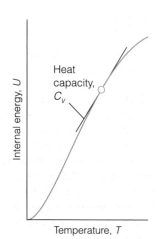

Fig. 1.11 The constant-volume heat capacity is the slope of a curve showing how the internal energy varies with temperature. The slope, and therefore the heat capacity, may be different at different temperatures.

Mathematical toolkit 1.1 *Differentiation*

Consider a function f with values $f(x)$ and $f(x+\delta x)$ at x and $x+\delta x$, respectively. The slope of this function at x is obtained by letting δx become zero, which we write as $\lim_{\delta x \to 0}$. In this limit, the δ is replaced by a d, and we write

$$\text{Slope at } x = \frac{df}{dx} = \lim_{\delta x \to 0} \frac{f(x+\delta x) - f(x)}{\delta x}$$

To work out the slope of any function, we develop the expression on the right: this process is called **differentiation**. The slope is called the **derivative** of the function. Examples of derivatives include:

$$\frac{dx^n}{dx} = nx^{n-1} \qquad \frac{de^{ax}}{dx} = ae^{ax} \qquad \frac{d\ln ax}{dx} = \frac{a}{x}$$

Most of the functions encountered in chemistry can be differentiated by using the following rules:

Rule 1. For two functions f and g:

$$\frac{d}{dx}(f+g) = \frac{df}{dx} + \frac{dg}{dx}$$

Rule 2 (the product rule). For two functions f and g:

$$\frac{d}{dx}fg = f\frac{dg}{dx} + g\frac{df}{dx}$$

Rule 3 (the quotient rule). For two functions f and g:

$$\frac{d}{dx}\frac{f}{g} = \frac{1}{g}\frac{df}{dx} - \frac{f}{g^2}\frac{dg}{dx}$$

Rule 4 (the chain rule). For a function $f = f(g)$, where $g = g(x)$,

$$\frac{df}{dx} = \frac{df}{dg}\frac{dg}{dx}$$

In the last rule, $f(g)$ is a 'function of a function', as in $\ln(1+x^2)$.

$$C_V = \frac{dU}{dT}$$

Definition of the constant-volume heat capacity   (1.9b)

(b) The internal energy as a state function

An important characteristic of the internal energy is that it is a **state function**, a physical property that depends only on the present state of the system and is independent of the path by which that state was reached. If we were to change the temperature of the system, then change the pressure, then adjust the temperature and pressure back to their original values, the internal energy would return to its original value too. A state function is very much like altitude: each point on the surface of the Earth can be specified by quoting its latitude and longitude, and (on land areas, at least) there is a unique property, the altitude, that has a fixed value at that point. In thermodynamics, the role of latitude and longitude is played by the pressure and temperature (and any other variables needed to specify the state of the system), and the internal energy plays the role of the altitude, with a single, fixed value for each state of the system.

The fact that U is a state function implies that *a change, ΔU, in the internal energy between two states of a system, is independent of the path between them* (Fig. 1.13). Once again, the altitude is a helpful analogy. If we climb a mountain between two fixed points, we make the same change in altitude regardless of the path we take between the two points. Likewise, if we compress a sample of gas until it reaches a certain pressure and then cool it to a certain temperature, the change in internal energy has a particular value. If, on the other hand, we changed the temperature and then the pressure but ensured that the two final values were the same as in the first experiment, then the overall change in internal energy would be exactly the same as before. This path independence of the value of ΔU is of the greatest importance in chemistry and for the study of bioenergetics, as we shall soon see.

A brief comment
More precisely, the constant-volume heat capacity is the *partial derivative* of the function U with respect to the variable T, denoted as

$$C_V = \left(\frac{\partial U}{\partial T}\right)_V$$

with the symbol ∂ replacing the symbol d, and the subscript V denoting that the variable V is held constant. Generally, a partial derivative of a function of more than one variable is the slope of the function with respect to one of the variables, all the other variables being held constant (Fig. 1.12). For more detail about partial derivatives, see *Mathematical toolkit 8.1*.

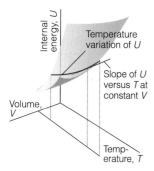

Fig. 1.12 The internal energy of a system varies with volume and temperature, perhaps as shown here by the surface. The variation of the internal energy with temperature at one particular constant volume is illustrated by the curve drawn parallel to *T*. The slope of this curve at any point is the partial derivative $(\partial U/\partial T)_V$.

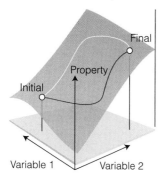

Fig. 1.13 The curved sheet shows how a property (for example, the altitude) changes as two variables (for example, latitude and longitude) are changed. The altitude is a state property because it depends only on the current state of the system. The change in the value of a state property is independent of the path between the two states. For example, the difference in altitude between the initial and final states shown in the diagram is the same whatever path (as depicted by the red and white lines) is used to travel between them.

(c) The First Law of thermodynamics

Suppose we now consider an isolated system. Because an isolated system can neither do work nor heat the surroundings, it follows that its internal energy cannot change. That is,

The internal energy of an isolated system is constant. `The First Law`

This statement is the **First Law of thermodynamics**. It is closely related to the law of conservation of energy but allows for transaction of energy by heating as well as by doing work. Unlike thermodynamics, mechanics does not deal with the concept of heat.

The experimental evidence for the First Law is the impossibility of making a 'perpetual motion machine', a device for producing work without consuming fuel. As we have already remarked, try as people might, they have never succeeded. No device has ever been made that creates internal energy to replace the energy drawn off as work. We cannot extract energy as work, leave the system isolated for some time, and hope that when we return the internal energy will have become restored to its original value. The same is true of organisms: energy required for the sustenance of life must be supplied continually in the form of food as work of motion, metabolism, and catabolism is done by the organism.

1.6 The enthalpy

Most biological processes take place in vessels that are open to the atmosphere and subjected to constant pressure and not maintained at constant volume, so we need to learn how to treat quantitatively the energy exchanges that take place at constant pressure.

In general, when a change takes place in a system open to the atmosphere, the volume of the system changes. For example, the thermal decomposition of 1.0 mol $CaCO_3(s)$ at 1 bar results in an increase in volume of 89 dm^3 at 800°C on account of the carbon dioxide gas produced. To create this large volume for the carbon dioxide to occupy, the surrounding atmosphere must be pushed back. That is, the system must perform expansion work. Therefore, although a certain quantity of heat may be supplied to bring about the endothermic decomposition, the increase in internal energy of the system is not equal to the energy supplied as heat because some energy has been used to do work of expansion (Fig. 1.14). In other words, because the volume has increased, some of the heat supplied to the system has leaked back into the surroundings as work.

Another example is the oxidation of a fat, such as tristearin, to carbon dioxide in the body. The overall reaction is

$$2\,C_{57}H_{110}O_6(s) + 163\,O_2(g) \rightarrow 114\,CO_2(g) + 110\,H_2O(l)$$

In this exothermic reaction there is a net *decrease* in volume equivalent to the elimination of (163 − 114) mol = 49 mol of gas molecules for every 2 mol of tristearin molecules that reacts. The decrease in volume at 25°C is about 600 cm^3 for the consumption of 1 g of fat. Because the volume of the system decreases, the atmosphere does work *on* the system as the reaction proceeds. That is, energy is transferred to the system as it contracts.[4] For this reaction, the decrease in the internal energy of the system is less than the energy released as heat because some energy has been restored by doing work.

---

[4] In effect, a weight has been lowered in the surroundings, so the surroundings can do less work after the reaction has occurred. Some of their energy has been transferred into the system.

(a) The definition of enthalpy

We can avoid the complication of having to take into account the work of expansion by introducing a new property that will be at the centre of our attention throughout the rest of the chapter and will recur throughout the book. The **enthalpy**, H, of a system is defined as

$$H = U + pV \qquad \text{Definition of the enthalpy} \qquad (1.10)$$

That is, the enthalpy differs from the internal energy by the addition of the product of the pressure, p, and the volume, V, of the system. This expression applies to *any* system or individual substance: don't be misled by the pV term into thinking that eqn 1.10 applies only to a perfect gas. A change in enthalpy (the only quantity we can measure in practice) arises from a change in the internal energy and a change in the product pV:

$$\Delta H = \Delta U + \Delta(pV) \qquad (1.11a)$$

where $\Delta(pV) = p_f V_f - p_i V_i$. If the change takes place at constant pressure p, the second term on the right simplifies to

$$\Delta(pV) = pV_f - pV_i = p(V_f - V_i) = p\Delta V$$

and we can write

$$\Delta H = \Delta U + p\Delta V \qquad \text{Enthalpy change at constant pressure} \qquad (1.11b)$$

We shall often make use of this important relation for processes occurring at constant pressure, such as chemical reactions taking place in containers open to the atmosphere.

Enthalpy is an extensive property. The **molar enthalpy**, $H_m = H/n$, of a substance, an intensive property, differs from the molar internal energy by an amount proportional to the molar volume, V_m, of the substance:

$$H_m = U_m + pV_m \qquad \text{Definition of the molar enthalpy} \qquad (1.12a)$$

This relation is valid for all substances. For a perfect gas we can go on to write $pV_m = RT$ and obtain

$$H_m = U_m + RT \qquad \text{Molar enthalpy of a perfect gas} \qquad (1.12b)$$

At 25°C, $RT = 2.5$ kJ mol^{-1}, so the molar enthalpy of a perfect gas differs from its molar internal energy by 2.5 kJ mol^{-1}. Because the molar volume of a solid or liquid is typically about 1000 times less than that of a gas, we can also conclude that the molar enthalpy of a solid or liquid is only about 2.5 J mol^{-1} (note: joules, not kilojoules) more than its molar internal energy, so the numerical difference is negligible. However, the conceptual importance is considerable, as we shall see.

(b) Changes in enthalpy

Although the enthalpy and internal energy of a sample may have similar values, the introduction of the enthalpy has very important consequences in thermodynamics. First, notice that because H is defined in terms of state functions (U, p, and V), *the enthalpy is a state function*. The implication is that the change in enthalpy, H, when a system changes from one state to another is independent of

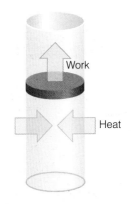

Fig. 1.14 The change in internal energy of a system that is free to expand or contract is not equal to the energy supplied by heating because some energy may escape back into the surroundings as work. However, the change in enthalpy of the system under these conditions *is* equal to the energy supplied by heating.

the path between the two states. Second, we show in the following *Justification* that the change in enthalpy of a system can be identified with the heat transferred to it at constant pressure:

$$\Delta H = q_p$$

Enthalpy change at constant pressure, no non-expansion work (1.13)

with the subscript p signifying that the pressure is held constant. Therefore, by imposing the constraint of constant pressure, we have identified an observable quantity (the energy transferred as heat) with a change in a state function, the enthalpy. Dealing with state functions greatly extends the power of thermo-dynamic arguments, because we don't have to worry about how we get from one state to another: all that matters is the initial and final states. For the particular case of the combustion of tristearin mentioned at the beginning of the section, in which 90 kJ of energy is released as heat at constant pressure, we would write $\Delta H = -90$ kJ regardless of how much expansion work is done.

Justification 1.2 *Heat transfers at constant pressure*

This *Justification* fulfils our promise that the calculation of expansion work can be done silently and automatically in the background and need not be done explicitly.

Consider a system open to the atmosphere, so that its pressure p is constant and equal to the external pressure p_{ex}. From eqn 1.11 we can write

$$\Delta H = \Delta U + p\Delta V = \Delta U + p_{ex}\Delta V$$

However, we know that the change in internal energy is given by eqn 1.6 ($\Delta U = w + q$) with $w = -p_{ex}\Delta V$ (provided the system does no other kind of work). When we substitute that expression into this one we obtain

$$\Delta H = (-p_{ex}\Delta V + q) + p_{ex}\Delta V = q$$

which is eqn 1.13.

An endothermic reaction ($q > 0$) taking place at constant pressure results in an increase in enthalpy ($\Delta H > 0$) because energy enters the system as heat. On the other hand, an exothermic process ($q < 0$) taking place at constant pressure corresponds to a decrease in enthalpy ($\Delta H < 0$) because energy leaves the system as heat. In summary:

exothermic process	endothermic process
$\Delta H < 0$	$\Delta H > 0$

Because all combustion reactions, including the controlled 'combustions' that contribute to respiration, are exothermic, they are accompanied by a decrease in enthalpy. These relations are consistent with the name *enthalpy*, which is derived from the Greek words meaning 'heat inside': the 'heat inside' the system is increased if the process is endothermic and absorbs energy as heat from the surroundings; it is decreased if the process is exothermic and releases energy as heat into the surroundings.[5]

---

[5] But heat does not actually 'exist' inside: only energy exists in a system; heat is a means of recover-ing that energy or increasing it. Heat is energy in transit, not a form in which energy is stored.

(c) The temperature dependence of the enthalpy

We have seen that the internal energy of a system rises as the temperature is increased. The same is true of the enthalpy, which also rises when the temperature is increased (Fig. 1.15). For example, the enthalpy of 100 g of water is greater at 80°C than at 20°C. We can measure the change by monitoring the energy that we must supply as heat to raise the temperature through 60°C when the sample is open to the atmosphere (or subjected to some other constant pressure); it is found that $\Delta H \approx +25$ kJ in this instance.

Just as we saw that the constant-volume heat capacity tells us about the temperature-dependence of the internal energy at constant volume, so the constant-pressure heat capacity tells us how the enthalpy of a system changes as its temperature is raised at constant pressure. To derive the relation, we combine the definition of heat capacity in eqn 1.5 ($C = q/\Delta T$) with eqn 1.13 and obtain

$$C_p = \frac{\Delta H}{\Delta T}$$

Definition of constant-pressure heat capacity   (1.14a)

That is, the constant-pressure heat capacity is the slope of a plot of enthalpy against temperature of a system kept at constant pressure. Because the plot might not be a straight line, in general we interpret C_p as the slope of the tangent to the curve at the temperature of interest (Fig. 1.16, Table 1.1). That is, the constant-pressure heat capacity is the derivative of the function H with respect to the variable T at a specified pressure or

$$C_p = \frac{\mathrm{d}H}{\mathrm{d}T}$$

(1.14b)

> **A brief illustration**
>
> Provided the heat capacity is constant over the range of temperatures of interest, we can write eqn 1.14a as $\Delta H = C_p \Delta T$. This relation means that when the temperature of 100 g of water (5.55 mol H_2O) is raised from 20°C to 80°C (so $\Delta T = +60$ K) at constant pressure, the enthalpy of the sample changes by
>
> $$\Delta H = C_p \Delta T = nC_{p,m}\Delta T = (5.55 \text{ mol}) \times (75.29 \text{ J K}^{-1} \text{ mol}^{-1}) \times (60 \text{ K}) = +25 \text{ kJ}$$
>
> The greater the temperature rise, the greater the change in enthalpy and therefore the greater the heating that is required to bring it about. Note that this calculation is only approximate because the heat capacity depends on the temperature and we have used an average value for the temperature range of interest.

The difference between $C_{p,m}$ and $C_{V,m}$ is significant for gases (for oxygen, $C_{V,m} = 20.8$ J K^{-1} mol^{-1} and $C_{p,m} = 29.1$ J K^{-1} mol^{-1}), which undergo large changes of volume when heated, but is negligible for most solids and liquids. For a perfect gas, you are invited to show in Exercise 1.19 that

$$C_{p,m} - C_{V,m} = R$$

Difference between the molar heat capacities of a perfect gas   (1.15)

The molar heat capacity of a substance at constant pressure is always greater than the molar heat capacity at constant volume. The reason is that when a system is

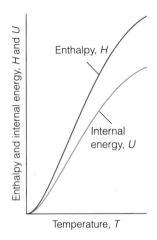

Fig. 1.15 The enthalpy of a system increases as its temperature is raised. Note that the enthalpy is always greater than the internal energy of the system and that the difference increases with temperature.

A brief comment
Again more precisely the constant-pressure heat capacity is the *partial derivative* of the function H with respect to the variable T, denoted as

$$C_p = \left(\frac{\partial H}{\partial T}\right)_p$$

and calculated by holding the variable p constant.

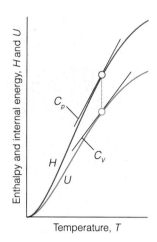

Enthalpy and internal energy, H and U

C_p

H

C_V

U

Temperature, T

Fig. 1.16 The heat capacity at constant pressure is the slope of the curve showing how the enthalpy varies with temperature; the heat capacity at constant volume is the corresponding slope of the internal energy curve. Note that the heat capacity varies with temperature (in general) and that C_p is greater than C_V.

A brief comment

Electrical charge is measured in *coulombs*, C. The motion of charge gives rise to an electric current, I, measured in coulombs per second, or *amperes*, A, where $1\ A = 1\ C\ s^{-1}$. If a constant current I flows through a potential difference $\mathcal{V}$ (measured in volts, V), the total energy supplied in an interval t is $I\mathcal{V}t$. Because $1\ A\ V\ s = 1\ (C\ s^{-1})\ V\ s = 1\ C\ V = 1\ J$, the energy is obtained in joules with the current in amperes, the potential difference in volts, and the time in seconds.

free to expand, some of the energy supplied as heat is free to escape back into the surroundings as work. Therefore, the rise in temperature at constant pressure is not as great as at constant volume (when no expansion work can be done), and the heat capacity is correspondingly greater.

In the laboratory 1.1 *Calorimetry*

Calorimetry is the study of heat transfer during physical and chemical processes. A **calorimeter**[6] is a device for measuring energy transferred as heat. Here we explore three common types of calorimeters used in investigations of nutrients, fuels, and biological processes.

(a) Bomb calorimeters

The most common device for measuring ΔU is an **adiabatic bomb calorimeter** (Fig. 1.17). The process under study is initiated inside a constant-volume container, the 'bomb'. The bomb is immersed in a stirred water bath, and the whole device is the calorimeter. The calorimeter is also immersed in an outer water bath. The water in the calorimeter and of the outer bath are both monitored and adjusted to the same temperature. This arrangement ensures that there is no net loss of heat from the calorimeter to the surroundings (the bath) and hence that the calorimeter is adiabatic.

The change in temperature, ΔT, of the calorimeter is proportional to the energy that the process releases or absorbs as heat. Therefore, by measuring ΔT we can determine q_V and hence find ΔU. The conversion of ΔT to q_V is best achieved by calibrating the calorimeter using a process of known energy output and determining the **calorimeter constant**, the constant C in the relation

$$q = C\Delta T \qquad \boxed{\text{Calorimeter constant}} \qquad (1.16)$$

The calorimeter constant may be measured electrically by passing a constant current, I, from a source of known potential difference, $\mathcal{V}$, through a heater for a known period of time, t, for then

$$q = I\mathcal{V}t \qquad (1.17)$$

A brief illustration

If we pass a current of 10.0 A from a 12 V supply for 300 s, then from eqn 1.17 the energy supplied as heat is

$$q = (10.0\ A) \times (12\ V) \times (300\ s) = 3.6 \times 10^4\ A\ V\ s = 36\ kJ$$

because $1\ A\ V\ s = 1\ J$. If the observed rise in temperature is 5.5 K, then the calorimeter constant is $C = (36\ kJ)/(5.5\ K) = 6.5\ kJ\ K^{-1}$.

Alternatively, C may be determined by using a reaction of known heat output, such as the combustion of benzoic acid (C_6H_5COOH), for which the heat output is 3227 kJ per mole of C_6H_5COOH consumed. With C known, it is simple to interpret an observed temperature rise as a release of energy as heat.

---

[6] The word *calorimeter* comes from 'calor', the Latin word for *heat*.

Bomb calorimetry is used in nutritional studies to determine the total energy content of a nutrient, also called its *gross energy* (G.E.) *content*. The results may be expressed in a number of units, but common in nutritional studies is the *large calorie* or *nutritional calorie* (abbreviation: Cal), which is defined as 1 Cal = 4.184 kJ exactly. The large calorie is the unit of energy used colloquially and on labels on packages of food products. This unit is distinct from the *calorie* (abbreviation: cal), or 'small calorie', still encountered in the scientific literature, 1 cal = 4.184 J exactly. It follows that 1 Cal = 1 kcal.

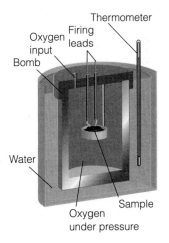

Fig. 1.17 A constant-volume adiabatic bomb calorimeter. The 'bomb' is the central vessel, which is strong enough to withstand high pressures. The calorimeter (for which the heat capacity must be known) is the entire assembly shown here. To ensure adiabaticity, the calorimeter is immersed in a water bath with a temperature continuously adjusted to that of the calorimeter at each stage of the combustion.

> **Example 1.2** *Calibrating a calorimeter and measuring the energy content of a nutrient*

In an experiment to measure the heat released by the combustion of a sample of nutrient, the compound was burned in a calorimeter and the temperature rose by 3.22°C. When a current of 1.23 A from a 12.0 V source flowed through a heater in the same calorimeter for 156 s, the temperature rose by 4.47°C. What is the energy content of the nutrient, taken as the heat released by the combustion reaction?

Strategy We calculate the heat supplied electrically by using eqn 1.17 and 1 A V s = 1 J. Then we use the observed rise in temperature to find the heat capacity of the calorimeter. Finally, we use this heat capacity to convert the temperature rise observed for the combustion into a heat output by writing $q = C\Delta T$ (or $q = C\Delta\theta$ if the temperature is given on the Celsius scale).

Solution The heat supplied during the calibration step is

$$q = IVt = (1.23 \text{ A}) \times (12.0 \text{ V}) \times (156 \text{ s})$$
$$= 1.23 \times 12.0 \times 156 \text{ A V s}$$
$$= 1.23 \times 12.0 \times 156 \text{ J}$$

This product works out as 2.30 kJ, but to avoid rounding errors we save the numerical work to the final stage. The heat capacity of the calorimeter is

$$C = \frac{q}{\Delta\theta} = \frac{1.23 \times 12.0 \times 156 \text{ J}}{4.47°C} = \frac{1.23 \times 12.0 \times 156}{4.47} \text{ J }°C^{-1}$$

The numerical value of C is 515 J °C^{-1}, but we don't evaluate it yet in the actual calculation. The heat output of the combustion is therefore

$$q = C\Delta\theta = \left(\frac{1.23 \times 12.0 \times 156}{4.47} \text{ J }°C^{-1}\right) \times 3.22°C = 1.66 \text{ kJ}$$
$$= 0.397 \text{ Cal}$$

A note on good practice
As well as keeping the numerical evaluation to the final stage (or at least not rounding intermediate values obtained with a calculator), show the units at each stage of the calculation.

> **Self-test 1.3** In an experiment to measure the heat released by the combustion of a sample of fuel, the compound was burned in an oxygen atmosphere inside a calorimeter and the temperature rose by 2.78°C. When a current of 1.12 A from an 11.5 V source flowed through a heater in the same calorimeter for 162 s, the temperature rose by 5.11°C. What is the heat released by the combustion reaction?
>
> *Answer:* 1.1 kJ

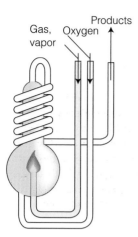

Fig. 1.18 An adiabatic flame calorimeter, an example of an isobaric calorimeter, consists of this component immersed in a stirred water bath. Combustion occurs as a known amount of reactant is passed through to fuel the flame and the rise of temperature is monitored.

A brief comment

The *rate of change of energy* is the power, expressed as joules per second, or *watts*, W: 1 W = 1 J s^{-1}. Because 1 J = 1 A V s, in terms of electrical units 1 W = 1 A V. We write the electrical power, P, as $P =$ (energy supplied)$/t = IVt/t = IV$.

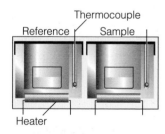

Fig. 1.19 A differential scanning calorimeter. The sample and a reference material are heated in separate but identical compartments. The output is the difference in power needed to maintain the compartments at equal temperatures as the temperature rises.

(b) Isobaric calorimeters

An enthalpy change can be measured calorimetrically by monitoring the temperature change that accompanies a physical or chemical change occurring at constant pressure. A calorimeter for studying processes at constant pressure is called an **isobaric calorimeter**. A simple example is a thermally insulated vessel open to the atmosphere: the heat released in the reaction is monitored by measuring the change in temperature of the contents. For a combustion reaction an **adiabatic flame calorimeter** may be used to measure ΔT when a given amount of substance burns in a supply of oxygen (Fig. 1.18). The relative efficiencies of fuels may be evaluated by this method (see Section 1.9).

(c) Differential scanning calorimeters

A **differential scanning calorimeter** (DSC) is more sophisticated than the calorimeters discussed so far. The term 'differential' refers to the fact that the behavior of the sample is compared to that of a reference material that does not undergo a physical or chemical change during the analysis. The term 'scanning' refers to the fact that the temperatures of the sample and reference material are increased, or scanned, systematically during the analysis.

A DSC consists of two small compartments that are heated electrically at a constant rate (Fig. 1.19). The temperature, T, at time t during a linear scan is $T = T_0 + \alpha t$, where T_0 is the initial temperature and α is the temperature scan rate (in kelvins per second, K s^{-1}). A computer controls the electrical power output in order to maintain the same temperature in the sample and reference compartments throughout the analysis.

The temperature of the sample changes significantly relative to that of the reference material if a chemical or physical process that involves heating occurs in the sample during the scan. To maintain the same temperature in both compartments, excess energy is transferred as heat to the sample during the process. For example, an endothermic process lowers the temperature of the sample relative to that of the reference and, as a result, the sample must be supplied with more energy (as heat) than the reference in order to maintain equal temperatures.

If no physical or chemical change occurs in the sample at temperature T, we can use eqn 1.5 to write $q_p = C_p \Delta T$, where $\Delta T = T - T_0 = \alpha t$ and we have assumed that C_p is independent of temperature. If an endothermic process occurs in the sample, we have to supply additional 'excess' energy by heating, $q_{p,\mathrm{ex}}$, to achieve the same change in temperature of the sample and can express this excess energy in terms of an additional contribution to the heat capacity, $C_{p,\mathrm{ex}}$, by writing $q_{p,\mathrm{ex}} = C_{p,\mathrm{ex}} \Delta T$. It follows that

$$C_{p,\mathrm{ex}} = \frac{q_{p,\mathrm{ex}}}{\Delta T} = \frac{q_{p,\mathrm{ex}}}{\alpha t} = \frac{P_{\mathrm{ex}}}{\alpha}$$

where $P_{\mathrm{ex}} = q_{p,\mathrm{ex}}/t$ is the excess electrical power necessary to equalize the temperature of the sample and reference compartments.

A DSC trace, which is called a **thermogram**, consists of a plot of P_{ex} or $C_{p,\mathrm{ex}}$ against T (Fig. 1.20). Broad peaks in the thermogram indicate processes requiring the transfer of energy by heating. We show in the following *Justification* that the enthalpy change of the process is

$$\Delta H = \int_{T_1}^{T_2} C_{p,\mathrm{ex}}\, dT \tag{1.18}$$

That is, the enthalpy change is the area under the curve of $C_{p,\text{ex}}$ against T between the temperatures at which the process begins and ends.

Justification 1.3 *The enthalpy change of a process from DSC data*

To calculate an enthalpy change from a thermogram, we begin by rewriting eqn 1.14b as

$$dH = C_{p,\text{ex}}dT$$

We proceed by integrating both sides of this expression from an initial temperature T_1 and initial enthalpy H_1 to a final temperature T_2 and enthalpy H_2.

$$\int_{H_1}^{H_2} dH = \int_{T_1}^{T_2} C_{p,\text{ex}}dT$$

Now we use the integral $\int dx = x + \text{constant}$ to write

$$\int_{H_1}^{H_2} dH = H_2 - H_1 = \Delta H$$

It follows that

$$\Delta H = \int_{T_1}^{T_2} C_{p,\text{ex}}dT$$

which is eqn 1.18.

A brief comment
Infinitesimally small quantities may be treated like any other quantity in algebraic manipulations, so the expression $dy/dx = a$ may be rewritten as $dy = adx$, $dx/dy = a^{-1}$, and so on.

To appreciate the utility of a DSC in biochemical investigations, we consider an important type of transformation that occurs in biological macromolecules, such as proteins and nucleic acids, and aggregates, such as biological membranes. Such large systems adopt complex three-dimensional structures as a result of intra- and inter-molecular interactions (*Fundamentals* F.1 and Chapter 11). **Denaturation**, the disruption of these interactions, can be achieved by adding chemical agents (such as urea, acids, or bases) or by changing the temperature, in which case the process is called **thermal denaturation**. Cooking is an example of thermal denaturation. For example, when eggs are cooked, the protein albumin is denatured irreversibly.

Differential scanning calorimetry is a powerful technique for the study of denaturation of biological macromolecules. Every biopolymer has a characteristic temperature, the **melting temperature**, T_m, at which the three-dimensional structure unravels and biological function is lost. For example, the thermogram shown in Fig. 1.20 indicates that the widely distributed protein ubiquitin retains its native structure up to about 45°C and 'melts' into a denatured state at higher temperatures. The area under the curve represents the heat absorbed in this process and can be identified with the enthalpy change. The thermogram also reveals the formation of new intermolecular interactions in the denatured form. The increase in heat capacity accompanying the native → denatured transition reflects the change from a more compact native conformation to one in which the more exposed amino acid side chains in the denatured form have more extensive interactions with the surrounding water molecules. Differential scanning calorimetry is a convenient method for such studies because it requires small samples, with masses as low as 0.5 mg.

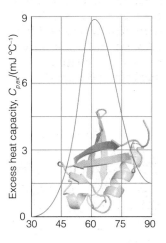

Fig. 1.20 A thermogram for the protein ubiquitin. The protein retains its native structure (shown as a green ribbon diagram) up to about 45°C and then undergoes an endothermic conformational change. (Adapted from B. Chowdhry and S. LeHarne, *J. Chem. Educ.* 74, 236 (1997).)

Physical and chemical change

We shall focus on the use of the enthalpy as a useful book-keeping property for tracing the flow of energy as heat during physical processes and chemical reactions at constant pressure. The discussion will lead naturally to a quantitative treatment of the factors that optimize the suitability of fuels, including 'biological fuels', the foods we ingest to meet the energy requirements of daily life.

1.7 Enthalpy changes accompanying physical processes

To begin to understand the complex structural changes that biological macromolecules undergo when heated or cooled, we need to understand how simpler physical changes occur.

To describe changes quantitatively, we need to keep track of the numerical value of a thermodynamic property with varying conditions, such as the states of the substances involved, the pressure, and the temperature. To simplify the calculations, chemists have found it convenient to report their data for a set of standard conditions at the temperature of their choice:

> The **standard state** of a substance is the pure substance at exactly 1 bar.[7]
>
> Definition of standard state

We denote the standard state value of a property by the superscript $^\circ$ on the symbol for the property, as in H_m° for the standard molar enthalpy of a substance and p° for the standard pressure of 1 bar. For example, the standard state of hydrogen gas is the pure gas at 1 bar and the standard state of solid calcium carbonate is the pure solid at 1 bar, with either the calcite or aragonite form specified. The physical state needs to be specified because we can speak of the standard states of the solid, liquid, and vapor forms of water, for instance, which are the pure solid, the pure liquid, and the pure vapor, respectively, at 1 bar in each case. The standard states of solutions, which are never 'pure', need to be treated differently (Section 3.8).

In older texts you might come across a standard state defined for 1 atm (101.325 kPa) in place of 1 bar. That is the old convention. In most cases, data for 1 atm differ only a little from data for 1 bar. You might also come across standard states defined as referring to 298.15 K. That is incorrect: temperature is not a part of the definition of standard state, and standard states may refer to any temperature (but it should be specified). Thus, it is possible to speak of the standard state of water vapor at 100 K, 273.15 K, or any other temperature. It is conventional, however, for data to be reported at the so-called **conventional temperature** of 298.15 K (25.00°C), and from now on, unless specified otherwise, all data will be for that temperature. For simplicity, we shall often refer to 298.15 K as '25°C'. Finally, a standard state need not be a stable state and need not be realizable in practice. Thus, the standard state of water vapor at 25°C is the vapor at 1 bar, but water vapor at that temperature and pressure would immediately condense to liquid water.

(a) Phase transitions

A **phase** is a specific state of matter that is uniform throughout in composition and physical state. The liquid and vapor states of water are two of its phases. The term 'phase' is more specific than 'state of matter' because a substance may exist in more than one solid form, each one of which is a solid phase. There are at least

---

[7] Remember that 1 bar = 10^5 Pa exactly.

12 forms of ice. No substance has more than one gaseous phase, so 'gas phase' and 'gaseous state' are effectively synonyms. The only substance that exists in more than one liquid phase is helium, although some evidence suggests that water might also have two liquid phases.

The conversion of one phase of a substance to another phase is called a **phase transition**. Thus, vaporization (liquid → gas) is a phase transition, as is a transition between solid phases (such as aragonite → calcite in geological processes). With a few exceptions, phase transitions are accompanied by a change of enthalpy, for the rearrangement of atoms or molecules usually requires or releases energy.

(b) Enthalpies of vaporization, fusion, and sublimation

The vaporization of a liquid, such as the conversion of liquid water to water vapor when a pool of water evaporates at 20°C or a kettle boils at 100°C, is an endothermic process ($\Delta H > 0$) because heating is required to bring about the change. At a molecular level, molecules are being driven apart from the grip they exert on one another, and this process requires energy. One of the body's strategies for maintaining its temperature at about 37°C is to use the endothermic character of the vaporization of water because the evaporation[8] of perspiration requires energy and withdraws it from the skin.

The energy that must be supplied as heat at constant pressure per mole of molecules that are vaporized under standard conditions (that is, pure liquid at 1 bar changing to pure vapor at 1 bar) is called the **standard enthalpy of vaporization** of the liquid and is denoted $\Delta_{vap}H^{\ominus}$ (Table 1.2). For example, 44 kJ of heat is required to vaporize 1 mol $H_2O(l)$ at 1 bar and 25°C, so $\Delta_{vap}H^{\ominus} = +44$ kJ mol^{-1}.

Alternatively, we can report the same information by writing the **thermochemical equation**[9]

$$H_2O(l) \rightarrow H_2O(g) \qquad \Delta H^{\ominus} = +44 \text{ kJ}$$

A note on good practice
The attachment of the subscript vap to the Δ is the modern convention; however, the older convention in which the subscript is attached to the H, as in ΔH_{vap}, is still widely used. All enthalpies of vaporization are positive, so the sign is not normally written explicitly in tables of data.

Table 1.2 Standard enthalpies of transition at the transition temperature*

Substance	Freezing point, T_{fus}/K	$\Delta_{fus}H^{\ominus}$/ (kJ mol^{-1})	Boiling point, T_b/K	$\Delta_{vap}H^{\ominus}$/ (kJ mol^{-1})
Ammonia, NH_3	195.3	5.65	239.7	23.4
Argon, Ar	83.8	1.2	87.3	6.5
Benzene, C_6H_6	278.7	9.87	353.3	30.8
Ethanol, C_2H_5OH	158.7	4.60	351.5	43.5
Helium, He	3.5	0.02	4.22	0.08
Hydrogen peroxide, H_2O_2	272.7	12.50	423.4	51.6
Mercury, Hg	234.3	2.292	629.7	59.30
Methane, CH_4	90.7	0.94	111.7	8.2
Methanol, CH_3OH	175.5	3.16	337.2	35.3
Propanone, CH_3COCH_3	177.8	5.72	329.4	29.1
Water, H_2O	273.15	6.01	373.2	40.7
				44.02 at 25°C
				45.07 at 0°C

*For values at 298.15 K, use the information in the *Resource section*.

[8] Evaporation is virtually synonymous with vaporization but commonly denotes vaporization to dryness.

[9] Unless otherwise stated, all data in this text are for 298.15 K.

A thermochemical equation shows the standard enthalpy change (including the sign) that accompanies the conversion of an amount of reactant equal to its stoichiometric coefficient in the accompanying chemical equation (in this case, 1 mol H_2O). If the stoichiometric coefficients in the chemical equation are multiplied through by 2, then the thermochemical equation would be written

$$2\,H_2O(l) \rightarrow 2\,H_2O(g) \qquad \Delta H^\ominus = +88\,kJ$$

This equation signifies that 88 kJ of heat is required to vaporize 2 mol $H_2O(l)$ at 1 bar and (recalling our convention) at 298.15 K.

There are some striking differences in standard enthalpies of vaporization: although the value for water is 44 kJ mol^{-1}, that for methane, CH_4, at its boiling point is only 8 kJ mol^{-1}. Even allowing for the fact that vaporization is taking place at different temperatures, the difference between the enthalpies of vaporization signifies that water molecules are held together in the bulk liquid much more tightly than methane molecules are in liquid methane. As should be recalled from introductory courses, the interaction responsible for the low volatility of water is the hydrogen bonding between neighboring H_2O molecules. The high enthalpy of vaporization of water has profound ecological consequences, for it is partly responsible for the survival of the oceans and the generally low humidity of the atmosphere. If only a small amount of heat had to be supplied to vaporize the oceans, the atmosphere would be much more heavily saturated with water vapor than is in fact the case.

Another common phase transition is **fusion**, or melting, as when ice melts to water. The change in molar enthalpy that accompanies fusion under standard conditions (pure solid at 1 bar changing to pure liquid at 1 bar) is called the **standard enthalpy of fusion**, $\Delta_{fus}H^\ominus$. Its value for water at 0°C is +6.01 kJ mol^{-1}. As for enthalpies of vaporization, all enthalpies of fusion are positive, and the sign is not written explicitly in tables. Notice that the enthalpy of fusion of water is much less than its enthalpy of vaporization. In vaporization the molecules become completely separated from each other, whereas in melting the molecules are merely loosened without separating completely (Fig. 1.21).

The reverse of vaporization is **condensation** and the reverse of fusion (melting) is **freezing**. The molar enthalpy changes are, respectively, the negative of the enthalpies of vaporization and fusion because the energy that is supplied (during heating) to vaporize or melt the substance is released when it condenses or

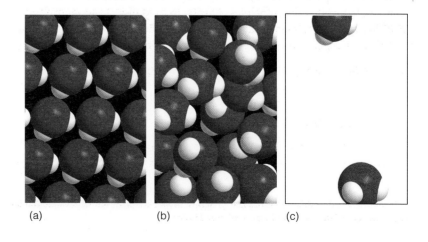

Fig. 1.21 When a solid (a) melts to a liquid (b), the molecules separate from one another only slightly, the intermolecular interactions are reduced only slightly, and there is only a small change in enthalpy. When a liquid vaporizes (c), the molecules are separated by a considerable distance, the intermolecular forces are reduced almost to zero, and the change in enthalpy is much greater.

(a)   (b)   (c)

freezes.[10] It is always the case that *the enthalpy change of a reverse transition is the negative of the enthalpy change of the forward transition* (under the same conditions of temperature and pressure):

$$H_2O(s) \rightarrow H_2O(l) \qquad \Delta H^{\ominus} = +6.01 \text{ kJ}$$

$$H_2O(l) \rightarrow H_2O(s) \qquad \Delta H^{\ominus} = -6.01 \text{ kJ}$$

and in general

$$\Delta_{forward}H^{\ominus} = -\Delta_{reverse}H^{\ominus} \qquad (1.19)$$

This relation follows from the fact that H is a state property, so it must return to the same value if a forward change is followed by the reverse of that change (Fig. 1.22). The high standard enthalpy of vaporization of water ($+44$ kJ mol^{-1}), signifying a strongly endothermic process, implies that the condensation of water (-44 kJ mol^{-1}) is a strongly exothermic process. That exothermicity is the origin of the ability of steam to scald severely because the energy is passed on to the skin.

The direct conversion of a solid to a vapor is called **sublimation**. The reverse process is called **vapor deposition**. Sublimation can be observed on a cold, frosty morning, when frost vanishes as vapor without first melting. The frost itself forms by vapor deposition from cold, damp air. The vaporization of solid carbon dioxide ('dry ice') is another example of sublimation. The standard molar enthalpy change accompanying sublimation is called the **standard enthalpy of sublimation**, $\Delta_{sub}H^{\ominus}$. Because enthalpy is a state property, the same change in enthalpy must be obtained both in the *direct* conversion of solid to vapor and in the *indirect* conversion, in which the solid first melts to the liquid and then that liquid vaporizes (Fig. 1.23):

$$\Delta_{sub}H^{\ominus} = \Delta_{fus}H^{\ominus} + \Delta_{vap}H^{\ominus} \qquad (1.20)$$

This result is an example of a more general statement that will prove useful time and again during our study of thermochemistry:

The enthalpy change of an overall process is the sum of the enthalpy changes for the steps (observed or hypothetical) into which it may be divided.

A brief illustration

To use eqn 1.20 correctly, the two enthalpies that are added together must be for the same temperature, so to get the enthalpy of sublimation of water at 0°C, we must add together the enthalpies of fusion (6.01 kJ mol^{-1}) and vaporization (45.07 kJ mol^{-1}) for this temperature. Adding together enthalpies of transition for different temperatures gives a meaningless result. It follows that

$$\Delta_{sub}H^{\ominus} = \Delta_{fus}H^{\ominus} + \Delta_{vap}H^{\ominus} = 6.01 \text{ kJ mol}^{-1} + 45.07 \text{ kJ mol}^{-1} = 51.08 \text{ kJ mol}^{-1}$$

1.8 Bond enthalpy

To understand bioenergetics at a molecular level we need to account for the flow of energy during chemical reactions as individual chemical bonds are broken and made.

---

[10] This relation is the origin of the obsolescent terms 'latent heat' of vaporization and fusion for what are now termed the enthalpy of vaporization and fusion.

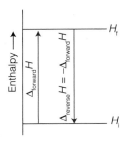

Fig. 1.22 An implication of the First Law is that the enthalpy change accompanying a reverse process is the negative of the enthalpy change for the forward process.

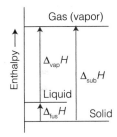

Fig. 1.23 The enthalpy of sublimation at a given temperature is the sum of the enthalpies of fusion and vaporization at that temperature. Another implication of the First Law is that the enthalpy change of an overall process is the sum of the enthalpy changes for the possibly hypothetical steps into which it may be divided.

A note on good practice
Molar quantities are expressed as a quantity per mole (as in kilojoules per mole, kJ mol^{-1}). Distinguish them from the magnitude of a property *for* 1 mol of substance, which is expressed as the quantity itself (as in kilojoules, kJ). All enthalpies of transition, denoted $\Delta_{trs}H$, are molar quantities.

Table 1.3 Selected bond enthalpies, $H(\text{A--B})/(\text{kJ mol}^{-1})$

Diatomic molecules

H–H	436	O=O	497	F–F	155	H–F	565
		N≡N	945	Cl–Cl	242	H–Cl	431
		O–H	428	Br–Br	193	H–Br	366
		C≡O	1074	I–I	151	H–I	299

Polyatomic molecules

H–CH₃	435	H–NH₂	431	H–OH	492
H–C₆H₅	469	O₂N–NO₂	57	HO–OH	213
H₃C–CH₃	368	O=CO	531	HO–CH₃	377
H₂C=CH₂	699			Cl–CH₃	352
HC≡CH	962			Br–CH₃	293
				I–CH₃	234

The thermochemical equation for the dissociation, or breaking, of a chemical bond can be written with the hydroxyl radical OH(g) as an example:

$$\text{HO(g)} \rightarrow \text{H(g)} + \text{O(g)} \qquad \Delta H^\ominus = +428 \text{ kJ}$$

The corresponding standard molar enthalpy change is called the **bond enthalpy**, so we would report the H–O bond enthalpy as 428 kJ mol⁻¹. All bond enthalpies are positive, so bond dissociation is an endothermic process.

Some bond enthalpies are given in Table 1.3. Note that the bond in molecular nitrogen, N_2, is very strong, at 945 kJ mol⁻¹, which helps to account for the chemical inertness of nitrogen and its ability to dilute the oxygen in the atmosphere without reacting with it. In contrast, the bond in molecular fluorine, F_2, is relatively weak, at 155 kJ mol⁻¹; the weakness of this bond contributes to the high reactivity of elemental fluorine. However, bond enthalpies alone do not account for reactivity because, although the bond in molecular iodine is even weaker, I_2 is less reactive than F_2, and the bond in CO is stronger than the bond in N_2, but CO forms many carbonyl compounds, such as $Ni(CO)_4$. The types and strengths of the bonds that the elements can make to other elements when a new substance is formed from reactants are additional factors.

A complication when dealing with bond enthalpies is that their values depend on the molecule in which the two linked atoms occur. For instance, the total standard enthalpy change for the atomization (the complete dissociation into atoms) of water:

$$\text{H}_2\text{O(g)} \rightarrow 2\,\text{H(g)} + \text{O(g)} \qquad \Delta H^\ominus = +927 \text{ kJ}$$

is not twice the O–H bond enthalpy in H_2O even though two O–H bonds are dissociated. There are in fact two different dissociation steps. In the first step, an O–H bond is broken in an H_2O molecule:

$$\text{H}_2\text{O(g)} \rightarrow \text{HO(g)} + \text{H(g)} \qquad \Delta H^\ominus = +492 \text{ kJ}$$

In the second step, the O–H bond is broken in an OH radical:

$$\text{HO(g)} \rightarrow \text{H(g)} + \text{O(g)} \qquad \Delta H^\ominus = +428 \text{ kJ}$$

A brief comment

Recall that a radical is a very reactive species containing one or more unpaired electrons. To emphasize the presence of an unpaired electron in a radical, it is common to use a dot (·) when writing the chemical formula. For example, the chemical formula of the hydroxyl radical may be written as ·OH. Hydroxyl radicals and other reactive species containing oxygen can be produced in organisms as undesirable by-products of electron transfer reactions and have been implicated in the development of cardiovascular disease, cancer, stroke, inflammatory disease, and other conditions.

Table 1.4 Mean bond enthalpies, $\Delta H_B/(\text{kJ mol}^{-1})$*

	H	C	N	O	F	Cl	Br	I	S	P	Si
H	436										
C	412	348 (1)									
		612 (2)									
		838 (3)									
		518 (a)†									
N	388	305 (1)	163 (1)								
		613 (2)	409 (2)								
		890 (3)	945 (3)								
O	463	360 (1)	157	146 (1)							
		743 (2)		497 (2)							
F	565	484	270	185	155						
Cl	431	338	200	203	254	242					
Br	366	276				219	193				
I	299	238				210	178	151			
S	338	259			496	250	212		264		
P	322									200	
Si	318		374		466						226

*Values are for single bonds except where otherwise stated (in parentheses).
†(a) Denotes aromatic.

The sum of the two steps is the atomization of the molecule. As can be seen from this example, the O–H bonds in H_2O and HO have similar but not identical bond enthalpies.

Although accurate calculations must use bond enthalpies for the molecule in question and its successive fragments, when such data are not available, there is no choice but to make estimates by using **mean bond enthalpies**, ΔH_B, which are the averages of bond enthalpies over a related series of compounds (Table 1.4). For example, the mean HO bond enthalpy, $\Delta H_B(\text{H–O}) = 463 \text{ kJ mol}^{-1}$, is the mean of the O–H bond enthalpies in H_2O and several other similar compounds, including methanol, CH_3OH.

Example 1.3   *Using mean bond enthalpies*

Use information from the *Resource section* and bond enthalpy data from Tables 1.3 and 1.4 to estimate the standard enthalpy change for the reaction

$$2 \; H_2O_2(l) \rightarrow 2 \; H_2O(l) + O_2(g)$$

in which liquid hydrogen peroxide decomposes into O_2 and water at 25°C. In the aqueous environment of biological cells, hydrogen peroxide—a very reactive species—is formed as a result of some processes involving O_2. The enzyme catalase helps rid organisms of toxic hydrogen peroxide by accelerating its decomposition. The enthalpy of vaporization of $H_2O_2(l)$ at 298 K is 51.5 kJ mol^{-1}.

Strategy In calculations of this kind, the procedure is to break the overall process down into a sequence of steps such that their sum is the chemical equation required.

Always ensure, when using bond enthalpies, that all the species are in the gas phase. That may mean including the appropriate enthalpies of vaporization or sublimation. One approach is to atomize all the reactants and then to build the products from the atoms so produced. When explicit bond enthalpies are available (that is, data are given in the tables available), use them; otherwise, use mean bond enthalpies to obtain estimates.

Solution The following steps are required:

	$\Delta H^{\ominus}/\text{kJ}$
Vaporization of 2 mol $H_2O_2(l)$, $2\,H_2O_2(l) \rightarrow 2\,H_2O_2(g)$	$2 \times (+51.5)$
Dissociation of 4 mol O–H bonds	$4 \times (+463)$
Dissociation of 2 mol O–O bonds in HO–OH	$2 \times (+213)$
Overall, so far: $2\,H_2O_2(l) \rightarrow 4\,H(g) + 4\,O(g)$	$+2381$

We have used the mean bond enthalpy value from Table 1.4 for the O–H bond and the exact bond enthalpy value for the O–O bond in HO–OH from Table 1.3. In the second step, four O–H bonds and one O=O bond are formed. The standard enthalpy change for bond formation (the reverse of dissociation) is the negative of the bond enthalpy. We can use exact values for the enthalpy of the O–H bond in $H_2O(g)$ and for the O=O bond in $O_2(g)$:

	$\Delta H^{\ominus}/\text{kJ}$
Formation of 4 mol O–H bonds	$4 \times (-492)$
Formation of 1 mol O_2	-497
Overall, in this step: $4\,O(g) + 4\,H(g) \rightarrow 2\,H_2O(g) + O_2(g)$	-2465

The final stage of the reaction is the condensation of 2 mol $H_2O(g)$

$$2\,H_2O(g) \rightarrow 2\,H_2O(l) \qquad \Delta H^{\ominus} = 2 \times (-44\,\text{kJ}) = -88\,\text{kJ}$$

The sum of the enthalpy changes is

$$\Delta H^{\ominus} = (+2381\,\text{kJ}) + (-2465\,\text{kJ}) + (-88\,\text{kJ}) = -172\,\text{kJ}$$

The experimental value is $-196\,\text{kJ}$.

Self-test 1.4 Estimate the enthalpy change for the reaction between 1 mol C_2H_5OH as liquid ethanol, a fuel made by fermenting corn, and $O_2(g)$ to yield $CO_2(g)$ and $H_2O(l)$ under standard conditions by using the bond enthalpies, mean bond enthalpies, and the appropriate standard enthalpies of vaporization.

Answer: $-1305\,\text{kJ}$; the experimental value is $-1368\,\text{kJ}$

1.9 Thermochemical properties of fuels

We need to understand the molecular origins of the energy content of biological fuels, the carbohydrates, fats, and proteins.

We saw in *Case study* 1.1 that photosynthesis and the oxidation of organic molecules are the most important processes that supply energy to organisms. In this section we begin our quantitative study of biological energy conversion by assessing the thermochemical properties of fuels.

Table 1.5 Standard enthalpies of combustion

Substance	$\Delta_c H^\ominus/(\text{kJ mol}^{-1})$
Carbon, C(s, graphite)	−394
Carbon monoxide, CO(g)	−283
Citric acid, $C_6H_8O_7$(s)	−1985
Ethanol, C_2H_5OH(l)	−1368
Glucose, $C_6H_{12}O_6$(s)	−2808
Glycine, $CH_2(NH_2)COOH$(s)	−969
Hydrogen, H_2(g)	−286
iso-Octane,* C_8H_{18}(l)	−5461
Methane, CH_4(g)	−890
Methanol, CH_3OH(l)	−726
Methylbenzene, $C_6H_5CH_3$(l)	−3910
Octane, C_8H_{18}(l)	−5471
Propane, C_3H_8(g)	−2220
Pyruvic acid, $CH_3(CO)COOH$(l)	−950
Sucrose, $C_{12}H_{22}O_{11}$(s)	−5645
Urea, $CO(NH_2)_2$(s)	−632

*2,2,4-Trimethylpentane.

The consumption of a fuel in a furnace or an engine is the result of a combustion. An example is the combustion of methane in a natural gas flame:

$$CH_4(g) + 2\,O_2(g) \rightarrow CO_2(g) + 2\,H_2O(l) \qquad \Delta H^\ominus = -890 \text{ kJ}$$

The **standard enthalpy of combustion**, $\Delta_c H^\ominus$, is the standard change in enthalpy per mole of combustible molecules. In this example, we would write $\Delta_c H^\ominus(CH_4, \text{g}) = -890 \text{ kJ mol}^{-1}$. Some typical values are given in Table 1.5. Note that $\Delta_c H^\ominus$ is a molar quantity and is obtained from the value of $\Delta H^\ominus$ by dividing by the amount of organic reactant consumed (in this case, by 1 mol CH_4).

According to the discussion in Sections 1.5 and 1.6, and the relation $\Delta U = q_V$, the energy transferred as heat at constant volume is equal to the change in internal energy, ΔU, not ΔH. To convert from ΔU to ΔH, we need to note that the molar enthalpy of a substance is related to its molar internal energy by $H_m = U_m + pV_m$ (eqn 1.12a). For condensed phases, pV_m is so small that it may be ignored. For example, the molar volume of liquid water is 18 cm³ mol⁻¹, and at 1.0 bar

$$pV_m = (1.0 \times 10^5 \text{ Pa}) \times (18 \times 10^{-6} \text{ m}^3 \text{ mol}^{-1}) = 1.8 \text{ Pa m}^3 \text{ mol}^{-1} = 1.8 \text{ J mol}^{-1}$$

However, the molar volume of a gas, and therefore the value of pV_m, is about 1000 times greater and cannot be ignored. For gases treated as perfect, pV_m may be replaced by RT. Therefore, if in the chemical equation the difference (products – reactants) in the stoichiometric coefficients of *gas phase* species is $\Delta\nu_{\text{gas}}$, we can write

$$\Delta_c H = \Delta_c U + \Delta\nu_{\text{gas}} RT \qquad (1.21)$$

Note that ν_{gas} (where ν is nu) is a dimensionless number.

> **A brief illustration**
>
> The energy released at constant volume as heat by the combustion of the amino acid glycine is −969.6 kJ mol⁻¹ at 298.15 K, so $\Delta_c U = -969.6$ kJ mol⁻¹. From the chemical equation
>
> $$NH_2CH_2COOH(s) + \tfrac{9}{4}O_2(g) \rightarrow 2\,CO_2(g) + \tfrac{5}{2}H_2O(l) + \tfrac{1}{2}N_2(g)$$
>
> we find that $\Delta v_{gas} = (2 + \tfrac{1}{2}) - \tfrac{9}{4} = \tfrac{1}{4}$. Therefore,
>
> $$\Delta_c H = \Delta_c U + \tfrac{1}{4}RT$$
> $$= -969.6 \text{ kJ mol}^{-1} + \tfrac{1}{4} \times (8.3145 \times 10^{-3} \text{ kJ K}^{-1} \text{ mol}^{-1}) \times (298.15 \text{ K})$$
> $$= -969.6 \text{ kJ mol}^{-1} + 0.62 \text{ kJ mol}^{-1} = -969.0 \text{ kJ mol}^{-1}$$

We shall see in Chapter 2 that the best assessment of the ability of a compound to act as a fuel to drive many of the processes occurring in the body makes use of the 'Gibbs energy'. However, a useful guide to the resources provided by a fuel, and the only one that matters when energy transferred as heat is being considered, is the enthalpy, particularly the enthalpy of combustion. The thermochemical properties of fuels and foods are commonly discussed in terms of their **specific enthalpy**, the magnitude of the enthalpy of combustion divided by the mass of the sample (typically in kilojoules per gram) or the **enthalpy density**, the magnitude of the enthalpy of combustion divided by the volume of the sample (typically in kilojoules per cubic decimeter). If the standard enthalpy of combustion is $\Delta_c H^{\ominus}$ and the molar mass of the compound is M, then the specific enthalpy is $\Delta_c H^{\ominus}/M$. Similarly, the enthalpy density is $\Delta_c H^{\ominus}/V_m$, where V_m is the molar volume of the material.

Table 1.6 lists the specific enthalpies and enthalpy densities of several fuels. The most suitable fuels are those with high specific enthalpies, as the advantage of a high molar enthalpy of combustion may be eliminated if a large mass of fuel is to be transported. We see that H_2 gas compares very well with more traditional fuels such as methane (natural gas), octane (gasoline), and methanol. Furthermore, the combustion of H_2 gas does not generate CO_2 gas, a pollutant implicated in the mechanism of global warming. As a result, H_2 gas has been proposed as an efficient, clean alternative to fossil fuels, such as natural gas and petroleum. However, we also see that H_2 gas has a very low enthalpy density, which arises from the fact

Table 1.6 Thermochemical properties of some fuels

Fuel	Combustion equation	$\Delta_c H^{\ominus}/(\text{kJ mol}^{-1})$	Specific enthalpy/ (kJ g⁻¹)	Enthalpy density*/ (kJ dm⁻³)
Hydrogen	$2\,H_2(g) + O_2(g) \rightarrow 2\,H_2O(l)$	−286	142	13
Methane	$CH_4(g) + 2\,O_2(g) \rightarrow CO_2(g) + 2\,H_2O(l)$	−890	55	40
iso-Octane†	$2\,C_8H_{18}(l) + 25\,O_2(g) \rightarrow 16\,CO_2(g) + 18\,H_2O(l)$	−5461	48	3.3×10^4
Methanol	$2\,CH_3OH(l) + 3\,O_2(g) \rightarrow 2\,CO_2(g) + 4\,H_2O(l)$	−726	23	1.8×10^4

*At atmospheric pressures and room temperature.
†2,2,4-Trimethylpentane.

that hydrogen is a very light gas. So, the advantage of a high specific enthalpy is undermined by the large volume of fuel to be transported and stored. Strategies are being developed to solve the storage problem. For example, the small H_2 molecules can travel through holes in the crystalline lattice of a sample of metal, such as titanium, where they bind as metal hydrides. In this way it is possible to increase the effective density of hydrogen atoms to a value that is higher than that of liquid H_2. Then the fuel can be released on demand by heating the metal.

We now assess the factors that optimize the enthalpy of combustion of carbon-based fuels, with an eye toward understanding such biological fuels as carbohydrates, fats, and proteins. The combustion of 1 mol $CH_4(g)$ releases 890 kJ of energy as heat per mole of C atoms:

$$CH_4(g) + 2\,O_2(g) \rightarrow CO_2(g) + 2\,H_2O(l) \qquad \Delta H^{\ominus} = -890\ kJ$$

Now consider the combustion of 1 mol $CH_3OH(g)$:

$$CH_3OH(g) + \tfrac{3}{2}O_2(g) \rightarrow CO_2(g) + 2\,H_2O(l) \qquad \Delta H^{\ominus} = -765\ kJ$$

This reaction is also exothermic, but now only 765 kJ of energy is released as heat per mole of C atoms. Much of the observed change in energy output between the reactions can be explained by noting that the replacement of a C–H bond by a C–O bond renders the carbon in methanol more oxidized than the carbon in methane, so it is reasonable to expect that less energy is released to complete the oxidation of carbon in methanol to CO_2. In general, the presence of partially oxidized C atoms (that is, carbon atoms bonded to oxygen atoms) in a material makes it a less suitable fuel than a similar material containing less oxidized C atoms.

Another factor that determines the enthalpy of combustion is the number of carbon atoms in hydrocarbon compounds. For example, whereas the enthalpy of combustion of methane is −890 kJ mol⁻¹, that of iso-octane (C_8H_{18}, 2,2,4-trimethylpentane **(1)**, a typical component of gasoline) is −5461 kJ mol⁻¹ (Table 1.6). The much larger value for iso-octane is a consequence of each molecule having eight C atoms to contribute to the formation of carbon dioxide, whereas methane has only one.

A brief comment
The concept of oxidation numbers, familiar from introductory chemistry, clarifies the point made in this paragraph. The formation of CO_2 from CH_4 involves an increase in the oxidation number—that is, an oxidation —of carbon from −4 in CH_4 to +4 in CO_2. By contrast, the carbon atom in CH_3OH has an oxidation number of −2 and is in a higher oxidation state than the carbon in methane.

Case study 1.2 | *Biological fuels*

A typical 18- to 20-year-old man requires a daily energy input of about 12 MJ (1 MJ = 10⁶ J) or about 2870 Cal; a woman of the same age needs about 9 MJ or about 2150 Cal. If the entire consumption were in the form of glucose, which has a specific enthalpy of 16 kJ g⁻¹, meeting energy needs would require the consumption of 750 g of glucose by a man and 560 g by a woman. In fact, the complex carbohydrates more commonly found in our diets have slightly higher specific enthalpies (17 kJ g⁻¹ = 4 Cal g⁻¹) than glucose itself, so a carbohydrate diet is slightly less daunting than a pure glucose diet, as well as being more appropriate in the form of fibre, the indigestible cellulose that helps move digestion products through the intestine.

1 *iso*-Octane
(2,2,4-trimethylpentane),
$(CH_3)_3CCH_2CH(CH_3)_2$

The specific enthalpy of fats, which are long-chain esters such as tristearin **(2)**, is much greater than that of carbohydrates, at around 38 kJ g⁻¹ (9 Cal g⁻¹), slightly less than the value for the hydrocarbon oils used as fuel (48 kJ g⁻¹ = 11 Cal g⁻¹). The reason for this difference lies in the fact that many of the carbon atoms in carbohydrates are bonded to oxygen atoms and are already partially oxidized, whereas most of the carbon atoms in fats are bonded to hydrogen and other carbon atoms and hence have lower oxidation numbers.

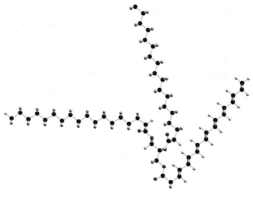

2 Tristearin $C_{57}H_{110}O_6$

As we have seen, the presence of partially oxidized carbons lowers the energy output of a fuel. Fats are commonly used as an energy store, to be used only when the more readily accessible carbohydrates have fallen into short supply. In Arctic species, the stored fat also acts as a layer of insulation; in desert species (such as the camel), the fat is also a source of water, one of its oxidation products.

Proteins are also used as a source of energy, but their components, the amino acids, are also used to construct other proteins. When proteins are oxidized (to urea, $CO(NH_2)_2$), the equivalent specific enthalpy is comparable to that of carbohydrates (about $17\ kJ\ g^{-1} = 4\ Cal\ g^{-1}$).

A brief illustration

A lunch consisting of a hamburger (about 350 Cal), potato chips (1 serving = 108 Cal), and a milk shake (about 502 Cal) would sum to about 960 Cal.[11,12] By contrast, a lighter lunch of halibut (about 205 Cal for a $\frac{1}{4}$-lb serving), a raw carrot (about 42 Cal), a large apple (101 Cal), and a glass of orange juice (about 120 Cal) would net only 468 Cal. The intake from these meals can be compared to the rates at which a 70-kg person can expend energy, depending on the nature of the activity:

Level of activity	Rate of energy expenditure/(Cal min^{-1})
Light (walking slowly)	2.5–5.0
Moderate (walking fast)	5.0–7.5
Heavy (running)	7.5–12.0

It follows that it would be necessary to walk slowly for about 3 to 6 hours (or run for 1 to 2 hours) to expend the energy taken in by eating the 960-Cal hamburger meal. Even though reading this textbook also requires energy, it would take about 16 hours for the hamburger meal to be 'burned off' by so sedentary an activity.

---

[11] Alarmingly, this single meal corresponds to about 33 per cent or 44 per cent of the daily energy requirements of a young man or woman, respectively.

[12] The data for this *brief illustration* are from C.H. Snyder, *The extraordinary chemistry of ordinary things*, Wiley (2002).

We have already remarked that not all the energy released by the oxidation of foods is used to perform work. The energy that is also released as heat needs to be discarded in order to maintain body temperature within its typical range of 35.6 to 37.8°C. A variety of mechanisms contribute to this aspect of **homeostasis**, the ability of an organism to counteract environmental changes with physiological responses. The general uniformity of temperature throughout the body is maintained largely by the flow of blood. When energy needs to be dissipated rapidly by heating, warm blood is allowed to flow through the capillaries of the skin, so producing flushing. Radiation is one means of heating the surroundings; another is evaporation and the energy demands of the enthalpy of vaporization of water.

A brief illustration

From the enthalpy of vaporization ($\Delta_{vap}H^{\ominus} = 44$ kJ mol^{-1} at 298 K), molar mass ($M = 18$ g mol^{-1}), and mass density ($\rho = 1.0$ g cm^{-3}, corresponding to 1.0×10^3 g dm^{-3}) of water, the energy removed as heat through evaporation per liter (cubic decimeter) of water perspired is

$$q = \frac{\rho \Delta_{vap}H^{\ominus}}{M} = \frac{(1.0 \times 10^3 \text{ g dm}^{-3}) \times (44 \text{ kJ mol}^{-1})}{18 \text{ g mol}^{-1}} = 2.4 \times 10^3 \text{ kJ dm}^{-3}$$

$$= 2.4 \text{ MJ dm}^{-3}$$

When vigorous exercise promotes sweating (through the influence of heat selectors on the hypothalamus), 1 to 2 dm^3 of perspired water can be produced per hour, corresponding to a loss of energy of approximately 2.4 to 4.8 MJ h^{-1}.

1.10 The combination of reaction enthalpies

To make progress in our study of bioenergetics, we need to develop methods for predicting the reaction enthalpies of complex biochemical reactions.

It is often the case that a reaction enthalpy is needed but is not available in tables of data. Now the fact that enthalpy is a state function comes in handy, because it implies that we can construct the required reaction enthalpy from the reaction enthalpies of known reactions. We have already seen a primitive example when we calculated the enthalpy of sublimation from the sum of the enthalpies of fusion and vaporization. The only difference is that we now apply the technique to a sequence of chemical reactions. The procedure is summarized by **Hess's law**, which in its modern form is:

The standard enthalpy of a reaction is the sum of the standard enthalpies of the reactions into which the overall reaction may be divided.

Although the procedure is given the status of a law, it hardly deserves the title because it is nothing more than a consequence of enthalpy being a state function, which implies that an overall enthalpy change can be expressed as a sum of enthalpy changes for each step in an indirect path. The individual steps need not be actual reactions that can be carried out in the laboratory—they may be entirely hypothetical reactions, the only requirement being that their equations should balance. Each step must correspond to the same temperature.

Example 1.4 *Using Hess's law*

In biological cells that have a plentiful supply of O_2, glucose is oxidized completely to CO_2 and H_2O (Section 1.9 and *Case study* 1.2). Muscle cells may be deprived of O_2 during vigorous exercise and, in that case, one molecule of glucose is converted to two molecules of lactic acid (Atlas C2) by the process of glycolysis (*Case study* 4.3). Given the thermochemical equations for the combustions of glucose and lactic acid:

$$C_6H_{12}O_6(s) + 6\,O_2(g) \rightarrow 6\,CO_2(g) + 6\,H_2O(l) \qquad \Delta H^{\circ} = -2808\;kJ$$
$$CH_3CH(OH)COOH(s) + 3\,O_2(g) \rightarrow 3\,CO_2(g) + 3\,H_2O(l) \quad \Delta H^{\circ} = -1344\;kJ$$

calculate the standard enthalpy for glycolysis:

$$C_6H_{12}O_6(s) \rightarrow 2\,CH_3CH(OH)COOH(s)$$

Is there a biological advantage of complete oxidation of glucose compared with glycolysis? Explain your answer.

Strategy We need to add or subtract the thermochemical equations so as to reproduce the thermochemical equation for the reaction required.

Solution We obtain the thermochemical equation for glycolysis from the following sum:

	$\Delta H^{\circ}/kJ$
$C_6H_{12}O_6(s) + 6\,O_2(g) \rightarrow 6\,CO_2(g) + 6\,H_2O(l)$	-2808
$6\,CO_2(g) + 6\,H_2O(l) \rightarrow 2\,CH_3CH(OH)COOH(s) + 6\,O_2(g)$	$2 \times (+1344\;kJ)$
Overall: $C_6H_{12}O_6(s) \rightarrow 2\,CH_3CH(OH)COOH(s)$	-120

It follows that the standard enthalpy for the conversion of glucose to lactic acid during glycolysis is $-120\;kJ\;mol^{-1}$, a mere 4 per cent of the enthalpy of combustion of glucose. Therefore, full oxidation of glucose is metabolically more useful than glycolysis because in the former process more energy becomes available for performing work.

Self-test 1.5 Calculate the standard enthalpy of the fermentation $C_6H_{12}O_6(s)$ $\rightarrow 2\,C_2H_5OH(l) + 2\,CO_2(g)$ from the standard enthalpies of combustion of glucose and ethanol (Table 1.5).

Answer: −72 kJ

1.11 Standard enthalpies of formation

We need to simplify even further the process of predicting reaction enthalpies of biochemical reactions.

The **standard reaction enthalpy**, $\Delta_r H^{\circ}$, is the difference between the standard molar enthalpies of the reactants and the products, with each term weighted by the stoichiometric coefficient, v (nu), in the chemical equation

$$\Delta_r H^{\circ} = \sum v H_m^{\circ}(products) - \sum v H_m^{\circ}(reactants) \qquad (1.22)$$

Definition of the standard reaction enthalpy

where $\sum$ (uppercase sigma) denotes a sum. Because the H_m° are molar quantities and the stoichiometric coefficients are pure numbers, the units of $\Delta_r H^{\circ}$ are kilojoules per mole. The standard reaction enthalpy is the change in enthalpy of the system when the reactants in their standard states (pure, 1 bar) are completely converted into products in their standard states (pure, 1 bar), with the change expressed in kilojoules per mole of reaction as written.

The problem with eqn 1.22 is that we have no way of knowing the absolute enthalpies of the substances. To avoid this problem, we can imagine the reaction as taking place by an indirect route, in which the reactants are first broken down into the elements and then the products are formed from the elements (Fig. 1.24). Specifically, the **standard enthalpy of formation**, $\Delta_f H^\circ$, of a substance is the standard enthalpy (per mole of the substance) for its formation from its elements in their reference states. The **reference state** of an element is its most stable form under the prevailing conditions (Table 1.7). Don't confuse 'reference state' with 'standard state': the reference state of carbon at 25°C is graphite (not diamond); the standard state of carbon is any specified phase of the element at 1 bar. For example, the standard enthalpy of formation of liquid water (at 25°C, as always in this text) is obtained from the thermochemical equation

$$H_2(g) + \tfrac{1}{2}O_2(g) \rightarrow H_2O(l) \qquad \Delta H^\circ = -286 \text{ kJ}$$

and is $\Delta_f H^\circ(H_2O, l) = -286 \text{ kJ mol}^{-1}$. Note that enthalpies of formation are molar quantities, so to go from ΔH° in a thermochemical equation to $\Delta_f H^\circ$ for that substance, divide by the amount of substance formed (in this instance, by 1 mol H_2O).

With the introduction of standard enthalpies of formation, we can write

$$\Delta_r H^\circ = \sum \nu \Delta_f H^\circ(\text{products}) - \sum \nu \Delta_f H^\circ(\text{reactants}) \quad \boxed{\begin{array}{c}\text{Calculation of}\\ \text{standard reaction}\\ \text{enthalpies}\end{array}} \quad (1.23)$$

The first term on the right is the enthalpy of formation of all the products from their elements; the second term on the right is the enthalpy of formation of all the reactants from their elements. The fact that the enthalpy is a state function means that a reaction enthalpy calculated in this way is identical to the value that would be calculated from eqn 1.22 if absolute enthalpies were available.

The values of some standard enthalpies of formation at 25°C are given in Table 1.8, and a longer list is given in the *Resource section*. The standard enthalpies of formation of elements in their reference states are zero by definition (because their formation is the null reaction: element → element). Note, however, that the standard enthalpy of formation of an element in a state other than its reference state is not zero:

$$C(s, \text{graphite}) \rightarrow C(s, \text{diamond}) \qquad \Delta H^\circ = +1.895 \text{ kJ}$$

Therefore, although $\Delta_f H^\circ(C, \text{graphite}) = 0$, $\Delta_f H^\circ(C, \text{diamond}) = +1.895 \text{ kJ mol}^{-1}$.

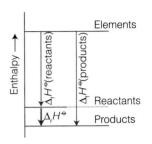

Fig. 1.24 An enthalpy of reaction may be expressed as the difference between the enthalpies of formation of the products and the reactants.

Table 1.7 Reference states of some elements at 298.15 K

Element	Reference state
Arsenic	Gray arsenic
Bromine	Liquid, $Br_2(l)$
Carbon	Graphite
Hydrogen	Gas, $H_2(g)$
Iodine	Solid, $I_2(s)$
Mercury	Liquid
Nitrogen	Gas, $N_2(g)$
Oxygen	Gas, $O_2(g)$
Phosphorus	White phosphorus, $P_4(s)$
Sulfur	Rhombic sulfur, $S_8(s)$

Example 1.5   *Using standard enthalpies of formation*

Glucose and fructose (Atlas S3) are simple carbohydrates with the molecular formula $C_6H_{12}O_6$. Sucrose (Atlas S5), or table sugar, is a complex carbohydrate with molecular formula $C_{12}H_{22}O_{11}$ that consists of a glucose unit covalently linked to a fructose unit (a water molecule is released as a result of the reaction between glucose and fructose to form sucrose). Estimate the standard enthalpy of combustion of sucrose from the standard enthalpies of formation of the reactants and products.

Table 1.8 Standard enthalpies of formation at 298.15 K*

Substance	$\Delta_f H^\ominus/(\text{kJ mol}^{-1})$	Substance	$\Delta_f H^\ominus/(\text{kJ mol}^{-1})$
Inorganic compounds		*Organic compounds*	
Ammonia, NH_3(g)	−46.11	Adenine, $C_5H_5N_5$(s)	+96.9
Carbon monoxide, CO(g)	−110.53	Alanine, $CH_3CH(NH_2)COOH$(s)	−604.0
Carbon dioxide, CO_2(g)	−393.51	Benzene, C_6H_6(l)	+49.0
Hydrogen sulfide, H_2S(g)	−20.63	Butanoic acid, $CH_3(CH_2)_2COOH$(l)	−533.8
Nitrogen dioxide, NO_2(g)	+33.18	Ethane, C_2H_6(g)	−84.68
Nitrogen monoxide, NO(g)	+90.25	Ethanoic acid, CH_3COOH(l)	−484.3
Sodium chloride, NaCl(s)	−411.15	Ethanol, C_2H_5OH(l)	−277.69
Water, H_2O(l)	−285.83	α-D-Glucose, $C_6H_{12}O_6$(s)	−1268
H_2O(g)	−241.82	Guanine, $C_5H_5N_5O$(s)	−183.9
		Glycine, $CH_2(NH_2)COOH$(s)	−528.5
		N-Glycylglycine, $C_4H_8N_2O_3$(s)	−747.7
		Hexadecanoic acid, $CH_3(CH_2)_{14}COOH$(s)	−891.5
		Leucine, $(CH_3)_2CHCH_2CH(NH_2)COOH$(s)	−637.4
		Methane, CH_4(g)	−74.81
		Methanol, CH_3OH(l)	−238.86
		Sucrose, $C_{12}H_{22}O_{11}$(s)	−2222
		Thymine, $C_5H_6N_2O_2$(s)	−462.8
		Urea, $(NH_2)_2CO$(s)	−333.1

*A longer list is given in the *Resource section*.

Strategy We write the chemical equation, identify the stoichiometric numbers of the reactants and products, and then use eqn 1.23. Note that the expression has the form 'products − reactants'. Numerical values of standard enthalpies of formation are given in the *Resource section*. The standard enthalpy of combustion is the enthalpy change per mole of substance, so we need to interpret the enthalpy change accordingly.

Solution The chemical equation is

$$C_{12}H_{22}O_{11}(s) + 12\,O_2(g) \rightarrow 12\,CO_2(g) + 11\,H_2O(l)$$

It follows that

$$
\begin{aligned}
\Delta_r H^\ominus &= \{12\Delta_f H^\ominus(CO_2,g) + 11\Delta_f H^\ominus(H_2O,l)\} \\
&\quad - \{\Delta_f H^\ominus(C_{12}H_{22}O_{11},g) + 12\Delta_f H^\ominus(O_2,g)\} \\
&= \{12 \times (-393.51\ \text{kJ mol}^{-1}) + 11 \times (-285.83\ \text{kJ mol}^{-1})\} \\
&\quad - \{(-2222\ \text{kJ mol}^{-1}) + 0\} \\
&= -5644\ \text{kJ mol}^{-1}
\end{aligned}
$$

A note on good practice
The standard enthalpy of formation of an element in its reference state (oxygen gas in this example) is written 0, not 0 kJ mol⁻¹, because it is zero whatever units we happen to be using.

Inspection of the chemical equation shows that, in this instance, the 'per mole' is per mole of sucrose, which is exactly what we need for an enthalpy of combustion. It follows that the estimate for the standard enthalpy of combustion of sucrose is −5644 kJ mol⁻¹. The experimental value is −5645 kJ mol⁻¹.

Self-test 1.6 Use standard enthalpies of formation to calculate the enthalpy of combustion of solid glycine to $CO_2(g)$, $H_2O(l)$, and $N_2(g)$.

Answer: $-973\ kJ\ mol^{-1}$, in agreement with the experimental value

(see the *Resource section*)

The reference states of the elements define a thermochemical 'sea level', and enthalpies of formation can be regarded as thermochemical 'altitudes' above or below sea level (Fig. 1.25). Compounds that have negative standard enthalpies of formation (such as water) are classified as **exothermic compounds**, for they lie at a lower enthalpy than their component elements (they lie below thermochemical sea level). Compounds that have positive standard enthalpies of formation (such as carbon disulfide) are classified as **endothermic compounds** and possess a higher enthalpy than their component elements (they lie above sea level).

1.12 Enthalpies of formation and computational chemistry

Table 1.4 is useful for many calculations, but its data cannot be used to estimate the differences between the standard enthalpies of formation of conformational isomers. For example, we would obtain the same enthalpy of formation for the equatorial and axial conformers of methylcyclohexane (**3** and **4**, respectively) if we were to use mean bond enthalpies. However, it has been observed experimentally that these conformers have different standard enthalpies of formation due to the steric repulsions in the axial conformer, which raise its energy relative to that of the equatorial conformer.

Computational chemistry is becoming the technique of choice for estimating standard enthalpies of formation of molecules with complex three-dimensional structures. Commercial software packages use the principles developed in Chapter 10 to calculate the standard enthalpy of formation of a conformer drawn on a computer screen. The difference between calculated standard enthalpies of formation of two conformers is then an estimate of the conformational energy difference. In the case of methylcyclohexane, the calculated conformational energy difference ranges from 5.9 to 7.9 kJ mol⁻¹, with the equatorial conformer having a lower standard enthalpy of formation than the axial conformer. These estimates compare favorably with the experimental value of 7.5 kJ mol⁻¹. However, good agreement between calculated and experimental values is relatively rare. Computational methods almost always predict correctly which conformer is more stable but do not always predict the correct magnitude of the conformational energy difference.

The computational approach also makes it possible to gain insight into the effect of solvation on the enthalpy of formation without conducting experiments. A calculation performed in the absence of solvent molecules estimates the properties of the molecule of interest in the gas phase. Computational methods are available that allow for the inclusion of several solvent molecules around a solute molecule, thereby taking into account the effect of molecular interactions with the solvent on the enthalpy of formation of the solute. Again, the numerical results are only estimates, and the primary purpose of the calculation is to predict whether interactions with the solvent increase or decrease the enthalpy of formation. As an example, consider the amino acid glycine, which can exist in a neutral (**5**) or zwitterionic (**6**) form, in which the amino group is protonated and the

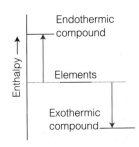

Fig. 1.25 The enthalpy of formation acts as a kind of thermochemical 'altitude' of a compound with respect to the 'sea level' defined by the elements from which it is made. Endothermic compounds have positive enthalpies of formation; exothermic compounds have negative energies of formation.

3 *eq*-Methylcyclohexane

4 *ax*-Methylcyclohexane

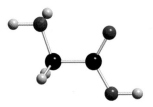

5 Glycine, NH_2CH_2COOH

6 Glycine, $^+NH_3CH_2CO_2^-$

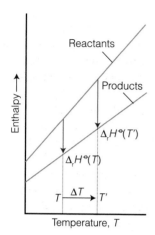

Fig. 1.26 The enthalpy of a substance increases with temperature, therefore if the total enthalpy of the reactants increases by a different amount from that of the products, the reaction enthalpy will change with temperature. The change in reaction enthalpy depends on the relative slopes of the two lines and hence on the heat capacities of the substances.

carboxyl group is deprotonated. It is possible to show computationally that in the gas phase the neutral form has a lower enthalpy of formation than the zwitterionic form. However, in water the opposite is true because of strong interactions between the polar solvent and the charges in the zwitterion.

1.13 The variation of reaction enthalpy with temperature

We need to know how to predict the reaction enthalpy of a biochemical reaction at one temperature from its value at other temperatures.

Suppose we want to know the enthalpy of a particular reaction at body temperature, 37°C, but have data available for 25°C, or suppose we want to know whether the oxidation of glucose is more exothermic when it takes place inside an Arctic fish that inhabits water at 0°C than when it takes place at mammalian body temperatures. In precise work, every attempt would be made to measure the reaction enthalpy at the temperature of interest, but it is useful to have a rapid way of estimating the sign and even a moderately reliable numerical value.

Figure 1.26 illustrates the technique. As we have seen, the enthalpy of a substance increases with temperature; therefore the total enthalpy of the reactants and the total enthalpy of the products increase, as shown in the illustration. Provided the two total enthalpy increases are different, the standard reaction enthalpy (their difference) will change as the temperature is changed. The change in the enthalpy of a substance depends on the slope of the graph and therefore on the constant-pressure heat capacities of the substances (recall Fig. 1.16). We can therefore expect the temperature dependence of the reaction enthalpy to be related to the difference in heat capacities of the products and the reactants. We show in the following *Justification* that this is indeed the case and that, when the heat capacities do not vary with temperature, the standard reaction enthalpy at a temperature T is related to the value at a different temperature T by a special formulation of **Kirchhoff's law**:

$$\Delta_r H^{\ominus}(T') = \Delta_r H^{\ominus}(T) + (T' - T)\Delta_r C_p^{\ominus} \qquad \boxed{\text{Kirchhoff's law}} \quad (1.24)$$

where $\Delta_r C_p^{\ominus}$ is the difference between the weighted sums of the standard molar heat capacities of the products and the reactants:

$$\Delta_r C_p^{\ominus} = \sum v C_{p,m}^{\ominus}(\text{products}) - \sum v C_{p,m}^{\ominus}(\text{reactants}) \qquad (1.25)$$

Values of standard molar constant-pressure heat capacities for a number of substances are given in the *Resource section*. Because eqn 1.24 applies only when the heat capacities are constant over the range of temperature of interest, its use is restricted to small temperature differences (of no more than 100 K or so).

Justification 1.4 *Kirchhoff's law*

To derive Kirchhoff's law, we consider the variation of the enthalpy with temperature. We begin by rewriting eqn 1.14b to calculate the change in the standard molar enthalpy H_m of each reactant and product as the temperature of the reaction mixture is increased:

$$dH_m^{\ominus} = C_{p,m}^{\ominus}dT$$

where $C_{p,m}^{\ominus}$ is the standard molar constant-pressure heat capacity, the molar heat capacity at 1 bar. We proceed by integrating both sides of the expression for dH_m° from an initial temperature T and initial enthalpy $H_m^\circ(T)$ to a final temperature T' and enthalpy $H_m^\circ(T')$:

$$\int_{H_m^\circ(T)}^{H_m^\circ(T')} dH = \int_T^{T'} C_{p,m}^\circ dT$$

It follows that for each reactant and product (assuming that no phase transition takes place in the temperature range of interest)

$$H_m^\circ(T') = H_m^\circ(T) + \int_T^{T'} C_{p,m}^\circ dT$$

Because this equation applies to each substance in the reaction, we use it and eqn 1.23 to write the following expression for $\Delta_r H^\circ(T')$:

$$\Delta_r H^\circ(T') = \Delta_r H^\circ(T) + \int_T^{T'} \Delta_r C_p^\circ dT$$

where $\Delta_r C_p^\circ$ is given by eqn 1.25. This equation is the exact form of Kirchhoff's law. The special case given by eqn 1.24 can be derived readily from it by making the approximation that $\Delta_r C_p^\circ$ is independent of temperature. Then the integral on the right evaluates to

$$\int_T^{T'} \Delta_r C_p^\circ dT = \Delta_r C_p^\circ \int_T^{T'} dT = \Delta_r C_p^\circ \times (T' - T)$$

and we obtain eqn 1.24.

A note on good practice
Because heat capacities can be measured more accurately than some reaction enthalpies, the exact form of Kirchhoff's law, with numerical integration of $\Delta_r C_p^\circ$ over the temperature range of interest, sometimes gives results more accurate than a direct measurement of the reaction enthalpy at the second temperature.

Example 1.6   *Using Kirchhoff's law*

The enzyme glutamine synthetase mediates the synthesis of the amino acid glutamine (Gln, **8**) from the amino acid glutamate (Glu, **7**) and ammonium ion:

$$\Delta_r H^\circ = +21.8 \text{ kJ mol}^{-1} \text{ at } 25°C$$

The process is endothermic and requires energy extracted from the oxidation of biological fuels and stored in ATP (*Case study* 1.1). Estimate the value of the reaction enthalpy at 60°C by using data found in this text (see the *Resource section*) and the following additional information: $C_{p,m}^\circ(\text{Gln, aq}) = 187.0 \text{ J K}^{-1} \text{ mol}^{-1}$ and $C_{p,m}^\circ(\text{Glu, aq}) = 177.0 \text{ J K}^{-1} \text{ mol}^{-1}$.

7 Glutamate ion

Strategy Calculate the value of $\Delta_r C_p^\circ$ from the available data and eqn 1.25 and use the result in eqn 1.24.

Solution From the *Resource section*, the standard molar constant-pressure heat capacities of $H_2O(l)$ and $NH_4^+(aq)$ are 75.3 J K^{-1} mol^{-1} and 79.9 J K^{-1} mol^{-1}, respectively. It follows that

$$\Delta_r C_p^\circ = \{C_{p,m}^\circ(\text{Gln, aq}) + C_{p,m}^\circ(H_2O, l)\} - \{C_{p,m}^\circ(\text{Glu, aq}) + C_{p,m}^\circ(NH_4^+, aq)\}$$
$$= \{(187.0 \text{ J K}^{-1} \text{ mol}^{-1}) + (75.3 \text{ J K}^{-1} \text{ mol}^{-1})\} - \{(177.0 \text{ J K}^{-1} \text{ mol}^{-1}) + (79.9 \text{ J K}^{-1} \text{ mol}^{-1})\}$$
$$= +5.4 \text{ J K}^{-1} \text{ mol}^{-1} = +5.4 \times 10^{-3} \text{ kJ K}^{-1} \text{ mol}^{-1}$$

8 Glutamine

Then, because $T' - T = +35$ K, from eqn 1.24 we find

$$\Delta_r H^\ominus(333 \text{ K}) = (+21.8 \text{ kJ mol}^{-1}) + (5.4 \times 10^{-3} \text{ kJ K}^{-1} \text{ mol}^{-1}) \times (35 \text{ K})$$
$$= (+21.8 \text{ kJ mol}^{-1}) + (0.19 \text{ kJ mol}^{-1})$$
$$= +22.0 \text{ kJ mol}^{-1}$$

Self-test 1.7 Estimate the standard enthalpy of combustion of solid glycine at 340 K from the data in Self-test 1.6 and the *Resource section*.

Answer: -9683 kJ mol^{-1}

The calculation in Example 1.6 shows that the standard reaction enthalpy at 60°C is only slightly different from that at 25°C. The reason is that the change in reaction enthalpy is proportional to the *difference* between the molar heat capacities of the products and the reactants, which is usually not very large. It is generally the case that provided the temperature range is not too wide, enthalpies of reactions vary only slightly with temperature. A reasonable first approximation is that standard reaction enthalpies are independent of temperature. However, notable exceptions are processes involving the unfolding of macromolecules, such as proteins (*In the laboratory* 1.1). The difference in molar heat capacities between the folded and unfolded states of proteins is usually rather large, in the order of a few kilojoules per mole, so the enthalpy of protein unfolding varies significantly with temperature.

Checklist of key concepts

1. A system is classified as open, closed, or isolated.

2. The surroundings remain at constant temperature and either constant volume or constant pressure when processes occur in the system.

3. An exothermic process releases energy as heat, q, to the surroundings; an endothermic process absorbs energy as heat.

4. Metabolism is the collection of chemical reactions that trap, store, and utilize energy in biological cells.

5. Catabolism is the collection of reactions associated with the oxidation of nutrients in the cell. Anabolism is the biosynthesis of small and large molecules.

6. Maximum expansion work is achieved in a reversible change.

7. The First Law of thermodynamics states that the internal energy of an isolated system is constant.

8. A change in internal energy is equal to the energy transferred as heat at constant volume ($\Delta U = q_V$); a change in enthalpy is equal to the energy transferred as heat at constant pressure ($\Delta H = q_p$).

9. The standard state of a substance is the pure substance at 1 bar.

10. Bomb calorimetry is a useful technique for the study of nutrients.

11. Isobaric calorimetry is a useful technique for the study of fuels.

12. Differential scanning calorimetry (DSC) is a useful technique for the investigation of phase transitions, especially those observed in biological macromolecules.

13. The standard enthalpy of transition, $\Delta_{trs}H^\ominus$, is the change in molar enthalpy when a substance in one phase changes into another phase, both phases being in their standard states.

14. The standard enthalpy of the reverse of a process is the negative of the standard enthalpy of the forward process, $\Delta_{reverse}H^\ominus = -\Delta_{forward}H^\ominus$.

15. The standard enthalpy of a process is the sum of the standard enthalpies of the individual processes into which it may be regarded as divided, as in $\Delta_{sub}H^\ominus = \Delta_{fus}H^\ominus + \Delta_{vap}H^\ominus$.

16. Hess's law states that the standard enthalpy of a reaction is the sum of the standard enthalpies of the reactions into which the overall reaction can be divided.

17. The standard enthalpy of formation of a compound, $\Delta_f H^\ominus$, is the standard reaction enthalpy for the formation of the compound from its elements in their reference states.

18. At constant pressure, exothermic compounds are those for which $\Delta_f H^\ominus < 0$; endothermic compounds are those for which $\Delta_f H^\ominus > 0$.

Checklist of key equations

Property or process	Equation	Comment
Work of expansion	$w = -p_{ex}\Delta V$	Constant pressure
Heat capacity	$C = q/\Delta T$	General definition
Change in internal energy	$\Delta U = w + q$	
Constant-volume heat capacity	$C_V = dU/dT$	Definition
Enthalpy	$H = U + pV$	Definition
Enthalpy change	$\Delta H = \Delta U + p\Delta V$	Constant pressure
Constant-pressure heat capacity	$C_p = dH/dT$	Definition
Difference between the molar heat capacities	$C_{p,m} - C_{V,m} = R$	Perfect gas
Standard reaction enthalpy	$\Delta_r H^\ominus = \sum v H_m^\ominus(\text{products}) - \sum v H_m^\ominus(\text{reactants})$	Definition
	$\Delta_r H^\ominus = \sum v \Delta_f H^\ominus(\text{products}) - \sum v \Delta_f H^\ominus(\text{reactants})$	Practical implementation
Kirchhoff's law	$\Delta_r H^\ominus(T') = \Delta_r H^\ominus(T) + \Delta_r C_p^\ominus(T' - T)$	Constant-pressure heat capacities are independent of temperature

Discussion questions

1.1 Provide molecular interpretations of work, heat, temperature, and heat capacity.

1.2 Suggest a reason why most molecules survive for long periods at room temperature.

1.3 Describe the general patterns of energy conversion in living organisms.

1.4 Explain the difference between the change in internal energy and the change in enthalpy of a chemical or physical process.

1.5 Explain the limitations of the following expressions:
(a) $\Delta H = \Delta U + p\Delta V$;
(b) $\Delta_r H^\ominus(T') = \Delta_r H^\ominus(T) + \Delta_r C_p^\ominus \times (T' - T)$.

1.6 A primitive air-conditioning unit for use in places where electrical power is not available can be made by hanging up strips of linen soaked in water. Explain why this strategy is effective.

1.7 In many experimental thermograms, such as that shown in Fig. 1.20, the baseline below T_1 is at a different level from that above T_2. Explain this observation.

1.8 Describe at least two calculational methods by which standard reaction enthalpies can be predicted. Discuss the advantages and disadvantages of each method.

1.9 Distinguish between (a) the standard state and the reference state of an element; (b) endothermic and exothermic compounds.

Exercises

Assume all gases are perfect unless stated otherwise. All thermochemical data are for 298.15 K.

1.10 The unit 1 electronvolt (1 eV) is defined as the energy acquired by an electron as it moves through a potential difference of 1 V. Suppose two states differ in energy by 1.0 eV. What is the ratio of their populations at (a) 300 K and (b) 3000 K?

1.11 How much metabolic energy must a bird of mass 200 g expend to fly to a height of 20 m? Neglect all losses due to friction, physiological imperfection, and the acquisition of kinetic energy.

1.12 Calculate the work of expansion accompanying the complete combustion of 1.0 g of glucose to carbon dioxide and (a) liquid water, and (b) water vapor at 20°C when the external pressure is 1.0 atm.

1.13 We are all familiar with the general principles of operation of an internal combustion reaction: the combustion of fuel drives out the piston. It is possible to imagine engines that use reactions other than combustions, and we need to assess the work they can do. A chemical reaction takes place in a container of cross-sectional area 100 cm²; the container has a piston at one end. As a result of the reaction, the piston is pushed out through 10.0 cm against a constant external pressure of 100 kPa. Calculate the work done by the system.

1.14 A sample of methane of mass 4.50 g occupies 12.7 dm³ at 310 K. Calculate the work done when the gas expands (a) isobarically against a constant external pressure of 30.0 kPa until its volume has increased by 3.3 dm³, and (b) isothermally by 3.3 dm³.

1.15 The heat capacity of air is much smaller than that of water, and relatively modest amounts of heat are needed to change its temperature. This is one of the reasons why desert regions, although very hot during the day, are bitterly cold at night. The heat capacity of air at room temperature and pressure is approximately 21 J K⁻¹ mol⁻¹. How much energy is required to raise the temperature of a room of dimensions 5.5 m × 6.5 m × 3.0 m by 10°C? If losses are neglected, how long will it take a heater rated at 1.5 kW to achieve that increase given that 1 W = 1 J s⁻¹?

1.16 The transfer of energy from one region of the atmosphere to another is of great importance in meteorology for it affects the weather. Calculate the heat needed to be supplied to a parcel of air containing 1.00 mol air molecules to maintain its temperature at 300 K when it expands reversibly and isothermally from 22 dm³ to 30.0 dm³ as it ascends.

1.17 A laboratory animal exercised on a treadmill, which, through pulleys, raised a mass of 200 g through 1.55 m. At the same time, the animal lost 5.0 J of energy as heat. Disregarding all other losses and regarding the animal as a closed system, what is its change in internal energy?

1.18 A sample of a serum of mass 25 g is cooled from 290 K to 275 K at constant pressure by the extraction of 1.2 kJ of energy as heat. Calculate q and ΔH and estimate the heat capacity of the sample.

1.19 (a) Show that for a perfect gas, $C_{p,m} - C_{V,m} = R$. (b) When 229 J of energy is supplied as heat at constant pressure to 3.00 mol $CO_2(g)$, the temperature of the sample increases by 2.06 K. Calculate the molar heat capacities at constant volume and constant pressure of the gas.

1.20 Use the information in Exercise 1.19 to calculate the change in (a) molar enthalpy and (b) molar internal energy when carbon dioxide is heated from 15°C (the temperature when air is inhaled) to 37°C (blood temperature, the temperature in our lungs).

1.21 Suppose that the molar internal energy of a substance over a limited temperature range could be expressed as a polynomial in T as $U_m(T) = a + bT + cT^2$. Find an expression for the constant-volume molar heat capacity at a temperature T.

1.22 The heat capacity of a substance is often reported in the form $C_{p,m} = a + bT + c/T^2$. Use this expression to make a more accurate estimate of the change in molar enthalpy of carbon dioxide when it is heated from 15°C to 37°C (as in Exercise 1.20), given $a = 44.22$ J K⁻¹ mol⁻¹, $b = 8.79 \times 10^{-3}$ J K⁻² mol⁻¹, and $c = -8.62 \times 10^5$ J K mol⁻¹. *Hint:* You will need to integrate $dH = C_p dT$.

1.23 Exercise 1.22 gives an expression for the temperature dependence of the constant-pressure molar heat capacity over a

limited temperature range. (a) How does the molar enthalpy of the substance change over that range? (b) Plot the molar enthalpy as a function of temperature using the data in Exercise 1.22.

1.24 Classify as endothermic or exothermic (a) a combustion reaction for which $\Delta_r H^{\oplus} = -2020$ kJ mol⁻¹, (b) a dissolution for which $\Delta H^{\oplus} = +4.0$ kJ mol⁻¹, (c) vaporization, (d) fusion, and (e) sublimation.

1.25 The pressures deep within the Earth are much greater than those on the surface, and to make use of thermochemical data in geochemical assessments we need to take the differences into account. (a) Given that the enthalpy of combustion of graphite is -393.5 kJ mol⁻¹ and that of diamond is -395.41 kJ mol⁻¹, calculate the standard enthalpy of the C(s, graphite) → C(s, diamond) transition. (b) Use the information in part (a) together with the densities of graphite (2.250 g cm⁻³) and diamond (3.510 g cm⁻³) to calculate the internal energy of the transition when the sample is under a pressure of 150 kbar.

1.26 A typical human produces about 10 MJ of energy transferred as heat each day through metabolic activity. (a) If a human body were an isolated system of mass 65 kg with the heat capacity of water, what temperature rise would the body experience? (b) Human bodies are actually open systems, and the main mechanism of heat loss is through the evaporation of water. What mass of water should be evaporated each day to maintain constant temperature?

1.27 Use the information in Tables 1.1 and 1.2 to calculate the total heat required to melt 100 g of ice at 0°C, heat it to 100°C, and then vaporize it at that temperature. Sketch a graph of temperature against time on the assumption that the sample is heated at a constant rate.

1.28 In preparation for a study of the metabolism of an organism, a small, sealed calorimeter was assembled. In the initial phase of the experiment, a current of 22.22 mA from an 11.8 V source was passed for 162 s through a heater inside the calorimeter. What is the change in internal energy of the calorimeter?

1.29 Water is heated to boiling under a pressure of 1.0 atm. When an electric current of 0.50 A from a 12 V supply is passed for 300 s through a resistance in thermal contact with it, it is found that 0.798 g of water is vaporized. Calculate the molar internal energy and enthalpy changes at the boiling point (373.15 K).

1.30 In an experiment to determine the energy content of a food, a sample of the food was burned in an oxygen atmosphere and the temperature rose by 2.89°C. When a current of 1.27 A from a 12.5 V source flowed through the same calorimeter for 157 s, the temperature rose by 3.88°C. What energy was released as heat by the combustion?

1.31 A sample of the sugar D-ribose ($C_5H_{10}O_5$) of mass 0.727 g was placed in a calorimeter and then ignited in the presence of excess oxygen. The temperature rose by 0.910 K. In a separate experiment in the same calorimeter, the combustion of 0.917 g of benzoic acid, for which the internal energy of combustion is -3226 kJ mol⁻¹, gave a temperature rise of 1.940 K. Calculate the enthalpy of formation of D-ribose.

1.32 Figure 1.27 shows the experimental DSC scan of hen white lysozyme (G. Privalov et al., *Anal. Biochem.* **79**, 232 (1995)) converted to kilojoules (from calories). Determine the enthalpy of unfolding of this protein by integration of the curve and the change in heat capacity accompanying the transition.

1.33 The mean bond enthalpies of C–C, C–H, C=O, and O–H bonds are 348, 412, 743, and 463 kJ mol⁻¹, respectively. The combustion of a fuel such as octane is exothermic because relatively weak bonds

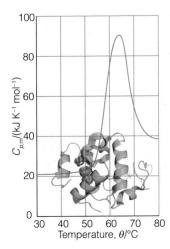

Fig. 1.27 Experimental DSC scan of hen white lysozyme.

break to form relatively strong bonds. Use this information to justify why glucose has a lower specific enthalpy than the lipid decanoic acid ($C_{10}H_{20}O_2$) even though these compounds have similar molar masses.

1.34 Use bond enthalpies and mean bond enthalpies to estimate (a) the enthalpy of the anaerobic breakdown of glucose to lactic acid in cells that are starved of O_2, $C_6H_{12}O_6(aq) \rightarrow 2\ CH_3CH(OH)COOH(aq)$, and (b) the enthalpy of combustion of glucose. Ignore the contributions of enthalpies of fusion and vaporization.

1.35 Glucose and fructose are simple sugars with the molecular formula $C_6H_{12}O_6$. Sucrose (table sugar) is a complex sugar with molecular formula $C_{12}H_{22}O_{11}$ that consists of a glucose unit covalently bound to a fructose unit (a water molecule is eliminated as a result of the reaction between glucose and fructose to form sucrose). (a) Calculate the energy released as heat when a typical table sugar cube of mass 1.5 g is burned in air. (b) To what height could you climb on the energy a table sugar cube provides assuming 25 per cent of the energy is available for work? (c) The mass of a typical glucose tablet is 2.5 g. Calculate the energy released as heat when a glucose tablet is burned in air. (d) To what height could you climb on the energy a tablet provides assuming 25 per cent of the energy released by the metabolism of glucose is available for work?

1.36 Camping gas is typically propane. The standard enthalpy of combustion of propane gas is -2220 kJ mol^{-1} and the standard enthalpy of vaporization of the liquid is $+15$ kJ mol^{-1}. Calculate (a) the standard enthalpy and (b) the standard internal energy of combustion of the liquid.

1.37 Ethane is flamed off in abundance from oil wells because it is unreactive and difficult to use commercially. But would it make a good fuel? The standard enthalpy of reaction for $2\ C_2H_6(g) + 7\ O_2(g) \rightarrow 4\ CO_2(g) + 6\ H_2O(l)$ is -3120 kJ. (a) What is the standard enthalpy of combustion of ethane? (b) What is the specific enthalpy of combustion of ethane? (c) Is ethane a more or less efficient fuel than methane?

1.38 Estimate the difference between the standard enthalpy of formation of $H_2O(l)$ as currently defined (at 1 bar) and its value using the former definition (at 1 atm).

1.39 Use information in the *Resource section* to calculate the standard enthalpies of the following reactions:

(a) the hydrolysis of a glycine–glycine dipeptide:

$$^+NH_3CH_2CONHCH_2CO_2^-(s) + H_2O(l) \rightarrow 2\ ^+NH_3CH_2CO_2^-(aq)$$

(b) the combustion of solid β-D-fructose

(c) the dissociation of nitrogen dioxide, which occurs in the atmosphere:

$$NO_2(g) \rightarrow NO(g) + O(g)$$

1.40 During glycolysis, glucose is partially oxidized to pyruvic acid, $CH_3COCOOH$, by NAD$^+$ (see Chapter 4) without the involvement of O_2. However, it is also possible to carry out the oxidation in the presence of O_2:

$$C_6H_{12}O_6(s) + O_2(g) \rightarrow 2\ CH_3COCOOH(s) + 2\ H_2O(l)$$
$$\Delta_r H^{\ominus} = -480.7 \text{ kJ mol}^{-1}$$

From these data and additional information in the *Resource section*, calculate the standard enthalpy of combustion and standard enthalpy of formation of pyruvic acid.

1.41 At 298 K, the enthalpy of denaturation of hen egg white lysozyme is $+217.6$ kJ mol^{-1} and the change in the constant-pressure molar heat capacity resulting from denaturation of the protein is $+6.3$ kJ K^{-1} mol^{-1}. (a) Estimate the enthalpy of denaturation of the protein at (i) 351 K, the 'melting' temperature of the macromolecule, and (ii) 263 K. State any assumptions in your calculations. (b) Based on your answers to part (a), is denaturation of hen egg white lysozyme always endothermic?

1.42 Estimate the enthalpy of vaporization of water at 100°C from its value at 25°C ($+44.01$ kJ mol^{-1}) given the constant-pressure heat capacities of 75.29 J K^{-1} mol^{-1} and 33.58 J K^{-1} mol^{-1} for liquid and gas, respectively.

1.43 Is the standard enthalpy of combustion of glucose likely to be higher or lower at blood temperature than at 25°C?

1.44 Using the fact that the enthalpy is a state function, derive a version of Kirchhoff's law (eqn 1.24) by adding contributions from the following processes: (a) the enthalpy change when the reactants are cooled from a temperature T to 298 K, (b) the reaction enthalpy at 298 K, and (c) the enthalpy change when the temperature of the products is increased from 298 K to T.

1.45 Derive a version of Kirchhoff's law (eqn 1.24) for the temperature dependence of the internal energy of reaction.

1.46 The formulation of Kirchhoff's law given in eqn 1.24 is valid when the difference in heat capacities is independent of temperature over the temperature range of interest. Suppose instead that $\Delta_r C_p^{\ominus} = a + bT + c/T^2$. Derive a more accurate form of Kirchhoff's law in terms of the parameters a, b, and c. *Hint:* The change in the reaction enthalpy for an infinitesimal change in temperature is $\Delta_r C_p^{\ominus} dT$. Integrate this expression between the two temperatures of interest.

Projects

1.47 The Boltzmann distribution can be used to calculate the average energy associated with each mode of motion of a molecule. However, for certain modes of motion, such as translation, there is a short cut, called the *equipartition theorem*:

> In a sample at a temperature T, all quadratic contributions to the total energy have the same mean value, namely $\frac{1}{2}kT$.

A 'quadratic contribution' means a contribution that depends on the square of the position or the velocity (or momentum).

(a) The kinetic energy of a particle of mass m free to undergo translation in three dimensions is $E_k = \frac{1}{2}mv_x^2 + \frac{1}{2}mv_y^2 + \frac{1}{2}mv_z^2$. What is the average kinetic energy of a particle free to move in three dimensions?

(b) Use the equipartition theorem to show that for a monatomic perfect gas:

$$U_m(T) = U_m(0) + \tfrac{3}{2}RT \qquad C_{V,m} = \tfrac{3}{2}R$$

where $U_m(0)$ is the molar internal energy at $T = 0$, when all translational motion has ceased.

(c) When the gas consists of molecules, we need to take into account the effect of rotation and vibration. A linear molecule, such as N_2 and CO_2, can rotate around two axes perpendicular to the line of the atoms, so it has two rotational modes of motion, each contributing a term $\frac{1}{2}kT$ to the internal energy. Show that

$$U_m(T) = U_m(0) + \tfrac{5}{2}RT \qquad C_{V,m} = \tfrac{5}{2}R$$
(linear molecule, translation and rotation only)

(d) A non-linear molecule, such as CH_4 or H_2O, can rotate around three axes and, again, each mode of motion contributes a term $\frac{1}{2}kT$ to the internal energy. Show that

$$U_m(T) = U_m(0) + 3RT \qquad C_{V,m} = 3R$$
(non-linear molecule, translation and rotation only)

(e) Molecules do not vibrate significantly at room temperature and, as a first approximation, the contribution of molecular vibrations to the internal energy is negligible except for very large molecules such as polymers and biological macromolecules. Use the information in this problem to justify the following statement: *the internal energy of a perfect gas does not change when the gas undergoes isothermal expansion.*

(f) Use the equipartition theorem to calculate the contribution of molecular motion to the total energy of a sample of 10.0 g of (i) argon, (ii) carbon dioxide, and (iii) methane at 20°C. *Hint:* For (ii) and (iii), take into account translation and rotation but not vibration.

(g) We saw in part (e) that the internal energy of a perfect gas does not change when the gas undergoes isothermal expansion. What is the change in enthalpy?

1.48 It is possible to see with the aid of a powerful microscope that a long piece of double-stranded DNA is flexible, with the distance between the ends of the chain adopting a wide range of values. This flexibility is important because it allows DNA to adopt very compact conformations as it is packaged in a chromosome (see Chapter 11). It is convenient to visualize a long piece of DNA as a *freely jointed chain*, a chain of N small, rigid units of length l that are free to make any angle with respect to each other. The length l, the *persistence length*, is approximately 45 nm, corresponding to approximately 130 base pairs. You will now explore the work associated with extending a DNA molecule.

(a) Suppose that a DNA molecule resists being extended from an equilibrium, more compact conformation with a *restoring force* $F = -k_f x$, where x is the difference in the end-to-end distance of the chain from an equilibrium value and k_f is the *force constant*. Systems showing this behavior are said to obey *Hooke's law*. (i) What are the limitations of this model of the DNA molecule? (ii) Using this model, write an expression for the work that must be done to extend a DNA molecule by x. Draw a graph of your conclusion.

(b) A better model of a DNA molecule is the *one-dimensional freely jointed chain*, in which a rigid unit of length l can make an angle of only 0° or 180° with an adjacent unit. In this case, the restoring force of a chain extended by $x = nl$ is given by

$$F = \frac{kT}{2l} \ln\left(\frac{1+v}{1-v}\right) \qquad v = n/N$$

where $k = 1.381 \times 10^{-23}$ J K^{-1} is Boltzmann's constant (not a force constant). (i) What are the limitations of this model? (ii) What is the magnitude of the force that must be applied to extend a DNA molecule with $N = 200$ by 90 nm? (iii) Plot the restoring force against v, noting that v can be either positive or negative. How is the variation of the restoring force with end-to-end distance different from that predicted by Hooke's law? (iv) Keeping in mind that the difference in end-to-end distance from an equilibrium value is $x = nl$ and, consequently, $dx = ldn = Nldv$, write an expression for the work of extending a DNA molecule. (v) Calculate the work of extending a DNA molecule from $v = 0$ to $v = 1.0$. *Hint:* You must integrate the expression for w. The task can be accomplished easily with mathematical software.

(c) Show that for small extensions of the chain, when $v \ll 1$, the restoring force is given by

$$F \approx \frac{vkT}{l} = \frac{nkT}{Nl}$$

Hint: See *Mathematical toolkit* 3.2 for a review of series expansions of functions.

(d) Is the variation of the restoring force with extension of the chain given in part (c) different from that predicted by Hooke's law? Explain your answer.

The Second Law

Some things happen; some things don't. A gas expands to fill the vessel it occupies; a gas that already fills a vessel does not suddenly contract into a smaller volume. A hot object cools to the temperature of its surroundings; a cool object does not suddenly become hotter than its surroundings. Hydrogen and oxygen combine explosively (once their ability to do so has been liberated by a spark) and form water; water left standing in oceans and lakes does not gradually decompose into hydrogen and oxygen. These everyday observations suggest that changes can be divided into two classes. A **spontaneous change** is a change that has a tendency to occur without work having to be done to bring it about. A spontaneous change has a natural tendency to occur. A non-spontaneous change is a change that can be brought about only by doing work. A non-spontaneous change has no natural tendency to occur. Non-spontaneous changes can be made to occur by doing work: a gas can be compressed into a smaller volume by pushing in a piston, the temperature of a cool object can be raised by forcing an electric current through a heater attached to it, and water can be decomposed by the passage of an electric current. However, in each case we need to act in some way on the system to bring about the non-spontaneous change. There must be some feature of the world that accounts for the distinction between the two types of change.

Throughout this chapter and the rest of the text we shall use the terms 'spontaneous' and 'non-spontaneous' in their thermodynamic sense. That is, we use them to signify that a change does or does not have a natural *tendency* to occur. In thermodynamics the term spontaneous has nothing to do with speed. Some spontaneous changes are very fast, such as the precipitation reaction that occurs when solutions of sodium chloride and silver nitrate are mixed. However, some spontaneous changes are so slow that there may be no observable change even after millions of years. For example, although the decomposition of benzene into carbon and hydrogen is spontaneous, it does not occur at a measurable rate under normal conditions, and benzene is a common laboratory commodity with a shelf life of (in principle) millions of years. Thermodynamics deals with the *tendency* to change; it is silent on the rate at which that tendency is realized.

We shall use the concepts introduced in this chapter as a basis for our study of bioenergetics and structure in biological systems. Our discussion of energy conversion in biological cells has focused on the chemical sources of energy that sustain life. We now begin an investigation—to be continued throughout the text—of the mechanisms by which energy in the form of radiation from the Sun or ingested as oxidizable molecules is converted into work of muscle contraction, neuronal activity, biosynthesis of essential molecules, and transport of material into and out of the cell. We shall also explain a remark made in Chapter 1, that only part of the energy of biological fuels leads to work, with the rest being dissipated in the surroundings as heat. We shall also see that the material discussed in this chapter is relevant not only to the transformations of energy but also to the structures of proteins.

Entropy

A few moments' thought is all that is needed to identify the reason why some changes are spontaneous and others are not. That reason is *not* the tendency of the system to move toward lower energy. This point is easily established by identifying an example of a spontaneous change in which there is no change in energy. The isothermal expansion of a perfect gas into a vacuum is spontaneous, but the total energy of the gas does not change because the molecules continue to travel at the same average speed and so keep their same total kinetic energy. Even in a process in which the energy of a system does decrease (as in the spontaneous cooling of a block of hot metal), the First Law requires the total energy to be constant. Therefore, in this case the energy of another part of the universe must increase if the energy decreases in the part that interests us. For instance, a hot block of metal in contact with a cool block cools and loses energy; however, the second block becomes warmer and increases in energy. It is equally valid to say that the second block has a tendency to go to higher energy as it is to say that the first block has a tendency to go to lower energy!

In the next few sections we shall develop the thermodynamic criteria for spontaneity by using an approach similar to that adopted in Chapter 1. At first sight the ideas, models, and mathematical expressions in our discussion may appear to be of no immediate concern to a biochemist. But in due course we shall see how they are of the greatest importance for an understanding of the flow of energy in biological systems, the reactions that sustain them, and the structures of biological macromolecules.

2.1 The direction of spontaneous change

To understand the processes occurring in organisms, we need to identify the factors that drive any physical or chemical change.

We shall now show that *the apparent driving force of spontaneous change is the tendency of energy and matter to disperse.* For example, the molecules of a gas may all be in one region of a container initially, but their ceaseless disorderly motion ensures that they spread rapidly throughout the entire volume of the container (Fig. 2.1). Because their motion is so random, there is a negligibly small probability that all the molecules will find their way back simultaneously into the region of the container they occupied initially. In this instance, the natural direction of change corresponds to the disorderly dispersal of matter.

A similar explanation accounts for spontaneous cooling, but now we need to consider the dispersal of energy rather than of matter. In a block of hot metal in which there is a temperature gradient, the atoms in the hot region are oscillating vigorously, and the hotter the region, the more vigorous their motion. The cooler region also consists of oscillating atoms, but their motion is less vigorous. The vigorously oscillating atoms jostle their neighbors in the cooler region, and the energy of the atoms in the block is handed on to the atoms in the cooler region (Fig. 2.2). The process continues until the vigor with which the atoms in the system are oscillating is uniform. The opposite flow of energy is very unlikely. It is highly improbable that there will be a net accumulation of energy in one

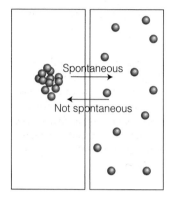

Fig. 2.1 One fundamental type of spontaneous process is the disorderly dispersal of matter. This tendency accounts for the spontaneous tendency of a gas to spread into and fill the container it occupies. It is extremely unlikely that all the particles will collect into one small region of the container. (In practice, the number of particles is of the order of 10^{23}.)

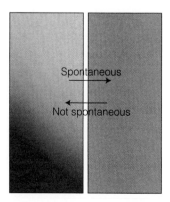

Fig. 2.2 Another fundamental type of spontaneous process is the disorderly dispersal of energy (represented by the color: red is high temperature, blue is low temperature).

region as a result of the jostling of molecules within the block. In this case, the natural direction of change corresponds to the disorderly dispersal of energy. Similar remarks apply to the cooling of a hot block to the temperature of its surroundings.

The tendency of energy to disperse also explains the fact that, despite countless attempts, it has proved impossible to construct an engine like that shown in Fig. 2.3, in which heat, perhaps from the combustion of a fuel, is drawn from a hot reservoir and completely converted into work, such as the work of moving an automobile. All actual heat engines have both a hot region, the 'source', and a cold region, the 'sink', and it has been found that for such engines to operate, some energy must be discarded into the cold sink as heat and not used to do work. In molecular terms, only some of the energy stored in the atoms and molecules of the hot source can be used to do work and transferred to the surroundings in an orderly way. For the engine to do work, some energy must be transferred to the cold sink as heat, to stimulate random motion of its atoms and molecules. Before long, we shall see that the heat engine, although it looks like an engineering concept, when appropriately interpreted is directly applicable to biochemical processes even though there is no actual 'engine' present.

In summary, we have identified two basic types of spontaneous physical process:

1) Matter tends to disperse in disorder.

2) Energy tends to disperse in disorder.

We now need to take the next step and see how these two fundamental processes result in some chemical reactions being spontaneous and others not. It may seem very puzzling that chaotic dispersal can account for the formation of such organized systems as proteins and biological cells. Nevertheless, in due course we shall see that change in all its forms, including the formation of organized structures, is driven by the tendency of energy and matter to disperse in disorder.

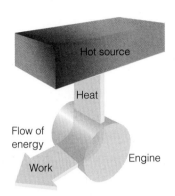

Fig. 2.3 The Second Law denies the possibility of the process illustrated here, in which heat is changed completely into work, there being no other change. The process is not in conflict with the First Law because the energy is conserved.

2.2 Entropy and the Second Law

> To make progress with our quantitative discussion of biological structure and reactivity, we need to associate the dispersal of energy and matter with the change in a state function.

The measure of the disorderly dispersal of energy or matter used in thermodynamics is called the **entropy**, S. We shall soon define entropy precisely and quantitatively, but for now all we need to know is that when matter and energy disperse in disorder, entropy increases. That being so, we can combine the two remarks above into a single statement known as the **Second Law of thermodynamics**:

The entropy of an isolated system tends to increase.   (The Second Law)

The 'isolated system' may consist of a system in which we have a special interest (a beaker containing reagents, a biological cell, or even an organelle within a cell) and that system's surroundings: the two components jointly form a little 'universe' in the thermodynamic sense.

(a) The definition of entropy

To make progress and turn the Second Law into a quantitatively useful statement, we shall use the following definition of a *change* in entropy:

$$\Delta S = \frac{q_{rev}}{T}$$

Definition of entropy change   (2.1)

That is, the change in entropy of a system is equal to the energy transferred as heat to it *reversibly* divided by the temperature at which the transfer takes place. This definition can be justified thermodynamically,[1] but we shall confine ourselves to showing that it is plausible and then show how to use it to obtain numerical values for a range of processes.

There are three points we need to understand about the definition in eqn 2.1:

The significance of the term 'reversible'

We met the concept of reversibility in Section 1.3, where we saw that it refers to the ability of an infinitesimal change in a control variable to change the direction of a process. Mechanical reversibility refers to the equality of pressure acting on either side of a movable wall. Thermal reversibility, the type involved in eqn 2.1, refers to the equality of temperature on either side of a thermally conducting wall. Reversible transfer of heat is smooth, careful, restrained transfer between two bodies at the same temperature. By making the transfer reversible, we ensure that there are no hot spots generated in the object that later disperse spontaneously and hence add to the entropy.

Why heat (not work) appears in the numerator

Recall from Section 1.2 that to transfer energy as heat, we make use of the random motion of molecules, whereas to transfer energy as work, we make use of orderly motion. It should be plausible that the change in entropy—the change in the degree of disorder of energy and matter—is proportional to the energy transfer that takes place by making use of random motion rather than orderly motion.

Why temperature appears in the denominator

The presence of the temperature in the denominator in eqn 2.1 takes into account the randomness of motion that is already present. If a given quantity of energy is transferred as heat to a hot object (one in which the atoms already undergo a significant amount of thermal motion), then the additional randomness of motion generated is less significant than if the same quantity of energy is transferred as heat to a cold object in which the atoms have less thermal motion. The difference is like sneezing in a busy street (an environment analogous to a high temperature), which adds little to the disorder already present, and sneezing in a quiet library (an environment analogous to a low temperature), which can be very disruptive.

> ### A brief illustration
>
> An organism inhabits a pond. In the course of its life, the organism transfers 100 kJ of heat to the pond water at 0°C (273 K). The resulting change in entropy of the water due to this transfer is
>
> $$\Delta S = \frac{q_{rev}}{T} = \frac{100 \times 10^3 \, J}{273 \, K} = +366 \, J \, K^{-1}$$
>
> The pond is large enough to ensure that the temperature of the water does not change as heat is transferred. The same transfer at 100°C (373 K) results in

---

[1] For a thermodynamic justification, see our *Physical chemistry* (2010).

$$\Delta S = \frac{100 \times 10^3 \, \text{J}}{373 \, \text{K}} = +268 \, \text{J K}^{-1}$$

The increase in entropy is greater at the lower temperature. Notice that the units of entropy are joules per kelvin (J K^{-1}). Entropy is an extensive property. When we deal with molar entropy, an intensive property, the units will be joules per kelvin per mole (J K^{-1} mol^{-1}).

The entropy (it can be proved) is a state function, a property with a value that depends only on the present state of the system.[2] The entropy is a measure of the current state of dispersal of energy and matter in the system, and how that change was achieved is not relevant to its current value. The implication of entropy being a state function is that a change in its value when a system undergoes a change of state is independent of how the change of state is brought about.

(b) The entropy change accompanying heating

We can often rely on intuition to judge whether the entropy increases or decreases when a substance undergoes a physical change. For instance, the entropy of a sample of gas increases as it expands because the molecules are able to move in a greater volume and so are more widely dispersed. Similarly, we should expect the entropy of a sample to increase as the temperature is raised because the thermal motion is greater at the higher temperature. As we show in the following *Justification*, provided the heat capacity is constant over the range of temperatures of interest,

$$\Delta S = C \ln \frac{T_f}{T_i}$$

> Change in entropy on heating   (2.2)

where C is the heat capacity of the system; if the pressure is held constant during the heating, we use the constant-pressure heat capacity, C_p, and if the volume is held constant, we use the constant-volume heat capacity, C_V.

Justification 2.1 *The change in entropy with temperature*

Equation 2.1 refers to the transfer of heat to a system at a temperature T. In general, the temperature changes as we heat a system, so we cannot use eqn 2.1 directly. Suppose, however, that we transfer only an infinitesimal energy as heat, dq, to the system; then there is only an infinitesimal change in temperature and we introduce negligible error if we keep the temperature in the denominator of eqn 2.1 equal to T during that transfer. As a result, the entropy increases by an infinitesimal amount dS given by

$$dS = \frac{dq_{rev}}{T}$$

To calculate dq, we recall from Section 1.4 that the heat capacity $C = q/\Delta T$, where ΔT is macroscopic change in temperature. For an infinitesimal change dT brought about by an infinitesimal transfer of heat dq we write $C = dq/dT$ and therefore d$q = CdT$, so we can write d$q_{rev} = CdT$ and therefore

---

[2] Again see our *Physical chemistry* (2010) for a proof.

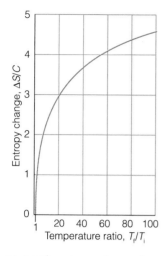

Fig. 2.4 The entropy of a sample with a heat capacity that can be regarded as independent of temperature in the range of interest increases as ln T as the temperature is increased. The increase is proportional to the heat capacity of the sample.

$$\Delta S = \frac{CdT}{T}$$

The total change in entropy, ΔS, when the temperature changes from T_i to T_f is the sum (integral) of all such infinitesimal terms:

$$\Delta S = \int_{T_i}^{T_f} \frac{CdT}{T}$$

For many substances and for small temperature ranges we may take C to be constant. (This is strictly true only for a monatomic perfect gas.) Then C may be taken outside the integral and the latter evaluated as follows:

$$\Delta S = \int_{T_i}^{T_f} \frac{CdT}{T} = C\int_{T_i}^{T_f} \frac{dT}{T} = C\ln\frac{T_f}{T_i}$$

Equation 2.2 is in line with what we expect. When $T_f > T_i$, $T_f/T_i > 1$, which implies that the logarithm is positive, that $\Delta S > 0$, and therefore that the entropy increases (Fig. 2.4). Note that the relation also shows a less obvious point, that the higher the heat capacity of the substance, the greater the change in entropy for a given rise in temperature. A moment's thought shows this conclusion to be reasonable too: a high heat capacity implies that a lot of heat is required to produce a given change in temperature, so the 'sneeze' (in terms of the analogy mentioned earlier) must be more powerful than when the heat capacity is low, and the entropy increase is correspondingly high.

Self-test 2.1 Calculate the change in molar entropy when water vapor is heated from 160°C to 170°C at constant volume. ($C_{V,m} = 26.92 \text{ J K}^{-1} \text{ mol}^{-1}$.)

Answer: +0.615 J K^{-1} mol^{-1}

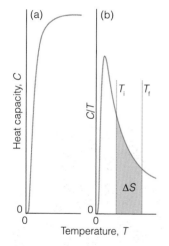

Fig. 2.5 The experimental determination of the change in entropy of a sample that has a heat capacity that varies with temperature, as shown in (a), involves measuring the heat capacity over the range of temperatures of interest, then plotting C/T against T and determining the area under the curve (the tinted area shown), as shown in (b). The heat capacity of all solids decreases toward zero as the temperature is reduced.

When we cannot assume that the heat capacity is constant over the temperature range of interest, which is the case for all solids at low temperatures, we have to allow for the variation of C with temperature. In *Justification* 2.1 we found, before making the assumption that the heat capacity is constant, that

$$\Delta S = \int_{T_i}^{T_f} \frac{CdT}{T} \tag{2.3}$$

All we need to recognize is the standard result from calculus, that the integral of a function between two limits is the area under the graph of the function between the two limits. In this case, the function is C/T, the heat capacity at each temperature divided by that temperature, and it follows that

ΔS = area under the graph of C/T plotted against T, between T_i and T_f
Experimental basis of determining an entropy change

This rule is illustrated in Fig. 2.5. To use eqn 2.3, we measure the heat capacity throughout the range of temperatures of interest. Then we divide each measurement by the corresponding temperature to get C/T at each temperature, plot these

Finally, we calculate the change in entropy for cooling the vapor from 100°C to 25°C (using eqn 2.2 again, but now with data for the vapor from Table 1.1):

$$\Delta S_3 = C_{p,m}(H_2O, \text{vapor}) \ln \frac{T_f}{T_i} = (33.58 \text{ J K}^{-1} \text{ mol}^{-1}) \times \ln \frac{298 \text{ K}}{373 \text{ K}}$$

$$= -7.5 \text{ J K}^{-1} \text{ mol}^{-1}$$

The sum of the three entropy changes is the entropy of transition at 25°C:

$$\Delta_{vap}S(298 \text{ K}) = \Delta S_1 + \Delta S_2 + \Delta S_3 = +118 \text{ J K}^{-1} \text{ mol}^{-1}$$

(d) Entropy changes in the surroundings

We can use the definition of entropy in eqn 2.1 to calculate the entropy change of the surroundings in contact with the system at the temperature T: $\Delta S_{sur} = q_{sur,rev}/T$. However, surroundings are so extensive that the spread of heat through them is effectively reversible, so the 'rev' subscript can be dropped and we can write $\Delta S_{sur} = q_{sur}/T$. Moreover, the heat entering the surroundings is lost from the system, so $q_{sur} = -q$. (For instance, if $q = +100$ J, an influx of 100 J into the system, then $q_{sur} = -100$ J, indicating that the surroundings have lost that 100 J.) Therefore, at this stage we can write $\Delta S_{sur} = -q/T$. Finally, if the change in the system is taking place at constant pressure, we can identify q with the change of enthalpy ΔH, and so obtain

for a process at constant pressure: $\Delta S_{sur} = -\dfrac{\Delta H}{T}$ Entropy change of the surroundings   (2.6)

This enormously important expression will lie at the heart of our discussion of bioenergetics and the structural consequences of the Second Law. We see that it is consistent with common sense: if the process is exothermic, ΔH is negative and therefore ΔS_{sur} is positive. The entropy of the surroundings increases if heat is released into them. If the process is endothermic ($\Delta H > 0$), then the entropy of the surroundings decreases.

A brief illustration

The enthalpy of vaporization of water at 20°C is 44 kJ mol^{-1}. When 10 cm^3 of water (corresponding to 10 g or 0.55 mol H_2O) in an open vessel evaporates at that temperature, the change in entropy of the surroundings is

$$\Delta S_{sur} = -\frac{(0.55 \text{ mol}) \times (44 \text{ kJ mol}^{-1})}{293 \text{ K}} = -83 \text{ J K}^{-1}$$

The entropy of the surroundings decreases because heat flows out of them into the water.

2.3 Absolute entropies and the Third Law of thermodynamics

To calculate the entropy changes associated with biological processes, we need to see how to compile tables that list the values of the entropies of substances.

The graphical procedure summarized by Fig. 2.5 and eqn 2.3 for the determination of the difference in entropy of a substance at two temperatures has a very

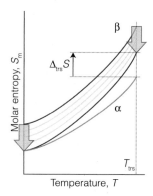

Fig. 2.8 The molar entropies of monoclinic and rhombic sulfur vary with temperature as shown here. Initially we do not know their values at $T = 0$. When we slide the two curves together by matching their separation to the measured entropy of transition at the transition temperature, we find that the entropies of the two forms are the same at $T = 0$.

important application. If $T_i = 0$, then the area under the graph between $T = 0$ and some temperature T gives us the value of $\Delta S = S(T) - S(0)$. However, at $T = 0$, all the motion of the atoms has been eliminated and there is no thermal disorder. Moreover, if the substance is perfectly crystalline, with every atom in a well-defined location, then there is no spatial disorder either. We can therefore suspect that at $T = 0$, the entropy is zero.

The thermodynamic evidence for this conclusion is based on observations like the following. Sulfur undergoes a phase transition from its rhombic form to its monoclinic polymorph at 96°C (369 K) and the enthalpy of transition is +402 J mol^{-1}. The entropy of transition is therefore +1.09 J K^{-1} mol^{-1} at this temperature. We can also measure the molar entropy of each phase relative to its value at $T = 0$ by determining the heat capacity from $T = 0$ up to the transition temperature (Fig. 2.8). At this stage, we do not know the values of the entropies at $T = 0$. However, as we see from the illustration, to match the observed entropy of transition at 369 K, *the molar entropies of the two crystalline forms must be the same at $T = 0$*. We cannot say that the entropies are zero at $T = 0$, but from the experimental data we do know that they are the same. This observation is generalized into the **Third Law of thermodynamics**:

> The entropies of all perfectly crystalline substances are the same at $T = 0$.
>
> The Third Law

For convenience (and in accordance with our understanding of entropy as a measure of disorder), we take this common value to be zero. Then, with this convention, according to the Third Law,

> $S(0) = 0$ for all perfectly ordered crystalline materials.

The **Third-Law entropy**, which is commonly called simply 'the entropy', at any temperature, $S(T)$, is based on setting $S(0) = 0$. The entropy of a substance depends on the pressure; we therefore select a standard pressure (1 bar) and report the **standard molar entropy**, $S_m^{\ominus}$, the molar entropy of a substance in its standard state at the temperature of interest. Some values at 298.15 K (the conventional temperature for reporting data) are given in Table 2.2.

It is worth taking a moment to look at the values in Table 2.2 to see that they are consistent with our understanding. All standard molar entropies are positive because raising the temperature of a sample above $T = 0$ invariably increases its entropy above the value $S(0) = 0$ because there is more thermal disorder. Another feature that we can understand in terms of disorder is illustrated by the standard molar entropy of diamond (2.4 J K^{-1} mol^{-1}), which is lower than that of graphite (5.7 J K^{-1} mol^{-1}). This difference is consistent with the atoms being linked less rigidly in graphite than in diamond and their thermal motion being correspondingly greater. The standard molar entropies of ice, water, and water vapor at 25°C are, respectively, 45, 70, and 189 J K^{-1} mol^{-1}, and the increase in values corresponds to the increasing molecular disorder on going from a solid to a liquid and then to a gas.

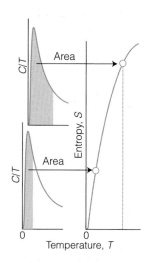

Fig. 2.9 The absolute entropy (or Third-Law entropy) of a substance is calculated by extending the measurement of heat capacities down to $T = 0$ (or as close to that value as possible) and then determining the area of the graph of C/T against T up to the temperature of interest. The area is equal to the absolute entropy at the temperature T.

| In the laboratory 2.1 | *The measurement of entropies* |

The Third-Law entropy at any temperature, $S(T)$, is equal to the area under the graph of C/T between $T = 0$ and the temperature T (Fig. 2.9). If there are any phase transitions (for example, melting) in that range, then the entropy of each transition at the transition temperature is calculated like that in eqn 2.4 and

Table 2.2 Standard molar entropies of some substances at 298.15 K*

Substance	$S_m^\circ/(J\ K^{-1}\ mol^{-1})$
Gases	
Ammonia, NH_3	192.5
Carbon dioxide, CO_2	213.7
Hydrogen, H_2	130.7
Nitrogen, N_2	191.6
Oxygen, O_2	205.1
Water vapor, H_2O	188.8
Liquids	
Acetic acid, CH_3COOH	159.8
Ethanol, CH_3CH_2OH	160.7
Water, H_2O	69.9
Solids	
Calcium carbonate, $CaCO_3$	92.9
Diamond, C	2.4
Glycine, $CH_2(NH_2)COOH$	103.5
Graphite, C	5.7
Sodium chloride, NaCl	72.1
Sucrose, $C_{12}H_{22}O_{11}$	360.2
Urea, $CO(NH_2)_2$	104.60

*See the *Resource section* for more values.

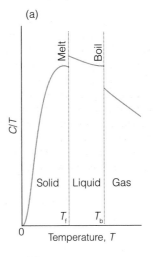

(a)

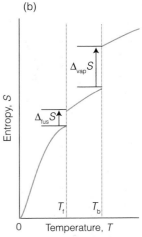

(b)

Fig. 2.10 The determination of entropy from heat capacity data. (a) Variation of C/T with the temperature of the sample. (b) The entropy, which is equal to the area beneath the upper curve up to the temperature of interest plus the entropy of each phase transition between $T = 0$ and the temperature of interest.

its contribution added to the contributions from each of the phases, as shown in Fig. 2.10. The entropies of gas-phase species may also be calculated from spectroscopic data about bond lengths and angles using the techniques of statistical thermodynamics, but few biologically interesting substances can be treated in this way.

To implement the calorimetric procedure the heat capacity of the substance is measured (for instance, by using a differential scanning calorimeter (DSC)) down to as low a temperature as feasible and then using eqn 2.3. In practice, a polynomial in T is fitted to the experimental data and then C_p/T is integrated from the lowest temperature attainable up to the temperature of interest. Thus, if the function $C_p(T) = a + bT + cT^2 + \cdots$ is fitted (for instance, by using a least-squares procedure in a software package) to the data between T_{lowest} and T_{trs}, where T_{trs} is the temperature of a phase transition, the entropy just before the phase transition is

$$S(T_{trs}) = S(T_{lowest}) + \int_{T_{lowest}}^{T_{trs}} \frac{C_p(T)}{T}\,dT$$

Then another polynomial is fitted to the heat capacities for the new phase up to the temperature of interest (or the next phase transition and a similar integral is evaluated). At each phase transition the enthalpy of transition is measured (once again, typically with a DSC), the entropy of transition is calculated as $\Delta_{trs}H(T_{trs})/T_{trs}$ by analogy with eqn 2.4, and this value is added to the value calculated by integrating the heat capacity.

There remains the experimental problem of determining $S(T_{lowest})$, the entropy at the lowest attainable temperature. If very low temperatures (within a few kelvins of $T = 0$) can be reached and reliable measurements of C_p made, it is possible to use an extrapolation based on the observation that many non-metallic substances have a heat capacity that obeys the **Debye T^3-law**:

> At temperatures close to $T = 0$, $C_p = aT^3$ Debye T^3-law   (2.7a)

where a is a constant that depends on the substance and is found by fitting this equation to a series of measurements of the heat capacity close to $T = 0$. With a determined, the entropy at low temperatures is simply

> At temperatures close to $T = 0$, $S(T) = \tfrac{1}{3}C_p(T)$ Entropy at low temperatures   (2.7b)

(See Exercise 2.21.) That is, the molar entropy at the low temperature T (which can be identified as T_{lowest}) is equal to one-third of the constant-pressure molar heat capacity at that temperature. Other extrapolation techniques have been developed that do not require reaching such low temperatures as those required for the Debye approximation to be reliable and are described in textbooks of laboratory procedures.

2.4 The molecular interpretation of the Second and Third Laws

To gain insight into the thermodynamic properties of biological assemblies and a deeper understanding of what drives a spontaneous change, we need to develop a molecular view of entropy.

The entry point into the molecular interpretation of the Second Law of thermodynamics is Boltzmann's insight into the manner in which molecules are distributed over their available energy levels, which we explored in *Fundamentals* F.3 and Section 1.2.

(a) The Boltzmann formula

Boltzmann made the link between the distribution of molecules over energy levels and the entropy. He proposed that the entropy of a system is given by

> $S = k \ln W$ Boltzmann formula for the entropy   (2.8)

where k is Boltzmann's constant and W is the number of **microstates**, the ways in which the molecules of a system can be arranged for the same total energy. At $T = 0$, all the molecules must be in the lowest energy state, and there is only one way of achieving that arrangement, so $W = 1$ and $S(0) = 0$ (because $\ln 1 = 0$), in accord with the Third Law. As the temperature is raised, more arrangements correspond to the same energy, so W increases and S rises.

Suppose we raise the temperature just enough for two molecules of a 100-molecule system to be able to leave their lowest energy state and occupy the first excited state. Two possible microstates are

$([3,4,\dots,100]_{\text{in state }0}[1,2]_{\text{in state }1})$ and
$([1,3,4\dots 42,44,\dots.100]_{\text{in state }0}[2,43]_{\text{in state }1})$

where molecules 1 and 2 are excited in the first microstate and molecules 2 and 43 are excited in the second. Each microstate lasts only for an instant and corresponds to a particular distribution of molecules over the available energy levels. These two microstates and a large number of others all correspond to the same **configuration**, in this case the configuration $\{98,2,0,\ldots\}$ we introduced in Section 1.2(c). In this case, there are $W = 4950$ possible microstates (that corresponds to the number of ways of choosing two molecules from 100).

As we saw in Section 1.2(c), there is a dominating configuration of the system—the one corresponding to the greatest number of microstates for a given total energy—and the properties of the system are those of this most probable configuration. That configuration is the one with populations given by the Boltzmann distribution. To use Boltzmann's formula for the entropy, we set the W that occurs in it equal to the W of this dominating configuration.

A brief illustration

Suppose that a protein molecule of 100 amino acid residues denatured into a random coil can adopt 1.0×10^{31} different conformations of the same energy. We set $W = 1.0 \times 10^{31}$ and calculate the entropy as

$$S = (1.38 \times 10^{-23}\ \mathrm{J\ K^{-1}}) \times \ln(1.0 \times 10^{31}) = 9.9 \times 10^{-22}\ \mathrm{J\ K^{-1}}$$

The corresponding molar entropy of the protein is 600 J K^{-1} mol^{-1} (to 2 significant figures; that is, 6.0×10^2 J K^{-1} mol^{-1}).

(b) The relation between thermodynamic and statistical entropy

The concept of the number of microstates makes quantitative the ill-defined qualitative concepts of 'disorder' and 'the dispersal of matter and energy' that we have used to introduce the concept of entropy: a more 'disorderly' distribution of energy and matter corresponds to a greater number of microstates associated with the same total energy. For instance, when a perfect gas expands, the available translational energy levels get closer together (Fig. 2.11), so it is possible to distribute the molecules over them in more ways than when the volume of the container is small and the energy levels are further apart. Therefore, as the container expands, W and therefore S increase, just as for thermodynamic entropy.

The Boltzmann approach also illuminates the thermodynamic definition itself (eqn 2.1) and in particular the role of the temperature. Molecules in a system at high temperature can occupy a large number of the available energy levels, so a small additional transfer of energy as heat will lead to a relatively small change in the number of accessible energy levels. Consequently, the number of microstates does not increase appreciably and neither does the entropy of the system. In contrast, the molecules in a system at low temperature have access to far fewer energy levels (at $T = 0$, only the lowest level is accessible), and the transfer of the same quantity of energy by heating will increase the number of accessible energy levels and the number of microstates significantly. Hence, the change in entropy on heating will be greater when the energy is transferred to a cold body than when it is transferred to a hot body. This argument suggests that the change in entropy should be inversely proportional to the temperature at which the transfer takes place, as in eqn 2.1.

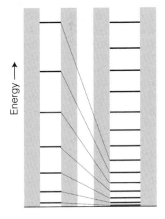

Fig. 2.11 When the size of a container is increased (shown here in two dimensions), the energy levels available to the molecules inside it move closer together so more are accessible at a given temperature (as indicated by the levels colored red).

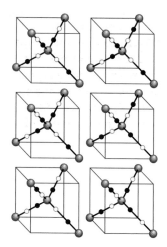

Fig. 2.12 The six possible arrangements of H atoms around a central O atom in ice. Occupied locations are indicated by black dots and unoccupied locations by a grey outline.

(c) The residual entropy

In most cases, $W = 1$ at $T = 0$ because there is only one way of putting all the molecules into the same, lowest state. Therefore, as we have seen, $S = 0$ at $T = 0$, in accord with the Third Law of thermodynamics. In certain cases, however, W may differ from 1 at $T = 0$. This is the case if disorder survives down to absolute zero because there is no energy advantage in adopting a particular orientation. For instance, there may be no energy difference between the arrangements ... AB AB AB ... and ... BA AB BA ..., so $W > 1$ even at $T = 0$. If $S > 0$ at $T = 0$ we say that the substance has a **residual entropy**. Ice has a residual entropy of 3.4 J K^{-1} mol^{-1}. It stems from the disorder in the hydrogen bonds between neighboring water molecules: a given O atom has two short O–H bonds and two long O···H bonds to its neighbors, but there is a degree of randomness in which two bonds are short and which two are long.

> **A brief illustration**
>
> Consider a sample of ice of N H_2O molecules. Each of the $2N$ H atoms can be either close to or relatively far from an O atom, resulting in 2^{2N} possible arrangements. However, of the $2^4 = 16$ possible arrangements around a single O atom, only 6 have two short and two long bonds (Fig. 2.12) and hence are acceptable. Therefore $W = 2^{2N}\left(\frac{6}{16}\right)^N = \left(\frac{3}{2}\right)^N$ and the residual entropy is
>
> $$S(0) = k \ln W = k \ln \left(\tfrac{3}{2}\right)^N = Nk \ln \tfrac{3}{2}$$
>
> The molar residual entropy (replace N by N_A and use $N_A k = R$) is therefore
>
> $$S_m(0) = R \ln \tfrac{3}{2} = 3.4 \text{ J K}^{-1} \text{ mol}^{-1}$$

2.5 Entropy changes accompanying chemical reactions

To move into the arena of biochemistry, where reactants are transformed into products, we need to establish procedures for using the tabulated values of absolute entropies to calculate entropy changes associated with chemical reactions; to assess the spontaneity of a biological process, we need to see how to take into account entropy changes in both the system and the surroundings.

Once again, we can sometimes use our intuition to predict the sign of the entropy change associated with a chemical reaction. When there is a net formation of a gas in a reaction, as in a combustion or the equivalent but controlled oxidations characteristic of organisms, we can usually anticipate that the entropy increases. When there is a net consumption of gas, as in the fixation of N_2 by certain micro-organisms, it is usually safe to predict that the entropy decreases. However, for a quantitative value of the change in entropy and to predict the sign of the change when no gases are involved, we need to do an explicit calculation.

(a) Standard reaction entropies

The difference in molar entropy between the products and the reactants in their standard states is called the **standard reaction entropy**, $\Delta_r S^\ominus$. It can be expressed in terms of the molar entropies of the substances in much the same way as we have already used for the standard reaction enthalpy:

$$\Delta_r S^\ominus = \sum \nu S_m^\ominus (\text{products}) - \sum \nu S_m^\ominus (\text{reactants}) \qquad \text{(2.9)}$$

> The standard entropy of reaction

where the ν are the stoichiometric coefficients in the chemical equation.

A brief illustration

The enzyme carbonic anhydrase catalyses the hydration of CO_2 gas in red blood cells: $CO_2(g) + H_2O(l) \rightarrow H_2CO_3(aq)$. We expect a negative entropy of reaction because a gas is consumed. To find the explicit value at 25°C, we use the information from the *Resource section* to write

$$\Delta_r S^\ominus = S_m^\ominus(H_2CO_3, aq) - \{S_m^\ominus(CO_2, g) + S_m^\ominus(H_2O, l)\}$$
$$= (187.4\ J\ K^{-1}\ mol^{-1}) - \{(213.74\ J\ K^{-1}\ mol^{-1}) + (69.91\ J\ K^{-1}\ mol^{-1})\}$$
$$= -96.3\ J\ K^{-1}\ mol^{-1}$$

Self-test 2.3   (a) Predict the sign of the entropy change associated with the complete oxidation of solid sucrose, $C_{12}H_{22}O_{11}(s)$, by O_2 gas to CO_2 gas and liquid H_2O. (b) Calculate the standard reaction entropy at 25°C.

Answer: (a) Positive; (b) $+512\ J\ K^{-1}\ mol^{-1}$

A note on good practice
Do not make the mistake of setting the standard molar entropies of elements equal to zero: they have non-zero values (provided $T > 0$), as we have already discussed.

(b) The spontaneity of chemical reactions

A process may be spontaneous even though the entropy change of the system itself is negative. Consider the binding of oxidized nicotinamide adenine dinucleotide (NAD^+; Atlas N4), an important electron carrier in metabolism (*Case studies* 1.1 and 1.2), to the enzyme lactate dehydrogenase, which plays a role in the catabolism and anabolism of carbohydrates. Experiments show that $\Delta_r S^\ominus = -16.8\ J\ K^{-1}\ mol^{-1}$ for binding at 25°C and pH = 7.0. The negative sign of the entropy change is expected because the association of two reactants gives rise to a more compact structure. The reaction results in a more organized structure, yet it is spontaneous!

The resolution of this apparent paradox underscores a feature of entropy that recurs throughout chemistry and biology: it is essential to consider the entropy of both the system and its surroundings when deciding whether or not a process is spontaneous. The reduction in entropy by $16.8\ J\ K^{-1}\ mol^{-1}$ relates only to the system, the reaction mixture. To apply the Second Law correctly, we need to calculate the total entropy, the sum of the changes in the system and the surroundings that jointly compose the entire 'isolated system' referred to in the Second Law. It may well be the case that the entropy of the system decreases when a change takes place, but there may be a more than compensating increase in entropy of the surroundings, so that overall the entropy change is positive. The opposite may also be true: a large decrease in the entropy of the surroundings may occur when the entropy of the system increases. In that case we would be wrong to conclude from the increase in the system alone that the change is spontaneous. Whenever considering the implications of entropy, we must always consider the total change of the system and its surroundings.

A brief illustration

To calculate the entropy change in the surroundings when a reaction takes place at constant pressure, we use eqn 2.6, interpreting the ΔH in that expression as the reaction enthalpy. For example, for the formation of the NAD^+-enzyme complex discussed above, with $\Delta_r H^\ominus = -24.2\ kJ\ mol^{-1}$, the change in entropy of the surroundings (which are maintained at 25°C, the same temperature as the reaction mixture) is

$$\Delta_r S_{sur} = -\frac{\Delta_r H}{T} = -\frac{(-24.2 \text{ kJ mol}^{-1})}{298 \text{ K}} = +81.2 \text{ J K}^{-1} \text{ mol}^{-1}$$

Now we can see that the total entropy change is positive:

$$\Delta_r S_{total} = (-16.8 \text{ J K}^{-1} \text{ mol}^{-1}) + (81.2 \text{ J K}^{-1} \text{ mol}^{-1}) = +64.4 \text{ J K}^{-1} \text{ mol}^{-1}$$

This calculation confirms that the reaction is spontaneous. In this case, the spontaneity is a result of the dispersal of energy that the reaction generates in the surroundings: the complex is dragged into existence, even though it has a lower entropy than the separated reactants, by the tendency of energy to disperse into the surroundings.

The Gibbs energy

One of the problems with entropy calculations is already apparent: we have to work out two entropy changes, the change in the system and the change in the surroundings, and then consider the sign of their sum. The great American theoretician J.W. Gibbs, who laid the foundations of chemical thermodynamics toward the end of the nineteenth century, discovered how to combine the two calculations into one. The combination of the two procedures in fact turns out to be of much greater relevance than just saving a little labor, and throughout this text we shall see consequences of the procedure he developed.

2.6 Focusing on the system

To simplify the discussion of the role of the total change in the entropy, we need to introduce a new state function, the Gibbs energy, which will be used extensively in our study of bioenergetics and biological structure.

The total entropy change that accompanies a process is

$$\Delta S_{total} = \Delta S + \Delta S_{sur} \qquad \boxed{\text{Total entropy change}} \qquad (2.10)$$

where ΔS is the entropy change for the system; for a spontaneous change, $\Delta S_{total} > 0$. If the process occurs at constant pressure and temperature, we can use eqn 2.6 to express the change in entropy of the surroundings in terms of the enthalpy change of the system, ΔH. When the resulting expression is inserted into this one, we obtain

$$\text{At constant temperature and pressure: } \Delta S_{total} = \Delta S - \frac{\Delta H}{T} \qquad (2.11)$$

The great advantage of this formula is that it expresses the total entropy change of the system and its surroundings in terms of the properties of the system alone. The only restriction is that the expression is confined to changes at constant pressure and temperature.

(a) The definition of the Gibbs energy

Now we take a very important step. First, we introduce the **Gibbs energy**, G, which is defined as[3]

---

[3] The Gibbs energy is still commonly referred to by its older name, the 'free energy'.

$$G = H - TS \qquad \boxed{\text{Definition of Gibbs energy}} \quad (2.12)$$

Because H, T, and S are state functions, G is a state function too. A change in Gibbs energy, ΔG, at constant temperature arises from changes in enthalpy and entropy and is

At constant temperature: $\Delta G = \Delta H - T\Delta S$ $\qquad \boxed{\text{Change in } G \text{ at constant } T}$ $\quad (2.13)$

By comparing eqns 2.11 and 2.13, we obtain

At constant temperature and pressure: $\Delta G = -T\Delta S_{\text{total}}$ $\qquad (2.14)$

We see that at constant temperature and pressure, the change in the Gibbs energy of a system is proportional to the overall change in the entropy of the system plus its surroundings.

(b) Spontaneity and the Gibbs energy

The difference in sign between ΔG and ΔS_{total} in eqn 2.14 implies that the condition for a process being spontaneous changes from $\Delta S_{\text{total}} > 0$ in terms of the total entropy (which is universally true) to $\Delta G < 0$ in terms of the Gibbs energy (for processes occurring at constant temperature and pressure). That is, *in a spontaneous change at constant temperature and pressure, the Gibbs energy decreases* (Fig. 2.13).

It may seem more natural to think of a system as falling to a lower value of some property. However, it must never be forgotten that to say that a system tends to fall toward lower Gibbs energy is only a modified way of saying that a system and its surroundings jointly tend toward a greater total entropy. The *only* criterion of spontaneous change is the total entropy of the system and its surroundings; the Gibbs energy merely contrives a way of expressing that total change in terms of the properties of the system alone and is valid only for processes that occur at constant temperature and pressure.

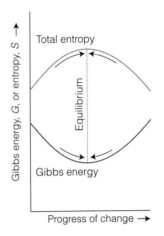

Fig. 2.13 The criterion of spontaneous change is the increase in total entropy of the system and its surroundings. Provided we accept the limitation of working at constant pressure and temperature, we can focus entirely on the properties of the system and express the criterion as a tendency to move to lower Gibbs energy.

$\boxed{\text{Case study 2.1}}$ *Life and the Second Law*

Every chemical reaction that is spontaneous under conditions of constant temperature and pressure, including those that drive the processes of growth, learning, and reproduction, is a reaction that proceeds in the direction of lower Gibbs energy, or—another way of expressing the same thing—results in the overall entropy of the system and its surroundings becoming greater. With these ideas in mind, it is easy to explain why life, which can be regarded as a collection of biological processes, proceeds in accord with the Second Law of thermodynamics.

It is not difficult to imagine conditions in the cell that may render spontaneous many of the reactions of catabolism described briefly in *Case study* 1.1. After all, the breakdown of large molecules, such as sugars and lipids, into smaller molecules leads to the dispersal of matter in the cell. Energy is also dispersed, as it is released on reorganization of bonds in foods when they are oxidized. More difficult to rationalize is life's requirement of the organization of a very large number of molecules into biological cells, which in turn assemble into organisms. To be sure, the entropy of the system—the organism—is very low because matter becomes less dispersed when molecules assemble to form cells, tissues, organs, and so on. However, the lowering of the system's entropy comes at the expense of an increase in the entropy of the surroundings.

To understand this point, recall from *Case study* 1.1 and *Case study* 1.2 that cells grow by converting energy from the Sun or oxidation of foods partially into work. The remaining energy is released as heat into the surroundings, so $q_{sur} > 0$ and $\Delta S_{sur} > 0$. As with any process, life is spontaneous and organisms thrive as long as the increase in the entropy of the organism's environment compensates for decreases in the entropy arising from the assembly of the organism. Alternatively, we may say that $\Delta G < 0$ for the overall sum of physical and chemical changes that we call life.

2.7 The hydrophobic interaction

To gain insight into the thermodynamic factors that contribute to the spontaneous assembly of biological macromolecules, we need to examine in detail some of the interactions that bring molecular building blocks together.

Throughout the text we shall see how concepts of physical chemistry can be used to establish some of the known 'rules' for the assembly of complex biological structures. Here, we describe how the Second Law can account for the formation of such organized assemblies as proteins and biological cell membranes.

As remarked in the *Prologue*, we do not know all the rules that govern the folding of proteins into well-defined three-dimensional structures. However, a number of general conclusions from experimental studies give some insight into the origin of tertiary and quaternary structure in proteins. Here we focus on the observation that, in an aqueous environment (including the interior of biological cells), the chains of a protein fold in such a way as to place hydrophobic groups (water-repelling, non-polar groups such as $-CH_2CH(CH_3)_2$) in the interior, which is often not very accessible to solvent, and hydrophilic groups (water-loving, polar or charged groups such as $-NH_3^+$) on the surface, which is in direct contact with the polar solvent. A species with both hydrophobic and hydrophilic regions is called **amphipathic**.[4] Phospholipids also are amphipathic molecules that can group together to form bilayer structures and cell membranes (recall Fig. F.1).

To understand the process in more detail, imagine a hypothetical initial state in which a polypeptide chain is immersed in water and has not acquired its final structure. Each hydrophobic group is surrounded by a cage of water molecules (Fig. 2.14). Now consider the actual final state in which hydrophobic groups are clustered together. Although the clustering together results in a negative contribution to the change in entropy of the system (the solution), fewer (albeit larger) cages are required and more solvent molecules are free to move. The net effect of the formation of clusters of hydrophobic groups is then a decrease in the organization of the solvent and a net *increase* in entropy of the system. This increase in entropy of the solvent is large enough to result in the association of hydrophobic groups in an aqueous environment being spontaneous. The process that drives the spontaneous clustering of hydrophobic groups in the presence of water is called the **hydrophobic interaction**.

Fig. 2.14 When a hydrophobic molecule (in shades of gray) is surrounded by water, the H_2O molecules (with their oxygen atoms shown in red) form a cage, of which a cross-section is shown here. As a result of this acquisition of structure, the entropy of water decreases, so the dispersal of the hydrophobic molecule into the water is entropy opposed; its coalescence is entropy favored.

[4] The *amphi-* part of the name is from the Greek word for 'both' and the *-pathic* part is from the same root (meaning 'feeling') as *sympathetic*.

Self-test 2.4 Two long-chain hydrophobic polypeptides can associate end-to-end so that only the ends meet or side-by-side so that the entire chains are in contact. Which arrangement would produce a larger entropy change when they come together?

Answer: The side-by-side arrangement

To understand the hydrophobic interaction more completely we need to know more about the energetics of the interaction of hydrophobic groups and water. Experiments indicate that the dissolution of a largely hydrophobic molecule in water is commonly endothermic ($\Delta_{diss}H > 0$) but that the entropy change is positive ($\Delta_{diss}S > 0$):

	$\Delta_{diss}G^{\oplus}/kJ\ mol^{-1}$	$\Delta_{diss}H^{\oplus}/kJ\ mol^{-1}$	$\Delta_{diss}S^{\oplus}/J\ K^{-1}\ mol^{-1}$
$CH_3CH_2CH_2CH_2OH$	−10	+9	+65
$CH_3CH_2CH_2CH_2CH_2OH$	−13	+8	+72

The positive entropy of dissolving is consistent with the solvent water becoming more disorganized in the presence of the hydrophobic tails of the alkanol molecules, as the hydrophobic interaction requires. These experimental values are consistent with a general rule that each additional $-CH_2-$ group contributes a further $-3\ kJ\ mol^{-1}$ to the Gibbs energy of dissolving.

An important consequence of this analysis is that low temperatures *disfavor* the hydrophobic interaction. Thus, from $\Delta G = \Delta H - T\Delta S$, lowering the temperature reduces the effect of ΔS and the ΔG can change from negative to positive. This is the reason why some proteins and viruses dissociate into their individual subunits as the temperature is lowered to 0°C.

A further aspect of this discussion is that we can set up a scale of hydrophobicities. The hydrophobicity of a small molecular group R is reported by defining the **hydrophobicity constant**, π, as

$$\pi = \log \frac{s(RX)}{s(HX)}$$

Definition of hydrophobicity constant   (2.15)

where $s(RX)$ is the ratio of the molar solubility (the maximum chemical amount that can be dissolved to form 1 dm³ of solution) of the compound RX in octan-1-ol, a non-polar solvent, to that in water, and $s(HX)$ is the ratio of the molar solubility of the compound HX in octan-1-ol to that in water. Therefore, positive values of π indicate hydrophobicity and negative values indicate hydrophilicity, the thermodynamic preference for water as a solvent. It is observed experimentally that the π values of most groups do not depend on the nature of X. However, measurements do suggest group additivity of π values:

-R	$-CH_3$	$-CH_2CH_3$	$-(CH_2)_2CH_3$	$-(CH_2)_3CH_3$	$-(CH_2)_4CH_3$
π	0.5	1	1.5	2	2.5

We see that acyclic saturated hydrocarbons become more hydrophobic as the carbon chain length increases. This trend can be rationalized by $\Delta_{diss}H$ becoming more positive and $\Delta_{diss}S$ more negative as the number of carbon atoms in the chain increases.

2.8 Work and the Gibbs energy change

To understand how biochemical reactions can be used to release energy as work in the cell, we need to gain deeper insight into the Gibbs energy.

An important feature of the Gibbs energy is that the value of ΔG for a process gives the maximum non-expansion work that can be extracted from the process at constant temperature and pressure. By **non-expansion work**, $w_{\text{non-exp}}$, we mean any work other than that arising from the expansion of the system. It may include electrical work, if the process takes place inside an electrochemical or biological cell, or other kinds of mechanical work, such as the winding of a spring or the contraction of a muscle. As we show in the following *Justification*,

At constant temperature and pressure: $\Delta G = w_{\text{max,non-exp}}$

$$\boxed{\text{Gibbs energy and non-expansion work}} \quad (2.16)$$

Justification 2.3 *Maximum non-expansion work*

We need to consider infinitesimal changes because dealing with reversible processes is then much easier. Our aim is to derive the relation between the infinitesimal change in Gibbs energy, dG, accompanying a process and the maximum amount of non-expansion work that the process can do, $dw_{\text{non-exp}}$. We start with the infinitesimal form of eqn 2.13,

at constant temperature: $dG = dH - TdS$

where, as usual, d denotes an infinitesimal difference. A good rule in the manipulation of thermodynamic expressions is to feed in definitions of the terms that appear. We do this twice. First, we use the expression for the change in enthalpy at constant pressure (eqn 1.11b, written as $dH = dU + pdV$) and obtain

at constant temperature and pressure: $dG = dU + pdV - TdS$

Then we replace dU in terms of infinitesimal contributions from work and heat ($dU = dw + dq$):

$dG = dw + dq + pdV - TdS$

The work done on the system consists of expansion work, $-p_{\text{ex}}dV$, and non-expansion work, $dw_{\text{non-exp}}$. Therefore,

$dG = -p_{\text{ex}}dV + dw_{\text{non-exp}} + dq + pdV - TdS$

This derivation is valid for any process taking place at constant temperature and pressure.

Now we specialize to a reversible change. For expansion work to be reversible, we need to match p and p_{ex}, in which case the first and fourth terms on the right cancel. Moreover, because the transfer of energy as heat is also reversible, we can replace dq by TdS, in which case the third and fifth terms also cancel. We are left with

at constant temperature and pressure, for a reversible process: $dG = dw_{\text{non-exp,rev}}$

Maximum work is done during a reversible change (Section 1.3(c)), so another way of writing this expression is at constant temperature and pressure: $dG = dw_{max,non\text{-}exp}$

Because this relation holds for each infinitesimal step between the specified initial and final states, it applies to the overall change too. Therefore, we obtain eqn 2.16.

Example 2.1   *Estimating a change in Gibbs energy for a metabolic process*

Suppose a certain small bird has a mass of 30 g. What is the minimum mass of glucose that it must consume to fly up to a branch 10 m above the ground? The change in Gibbs energy that accompanies the oxidation of 1.0 mol $C_6H_{12}O_6(s)$ to carbon dioxide gas and liquid water at 25°C is −2808 kJ.

Strategy First, we need to calculate the work needed to raise a mass m through a height h on the surface of the Earth. As we saw in eqn 1.2, this work is equal to mgh, where g is the acceleration of free fall. This work, which is non-expansion work, can be identified with ΔG. We need to determine the amount of substance that corresponds to the required change in Gibbs energy and then convert that amount to a mass by using the molar mass of glucose.

Solution The non-expansion work to be done is

$$w_{non\text{-}exp} = (30 \times 10^{-3}\,kg) \times (9.81\,m\,s^{-2}) \times (10\,m) = 3.0 \times 9.81 \times 1.0 \times 10^{-1}\,J$$

(because $1\,kg\,m^2\,s^{-2} = 1\,J$). The amount, n, of glucose molecules required for oxidation to give a change in Gibbs energy of this value given that 1 mol provides 2808 kJ is

$$n = \frac{3.0 \times 9.81 \times 1.0 \times 10^{-1}\,J}{2.808 \times 10^6\,J\,mol^{-1}} = \frac{3.0 \times 9.81 \times 1.0 \times 10^{-7}}{2.808}\,mol$$

Therefore, because the molar mass, M, of glucose is 180 g mol^{-1}, the mass, m, of glucose that must be oxidized is

$$m = nM = \left(\frac{3.0 \times 9.81 \times 1.0 \times 10^{-7}}{2.808}\,mol \right) \times (180\,g\,mol^{-1}) = 1.9 \times 10^{-4}\,g$$

That is, the bird must consume at least 0.19 mg of glucose for the mechanical effort (and more if it thinks about it).

Self-test 2.5   A hardworking human brain, perhaps one that is grappling with physical chemistry, operates at about 25 J s^{-1}. What mass of glucose must be consumed to sustain that metabolic rate for an hour?

Answer: 5.8 g

The great importance of the Gibbs energy in chemistry is becoming apparent. At this stage, we see that it is a measure of the non-expansion work resources of chemical reactions: if we know ΔG, then we know the maximum non-expansion work that we can obtain by harnessing the reaction in some way. In some cases, the non-expansion work is extracted as electrical energy. This is the case when electrons are transferred across cell membranes in some key reactions of photosynthesis and respiration (see Sections 5.10 and 5.11).

Some insight into the physical significance of G itself comes from its definition as $H - TS$. The enthalpy is a measure of the energy that can be obtained from the system as heat. The term TS is a measure of the quantity of energy stored in the *random* motion of the molecules making up the sample. Work, as we have seen, is energy transferred in an orderly way, so we cannot expect to obtain work from the energy stored randomly. The difference between the total stored energy and the energy stored randomly, $H - TS$, is available for doing work, and we recognize that difference as the Gibbs energy. In other words, the Gibbs energy is the energy stored in the uniform motion and arrangement of the molecules in the system.

Case study 2.2   *The action of adenosine triphosphate*

In biological cells, the energy released by the oxidation of foods (*Case study* 1.1) is stored in adenosine triphosphate (ATP or ATP^{4-}, Atlas N3). The essence of ATP's action is its ability to lose its terminal phosphate group by hydrolysis and to form adenosine diphosphate (ADP or ADP^{3-}, Atlas N2):

$$ATP^{4-}(aq) + H_2O(l) \rightarrow ADP^{3-}(aq) + HPO_4^{2-}(aq) + H_3O^+(aq)$$

At pH = 7.0 and 37°C (310 K, blood temperature) the enthalpy and Gibbs energy of hydrolysis are $\Delta_r H = -20$ kJ mol^{-1} and $\Delta_r G = -31$ kJ mol^{-1}, respectively. Under these conditions, the hydrolysis of 1 mol ATP^{4-}(aq) results in the extraction of up to 31 kJ of energy that can be used to do non-expansion work, such as the synthesis of proteins from amino acids, muscular contraction, and the activation of neuronal circuits in our brains, as we shall see in Chapter 5. If no attempt is made to extract any energy as work, then 20 kJ (in general, ΔH) of heat will be produced.

Checklist of key concepts

☐ 1. A spontaneous change is a change that has a tendency to occur without work having to be done to bring it about.

☐ 2. Matter and energy tend to disperse.

☐ 3. The Second Law states that the entropy of an isolated system tends to increase.

☐ 4. In general, the entropy change accompanying the heating of a system is equal to the area under the graph of C/T against T between the two temperatures of interest.

☐ 5. The Third Law of thermodynamics states that the entropies of all perfectly crystalline substances are the same at $T = 0$ (and may be taken to be zero).

☐ 6. The Boltzmann formula expresses the statistical entropy in terms of the number of microstates of a system.

☐ 7. The Gibbs energy is defined as $G = H - TS$ and is a state function.

☐ 8. At constant temperature and pressure, a system tends to change in the direction of decreasing Gibbs energy.

☐ 9. The hydrophobic interaction is a process that leads to the organization of solute molecules and is driven by a tendency toward greater dispersal of solvent molecules.

☐ 10. At constant temperature and pressure, the change in Gibbs energy accompanying a process is equal to the maximum non-expansion work the process can do.

Checklist of key equations

Property or process	Equation	Comment
Entropy change	$\Delta S = q_{rev}/T$	Definition
Entropy change of surroundings	$\Delta S_{sur} = -q/T$	q is heat supplied to the system
	$\Delta S_{sur} = -\Delta H/T$	ΔH is the enthalpy change of the system; constant pressure process
Boltzmann formula for the entropy	$S = k \ln W$	W is the number of microstates
Entropy of transition	$\Delta_{trs}S\,(T_{trs}) = \Delta_{trs}H(T_{trs})/T_{trs}$	At the transition temperature
Entropy change due to a change in temperature	$\Delta S = C\ln(T_2/T_1)$	Heat capacity constant in the range of interest
Standard reaction entropy	$\Delta_r S^\oplus = \sum \nu S_m^\oplus(\text{products}) - \sum \nu S_m^\oplus(\text{reactants})$	Definition
Gibbs energy	$G = H - TS$	Definition
Change in Gibbs energy	$\Delta G = \Delta H - T\Delta S$	At constant temperature
Relation to maximum non-expansion work	$\Delta G = w_{max,non-exp}$	At constant temperature and pressure

Discussion questions

2.1 The following expressions have been used to establish criteria for spontaneous change: $\Delta S_{isolated\,system} > 0$ and $\Delta G < 0$. Discuss the origin, significance, and applicability of each criterion.

2.2 Explain the limitations of the following expressions: (a) $\Delta S = C\ln(T_f/T_i)$, (b) $\Delta G = \Delta H - T\Delta S$, and (c) $\Delta G = w_{max,non-exp}$.

2.3 Suggest a procedure for the measurement of the entropy of unfolding of a protein with differential scanning calorimetry (see *In the laboratory* 1.1).

2.4 Justify the identification of the statistical entropy with the thermodynamic entropy.

2.5 Explain the origin of the residual entropy.

2.6 Without performing a calculation, predict whether the standard entropies of the following reactions are positive or negative:

(a) Ala–Ser–Thr–Lys–Gly–Arg–Ser $\xrightarrow{\text{trypsin}}$ Ala–Ser–Thr–Lys– + Gly–Arg

(b) $N_2(g) + 3\,H_2(g) \xrightarrow{\text{trypsin}} 2\,NH_3(g)$

(c) $ATP^{4-}(aq) + H_2O(1) \xrightarrow{\text{trypsin}} ADP^{3-}(aq) + HPO_4^{2-}(aq) + H_3O^+(aq)$

2.7 Provide a molecular interpretation of the hydrophobic interaction.

Exercises

2.8 A goldfish swims in a bowl of water at 20°C. Over a period of time, the fish transfers 120 J to the water as a result of its metabolism. What is the change in entropy of the water?

2.9 Suppose that when you exercise, you consume 100 g of glucose and that all the energy released as heat remains in your body at 37°C. What is the change in entropy of your body?

2.10 Suppose you put a cube of ice of mass 100 g into a glass of water at just above 0°C. When the ice melts, about 33 kJ of energy is absorbed from the surroundings as heat. What is the change in entropy of (a) the sample (the ice) and (b) the surroundings (the glass of water)?

2.11 Calculate the change in entropy of 100 g of ice at 0°C as it is melted, heated to 100°C, and then vaporized at that temperature. Suppose that the changes are brought about by a heater that supplies energy at a constant rate, and sketch a graph showing (a) the change in temperature of the system, (b) the enthalpy of the system, and (c) the entropy of the system as a function of time.

2.12 What is the change in entropy of 100 g of water when it is heated from room temperature (20°C) to body temperature (37°C)? Use $C_{p,m} = 75.5$ J K^{-1} mol^{-1}.

2.13 Estimate the molar entropy of potassium chloride at 5.0 K given that its molar heat capacity at that temperature is 1.2 mJ K^{-1} mol^{-1}.

2.14 Equation 2.2 is based on the assumption that the heat capacity is independent of temperature. Suppose, instead, that the heat capacity depends on temperature as $C = a + bT + a/T^2$. Find an expression for the change of entropy accompanying heating from T_i to T_f. *Hint*: See *Justification* 2.1.

2.15 Calculate the change in entropy when 100 g of water at 80°C is poured into 100 g of water at 10°C in an insulated vessel given that $C_{p,m} = 75.5 \, J \, K^{-1} \, mol^{-1}$.

2.16 The protein lysozyme unfolds at a transition temperature of 75.5°C, and the standard enthalpy of transition is 509 kJ mol^{-1}. Calculate the entropy of unfolding of lysozyme at 25.0°C, given that the difference in the constant-pressure heat capacities on unfolding is 6.28 kJ K^{-1} mol^{-1} and can be assumed to be independent of temperature. *Hint*: Imagine that the transition at 25.0°C occurs in three steps: (i) heating of the folded protein from 25.0°C to the transition temperature, (ii) unfolding at the transition temperature, and (iii) cooling of the unfolded protein to 25.0°C. Because the entropy is a state function, the entropy change at 25.0°C is equal to the sum of the entropy changes of the steps.

2.17 The enthalpy of the graphite → diamond phase transition, which under 100 kbar occurs at 2000 K, is +1.9 kJ mol^{-1}. Calculate the entropy of transition at that temperature.

2.18 The enthalpy of vaporization of methanol is 35.27 kJ mol^{-1} at its normal boiling point of 64.1°C. Calculate (a) the entropy of vaporization of methanol at this temperature and (b) the entropy change of the surroundings.

2.19 Trouton's rule summarizes the results of experiments showing that the entropy of vaporization measured at the boiling point, $\Delta_{vap}S = \Delta_{vap}H(T_b)/T_b$, is approximately the same and equal to about 85 J K^{-1} mol^{-1} for all liquids except when hydrogen bonding or some other kind of specific molecular interaction is present. (a) Provide a molecular interpretation for Trouton's rule. (b) Estimate the entropy of vaporization and the enthalpy of vaporization of octane, which boils at 126°C. (c) Trouton's rule does not apply to water because in the liquid, water molecules are held together by an extensive network of hydrogen bonds. Provide a molecular interpretation for the observation that Trouton's rule underestimates the value of the entropy of vaporization of water.

2.20 Calculate the entropy of fusion of a compound at 25°C given that its enthalpy of fusion is 32 kJ mol^{-1} at its melting point of 146°C and the molar heat capacities (at constant pressure) of the liquid and solid forms are 28 J K^{-1} mol^{-1} and 19 J K^{-1} mol^{-1}, respectively.

2.21 Show that at temperatures close to $T = 0$, $S(T) = \frac{1}{3}C_p(T)$.

2.22 Calculate the residual molar entropy of a solid in which the molecules can adopt (a) three, (b) five, and (c) six orientations of equal energy at $T = 0$.

2.23 Calculate the standard reaction entropy at 298 K of the fermentation of glucose to ethanol:
$C_6H_{12}O_6(s) \rightarrow 2 \, C_2H_5OH(l) + 2 \, CO_2(g)$.

2.24 The constant-pressure molar heat capacities of linear gaseous molecules are approximately $\frac{7}{2}R$ and those of non-linear gaseous molecules are approximately $4R$. Estimate the change in standard reaction entropy of the following two reactions when the temperature is increased by 10 K at constant pressure:

(a) $2 \, H_2(g) + O_2(g) \rightarrow 2 \, H_2O(l)$

(b) $CH_4(g) + 2 \, O_2(g) \rightarrow CO_2(g) + 2 \, H_2O(g)$

2.25 Use the information in Exercise 2.24 to calculate the standard Gibbs energy of reaction of $N_2 + 3 \, H_2(g) \rightarrow 2 \, NH_3(g)$.

2.26 In a particular biological reaction taking place in the body at 37°C, the change in enthalpy was −125 kJ mol^{-1} and the change in entropy was −126 J K^{-1} mol^{-1}. (a) Calculate the change in Gibbs energy. (b) Is the reaction spontaneous? (c) Calculate the total change in entropy of the system and the surroundings.

2.27 The change in Gibbs energy that accompanies the oxidation of $C_6H_{12}O_6(s)$ to carbon dioxide and water vapor at 25°C is −2808 kJ mol^{-1}. How much glucose does a person of mass 65 kg need to consume to climb through 10 m?

2.28 A non-spontaneous reaction may be driven by coupling it to a reaction that is spontaneous. The formation of glutamine from glutamate and ammonium ions requires 14.2 kJ mol^{-1} of energy input. It is driven by the hydrolysis of ATP to ADP mediated by the enzyme glutamine synthetase. (a) Given that the change in Gibbs energy for the hydrolysis of ATP corresponds to $\Delta G = -31$ kJ mol^{-1} under the conditions prevailing in a typical cell, can the hydrolysis drive the formation of glutamine? (b) How many moles of ATP must be hydrolyzed to form 1 mol glutamine?

2.29 The hydrolysis of acetyl phosphate has $\Delta G = -42$ kJ mol^{-1} under typical biological conditions. If the phosphorylation of acetic acid were to be coupled to the hydrolysis of ATP, what is the minimum number of ATP molecules that would need to be involved?

2.30 Suppose that the radius of a typical cell is 10 μm and that inside it 10^6 ATP molecules are hydrolyzed each second. What is the power density of the cell in watts per cubic meter (1 W = 1 J s^{-1})? A computer battery delivers about 15 W and has a volume of 100 cm^3. Which has the greater power density, the cell or the battery? (For data, see Exercise 2.28.)

Projects

2.31 The following is an example of a structure–activity relation (SAR), in which it is possible to correlate the effect of a structural change in a compound with its biological function. The use of SARs can improve the design of drugs for the treatment of disease because it facilitates the prediction of the biological activity of a compound before it is synthesized. The binding of non-polar groups of amino acid to hydrophobic sites in the interior of proteins is governed largely by hydrophobic interactions.

(a) Consider a family of hydrocarbons R–H. The hydrophobicity constants, π, for R = CH$_3$, CH$_2$CH$_3$, (CH$_2$)$_2$CH$_3$, (CH$_2$)$_3$CH$_3$, and (CH$_2$)$_4$CH$_3$ are, respectively, 0.5, 1.0, 1.5, 2.0, and 2.5. Use these data to predict the π value for (CH$_2$)$_6$CH$_3$.

(b) The equilibrium constants K_I for the dissociation of inhibitors (**1**) from the enzyme chymotrypsin (Atlas P3) were measured for different substituents R:

1

R	CH₃CO	CN	NO₂	CH₃	Cl
π	−0.20	−0.025	0.33	0.50	0.90
$\log K_I$	−1.73	−1.90	−2.43	−2.55	−3.40

Plot $\log K_I$ against π. Does the plot suggest a linear relationship? If so, what are the slope and intercept to the $\log K_I$ axis of the line that best fits the data?

(c) Predict the value of K_I for the case R = H.

2.32 An *exergonic reaction* is a reaction for which $\Delta G < 0$, and an *endergonic reaction* is a reaction for which $\Delta G > 0$. Here we investigate the molecular basis for the observation first discussed in *Case study* 2.2 that the hydrolysis of ATP is exergonic at pH = 7.0 and 310 K:

$$ATP^{4-}(aq) + H_2O(l) \rightarrow ADP^{3-}(aq) + HPO_4^{2-}(aq) + H_3O^+(aq)$$
$$\Delta_r G = -31 \text{ kJ mol}^{-1}$$

(a) It is thought that the exergonicity of ATP hydrolysis is due in part to the fact that the standard entropies of hydrolysis of polyphosphates are positive. Why would an increase in entropy accompany the hydrolysis of a triphosphate group into a diphosphate and a phosphate group?

(b) Under identical conditions, the Gibbs energies of hydrolysis of H₄ATP and MgATP²⁻, a complex between the Mg^{2+} ion and ATP^{4-}, are less negative than the Gibbs energy of hydrolysis of ATP^{4-}. This observation has been used to support the hypothesis that electrostatic repulsion between adjacent phosphate groups is a factor that controls the exergonicity of ATP hydrolysis. Provide a rationale for the hypothesis and discuss how the experimental evidence supports it. Do these electrostatic effects contribute to the $\Delta_r H$ or $\Delta_r S$ terms that determine the exergonicity of the reaction? *Hint*: In the MgATP²⁻ complex, the Mg^{2+} ion and ATP^{4-} anion form two bonds: one that involves a negatively charged oxygen belonging to the terminal phosphate group of ATP^{4-} and another that involves a negatively charged oxygen belonging to the phosphate group adjacent to the terminal phosphate group of ATP^{4-}.

(c) Stabilization due to resonance in ATP^{4-} and the HPO_4^{2-} ion is thought to be one of the factors that controls the exergonicity of ATP hydrolysis. Provide a rationale for the hypothesis. Does stabilization through resonance contribute to the $\Delta_r H$ or $\Delta_r S$ terms that determine the exergonicity of the reaction?

3 Phase equilibria

Boiling, freezing, the unfolding of proteins, and the unzipping of a DNA double helix are all examples of **phase transitions**, or changes of phase without change of chemical composition. Many phase changes are common everyday phenomena, and their description is an important part of physical chemistry. They occur whenever a solid changes into a liquid, as in the melting of ice, or a liquid changes into a vapor, as in the vaporization of water in our lungs. They also occur when one solid phase changes into another, as in the conversion of one phase of a biological membrane into another as it is heated.

The thermodynamics of phase changes of pure materials is also important because it prepares us first for the study of mixtures and then for the study of chemical equilibria (Chapter 4). Some of the thermodynamic concepts developed in this chapter also form the basis of important experimental techniques in biochemistry, such as the measurement of molar masses of proteins and nucleic acids and the investigation of the binding of small molecules to proteins.

The thermodynamics of transition

Because the Gibbs energy, $G = H - TS$, provides a signpost of spontaneous change when the pressure and temperature are constant, and we need to know the conditions under which a transition from one state to another becomes spontaneous, it is at the centre of all that follows. In particular, we need to know how G depends on the pressure and temperature. As we work out these dependencies, we shall acquire deep insight into the thermodynamic properties of biologically important substances and the transitions they can undergo.

3.1 The condition of stability

To understand processes ranging from the melting of ice to the denaturation of biopolymers, we need to understand the relative thermodynamic stabilities of the phases of a substance.

First, we need to establish the importance of the *molar* Gibbs energy, $G_m = G/n$, in the discussion of phase transitions of a pure substance. The molar Gibbs energy, an intensive property, is characteristic of the phase of the substance. For instance, the molar Gibbs energy of liquid water is in general different from that of water vapor at the same temperature and pressure. When an amount n of the substance changes from phase 1 (for instance, liquid) with molar Gibbs energy $G_m(1)$ to phase 2 (for instance, vapor) with molar Gibbs energy $G_m(2)$, the change in Gibbs energy is

$$\Delta G = nG_{\mathrm{m}}(2) - nG_{\mathrm{m}}(1) = n\{G_{\mathrm{m}}(2) - G_{\mathrm{m}}(1)\}$$

We know that a negative value of ΔG indicates that the change from phase 1 to phase 2 is spontaneous at constant temperature and pressure. It follows that the change from phase 1 to phase 2 is spontaneous if the molar Gibbs energy of phase 2 is lower than that of phase 1. In other words, *a substance has a spontaneous tendency to change into the phase with the lower molar Gibbs energy.*

If at a certain temperature and pressure the solid phase of a substance has a lower molar Gibbs energy than its liquid phase, then the solid phase is thermodynamically more stable and the liquid will (or at least has a tendency to) freeze. If the opposite is true, the liquid phase is thermodynamically more stable and the solid will melt. For example, at 1 atm, ice has a lower molar Gibbs energy than liquid water when the temperature is below 0°C, and under these conditions water converts spontaneously to ice.

3.2 The variation of Gibbs energy with pressure

To discuss how phase transitions depend on the pressure and to lay the foundation for understanding the behavior of solutions of biological macromolecules, we need to know how the molar Gibbs energy varies with pressure.

Why should biologists be interested in the variation of the Gibbs energy with the pressure since in most cases their systems are at pressures close to 1 atm? You should recall the discussion in Chapter 1, where we pointed out that to study the thermodynamic properties of a liquid (in which biochemists do have an interest), we can explore the properties of a vapor which, as a gas, are easy to formulate, and then imagine bringing the vapor into equilibrium with the liquid. Then the properties of the liquid mirror those of the vapor. That is the strategy we adopt throughout this chapter. First we establish equations that apply to gases. Then we consider equilibria between gases and liquids and adapt the gas-phase expressions to describe what really interests us, the properties of liquids.

We show in the following *Justification* that when the temperature is held constant and the pressure is changed from p_{i} to p_{f}, the molar Gibbs energy of an incompressible liquid becomes

$$G_{\mathrm{m}}(p_{\mathrm{f}}) = G_{\mathrm{m}}(p_{\mathrm{i}}) + (p_{\mathrm{f}} - p_{\mathrm{i}})V_{\mathrm{m}} \qquad \text{Variation of the Gibbs energy with pressure (for an incompressible liquid)} \qquad (3.1)$$

where V_{m} is the molar volume of the substance. This expression is valid when the molar volume is constant in the pressure range of interest, which is true of most liquids (and solids) under normal circumstances. Even though gases are far from incompressible, we can also use eqn 3.1 for the *qualitative* discussion of the pressure dependence of G_{m} of a gas provided the change in pressure is small.

Justification 3.1   *The variation of G of an incompressible liquid with pressure*

When the temperature, volume, and pressure of a substance are changed by infinitesimal amounts, H changes to $H + \mathrm{d}H$, T changes to $T + \mathrm{d}T$, and S changes to $S + \mathrm{d}S$. As a result, G changes to $G + \mathrm{d}G$, where

$$G + \mathrm{d}G = (H + \mathrm{d}H) - (T + \mathrm{d}T)(S + \mathrm{d}S) = H + \mathrm{d}H - TS - T\mathrm{d}S - S\mathrm{d}T - \mathrm{d}T\mathrm{d}S$$

The G on the left cancels the $H - TS$ on the right, the doubly infinitesimal $\mathrm{d}T\mathrm{d}S$ can be neglected, and we are left with

$$dG = dH - TdS - SdT$$

In a similar way, from the definition $H = U + pV$, letting U change to $U + dU$, and so on, and neglecting the doubly infinitesimal term $dpdV$, we can write

$$dH = dU + pdV + Vdp$$

At this point we need to know how the internal energy changes, and write

$$dU = dq + dw$$

If initially we consider only reversible changes, we can replace dq by TdS (because $dS = dq_{rev}/T$) and dw by $-pdV$ (because $dw = -p_{ex}dV$ and $p_{ex} = p$ for a reversible change) and obtain

$$dU = TdS - pdV$$

Now we substitute this expression into the expression for dH and that expression into the expression for dG and obtain

$$dG = TdS - pdV + pdV + Vdp - TdS - SdT$$

It follows that

$$dG = Vdp - SdT \qquad \boxed{\text{Variation of the Gibbs energy with pressure and temperature}} \qquad (3.2)$$

Now here is a subtle but important point. To derive this result we have supposed that the changes in conditions have been made reversibly. However, G is a state function and so the change in its value is independent of path. Therefore, the expression is valid for any change within a system of known composition, not just a reversible change.

At this point we decide to keep the temperature constant and set $dT = 0$; this leaves

$$dG = Vdp$$

and, for molar quantities, $dG_m = V_m dp$. This expression is exact but applies only to an infinitesimal change in the pressure. For an observable change, we replace dG_m by $G_m(p_f) - G_m(p_i)$ and dp by $p_f - p_i$, respectively, and obtain eqn 3.1, provided the molar volume is constant over the range of interest.

A note on good practice
When confronted with a proof in thermodynamics, go back to fundamental definitions (as we did three times in succession in this derivation: first of G, then of H, and finally of U).

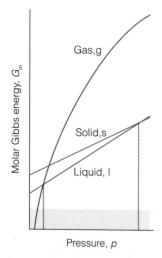

Fig. 3.1 The variation of molar Gibbs energy with pressure. The region of stability of each phase is indicated in the band at the bottom of the illustration.

Equation 3.1 tells us that, because all molar volumes are positive, *the molar Gibbs energy increases* $(G_m(p_f) > G_m(p_i))$ *when the pressure increases* $(p_f > p_i)$. We also see that, for a given change in pressure, the resulting change in molar Gibbs energy is greatest for substances with large molar volumes. Again bearing in mind that we can apply eqn 3.1 *qualitatively* to gases over small changes in pressure, we see that because the molar volume of a gas is much larger than that of a condensed phase (a liquid or a solid), the dependence of G_m on p is much greater for a gas than for a condensed phase. For most substances (water is an important exception), the molar volume of the liquid phase is greater than that of the solid phase. Therefore, for most substances, the slope of a graph of G_m against p is greater for a liquid than for a solid. These characteristics are illustrated in Fig. 3.1.

As we see from Fig. 3.1, when we increase the pressure on a substance, the molar Gibbs energy of the gas phase rises above that of the liquid, then the molar Gibbs energy of the liquid rises above that of the solid. Because the system has a

Mathematical toolkit 3.1 *Integration*

The area under a graph of any function f is found by the techniques of integration. For instance, the area under the graph of the function $f(x)$ between $x = a$ and $x = b$ is denoted by

$$\text{area between } a \text{ and } b = \int_a^b f(x)dx$$

The elongated S symbol on the right is called the **integral** of the function f. When written as $\int$ alone, it is the **indefinite integral** of the function. When written with limits (as in the expression above), it becomes the **definite integral** of the function. The definite integral is the indefinite integral evaluated at the upper limit (b) minus the indefinite integral evaluated at the lower limit (a).

A very useful integral in physical chemistry is

$$\int \frac{dx}{x} = \ln x + \text{constant}$$

where $\ln x$ is the natural logarithm of x. To evaluate the integral between the limits $x = a$ and $x = b$, we write

$$\int_a^b \frac{dx}{x} = (\ln x + \text{constant})\Big|_a^b$$

$$= (\ln b + \text{constant}) - (\ln a + \text{constant})$$

$$= \ln b - \ln a = \ln \frac{b}{a}$$

We see that the constant cancels. For instance, the area under the graph of $1/x$ lying between $a = 2$ and $b = 3$ is $\ln(\frac{3}{2}) = 0.41$.

---

tendency to convert into the state of lowest molar Gibbs energy, the graphs show that at low pressures the gas phase is the most stable, then at higher pressures the liquid phase becomes the most stable, followed by the solid phase. In other words, under pressure the substance condenses to a liquid, and then further pressure can result in the formation of a solid.

We can use eqn 3.1 to predict the actual shape of graphs like those in Fig. 3.1. For a solid or liquid, the molar volume is almost independent of pressure, so eqn 3.1 is an excellent approximation to the change in molar Gibbs energy. It shows that the molar Gibbs energy of a solid or liquid increases linearly with pressure. However, because the molar volume of a condensed phase is so small, the dependence is very weak, and for typical ranges of pressure of interest to us, we can ignore the pressure dependence of G. The molar Gibbs energy of a gas, however, does depend on the pressure, and because the molar volume of a gas is large, the dependence is significant. To find a quantitative expression for the pressure dependence that is valid over a substantial pressure range, we have to take into account the fact that a gas is compressible and that the molar volume decreases as pressure is applied. We therefore expect the Gibbs energy to increase with pressure, but for it to become less sensitive to pressure as the pressure rises (because the molar volume is decreasing). We show in the following *Justification* that

$$G_m(p_f) = G_m(p_i) + RT \ln \frac{p_f}{p_i} \tag{3.3}$$

This equation shows that the molar Gibbs energy increases logarithmically (as $\ln p$) with the pressure (Fig. 3.2). The flattening of the curve at high pressures reflects the fact that, as we anticipated, as V_m gets smaller, G_m becomes less responsive to pressure.

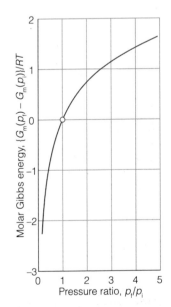

Fig. 3.2 The variation of the molar Gibbs energy of a perfect gas with pressure.

Justification 3.2 *The pressure variation of the Gibbs energy of a perfect gas*

We start with the exact expression for the effect of an infinitesimal change in pressure obtained in *Justification 3.1*, that $dG_m = V_m dp$. For a change in pressure from p_i to p_f, we need to add together (integrate) all these infinitesimal changes and write

$$\Delta G_m = \int_{p_i}^{p_f} V_m dp$$

To evaluate the integral, we must know how the molar volume depends on the pressure. The easiest case to consider is a perfect gas, for which $V_m = RT/p$. Then

$$\Delta G_m = \int_{p_i}^{p_f} \frac{RT}{p} dp = RT \int_{p_i}^{p_f} \frac{dp}{p} = RT \ln \frac{p_f}{p_i}$$

We have used the standard integral described in *Mathematical toolkit 3.1*. Finally, with $\Delta G_m = G_m(p_f) - G_m(p_i)$, we get eqn 3.3.

3.3 The variation of Gibbs energy with temperature

To understand why phase transitions, including the denaturation of a biopolymer, occur at a specific temperature, we need to know how molar Gibbs energy varies with temperature.

For small changes in temperature, we show in the following *Justification* that the change in molar Gibbs energy at constant pressure may be written as

$$G_m(T_f) = G_m(T_i) - (T_f - T_i)S_m$$

Variation of the Gibbs energy with temperature   (3.4)

This expression is valid provided the entropy of the substance is unchanged over the range of temperatures of interest.

Justification 3.3 *The variation of the Gibbs energy with temperature*

The starting point for this short derivation is eqn 3.2 in *Justification* 3.1, which we rewrite as the change in molar Gibbs energy when both the pressure and the temperature are changed by infinitesimal amounts:

$$dG_m = V_m dp - S_m dT$$

If we hold the pressure constant, $dp = 0$, and

$$dG_m = -S_m dT$$

This expression is exact. If we suppose that the molar entropy is unchanged in the range of temperatures of interest, we can replace the infinitesimal changes by observable changes and so obtain eqn 3.4.

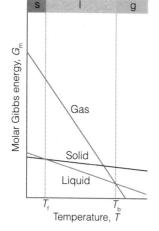

Fig. 3.3 The variation of molar Gibbs energy with temperature. All molar Gibbs energies decrease with increasing temperature. The regions of temperature over which the solid, liquid, and gaseous forms of a substance have the lowest molar Gibbs energy are indicated in the band at the top of the illustration.

Equation 3.4 tells us that, because molar entropy is positive, *an increase in temperature* ($T_f > T_i$) *results in a decrease in* G_m ($G_m(T_f) < G_m(T_i)$). Moreover, for a given change of temperature, the change in molar Gibbs energy is proportional to the molar entropy. For a given substance, because the molar entropy of the gas phase is greater than that for a condensed phase, the molar Gibbs energy falls more steeply with temperature for a gas than for a condensed phase. The molar entropy of the liquid phase of a substance is greater than that of its solid phase, so the slope is least steep for a solid. Figure 3.3 summarizes these characteristics.

Figure 3.3 reveals the thermodynamic reason why substances melt and vaporize as the temperature is raised. At low temperatures, the solid phase has the lowest molar Gibbs energy and is therefore the most stable. However, as the temperature is raised, the molar Gibbs energy of the liquid phase falls below that of the solid phase and the substance melts. At even higher temperatures, the molar Gibbs energy of the vapor plunges down below that of the liquid phase, and the vapor becomes the most stable phase. In other words, above a certain temperature, the liquid vaporizes.

We can also start to understand why some substances, such as solid carbon dioxide, sublime to a vapor without first forming a liquid. There is no fundamental requirement for the three lines to lie exactly in the positions we have drawn them in Fig. 3.3: the liquid line, for instance, could lie where we have drawn it in Fig. 3.4. Now we see that at no temperature (at the given pressure) does the liquid phase have the lowest molar Gibbs energy. Such a substance converts spontaneously directly from the solid to the vapor. That is, the substance sublimes.

The **transition temperature** between two phases, such as between liquid and solid or between conformations of a protein, is the temperature, at a given pressure, at which the two phases are in equilibrium and therefore their molar Gibbs energies are equal. At 1 atm, for instance, ice and liquid water are in equilibrium at 0°C and $G_m(H_2O,l) = G_m(H_2O,s)$.

As always when using thermodynamic arguments, it is important to keep in mind the distinction between the spontaneity of a phase transition and its rate. *Spontaneity is a tendency, not necessarily an actuality.* A phase transition predicted to be spontaneous may occur so slowly as to be unimportant in practice. For instance, at normal temperatures and pressures the molar Gibbs energy of graphite is 3 kJ mol^{-1} lower than that of diamond, so there is a thermodynamic tendency for diamond to convert into graphite. However, for this transition to take place, the carbon atoms of diamond must change their locations, and because the bonds between the atoms are so strong and large numbers of bonds must change simultaneously, this process is immeasurably slow except at high temperatures. In gases and liquids the mobilities of the molecules normally allow phase transitions to occur rapidly, but in solids thermodynamic instability may be frozen in and a thermodynamically unstable phase may persist for thousands of years. The molecules of liquids are mobile, so this 'metastability' is much less likely to occur. Nevertheless, even a liquid may persist above its boiling point as a **superheated liquid** if it is heated carefully and there are no so-called **nucleation centers**, such as scratches on the interior of the containing vessel, at which the vapor can form.

3.4 Phase diagrams

To prepare for being able to describe phase transitions in biological macromolecules, first we need to explore the conditions for equilibrium between phases of simpler substances.

The **phase diagram** of a substance is a map showing the conditions of temperature and pressure at which its various phases are thermodynamically most stable (Fig. 3.5). For example, at point A in the illustration, the vapor phase of the substance is thermodynamically the most stable, but at C the liquid phase is the most stable.

The boundaries between regions in a phase diagram, which are called **phase boundaries**, show the values of p and T at which the two neighboring phases are in equilibrium. For example, if the system is arranged to have a pressure and

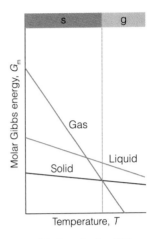

Fig. 3.4 If the line for the Gibbs energy of the liquid phase does not cut through the line for the solid phase (at a given pressure) before the line for the gas phase cuts through the line for the solid, the liquid is not stable at any temperature at that pressure. Such a substance sublimes.

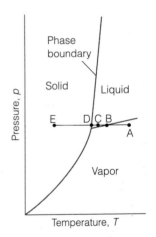

Fig. 3.5 A typical phase diagram, showing the regions of pressure and temperature at which each phase is the most stable. The phase boundaries (three are shown here) show the values of pressure and temperature at which the two phases separated by the line are in equilibrium. The significance of the letters A, B, C, D, and E (also referred to in Fig. 3.8) is explained in the text.

temperature represented by point B, then the liquid and its vapor are in equilibrium (like liquid water and water vapor at 1 atm and 100°C). If the temperature is reduced at constant pressure, the system moves to point C, where the liquid is stable (like water at 1 atm and at temperatures between 0°C and 100°C). If the temperature is reduced still further to D, then the solid and liquid phases are in equilibrium (like ice and water at 1 atm and 0°C). A further reduction in temperature to E takes the system into the region where the solid is the stable phase.

Any point lying on a phase boundary represents a pressure and temperature at which there is a 'dynamic equilibrium' between the two adjacent phases. A state of **dynamic equilibrium** is one in which a reverse process is taking place at the same rate as the forward process. Although there may be a great deal of activity at a molecular level, there is no net change in the bulk properties or appearance of the sample. For example, any point on the liquid–vapor boundary represents a state of dynamic equilibrium in which vaporization and condensation continue at matching rates. Molecules are leaving the surface of the liquid at a certain rate, and molecules already in the gas phase are returning to the liquid at the same rate; as a result, there is no net change in the number of molecules in the vapor and hence no net change in its pressure. Similarly, a point on the solid–liquid curve represents conditions of pressure and temperature at which molecules are ceaselessly breaking away from the surface of the solid and contributing to the liquid. However, they are doing so at a rate that exactly matches that at which molecules already in the liquid are settling onto the surface of the solid and contributing to the solid phase.

(a) Phase boundaries

The pressure of a vapor that is in equilibrium with its condensed phase is called the **vapor pressure** of the substance. Vapor pressure increases with temperature because, as the temperature is raised, more molecules have sufficient energy to leave their neighbors in the liquid. To determine the vapor pressure, a small amount of liquid can be introduced into the near-vacuum at the top of a mercury barometer and the depression of the column measured (Fig. 3.6). To ensure that the pressure exerted by the vapor is truly the vapor pressure, enough liquid must be added for some to remain after the vapor forms, for only then are the liquid and vapor phases in equilibrium. The temperature can be changed to determine another point on the curve, and so on (Fig. 3.7).

The plot of the vapor pressure against temperature is also the liquid–vapor boundary in a phase diagram. To appreciate that interpretation, suppose we have a liquid in a cylinder fitted with a piston. If at some temperature we apply a pressure greater than the vapor pressure of the liquid, the vapor is eliminated, the piston rests on the surface of the liquid, and the system moves to one of the points in the 'liquid' region of the phase diagram. If instead we reduce the pressure on the system to a value below the vapor pressure at that temperature, the system moves to one of the points in the 'vapor' region of the diagram. At the vapor pressure itself, vapor and liquid are in equilibrium, and the state of the system is represented by a point on the phase boundary.

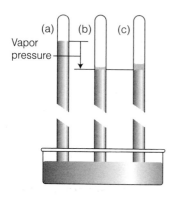

Fig. 3.6 When a small volume of water is introduced into the vacuum above the mercury in a barometer (a), the mercury is depressed (b) by an amount that is proportional to the vapor pressure of the liquid. (c) The same pressure is observed however much liquid is present (provided some is present).

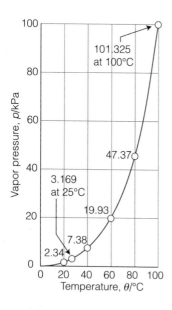

Fig. 3.7 The experimental variation of the vapor pressure of water with temperature.

Self-test 3.1) What would be observed when a pressure of 50 Torr is applied to a sample of water in equilibrium with its vapor at 25°C, when its vapor pressure is 23.8 Torr?

Answer: The sample condenses entirely to liquid.

The same approach can be used to plot the solid–vapor boundary, which is a graph of the vapor pressure of the solid against temperature. The **sublimation vapor pressure** of a solid, the pressure of the vapor in equilibrium with a solid at a particular temperature, is usually much lower than that of a liquid because the molecules are more strongly bound together in the solid than in the liquid.

A more sophisticated procedure is needed to determine the locations of solid–solid phase boundaries like that between the different forms of ice, for instance, because the transition between two solid phases is more difficult to detect. One approach is to use **thermal analysis**, which takes advantage of the heat released during a transition. In a typical thermal analysis experiment, a sample is allowed to cool and its temperature is monitored. When the transition occurs, energy is released as heat and the cooling stops until the transition is complete (Fig. 3.8). The transition temperature is obvious from the shape of the graph and is used to mark a point on the phase diagram. The pressure can then be changed and the corresponding transition temperature determined.

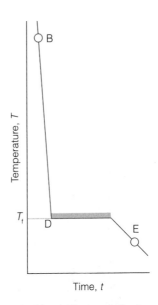

Fig. 3.8 The cooling curve for the B–E section of the horizontal line in Fig. 3.5. The halt at D corresponds to the pause in cooling while the liquid freezes and releases its enthalpy of transition. The halt lets us locate T_f even if the transition cannot be observed visually.

(b) The location of phase boundaries

Thermodynamics provides us with a way of predicting the location of the phase boundaries and relating their location and shape to the thermodynamic properties of the system. For instance, the shape of the vapor pressure curve (the liquid–vapor boundary) is related to the enthalpy of vaporization of the liquid.

Suppose two phases, such as liquid and vapor, are in equilibrium at a given pressure and temperature. As we have seen, at a given temperature, the pressure corresponding to equilibrium is the vapor pressure of the liquid. If we change the temperature, the vapor pressure changes to a different value. That is, there is a relation between the change in temperature, dT, and the accompanying change in vapor pressure, dp. If we were considering the equilibrium between a solid and a liquid, the focus would be different: in this case we would typically be interested in the change in melting point as the pressure is increased. We show in the following *Justification* that the relation between dT and dp that ensures that in either case the two phases remain in equilibrium is given by the **Clapeyron equation** for the slope of the phase boundary at any temperature

$$\frac{dp}{dT} = \frac{\Delta_{trs}H}{T\Delta_{trs}V}$$

$\boxed{\text{Clapeyron equation}}$ (3.5)

where $\Delta_{trs}H$ is the enthalpy of transition and $\Delta_{trs}V$ is the volume of transition (the change in molar volume that accompanies the transition) at the temperature of interest. For the liquid–vapor equilibrium, the equation in the form $dp = (\Delta_{vap}H/T\Delta_{vap}V)dT$ gives the change in vapor pressure when the temperature is changed; for the solid–liquid equilibrium, the equation in the form $dT = (T\Delta_{fus}V/\Delta_{fus}H)dp$ gives the change in melting point caused by a change in pressure.

For the solid–liquid phase boundary, the enthalpy of fusion is positive because melting is endothermic for all substances of interest in biology. For most substances, the molar volume increases slightly on melting, so $\Delta_{fus}V$ is positive but small. It follows that the melting temperature changes very little when the pressure is changed. In other words, the slope of the phase boundary is large and positive (up from left to right). Water, however, is quite different, for although its melting is endothermic, its molar volume decreases on melting (liquid water is denser than ice at 0°C, which is why ice floats on water), so $\Delta_{fus}V$ is small but negative. Consequently, an increase in pressure brings about a decrease in the melting point of ice.

A brief illustration

For water at 1 bar, $\Delta_{fus}H^{\ominus} = 6.008 \times 10^3$ J mol^{-1} and $\Delta_{fus}V^{\ominus} = -1.634 \times 10^{-6}$ m^3 mol^{-1}. It follows from eqn 3.5 that at $T = 273.15$ K, the melting point of ice,

$$\frac{dp}{dT} = \frac{6.008 \times 10^3 \text{ J mol}^{-1}}{(273.15 \text{ K}) \times (-1.634 \times 10^{-6} \text{ m}^3 \text{ mol}^{-1})} = -1.346 \times 10^7 \text{ Pa K}^{-1}$$

where we have used 1 Pa = 1 N m^{-2} and 1 J = 1 N m to write 1 Pa = 1 J m^{-3}. In other words, the slope of the ice–water phase boundary is steep but negative (down from left to right).

For the liquid–vapor boundary (the vapor pressure curve), both the enthalpy and volume of vaporization are invariably positive, so the vapor pressure invariably increases with temperature (dp is positive if dT is positive). However, we have to be cautious because although the enthalpy of vaporization is not very sensitive to temperature, the volume of vaporization depends strongly on the temperature (through the effect of temperature on the volume of a gas). If we suppose that the vapor behaves as a perfect gas, then we show in the following *Justification* that the relation between a change in temperature and a change in vapor pressure is given by the **Clausius–Clapeyron equation**:

$$\frac{d \ln p}{dT} = \frac{\Delta_{vap}H}{RT^2}$$

Clausius–Clapeyron equation   (3.6)

A brief illustration

For water at 1 bar and 373.2 K (the boiling point of water), $\Delta_{vap}H^{\ominus} = 4.07 \times 10^4$ J mol^{-1}, and it follows that

$$\frac{d \ln p}{dT} = \frac{4.07 \times 10^4 \text{ J mol}^{-1}}{(8.314 \text{ J K}^{-1} \text{ mol}^{-1}) \times (373.2 \text{ K})^2} = 3.51 \times 10^{-2} \text{ K}^{-1}$$

The liquid–vapor boundary of the phase diagram for water has a positive slope, and we shall see in Section 3.4(d) that the slope of the liquid–vapor phase boundary is much less steep than the slope of the ice–water phase boundary.

Justification 3.4   *The Clapeyron and Clausius–Clapeyron equations*

The derivation of the Clapeyron equations is based on eqn 3.2, written as $dG_m = V_m dp - S_m dT$. At a certain pressure and temperature two phases, which we call 1 and 2 but can imagine to be a liquid and a vapor, respectively, are in equilibrium and $G_m(1) = G_m(2)$ (Fig. 3.9). When the temperature is changed by dT and the pressure changes by dp, the molar Gibbs energies change as follows:

$$dG_m(1) = V_m(1)dp - S_m(1)dT \qquad dG_m(2) = V_m(2)dp - S_m(2)dT$$

The two phases are in equilibrium before the change and remain in equilibrium after the change, so the two changes in molar Gibbs energy must be equal, $dG_m(1) = dG_m(2)$. It follows that

$$V_m(1)dp - S_m(1)dT = V_m(2)dp - S_m(2)dT$$

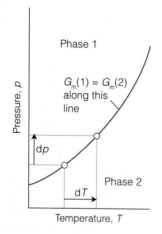

Fig. 3.9 At equilibrium, two phases have the same molar Gibbs energy. When the temperature is changed by dT, for the two phases to remain in equilibrium, the pressure must be changed by dp so that the Gibbs energies of the two phases remain equal.

and therefore

$$\{V_m(2) - V_m(1)\}dp = \{S_m(2) - S_m(1)\}dT$$

With $\Delta_{trs}V = V_m(2) - V_m(1)$ and $\Delta_{trs}S = S_m(2) - S_m(1)$, this equation becomes

$$\Delta_{trs}V dp = \Delta_{trs}S dT$$

or

$$\frac{dp}{dT} = \frac{\Delta_{trs}S}{\Delta_{trs}V}$$

We saw in Section 2.2 that the transition entropy is related to the enthalpy of transition by $\Delta_{trs}S = \Delta_{trs}H/T_{trs}$, so eqn 3.5 follows immediately. We have dropped the 'trs' subscript from the temperature in eqn 3.5 because all the points on the phase boundary—the only points we are considering in eqn 3.5—are transition temperatures.

To move on to the Clausius–Clapeyron equation, we consider the case of vaporization. Because the molar volume of a gas is much larger than the molar volume of a liquid, we can replace $\Delta_{vap}V = V_m(g) - V_m(l)$ by $V_m(g)$ alone and write

$$\frac{dp}{dT} \approx \frac{\Delta_{vap}H}{TV_m(g)}$$

Next, we suppose that the vapor behaves as a perfect gas and write its molar volume as $V_m(g) = RT/p$. Then

$$\frac{dp}{dT} = \frac{\Delta_{vap}H}{T(RT/p)} = \frac{p\Delta_{vap}H}{RT^2}$$

and therefore

$$\frac{1}{p}\frac{dp}{dT} = \frac{\Delta_{vap}H}{RT^2}$$

A standard result of calculus is $d \ln x/dx = 1/x$, and therefore (by multiplying both sides by dx), $dx/x = d \ln x$. In this case, $dp/p = d \ln p$, and eqn 3.6 follows.

A note on good practice
Keep a note of any approximations made in a derivation, for they limit the range of applicability of an expression. We have made two approximations in the derivation of the Clausius–Clapeyron equation: (1) the molar volume of a gas is much greater than that of a liquid and (2) the vapor behaves as a perfect gas.

(c) Characteristic points

We have seen that as the temperature of a liquid is raised, its vapor pressure increases. What we observe, however, depends on whether the heating takes place in a closed or an open container.

First, consider what we would observe when we heat a liquid in an open vessel. At a certain temperature, the vapor pressure becomes equal to the external pressure. At this temperature, the vapor can drive back the surrounding atmosphere and expand indefinitely. Moreover, because there is no constraint on expansion, bubbles of vapor can form throughout the body of the liquid, the condition known as **boiling**. The temperature at which the vapor pressure of a liquid is equal to the external pressure is called the **boiling temperature**. When the external pressure is 1 atm, the boiling temperature is called the **normal boiling point**, T_b. It follows that we can predict the normal boiling point of a liquid by noting the temperature on the phase diagram at which its vapor pressure is 1 atm.

Now consider what happens when we heat the liquid in a closed vessel. Because the vapor cannot escape, its density increases as the vapor pressure rises and in due course the density of the vapor becomes equal to that of the remaining liquid. At this stage the surface between the two phases disappears (Fig. 3.10).

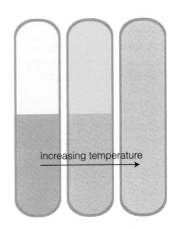

Fig. 3.10 When a liquid is heated in a sealed container, the density of the vapor phase increases and that of the liquid phase decreases, as depicted here by the changing density of shading. There comes a stage at which the two densities are equal and the interface between the two fluids disappears. This disappearance occurs at the critical temperature. The container needs to be strong: the critical temperature of water is at 373°C and the vapor pressure is then 218 atm.

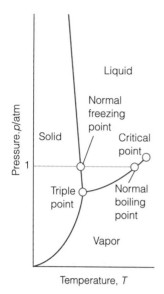

Fig. 3.11 The significant points of a phase diagram. The liquid–vapor phase boundary terminates at the *critical point*. At the *triple point*, solid, liquid, and vapor are in dynamic equilibrium. The *normal freezing point* is the temperature at which the liquid freezes when the pressure is 1 atm; the *normal boiling point* is the temperature at which the vapor pressure of the liquid is 1 atm.

Table 3.1 Critical constants*

	p_c/atm	V_c/(cm³ mol⁻¹)	T_c/K
Ammonia, NH_3	111	73	406
Argon, Ar	48	75	151
Benzene, C_6H_6	49	260	563
Carbon dioxide, CO_2	73	94	304
Hydrogen, H_2	13	65	33
Methane, CH_4	46	99	191
Oxygen, O_2	50	78	155
Water, H_2O	218	55	647

*The critical volume, V_c, is the molar volume at the critical pressure and critical volume.

The temperature at which the surface disappears is the **critical temperature**, T_c. The vapor pressure at the critical temperature is called the **critical pressure**, p_c, and the critical temperature and critical pressure together identify the **critical point** of the substance (see Table 3.1). If we exert pressure on a sample that is above its critical temperature, we produce a denser fluid. However, no surface appears to separate the two parts of the sample and a single uniform phase, a **supercritical fluid**, continues to fill the container. That is, we have to conclude that *a liquid cannot be produced by the application of pressure to a substance if it is at or above its critical temperature*. That is why the liquid–vapor boundary in a phase diagram terminates at the critical point (Fig. 3.11). A supercritical fluid is not a true liquid, but it behaves like a liquid in many respects—for example, it has a density similar to that of a liquid.

> **A brief illustration**
>
> Supercritical carbon dioxide, $scCO_2$, is the center of attention for an increasing number of solvent-based processes. The critical temperature of CO_2, 304.2 K (31.0°C) and its critical pressure, 72.9 atm, are readily accessible, it is cheap, and it can readily be recycled. A great advantage of $scCO_2$ is that there are no noxious residues once the solvent has been allowed to evaporate, so, coupled with its low critical temperature, $scCO_2$ is ideally suited to food processing and the production of pharmaceuticals. It is used, for instance, to remove caffeine from coffee or fats from milk. The supercritical fluid is also increasingly being used for dry cleaning, which avoids the use of carcinogenic and environmentally damaging chlorinated hydrocarbons.

The temperature at which the liquid and solid phases of a substance coexist in equilibrium at a specified pressure is called the **melting temperature** of the substance. Because a substance melts at the same temperature as it freezes, the melting temperature is the same as the **freezing temperature**. The solid–liquid boundary therefore shows how the melting temperature of a solid varies with pressure.

The melting temperature when the pressure on the sample is 1 atm is called the **normal melting point** or the **normal freezing point**, T_f. A liquid freezes when the energy of the molecules in the liquid is so low that they cannot escape from the attractive forces of their neighbors and lose their mobility.

There is a set of conditions under which three different phases (typically solid, liquid, and vapor) all simultaneously coexist in equilibrium. It is represented by the **triple point**, where the three phase boundaries meet. The triple point of a pure substance is a characteristic, unchangeable physical property of the substance. For water the triple point lies at 273.16 K and 611 Pa, and ice, liquid water, and water vapor coexist in equilibrium at no other combination of pressure and temperature.[1] At the triple point, the rates of each forward and reverse process are equal (but the three individual rates are not necessarily the same).

The triple point and the critical point are important features of a substance because they act as frontier posts for the existence of the liquid phase. As we see from Fig. 3.12a, if the slope of the solid-liquid phase boundary is as shown in the diagram:

The triple point marks the lowest temperature at which the liquid can exist.

The critical point marks the highest temperature at which the liquid can exist.

We shall see in the following section that for water, the solid–liquid phase boundary slopes in the opposite direction, and then only the second of these conclusions is relevant (see Fig. 3.12b).

(d) The phase diagram of water

Figure 3.13 is the phase diagram for water. The liquid–vapor phase boundary shows how the vapor pressure of liquid water varies with temperature. We can use this curve to decide how the boiling temperature varies with changing external pressure. For example, when the external pressure is 149 Torr (at an altitude of 12 km), water boils at 60°C because that is the temperature at which the vapor pressure is 149 Torr (19.9 kPa).

The solid–liquid boundary line in Fig. 3.14 shows how the melting temperature of water depends on the pressure. For example, although ice melts at 0°C at 1 atm, it melts at −1°C when the pressure is 130 atm. The very steep slope of the boundary indicates that enormous pressures are needed to bring about significant changes. Notice that the line slopes down from left to right, which—as we anticipated—means that the melting temperature of ice falls as the pressure is raised. We can trace the reason for this unusual behavior to the decrease in volume that occurs when ice melts: it is favorable for the solid to transform into the denser liquid as the pressure is raised. The decrease in volume is a result of the very open structure of the crystal structure of ice: as shown in Fig. 3.15, the water molecules are held apart, as well as together, by the hydrogen bonds between them, but the structure partially collapses on melting and the liquid is denser than the solid.

Figure 3.13 shows that water has one liquid phase but many different solid phases other than ordinary ice ('ice I', shown in Fig. 3.15). These solid phases differ in the arrangement of the water molecules: under the influence of very high pressures, hydrogen bonds buckle and the H_2O molecules adopt different

---

[1] The triple point of water is used to define the Kelvin scale of temperatures: the triple point is defined as lying at 273.16 K exactly. The normal freezing point of water is found experimentally to lie approximately 0.01 K below the triple point, at very close to 273.15 K.

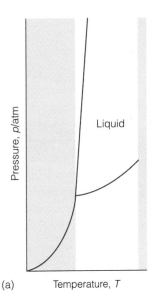

(a)   Temperature, T

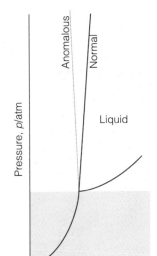

(b)   Temperature, T

Fig. 3.12 (a) For substances that have phase diagrams resembling the one shown here (which is common for most substances, with the important exception of water), the triple point and the critical point mark the range of temperatures over which the substance can exist as a liquid. The shaded areas show the regions of temperature in which a liquid cannot exist as a stable phase. (b) A liquid cannot exist as a stable phase if the pressure is below that of the triple point for normal or anomalous liquids.

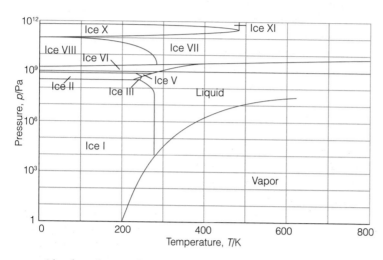

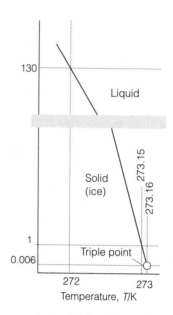

Fig. 3.13 The phase diagram for water showing the different solid phases.

Fig. 3.14 The solid–liquid boundary of water in more detail. The graph is schematic and not to scale.

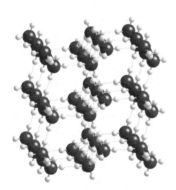

Fig. 3.15 The structure of ice I. Each O atom is at the center of a tetrahedron of four O atoms at a distance of 276 pm. The central O atom is attached by two short O–H bonds to two H atoms and by two long hydrogen bonds to the H atoms of two of the neighboring molecules. Overall, the structure consists of planes of puckered hexagonal rings of H_2O molecules (like the chair form of cyclohexane). This structure collapses partially on melting, leading to a liquid that is denser than the solid.

arrangements. These **polymorphs**, or different solid phases, of ice may be responsible for the advance of glaciers, for ice at the bottom of glaciers experiences very high pressures where it rests on jagged rocks. The sudden apparent explosion of Halley's comet in 1991 may have been due to the conversion of one form of ice into another in its interior. Figure 3.13 also shows that four or more phases of water (such as two solid forms, liquid, and vapor) are never in equilibrium. This observation is justified and generalized to all substances by the *phase rule*, which is derived in *Further information 3.1*.

Phase transitions in biopolymers and aggregates

In *Fundamentals* F.1 and Chapter 2 we saw that proteins and biological membranes can exist in ordered structures stabilized by a variety of molecular interactions, such as hydrogen bonds and hydrophobic interactions. However, when certain conditions are changed, the helical and sheet structures of a polypeptide chain may collapse into a random coil and the hydrocarbon chains in the interior of bilayer membranes may become more or less flexible. These structural changes may be regarded as phase transitions in which molecular interactions in compact phases are disrupted at characteristic transition temperatures to yield phases in which the atoms can move more randomly.

3.5 The stability of nucleic acids and proteins

To understand melting of proteins and nucleic acids at specific transition temperatures, we need to explore quantitatively the effect of intermolecular interactions on the stability of compact conformations of biopolymers.

From *In the laboratory* 1.1 we learned that the thermal denaturation of a biopolymer may be thought of as a kind of intramolecular melting from an organized structure to a flexible coil. This melting occurs at a specific **melting temperature**, T_m, which increases with the strength and number of intramolecular and intermolecular interactions in the material. Denaturation is a **cooperative process** in the sense that the biopolymer becomes increasingly more susceptible to denaturation once the process begins. This cooperativity is observed as a sharp step in a plot of fraction of unfolded polymer against temperature (Fig. 3.16). The melting temperature, T_m, is the temperature at which the fraction of unfolded polymer is 0.5.

Closer examination of thermal denaturation reveals some of the chemical factors that determine protein and nucleic acid stability. For example, the thermal stability of DNA increases with the number of C–G base pairs in the sequence because each C–G base pair has three hydrogen bonds (**1**), whereas each T–A base pair has only two (**2**). More energy is required to unravel a double helix that has a higher proportion of hydrogen bonding interactions per base pair.

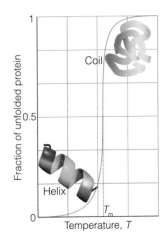

Fig. 3.16 A protein unfolds as the temperature of the sample increases. The sharp step in the plot of fraction of unfolded protein against temperature indicates that the transition is cooperative. The melting temperature, T_m, is the temperature at which the fraction of unfolded polymer is 0.5.

1 C-G base pair

2 T-A base pair

Example 3.1   *Predicting the melting temperature of DNA*

The melting temperature of a DNA molecule can be determined by differential scanning calorimetry (*In the laboratory* 1.1). The following data were obtained in 0.010 M Na_3PO_4(aq) for a series of DNA molecules with varying base pair composition, with f the fraction of C–G base pairs:

f	0.375	0.509	0.589	0.688	0.750
T_m/K	339	344	348	351	354

Estimate the melting temperature of a DNA molecule containing 40.0 per cent C–G base pairs.

Strategy We need to look for a quantitative relation between the melting temperature and the composition of DNA. We can begin by plotting T_m against fraction of C–G base pairs and examining the shape of the curve. If visual inspection of the plot suggests a linear relation, then the melting point at any composition can be predicted from the equation of the line that fits the data.

Solution Figure 3.17 shows that T_m varies linearly with the fraction of C–G base pairs, at least in this range of composition. The equation of the line that fits the data is

$$T_m/K = 325 + 39.7f$$

It follows that $T_m = 341$ K for 40.0 per cent C–G base pairs (at $f = 0.400$).

A note on good practice
In this example we do not have a good theory to guide us in the choice of a mathematical model to describe the behavior of the system over a wide range of parameters. We are limited to finding a purely empirical relation—in this case a simple first-order polynomial equation—that fits the available data. It follows that we should not attempt to predict the property of a system that falls outside the narrow range of the data used to generate the fit because the mathematical model may have to be enhanced (for example, by using higher-order polynomial equations) to describe the system over a wider range of conditions. In the present case, we should not attempt to predict the T_m of DNA molecules outside the range $0.375 < f < 0.750$.

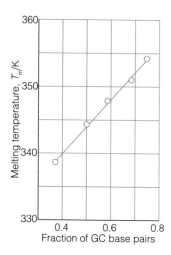

Fig. 3.17 Data for Example 3.1 showing the variation of the melting temperature of DNA molecules with the fraction of C–G base pairs. All the samples also contain 1.0×10^{-2} mol dm^{-3} Na$_3$PO$_4$.

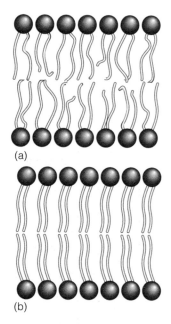

(a)

(b)

Fig. 3.18 A depiction of the variation with temperature of the flexibility of hydrocarbon chains in a lipid bilayer. (a) At physiological temperature, the bilayer exists as a liquid crystal, in which some order exists but the chains writhe. (b) At a specific temperature, the chains are largely frozen and the bilayer is said to exist as a gel.

Self-test 3.2 The following calorimetric data were obtained in solutions containing 0.15 M NaCl(aq) for the same series of DNA molecules studied in *Example* 3.1. Estimate the melting temperature of a DNA molecule containing 40.0 per cent C–G base pairs under these conditions.

f	0.375	0.509	0.589	0.688	0.750
T_m/K	359	364	368	371	374

Answer: 360 K

Example 3.1 and *Self-test* 3.2 reveal that DNA is rather stable toward thermal denaturation, with T_m values ranging from about 340 K to 375 K, which is significantly higher than body temperature (310 K). The data also show that increasing the concentration of ions in solution increases the melting temperature of DNA. The stabilizing effect of ions can be traced to the fact that DNA has negatively charged phosphate groups decorating its surface. When the concentration of ions in solution is low, repulsive Coulomb interactions between neighboring phosphate groups destabilize the double helix and lower the melting temperature. On the other hand, positive ions, such as the Na$^+$ ions in Self-test 3.2, bind electrostatically to the surface of DNA and mitigate repulsive interactions between phosphate groups. The result is stabilization of the double helical conformation and an increase in T_m.

In contrast to DNA, proteins are relatively unstable toward thermal denaturation. For example, $T_m = 320$ K for ribonuclease T$_1$ (an enzyme that cleaves RNA in the cell), which is close to body temperature. More surprisingly, the Gibbs energy for the unfolding of ribonuclease T$_1$ at pH = 7.0 and 298 K is only +22.5 kJ mol^{-1}, which is comparable to the energy required to break a single hydrogen bond (about 20 kJ mol^{-1}) despite the fact that the formation of helices and sheets in proteins requires many hydrogen bonds. Therefore, unlike DNA, the stability of a protein does not increase in a simple way with the number of hydrogen bonding interactions. Although the reasons for the low stability of proteins are not known, the answer probably lies in a delicate balance of all intra- and intermolecular interactions that allow a protein to fold into its active conformation (Chapter 11).

3.6 Phase transitions of biological membranes

To understand why cell membranes are sufficiently rigid to encase life's molecular machines while being flexible enough to allow for cell division, we need to explore the factors that determine the melting temperatures of lipid bilayers.

All lipid bilayers undergo a transition from a state of high to low chain mobility at a temperature that depends on the structure of the lipid. To visualize the transition, we consider what happens to a membrane as we lower its temperature (Fig. 3.18). There is sufficient energy available at normal temperatures for limited bond rotation to occur and the flexible chains to writhe around. However, the membrane is still highly organized in the sense that the bilayer structure does not come apart and the system is best described as a *liquid crystal*, a substance having liquid-like, imperfect long-range order in at least one direction in space but positional or orientational order in at least one other direction (Fig. 3.18a). At lower temperatures, the amplitudes of the writhing motion decrease until a specific temperature is reached at which motion is largely frozen. The membrane is then said to exist as a *gel* (Fig. 3.18b). Biological membranes exist as liquid crystals at physiological temperatures.

Phase transitions in membranes are often observed as 'melting' from gel to liquid crystal by differential scanning calorimetry (*In the laboratory* 1.1). The data show relations between the structure of the lipid and the melting temperature. For example, the melting temperature increases with the length of the hydrophobic chain of the lipid. This correlation is reasonable, as we expect longer chains to be held together more strongly by hydrophobic interactions than shorter chains (Section 2.7). It follows that stabilization of the gel phase in the membranes of lipids with long chains results in relatively high melting temperatures. On the other hand, any structural elements that prevent alignment of the hydrophobic chains in the gel phase lead to low melting temperatures. Indeed, lipids containing unsaturated chains, those containing some C=C bonds, form membranes with lower melting temperatures than those formed from lipids with fully saturated chains, those consisting of C–C bonds only.

Interspersed among the phospholipids of biological membranes are sterols, such as cholesterol (Atlas L1), which is largely hydrophobic but does contain a hydrophilic –OH group. Sterols, which are present in different proportions in different types of cells, prevent the hydrophobic chains of lipids from 'freezing' into a gel and, by disrupting the packing of the chains, spread the melting point of the membrane over a range of temperatures.

Self-test 3.3) Organisms are capable of biosynthesizing lipids of different composition so that cell membranes have melting temperatures close to the ambient temperature. Why do bacterial and plant cells grown at low temperatures synthesize more phospholipids with unsaturated chains than do cells grown at higher temperatures?

> **Answer:** Insertion of lipids with unsaturated chains lowers the plasma membrane's melting temperature to a value that is close to the lower ambient temperature.

Case study 3.1) *The use of phase diagrams in the study of proteins*

As in the discussion of pure substances, the phase diagram of a mixture shows which phase is most stable for the given conditions. However, composition is now a variable in addition to the pressure and temperature. Phase equilibria in binary mixtures may be explored by collecting data at constant pressure and displaying the results as a **temperature–composition diagram**, in which one axis is the temperature and the other axis is the mole fraction or concentration.

Temperature–composition diagrams may be used to characterize intermediates in the unfolding of a protein caused by denaturation with a chemical agent. For example, urea, $CO(NH_2)_2$, competes for NH and CO groups, interferes with hydrogen bonding in a polypeptide, and disrupts the intramolecular interactions responsible for its native three-dimensional conformation. A temperature–composition diagram, such as the idealized form shown in Fig. 3.19, can reveal conditions under which different forms of the polypeptide can exist. The idealized diagram shows three structural regions, or phases: the native form, the unfolded form, and a 'molten globule' form, a partially unfolded but still compact form of the protein. As usual, two phases in equilibrium

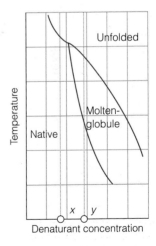

Fig. 3.19 An example of a temperature–composition diagram showing denaturation of a protein in a native phase into molten globule and fully unfolded phases. The concentrations marked *x* and *y* will be used in Exercise 3.39.

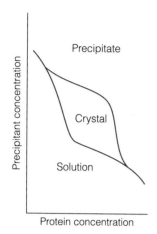

Fig. 3.20 An example of a phase diagram in which the mole fraction of a precipitant, a substance that causes precipitation, is plotted against the mole fraction of a protein. The data help biochemists find conditions under which a protein crystallizes.

define a line in the diagram, and a point represents a unique set of conditions under which the three phases are in equilibrium.

In another type of phase diagram, the mole fraction or concentration of one component of a mixture is plotted against the mole fraction or concentration of another component, and experiments are conducted at constant temperature and pressure. Phase diagrams so constructed help biochemists find conditions under which a protein may form an ordered crystal amenable to study by X-ray diffraction techniques, which can reveal the three-dimensional arrangement of atoms in biological assemblies (see the *Prologue* and Chapter 11). A common crystallization technique for charged proteins consists of adding large amounts of a salt, such as $(NH_4)_2SO_4$, to a buffer solution containing the biopolymer. The increase in the ionic strength of the solution decreases the solubility of the protein to such an extent that the protein precipitates. The idealized phase diagram in Fig. 3.20 shows that ordered crystals precipitate over a relatively narrow range of protein and salt concentrations. Precise knowledge of crystallization conditions is a key to the reproducibility of X-ray diffraction experiments.

The thermodynamic description of mixtures

We now leave pure materials and the limited but important changes they can undergo and examine mixtures. We shall consider only **homogeneous mixtures**, or solutions, in which the composition is uniform however small the sample. The component in smaller abundance is called the **solute** and that in larger abundance is the **solvent**. These terms, however, are normally but not invariably reserved for solids dissolved in liquids; one liquid mixed with another is normally called simply a 'mixture' of the two liquids. In this chapter we consider mainly **nonelectrolyte solutions**, where the solute is not present as ions. Examples are sucrose dissolved in water, sulfur dissolved in carbon disulfide, and a mixture of ethanol and water. Although we also consider some of the special problems of **electrolyte solutions**, in which the solute consists of ions that interact strongly with one another, we defer a full study until Chapter 5. The measures of concentration commonly encountered in physical chemistry are reviewed in *Further information 3.2*.

3.7 The chemical potential

To assess the spontaneity of a biological process involving a mixture, we need to know how to compute the contribution of each substance to the total Gibbs energy of the mixture.

A **partial molar property** is the contribution (per mole) that a substance makes to an overall property of a mixture. The most important partial molar property for our purposes is the **partial molar Gibbs energy**, $G_{J,m}$, of a substance J, which is the contribution of J (per mole of J) to the total Gibbs energy of a mixture. It follows that if we know the partial molar Gibbs energies of two substances A and B in a mixture of a given composition, then we can calculate the total Gibbs energy of the mixture by using

$$G = n_A G_{A,m} + n_B G_{B,m}$$

(3.7)

To gain insight into the significance of the partial molar Gibbs energy, consider a mixture of ethanol and water. Ethanol has a particular partial molar Gibbs energy when it is pure (and every molecule is surrounded by other ethanol molecules), and it has a different partial molar Gibbs energy when it is in an aqueous solution of a certain composition (because then each ethanol molecule is surrounded by a mixture of ethanol and water molecules).

The partial molar Gibbs energy is so important in chemistry that it is given a special name and symbol. From now on, we shall call it the **chemical potential** and denote it μ (mu). Then eqn 3.7 becomes

$$G = n_A \mu_A + n_B \mu_B \tag{3.8}$$

where μ_A is the chemical potential of A in the mixture and μ_B is the chemical potential of B. In the course of this chapter and the next we shall see that the name 'chemical potential' is very appropriate, for it will become clear that μ_J is a measure of the ability of J to bring about physical and chemical change. A substance with a high chemical potential has a high ability, in a sense we shall explore, to drive a reaction or some other physical process forward.

We saw in Section 3.1 that the molar Gibbs energy of a pure substance is the same in all the phases at equilibrium. We can use the same argument to show in the following *Justification* that *a system is at equilibrium when the chemical potential of each substance has the same value in every phase in which it occurs.* We can think of the chemical potential as the pushing power of each substance, and equilibrium is reached only when each substance pushes with the same strength in any phase it occupies.

Justification 3.5   *The uniformity of chemical potential*

Suppose a substance J occurs in different phases in different regions of a system. For instance, we might have a liquid mixture of ethanol and water and a mixture of their vapors. Let the substance J have chemical potential $\mu_J(l)$ in the liquid mixture and $\mu_J(g)$ in the vapor. We could imagine an infinitesimal amount, dn_J, of J migrating from the liquid to the vapor. As a result, the Gibbs energy of the liquid phase falls by $\mu_J(l)dn_J$ and that of the vapor rises by $\mu_J(g) dn_J$. The net change in Gibbs energy is

$$dG = \mu_J(g)dn_J - \mu_J(l)dn_J = \{\mu_J(g) - \mu_J(l)\}dn_J$$

There is no tendency for this migration (and the reverse process, migration from the vapor to the liquid) to occur, and the system is at equilibrium if $dG = 0$, which requires that $\mu_J(g) = \mu_J(l)$. The argument applies to each component of the system. Therefore, for a substance to be at equilibrium throughout the system, its chemical potential must be the same everywhere, as asserted in the text.

3.8 Ideal and ideal-dilute solutions

Because in biochemistry we are concerned primarily with liquid solutions, we need expressions for the chemical potentials of solutes and solvents.

We need an explicit formula for the variation of the chemical potential of a substance with the composition of the mixture. Here we use the strategy mentioned at the start of the chapter: we begin by considering the chemical potential of a gas, not because gases are particularly interesting in biology but because we can use the resulting expression to derive results for solutions.

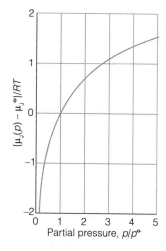

Fig. 3.21 The variation with partial pressure of the chemical potential of a perfect gas. Note that the chemical potential increases with pressure.

(a) The chemical potential of a gas

Our starting point is eqn 3.3, $G_m(p_f) = G_m(p_i) + RT \ln(p_f/p_i)$, which shows how the molar Gibbs energy of a perfect gas depends on pressure. First, we set $p_f = p$, the pressure of interest, and $p_i = p^\ominus$, the standard pressure (1 bar). At the latter pressure, the molar Gibbs energy has its standard value, $G_m^\ominus$, so we can write

$$G_m(p) = G_m^\ominus + RT \ln(p/p^\ominus) \tag{3.9}$$

Next, for a *mixture* of perfect gases, we interpret p as the *partial* pressure of the gas, and G_m is the *partial* molar Gibbs energy, the chemical potential. Therefore, for a mixture of perfect gases, for each component J present at a partial pressure p_J,

$$\mu_J = \mu_J^\ominus + RT \ln(p_J/p^\ominus) \tag{3.10a}$$

In this expression, $\mu_J^\ominus$ is the **standard chemical potential** of the gas J, which is identical to its standard molar Gibbs energy, the value of G_m for the pure gas at 1 bar. If we adopt the convention that, whenever p_J appears in a formula, it is to be interpreted as $p_J/p^\ominus$ (so, if the pressure is 2.0 bar, $p_J = 2.0$), we can write eqn 3.10a more simply as

$$\mu_J = \mu_J^\ominus + RT \ln p_J \tag{3.10b}$$

Figure 3.21 illustrates the pressure dependence of the chemical potential of a perfect gas predicted by this equation. Note that the chemical potential becomes negatively infinite as the pressure tends to zero, rises to its standard value at 1 bar (because ln 1 = 0), and then increases slowly (logarithmically, as ln p) as the pressure is increased further.

Equation 3.10 tells us that *the higher the partial pressure of a gas, the higher its chemical potential*. This conclusion is consistent with the interpretation of the chemical potential as an indication of the potential of a substance to be active chemically: the higher the partial pressure, the more active chemically the species. In this instance the chemical potential represents the tendency of the substance to react when it is in its standard state (the significance of the term $\mu^\ominus$) plus an additional tendency that reflects whether it is at a different pressure. A higher partial pressure gives a substance more chemical 'punch', just like winding a spring gives a spring more physical punch (that is, enables it to do more work).

Self-test 3.4　Suppose that the partial pressure of a perfect gas falls from 1.00 bar to 0.50 bar as it is consumed in a reaction at 25°C. What is the change in chemical potential of the substance?

Answer: -1.7 kJ mol^{-1}

(b) The chemical potential of a solvent

We can anticipate that the chemical potential of a species ought to increase with concentration because the higher its concentration, the greater its chemical 'punch'. In the following, we use J to denote a substance in general, A to denote a solvent, and B to denote a solute.

The key to linking the properties of a solution to those of a gas and setting up an expression for the chemical potential of a solute is the work done by the French chemist François Raoult (1830–1901), who spent most of his life measuring the vapor pressures of solutions. He measured the **partial vapor pressure**, p_J, of each component in the mixture, the partial pressure of the vapor of each component in dynamic equilibrium with the liquid mixture, and established what is now called **Raoult's law**:

> The partial vapor pressure of a substance in a liquid mixture is proportional to its mole fraction in the mixture and its vapor pressure when pure: $p_J = x_J p_J^*$ | Raoult's law | (3.11)

In this expression, p_J^* is the vapor pressure of the pure substance.

> **A brief illustration**
>
> When the mole fraction of water in an aqueous solution is 0.90, then, provided Raoult's law is obeyed, the partial vapor pressure of the water in the solution is 90 per cent that of pure water. This conclusion is approximately true whatever the identity of the solute and the solvent (Fig. 3.22).

The molecular origin of Raoult's law is the effect of the solute on the entropy of the solution. The entropy of the solvent arises from the random locations and the thermal motion of its molecules. The vapor pressure then represents the tendency of the system and its surroundings to reach a higher entropy. When a solute is present, the molecules in the solution are more dispersed than in the pure solvent, so we cannot be sure that a molecule chosen at random will be a solvent molecule (Fig. 3.23). Because the entropy of the solution is higher than that of the pure solvent, the solution has a lower tendency to acquire an even higher entropy by the solvent vaporizing. In other words, the vapor pressure of the solvent in the solution is lower than that of the pure solvent.

A hypothetical solution of a solute B in a solvent A that obeys Raoult's law throughout the composition range from pure A to pure B is called an **ideal solution**. The law is most reliable when the components of a mixture have similar molecular shapes and are held together in the liquid by similar types and strengths of intermolecular forces. An example is a mixture of two structurally similar hydrocarbons. A mixture of benzene and methylbenzene (toluene) is a good approximation to an ideal solution, for the partial vapor pressure of each component satisfies Raoult's law reasonably well throughout the composition range from pure benzene to pure methylbenzene (Fig. 3.24).

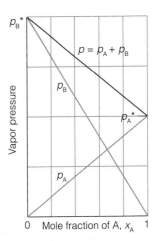

Fig. 3.22 The partial vapor pressures of the two components of an ideal binary mixture are proportional to the mole fractions of the components in the liquid. The total pressure of the vapor is the sum of the two partial vapor pressures.

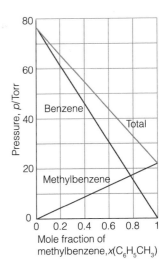

Fig. 3.24 Two similar substances, in this case benzene and methylbenzene (toluene), behave almost ideally and have vapor pressures that closely resemble those for the ideal case depicted in Fig. 3.22.

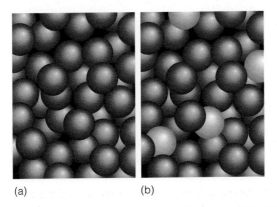

Fig. 3.23 (a) In a pure liquid, we can be confident that any molecule selected from the sample is a solvent molecule. (b) When a solute is present, we cannot be sure that blind selection will give a solvent molecule, so the entropy of the system is greater than in the absence of the solute.

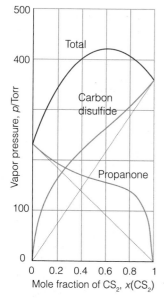

Fig. 3.25 Strong deviations from ideality are shown by dissimilar substances, in this case carbon disulfide and acetone (propanone). Note, however, that Raoult's law is obeyed by propanone when only a small amount of carbon disulfide is present (on the left) and by carbon disulfide when only a small amount of propanone is present (on the right).

A note on good practice
An asterisk (*) denotes a pure substance, but not one that is necessarily in its standard state. Only if the pressure is 1 bar would $\mu_A^\star$ be the standard chemical potential of A, and it would then be written as $\mu_A^\ominus$.

No mixture is perfectly ideal, and all real mixtures show deviations from Raoult's law. However, the deviations are small for the component of the mixture that is in large excess (the solvent) and become smaller as the concentration of solute decreases (Fig. 3.25). We can usually be confident that Raoult's law is reliable for the solvent when the solution is very dilute. More formally, Raoult's law is a *limiting law* (like the perfect gas law) and is strictly valid only at the limit of zero concentration of solute.

The theoretical importance of Raoult's law is that, because it relates vapor pressure to composition and we know how to relate pressure to chemical potential, we can use the law to relate chemical potential to the composition of a solution. As we show in the following *Justification*, the chemical potential of a solvent A present in solution at a mole fraction x_A is

$$\mu_A = \mu_A^\star + RT \ln x_A$$

Chemical potential of the solvent in an ideal solution (3.12)

where $\mu_A^\star$ is the chemical potential of pure A. This expression is valid throughout the concentration range for either component of a binary ideal solution. It is valid for the solvent of a real solution the closer the composition approaches pure solvent (pure A).

Justification 3.6 *The chemical potential of a solvent*

When a solvent A in a solution is in equilibrium with its vapor at a partial pressure p_A, the chemical potentials of the two phases are equal and we can write $\mu_A(l) = \mu_A(g)$ (Fig. 3.26). However, we have just derived an expression for the chemical potential of a vapor, eqn 3.10, so at equilibrium

$$\mu_A(l) = \mu_A^\ominus(g) + RT \ln p_A$$

According to Raoult's law, $p_A = x_A p_A^\star$, so we can use the relation $\ln(xy) = \ln x + \ln y$ to write

$$\mu_A(l) = \mu_A^\ominus(g) + RT \ln (x_A p_A^\star) = \mu_A^\ominus(g) + RT \ln p_A^\star + RT \ln x_A$$

The first two terms on the right, $\mu_A^\ominus(g)$ and $RT \ln p_A^\star$, are independent of the composition of the mixture and can be combined into the constant $\mu_A^\star$, the chemical potential of pure liquid A. Equation 3.12 then follows.

Figure 3.27 shows the variation of the chemical potential of the solvent predicted by this expression. Note that the chemical potential has its pure value at $x_A = 1$ (when only A is present). The essential feature of eqn 3.12 is that because $x_A < 1$ implies that $\ln x_A < 0$, *the chemical potential of a solvent is lower in a solution than when it is pure*. Provided the solution is almost ideal, a solvent in which a solute is present has less chemical 'punch' (including a lower ability to generate a vapor pressure) than when it is pure.

(**Self-test 3.5**) By how much is the chemical potential of benzene reduced at 25°C by a solute that is present at a mole fraction of 0.10?

Answer: 0.26 kJ mol⁻¹

(c) The chemical potential of a solute

Raoult's law provides a good description of the vapor pressure of the *solvent* in a very dilute solution, when the solvent A is almost pure. However, we cannot in

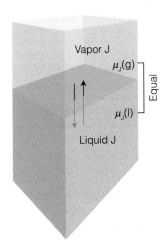

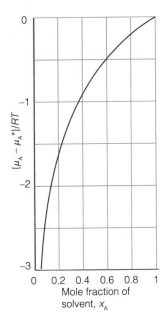

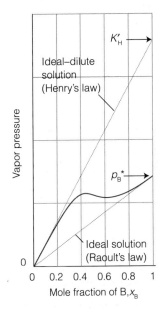

Fig. 3.26 At equilibrium, the chemical potential of a substance in its liquid phase is equal to the chemical potential of the substance in its vapor phase.

Fig. 3.27 The variation of the chemical potential of the solvent with the composition of the solution. Note that the chemical potential of the solvent is lower in the mixture than for the pure liquid (for an ideal system). This behavior is likely to be shown by a dilute solution in which the solvent is almost pure (and obeys Raoult's law).

Fig. 3.28 When a component (the solvent) is almost pure, it behaves in accord with Raoult's law and has a vapor pressure that is proportional to the mole fraction in the liquid mixture and a slope p^*, the vapor pressure of the pure substance. When the same substance is the minor component (the solute), its vapor pressure is still proportional to its mole fraction, but the constant of proportionality is now K'_H.

general expect it to be a good description of the vapor pressure of the solute B because a solute in dilute solution is very far from being pure. In a dilute solution, each solute molecule is surrounded by nearly pure solvent, so its environment is quite unlike that in the pure solute, and except when solute and solvent are very similar (such as benzene and methylbenzene), it is very unlikely that the vapor pressure of the solute will be related in a simple manner to the vapor pressure of the pure solute. However, it is found experimentally that in dilute solutions, the vapor pressure of the solute is in fact proportional to its mole fraction, just as for the solvent. Unlike the solvent, however, the constant of proportionality is not in general the vapor pressure of the pure solute. This linear but different dependence was discovered by the English chemist William Henry (1774–1836) and is summarized as **Henry's law**:

The vapor pressure of a volatile solute B is proportional to its mole fraction in a solution: $p_B = K'_H x_B$ (Henry's law)   (3.13)

Here K'_H, which is called **Henry's law constant**, is characteristic of the solute and chosen so that the straight line predicted by eqn 3.13 is tangent to the experimental curve at $x_B = 0$ (Fig. 3.28). Henry's law is usually obeyed only at low concentrations of the solute (close to $x_B = 0$). Solutions that are dilute enough for the solute to obey Henry's law are called **ideal–dilute solutions**.

The Henry's law constants of some gases are listed in Table 3.2. The values given there are for the law rewritten to show how the molar concentration depends on the partial pressure, rather than vice versa:

$[J] = K_H p_J$ (Another version of Henry's law)   (3.14)

Table 3.2 Henry's law constants for gases dissolved in water at 25°C

	$K_H/$ (mol m^{-3} kPa^{-1})
Carbon dioxide, CO$_2$	3.39×10^{-1}
Hydrogen, H$_2$	7.78×10^{-3}
Methane, CH$_4$	1.48×10^{-2}
Nitrogen, N$_2$	6.48×10^{-3}
Oxygen, O$_2$	1.30×10^{-2}

The Henry's law constant, K_H, is commonly reported in moles per cubic metre per kilopascal (mol m^{-3} kPa^{-1}). This form of the law and these units make it very easy to calculate the molar concentration of the dissolved gas, simply by multiplying the partial pressure of the gas (in kilopascals) by the appropriate constant. Equation 3.14 is used, for instance, to estimate the concentration of O_2 in natural waters or the concentration of carbon dioxide in blood plasma.

Example 3.2   *Determining whether a natural water can support aquatic life*

The concentration of O_2 in water required to support aerobic aquatic life is about 4.0 mg dm^{-3}. What is the minimum partial pressure of oxygen in the atmosphere that can achieve this concentration?

Strategy The strategy of the calculation is to determine the partial pressure of oxygen that, according to Henry's law (written as eqn 3.14), corresponds to the concentration specified.

Solution Equation 3.14 becomes

$$p_{O_2} = \frac{[O_2]}{K_H}$$

We note that the molar concentration of O_2 is

$$[O_2] = \frac{4.0 \times 10^{-3}\,\text{g dm}^{-3}}{32\,\text{g mol}^{-1}} = \frac{4.0 \times 10^{-3}\,\text{mol}}{32\,\text{dm}^3} = \frac{4.0 \times 10^{-3}\,\text{mol}}{32 \times 10^{-3}\,\text{m}^3} = \frac{4.0}{32}\,\text{mol m}^{-3}$$

From Table 3.2, K_H for oxygen in water is 1.30×10^{-2} mol m^{-3} kPa^{-1}, therefore the partial pressure needed to achieve the stated concentration is

$$p_{O_2} = \frac{(4.0/32)\,\text{mol m}^{-3}}{1.30 \times 10^{-2}\,\text{mol m}^{-3}\,\text{kPa}^{-1}} = 9.6\,\text{kPa}$$

The partial pressure of oxygen in air at sea level is 21 kPa (158 Torr), which is greater than 9.6 kPa (72 Torr), so the required concentration can be maintained under normal conditions.

Self-test 3.6   What partial pressure of methane is needed to dissolve 21 mg of methane in 100 g of benzene at 25°C ($K_H' = 5.69 \times 10^4$ kPa, for Henry's law in the form given in eqn 3.13)?

Answer: 57 kPa (4.3×10^2 Torr)

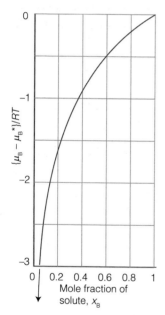

Fig. 3.29 The variation of the chemical potential of the solute with the composition of the solution expressed in terms of the mole fraction of solute. Note that the chemical potential of the solute is lower in the mixture than for the pure solute (for an ideal system). This behavior is likely to be shown by a dilute solution in which the solvent is almost pure and the solute obeys Henry's law.

Henry's law lets us write an expression for the chemical potential of a solute in a solution. We show in the following *Justification* that the chemical potential of the solute when it is present at a mole fraction x_B is

$$\mu_B = \mu_B^* + RT \ln x_B \qquad \boxed{\text{Chemical potential of the solute in terms of the mole fraction}} \quad (3.15)$$

This expression, which is illustrated in Fig. 3.29, applies when Henry's law is valid, in very dilute solutions. The chemical potential of the solute has its pure value when it is present alone ($x_B = 1$, $\ln 1 = 0$) and a smaller value when dissolved (when $x_B < 1$, $\ln x_B < 0$).

Justification 3.7   *The chemical potential of the solute*

We apply the same reasoning as in *Justification* 3.6. When a solute B in a solution is in equilibrium with its vapor at a partial pressure p_B, we can write $\mu_B(l) = \mu_B(g)$ and (from eqn 3.10)

$$\mu_B(l) = \mu_B^\ominus(g) + RT \ln p_B$$

According to Henry's law, $p_B = K_H' x_B$, so it follows that

$$\mu_B(l) = \mu_B^\ominus(g) + RT \ln K_H' x_B = \mu_B^\ominus(g) + RT \ln K_H' + RT \ln x_B$$

The terms $\mu_B^\ominus(g)$ and $RT \ln K_H$ are independent of the composition of the mixture and can be combined into the constant $\mu_B^\star$, the chemical potential of pure liquid B. Equation 3.15 then follows.

We often express the composition of a solution in terms of the molar concentration of the solute, [B], rather than as a mole fraction. The mole fraction and the molar concentration are proportional to each other in dilute solutions, so we write $x_B = \text{constant} \times [B]/c^\ominus$, where $c^\ominus = 1$ mol dm^{-3} is introduced to ensure that the constant is dimensionless. We shall call $c^\ominus$ the **standard molar concentration**. Then eqn 3.15 becomes

$$\mu_B = \mu_B^\star + RT \ln(\text{constant}) + RT \ln([B]/c^\ominus)$$

We can combine the first two terms into a single constant, which we denote $\mu_B^\ominus$, and write this relation as

$$\mu_B = \mu_B^\ominus + RT \ln([B]/c^\ominus) \qquad \boxed{\text{Chemical potential of the solute in terms of the molar concentration}} \quad (3.16a)$$

This equation is the best way to write the relation, but it is cumbersome, and for the rest of the chapter we shall write $[B]/c^\ominus$ simply as [B] and—to conform to the requirement stated in the *note on good practice*—interpret [B] as the molar concentration with the units deleted (we treated pressure similarly earlier in the chapter). Thus, if in fact [B] = 0.1 mol dm^{-3}, so $[B]/c^\ominus = 0.1$, from now on we shall write [B] = 0.1 and use eqn 3.16a in the form

$$\mu_B = \mu_B^\ominus + RT \ln[B] \qquad \boxed{\text{Simplified form of eqn 3.16a}} \quad (3.16b)$$

Figure 3.30 illustrates the variation of chemical potential with concentration predicted by this equation. The chemical potential of the solute has its standard value when the molar concentration of the solute is $c^\ominus = 1$ mol dm^{-3}.

At this stage a summary of the results so far might be helpful:

Species	Chemical potential	Comment
Gas, J	$\mu_J = \mu_J^\ominus + RT \ln p_J$	Perfect gas
Solvent, A	$\mu_A = \mu_A^\star + RT \ln x_A$	Dilute solution
Solute, B	$\mu_B = \mu_B^\star + RT \ln x_B$	Dilute solution
	$\mu_B = \mu_B^\ominus + RT \ln[B]$	

Case study 3.2   *Gas solubility and breathing*

We inhale about 500 cm^3 of air with each breath we take. The influx of air is a result of changes in volume of the lungs as the diaphragm is depressed and the chest expands, which results in a decrease in pressure of about 100 Pa

A note on good practice
It is meaningless to take logarithms of quantities with units, so always ensure that the x of $\ln x$ is a pure number.

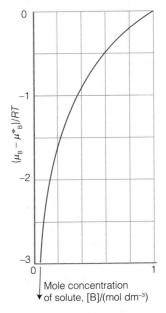

Fig. 3.30 The variation of the chemical potential of the solute with the composition of the solution that obeys Henry's law expressed in terms of the molar concentration of solute. The chemical potential has its standard value at [B] = 1 mol dm^{-3}.

relative to atmospheric pressure. Expiration occurs as the diaphragm rises and the chest contracts, and gives rise to a differential pressure of about 100 Pa above atmospheric pressure. The total volume of air in the lungs is about 6 dm^3, and the additional volume of air that can be exhaled forcefully after normal expiration is about 1.5 dm^3. Some air remains in the lungs at all times to prevent the collapse of the alveoli.

The effect of gas exchange between blood and air inside the alveoli of the lungs means that the composition of the air in the lungs is different from that in the atmosphere, and changes throughout the breathing cycle. Alveolar gas is in fact a mixture of newly inhaled air and air about to be exhaled. The concentration of oxygen present in arterial blood is equivalent to a partial pressure of about 40 Torr (5.3 kPa), whereas the partial pressure of freshly inhaled air in the alveoli of the lungs is about 100 Torr (13.3 kPa). Arterial blood remains in the capillary passing through the wall of an alveolus for about 0.75 s, but such is the steepness of the pressure gradient that it becomes fully saturated with oxygen in about 0.25 s. If the lungs collect fluids (as in pneumonia), then the respiratory membrane thickens, diffusion is greatly slowed, and body tissues begin to suffer from oxygen starvation. Carbon dioxide moves in the opposite direction across the respiratory tissue, but the partial pressure gradient is much less, corresponding to about 5 Torr (0.7 kPa) in blood and 40 Torr (5.3 kPa) in air at equilibrium in the alveoli of the lungs. However, because carbon dioxide is much more soluble in the alveolar fluid than oxygen is, equal amounts of oxygen and carbon dioxide are exchanged in each breath.

A hyperbaric oxygen chamber, in which oxygen is at an elevated partial pressure, is used to treat certain types of disease. Carbon monoxide poisoning can be treated in this way, as can the consequences of shock. Diseases that are caused by anaerobic bacteria, such as gas gangrene and tetanus, can also be treated because the bacteria cannot thrive in high oxygen concentrations.

(d) Real solutions: activities

No actual solutions are ideal, and many solutions deviate from ideal–dilute behavior as soon as the concentration of solute rises above a small value. In thermodynamics we try to preserve the form of equations developed for ideal systems so that it becomes easy to step between the two types of system.[2] This is the thought behind the introduction of the **activity**, a_J, of a substance, which is a kind of effective concentration. The activity is defined so that the expression

$$\mu_J = \mu_J^{\ominus} + RT \ln a_J$$

| The chemical potential in terms of the activity | (3.17) |

is true at *all* concentrations and for both the solvent and the solute.

For ideal solutions, $a_J = x_J$, and the activity of each component is equal to its mole fraction. For ideal–dilute solutions using the definition in eqn 3.17, $a_B = [B]/c^{\ominus}$, and the activity of the solute is equal to the numerical value of its molar concentration. For *non*-ideal solutions we write

For the solvent: $a_A = \gamma_A x_A$

For the solute: $a_B = \gamma_B [B]/c^{\ominus}$

| The activity in terms of the activity coefficient | (3.18) |

---

[2] An added advantage is that there are fewer equations to remember!

Table 3.3 Activities and standard states*

Substance	Standard state	Activity, a
Solid	Pure solid, 1 bar	1
Liquid	Pure liquid, 1 bar	1
Gas	Pure gas, 1 bar	$p/p^{\ominus}$
Solute	Molar concentration of 1 mol dm^{-3}	$[J]/c^{\ominus}$

$p^{\ominus} = 1$ bar ($= 10^5$ Pa), $c^{\ominus} = 1$ mol dm^{-3}.
*Activities are for perfect gases and ideal–dilute solutions; all activities are dimensionless.

where the γ (gamma) in each case is the **activity coefficient**. Activity coefficients depend on the composition of the solution, and we should note the following:

Because the solvent behaves more in accord with Raoult's law as it becomes pure, $\gamma_A \rightarrow 1$ as $x_A \rightarrow 1$.

Because the solute behaves more in accord with Henry's law as the solution becomes very dilute, $\gamma_B \rightarrow 1$ as $[B] \rightarrow 0$.

Conventions concerning standard states and activities of ideal systems are summarized in Table 3.3.

Activities and activity coefficients are often branded as 'fudge factors'. To some extent that is true. However, their introduction does allow us to derive thermodynamically exact expressions for the properties of nonideal solutions. Moreover, in a number of cases it is possible to calculate or measure the activity coefficient of a species in solution. In this text we shall normally derive thermodynamic relations in terms of activities, but when we want to make contact with actual measurements, we shall set the activities equal to the 'ideal' values in Table 3.3.

Case study 3.3   *The Donnan equilibrium*

The term **Donnan equilibrium** refers to the distribution of ions between two solutions in contact through a semipermeable membrane, in one of which there is a polyelectrolyte, such as $Na_\nu P$ (with $P^{\nu-}$ a polyanion), and where the membrane is not permeable to the large charged macromolecule. This arrangement is one that actually occurs in living systems, where we have seen that osmosis is an important feature of cell operation. The thermodynamic consequences of the distribution and transfer of charged species across cell membranes is explored further in Chapter 5.

Consider a situation in which a high concentration of a salt such as NaCl is added to the solution on both sides of the membrane so that the number of cations that $P^{\nu-}$ provides is insignificant in comparison with the number supplied by the additional salt. Apart from small imbalances of charge close to the membrane (which have important consequences, as we shall see in Chapter 5), electrical neutrality must be preserved in the bulk on both sides of the membrane: if an anion migrates, a cation must accompany it. For simplicity, we take the volumes of the solutions on each side of the membrane to be equal.

On one side of the membrane—call it the 'left-hand' side—there are $P^{\nu-}$, Na^+, and Cl^- ions. In the 'right-hand' side there are Na^+ and Cl^- ions. The condition for equilibrium is that the chemical potentials of the Na^+ and Cl^- ions in solution are the same in both sides, so a net flow of Na^+ and Cl^- ions occurs until the chemical potentials are equalized. This equality occurs when

$$\mu^{\ominus}(Na^+) + \mu^{\ominus}(Cl^-) + RT \ln a_L(Na^+) + RT \ln a_L(Cl^-)$$
$$= \mu^{\ominus}(Na^+) + \mu^{\ominus}(Cl^-) + RT \ln a_R(Na^+) + RT \ln a_R(Cl^-)$$

where the subscripts L and R refer to the left-hand and right-hand sides, respectively, separated by the membrane. It follows that

$$RT \ln a_L(Na^+) a_L(Cl^-) = RT \ln a_R(Na^+) a_R(Cl^-)$$

If we ignore activity coefficients and interpret $[Na^+]/c^{\ominus}$ and $[Cl^-]/c^{\ominus}$ as $[Na^+]$ and $[Cl^-]$, respectively, the two expressions are equal when $[Na^+]_L[Cl^-]_L = [Na^+]_R[Cl^-]_R$. As the Na^+ ions are supplied by the polyelectrolyte as well as the added salt, the conditions for bulk electrical neutrality lead to the charge-balance equations $[Na^+]_L = [Cl^-]_L + \nu[P^{\nu-}]$ and $[Na^+]_R = [Cl^-]_R$. We can now combine these three conditions to obtain expressions for the differences in ion concentrations across the membrane. For example, we write

$$[Na^+]_L = \frac{[Na^+]_R[Cl^-]_R}{[Cl^-]_L} = \frac{[Na^+]_R^2}{[Na^+]_L - \nu[P^{\nu-}]}$$

which rearranges to

$$[Na^+]_L^2 - [Na^+]_R^2 = \nu[P^{\nu-}][Na^+]_L$$

After applying the relation $a^2 - b^2 = (a + b)(a - b)$ and rearranging, we obtain

$$[Na^+]_L - [Na^+]_R = \frac{\nu[P^{\nu-}][Na^+]_L}{[Na^+]_L + [Na^+]_R}$$

It follows from the definition $[Cl^-] = \frac{1}{2}([Cl^-]_L + [Cl^-]_R)$ and the charge–balance equations that

$$[Na^+]_L + [Na^+]_R = [Cl^-]_L + [Cl^-]_R + \nu[P^{\nu-}] = 2[Cl^-] + \nu[P^{\nu-}]$$

Substitution of this result into the equation for $[Na^+]_L - [Na^+]_R$ leads to

$$[Na^+]_L - [Na^+]_R = \frac{\nu[P^{\nu-}][Na^+]_L}{2[Cl^-] + \nu[P^{\nu-}]} \tag{3.19a}$$

Similar manipulations lead to an equation for the difference in chloride ion concentration:

$$[Cl^-]_L - [Cl^-]_R = -\frac{\nu[P^{\nu-}][Cl^-]_L}{[Cl^-]_L + [Cl^-]_R}$$

which becomes

$$[Cl^-]_L - [Cl^-]_R = -\frac{\nu[P^{\nu-}][Cl^-]_L}{2[Cl^-]} \tag{3.19b}$$

Note that cations will dominate the anions in the compartment that contains the polyanion because the concentration difference is positive for Na^+ and negative for Cl^-. It also follows that from a measurement of the ion concentrations, it is possible to determine the net charge of the polyanion, which may be unknown.

Example 3.3   *Analyzing a Donnan equilibrium*

Suppose that two equal volumes of 0.200 M NaCl(aq) solution are separated by a membrane and that the left-hand side of the experimental arrangement contains a polyelectrolyte Na_6P at a concentration of 50 g dm^{-3}. Assuming that the membrane is not permeable to the polyanion, which has a molar mass of 55 kg mol^{-1}, calculate the molar concentrations of Na$^+$ and Cl$^-$ in each compartment.

Strategy We saw above that the sum of the equilibrium concentrations of Na$^+$ in both compartments is

$$[Na^+]_L + [Na^+]_R = 2[Cl^-] + \nu[P^{\nu-}]$$

with $[Cl^-] = 0.200$ mol dm^{-3}, and $[P^{\nu-}]$ being calculated from the mass concentration and the molar mass of the polyanion. At this point, we have one equation and two unknowns, $[Na^+]_L$ and $[Na^+]_R$, so we use a second equation, eqn 3.19a, to solve for both Na$^+$ ion concentrations. To calculate the Cl$^-$ ion concentrations, we use $[Cl^-]_R = [Na^+]_R$ and $[Cl^-]_L = [Na^+]_L - \nu[P^{\nu-}]$, with $\nu = 6$.

Solution The molar concentration of the polyanion is $[P^{\nu-}] = 9.1 \times 10^{-4}$ mol dm^{-3}. It follows from eqn 3.19a that

$$[Na^+]_L - [Na^+]_R = \frac{6 \times (9.1 \times 10^{-4} \text{ mol dm}^{-3}) \times [Na^+]_L}{2 \times (0.200 \text{ mol dm}^{-3}) + 6 \times (9.1 \times 10^{-4} \text{ mol dm}^{-3})}$$

The sum of Na$^+$ concentrations is

$$[Na^+]_L + [Na^+]_R = 2 \times (0.200 \text{ mol dm}^{-3}) + 6 \times (9.1 \times 10^{-4} \text{ mol dm}^{-3})$$
$$= 0.405 \text{ mol dm}^{-3}$$

The solutions of these two equations are

$$[Na^+]_L = 0.204 \text{ mol dm}^{-3} \qquad [Na^+]_R = 0.201 \text{ mol dm}^{-3}$$

Then

$$[Cl^-]_R = [Na^+]_R = 0.201 \text{ mol dm}^{-3}$$
$$[Cl^-]_L = [Na^+]_L - 6[P^{\nu-}] = 0.199 \text{ mol dm}^{-3}$$

Self-test 3.7   Repeat the calculation for 0.300 M NaCl(aq), a polyelectrolyte $Na_{10}P$ of molar mass 33 kg mol^{-1} at a mass concentration of 50.0 g dm^{-3}.

Answer: $[Na^+]_L = 0.31$ mol dm^{-3}, $[Na^+]_R = 0.30$ mol dm^{-3}

(e) The thermodynamics of dissolving

We now have enough information to formulate a thermodynamic description of dissolving to form an ideal solution. As we see in the following *Justification*, when an amount n_B of a solute B dissolves in an amount n_A of a solvent A at a temperature T,

$$\Delta G = nRT\{x_A \ln x_A + x_B \ln x_B\} \qquad \boxed{\text{Gibbs energy of dissolving}} \quad (3.20)$$

with $n = n_A + n_B$ and the x_J the mole fractions in the mixture.

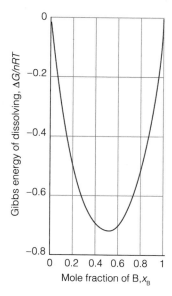

Fig. 3.31 The variation of the Gibbs energy of dissolving with composition for two components at constant temperature and pressure. Note that $\Delta G < 0$ for all compositions, which indicates that two components mix spontaneously in all proportions.

Justification 3.8 *The Gibbs energy of dissolving*

The Gibbs energy of the two unmixed components is the sum of their individual Gibbs energies:

$$G_i = n_A \mu_A^\star + n_B \mu_B^\star$$

When B is dissolved in A to form an ideal solution, the Gibbs energy becomes

$$G_f = n_A \mu_A + n_B \mu_B = n_A \{\mu_A^\star + RT \ln x_A\} + n_B \{\mu_B^\star + RT \ln x_B\}$$
$$= n_A \mu_A^\star + n_A RT \ln x_A + n_B \mu_B^\star + n_B RT \ln x_B$$

where the x_J are the mole fractions of the two components in the solution. The difference $G_f - G_i$ is the change in Gibbs energy that accompanies dissolving. The pure chemical potentials cancel, so

$$\Delta G = RT\{n_A \ln x_A + n_B \ln x_B\}$$

Because $x_J = n_J/n$, we can substitute $n_A = x_A n$ and $n_B = x_B n$ into the expression above and obtain

$$\Delta G = nRT\{x_A \ln x_A + x_B \ln x_B\}$$

which is eqn 3.20.

Equation 3.20 tells us the change in Gibbs energy when a solute dissolves to give an ideal solution (Fig. 3.31). The crucial feature is that because x_A and x_B are both less than 1, the two logarithms are negative ($\ln x < 0$ if $x < 1$), so $\Delta G < 0$ at all compositions. Therefore, *dissolving to form an ideal solution is spontaneous in all proportions*. Furthermore, if we compare eqn 3.20 with $\Delta G = \Delta H - T\Delta S$, we can conclude that:

$$\Delta H = 0 \qquad \boxed{\text{Enthalpy of dissolving}} \quad (3.21a)$$

$$\Delta S = -nR\{x_A \ln x_A + x_B \ln x_B\} \qquad \boxed{\text{Entropy of dissolving}} \quad (3.21b)$$

The value of ΔH indicates that although there are interactions between the molecules, the solute–solute, solvent–solvent, and solute–solvent interactions are all the same, so the solute slips into solution without a change in enthalpy. There is an increase in entropy because the molecules are more dispersed in the solution than in the unmixed components. The entropy of the surroundings is unchanged because the enthalpy of the system is constant, so no energy escapes as heat into the surroundings. It follows that the increase in entropy of the system is the 'driving force' of the dissolving.

Colligative properties

An ideal solute has no effect on the enthalpy of a solution in the sense that the enthalpy of mixing is zero. However, it does affect the entropy, and we found in eqn 3.21 that $\Delta S > 0$ when a solute dissolves in a solvent to give an ideal solution. We can therefore expect a solute to modify the physical properties of the solution. Apart from lowering the vapor pressure of the solvent, which we have already considered, a nonvolatile solute has three main effects: it raises the boiling point of a solution, it lowers the freezing point, and it gives rise to an osmotic pressure.

Table 3.4 Cryoscopic and ebullioscopic constants

Solvent	$K_f/(K\ kg\ mol^{-1})$	$K_b/(K\ kg\ mol^{-1})$
Acetic acid	3.90	3.07
Benzene	5.12	2.53
Camphor	40	
Carbon disulfide	3.8	2.37
Naphthalene	6.94	5.8
Phenol	7.27	3.04
Tetrachloromethane	30	4.95
Water	1.86	0.51

(The meaning of the last will be explained shortly.) These properties, which are called **colligative properties**, stem from a change in the dispersal of solvent molecules that depends on the number of solute particles present but is independent of the identity of the species we use to bring it about.[3] Thus, a 0.01 mol kg^{-1} aqueous solution of any nonelectrolyte should have the same boiling point, freezing point, and osmotic pressure.

3.9 The modification of boiling and freezing points

To understand the origins of the colligative properties and their effect on biological processes, it is useful to explore the modification of the boiling and freezing points of a solvent in a solution.

It is found empirically, and can be justified thermodynamically, that the **elevation of boiling point**, T_b, and the **depression of freezing point**, T_f, are both proportional to the molality, b_B, of the solute:

$$\Delta T_b = K_b b_B \qquad \boxed{\text{Elevation of the boiling point}}$$

$$\Delta T_f = K_f b_B \qquad \boxed{\text{Depression of the freezing point}} \qquad (3.22)$$

where K_b is the **ebullioscopic constant** and K_f is the **cryoscopic constant** of the solvent.[4] The two constants can be estimated from other properties of the solvent, but both are best treated as empirical constants (Table 3.4).

Self-test 3.8   Estimate the lowering of the freezing point of the solution made by dissolving 3.0 g (about one cube) of sucrose in 100 g of water.

Answer: 0.16 K

To understand the origin of these effects, we shall make two simplifying assumptions:

1)   The solute is not volatile and therefore does not appear in the vapor phase.

2)   The solute is insoluble in the solid solvent and therefore does not appear in the solid phase.

---

[3] Hence, the name *colligative*, meaning 'depending on the collection'.

[4] They are also called the 'boiling-point constant' and the 'freezing-point constant'.

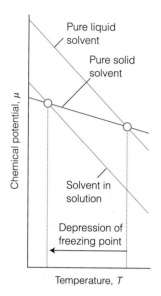

Fig. 3.32 The chemical potentials of pure solid solvent and pure liquid solvent also decrease with temperature, and the point of intersection, where the chemical potential of the liquid rises above that of the solid, marks the freezing point of the pure solvent. A solute lowers the chemical potential of the solvent but leaves that of the solid unchanged. As a result, the intersection point lies farther to the left and the freezing point is therefore lowered.

For example, a solution of sucrose in water consists of a solute (sucrose, $C_{12}H_{22}O_{11}$) that is not volatile and therefore never appears in the vapor, which is therefore pure water vapor. The sucrose is also left behind in the liquid solvent when ice begins to form, so the ice remains pure.

The origin of colligative properties is the lowering of chemical potential of the solvent by the presence of a solute, as expressed by eqn 3.12. We saw in Section 3.3 that the freezing and boiling points correspond to the temperatures at which the graph of the molar Gibbs energy of the liquid intersects the graphs of the molar Gibbs energy of the solid and vapor phases, respectively. Because we are now dealing with mixtures, we have to think about the *partial* molar Gibbs energy (the chemical potential) of the solvent. The presence of a solute lowers the chemical potential of the liquid, but because the vapor and solid remain pure, their chemical potentials remain unchanged. As a result, we see from Fig. 3.32 that the freezing point moves to lower values; likewise, from Fig. 3.33 we see that the boiling point moves to higher values. In other words, the freezing point is depressed, the boiling point is elevated, and the liquid phase exists over a wider range of temperatures.

The elevation of boiling point is too small to have any practical significance. A practical consequence of the lowering of freezing point, and hence the lowering of the melting point of the pure solid, is its employment in organic chemistry to judge the purity of a sample, for any impurity lowers the melting point of a substance from its accepted value. The salt water of the oceans freezes at temperatures lower than that of fresh water, and salt is spread on highways to delay the onset of freezing. The addition of 'antifreeze' to car engines and, by natural processes, to arctic fish, is commonly held up as an example of the lowering of freezing point, but the concentrations are far too high for the arguments we have used here to be applicable. The 1,2-ethanediol ('glycol') used as antifreeze probably just interferes with bonding between water molecules. Likewise, the antifreeze proteins of arctic fish act by binding to small ice crystals and preventing larger crystals from forming.

Mathematical toolkit 3.2   *Power series and expansions*

A **power series** has the form

$$c_0 + c_1(x - a) + c_2(x - a)^2 + \cdots + c_n(x - a)^n + \cdots$$

$$= \sum_{n=0}^{\infty} c_n(x - a)^n$$

where c_n and a are constants. It is often useful to express a function $f(x)$ in the vicinity of $x = a$ as a special power series called the **Taylor series**, or **Taylor expansion**, which has the form

$$f(x) = f(a) + \left(\frac{df}{dx}\right)_a (x - a) + \frac{1}{2!}\left(\frac{d^2f}{dx^2}\right)_a (x - a)^2 +$$

$$\cdots + \frac{1}{n!}\left(\frac{d^n f}{dx^n}\right)_a (x - a)^n + \cdots$$

$$= \sum_{n=0}^{\infty} \frac{1}{n!}\left(\frac{d^n f}{dx^n}\right)_a (x - a)^n$$

where $n!$ denotes a **factorial** given by $n! = n(n - 1)(n - 2)\ldots 1$.

The following Taylor expansions are often useful:

$$(1 + x)^{-1} = 1 - x + x^2 - \cdots$$

$$e^x = 1 + x + \tfrac{1}{2}x^2 + \cdots$$

$$\ln x = (x - 1) - \tfrac{1}{2}(x - 1)^2 + \tfrac{1}{3}(x - 1)^3 - \tfrac{1}{4}(x - 1)^4 + \cdots$$

$$\ln(1 + x) = x - \tfrac{1}{2}x^2 + \tfrac{1}{3}x^3 - \cdots$$

If $x \ll 1$, then

$$(1 + x)^{-1} \approx 1 - x$$

$$e^x \approx 1 + x$$

$$\ln(1 + x) \approx x.$$

3.10 Osmosis

To understand why cells neither collapse nor burst easily, we need to explore the thermodynamics of the transfer of water through cell membranes.

The phenomenon of **osmosis** is the passage of a pure solvent into a solution separated from it by a **semipermeable membrane**,[5] a membrane that is permeable to the solvent but not to the solute (Fig. 3.34). The membrane might have microscopic holes that are large enough to allow water molecules to pass through, but not ions or carbohydrate molecules with their bulky coating of hydrating water molecules. The **osmotic pressure**, Π (uppercase pi), is the pressure that must be applied to the solution to stop the inward flow of solvent.

In the simple arrangement shown in Fig. 3.34, the pressure opposing the passage of solvent into the solution arises from the hydrostatic pressure of the column of solution that the osmosis itself produces. This column is formed when the pure solvent flows through the membrane into the solution and pushes the column of solution higher up the tube. Equilibrium is reached when the downward pressure exerted by the column of solution is equal to the upward osmotic pressure. A complication of this arrangement is that the entry of solvent into the solution results in dilution of the latter, so it is more difficult to treat mathematically than an arrangement in which an externally applied pressure opposes any flow of solvent into the solution.

The osmotic pressure of a solution is proportional to the concentration of solute. In fact, we show in the following *Justification* that the expression for the osmotic pressure of an ideal solution, which is called the **van 't Hoff equation**, bears an uncanny resemblance to the expression for the pressure of a perfect gas:

$$\Pi V \approx n_B RT \qquad \text{van 't Hoff equation} \qquad (3.23a)$$

Because $n_B/V = [B]$, the molar concentration of the solute, a simpler form of this equation is

$$\Pi \approx [B]RT \qquad \begin{array}{c}\text{Another version of the}\\\text{van 't Hoff equation}\end{array} \qquad (3.23b)$$

This equation applies only to solutions that are sufficiently dilute to behave as ideal–dilute solutions.

Justification 3.9 *The van 't Hoff equation*

The thermodynamic treatment of osmosis makes use of the fact that, at equilibrium, the chemical potential of the solvent A is the same on each side of the membrane (Fig. 3.35). The starting relation is therefore

$$\mu_A(\text{pure solvent at pressure } p) = \mu_A(\text{solvent in the solution at pressure } p + \Pi)$$

The pure solvent is at atmospheric pressure, p, and the solution is at a pressure $p + \Pi$ on account of the additional pressure, Π, that has to be exerted on the solution to establish equilibrium. We shall write the chemical potential of the pure solvent at the pressure p as $\mu_A^\star(p)$. The chemical potential of the solvent in the solution is lowered by the solute, but it is raised on account of the greater pressure, $p + \Pi$, acting on the solution. We denote this chemical potential by $\mu_A(x_A, p + \Pi)$. Our task is to find the extra pressure Π needed to balance the lowering of chemical potential caused by the solute.

[5] The name *osmosis* is derived from the Greek word for 'push'.

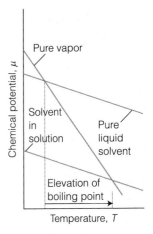

Fig. 3.33 The chemical potentials of pure solvent vapor and pure liquid solvent decrease with temperature, and the point of intersection, where the chemical potential of the vapor falls below that of the liquid, marks the boiling point of the pure solvent. A solute lowers the chemical potential of the solvent but leaves that of the vapor unchanged. As a result, the intersection point lies farther to the right and the boiling point is therefore raised.

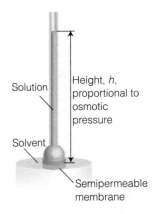

Fig. 3.34 In a simple osmosis experiment, a solution is separated from the pure solvent by a semipermeable membrane. Pure solvent passes through the membrane and the solution rises in the inner tube. The net flow ceases when the pressure exerted by the column of liquid is equal to the osmotic pressure of the solution.

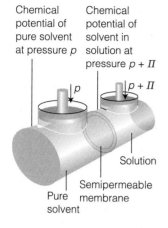

Chemical potential of pure solvent at pressure p

Chemical potential of solvent in solution at pressure $p + \Pi$

p $p + \Pi$

Solution

Semipermeable membrane

Pure solvent

Fig. 3.35 The basis of the calculation of osmotic pressure. The presence of a solute lowers the chemical potential of the solvent in the right-hand compartment, but the application of pressure raises it. The osmotic pressure is the pressure needed to equalize the chemical potential of the solvent in the two compartments.

The condition for equilibrium written above is

$$\mu_A^\star(p) = \mu_A(x_A, p + \Pi)$$

We take the effect of the solute into account using eqn 3.12:

$$\mu_A(x_A, p + \Pi) = \mu_A^\star(p + \Pi) + RT \ln x_A$$

The effect of pressure on an (assumed incompressible) liquid is given by eqn 3.1 ($G_m(p_f) = G_m(p_i) + (p_f - p_i)V_m$) but now expressed in terms of the chemical potential and the partial molar volume of the solvent:

$$\mu_A^\star(p + \Pi) = \mu_A^\star(p) + (p + \Pi - p)V_A = \mu_A^\star(p) + \Pi V_A$$

When the last three equations are combined, we get

$$\mu_A^\star(p) = \mu_A^\star(p) + \Pi V_A + RT \ln x_A$$

and therefore

$$-RT \ln x_A = \Pi V_A$$

The mole fraction of the solvent is equal to $1 - x_B$, where x_B is the mole fraction of solute molecules. In dilute solution, $\ln(1 - x_B)$ is approximately equal to $-x_B$ (see *Mathematical toolkit 3.2*), so this equation becomes

$$RT x_B \approx \Pi V_A$$

When the solution is dilute, $x_B = n_B/n \approx n_B/n_A$. Moreover, because $n_A V_A \approx V$, the total volume of the solution, this equation becomes eqn 3.23.

Osmosis helps biological cells maintain their structure. Cell membranes are semipermeable and allow water, small molecules, and hydrated ions to pass, while blocking the passage of biopolymers synthesized inside the cell. The difference in concentrations of solutes inside and outside the cell gives rise to an osmotic pressure, and water passes into the more concentrated solution in the interior of the cell, carrying small nutrient molecules. The influx of water also keeps the cell swollen, whereas dehydration causes the cell to shrink. These effects are important in everyday medical practice. To maintain the integrity of blood cells, solutions that are injected into the bloodstream for blood transfusions and intravenous feeding must be *isotonic* with the blood, meaning that they must have the same osmotic pressure as blood. If the injected solution is too dilute, or *hypotonic*, the flow of solvent into the cells, required to equalize the osmotic pressure, causes the cells to burst and die by a process called *hemolysis*. If the solution is too concentrated, or *hypertonic*, equalization of the osmotic pressure requires flow of solvent out of the cells, which shrink and die.

Osmosis also forms the basis of **dialysis**, a common technique for the removal of impurities from solutions of biological macromolecules. In a dialysis experiment, a solution of macromolecules containing impurities, such as ions or small molecules (including small proteins or nucleic acids), is placed in a bag made of a material that acts as a semipermeable membrane and the filled bag is immersed in a solvent. The membrane permits the passage of the small ions and molecules but not the larger macromolecules, so the former migrate through the membrane, leaving the macromolecules behind. In practice, purification of the sample requires several changes of solvent to coax most of the impurities out of the dialysis bag.

In the laboratory 3.1 Osmometry

Osmometry is the determination of molar mass by the measurement of osmotic pressure. Biological macromolecules dissolve to produce solutions that are far from ideal, but we can still calculate the osmotic pressure by assuming that the van't Hoff equation is only the first term of a lengthier expression:

$$\Pi = [B]RT\{1 + B[B] + \cdots\}$$

Expanded van 't Hoff equation   (3.24a)

The empirical parameter B in this expression is called the **osmotic virial coefficient**. To use eqn 3.24a, we rearrange it into a form that gives a straight line by dividing both sides by $[B]$:

$$\Pi/[B] = RT + BRT[B] + \cdots \tag{3.24b}$$

As we illustrate in the following example, the molar mass of the solute B can be found by measuring the osmotic pressure at a series of mass concentrations and making a plot of $\Pi/[B]$ against $[B]$ (Fig. 3.36).

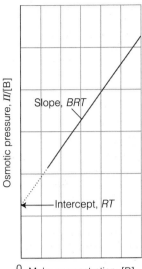

Fig. 3.36 The plot and extrapolation made to analyze the results of an osmometry experiment.

Example 3.4 Determining the molar mass of an enzyme from measurements of the osmotic pressure

The osmotic pressures of solutions of an enzyme in water at 298 K are given below. Find the molar mass of the enzyme.

$c/(\text{g dm}^{-3})$	1.00	2.00	4.00	7.00	9.00
Π/Pa	27.5	69.6	197	500	785

Strategy First, we need to express eqn 3.24b in terms of the mass concentration, c, so that we can use the data. The molar concentration $[B]$ of the solute is related to the mass concentration $c_B = m_B/V$ by

$$c_B = \frac{m_B}{V} = \frac{m_B}{n_B} \times \frac{n_B}{V} = M \times [B]$$

where M is the molar mass of the solute ($M = m_B/n_B$), so $[B] = c_B/M$. With this substitution, eqn 3.24b becomes

$$\frac{\Pi}{c_B/M} = RT + \frac{BRTc_B}{M} + \cdots$$

Division through by M gives

$$\frac{\Pi}{c_B} = \frac{RT}{M} + \left(\frac{BRT}{M^2}\right)c_B + \cdots$$

It follows that, by plotting Π/c_B against c_B, the results should fall on a straight line with intercept RT/M on the vertical axis at $c_B = 0$. Therefore, by locating the intercept by extrapolation of the data to $c_B = 0$, we can find the molar mass of the solute.

Solution The following values of Π/c_B can be calculated from the data:

$c_B/(\text{g dm}^{-3})$	1.00	2.00	4.00	7.00	9.00
$(\Pi/\text{Pa})/(c_B/\text{g dm}^{-3})$	27.5	34.8	49.3	71.5	87.2

The points are plotted in Fig. 3.37. The intercept with the vertical axis at $c_B = 0$ is at

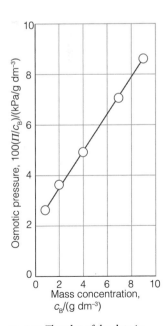

Fig. 3.37 The plot of the data in *Example* 3.4.

A note on good practice
Graphs should be plotted on axes labeled with pure numbers. Note how the plotted quantities are divided by their units, so that $c_B/(\text{g dm}^{-3})$, for instance, is a dimensionless number. By carrying the units through every stage of the calculation, we end up with the correct units for M. It is far better to proceed systematically in this way than to try to guess the units at the end of the calculation.

$$\frac{\Pi/\text{Pa}}{c_B/(\text{g dm}^{-3})} = 19.8$$

which we can rearrange into

$$\Pi/c_B = 19.8 \text{ Pa g}^{-1} \text{ dm}^3$$

Therefore, because this intercept is equal to RT/M, we can write

$$M = \frac{RT}{19.8 \text{ Pa g}^{-1} \text{ dm}^3}$$

It follows that

$$M = \frac{(8.314\,47 \times 10^3 \text{ Pa dm}^3 \text{ K}^{-1} \text{ mol}^{-1}) \times (298 \text{ K})}{19.8 \text{ Pa g}^{-1} \text{ dm}^3} = 1.25 \times 10^5 \text{ g mol}^{-1}$$

The molar mass of the enzyme is therefore close to 125 kg mol^{-1} (corresponding to a molecular mass of 125 kDa).

Self-test 3.9 The osmotic pressures of solutions of a protein at 25°C were as follows:

$c/(\text{g dm}^{-3})$	0.50	1.00	1.50	2.00	2.50
Π/Pa	40.0	110	200	330	490

What is the molar mass of the protein?

Answer: 49 kg mol^{-1}

Checklist of key concepts

1. A phase diagram of a substance shows the conditions of pressure and temperature at which its various phases are most stable.

2. A phase boundary depicts the pressures and temperatures at which two phases are in equilibrium.

3. The boiling temperature is the temperature at which the vapor pressure is equal to the external pressure; the normal boiling point is the temperature at which the vapor pressure is 1 atm. The triple point is the condition of pressure and temperature at which three phases are in mutual equilibrium.

4. A partial molar quantity is the contribution of a component (per mole) to the overall property of a mixture.

5. The chemical potential of a component is the partial molar Gibbs energy of that component in a mixture.

6. An ideal solution is one in which both components obey Raoult's law over the entire composition range.

7. An ideal–dilute solution is one in which the solute obeys Henry's law.

8. The activity of a substance is an effective concentration; see Table 3.3.

9. The Donnan equilibrium determines the distribution of ions between two solutions in contact through a membrane, in one of which there is a polyelectrolyte and where the membrane is not permeable to the large charged macromolecule.

10. A colligative property is a property that depends on the number of solute particles, not their chemical identity; it arises from the effect of a solute on the entropy of the solution.

11. Colligative properties include lowering of vapor pressure, depression of freezing point, elevation of boiling point, and osmotic pressure.

12. The molar masses of biological polymers can be determined by measurements of the osmotic pressure of their solutions.

Checklist of key equations

Property or process	Equation	Comment
Dependence of the Gibbs energy on pressure	$\Delta G_m = V_m \Delta p$	Liquids and solids
	$\Delta G_m = RT \ln(p_f/p_i)$	Perfect gases
Dependence of the Gibbs energy on temperature	$\Delta G_m = -S_m \Delta T$	
Clapeyron equation	$dp/dT = \Delta_{trs}H/T\Delta_{trs}V$	
Clausius–Clapeyron equation	$d \ln p/dT = \Delta_{vap}H/RT^2$	Vapor behaves as a perfect gas
Gibbs energy of a binary mixture	$G = n_A\mu_A + n_B\mu_B$	
Chemical potential	$\mu_J = \mu_J^{\ominus} + RT \ln p_J$	Perfect gas
	$\mu_J = \mu_J^{*} + RT \ln x_J$	Solvent and solute in a dilute solution
Raoult's law	$p_J = x_J p_J^{*}$	For an ideal solution
Henry's law	$p_B = x_B K'_H$, $[B] = K_H p_B$	For an ideal–dilute solution
Elevation of the boiling point	$\Delta T_b = K_b b_B$	
Depression of the freezing point	$\Delta T_f = K_f b_B$	
van 't Hoff equation	$\Pi V \approx n_B RT$	Ideal solution

Further information

Further information 3.1 *The phase rule*

To explore whether *four* phases of a single substance could ever be in equilibrium (such as four of the many phases of ice), we think about the thermodynamic criterion for four phases to be in equilibrium. For equilibrium, the four molar Gibbs energies would all have to be equal, and we could write

$$G_m(1) = G_m(2) \qquad G_m(2) = G_m(3) \qquad G_m(3) = G_m(4)$$

(The other equalities, $G_m(1) = G_m(4)$, and so on, are implied by these three equations.) Each Gibbs energy is a function of the pressure and temperature, so we should think of these three relations as three equations for the two unknowns p and T. In general, three equations for two unknowns have no solution. For instance, the three equations $5x + 3y = 4$, $2x + 6y = 5$, and $x + y = 1$ have no solutions (try it). Therefore, we have to conclude that the four molar Gibbs energies cannot all be equal. In other words, *four phases of a single substance cannot coexist in mutual equilibrium.*

The conclusion we have reached is a special case of one of the most elegant results of chemical thermodynamics. The **phase rule** was derived by Gibbs and states that, for a system at equilibrium,

$$F = C - P + 2$$

Here F is the number of degrees of freedom, C is the number of components, and P is the number of phases. The **number of components**, C, in a system is the minimum number of independent species necessary to define the composition of all the phases present in the system. The definition is easy to apply when the species present in a system do not react, for then we simply count their number. For instance, pure water is a one-component system ($C = 1$), and a mixture of ethanol and water is a two-component system ($C = 2$). The **number of degrees of freedom**, F, of a system is the number of intensive variables (such as the pressure, temperature, or mole fractions) that can be changed independently without disturbing the number of phases in equilibrium.

For a one-component system, such as pure water, we set $C = 1$ and the phase rule simplifies to $F = 3 - P$. When only one phase is present, $F = 2$, which implies that p and T can be varied independently. In other words, a single phase is represented by an *area* on a phase diagram. When two phases are in equilibrium, $F = 1$, which implies that pressure is not freely variable if we have set the temperature. That is, the equilibrium of two phases is represented by a *line* in a phase diagram: a line in a graph shows how one variable must change if another variable is varied (Fig. 3.38). Instead of selecting the temperature, we can select the pressure, but having done so, the two phases come into equilibrium at a single definite temperature. Therefore, freezing (or any other phase transition of a single substance) occurs at a definite temperature at a given pressure. When three phases are in equilibrium, $F = 0$. This special 'invariant condition' can

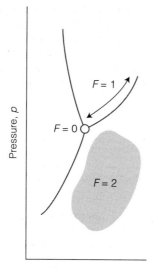

Fig. 3.38 The features of a phase diagram represent different degrees of freedom. When only one phase is present, $F = 2$ and the pressure and temperature can be varied at will. When two phases are present in equilibrium, $F = 1$: now if the temperature is changed, the pressure must be changed by a specific amount. When three phases are present in equilibrium, $F = 0$ and there is no freedom to change either variable.

therefore be established only at a definite temperature and pressure. The equilibrium of three phases is therefore represented by a *point*, the triple point, on the phase diagram. If we set $P = 4$, we get the absurd result that F is negative; that result is in accord with the conclusion at the start of this section that four phases cannot be in equilibrium in a one-component system.

Further information 3.2 *Measures of concentration*

A useful measure of concentration of component J of a mixture is its **mole fraction**, the amount of J molecules expressed as a fraction of the total amount of molecules in the mixture. In a mixture that consists of n_A A molecules, n_B B molecules, and so on (where the n_J are amounts in moles), the mole fraction of J (where J = A, B, . . .) is

$$x_J = \frac{n_J}{n}$$
Definition of the mole fraction (3.25a)

where $n = n_A + n_B + \dots$. For a **binary mixture**, one that consists of two species, this general expression becomes

$$x_A = \frac{n_A}{n_A + n_B} \qquad x_B = \frac{n_B}{n_A + n_B} \qquad x_A + x_B = 1 \qquad (3.25b)$$

When only A is present, $x_A = 1$ and $x_B = 0$. When only B is present, $x_B = 1$ and $x_A = 0$. When both are present in the same amounts, $x_A = \frac{1}{2}$ and $x_B = \frac{1}{2}$ (Fig. 3.39).

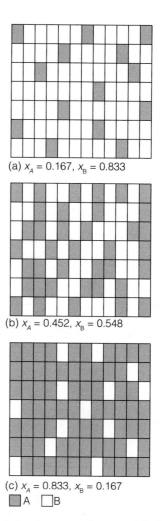

(a) $x_A = 0.167$, $x_B = 0.833$

(b) $x_A = 0.452$, $x_B = 0.548$

(c) $x_A = 0.833$, $x_B = 0.167$

A   B

Fig. 3.39 The mole fraction is an indication of the fraction of molecules in a sample that are of the specified identity. In this figure, the mole fraction x_A of component A (in blue) of a binary mixture increases from (a) 0.167, to (b) 0.452, to (c) 0.833.

Self-test 3.10   Calculate the mole fractions of N_2, O_2, and Ar in dry air at sea level, given that 100.0 g of air consists of 75.5 g of N_2, 23.2 g of O_2, and 1.3 g of Ar. (*Hint:* Begin by converting each mass to an amount in moles.)

Answer: 0.780, 0.210, 0.009

For any type of gas (perfect or not) in a gaseous mixture, the **partial pressure**, p_J, of the gas J is defined as

$$p_J = x_J p$$
Definition of the partial pressure (3.26)

where x_J is the mole fraction of the gas J in the mixture. For perfect gases, the partial pressure of a gas defined in this way is also the pressure that the gas would exert if it were alone in the container at the same temperature.

From *Self-test* 3.10, we have $x_{N_2} = 0.780$, $x_{O_2} = 0.210$, and $x_{Ar} = 0.009$ for dry air at sea level. It then follows from eqn 3.26 that when the total atmospheric pressure is 100 kPa, the partial pressure of nitrogen is $p_{N_2} = x_{N_2}p = 0.780 \times (100 \text{ kPa}) = 78.0$ kPa. Similarly, for the other two components we find $p_{O_2} = 21.0$ kPa and $p_{Ar} = 0.9$ kPa.

The **molar concentration**, [J] or c_J, of a solute J in a solution (more formally, the 'amount of substance concentration') is the chemical amount of J divided by the volume of the solution:[6]

$$[J] = \frac{n_J}{V} \qquad \text{Definition of the molar concentration} \qquad (3.27)$$

Molar concentration is typically reported in moles per cubic decimeter (and commonly as moles per liter, mol L^{-1}). The unit 1 mol dm^{-3} is commonly denoted 1 M (and read 'molar'). Once we know the molar concentration of a solute, we can calculate the amount of that substance in a given volume, V, of solution by writing

$$n_J = [J]V \qquad (3.28)$$

The **molality**, b_J, of a solute J in a solution is the amount of substance divided by the mass of solvent used to prepare the solution:

$$b_J = \frac{n_J}{m_{solvent}} \qquad \text{Definition of molality} \qquad (3.29)$$

Molality is typically reported in moles of solute per kilogram of solvent (mol kg^{-1}). This unit is sometimes (but unofficially) denoted m, with 1 m = 1 mol kg^{-1}. An important distinction between molar concentration and molality is that whereas the former is defined in terms of the volume of the solution, the molality is defined in terms of the mass of solvent used to prepare the solution. A distinction to remember is that molar concentration varies with temperature as the solution expands and contracts, but the molality does not. For dilute solutions in water, the numerical values of the molality and molar concentration differ very little because 1 dm^3 of solution is mostly water and has a mass close to 1 kg; for concentrated aqueous solutions and for all nonaqueous solutions with densities different from 1 g cm^{-3}, the two values are very different.

As we have indicated, we use molality when we need to emphasize the relative amounts of solute and solvent molecules. To see why this is so, we note that the mass of

[6] Molar concentration is still widely called 'molarity'.

solvent is proportional to the amount of solvent molecules present, so from eqn 3.29 we see that the molality is proportional to the ratio of the amounts of solute and solvent molecules. For example, any 1.0 m aqueous nonelectrolyte solution contains 1.0 mol solute particles per 55.5 mol H_2O molecules, so in each case there is 1 solute molecule per 55.5 solvent molecules.

Example 3.5 *Relating mole fraction and molality*

What is the mole fraction of glycine molecules in 0.140 m $NH_2CH_2COOH(aq)$? Disregard the effects of protonation and deprotonation.

Strategy We consider a sample that contains (exactly) 1 kg of solvent and hence an amount $n_J = b_J \times (1 \text{ kg})$ of solute molecules. The amount of solvent molecules in exactly 1 kg of solvent is

$$n_{solvent} = \frac{1 \text{ kg}}{M}$$

where M is the molar mass of the solvent. Once these two amounts are available, we can calculate the mole fraction by using eqn 3.25 with $n = n_J + n_{solvent}$.

Solution It follows from the discussion in the Strategy that the amount of glycine (gly) molecules in exactly 1 kg of solvent is

$$n_{gly} = (0.140 \text{ mol kg}^{-1}) \times (1 \text{ kg}) = 0.140 \text{ mol}$$

The amount of water molecules in exactly 1 kg (10^3 g) of water is

$$n_{water} = \frac{10^3 \text{ g}}{18.02 \text{ g mol}^{-1}} = \frac{10^3}{18.02} \text{ mol}$$

The total amount of molecules present is

$$n = 0.140 \text{ mol} + \frac{10^3}{18.02} \text{ mol}$$

The mole fraction of glycine molecules is therefore

$$x_{gly} = \frac{0.140 \text{ mol}}{\{0.140 + (10^3/18.02)\} \text{ mol}} = 2.52 \times 10^{-3}$$

A note on good practice We refer to *exactly* 1 kg of solvent to avoid problems with significant figures.

Self-test 3.11 Calculate the mole fraction of sucrose molecules in 1.22 m $C_{12}H_{22}O_{11}(aq)$.

Answer: 2.15×10^{-2}

Discussion questions

3.1 Why does the chemical potential vary with **(a)** temperature and **(b)** pressure?

3.2 Discuss the implications for phase stability of the variation of chemical potential with temperature and pressure.

3.3 State and justify the thermodynamic criterion for solution–vapor equilibrium.

3.4 Explain the significance of the Clapeyron equation and the Clausius–Clapeyron equation.

3.5 How would you expect the shape of the curve shown in Fig. 3.16 to change if the degree of cooperativity of denaturation of a protein were to increase or decrease for a constant value of the melting temperature?

3.6 What is meant by the activity of a solute?

3.7 Explain the origin of colligative properties. Why do they not depend on the chemical identity of the solute?

3.8 Explain how osmometry can be used to determine the molar mass of a biological macromolecule.

Exercises

3.9 What is the difference in molar Gibbs energy due to pressure alone of **(a)** water (density 1.03 g cm^{-3}) at the ocean surface and in the Mindañao trench (depth 11.5 km) and **(b)** mercury (density 13.6 g cm^{-3}) at the top and bottom of the column in a barometer? *Hint:* At the very top, the pressure on the mercury is equal to the vapor pressure of mercury, which at 20°C is 160 mPa.

3.10 The density of the fat tristearin is 0.95 g cm^{-3}. Calculate the change in molar Gibbs energy of tristearin when a deep-sea creature is brought to the surface ($p = 1.0$ atm) from a depth of 2.0 km. To calculate the hydrostatic pressure, take the mean density of water to be 1.03 g cm^{-3}.

3.11 Calculate the change in molar Gibbs energy of carbon dioxide (treated as a perfect gas) at 20°C when its pressure is changed isothermally from 1.0 bar to **(a)** 2.0 bar and **(b)** 0.000 27 atm, its partial pressure in air.

3.12 The standard molar entropies of water ice, liquid, and vapor are 37.99, 69.91, and 188.83 J K^{-1} mol^{-1}, respectively. On a single graph, show how the Gibbs energies of each of these phases vary with temperature.

3.13 An open vessel containing **(a)** water, **(b)** benzene, and **(c)** mercury stands in a laboratory measuring 6.0 m × 5.3 m × 3.2 m at 25°C. What mass of each substance will be found in the air if there is no ventilation? (The vapor pressures are **(a)** 2.3 kPa, **(b)** 10 kPa, and **(c)** 0.30 Pa.)

3.14 **(a)** Use the Clapeyron equation to estimate the slope of the solid–liquid phase boundary of water given that the enthalpy of fusion is 6.008 kJ mol^{-1} and the densities of ice and water at 0°C are 0.916 71 and 0.999 84 g cm^{-3}, respectively. *Hint:* Express the entropy of fusion in terms of the enthalpy of fusion and the melting point of ice. **(b)** Estimate the pressure required to lower the melting point of ice by 1°C.

3.15 **(a)** Use the Clausius–Clapeyron equation to show that the vapor pressure p' at a temperature T' is related to the vapor pressure p at a temperature T by

$$\ln p' = \ln p + \frac{\Delta_{vap}H}{R}\left(\frac{1}{T} - \frac{1}{T'}\right)$$

(b) The vapor pressure of mercury at 20°C is 160 mPa. What is its vapor pressure at 40°C given that its enthalpy of vaporization is 59.30 kJ mol^{-1}?

3.16 **(a)** The vapor pressures of substances are commonly reported as $\log(p/\text{kPa}) = A - B/T$, where A and B are constants. Show that the expression you derived in Exercise 3.15 reduces to this form, and write expressions for the constants A and B. **(b)** For benzene in the range 0–42°C, $A = 7.0871$ and $B = 1785$ K. What is the enthalpy of vaporization of benzene?

3.17 On a cold, dry morning after a frost, the temperature was −5°C and the partial pressure of water in the atmosphere fell to 2 Torr. Will the frost sublime? What partial pressure of water would ensure that the frost remained?

3.18 **(a)** Refer to Fig. 3.13 and describe the changes that would be observed when water vapor at 1.0 bar and 400 K is cooled at constant pressure to 260 K. **(b)** Suggest the appearance of a plot of temperature against time if energy is removed at a constant rate. To judge the relative slopes of the cooling curves, you need to know that the constant-pressure molar heat capacities of water vapor, liquid, and solid are approximately $4R$, $9R$, and $4.5R$; the enthalpies of transition are given in Table 1.2.

3.19 Refer to Fig. 3.13 and describe the changes that would be observed when cooling takes place at the pressure of the triple point.

3.20 A thermodynamic treatment allows predictions to be made of the temperature T_m for the unfolding of a helical polypeptide into a random coil. If a polypeptide has n amino acids, $n - 4$ hydrogen bonds are formed to form an α-helix, the most common type of helix in naturally occurring proteins (see Chapter 11). Because the first and last residues in the chain are free to move, it follows that $n - 2$ residues form the compact helix and have restricted motion. Based on these ideas, the molar Gibbs energy of unfolding of a polypeptide with $n \geq 5$ may be written as

$$\Delta G_m = (n - 4)\Delta_{hb}H_m - (n - 2)T\Delta_{hb}S_m$$

where $\Delta_{hb}H_m$ and $\Delta_{hb}S_m$ are, respectively, the molar enthalpy and entropy of dissociation of hydrogen bonds in the polypeptide. **(a)** Justify the form of the equation for the Gibbs energy of unfolding. That is, why are the enthalpy and entropy terms written as $(n - 4)\Delta_{hb}H_m$ and $(n - 2)\Delta_{hb}S_m$, respectively? **(b)** Show that T_m may be written as

$$T_m = (n - 4)\Delta_{hb}H_m/(n - 2)\Delta_{hb}S_m$$

(c) Plot $T_m/(\Delta_{hb}H_m/\Delta_{hb}S_m)$ for $5 \leq n \leq 20$. At what value of n does T_m change by less than 1 per cent when n increases by one?

3.21 A thermodynamic treatment allows predictions of the stability of DNA. The table below lists the standard Gibbs energies, enthalpies, and entropies of formation at 298 K of short sequences of base pairs as two polynucleotide chains come together:

Sequence	5′–A–G 3′–T–C	5′–G–C 3′–C–G	5′–T–G 3′–A–C
$\Delta_{seq}G^\circ/(\text{kJ mol}^{-1})$	−5.4	−10.5	−6.7
$\Delta_{seq}H^\circ/(\text{kJ mol}^{-1})$	−25.5	−46.4	−31.0
$\Delta_{seq}S^\circ/(\text{kJ mol}^{-1})$	−67.4	−118.8	−80.8

To estimate the standard Gibbs energy of formation of a double-stranded piece of DNA, $\Delta_{DNA}G^\circ$, we sum the contributions from the formation of the sequences and add to that quantity the standard Gibbs energy of initiation of the process, which in the case treated in this exercise may be set equal to $\Delta_{init}G^\circ = +14.2 \text{ kJ mol}^{-1}$:

$$\Delta_{DNA}G^\circ = \Delta_{init}G^\circ + \Delta_{seq}G^\circ(\text{sequences})$$

Similar procedures lead to $\Delta_{DNA}H^\circ$ and $\Delta_{DNA}S^\circ$. **(a)** Provide a molecular explanation for the fact that $\Delta_{init}G^\circ$ is positive and $\Delta_{seq}G^\circ$ negative. **(b)** Estimate the standard Gibbs energy, enthalpy, and entropy changes for the following reaction:

5′–A–G–C–T–G–3′ + 5′–C–A–G–C–T– 3′ →
5′–A–G–C–T–G–3′
3′–T–C–G–A–C–5′

Use $\Delta_{init}H^\circ = +2.5 \text{ kJ mol}^{-1}$ and $\Delta_{init}S^\circ = −37.7 \text{ J K}^{-1} \text{ mol}^{-1}$. **(c)** Estimate the 'melting' temperature of the piece of DNA shown in part **(b)**.

3.22 The vapor pressure of water at blood temperature is 47 Torr. What is the partial pressure of dry air in our lungs when the total pressure is 760 Torr?

3.23 A gas mixture being used to simulate the atmosphere of another planet consists of 320 mg of methane, 175 mg of argon, and 225 mg of nitrogen. The partial pressure of nitrogen at 300 K is 15.2 kPa. Calculate **(a)** the volume and **(b)** the total pressure of the mixture.

3.24 Calculate the mass of glucose you should use to prepare **(a)** 250.0 cm³ of 0.112 M $C_6H_{12}O_6$(aq) and **(b)** 0.112 m $C_6H_{12}O_6$(aq) using 250.0 g of water.

3.25 What is the mole fraction of alanine in 0.134 m $CH_3CH(NH_2)COOH$(aq)?

3.26 What mass of sucrose, $C_{12}H_{22}O_{11}$, should you dissolve in 100.0 g of water to obtain a solution in which the mole fraction of $C_{12}H_{22}O_{11}$ is 0.124?

3.27 Calculate **(a)** the (molar) Gibbs energy of mixing and **(b)** the (molar) entropy of mixing when the two major components of air (nitrogen and oxygen) are mixed to form air. The mole fractions of N_2 and O_2 are 0.78 and 0.22, respectively. Is the mixing spontaneous?

3.28 Suppose now that argon is added to the mixture in Exercise 3.27 to bring the composition closer to real air, with mole fractions 0.780, 0.210, and 0.0096, respectively. What is the additional change in molar Gibbs energy and entropy? Is the mixing spontaneous?

3.29 Estimate the vapor pressure of seawater at 20°C given that the vapor pressure of pure water is 2.338 kPa at that temperature and the solute is largely Na^+ and Cl^- ions, each present at about 0.50 mol dm⁻³.

3.30 Hemoglobin, the red blood protein responsible for oxygen transport, binds about 1.34 cm³ of oxygen per gram. Normal blood has a hemoglobin concentration of 150 g dm⁻³. Hemoglobin in the lungs is about 97 per cent saturated with oxygen but in the capillary is only about 75 per cent saturated. What volume of oxygen is given up by 100 cm³ of blood flowing from the lungs in the capillary?

3.31 In scuba diving (where *scuba* is an acronym formed from 'self-contained underwater breathing apparatus'), air is supplied at a higher pressure so that the pressure within the diver's chest matches the pressure exerted by the surrounding water. The latter increases by about 1 atm for each 10 m of descent. One unfortunate consequence of breathing air at high pressures is that nitrogen is much more soluble in fatty tissues than in water, so it tends to dissolve in the central nervous system, bone marrow, and fat reserves. The result is *nitrogen narcosis*, with symptoms like intoxication. If the diver rises too rapidly to the surface, the nitrogen comes out of its lipid solution as bubbles, which causes the painful and sometimes fatal condition known as *the bends*. Many cases of scuba drowning appear to be consequences of arterial embolisms (obstructions in arteries caused by gas bubbles) and loss of consciousness as the air bubbles rise into the head. The Henry's law constant in the form $c = Kp$ for the solubility of nitrogen is 0.18 µg/(g H_2O atm). **(a)** What mass of nitrogen is dissolved in 100 g of water saturated with air at 4.0 atm and 20°C? Compare your answer to that for 100 g of water saturated with air at 1.0 atm. (Air is 78.08 mole per cent N_2.) **(b)** If nitrogen is four times as soluble in fatty tissues as in water, what is the increase in nitrogen concentration in fatty tissue in going from 1 atm to 4 atm?

3.32 Calculate the concentration of carbon dioxide in fat given that the Henry's law constant is 8.6×10^4 Torr and the partial pressure of carbon dioxide is 55 kPa.

3.33 The rise in atmospheric carbon dioxide results in higher concentrations of dissolved carbon dioxide in natural waters. Use Henry's law and the data in Table 3.2 to calculate the solubility of CO_2 in water at 25°C when its partial pressure is **(a)** 4.0 kPa and **(b)** 100 kPa.

3.34 The mole fractions of N_2 and O_2 in air at sea level are approximately 0.78 and 0.21. Calculate the molalities of the solution formed in an open flask of water at 25°C.

3.35 Estimate the freezing point of 150 cm³ of water sweetened with 7.5 g of sucrose.

3.36 A compound A existed in equilibrium with its dimer, A_2, in an aqueous solution. Derive an expression for the equilibrium constant $K = [A_2]/[A]^2$ in terms of the depression in vapor pressure caused by a given concentration of compound. *Hint:* Suppose that a fraction f of the A molecules are present as the dimer. The depression of vapor pressure is proportional to the total concentration of A and A_2 molecules regardless of their chemical identities.

3.37 The osmotic pressure of an aqueous solution of urea at 300 K is 120 kPa. Calculate the freezing point of the same solution.

3.38 The molar mass of an enzyme was determined by dissolving it in water, measuring the osmotic pressure at 20°C and extrapolating the data to zero concentration. The following data were used:

$c/(\text{mg cm}^{-3})$	3.221	4.618	5.112	6.722
h/cm	5.746	8.238	9.119	11.990

Calculate the molar mass of the enzyme. *Hint:* Begin by expressing eqn 3.24 in terms of the height of the solution by using $\Pi = \rho gh$; take $\rho = 1.000$ g cm⁻³.

Projects

3.39 We now explore further the use of temperature–composition diagrams in the study of biological systems.

(a) Use the phase rule described in *Further information* 3.1 to justify the statement that in a temperature–composition diagram for a binary mixture, two-phase equilibria define a line and a three-phase equilibrium is represented by a point.

(b) Consider Fig. 3.19. (i) Is the molten globule form ever stable when the denaturant concentration is below the level marked x? (ii) Describe what happens to the polymer as the native form is heated in the presence of denaturant at concentration y.

(c) In an experimental study of membrane-like assemblies, a phase diagram like that shown in Fig. 3.40 was obtained. The two components are dielaidoylphosphatidylcholine (DEL) and dipalmitoylphosphatidylcholine (DPL). Explain what happens as a liquid mixture of composition $x_{DEL} = 0.5$ is cooled from 45°C.

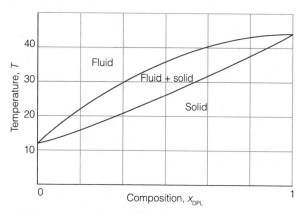

Fig. 3.40

3.40 Dialysis may also be used to study the binding of small molecules to macromolecules, such as an inhibitor to an enzyme, an antibiotic to DNA, and any other instance of cooperation or inhibition by small molecules attaching to large ones. To see how this is possible, suppose inside the dialysis bag the molar concentration of the macromolecule M is $[M]$ and the total concentration of small molecule A is $[A]_{in}$. This total concentration is the sum of the concentrations of free A and bound A, which we write $[A]_{free}$ and $[A]_{bound}$, respectively. At equilibrium, $\mu_{A,free} = \mu_{A,out}$, which implies that $[A]_{free} = [A]_{out}$, provided the activity coefficient of A is the same in both solutions. Therefore, by measuring the concentration of A in the solution outside the bag, we can find the concentration of unbound A in the macromolecule solution and, from the difference $[A]_{in} - [A]_{free} = [A]_{in} - [A]_{out}$, the concentration of bound A. Now we explore the quantitative consequences of the experimental arrangement just described.

(a) The average number of A molecules bound to M molecules, v, is

$$v = \frac{[A]_{bound}}{[M]} = \frac{[A]_{in} - [A]_{out}}{[M]}$$

The bound and unbound A molecules are in equilibrium, $M + A \rightleftharpoons MA$. Recall from introductory chemistry that we may write the equilibrium constant for binding, K, as

$$K = \frac{[MA]}{[M]_{free}[A]_{free}}$$

Now show that

$$K = \frac{v}{(1-v)[A]_{out}}$$

(b) If there are N *identical* and *independent* binding sites on each macromolecule, each macromolecule behaves like N separate smaller macromolecules, with the same value of K for each site. It follows that the average number of A molecules per site is v/N. Show that, in this case, we may write the *Scatchard equation*:

$$\frac{v}{[A]_{out}} = KN - Kv$$

(c) The Scatchard equation implies that a plot of $v/[A]_{out}$ against v should be a straight line of slope K and intercept KN at $v = 0$. To apply the Scatchard equation, consider the binding of ethidium bromide (EB) to a short piece of DNA by a process called *intercalation*, in which the aromatic ethidium cation fits between two adjacent DNA base pairs. A 1.00×10^{-6} mol dm^{-3} aqueous solution of the DNA sample was dialyzed against an excess of EB. The following data were obtained for the total concentration of EB:

[EB]/(μmol dm^{-3})					
Side without DNA	0.042	0.092	0.204	0.526	1.150
Side with DNA	0.292	0.590	1.204	2.531	4.150

From these data, make a Scatchard plot and evaluate the equilibrium constant, K, and total number of sites per DNA molecule. Is the identical and independent sites model for binding applicable?

(d) For nonidentical independent binding sites, the Scatchard equation is

$$\frac{v}{[A]_{out}} = \sum_i \frac{N_i K_i}{1 + K_i[A]_{out}}$$

Plot $v/[A]$ for the following cases. (i) There are four independent sites on an enzyme molecule and the equilibrium constant is $K = 1.0 \times 10^7$. (ii) There are a total of six sites per enzyme molecule. Four of the sites are identical and have an equilibrium constant of 1×10^5. The binding constants for the other two sites are 2×10^6.

Chemical equilibrium

4

Now we arrive at the point where real chemistry begins. Chemical thermodynamics is used to predict whether a mixture of reactants has a spontaneous tendency to change into products, to predict the composition of the reaction mixture at equilibrium, and to predict how that composition will be modified by changing the conditions. In biology, life is the avoidance of equilibrium; the attainment of equilibrium is death. Knowing whether equilibrium lies in favor of reactants or products under certain conditions is a good indication of the feasibility of a biochemical reaction. Indeed, the material we cover in this chapter is of crucial importance for understanding the mechanisms of oxygen transport in blood, metabolism, and all the processes going on inside organisms. One very important feature of life is that one reaction can drive another in a vast web of processes. Broadly speaking, that is why we have to eat. We shall see later in the chapter how thermodynamics can be used to assess the ability of one reaction to drive another.

There is one word of warning that is essential to remember: *thermodynamics is silent about the rates of reaction*. All it can do is to identify whether a particular reaction mixture has a tendency to form products; it cannot say whether that tendency will ever be realized. We explore what determines the rates of chemical reactions in Chapters 6 through 8.

Thermodynamic background

The thermodynamic criterion for spontaneous change at constant temperature and pressure is $\Delta G < 0$. The principal idea behind this chapter, therefore, is that, *at constant temperature and pressure, a reaction mixture tends to adjust its composition until its Gibbs energy is a minimum*. If the Gibbs energy of a mixture varies as shown in Fig. 4.1a, very little of the reactants convert into products before G has reached its minimum value, and the reaction 'does not go'. If G varies as shown in Fig. 4.1c, then a high proportion of products must form before G reaches its minimum and the reaction 'goes'. In many cases, the equilibrium mixture contains almost no reactants or almost no products. Many reactions have a Gibbs energy that varies as shown in Fig. 4.1b, and at equilibrium the reaction mixture contains substantial amounts of both reactants and products.

4.1 The reaction Gibbs energy

To explore metabolic processes, we need a measure of the driving power of a chemical reaction, and to understand the chemical composition of cells, we need to know what those compositions would be if the reactions taking place in them had reached equilibrium.

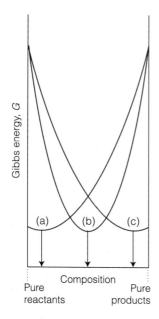

Fig. 4.1 The variation of Gibbs energy of a reaction mixture with progress of the reaction, pure reactants on the left and pure products on the right. (a) This reaction 'does not go': the minimum in the Gibbs energy occurs very close to the reactants. (b) This reaction reaches equilibrium with approximately equal amounts of reactants and products present in the mixture. (c) This reaction goes almost to completion, as the minimum in Gibbs energy lies very close to pure products.

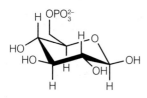

1 Glucose-6-phosphate

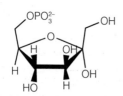

2 Fructose-6-phosphate

To keep our ideas in focus, we consider two important processes. One is the isomerism of glucose-6-phosphate (**1**, G6P) to fructose-6-phosphate (**2**, F6P), which is an early step in the anaerobic breakdown of glucose (*Case study* 4.3):

$$G6P(aq) \rightarrow F6P(aq) \tag{A}$$

The second is the binding of $O_2(g)$ to the protein hemoglobin, Hb, in blood:

$$Hb(aq) + 4\,O_2(g) \rightarrow Hb(O_2)_4(aq) \tag{B}$$

These two reactions are specific examples of a general reaction of the form

$$a\,A + b\,B \rightarrow c\,C + d\,D \tag{C}$$

with arbitrary physical states.

First, consider reaction A. Suppose that in a short interval while the reaction is in progress, the amount of G6P changes infinitesimally by $-dn$. As a result of this change in amount, the contribution of G6P to the total Gibbs energy of the system changes by $-\mu_{G6P}dn$, where μ_{G6P} is the chemical potential (the partial molar Gibbs energy) of G6P in the reaction mixture. In the same interval, the amount of F6P changes by $+dn$, so its contribution to the total Gibbs energy changes by $+\mu_{F6P}dn$, where μ_{F6P} is the chemical potential of F6P. The change in Gibbs energy of the system is

$$dG = \mu_{F6P}dn - \mu_{G6P}dn$$

On dividing through by dn, we obtain the **reaction Gibbs energy**, $\Delta_r G$:

$$\frac{dG}{dn} = \mu_{F6P} - \mu_{G6P} = \Delta_r G \tag{4.1a}$$

There are two ways to interpret $\Delta_r G$. First, it is the difference of the chemical potentials of the products and reactants *at the current composition of the reaction mixture*. Second, we can think of $\Delta_r G$ as the derivative of G with respect to n, which is the slope of the graph of G plotted against the changing composition of the system (Fig. 4.2). As we see from the illustration, that slope changes as the reaction proceeds because the two chemical potentials change as the composition of the reaction mixture changes.

The binding of oxygen to hemoglobin provides a slightly more complicated example. When the amount of Hb changes by $-dn$, from the reaction stoichiometry the change in the amount of O_2 is $-4dn$ and the change in the amount of $Hb(O_2)_4$ is $+dn$. The overall change in the Gibbs energy of the mixture is therefore

$$dG = \mu_{Hb(O_2)_4} \times dn - \mu_{Hb} \times dn - \mu_{O_2} \times 4dn$$
$$= (\mu_{Hb(O_2)_4} - \mu_{Hb} - 4\mu_{O_2})dn$$

where the μ_J are the chemical potentials of the species in the reaction mixture. In this case, therefore, the reaction Gibbs energy is

$$\Delta_r G = \frac{dG}{dn} = \mu_{Hb(O_2)_4} - \mu_{Hb} - 4\mu_{O_2} \tag{4.1b}$$

Note that each chemical potential is multiplied by the corresponding stoichiometric coefficient and that reactants are subtracted from products. For the general reaction C,

$$\Delta_r G = (c\mu_C + d\mu_D) - (a\mu_A + b\mu_B) \qquad \boxed{\text{Gibbs energy of reaction}} \quad (4.1c)$$

where the chemical potentials are those for each substance at the current composition of the mixture.

The chemical potential of a substance depends on the composition of the mixture in which it is present and is high when its concentration or partial pressure is high. Therefore, $\Delta_r G$ changes as the composition changes (Fig. 4.3). Remember that $\Delta_r G$ is the *slope* of G plotted against composition. We see that $\Delta_r G < 0$ and the slope of G is negative (down from left to right) when the mixture is rich in the reactants A and B because μ_A and μ_B are then high. Conversely, $\Delta_r G > 0$ and the slope of G is positive (up from left to right) when the mixture is rich in the products C and D because μ_C and μ_D are then high. At compositions corresponding to $\Delta_r G < 0$ the reaction tends to form more products; where $\Delta_r G > 0$, the *reverse* reaction is spontaneous, and the products tend to decompose into reactants. Where $\Delta_r G = 0$ (at the minimum of the graph where the slope is zero), the reaction has no tendency to form either products or reactants. In other words, the reaction is at equilibrium. That is, *the criterion for chemical equilibrium at constant temperature and pressure is*

$$\Delta_r G = 0 \qquad \boxed{\text{Criterion of chemical equilibrium}} \quad (4.2)$$

4.2 The variation of $\Delta_r G$ with composition

The reactants and products in a biological cell are rarely at equilibrium, so we need to know how the reaction Gibbs energy depends on their concentrations.

Our starting point is the general expression for the composition dependence of the chemical potential derived in Section 3.8:

$$\mu_J = \mu_J^\ominus + RT \ln a_J \qquad \boxed{\text{Chemical potential of a species J}} \quad (4.3)$$

where a_J is the activity of the species J. When we are dealing with systems that may be treated as ideal, which will be the case in this chapter, we use the identifications given in Table 3.3:

For solutes in an ideal solution, $a_J = [\text{J}]/c^\ominus$, the molar concentration of J relative to the standard value $c^\ominus = 1\ \text{mol dm}^{-3}$.

For perfect gases, $a_J = p_J/p^\ominus$, the partial pressure of J relative to the standard pressure $p^\ominus = 1$ bar.

For pure solids and liquids, $a_J = 1$.

As in Chapter 3, to simplify the appearance of expressions in what follows, we shall not write $c^\ominus$ and $p^\ominus$ explicitly.

(a) The reaction quotient
Substitution of eqn 4.3 into eqn 4.1c gives

$$\Delta_r G = \{c(\mu_C^\ominus + RT \ln a_C) + d(\mu_D^\ominus + RT \ln a_D)\}$$
$$- \{a(\mu_A^\ominus + RT \ln a_A) + b(\mu_B^\ominus + RT \ln a_B)\}$$
$$= \{(c\mu_C^\ominus + d\mu_D^\ominus) - (a\mu_A^\ominus + b\mu_A^\ominus)\}$$
$$+ RT\{c \ln a_C + d \ln a_D - a \ln a_A - b \ln a_B\}$$

The first term on the right in the second equality is the **standard reaction Gibbs energy**, $\Delta_r G^\ominus$:

$$\Delta_r G^\ominus = \{c\mu_C^\ominus + d\mu_D^\ominus\} - \{a\mu_A^\ominus + b\mu_B^\ominus\} \qquad (4.4a)$$

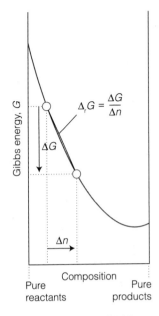

Fig. 4.2 The variation of Gibbs energy with progress of reaction showing how the reaction Gibbs energy, $\Delta_r G$, is related to the slope of the curve at a given composition. When ΔG and Δn are both infinitesimal, the slope is written dG/dn.

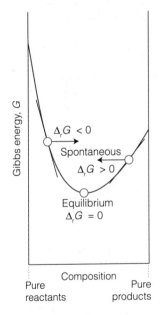

Fig. 4.3 At the minimum of the curve, corresponding to equilibrium, $\Delta_r G = 0$. To the left of the minimum, $\Delta_r G < 0$, and the forward reaction is spontaneous. To the right of the minimum, $\Delta_r G > 0$, and the reverse reaction is spontaneous.

Because the standard states refer to the pure materials, the standard chemical potentials in this expression are the standard molar Gibbs energies of the (pure) species. Therefore, eqn 4.4a is the same as

$$\Delta_r G^\ominus = \{cG_m^\ominus(C) + dG_m^\ominus(D)\} - \{aG_m^\ominus(A) + bG_m^\ominus(B)\} \qquad \boxed{\text{Standard Gibbs energy of reaction}} \qquad (4.4b)$$

We consider this important quantity in more detail shortly. At this stage, therefore, we know that

$$\Delta_r G = \Delta_r G^\ominus + RT\{c \ln a_C + d \ln a_D - a \ln a_A - b \ln a_B\}$$

and the expression for $\Delta_r G$ is beginning to look much simpler.

To make further progress, we rearrange the remaining terms on the right as follows:

$$c \ln a_C + d \ln a_D - a \ln a_A - b \ln a_B = \ln a_C^c + \ln a_D^d - \ln a_A^a - \ln a_B^b$$

$$= \ln a_C^c a_D^d - \ln a_A^a a_B^b = \ln \frac{a_C^c a_D^d}{a_A^a a_B^b}$$

At this point, we have deduced that

$$\Delta_r G = \Delta_r G^\ominus + RT \ln \frac{a_C^c a_D^d}{a_A^a a_B^d} \qquad (4.5)$$

To simplify the appearance of this expression still further, we introduce the (dimensionless) **reaction quotient**, Q, for reaction C:

$$Q = \frac{a_C^c a_D^d}{a_A^a a_B^b} \qquad \boxed{\text{Definition of reaction quotient}} \qquad (4.6)$$

Note that Q has the form of products divided by reactants, with the activity of each species raised to a power equal to its stoichiometric coefficient in the reaction; because activities are dimensionless quantities, Q is a dimensionless quantity. We can now write the overall expression for the reaction Gibbs energy at any composition of the reaction mixture as

$$\Delta_r G = \Delta_r G^\ominus + RT \ln Q \qquad \boxed{\text{Reaction Gibbs energy}} \qquad (4.7)$$

This simple but hugely important equation will occur several times in different disguises.

| Example 4.1 | *Formulating a reaction quotient* |

Formulate the reaction quotients for reactions A (the isomerism of glucose-6-phosphate) and B (the binding of oxygen to hemoglobin).

Strategy Use Table 3.3 to express activities in terms of molar concentrations or pressures. Then use eqn 4.6 to write an expression for the reaction quotient Q. In reactions involving gases and solutes, the expression for Q will contain pressures and molar concentrations.

Solution The reaction quotient for reaction A is

$$Q = \frac{a_{F6P}}{a_{G6P}} = \frac{[F6P]/c^{\ominus}}{[G6P]/c^{\ominus}} = \frac{[F6P]}{[G6P]}$$

For reaction B, the binding of oxygen to hemoglobin, the reaction quotient is

$$Q = \frac{a_{Hb(O_2)_4}}{a_{Hb}a_{O_2}^4} = \frac{[Hb(O_2)_4]/c^{\ominus}}{([Hb]/c^{\ominus})(p_{O_2}/p^{\ominus})^4}$$

Because we are not writing the standard concentration and pressure explicitly, this expression simplifies to

$$Q = \frac{[Hb(O_2)_4]}{[Hb]p_{O_2}^4}$$

with p_J the numerical value of the partial pressure of J in bar (so if $p_{O_2} = 2.0$ bar, we just write $p_{O_2} = 2.0$ when using this expression).

Self-test 4.1   Write the reaction quotient for the esterification reaction $CH_3COOH + C_2H_5OH \rightarrow CH_3COOC_2H_5 + H_2O$. (All four components are present in the reaction mixture as liquids: the mixture is not an aqueous solution.)

Answer: $Q \approx [CH_3COOC_2H_5][H_2O]/[CH_3COOH][C_2H_5OH]$

(b) Biological standard states

The thermodynamic definition of standard states of solutes takes them as being at unit activity (in elementary work, at $c^{\ominus} = 1$ mol dm^{-3}). The conventional standard state of hydrogen ions ($a_{H_3O^+} = 1$, corresponding to pH = 0, a strongly acidic solution) is not appropriate to normal biological conditions inside cells, where the pH is close to 7. Therefore, in biochemistry it is common to adopt the **biological standard state**, in which pH = 7, a neutral solution. When we adopt this convention we label the corresponding standard quantities as $G^{\oplus}$, $H^{\oplus}$, and $S^{\oplus}$.[1] Equation 4.8 allows us to relate the two standard Gibbs energies of formation.

For a reaction of the form

reactants $+ v\,H_3O^+(aq) \rightarrow$ products

the biological and thermodynamic standard states are related by

$$\Delta_r G^{\oplus} = \Delta_r G^{\ominus} - RT\ln(10^{-7})^v = \Delta_r G^{\ominus} + 7vRT\ln 10 \qquad \boxed{\text{Relation between standard values}} \quad (4.8)$$

where we have used the relations $(x^a)^b = x^{ab}$ and $\ln x^{ab} = ab\ln x$. It follows that

at 298.15 K: $\Delta_r G^{\oplus} = \Delta_r G^{\ominus} + v(39.96\text{ kJ mol}^{-1})$

at 37°C (310 K, body temperature): $\Delta_r G^{\oplus} = \Delta_r G^{\ominus} + v(41.5\text{ kJ mol}^{-1})$

There is no difference between thermodynamic and biological standard values if hydrogen ions are not involved in the reaction ($v = 0$).

---

[1] Another convention to denote the biological standard state is to write $X^{\circ\prime}$ or $X^{\oplus\prime}$.

> **Example 4.2** *Converting between thermodynamic and biological standard states*

The standard reaction Gibbs energy for the hydrolysis of ATP is $+10\ \text{kJ mol}^{-1}$ at 298 K. What is the biological standard state value?

Strategy Because protons occur as products, lowering their concentration (from $1\ \text{mol dm}^{-3}$ to $10^{-7}\ \text{mol dm}^{-3}$) suggests that the reaction will have a higher tendency to form products. Therefore, we expect a more negative value of the reaction Gibbs energy for the biological standard than for the thermodynamic standard. The two types of standard are related by eqn 4.8, with the activity of hydrogen ions 10^{-7} in place of 1.

Solution The reaction quotient for the hydrolysis reaction

$$\text{ATP}^{4-}(\text{aq}) + \text{H}_2\text{O}(\text{l}) \rightarrow \text{ADP}^{3-}(\text{aq}) + \text{HPO}_4^{2-}(\text{aq}) + \text{H}_3\text{O}^+(\text{aq})$$

when all the species are in their standard states except the hydrogen ions, which are present at $10^{-7}\ \text{mol dm}^{-3}$, is

$$Q = \frac{a_{\text{ADP}^{3-}} a_{\text{HPO}_4^{2-}} a_{\text{H}_3\text{O}^+}}{a_{\text{ATP}^{4-}} - a_{\text{H}_2\text{O}}} = \frac{1 \times 1 \times 10^{-7}}{1 \times 1} = 10^{-7}$$

The thermodynamic and biological standard values are therefore related by eqn 4.8 in the form

$$\Delta_r G^\oplus = \Delta_r G^\ominus + RT \ln(10^{-7})$$

At 298 K

$$\begin{aligned}\Delta_r G^\oplus &= 10\ \text{kJ mol}^{-1} + (8.3145\ \text{J K}^{-1}\ \text{mol}^{-1}) \times (298\ \text{K}) \times \ln(10^{-7}) \\ &= 10\ \text{kJ mol}^{-1} - 40\ \text{kJ mol}^{-1} \\ &= -30\ \text{kJ mol}^{-1}\end{aligned}$$

Note how the large change in pH changes the sign of the standard reaction Gibbs energy.

Self-test 4.2 The overall reaction for the glycolysis reaction (*Case study* 4.3) is $\text{C}_6\text{H}_{12}\text{O}_6(\text{aq}) + 2\ \text{NAD}^+(\text{aq}) + 2\ \text{ADP}^{3-}(\text{aq}) + 2\ \text{HPO}_4^{2-}(\text{aq}) + 2\ \text{H}_2\text{O}(\text{l}) \rightarrow 2\ \text{CH}_3\text{COCO}_2^-(\text{aq}) + 2\ \text{NADH}(\text{aq}) + 2\ \text{ATP}^{4-}(\text{aq}) + 2\ \text{H}_3\text{O}^+(\text{aq})$. For this reaction, $\Delta_r G^\oplus = -80.6\ \text{kJ mol}^{-1}$ at 298 K. What is the value of $\Delta_r G^\ominus$?

Answer: $-0.7\ \text{kJ mol}^{-1}$

4.3 Reactions at equilibrium

> We need to be able to identify the equilibrium composition of a reaction so that we can discuss the approach to equilibrium systematically.

At equilibrium, the reaction quotient has a certain (dimensionless) value called the **equilibrium constant**, K, of the reaction:

$$K = \left(\frac{a_C^c a_D^d}{a_A^a a_B^b} \right)_{\text{equilibrium}}$$

| Definition of equilibrium constant | (4.9) |

We shall not normally write *equilibrium*; the context will always make it clear that Q refers to an *arbitrary* stage of the reaction, whereas K, the value of Q at equilibrium, is calculated from the equilibrium composition. It now follows from eqn 4.7 that at equilibrium

$$0 = \Delta_r G^{\ominus} + RT \ln K$$

and therefore that

$$\Delta_r G^{\ominus} = -RT \ln K \qquad \boxed{\text{Expression for calculating an equilibrium constant}} \qquad (4.10a)$$

This is one of the most important equations in the whole of chemical thermodynamics. Its principal use is to predict the value of the equilibrium constant of any reaction from tables of thermodynamic data, like those in the *Resource section*. Alternatively, we can use it to determine $\Delta_r G^{\ominus}$ by measuring the equilibrium constant of a reaction.

A brief illustration

The first step in the metabolic breakdown of glucose is its phosphorylation to G6P:

$$\text{glucose(aq)} + \text{ATP(aq)} \rightarrow \text{G6P(aq)} + \text{ADP(aq)} + \text{H}^+\text{(aq)}$$

The standard reaction Gibbs energy for the reaction is -34 kJ mol^{-1} at 37°C, so it follows from eqn 4.10a that

$$\ln K = -\frac{\Delta_r G^{\ominus}}{RT} = -\frac{(-3.4 \times 10^4 \text{ J mol}^{-1})}{(8.3145 \text{ J K}^{-1} \text{ mol}^{-1}) \times (310 \text{ K})} = \frac{3.4 \times 10^4}{8.3145 \times 310}$$

To calculate the equilibrium constant of the reaction, which (like the reaction quotient) is a dimensionless number, we use the relation $e^{\ln x} = x$ with $x = K$:

$$K = e^{3.4 \times 10^4/8.3145 \times 310} = 5.4 \times 10^5$$

A note on good practice
The exponential function (e^x) is very sensitive to the value of x, so evaluate it only at the end of a numerical calculation.

Self-test 4.3   Calculate the equilibrium constant of the reaction $N_2(g) + 3 H_2(g) \rightleftharpoons 2 NH_3(g)$ at 25°C, given that $\Delta_r G^{\ominus} = -32.90 \text{ kJ mol}^{-1}$.

Answer: 5.8×10^5

If the biological standard state is used in place of the thermodynamic standard state, we write eqn 4.10a in the same way,

$$\Delta_r G^{\oplus} = -RT \ln K \qquad \boxed{\text{Expression for calculating an equilibrium constant}} \qquad (4.10b)$$

but interpret the concentration of any hydronium ions that occurs in K as relative to $c^{\oplus} = 10^{-7} \text{ mol dm}^{-3}$ rather than relative to $c^{\ominus} = 1 \text{ mol dm}^{-3}$. That is, we interpret the activity of hydronium ions in the expression for K as $a_{\text{H}_3\text{O}^+} = [\text{H}_3\text{O}^+]/c^{\oplus}$.

A brief illustration

The biological standard reaction Gibbs energy for the reaction in the preceding brief illustration is $-75.5 \text{ kJ mol}^{-1}$. The same calculation illustrated there but with this value gives $K' = 5.3 \times 10^{12}$. This value is for

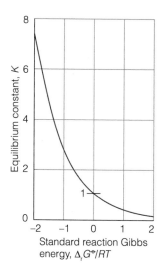

Fig. 4.4 The relation between standard reaction Gibbs energy and the equilibrium constant of the reaction.

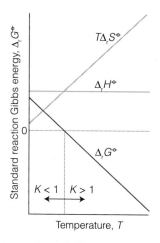

Fig. 4.5 An endothermic reaction may have $K > 1$ provided the temperature is high enough for $T\Delta_r S^\ominus$ to be large enough that, when subtracted from $\Delta_r H^\ominus$, the result is negative.

$$K' = \frac{[\text{G6P}][\text{ADP}]([\text{H}_3\text{O}^+]/10^{-7})}{[\text{glucose}][\text{ATP}]} = 5.3 \times 10^{12}$$

whereas the former value was for

$$K = \frac{[\text{glucose}][\text{ADP}][\text{H}_3\text{O}^+]}{[\text{F6P}][\text{ATP}]} = 5.4 \times 10^5$$

Multiplication of both sides of the first equation of this pair by 10^{-7} gives a result in accord with the second equation.

(a) The significance of the equilibrium constant

An important feature of eqn 4.10a is that it tells us that $K > 1$ if $\Delta_r G^\ominus < 0$ (and correspondingly that $K' > 1$ if $\Delta_r G^\oplus < 0$). Broadly speaking, $K > 1$ implies that products are dominant at equilibrium, so we can conclude that *a reaction is thermodynamically feasible if $\Delta_r G^\ominus < 0$* (Fig. 4.4). Conversely, because eqn 4.10a tells us that $K < 1$ if $\Delta_r G^\ominus > 0$, then we know that the reactants will be dominant in a reaction mixture at equilibrium if $\Delta_r G^\ominus > 0$. In other words, *a reaction with $\Delta_r G^\ominus > 0$ is not thermodynamically feasible*. Some care must be exercised with these rules, however, because the products will be significantly more abundant than reactants only if $K \gg 1$ (more than about 10^3), and even a reaction with $K < 1$ may have a reasonable abundance of products at equilibrium.

Table 4.1 summarizes the conditions under which $\Delta_r G^\ominus < 0$ and $K > 1$. Because $\Delta_r G^\ominus = \Delta_r H^\ominus - T\Delta_r S^\ominus$, the standard reaction Gibbs energy is certainly negative if both $\Delta_r H^\ominus < 0$ (an exothermic reaction) and $\Delta_r S^\ominus > 0$ (a reaction system that becomes more disorderly, such as by forming a gas). The standard reaction Gibbs energy is also negative if the reaction is endothermic ($\Delta_r H^\ominus > 0$) and $T\Delta_r S^\ominus$ is sufficiently large and positive. Note that for an endothermic reaction to have $\Delta_r G^\ominus < 0$, its standard reaction entropy *must* be positive. Moreover, the temperature must be high enough for $T\Delta_r S^\ominus$ to be greater than $\Delta_r H^\ominus$ (Fig. 4.5). The switch of $\Delta_r G^\ominus$ from positive to negative, corresponding to the switch from $K < 1$ (the reaction 'does not go') to $K > 1$ (the reaction 'goes'), occurs at a temperature given by equating $\Delta_r H^\ominus - T\Delta_r S^\ominus$ to 0, which gives

$$T = \frac{\Delta_r H^\ominus}{\Delta_r S^\ominus}$$

| Temperature at which an endothermic reaction becomes spontaneous | (4.11) |

Table 4.1 Thermodynamic criteria of spontaneity

1. If the reaction is exothermic ($\Delta_r H^\ominus < 0$) and $\Delta_r S^\ominus > 0$

   $\Delta_r G^\ominus < 0$ and $K > 1$ at all temperatures

2. If the reaction is exothermic ($\Delta_r H^\ominus < 0$) and $\Delta_r S^\ominus < 0$

   $\Delta_r G^\ominus < 0$ and $K > 1$ provided that $T < \Delta_r H^\ominus/\Delta_r S^\ominus$

3. If the reaction is endothermic ($\Delta_r H^\ominus > 0$) and $\Delta_r S^\ominus > 0$

   $\Delta_r G^\ominus < 0$ and $K > 1$ provided that $T > \Delta_r H^\ominus/\Delta_r S^\ominus$

4. If the reaction is endothermic ($\Delta_r H^\ominus > 0$) and $\Delta_r S^\ominus < 0$

   $\Delta_r G^\ominus < 0$ and $K > 1$ at no temperature

Self-test 4.4 Suppose that the enthalpy change accompanying the dissociation of base pairs is of the order of +15 kJ per mole of base pairs and the corresponding entropy change is 45 J K^{-1} mol^{-1}. At what temperature can you expect a DNA chain to denature spontaneously?

Answer: 60°C

(b) The composition at equilibrium

An equilibrium constant expresses the composition of an equilibrium mixture as a ratio of products of activities. Even if we confine our attention to ideal systems, it is still necessary to do some work to extract the actual equilibrium concentrations or partial pressures of the reactants and products given their initial values (see, for example, *Example* 4.5).

Example 4.3 *Calculating an equilibrium composition*

Consider reaction A, for which $\Delta_r G^\oplus = +1.7$ kJ mol^{-1} at 25°C. Estimate the fraction f of F6P in equilibrium with G6P at 25°C, where f is defined as

$$f = \frac{[\text{F6P}]}{[\text{F6P}] + [\text{G6P}]}$$

Strategy Express f in terms of K. To do so, recognize that if the numerator and denominator in the expression for f are both divided by [G6P]; then the ratios [F6P]/[G6P] can be replaced by K. Calculate the value of K by using eqn 4.10a.

Solution Division of the numerator and denominator by [G6P] gives

$$f = \frac{[\text{F6P}]/[\text{G6P}]}{[\text{F6P}]/[\text{G6P}] + 1} = \frac{K}{K+1}$$

We find the equilibrium constant by rearranging eqn 4.10a into

$$K = e^{-\Delta_r G^\oplus / RT}$$

with

$$\frac{\Delta_r G^\oplus}{RT} = \frac{1.7 \times 10^3 \, \text{J mol}^{-1}}{(8.3145 \, \text{J K}^{-1} \, \text{mol}^{-1}) \times (298 \, \text{K})} = \frac{1.7 \times 10^3}{8.3145 \times 298}$$

Therefore,

$$K = e^{-1.7 \times 10^3 / 8.3145 \times 298} = 0.50$$

and

$$f = \frac{0.50}{0.50 + 1} = 0.33$$

That is, at equilibrium, 33 per cent of the solute is F6P and 67 per cent is G6P.

Self-test 4.5 Estimate the composition of a solution in which two isomers A and B are in equilibrium (A $\rightleftharpoons$ B) at 37°C and $\Delta_r G^\oplus = -2.2$ kJ mol^{-1}.

Answer: The fraction of B at equilibrium is $f_{eq} = 0.70$.

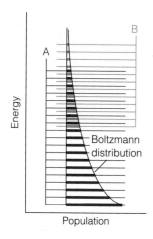

Fig. 4.6 The Boltzmann distribution of populations over the energy levels of two species A and B with similar densities of energy levels; the reaction A → B is endothermic in this example. The bulk of the population is associated with the species A, so that species is dominant at equilibrium.

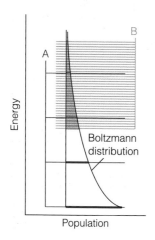

Fig. 4.7 Even though the reaction A → B is endothermic, the density of energy levels in B is so much greater than that in A, the population associated with B is greater than that associated with A; hence B is dominant at equilibrium.

(c) The molecular origin of chemical equilibrium

We can obtain a deeper insight into the origin and significance of the equilibrium constant by considering the Boltzmann distribution of molecules over the available states of a system composed of reactants and products. When atoms can exchange partners, as in a chemical reaction, the available states of the system include arrangements in which the atoms are present in the form of reactants and other atoms are present in the form of products. These arrangements have their characteristic sets of energy levels, but the Boltzmann distribution does not distinguish between their identities, only their energies. The atoms distribute themselves over both sets of energy levels in accord with the Boltzmann distribution (Fig. 4.6). At a given temperature, there is a specific distribution of populations and hence a specific composition of the reaction mixture.

It can be appreciated from Fig. 4.6 that if the reactants and products both have similar arrays of molecular energy levels, then the dominant species in a reaction mixture at equilibrium will be the species with the lower set of energy levels. However, the fact that the equilibrium constant is related to the *Gibbs* energy (through $\ln K = -\Delta_r G^{\ominus}/RT$) is a signal that entropy plays a role as well as energy. Its role can be appreciated by referring to Fig. 4.7. We see that although the B energy levels lie higher than the A energy levels, in this instance they are much more closely spaced. As a result, their total population may be considerable and B could even dominate in the reaction mixture at equilibrium. Closely spaced energy levels correlate with a high entropy, so in this case we see that entropy effects dominate adverse energy effects. That is, a positive reaction enthalpy results in a lowering of the equilibrium constant (that is, an endothermic reaction can be expected to have an equilibrium composition that favors the reactants). However, if there is positive reaction entropy, then the equilibrium composition may favor products, despite the endothermic character of the reaction.

Case study 4.1 *Binding of oxygen to myoglobin and hemoglobin*

Biochemical equilibria can be far more complex than those we have considered so far, but exactly the same principles apply. An example of a complex process is the binding of O_2 by hemoglobin in blood, which is described only approximately by reaction B. The protein myoglobin (Mb, Atlas P10) stores O_2 in muscle, and the protein hemoglobin (Hb, Atlas P7) transports O_2 in blood. These two proteins are related, for hemoglobin is a tetramer of four myoglobin-like molecules. In each protein, the O_2 molecule attaches to an iron ion in a heme group (Atlas R2) (Fig. 4.8).

First, consider the equilibrium between Mb and O_2:

$$Mb(aq) + O_2(g) \rightleftharpoons MbO_2(aq) \qquad K = \frac{[MbO_2]}{[Mb]p}$$

where p is the numerical value of the partial pressure of O_2 gas in bar. It follows that the *fractional saturation*, s, the fraction of Mb molecules that are oxygenated, is

$$s = \frac{[MbO_2]}{[Mb]_{total}} = \frac{[MbO_2]}{[Mb] + [MbO_2]} = \frac{Kp}{1 + Kp} \qquad \boxed{\text{Fractional saturation of myoglobin}} \qquad (4.12)$$

The dependence of s on p is shown in Fig. 4.9.

Now consider the equilibrium between Hb and O_2:

$$Hb(aq) + O_2(g) \rightleftharpoons HbO_2(aq) \qquad K_1 = \frac{[HbO_2]}{[Hb]p}$$

$$HbO_2(aq) + O_2(g) \rightleftharpoons Hb(O_2)_2(aq) \qquad K_2 = \frac{[Hb(O_2)_2]}{[HbO_2]p}$$

$$Hb(O_2)_2(aq) + O_2(g) \rightleftharpoons Hb(O_2)_3(aq) \qquad K_3 = \frac{[Hb(O_2)_3]}{[Hb(O_2)_2]p}$$

$$Hb(O_2)_3(aq) + O_2(g) \rightleftharpoons Hb(O_2)_4(aq) \qquad K_4 = \frac{[Hb(O_2)_4]}{[Hb(O_2)_3]p}$$

To develop an expression for s, we express $[Hb(O_2)_2]$ in terms of $[HbO_2]$ by using K_2, then express $[HbO_2]$ in terms of $[Hb]$ by using K_1, and likewise for all the other concentrations of $Hb(O_2)_3$ and $Hb(O_2)_4$. It follows that

$$[HbO_2] = K_1[Hb]p \qquad\qquad [Hb(O_2)_2] = K_1K_2[Hb]p^2$$
$$[Hb(O_2)_3] = K_1K_2K_3[Hb]p^3 \qquad [Hb(O_2)_4] = K_1K_2K_3K_4[Hb]p^4$$

The total concentration of bound O_2 is

$$[O_2]_{bound} = [HbO_2] + 2[Hb(O_2)_2] + 3[Hb(O_2)_3] + 4[Hb(O_2)_4]$$
$$= (1 + 2K_2p + 3K_2K_3p^2 + 4K_2K_3K_4p^3)K_1[Hb]p$$

where we have used the fact that n O_2 molecules are bound in $Hb(O_2)_n$, so the concentration of bound O_2 in $Hb(O_2)_2$ is $2[Hb(O_2)_2]$, and so on. The total concentration of hemoglobin is

$$[Hb]_{total} = (1 + K_1p + K_1K_2p^2 + K_1K_2K_3p^3 + K_1K_2K_3K_4p^4)[Hb]$$

Because each Hb molecule has four sites at which O_2 can attach, the fractional saturation is

$$s = \frac{[O_2]_{bound}}{4[Hb]_{total}}$$
$$= \frac{(1 + 2K_2p + 3K_2K_3p^2 + 4K_2K_3K_4p^3)K_1p}{4(1 + K_1p + K_1K_2p^2 + K_1K_2K_3p^3 + K_1K_2K_3K_4p^4)} \qquad \begin{array}{l}\text{Fractional} \\ \text{saturation of} \\ \text{hemoglobin}\end{array} \quad (4.13)$$

A reasonable fit of the experimental data can be obtained with $K_1 = 0.01$, $K_2 = 0.02$, $K_3 = 0.04$, and $K_4 = 0.08$ when p is expressed in torr.

The binding of O_2 to hemoglobin is an example of **cooperative binding**, in which the binding of a ligand (in this case O_2) to a biopolymer (in this case Hb) becomes more favorable thermodynamically (that is, the equilibrium constant increases) as the number of bound ligands increases up to the maximum number of binding sites. We see the effect of cooperativity in Fig. 4.9. Unlike the myoglobin saturation curve, the hemoglobin saturation curve is *sigmoidal* (S shaped): the fractional saturation is small at low ligand concentrations, increases sharply at intermediate ligand concentrations, and then levels off at high ligand concentrations. Cooperative binding of O_2 by hemoglobin is explained by an **allosteric effect**, in which an adjustment of the conformation of a molecule when one substrate binds affects the ease with which a subsequent substrate molecule binds. The details of the allosteric effect in hemoglobin will be explored in *Case study* 10.4.

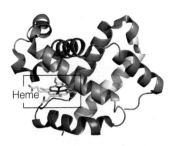

Fig. 4.8 One of the four polypeptide chains that make up the human hemoglobin molecule. The chains, which are similar to the oxygen storage protein myoglobin, consist of helical and sheet-like regions. The heme group is at the lower left.

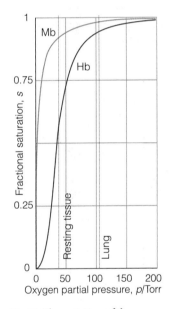

Fig. 4.9 The variation of the fractional saturation of myoglobin and hemoglobin molecules with the partial pressure of oxygen. The different shapes of the curves account for the different biological functions of the two proteins.

The differing shapes of the saturation curves for myoglobin and hemoglobin have important consequences for the way O_2 is made available in the body: in particular, the greater sharpness of the Hb saturation curve means that Hb can load O_2 more fully in the lungs and unload it more fully in different regions of the organism. In the lungs, where $p \approx 105$ Torr (14 kPa), $s \approx 0.98$, representing almost complete saturation. In resting muscular tissue, p is equivalent to about 38 Torr (5 kPa), corresponding to $s \approx 0.75$, implying that sufficient O_2 is still available should a sudden surge of activity take place. If the local partial pressure falls to 22 Torr (3 kPa), s falls to about 0.1. Note that the steepest part of the curve falls in the range of typical tissue oxygen partial pressure. Myoglobin, on the other hand, begins to release O_2 only when p has fallen below about 22 Torr, so it acts as a reserve to be drawn on only when the Hb oxygen has been used up.

4.4 The standard reaction Gibbs energy

The standard reaction Gibbs energy is central to the discussion of chemical equilibria and the calculation of equilibrium constants. It is also a useful indicator of the energy available from catabolism to drive anabolic processes, such as the synthesis of proteins.

We have seen that standard reaction Gibbs energy, $\Delta_r G^{\ominus}$, is defined as the difference in standard molar Gibbs energies of the products and the reactants weighted by the stoichiometric coefficients, ν, in the chemical equation

$$\Delta_r G^{\ominus} = \sum_{\text{Products}} \nu G_m^{\ominus} - \sum_{\text{Reactants}} \nu G_m^{\ominus} \qquad \boxed{\text{Definition of standard Gibbs energy of reaction}} \quad (4.14)$$

For example, the standard reaction Gibbs energy for reaction A is the difference between the molar Gibbs energies of fructose-6-phosphate and glucose-6-phosphate in solution at 1 mol dm^{-3} and 1 bar.

We cannot calculate $\Delta_r G^{\ominus}$ from the standard molar Gibbs energies themselves because these quantities are not known. One practical approach is to calculate the standard reaction enthalpy from standard enthalpies of formation (Section 1.11), the standard reaction entropy from Third-Law entropies (Section 2.5), and then to combine the two quantities by using

$$\Delta_r G^{\ominus} = \Delta_r H^{\ominus} - T\Delta_r S^{\ominus} \qquad \boxed{\text{Construction of } \Delta_r G^{\ominus}} \quad (4.15)$$

Example 4.4   *Calculating the standard reaction Gibbs energy of an enzyme-catalyzed reaction*

Evaluate the standard reaction Gibbs energy at 25°C for the reaction $CO_2(g)$ + $H_2O(l) \rightarrow H_2CO_3(aq)$ catalyzed by the enzyme carbonic anhydrase in red blood cells.

Strategy Obtain the relevant standard enthalpies of formation and standard entropies from the *Resource section*. Then calculate the standard reaction enthalpy and the standard reaction entropy from

$$\Delta_r H^{\ominus} = \sum_{\text{Products}} \nu\Delta_f H^{\ominus} - \sum_{\text{Reactants}} \nu\Delta_f H^{\ominus}$$

$$\Delta_r S^{\ominus} = \sum_{\text{Products}} \nu S_m^{\ominus} - \sum_{\text{Reactants}} \nu S_m^{\ominus}$$

and the standard reaction Gibbs energy from eqn 4.14.

Solution The standard reaction enthalpy is

$$\Delta_r H^{\ominus} = \Delta_f H^{\ominus}(H_2CO_3,aq) - \{\Delta_f H^{\ominus}(CO_2,g) + \Delta_f H^{\ominus}(H_2O,l)\}$$
$$= -699.65 \text{ kJ mol}^{-1} - \{(-393.51 \text{ kJ mol}^{-1}) + (-285.83 \text{ kJ mol}^{-1})\}$$
$$= -20.31 \text{ kJ mol}^{-1}$$

The standard reaction entropy was calculated in the *brief illustration* in Section 2.5:

$$\Delta_r S^{\ominus} = -96.3 \text{ J K}^{-1} \text{ mol}^{-1}$$

which, because 96.3 J is the same as 9.63×10^{-2} kJ, corresponds to -9.63×10^{-2} kJ K^{-1} mol^{-1}. Therefore, from eqn 4.15,

$$\Delta_r G^{\ominus} = (-20.31 \text{ kJ mol}^{-1}) - (298.15 \text{ K}) \times (-9.63 \times 10^{-2} \text{ kJ K}^{-1} \text{ mol}^{-1})$$
$$= +8.40 \text{ kJ mol}^{-1}$$

Self-test 4.6 Use the information in the *Resource section* to determine the standard reaction Gibbs energy for $3 O_2(g) \rightarrow 2 O_3(g)$ from standard enthalpies of formation and standard entropies.

Answer: +326.4 kJ mol^{-1}

(a) Standard Gibbs energies of formation

We saw in Section 1.11 how to use standard enthalpies of formation of substances to calculate standard reaction enthalpies. We can use the same technique for standard reaction Gibbs energies. To do so, we list the **standard Gibbs energy of formation**, $\Delta_f G^{\ominus}$, of a substance, which is the standard reaction Gibbs energy (per mole of the species) for its formation from the elements in their reference states. The concept of reference state was introduced in Section 1.11 (a reminder: it is the most stable form of the element under the prevailing conditions; do not confuse 'reference state' with 'standard state', but be aware that a reference state of an element will also be in its standard state if the pressure is 1 bar); the temperature is arbitrary, but we shall almost always take it to be 25°C (298 K). For example, the standard Gibbs energy of formation of liquid water, $\Delta_f G^{\ominus}(H_2O,l)$, is the standard reaction Gibbs energy for

$$H_2(g) + \tfrac{1}{2} O_2(g) \rightarrow H_2O(l)$$

and is −237 kJ mol^{-1} at 298 K. Some standard Gibbs energies of formation are listed in Table 4.2 and more can be found in the *Resource section*. It follows from the definition that the standard Gibbs energy of formation of an element in its reference state is zero because reactions such as C(s, graphite) → C(s, graphite) are null (that is, nothing happens). The standard Gibbs energy of formation of an element in a phase different from its reference state is nonzero:

$$C(s, \text{graphite}) \rightarrow C(s, \text{diamond}) \qquad \Delta_f G^{\ominus}(C, \text{diamond}) = +2.90 \text{ kJ mol}^{-1}$$

Many of the values in the tables have been compiled by combining the standard enthalpy of formation of the species with the standard entropies of the compound and the elements, as illustrated in *Example* 4.4, but there are other sources of data and we encounter some of them later.

Standard Gibbs energies of formation can be combined to obtain the standard Gibbs energy of almost any reaction. We use the now familiar expression

Table 4.2 Standard Gibbs energies of formation at 298.15 K*

Substance	$\Delta_f G^{\ominus}/(kJ\ mol^{-1})$
Gases	
Carbon dioxide, CO_2	−394.36
Methane, CH_4	−50.72
Nitrogen oxide, NO	+86.55
Water, H_2O	−228.57
Liquids	
Ethanol, CH_3CH_2OH	−174.78
Hydrogen peroxide, H_2O_2	−120.35
Water, H_2O	−237.13
Solids	
α-D-Glucose $C_6H_{12}O_6$	−917.2
Glycine, $CH_2(NH_2)COOH$	−532.9
Sucrose, $C_{12}H_{22}O_{11}$	−1543
Urea, $CO(NH_2)_2$	−197.33
Solutes in aqueous solution	
Carbon dioxide, CO_2	−385.98
Carbonic acid, H_2CO_3	−623.08
Phosphoric acid, H_3PO_4	−1018.7

*Additional values are given in the *Data* section.

$$\Delta_r G^{\ominus} = \sum_{\text{Products}} v\Delta_f G^{\ominus} - \sum_{\text{Reactants}} v\Delta_f G^{\ominus}$$

Calculation of standard Gibbs energy of reaction   (4.16)

If we need the biological standard reaction Gibbs energy, we convert $\Delta_r G^{\ominus}$ to $\Delta_r G^{\oplus}$ by using eqn 4.8.

A brief illustration

To determine the standard reaction Gibbs energy for the complete oxidation of solid sucrose, $C_{12}H_{22}O_{11}(s)$, by oxygen gas to carbon dioxide gas and liquid water,

$$C_{12}H_{22}O_{11}(s) + 12\ O_2(g) \rightarrow 12\ CO_2(g) + 11\ H_2O(l)$$

we carry out the following calculation:

$$\Delta_r G^{\ominus} = \{12\Delta_f G^{\ominus}(CO_2,g) + 11\Delta_f G^{\ominus}(H_2O,l)\} - \{\Delta_f G^{\ominus}(C_{12}H_{22}O_{11},s) + 12\Delta_f G^{\ominus}(O_2,g)\}$$
$$= \{12(-394\ kJ\ mol^{-1}) + 11(-237\ kJ\ mol^{-1})\} - \{-1543\ kJ\ mol^{-1} + 0\}$$
$$= -5.79 \times 10^3\ kJ\ mol^{-1}$$

Self-test 4.7   Calculate the standard reaction Gibbs energy of the oxidation of ammonia to nitric oxide according to the equation $4\ NH_3(g) + 5\ O_2(g) \rightarrow 4\ NO(g) + 6\ H_2O(g)$.

Answer: $-959.42\ kJ\ mol^{-1}$

(b) Stability and instability

Standard Gibbs energies of formation of compounds have their own significance as well as being useful in calculations of K. They are a measure of the 'thermodynamic altitude' of a compound above or below a 'sea level' of stability represented by the elements in their reference states (Fig. 4.10). If the standard Gibbs energy of formation is positive and the compound lies above 'sea level', then the compound has a spontaneous tendency to sink toward thermodynamic sea level and decompose into the elements. That is, $K < 1$ for their formation reaction. We say that a compound with $\Delta_f G^{\ominus} > 0$ is **thermodynamically unstable** with respect to its elements or that it is **endergonic**. Thus, the endergonic substance ozone, for which $\Delta_f G^{\ominus} = +163$ kJ mol^{-1}, has a spontaneous tendency to decompose into oxygen under standard conditions at 25°C. More precisely, the equilibrium constant for the reaction $\frac{3}{2} O_2(g) \rightleftharpoons O_3(g)$ is less than 1 (much less, in fact: $K = 2.7 \times 10^{-29}$). However, although ozone is thermodynamically unstable, it can survive if the reactions that convert it into oxygen are slow. That is the case in the upper atmosphere, and the O_3 molecules in the ozone layer survive for long periods. Benzene ($\Delta_f G^{\ominus} = +124$ kJ mol^{-1}) is also thermodynamically unstable with respect to its elements ($K = 1.8 \times 10^{-22}$). However, the fact that bottles of benzene are everyday laboratory commodities also reminds us of the point made at the start of the chapter, that *spontaneity is a thermodynamic tendency that might not be realized at a significant rate in practice*.

Another useful point that can be made about standard Gibbs energies of formation is that there is no point in searching for *direct* syntheses of a thermodynamically unstable compound from its elements (under standard conditions, at the temperature to which the data apply) because the reaction does not occur in the required direction: the *reverse* reaction, decomposition, is spontaneous. Endergonic compounds must be synthesized by alternative routes or under conditions for which their Gibbs energy of formation is negative and they lie beneath thermodynamic sea level.

Compounds with $\Delta_f G^{\ominus} < 0$ (corresponding to $K > 1$ for their formation reactions) are said to be **thermodynamically stable** with respect to their elements or **exergonic**. Exergonic compounds lie below the thermodynamic sea level of the elements (under standard conditions). An example is the exergonic compound ethane, with $\Delta_f G^{\ominus} = -33$ kJ mol^{-1}: the negative sign shows that the formation of ethane gas from its elements is spontaneous in the sense that $K > 1$ (in fact, $K = 7.1 \times 10^5$ at 25°C).

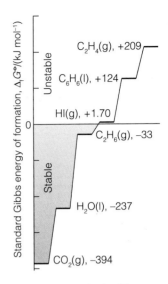

Fig. 4.10 The standard Gibbs energy of formation of a compound is like a measure of the compound's altitude above or below sea level: compounds that lie above sea level have a spontaneous tendency to decompose into the elements (and to revert to sea level). Compounds that lie below sea level are stable with respect to decomposition into the elements.

The response of equilibria to the conditions

In introductory chemistry, we meet the empirical rule of thumb known as **Le Chatelier's principle**:

> When a system at equilibrium is subjected to a disturbance, the composition of the system adjusts so as to tend to minimize the effect of the disturbance.

Le Chatelier's principle is only a rule of thumb, and to understand why reactions respond as they do and to calculate the new equilibrium composition, we need to use thermodynamics. We need to keep in mind that some changes in conditions affect the value of $\Delta_r G^{\ominus}$ and therefore of K (temperature is the only instance), whereas others change the consequences of K having a particular fixed value without changing the value of K (the pressure, for instance).

4.5 The presence of a catalyst

Enzymes are biological versions of catalysts and are so ubiquitous that we need to know how their action affects chemical equilibria.

We study the action of catalysts (a substance that accelerates a reaction without itself appearing in the overall chemical equation), especially enzymes, in Chapter 8 and at this stage do not need to know in detail how they work other than that they provide an alternative, faster route from reactants to products. Although the new route from reactants to products is faster, the initial reactants and the final products are the same. The quantity $\Delta_r G^\oplus$ is defined as the difference of the standard molar Gibbs energies of the reactants and products, so it is independent of the path linking the two. It follows that an alternative pathway between reactants and products leaves $\Delta_r G^\oplus$ and therefore K unchanged. That is, *the presence of a catalyst does not change the equilibrium constant of a reaction.*

4.6 The effect of temperature

In organisms, biochemical reactions occur over a very narrow range of temperatures, and changes by only a few degrees can have serious consequences, including death. Therefore, it is important to know how changes in temperature, such as those brought about by infections, affect biological processes.

According to Le Chatelier's principle, we can expect a reaction to respond to a lowering of temperature by releasing heat and to respond to an increase of temperature by absorbing heat. That is:

When the temperature is raised, the equilibrium composition of an exothermic reaction will tend to shift toward reactants; the equilibrium composition of an endothermic reaction will tend to shift toward products.

In each case, the response tends to minimize the effect of raising the temperature. But *why* do reactions at equilibrium respond in this way? Le Chatelier's principle is only a rule of thumb and gives no clue to the reason for this behavior. As we shall now see, the origin of the effect is the dependence of $\Delta_r G^\oplus$, and therefore of K, on the temperature.

First, we consider the effect of temperature on $\Delta_r G^\oplus$. We use the relation $\Delta_r G^\oplus = \Delta_r H^\oplus - T\Delta_r S^\oplus$ and make the assumption that neither the reaction enthalpy nor the reaction entropy varies much with temperature (over small ranges, at least). It follows that

$$\text{change in } \Delta_r G^\oplus = -(\text{change in } T) \times \Delta_r S^\oplus \tag{4.17}$$

This expression is easy to apply when there is a consumption or formation of gas because, as we have seen (Section 2.5), gas formation dominates the sign of the reaction entropy.

Now consider the effect of temperature on K itself. At first, this problem looks troublesome because both T and $\Delta_r G^\oplus$ appear in the expression for K. However, as we show in the following *Justification*, the effect of temperature can be expressed very simply as the **van 't Hoff equation**.[2]

$$\ln K_2 = \ln K_1 + \frac{\Delta_r H^\oplus}{R}\left(\frac{1}{T_1} - \frac{1}{T_2}\right) \qquad \boxed{\text{van 't Hoff equation}} \tag{4.18}$$

---

[2] There are several 'van 't Hoff equations'. To distinguish them, this one is sometimes called the *van 't Hoff isochore*.

where K_1 is the equilibrium constant at the temperature T_1 and K_2 is its value when the temperature is T_2. All we need to know to calculate the temperature dependence of an equilibrium constant, therefore, is the standard reaction enthalpy.

Justification 4.1   *The van 't Hoff equation*

As before, we use the approximation that the standard reaction enthalpy and entropy are independent of temperature over the range of interest, so the entire temperature dependence of $\Delta_r G^{\ominus}$ stems from the T in $\Delta_r G^{\ominus} = \Delta_r H^{\ominus} - T\Delta_r S^{\ominus}$. At a temperature T_1,

$$\ln K_1 = -\frac{\Delta_r G^{\ominus}}{RT_1} = -\frac{\Delta_r H^{\ominus}}{RT_1} + \frac{\Delta_r S^{\ominus}}{R}$$

At another temperature T_2, when $\Delta_r G^{\ominus\prime} = \Delta_r H^{\ominus} - T_2\Delta_r S^{\ominus}$ and the equilibrium constant is K_2, a similar expression holds:

$$\ln K_2 = -\frac{\Delta_r H^{\ominus}}{RT_2} + \frac{\Delta_r S^{\ominus}}{R}$$

The difference between the two is eqn 4.18.

Let's explore the information in the van 't Hoff equation. Consider the case when $T_2 > T_1$. Then the term in parentheses in eqn 4.18 is positive. If $\Delta_r H^{\ominus} > 0$, corresponding to an endothermic reaction, the entire term on the right is positive. In this case, therefore, $\ln K_2 > \ln K_1$. That being so, we conclude that $K_2 > K_1$ for an endothermic reaction. In general, *the equilibrium constant of an endothermic reaction increases with temperature*. The opposite is true when $\Delta_r H^{\ominus} < 0$, so we can conclude that *the equilibrium constant of an exothermic reaction decreases with an increase in temperature*.

Statistical principles also give us insight into the temperature dependence of the equilibrium constant. The typical arrangement of energy levels for an endothermic reaction is shown in Fig. 4.11a. When the temperature is increased, the Boltzmann distribution adjusts and the populations change as shown. The change corresponds to an increased population of the higher energy states at the expense of the population of the lower-energy states. We see that the states that arise from the B molecules become more populated at the expense of the A molecules. Therefore, the total population of B states increases, and B becomes more abundant in the equilibrium mixture. Conversely, if the reaction is exothermic (Fig. 4.11b), then an increase in temperature increases the population of the A states (which start at higher energy) at the expense of the B states, so the reactants become more abundant.

Coupled reactions in bioenergetics

We remarked in the introduction to this chapter that thermodynamics enables us to determine whether one reaction can drive another forward. We now have enough information to take this step. A simple mechanical analogy is a pair of weights joined by a string (Fig. 4.12): the lighter of the pair of weights will be pulled up as the heavier weight falls down. Although the lighter weight has a

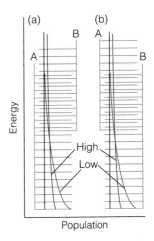

Fig. 4.11 The effect of temperature on a chemical equilibrium can be interpreted in terms of the change in the Boltzmann distribution with temperature and the effect of that change in the population of the species. (a) In an endothermic reaction, the population of B increases at the expense of A as the temperature is raised. (b) In an exothermic reaction, the opposite happens.

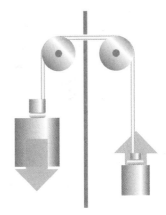

Fig. 4.12 If two weights are coupled as shown here, then the heavier weight will move the lighter weight in its nonspontaneous direction: overall, the process is still spontaneous. The weights are the analogs of two chemical reactions: a reaction with a large negative $\Delta_r G$ can force another reaction with a smaller $\Delta_r G$ to run in its nonspontaneous direction.

natural tendency to move downward, its coupling to the heavier weight results in it being raised. The thermodynamic analog is an **endergonic reaction**, a reaction with a positive Gibbs energy, $\Delta_r G$ (the analog of the lighter weight moving up), being forced to occur by coupling it to an **exergonic reaction**, a reaction with a negative Gibbs energy, $\Delta_r G'$ (the analog of the heavier weight falling down). The overall reaction is spontaneous because the sum $\Delta_r G + \Delta_r G'$ is negative. The whole of life's activities depend on couplings of this kind, for the oxidation reactions of food act as the heavy weights that drive other reactions forward and result in the formation of proteins from amino acids, the actions of muscles for propulsion, and even the activities of the brain for reflection, learning, and imagination.

Case study 4.2) *ATP and the biosynthesis of proteins*

The function of adenosine triphosphate, ATP^{4-} (Atlas N3) or (more succinctly) ATP, is to store the energy made available when food is oxidized and then to supply it on demand to a wide variety of processes, including muscular contraction, reproduction, and vision. We saw in *Case study* 2.2 that the essence of ATP's action is its ability to lose its terminal phosphate group by hydrolysis and to form adenosine diphosphate, ADP^{3-} (Atlas N2):

$$ATP^{4-}(aq) + H_2O(l) \rightarrow ADP^{3-}(aq) + HPO_4^{2-}(aq) + H_3O^+(aq)$$

This reaction is exergonic under the conditions prevailing in cells and can drive an endergonic reaction forward if suitable enzymes are available to couple the reactions. One reason why ATP is so potent is that its concentration in cells is high, so its chemical potential is also high.

The biological standard values for the hydrolysis of ATP at 37°C are

$$\Delta_r G^{\oplus} = -31 \text{ kJ mol}^{-1} \qquad \Delta_r H^{\oplus} = -20 \text{ kJ mol}^{-1} \qquad \Delta_r S^{\oplus} = +34 \text{ J K}^{-1} \text{ mol}^{-1}$$

The hydrolysis is therefore exergonic ($\Delta_r G < 0$) under these conditions, and 31 kJ mol^{-1} is available for driving other reactions. On account of its exergonic character, the ADP–phosphate bond has been called a 'high-energy phosphate bond'. The name is intended to signify a high tendency to undergo reaction and should not be confused with 'strong' bond in its normal chemical sense (that of a high bond enthalpy). In fact, even in the biological sense it is not of very 'high energy'. The action of ATP depends on the bond being intermediate in strength. Thus ATP acts as a phosphate donor to a number of acceptors (such as glucose) but is recharged with a new phosphate group by more powerful phosphate donors in the phosphorylation steps in the respiration cycle.

In the cell, each ATP molecule can be used to drive an endergonic reaction for which $\Delta_r G^{\oplus}$ does not exceed 31 kJ mol^{-1}. For example, the biosynthesis of sucrose from glucose and fructose can be driven by enzyme-catalyzed processes in plants because the reaction is endergonic to the extent $\Delta_r G^{\oplus} = +23$ kJ mol^{-1}. The biosynthesis of proteins is strongly endergonic, not only on account of the enthalpy change but also on account of the large decrease in entropy that occurs when many amino acid residues are assembled into a precisely determined sequence. For instance, the formation of a peptide link is endergonic, with $\Delta_r G^{\oplus} = +17$ kJ mol^{-1}, but the biosynthesis occurs indirectly and is equivalent to the consumption of three ATP molecules for each link.

In a moderately small protein such as myoglobin, with about 150 peptide links, the construction alone requires 450 ATP molecules and therefore about 12 mol of glucose molecules for 1 mol of protein molecules.

Self-test 4.8   Fats yield almost twice as much energy per gram as carbohydrates. What mass of fat would need to be metabolized to synthesize 1.0 mol of myoglobin molecules?

Answer: 1.1 kg

Adenosine triphosphate is not the only phosphate species capable of driving other less exergonic reactions. For instance, creatine phosphate (3) can release its phosphate group in a hydrolysis reaction, and $\Delta_r G^\oplus = -43$ kJ mol^{-1}. These different exergonicities give rise to the concept of **transfer potential**, which is the negative of the value of $\Delta_r G^\oplus$ for the hydrolysis reaction. Thus, the transfer potential of creatine phosphate is +43 kJ mol^{-1}. Just as one exergonic reaction can drive a less exergonic reaction, so the hydrolysis of a species with a high transfer potential can drive the phosphorylation of a species with a lower transfer potential (Table 4.3).

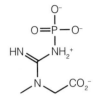

3 Creatine phosphate

Table 4.3 Transfer potentials at 298.15 K

Substance	Transfer potential, $-\Delta_r G^\circ$/(kJ mol^{-1})
AMP	14
ATP, ADP	31
1,3-Bis(phospho)glycerate	49
Creatine phosphate	43
Glucose-6-phosphate	14
Glycerol-1-phosphate	10
Phosphoenolpyruvate	62
Pyrophosphate, HP$_2$O$_7^{3-}$	33

Case study 4.3   *The oxidation of glucose*

The breakdown of glucose in the cell begins with *glycolysis*, a partial oxidation of glucose by nicotinamide adenine dinucleotide (NAD$^+$, Atlas N4) to pyruvate ion, CH$_3$COCO$_2^-$ (4). Metabolism continues in the form of the citric acid cycle, in which pyruvate ions are oxidized to CO$_2$, and ends with oxidative phosphorylation, in which O$_2$ is reduced to H$_2$O. Glycolysis is the main source of energy during *anaerobic metabolism*, a form of metabolism in which inhaled O$_2$ does not play a role. The citric acid cycle and oxidative phosphorylation are the main mechanisms for the extraction of energy from carbohydrates during *aerobic metabolism*, a form of metabolism in which inhaled O$_2$ does play a role.

Glycolysis occurs in the *cytosol*, the aqueous material encapsulated by the cell membrane, and consists of 10 enzyme-catalyzed reactions (Fig. 4.13).

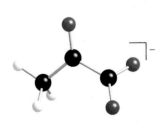

4 Pyruvate ion

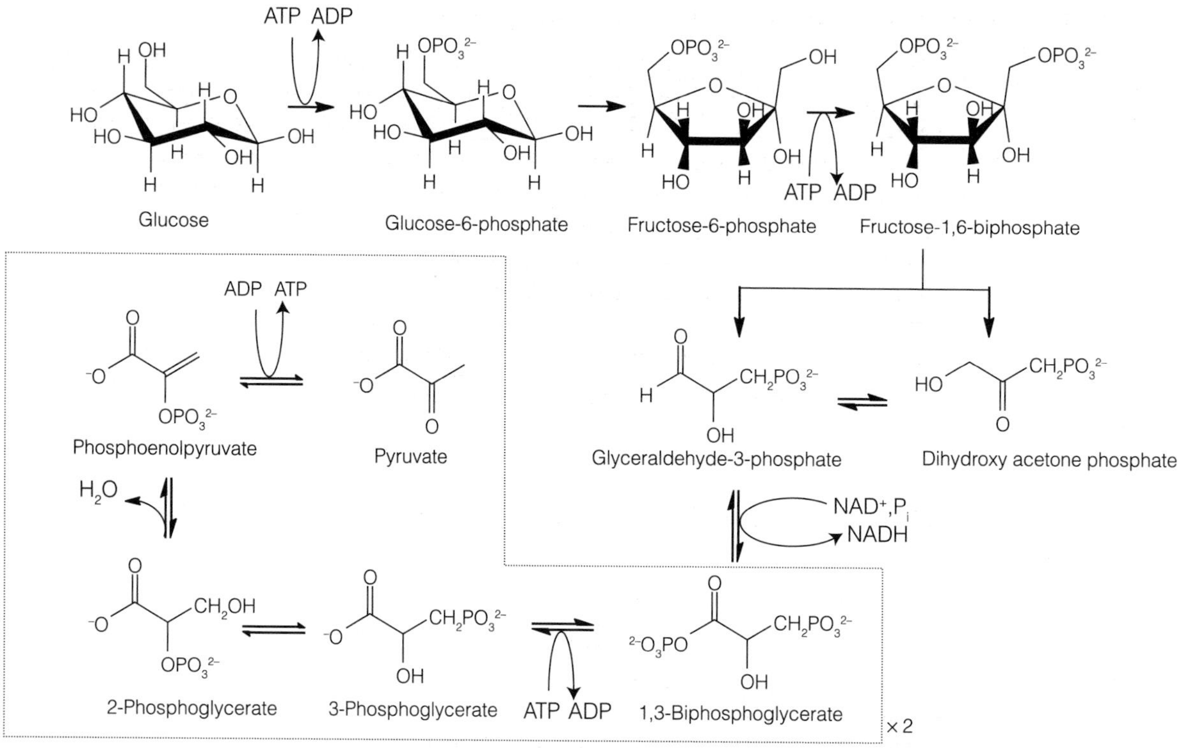

Fig. 4.13 The reactions of glycolysis, in which glucose is partially oxidized by nicotinamide adenine dinucleotide (NAD⁺, Atlas N4) to pyruvate ion.

The process needs to be initiated by consumption of two molecules of ATP per molecule of glucose. The first ATP molecule is used to drive the phosphorylation of glucose to glucose-6-phosphate (G6P):

$$\text{glucose(aq)} + \text{ATP(aq)} \rightarrow \text{G6P(aq)} + \text{ADP(aq)} + \text{H}^+(\text{aq}) \quad \Delta_r G^\oplus = -17\ \text{kJ mol}^{-1}$$

(Note that ATP, ADP, and G6P denote charged species, so charges are in fact balanced in this and similar equations, but charge balance is not displayed explicitly.) As we saw in Section 4.1, the next step is the isomerization of G6P to fructose-6-phosphate (F6P). The second ATP molecule consumed during glycolysis drives the phosphorylation of F6P to fructose-1,6-diphosphate (FDP):

$$\text{F6P(aq)} + \text{ATP(aq)} \rightarrow \text{FDP(aq)} + \text{ADP(aq)} + \text{H}^+(\text{aq}) \quad \Delta_r G^\oplus = -14\ \text{kJ mol}^{-1}$$

In the next step, FDP is broken into two three-carbon units, dihydroxyacetone phosphate (1,3-dihydroxypropanone phosphate, $\text{CH}_2\text{OHCOCH}_2\text{OPO}_3^{2-}$, **5**) and glyceraldehyde-3-phosphate (**6**), which exist in mutual equilibrium. Only the glyceraldehyde-3-phosphate is oxidized by NAD⁺ to pyruvate ion, with formation of two ATP molecules. As glycolysis proceeds, all the dihydroxyacetone phosphate is converted to glyceraldehyde-3-phosphate, so the result is the consumption of two NAD⁺ molecules and the formation of four ATP molecules per molecule of glucose.

A brief comment

From now on, we shall represent biochemical reactions with chemical equations written with a shorthand method, in which some substances are given 'nicknames' and charges are not always given explicitly. For example, $\text{H}_2\text{PO}_4^{2-}$ is written as P_i, ATP⁴⁻ as ATP, and so on. We need to show hydrogen ions explicitly (because they account for differences between thermodynamic and biological standard states and for the role of pH); in such cases charges will not seem to be balanced.

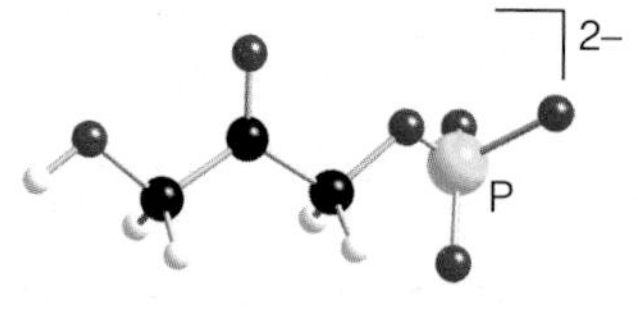

5 Dihydroxypropanone phosphate 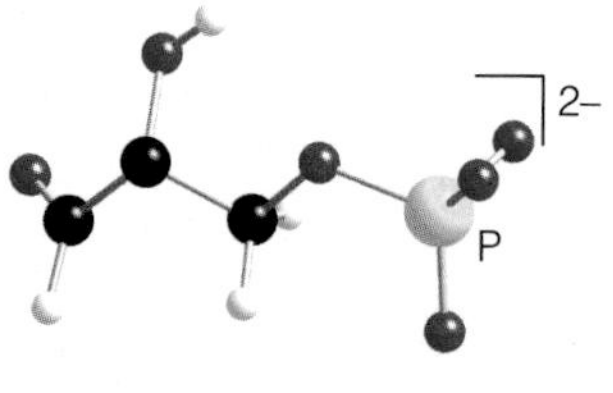   **6** Glyceraldehyde-3-phosphate

The oxidation of glucose by NAD^+ to pyruvate ions has $\Delta_r G^\oplus = -147$ kJ mol^{-1} at blood temperature. In glycolysis, the oxidation of one glucose molecule is coupled to the *net* conversion of two ADP molecules to two ATP molecules ('net' because two ATP molecules are consumed and four are formed), so the net reaction of glycolysis is

$$\text{glucose(aq)} + 2\ NAD^+(aq) + 2\ ADP(aq) + 2\ P_i(aq) + 2\ H_2O(l) \rightarrow$$
$$2\ CH_3COCO_2^-(aq) + 2\ NADH(aq) + 2\ ATP(aq) + 2\ H_3O^+(aq)$$

The biological standard reaction Gibbs energy is $(-147) - 2(-31)$ kJ mol^{-1} = -85 kJ mol^{-1}. The reaction is exergonic and therefore spontaneous under biological standard conditions: the oxidation of glucose is used to 'recharge' the ATP.

In cells that are deprived of O_2, pyruvate ion is reduced to lactate ion, $CH_3CH(OH)CO_2^-$ (7) by NADH.[3] Very strenuous exercise, such as bicycle racing, can decrease sharply the concentration of O_2 in muscle cells, and the condition known as muscle fatigue results from increased concentrations of lactate ion.

The standard Gibbs energy of combustion of glucose is -2880 kJ mol^{-1}, so terminating its oxidation at pyruvate is a poor use of resources, akin to the partial combustion of hydrocarbon fuels in a badly tuned engine. In the presence of O_2, pyruvate is oxidized further during the citric acid cycle and oxidative phosphorylation, which occur in the mitochondria of cells.

The further oxidation of carbon derived from glucose begins with a reaction between pyruvate ion, NAD^+, and coenzyme A (CoA, Atlas N6) to give acetyl CoA, NADH, and CO_2. Acetyl CoA is then oxidized by NAD^+ and flavin adenine dinucleotide (FAD, Atlas N7) in the citric acid cycle (Fig. 4.14), which

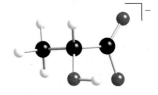

7 Lactate ion

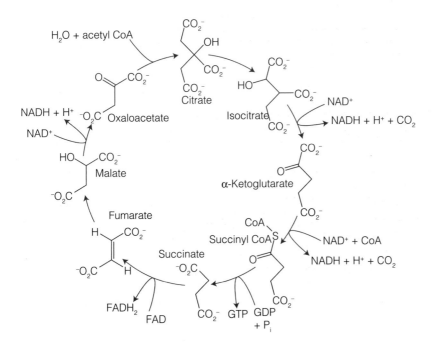

Fig. 4.14 The reactions of the citric acid cycle, in which acetyl CoA is oxidized by NAD^+ and FAD, resulting in the synthesis of GTP (shown) or ATP, depending on the type of cell. The GTP molecules are eventually converted to ATP.

[3] In yeast, the terminal products are ethanol and CO_2.

requires eight enzymes and results in the synthesis of guanosine triphosphate (GTP, Atlas N8) from guanosine diphosphate, GDP, or of ATP from ADP:

$$\text{Acetyl CoA(aq)} + 3\,\text{NAD}^+(\text{aq}) + \text{FAD(aq)} + \text{GDP(aq)} + \text{P}_i(\text{aq}) + 2\,\text{H}_2\text{O(l)} \rightarrow$$
$$2\,\text{CO}_2(\text{g}) + 3\,\text{NADH(aq)} + 2\,\text{H}_3\text{O}^+(\text{aq}) + \text{FADH}_2(\text{aq}) + \text{GTP(aq)} + \text{CoA(aq)}$$
$$\Delta_r G^\ominus = -57\ \text{kJ mol}^{-1}$$

In cells that produce GTP, the enzyme nucleoside diphosphate kinase catalyzes the transfer of a phosphate group to ADP to form ATP:

$$\text{GTP(aq)} + \text{ADP(aq)} \rightarrow \text{GDP(aq)} + \text{ATP(aq)}$$

For this reaction, $\Delta_r G^\ominus = 0$ because the phosphate group transfer potentials for GTP and ATP are essentially identical. Overall, we write the oxidation of glucose as a result of glycolysis and the citric acid cycle as

$$\text{glucose(aq)} + 10\,\text{NAD}^+(\text{aq}) + 2\,\text{FAD(aq)} + 4\,\text{ADP(aq)} + 4\,\text{P}_i(\text{aq}) + 2\,\text{H}_2\text{O(l)} \rightarrow$$
$$6\,\text{CO}_2(\text{g}) + 10\,\text{NADH(aq)} + 6\,\text{H}_3\text{O}^+(\text{aq}) + 2\,\text{FADH}_2(\text{aq}) + 4\,\text{ATP(aq)}$$

The NADH and FADH_2 go on to reduce O_2 during oxidative phosphorylation (Section 5.10b), which also produces ATP. The citric acid cycle and oxidative phosphorylation generate as many as 38 ATP molecules for each glucose molecule consumed. Each mole of ATP molecules extracts 31 kJ from the 2880 kJ supplied by 1 mol $C_6H_{12}O_6$ (180 g of glucose), so 1178 kJ is stored for later use. Therefore, aerobic oxidation of glucose is much more efficient than glycolysis.

Proton transfer equilibria

An enormously important biological aspect of chemical equilibrium is that involving the transfer of protons (hydrogen ions, H^+) between species in aqueous environments, such as living cells. Even small drifts in the equilibrium concentration of hydrogen ions can result in disease, cell damage, and death. In this section we see how the general principles outlined earlier in the chapter are applied to proton transfer equilibria.

4.7 Brønsted–Lowry theory

Cells have elaborate procedures for using proton transfer equilibria, and this function cannot be understood without knowing which species provide protons and which accept them and how to express the concentration of hydrogen ions in solution.

According to the **Brønsted–Lowry theory** of acids and bases, an **acid** is a proton donor and a **base** is a proton acceptor. The proton, which in this context means a hydrogen ion, H^+, is highly mobile and acids and bases in water are always in equilibrium with their deprotonated and protonated counterparts and hydronium ions (H_3O^+, 8). Thus, an acid HA, such as HCN, immediately establishes the equilibrium

$$\text{HA(aq)} + \text{H}_2\text{O(l)} \rightleftharpoons \text{H}_3\text{O}^+(\text{aq}) + \text{A}^-(\text{aq}) \qquad K = \frac{a_{\text{H}_3\text{O}^+}a_{\text{A}^-}}{a_{\text{HA}}a_{\text{H}_2\text{O}}} \qquad (4.19\text{a})$$

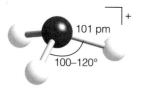

8 Hydronium ion, H_3O^+

101 pm

100–120°

A base B, such as NH_3, immediately establishes the equilibrium

$$B(aq) + H_2O(l) \rightleftharpoons HB^+(aq) + OH^-(aq) \qquad K = \frac{a_{HB^+} a_{OH^-}}{a_B a_{H_2O}} \qquad (4.19b)$$

In these equilibria, A^- is the **conjugate base** of the acid HA, and BH^+ is the **conjugate acid** of the base B. Even in the absence of added acids and bases, proton transfer occurs between water molecules, and the **autoprotolysis equilibrium**[4]

$$2\,H_2O(l) \rightleftharpoons H_3O^+(aq) + OH^-(aq) \qquad K = \frac{a_{H_3O^+} a_{OH^-}}{a_{H_2O}^2} \qquad \boxed{\text{Autoprotolysis equilibrium}} \qquad (4.20)$$

is always present.

As will be familiar from introductory chemistry, the hydronium ion concentration is commonly expressed in terms of the pH, which is defined formally as

$$pH = -\log a_{H_3O^+} \qquad \boxed{\text{Definition of pH}} \qquad (4.21)$$

where the logarithm is to base 10. In elementary work, the hydronium ion activity is replaced by the numerical value of its molar concentration, $[H_3O^+]$, which is equivalent to setting the activity coefficient γ equal to 1.

A brief illustration

If the molar concentration of H_3O^+ is 2.0 mmol dm^{-3} (where 1 mmol = 10^{-3} mol), then

$$pH \approx -\log(2.0 \times 10^{-3}) = 2.70$$

If the molar concentration were 10 times less, at 0.20 mmol dm^{-3}, then the pH would be 3.70.

Notice that *the higher the pH, the lower the concentration of hydronium ions in the solution* and that a change in pH by 1 unit corresponds to a 10-fold change in their molar concentration. However, it should never be forgotten that the replacement of activities by molar concentration is invariably hazardous. Because ions interact over long distances, the replacement is unreliable for all but the most dilute solutions.

Self-test 4.9   Death is likely if the pH of human blood plasma changes by more than ±0.4 from its normal value of 7.4. What is the approximate range of molar concentrations of hydrogen ions for which life can be sustained?

Answer: 16 nmol dm^{-3} to 100 nmol dm^{-3} (1 nmol = 10^{-9} mol)

4.8 Protonation and deprotonation

The protonation and deprotonation of molecules are key steps in many biochemical reactions, and we need to be able to describe procedures for treating protonation and deprotonation processes quantitatively.

All the solutions we consider are so dilute that we can regard the water present as being a nearly pure liquid and therefore as having unit activity (see Table 3.3).

---

[4] Autoprotolysis is also called *autoionization*.

This feature leads to convenient expressions for quantities that measure the strengths of acids and bases and the extent of protonation of bases and deprotonation of acids.

(a) The strengths of acids and bases

When we set $a_{H_2O} = 1$ for all the solutions we consider, the resulting equilibrium constant is called the **acidity constant**, K_a, of the acid HA:[5]

$$HA(aq) + H_2O(l) \rightleftharpoons H_3O^+(aq) + A^-(aq)$$

$$K_a = \frac{a_{H_3O^+}a_{A^-}}{a_{HA}} \qquad \boxed{\text{Definition of acidity constant}} \qquad (4.22a)$$

In elementary applications, the activities are replaced by the numerical values of the molar concentrations, and we write

$$K_a = \frac{[H_3O^+][A^-]}{[HA]} \qquad (4.22b)$$

Data are widely reported in terms of the negative common (base 10) logarithm of this quantity:

$$pK_a = -\log K_a \qquad \boxed{\text{Definition of } pK_a} \qquad (4.23)$$

It follows from eqn 4.10a ($\Delta_r G^\ominus = -RT \ln K$) that pK_a is proportional to $\Delta_r G^\ominus$ for the proton transfer reaction. More explicitly, $pK_a = \Delta_r G^\ominus/(RT \ln 10)$, with $\ln 10 = 2.303\ldots$. Therefore, manipulations of pK_a and related quantities are actually manipulations of standard reaction Gibbs energies in disguise.

$\boxed{\text{Self-test 4.10}}$ Show that $pK_a = \Delta_r G^\ominus/(RT \ln 10)$. *Hint:* $\ln x = \ln 10 \times \log x$.

The value of the acidity constant indicates the extent to which proton transfer occurs at equilibrium in aqueous solution. The smaller the value of K_a (for instance 10^{-8} compared with 10^{-6}) and therefore the larger the value of pK_a (for instance, 8 compared with 6), the lower is the concentration of deprotonated molecules. Most acids have $K_a < 1$ (and usually much less than 1), with $pK_a > 0$, indicating only a small extent of deprotonation in water. These acids are classified as **weak acids**. A few acids, most notably, in aqueous solution, HCl, HBr, HI, HNO_3, H_2SO_4, and $HClO_4$, are classified as **strong acids** and are commonly regarded as being completely deprotonated in aqueous solution.[6]

The corresponding expression for a base is called the **basicity constant**, K_b:

$$B(aq) + H_2O(l) \rightleftharpoons HB^+(aq) + OH^-(aq)$$

$$K_b = \frac{a_{HB^+}a_{OH^-}}{a_B} \qquad \boxed{\text{Definition of basicity constant}} \qquad (4.24a)$$

and the corresponding value of $pK_b = -\log K_b$. As for acids, in elementary applications the activities are replaced by the numerical values of the molar concentrations and we use

---

[5] Acidity constants are also called *acid ionization constants* and, less appropriately, *dissociation constants*.

[6] Sulfuric acid, H_2SO_4, is strong with respect only to its first deprotonation; HSO_4^- is weak.

$$K_b = \frac{[HB^+][OH^-]}{[B]} \qquad (4.24b)$$

A **strong base** is fully protonated in solution in the sense that $K_b > 1$. One example is the oxide ion, O^{2-}, which cannot survive in water but is immediately and fully converted into its conjugate acid OH^-. A **weak base** is not fully protonated in water, in the sense that $K_b < 1$ (and usually much less than 1). Ammonia, NH_3, and its organic derivatives the amines are all weak bases in water, and only a small proportion of their molecules exist as the conjugate acid (NH_4^+ or RNH_3^+).

The **autoprotolysis constant** for water, K_w, is obtained in a similar way by setting the activity of water in eqn 4.19 to its 'pure' value:

$$K_w = a_{H_3O^+} a_{OH^-} \approx [H_3O^+][OH^-] \qquad \boxed{\begin{array}{l}\text{Definition of the} \\ \text{autoprotolysis} \\ \text{constant of water}\end{array}} \qquad (4.25)$$

At 25°C, $K_w = 1.0 \times 10^{-14}$ and $pK_w = -\log K_w = 14.00$.

As may be confirmed by multiplying the two constants together, the basicity constant of a base B and the acidity constant of its conjugate acid, HB^+,

$$B(aq) + H_2O(l) \rightleftharpoons HB^+(aq) + OH^-(aq) \qquad K_b = \frac{a_{HB^+} a_{OH^-}}{a_B}$$

$$HB^+(aq) + H_2O(l) \rightleftharpoons H_3O^+(aq) + B(aq) \qquad K_a = \frac{a_{H_3O^+} a_B}{a_{HB^+}}$$

are related by

$$K_a K_b = \frac{a_{H_3O^+} a_B}{a_{HB^+}} \times \frac{a_{HB^+} a_{OH^-}}{a_B} = a_{H_3O^+} a_{OH^-} = K_w \qquad \boxed{\begin{array}{l}\text{Relation between} \\ K_a \text{ and } K_b\end{array}} \qquad (4.26a)$$

The implication of this relation is that K_a increases as K_b decreases to maintain a product equal to the constant K_w. That is, *as the strength of a base decreases, the strength of its conjugate acid increases* and vice versa. On taking the negative common logarithm of both sides of eqn 4.26a, we obtain

$$pK_a + pK_b = pK_w \qquad \boxed{\begin{array}{l}\text{Relation between} \\ pK_a \text{ and } pK_b\end{array}} \qquad (4.26b)$$

The great advantage of this relation is that the pK_b values of bases may be expressed as the pK_a of their conjugate acids, so the strengths of all weak acids and bases may be listed in a single table (Table 4.4).

(**A brief illustration**)

If the acidity constant of the conjugate acid ($CH_3NH_3^+$) of the base methylamine (CH_3NH_2) is reported as $pK_a = 10.56$, we can infer that the basicity constant of methylamine itself is

$$pK_b = pK_w - pK_a = 14.00 - 10.56 = 3.44$$

Another useful relation is obtained by taking the negative common logarithm of both sides of the definition of K_w in eqn 4.24, which gives

$$pH + pOH = pK_w \qquad \boxed{\begin{array}{l}\text{Relation between} \\ pH \text{ and } pOH\end{array}} \qquad (4.27)$$

Table 4.4 Acidity and basicity constants* at 298.15 K

Acid/base	K_b	pK_b	K_a	pK_a
Strongest weak acids				
Trichloroacetic acid, CCl_3COOH	3.3×10^{-14}	13.48	3.0×10^{-1}	0.52
Benzenesulfonic acid, $C_6H_5SO_3H$	5.0×10^{-14}	13.30	2×10^{-1}	0.70
Iodic acid, HIO_3	5.9×10^{-14}	13.23	1.7×10^{-1}	0.77
Sulfurous acid, H_2SO_3	6.3×10^{-13}	12.19	1.6×10^{-2}	1.81
Chlorous acid, $HClO_2$	1.0×10^{-12}	12.00	1.0×10^{-2}	2.00
Phosphoric acid, H_3PO_4	1.3×10^{-12}	11.88	7.6×10^{-3}	2.12
Chloroacetic acid, $CH_2ClCOOH$	7.1×10^{-12}	11.15	1.4×10^{-3}	2.85
Lactic acid, $CH_3CH(OH)COOH$	1.2×10^{-11}	10.92	8.4×10^{-4}	3.08
Nitrous acid, HNO_2	2.3×10^{-11}	10.63	4.3×10^{-4}	3.37
Hydrofluoric acid, HF	2.9×10^{-11}	10.55	3.5×10^{-4}	3.45
Formic acid, HCOOH	5.6×10^{-11}	10.25	1.8×10^{-4}	3.75
Benzoic acid, C_6H_5COOH	1.5×10^{-10}	9.81	6.5×10^{-4}	4.19
Acetic acid, CH_3COOH	5.6×10^{-10}	9.25	1.8×10^{-4}	4.75
Carbonic acid, H_2CO_3	2.3×10^{-8}	7.63	4.3×10^{-7}	6.37
Hypochlorous acid, HClO	3.3×10^{-7}	6.47	3.0×10^{-8}	7.53
Hypobromous acid, HBrO	5.0×10^{-6}	5.31	2.0×10^{-9}	8.69
Boric acid, $B(OH)_3H^†$	1.4×10^{-5}	4.86	7.2×10^{-10}	9.14
Hydrocyanic acid, HCN	2.0×10^{-5}	4.69	4.9×10^{-10}	9.31
Phenol, C_6H_5OH	7.7×10^{-5}	4.11	1.3×10^{-10}	9.89
Hypoiodous acid, HIO	4.3×10^{-4}	3.36	2.3×10^{-11}	10.64
Weakest weak bases				
Urea, $CO(NH_2)_2$	1.3×10^{-14}	13.90	7.7×10^{-1}	0.10
Aniline, $C_6H_5NH_2$	4.3×10^{-10}	9.37	2.3×10^{-5}	4.63
Pyridine, C_5H_5N	1.8×10^{-9}	8.75	5.6×10^{-6}	5.35
Hydroxylamine, NH_2OH	1.1×10^{-8}	7.97	9.1×10^{-7}	6.03
Nicotine, $C_{10}H_{11}N_2$	1.0×10^{-6}	5.98	1.0×10^{-8}	8.02
Morphine, $C_{17}H_{19}O_3N$	1.6×10^{-6}	5.79	6.3×10^{-9}	8.21
Hydrazine, NH_2NH_2	1.7×10^{-6}	5.77	5.9×10^{-9}	8.23
Ammonia, NH_3	1.8×10^{-5}	4.75	5.6×10^{-10}	9.25
Trimethylamine, $(CH_3)_3N$	6.5×10^{-5}	4.19	1.5×10^{-10}	9.81
Methylamine, CH_3NH_2	3.6×10^{-4}	3.44	2.8×10^{-11}	10.56
Dimethylamine, $(CH_3)_2NH$	5.4×10^{-4}	3.27	1.9×10^{-11}	10.73
Ethylamine, $C_2H_5NH_2$	6.5×10^{-4}	3.19	1.5×10^{-11}	10.81
Triethylamine, $(C_2H_5)_3N$	1.0×10^{-3}	2.99	1.0×10^{-11}	11.01
Strongest weak bases				

*Values for polyprotic acids—those capable of donating more than one proton—refer to the first deprotonation.
†The proton transfer equilibrium is $B(OH)_3(aq) + 2 H_2O(l) \rightleftharpoons H_3O^+(aq) + B(OH)_4^-(aq)$.

where pOH $= -\log a_{OH^-}$. This enormously important relation means that the activities (in elementary work, the molar concentrations) of hydronium and hydroxide ions of a given solution are related by a seesaw relation: as one goes up, the other goes down to preserve the value of pK_w.

Self-test 4.11 The molar concentration of OH⁻ ions in a certain solution is 0.010 mmol dm⁻³. What is the pH of the solution?

Answer: 9.00

(b) The pH of a solution of a weak acid

The most reliable way to estimate the pH of a solution of a weak acid is to consider the contributions from deprotonation of the acid and autoprotolysis of water to the total concentration of hydronium ion in solution (see *Further information* 4.1). Autoprotolysis may be ignored if the weak acid is the main contributor of hydronium ions, a condition that is satisfied if the acid is not very weak and is present at not too low a concentration. Then we can estimate the pH of a solution of a weak acid and calculate either of these fractions by using the following strategy:

Organize the necessary work into a table with columns headed by the species present in the mixture (ignoring H_2O) and, in successive rows write:

1. The initial molar concentrations of the species, ignoring any contributions to the concentration of H_3O^+ or OH⁻ from the autoprotolysis of water.
2. The changes in these quantities that must take place for the system to reach equilibrium.
3. The resulting equilibrium values.

In most cases, we do not know the change that must occur for the system to reach equilibrium, so the change in the concentration of H_3O^+ is written as x and the reaction stoichiometry is used to write the corresponding changes in the other species. When the values at equilibrium (the last row of the table) are substituted into the expression for the acidity constant, we obtain an equation for x in terms of K_a. This equation can be solved for x. In general, solution of the equation for x results in several mathematically possible values of x. We select the chemically acceptable solution by considering the signs of the predicted concentrations: they must be positive.

Example 4.5 *Estimating the pH of a solution of a weak acid*

Acetic acid lends a sour taste to vinegar and is produced by aerobic oxidation of ethanol by bacteria in fermented beverages, such as wine and cider: $CH_3CH_2OH(aq) + O_2(g) \rightarrow CH_3COOH(aq) + H_2O(l)$. Estimate the pH of (a) 0.15 M $CH_3COOH(aq)$ and (b) 1.5×10^{-4} M $CH_3COOH(aq)$.

Strategy Proceed as outlined above.

Solution We draw up the following equilibrium table based on the proton transfer equilibrium $CH_3COOH(aq) + H_2O(l) \rightleftharpoons H_3O^+(aq) + CH_3CO_2^-(aq)$.

Species	CH_3COOH	H_3O^+	$CH_3CO_2^-$
Initial concentration/(mol dm⁻³)	0.15	0	0
Change to reach equilibrium/(mol dm⁻³)	$-x$	$+x$	$+x$
Equilibrium concentration/(mol dm⁻³)	$0.15 - x$	x	x

(a) The value of x is found by inserting the equilibrium concentrations into the expression for the acidity constant:

$$K_a = \frac{[H_3O^+][CH_3CO_2^-]}{[CH_3COOH]} = \frac{x \times x}{0.15 - x}$$

We could arrange the expression into a quadratic equation. However, it is more instructive to make use of the smallness of x to replace $0.15 - x$ by 0.15 (this approximation is valid if $x \ll 0.15$). Then the simplified equation rearranges first to $0.15 \times K_a = x^2$ and then to

$$x = (0.15 \times K_a)^{1/2} = (0.15 \times 1.8 \times 10^{-5})^{1/2} = 1.6 \times 10^{-3}$$

where we have used $K_a = 1.8 \times 10^{-5}$ (Table 4.4). Therefore, pH = 2.80. Calculations of this kind are rarely accurate to more than one decimal place in the pH (and even that may be too optimistic) because the effects of ion–ion interactions have been ignored, so this answer would be reported as pH = 2.8.

(b) Had we proceeded in the same way with the new concentration, we would calculate $x = 5.2 \times 10^{-5}$, which although less than the initial concentration is not much less. Therefore, we must solve the quadratic equation

$$x^2 + K_a x - (1.5 \times 10^{-4})K_a = 0$$

by setting $a = 1$, $b = 1.8 \times 10^{-5}$, and $c = -1.5 \times 1.8 \times 10^{-9}$:

$$x = \frac{-1.8 \times 10^{-5} \pm ((-1.8 \times 10^{-5})^2 - 4(-1.5 \times 1.8 \times 10^{-9}))^{1/2}}{2}$$

$$= 4.4 \times 10^{-5} \text{ or } -6.2 \times 10^{-5}$$

Because x is equal to the concentration of H_3O^+, it cannot be negative, so we select $x = 4.4 \times 10^{-5}$. It follows that pH = 4.4. (The illegal calculation would have given 4.3.)

Self-test 4.12 Estimate the pH of 0.010 M $CH_3CH(OH)COOH(aq)$ (lactic acid) from the data in Table 4.4. Before carrying out the numerical calculation, decide whether you expect the pH to be higher or lower than that calculated for the same concentration of acetic acid.

Answer: 2.6

A note on good practice
When an approximation has been made, verify at the end of the calculation that the approximation is consistent with the result obtained. In this case, we assumed that $x \ll 0.15$ and have found that $x = 1.6 \times 10^{-3}$, which is consistent.

Another note on good practice
Acetic acid (ethanoic acid) is written CH_3COOH because the two O atoms are inequivalent; its conjugate base, the acetate ion (ethanoate ion), is written $CH_3CO_2^-$ because the two O atoms are now equivalent (by resonance).

Mathematical toolbox 4.1 *Quadratic equations*

A quadratic equation is an equation of the form

$$ax^2 + bx + c = 0$$

where a, b, and c are constants. The two roots (solutions) of the equation are given by the expression

$$x = \frac{-b \pm (b^2 - 4ac)^{1/2}}{2a}$$

The graph of the function $ax^2 + bx + c$ is a parabola, which cuts the x-axis at the two roots of the corresponding quadratic equation:

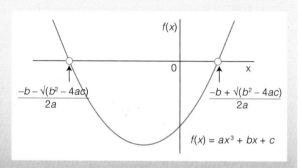

(c) The pH of a solution of a weak base

The calculation of the pH of a solution of a base involves an additional step. The first step is to calculate the concentration of OH^- ions in the solution from the value of K_b by using the equilibrium-table technique and to express it as the pOH of the solution. The additional step is to convert that pOH into a pH by using the water autoprotolysis equilibrium, eqn 4.27, in the form $pH = pK_w - pOH$, with $pK_w = 14.00$ at 25°C.

Example 4.6   *Estimating the pH of a solution of a weak base*

Calculate the pH of an 0.20 M aqueous solution of methylamine, CH_3NH_2, for which $pK_b = 3.44$.

Strategy Proceed as outlined above.

Solution We draw up the following equilibrium table based on the proton transfer equilibrium $CH_3NH_2(aq) + H_2O(l) \rightleftharpoons CH_3NH_3^+(aq) + OH^-(aq)$.

Species	CH_3NH_2	$CH_3NH_3^+$	OH^-
Initial concentration/(mol dm^{-3})	0.20	0	0
Change to reach equilibrium/(mol dm^{-3})	$-x$	$+x$	$+x$
Equilibrium concentration/(mol dm^{-3})	$0.20 - x$	x	x

Then, because $K_b = 10^{-3.44} = 3.6 \times 10^{-4}$, and anticipating that $x \ll 0.2$,

$$\frac{[CH_3NH_3^+][OH^-]}{[CH_3NH_2]} = \frac{x \times x}{0.20 - x} \approx \frac{x^2}{0.2} = 3.6 \times 10^{-4}$$

It follows that $[OH^-] = x = 0.0085$, so $pOH = -\log(0.0085) = 2.07$. Therefore, $pH = 14.00 - 2.07 = 11.93$ (more realistically, 11.9).

Self-test 4.13 The base quinoline has $pK_b = 9.12$. Estimate the pH of an 0.010 M aqueous solution of quinoline.

Answer: 8.4

(d) The extent of protonation and deprotonation

The extent of deprotonation of a weak acid in solution depends on the acidity constant and the initial concentration of the acid, its concentration as prepared. The **fraction deprotonated**, the fraction of acid molecules HA that have donated a proton, is

$$f_{deprotonated} = \frac{[A^-]_{at\ equilibrium}}{[HA]_{as\ prepared}} \qquad \boxed{\text{Fraction of HA deprotonated}} \quad (4.28a)$$

The extent to which a weak base B is protonated is reported in terms of the **fraction protonated**:

$$f_{protonated} = \frac{[HB^+]_{at\ equilibrium}}{[B]_{as\ prepared}} \qquad \boxed{\text{Fraction of B protonated}} \quad (4.28b)$$

The calculation of either f proceeds in the same way as for the calculation of the pH of a solution, the only difference being how the calculated value of x in the equilibrium table is used.

The fraction of acetic acid molecules deprotonated in the solution referred to in Example 4.5a is

$$f_{\text{deprotonated}} = \frac{[CH_3CO_2^-]_{\text{at equilibrium}}}{[CH_3COOH]_{\text{as prepared}}} = \frac{x}{0.15} = \frac{1.6 \times 10^{-3}}{0.15} = 0.011$$

That is, 1.1 per cent of CH_3COOH molecules have lost their acidic proton.

Self-test 4.14 Estimate the fraction of quinoline molecules protonated in a 0.010 M aqueous solution of quinoline.

Answer: 1/3600

(e) The pH of solutions of salts

The ions present when a salt is added to water may themselves be either acids or bases and consequently affect the pH of the solution. For example, when ammonium chloride is added to water, it provides both an acid (NH_4^+) and a base (Cl^-). The solution consists of a weak acid (NH_4^+) and a very weak base (Cl^-). The net effect is that the solution is acidic. Similarly, a solution of sodium acetate consists of an ion that is neither acidic nor basic (the Na^+ ion) and a base ($CH_3CO_2^-$). The net effect is that the solution is basic, and its pH is greater than 7.

To estimate the pH of the solution, we proceed in exactly the same way as for the addition of a 'conventional' acid or base, for in the Brønsted–Lowry theory there is no distinction between 'conventional' acids such as acetic acid and the conjugate acids of bases (such as NH_4^+).

A brief illustration

To calculate the pH of 0.010 M $NH_4Cl(aq)$ at 25°C, we proceed exactly as in *Example* 4.5, taking the initial concentration of the acid (NH_4^+) to be 0.010 mol dm^{-3}. The K_a to use is the acidity constant of the acid NH_4^+, which is listed in Table 4.4. Alternatively, we use K_b for the conjugate base (NH_3) of the acid and convert that quantity to K_a by using eqn 4.26a ($K_a K_b = K_w$). We find pH = 5.63, which is on the acid side of neutral.

Exactly the same procedure is used to find the pH of a solution of a salt of a weak acid, such as sodium acetate. The equilibrium table is set up by treating the anion $CH_3CO_2^-$ as a base (which it is) and using for K_b the value obtained from the value of K_a for its conjugate acid (CH_3COOH).

Self-test 4.15 Estimate the pH of 0.0025 M $NH(CH_3)_3Cl(aq)$ at 25°C.

Answer: 6.2

4.9 Polyprotic acids

Many biological macromolecules, such as the nucleic acids, contain multiple proton donor sites, and we need to see how to handle this complication quantitatively.

Table 4.5 Successive acidity constants of polyprotic acids at 298.15 K

Acid	K_{a1}	pK_{a1}	K_{a2}	pK_{a2}	K_{a3}	pK_{a3}
Carbonic acid, H_2CO_3	4.3×10^{-7}	6.37	5.6×10^{-11}	10.25		
Hydrosulfuric acid, H_2S	1.3×10^{-7}	6.89	7.1×10^{-15}	14.15		
Oxalic acid, $(COOH)_2$	5.9×10^{-2}	1.23	6.5×10^{-5}	4.19		
Phosphoric acid, H_3PO_4	7.6×10^{-3}	2.12	6.2×10^{-8}	7.21	2.1×10^{-13}	12.67
Phosphorous acid, H_2PO_3	1.0×10^{-2}	2.00	2.6×10^{-7}	6.59		
Sulfuric acid, H_2SO_4	Strong		1.2×10^{-2}	1.92		
Sulfurous acid, H_2SO_3	1.5×10^{-2}	1.81	1.2×10^{-7}	6.91		
Tartaric acid, $C_2H_4O_2(COOH)_2$	6.0×10^{-4}	3.22	1.5×10^{-5}	4.82		

A **polyprotic acid** is a molecular compound that can donate more than one proton. Two examples are sulfuric acid, H_2SO_4, which can donate up to two protons, and phosphoric acid, H_3PO_4, which can donate up to three. A polyprotic acid is best considered to be a molecular species that can give rise to a series of Brønsted acids as it donates its succession of protons. Thus, sulfuric acid is the parent of two Brønsted acids, H_2SO_4 itself and HSO_4^-, and phosphoric acid is the parent of three Brønsted acids, namely H_3PO_4, $H_2PO_4^-$, and HPO_4^{2-}.

For a species H_2A with two acidic protons (such as H_2SO_4), the successive equilibria we need to consider are

$$H_2A(aq) + H_2O(l) \rightleftharpoons H_3O^+(aq) + HA^-(aq)$$

$$K_{a1} = \frac{a_{H_3O^+} \cdot a_{HA^-}}{a_{H_2A}} \qquad \boxed{\text{First deprotonation}} \quad (4.29a)$$

$$HA^-(aq) + H_2O(l) \rightleftharpoons H_3O^+(aq) + A^{2-}(aq)$$

$$K_{a2} = \frac{a_{H_3O^+} \cdot a_{A^{2-}}}{a_{HA^-}} \qquad \boxed{\text{Second deprotonation}} \quad (4.29b)$$

In the first of these equilibria, HA^- is the conjugate base of H_2A. In the second, HA^- acts as the acid and A^{2-} is its conjugate base. Values are given in Table 4.5. In all cases, K_{a2} is smaller than K_{a1}, typically by three orders of magnitude for small molecular species, because the second proton is more difficult to remove, partly on account of the negative charge on HA^-. Enzymes are polyprotic acids, for they possess many protons that can be donated to a substrate molecule or to the surrounding aqueous medium of the cell. For them, successive acidity constants vary much less because the molecules are so large that the loss of a proton from one part of the molecule has little effect on the ease with which another some distance away may be lost.

| **Example 4.7** | *Calculating the concentration of carbonate ion in carbonic acid* |

Groundwater contains dissolved carbon dioxide, carbonic acid, hydrogencarbonate ions, and a very low concentration of carbonate ions. Estimate the molar concentration of CO_3^{2-} ions in a solution in which water and $CO_2(g)$ are in equilibrium. We must be very cautious in the interpretation of calculations involving carbonic acid because equilibrium between dissolved CO_2 and

H_2CO_3 is achieved only very slowly. In organisms, attainment of equilibrium is facilitated by the enzyme carbonic anhydrase.

Strategy We start with the equilibrium that produces the ion of interest (such as A^{2-}) and write its activity in terms of the acidity constant for its formation (K_{a2}). That expression will contain the activity of the conjugate acid (HA^-), which we can express in terms of the activity of *its* conjugate acid (H_2A) by using the appropriate acidity constant (K_{a1}). This equilibrium dominates all the rest provided the molecule is small and there are marked differences between its acidity constants, so it may be possible to make an approximation at this stage.

Solution The CO_3^{2-} ion, the conjugate base of the acid HCO_3^- is produced in the equilibrium

$$HCO_3^-(aq) + H_2O(l) \rightleftharpoons H_3O^+(aq) + CO_3^{2-}(aq) \qquad K_{a2} = \frac{a_{H_3O^+} \cdot a_{CO_3^{2-}}}{a_{HCO_3^-}}$$

Hence,

$$a_{CO_3^{2-}} = \frac{a_{HCO_3^-} K_{a2}}{a_{H_3O^+}}$$

The HCO_3^- ions are produced in the equilibrium

$$H_2CO_3(aq) + H_2O(l) \rightleftharpoons H_3O^+(aq) + HCO_3^-(aq)$$

One H_3O^+ ion is produced for each HCO_3^- ion produced. These two concentrations are not exactly the same because a little HCO_3^- is lost in the second deprotonation and the amount of H_3O^+ has been increased by it. Also, HCO_3^- is a weak base and abstracts a proton from water to generate H_2CO_3 (see Section 4.10). However, those secondary changes can safely be ignored in an approximate calculation. Because the molar concentrations of HCO_3^- and H_3O^+ are approximately the same, we can suppose that their activities are also approximately the same and set $a_{HCO_3^-} \approx a_{H_3O^+}$. When this equality is substituted into the expression for $a_{CO_3^{2-}}$, we obtain

$$[CO_3^{2-}] \approx K_{a2}$$

Because we know from Table 4.5 that $pK_{a2} = 10.25$, it follows that $[CO_3^{2-}] = 5.6 \times 10^{-11}$ and therefore that the molar concentration of CO_3^{2-} ions is 56 pmol dm^{-3}.

Self-test 4.16 Calculate the molar concentration of S^{2-} ions in $H_2S(aq)$.

Answer: 7.1 fmol dm^{-3}

Case study 4.4   *The fractional composition of a solution of lysine*

The amino acid lysine (HLys, Atlas A12) can accept two protons on its nitrogen atoms and donate one from its carboxyl group. The neutral acid is HLys and the fully protonated form is H_3Lys^{2+}. Let's see how the composition of an aqueous solution that contains 0.010 mol dm^{-3} of lysine varies with pH. The pK_a values of amino acids are given in Table 4.6.

We expect the fully protonated species (H_3Lys^{2+}) at low pH (high H_3O^+ concentration), the partially protonated species (H_2Lys^+ and HLys) at intermediate pH, and the fully deprotonated species (Lys^-) at high pH. The three acidity constants (using the notation in Table 4.6 and replacing activities by molar concentrations) are

$$H_3Lys^{2+}(aq) + H_2O(l) \rightleftharpoons H_3O^+(aq) + H_2Lys^+(aq) \qquad K_{a1} = \frac{[H_3O^+][H_2Lys^+]}{[H_3Lys^{2+}]}$$

$$H_2Lys^+(aq) + H_2O(l) \rightleftharpoons H_3O^+(aq) + HLys(aq) \qquad K_{a2} = \frac{[H_3O^+][HLys]}{[H_2Lys^+]}$$

$$HLys(aq) + H_2O(l) \rightleftharpoons H_3O^+(aq) + Lys^-(aq) \qquad K_{a3} = \frac{[H_3O^+][Lys^-]}{[HLys]}$$

We also know that the total concentration of lysine in all its forms is

$$L = [H_3Lys^{2+}] + [H_2Lys^+] + [HLys] + [Lys^-] \qquad \boxed{\text{Total concentration of lysine}} \qquad (4.30)$$

We now have four equations for four unknown concentrations. To solve the equations, we proceed systematically, using K_{a3} to express $[Lys^-]$ in terms of [HLys], then K_{a2} to express [HLys] in terms of $[H_2Lys^+]$, and so on:

$$[Lys^-] = \frac{K_{a3}[HLys]}{[H_3O^+]} = \frac{K_{a2}K_{a3}[H_2Lys^+]}{[H_3O^+]^2} = \frac{K_{a1}K_{a2}K_{a3}[H_3Lys^{2+}]}{[H_3O^+]^3}$$

$$[HLys] = \frac{K_{a2}[H_2Lys^+]}{[H_3O^+]} = \frac{K_{a1}K_{a2}[H_3Lys^{2+}]}{[H_3O^+]^2}$$

$$[H_2Lys^+] = \frac{K_{a1}[H_3Lys^{2+}]}{[H_3O^+]}$$

Then the expression for the total concentration L can be written in terms of $[H_3Lys^{2+}]$:

$$L = \frac{H[H_3Lys^{2+}]}{[H_3O^+]^3}$$

where

$$H = [H_3O^+]^3 + K_{a1}[H_3O^+]^2 + K_{a1}K_{a2}[H_3O^+] + K_{a1}K_{a2}K_{a3} \qquad (4.31)$$

It then it follows that the fractions of each species present in the solution are

$$f(H_3Lys^{2+}) = \frac{[H_3Lys^{2+}]}{L} = \frac{[H_3O^+]^3}{H}$$

$$f(H_2Lys^+) = \frac{[H_2Lys^+]}{L} = \frac{K_{a1}[H_3O^+]^2}{H}$$

$$f(HLys) = \frac{[HLys]}{L} = \frac{K_{a1}K_{a2}[H_3O^+]}{H} \qquad \boxed{\text{Fractional composition}} \qquad (4.32)$$

$$f(Lys^-) = \frac{[Lys^-]}{L} = \frac{K_{a1}K_{a2}K_{a3}}{H}$$

These fractions are plotted against pH (by using $[H_3O^+] = 10^{-pH}$) in Fig. 4.15. Note that:

Table 4.6 Acidity constants of amino acids at 298.15 K*

Acid	pK_{a1}	pK_{a2}	pK_{a3}
Ala	2.33	9.71	
Arg	2.03	9.00	12.10
Asn	2.16	8.73	
Asp	1.95	3.71	9.66
Cys	1.91	8.14	10.28
Gln	2.18	9.00	
Glu	2.16	4.15	9.58
Gly	2.34	9.58	
His	1.70	6.04	9.09
Ile	2.26	9.60	
Leu	2.32	9.58	
Lys	2.15	9.16	10.67
Met	2.16	9.08	
Phe	2.18	9.09	
Pro	1.95	10.47	
Ser	2.13	9.05	
Thr	2.20	9.96	
Trp	2.38	9.34	
Tyr	2.24	9.04	10.10
Val	2.27	9.52	

*For the identities of the acids, see the *Atlas* of structures. The acidity constants refer, respectively, to the most highly protonated form, the next most, and so on. So the values for Lys, for instance, refer to H_3Lys^{2+}, H_2Lys^+, and HLys (the electrically neutral molecule).

Fig. 4.15 The fractional composition of the protonated and deprotonated forms of lysine (Lys) in aqueous solution as a function of pH. Note that conjugate pairs are present at equal concentrations when the pH is equal to the pK_a of the acid member of the pair.

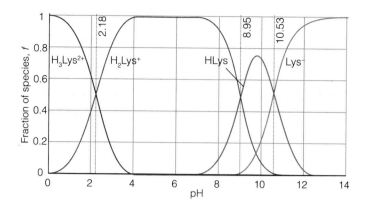

Fig. 4.16 The fractional composition of the protonated and deprotonated forms of histidine (His) in aqueous solution as a function of pH.

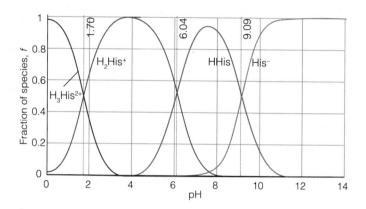

A note on good practice
Take note of the symmetry of the expressions in eqn 4.31. By doing so, it is easy to write down the corresponding expressions for species with different numbers of acidic protons without repeating the lengthy calculation.

- H_3Lys^{2+} is dominant for $pH < pK_{a1}$
- H_3Lys^{2+} and H_2Lys^+ have the same concentration at $pH = pK_{a1}$
- H_2Lys^+ is dominant for $pH > pK_{a1}$ until HLys becomes dominant, and so on.

In a neutral solution at pH = 7, the dominant species is H_2Lys^+, for pH = 7 lies between pK_{a1} and pK_{a2}: below pK_{a1}, H_3Lys^{2+} is dominant and above pK_{a2}, HLys is dominant.

Self-test 4.17 Construct the diagram for the fraction of protonated species in an aqueous solution of histidine (Atlas A9).

Answer: Fig. 4.16

We can summarize the behavior discussed in *Case study* 4.4 and illustrated in Figs 4.15 and 4.16 as follows. Consider each conjugate acid–base pair, with acidity constant K_a; then:

The acid form is dominant for $pH < pK_a$.

The conjugate pair have equal concentrations at $pH = pK_a$.

The base form is dominant for $pH > pK_a$.

In each case, the other possible forms of a polyprotic system can be ignored, provided the pK_a values are not too close together.

4.10 Amphiprotic systems

Many molecules of biochemical significance (including the amino acids) can act as both proton donors and proton acceptors, and we need to be able to treat this dual function quantitatively.

An **amphiprotic** species is a molecule or ion that can both accept and donate protons. For instance, HCO_3^- can act as an acid (to form CO_3^{2-}) and as a base (to form H_2CO_3). Among the most important amphiprotic compounds are the amino acids, which can act as proton donors by virtue of their carboxyl groups and as bases by virtue of their amino groups.

(a) The fractional composition of amino acid solutions

In solution, amino acids are present largely in their **zwitterionic** ('double ion') form, in which the amino group is protonated and the carboxyl group is deprotonated: the acidic proton of the carboxyl group has been donated to the basic amino group (but not necessarily of the same molecule). The zwitterionic form of glycine, NH_2CH_2COOH, for instance, is $^+H_3NCH_2CO_2^-$. We can suppose that in an aqueous solution of glycine, the species present are NH_2CH_2COOH (and in general BAH, where B represents the basic amino group and AH the carboxylic acid group), $NH_2CH_2CO_2^-$ (BA^-), $^+NH_3CH_2COOH$ (^+HBAH), and the zwitterion $^+NH_3CH_2CO_2^-$ ($^+HBA^-$). The proton transfer equilibria in water are

$$BAH(aq) + H_2O(l) \rightleftharpoons H_3O^+(aq) + BA^-(aq) \qquad K_1$$

$$^+HBAH(aq) + H_2O(l) \rightleftharpoons H_3O^+(aq) + BAH(aq) \qquad K_2$$

$$^+HBA^-(aq) + H_2O(l) \rightleftharpoons H_3O^+(aq) + BA^-(aq) \qquad K_3$$

By following the same procedure as in *Case study* 4.4, we find the following expressions for the composition of the solution:

$$f(BA^-) = \frac{K_1K_2K_3}{H}$$

$$f(BAH) = \frac{K_2K_3[H_3O^+]}{H}$$

$$f(^+HBA^-) = \frac{K_1K_2[H_3O^+]}{H}$$

$$f(^+HBAH) = \frac{K_3[H_3O^+]^2}{H} \qquad \boxed{\text{Fractional composition}} \quad (4.33)$$

with $H = [H_3O^+]^2K_3 + [H_3O^+](K_1 + K_3)K_2 + K_1K_2K_3$. The variation of composition with pH is shown in Fig. 4.17. Because we can expect the zwitterion to be a much weaker acid than the neutral molecule (because the negative charge on the carboxylate group hinders the escape of the proton from the conjugate acid of the amino group), we can anticipate that $K_3 \ll K_1$ and therefore that $f(BAH) \ll f(^+HBA^-)$ at all values of pH.

(b) The pH of solutions of amphiprotic anions

The further question we need to tackle is the pH of a solution of a salt with an amphiprotic anion, such as a solution of $NaHCO_3$. Is the solution acidic on account of the acid character of HCO_3^- or is it basic on account of the anion's basic character? As we show in *Further information* 4.2, under the circumstances specified there ($K_{a2} \ll K_{a1}$, $S \gg K_w/K_{a2}$, and $S \gg K_{a1}$), where S is the numerical

Fig. 4.17 The fractional composition of the protonated and deprotonated forms of an amino acid NH₂CHRCOOH (with arbitrarily chosen values of pK), in which the group R does not participate in proton transfer reactions.

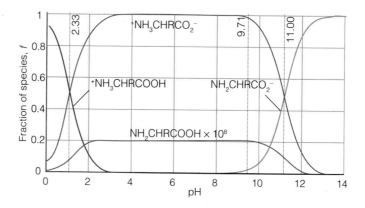

value of the molar concentration of the salt providing the anion), the pH of such a solution is given by

$$pH = \tfrac{1}{2}(pK_{a1} + pK_{a2})$$

> The pH of the solution of an amphiprotic anion

(4.34)

A brief illustration

Using values from Table 4.5, we can immediately conclude that the pH of the solution of sodium hydrogencarbonate *of any concentration* (subject to the conditions just quoted) is

$$pH = \tfrac{1}{2}(6.37 + 10.25) = 8.31$$

The solution is basic. We can treat a solution of potassium hydrogenphosphate in the same way, taking into account only the second and third acidity constants of H_3PO_4 because protonation as far as H_3PO_4 is negligible (see Table 4.5):

$$pH = \tfrac{1}{2}(7.21 + 12.67) = 9.94$$

4.11 Buffer solutions

> Cells cease to function and may be damaged irreparably if the pH changes significantly, so we need to understand how the pH is stabilized by a buffer.

Suppose that we make an aqueous solution by dissolving known amounts of a weak acid (which provides the species HA) and its conjugate base (which provides the species A⁻). To calculate the pH of this solution, we make use of the expression for K_a of the weak acid, eqn 4.22, with [HA] = [acid] and [A⁻] = [base], and write

$$K_a = \frac{a_{H_3O^+}a_{base}}{a_{acid}} \approx \frac{a_{H_3O^+}[base]}{[acid]}$$

which rearranges first to

$$a_{H_3O^+} \approx \frac{K_a[acid]}{[base]}$$

and then, by taking negative common logarithms, to the **Henderson–Hasselbalch equation**:

$$pH = pK_a - \log \frac{[acid]}{[base]}$$

Henderson–Hasselbalch equation: (4.35)

When the concentrations of the conjugate acid and base are equal, the second term on the right of eqn 4.35 is $\log 1 = 0$, so under these conditions $pH = pK_a$. Although the equation has been derived without making any assumptions about [acid] and [base], it is common to suppose that, because the acid is weak, [acid] and [base] are unchanged from the values used to make up the solution; that is, we disregard the small amount of deprotonation of the added acid and the small amount of protonation of the added base.

A brief illustration

To calculate the pH of a solution formed from equal amounts of $CH_3COOH(aq)$ and $NaCH_3CO_2(aq)$, we note that the latter dissociates (in the sense that the ions separate) fully in water, yielding $Na^+(aq)$ and $CH_3CO_2^-(aq)$, the conjugate base of $CH_3COOH(aq)$. Because $[CH_3COOH] = [CH_3CO_2^-]$ (that is, [acid] = [base]), for this solution provided we disregard protonation and deprotonation, $pH \approx pK_a$. Because the pK_a of $CH_3COOH(aq)$ is 4.75 (Table 4.4), it follows that $pH = 4.8$ (more realistically, $pH = 5$).

Self-test 4.18 Calculate the pH of an aqueous solution that contains equal amounts of NH_3 and NH_4Cl.

Answer: 9.25; more realistically: 9

It is observed that solutions containing known amounts of an acid and that acid's conjugate base show **buffer action**, the ability of a solution to oppose changes in pH when small amounts of strong acids and bases are added. An **acid buffer** solution, one that stabilizes the solution at a pH below 7, is typically prepared by making a solution of a weak acid (such as acetic acid) and a salt that supplies its conjugate base (such as sodium acetate). A **base buffer**, one that stabilizes a solution at a pH above 7, is prepared by making a solution of a weak base (such as ammonia) and a salt that supplies its conjugate acid (such as ammonium chloride). Physiological buffers are responsible for maintaining the pH of blood within a narrow range of 7.37 to 7.43, thereby stabilizing the active conformations of biological macromolecules and optimizing the rates of biochemical reactions.

An acid buffer stabilizes the pH of a solution because the abundant supply of A^- ions (from the salt) can remove any H_3O^+ ions brought by additional strong acid; furthermore, the abundant supply of HA molecules (from the acid component of the buffer) can provide H_3O^+ ions to react with any strong base that is added. Similarly, in a base buffer the weak base B can accept protons when a strong acid is added and its conjugate acid BH^+ can supply protons if a strong base is added. The following example explores the quantitative basis of buffer action.

Example 4.8   *Assessing buffer action*

Estimate the effect of addition of 0.020 mol of hydronium ions (from a solution of a strong acid, such as hydrochloric acid) on the pH of 1.0 dm³ of (a) 0.15 M $CH_3COOH(aq)$ and (b) a buffer solution containing 0.15 M $CH_3COOH(aq)$ and 0.15 M $NaCH_3CO_2(aq)$.

Strategy Before addition of hydronium ions, the pH of solutions (a) and (b) is 2.8 (*Example* 4.5) and 4.8 (see the preceding *brief illustration*). After addition to solution (a) the initial molar concentration of $CH_3COOH(aq)$ is 0.15 M and that of $H_3O^+(aq)$ is (0.020 mol)/(1.0 dm³) = 0.020 M. After addition to solution (b), the initial molar concentrations of $CH_3COOH(aq)$, $CH_3CO_2^-(aq)$, and $H_3O^+(aq)$ are 0.15 M, 0.15 M, and 0.020 M, respectively. The weak base already present in solution, $CH_3CO_2^-(aq)$, reacts immediately with the added hydronium ions:

$$CH_3CO_2^-(aq) + H_3O^+(aq) \rightarrow CH_3COOH(aq) + H_2O(l)$$

We use the adjusted concentrations of $CH_3COOH(aq)$ and $CH_3CO_2^-(aq)$ and eqn 4.45 to calculate a new value of the pH of the buffer solution.

Solution For addition of a strong acid to solution (a), we draw up the following equilibrium table to show the effect of the addition of hydronium ions:

Species	CH_3COOH	H_3O^+	$CH_3CO_2^-$
Initial concentration/(mol dm⁻³)	0.15	0.020	0
Change to reach equilibrium/(mol dm⁻³)	$-x$	$+x$	$+x$
Equilibrium concentration/(mol dm⁻³)	$0.15-x$	$0.020+x$	x

The value of x is found by inserting the equilibrium concentrations into the expression for the acidity constant:

$$K_a = \frac{[H_3O^+][CH_3CO_2^-]}{[CH_3COOH]} = \frac{(0.020+x)x}{0.15-x}$$

As in *Example* 4.5, we assume that x is very small; in this case $x \ll 0.020$, and write

$$K_a \approx \frac{0.020x}{0.15}$$

Then

$$x = (0.15/0.020) \times K_a = 7.5 \times 1.8 \times 10^{-5} = 1.4 \times 10^{-4}$$

We see that our approximation is valid and, therefore, $[H_3O^+] = 0.020 + x \approx 0.020$ and pH = 1.7. It follows that the pH of the unbuffered solution (a) changes from 2.8 to 1.7 on addition of 0.020 M H_3O^+ (aq).

Now we consider the addition of 0.020 M $H_3O^+(aq)$ to solution (b). Reaction between the strong acid and weak base consumes the added hydronium ions and changes the concentration of $CH_3CO_2^-(aq)$ to 0.13 M and the concentration of $CH_3COOH(aq)$ to 0.17 M. It follows from eqn 4.35 that

$$pH = pK_a - \log\frac{[CH_3COOH]}{[CH_3CO_2^-]} = 4.75 - \log\frac{0.17}{0.13} = 4.6$$

The pH of the buffer solution (b) changes only slightly from 4.8 to 4.6 on addition of 0.020 M $H_3O^+(aq)$.

Self-test 4.19 Estimate the change in pH of solution (b) from *Example* 4.8 after addition of 0.020 mol of OH⁻(aq).

Answer: 4.9

Case study 4.5   *Buffer action in blood*

The pH of blood in a healthy human being varies from 7.37 to 7.43. There are two buffer systems that help maintain the pH of blood relatively constant: one arising from a carbonic acid/bicarbonate (hydrogencarbonate) ion equilibrium and another involving protonated and deprotonated forms of hemoglobin.

Carbonic acid forms in blood from the reaction between water and CO_2 gas, which comes from inhaled air and is also a by-product of metabolism (*Case study* 4.3):

$$CO_2(g) + H_2O(l) \rightleftharpoons H_2CO_3(aq)$$

In red blood cells, this reaction is catalyzed by the enzyme carbonic anhydrase. Aqueous carbonic acid then deprotonates to form bicarbonate (hydrogencarbonate) ion:

$$H_2CO_3(aq) \rightleftharpoons H^+(aq) + HCO_3^-(aq)$$

The fact that the pH of normal blood is approximately 7.4 implies that $[HCO_3^-]/[H_2CO_3] \approx 20$. The body's control of the pH of blood is an example of **homeostasis**, the ability of an organism to counteract environmental changes with physiological responses. For instance, the concentration of carbonic acid can be controlled by respiration: exhaling air depletes the system of $CO_2(g)$ and $H_2CO_3(aq)$ so the pH of blood rises when air is exhaled. The kidneys also play a role in the control of the concentration of hydronium ions. There, ammonia formed by the release of nitrogen from some amino acids (such as glutamine) combines with excess hydronium ions and the ammonium ion is excreted through urine.

The condition known as *alkalosis* occurs when the pH of blood rises above about 7.45. *Respiratory alkalosis* is caused by hyperventilation, or excessive respiration. The simplest remedy consists of breathing into a paper bag in order to increase the levels of inhaled CO_2. *Metabolic alkalosis* may result from illness, poisoning, repeated vomiting, and overuse of diuretics. The body may compensate for the increase in the pH of blood by decreasing the rate of respiration.

Acidosis occurs when the pH of blood falls below about 7.35. In *respiratory acidosis*, impaired respiration increases the concentration of dissolved CO_2 and lowers the blood's pH. The condition is common in victims of smoke inhalation and patients with asthma, pneumonia, and emphysema. The most efficient treatment consists of placing the patient in a ventilator. *Metabolic acidosis* is caused by the release of large amounts of lactic acid or other acidic by-products of metabolism (*Case study* 4.3), which react with bicarbonate ion to form carbonic acid, thus lowering the blood's pH. The condition is common in patients with diabetes and severe burns.

The concentration of hydronium ion in blood is also controlled by hemoglobin, which can exist in deprotonated (basic) or protonated (acidic) forms, depending on the state of protonation of several histidines on the protein's surface (see Fig. 4.16 for a diagram of the fraction of protonated species in an aqueous solution of histidine). The carbonic acid/bicarbonate ion equilibrium and proton equilibria in hemoglobin also regulate the oxygenation of blood. The key to this regulatory mechanism is the **Bohr effect**, the observation that hemoglobin binds O_2 strongly when it is deprotonated and releases O_2 when it is protonated. It follows that when dissolved CO_2 levels are high and the pH of blood falls slightly, hemoglobin becomes protonated and releases bound O_2 to tissue. Conversely, when CO_2 is exhaled and the pH rises slightly, hemoglobin becomes deprotonated and binds O_2.

Checklist of key concepts

1. The reaction Gibbs energy, $\Delta_r G$, is the slope of a plot of Gibbs energy against composition.

2. The condition of chemical equilibrium at constant temperature and pressure is $\Delta_r G = 0$.

3. The standard reaction Gibbs energy is the difference of the standard Gibbs energies of formation of the products and reactants weighted by the stoichiometric coefficients in the chemical equation.

4. The equilibrium constant is the value of the reaction quotient at equilibrium.

5. A compound is thermodynamically stable with respect to its elements if $\Delta_f G^\ominus < 0$.

6. The equilibrium constant of a reaction is independent of the presence of a catalyst.

7. The equilibrium constant K increases with temperature if $\Delta_r H^\ominus > 0$ (an endothermic reaction) and decreases if $\Delta_r H^\ominus < 0$ (an exothermic reaction).

8. An endergonic reaction has a positive Gibbs energy; an exergonic reaction has a negative Gibbs energy.

9. The biological standard state corresponds to pH = 7.

10. An endergonic reaction may be driven forward by coupling it to an exergonic reaction.

11. The strength of an acid HA is reported in terms of its acidity constant and that of a base B in terms of its basicity constant.

12. The acid form of a species is dominant if $pH < pK_a$, and the base form is dominant if $pH > pK_a$.

13. The pH of a buffer solution containing equal concentrations of a weak acid and its conjugate base is $pH = pK_a$.

Checklist of key equations

Property	Equation	Comment
Reaction quotient	$Q = a_C^c a_D^d / a_A^a a_B^b$	Dimensionless
Equilibrium constant	$K = (a_C^c a_D^d / a_A^a a_B^b)_{equilibrium}$	Dimensionless
Reaction Gibbs energy	$\Delta_r G = \Delta_r G^{\ominus} + RT \ln Q$	
Standard reaction Gibbs energy	$\Delta_r G^{\ominus} = \sum_{Products} v \Delta_f G^{\ominus} - \sum_{Reactants} v \Delta_f G^{\ominus}$	Procedure for calculation
Relation to K	$\Delta_r G^{\ominus} = -RT \ln K$	
van 't Hoff equation	$\ln K_2 = \ln K_1 + (\Delta_r H^{\ominus}/R)(1/T_1 - 1/T_2)$	Assumes $\Delta_r H^{\ominus}$ constant over range
Relation between standard states	$\Delta_r G^{\oplus} = \Delta_r G^{\ominus} + 7vRT \ln 10$	
pH	$pH = -\log a_{H_3O^+}$	Definition
Relation between pH and pOH	$pH + pOH = pK_w$	
Acidity constant	$K_a = a_{H_3O^+} a_{A^-} / a_{HA}$	Dilute solutions ($a_{H_2O} = 1$)
Basicity constant	$K_b = a_{HB^+} a_{OH^-} / a_B$	Dilute solutions ($a_{H_2O} = 1$)
Autoprotolysis constant	$K_w = a_{H_3O^+} a_{OH^-}$	
Relation between pK_a and pK_b	$pK_a + pK_b = pK_w$	
pH of amphiprotic anion solution	$pH = \frac{1}{2}(pK_{a1} + pK_{a2})$	$K_{a2} \ll K_{a1}$, $S \gg K_w/K_{a2}$, and $S \gg K_{a1}$
Henderson–Hasselbalch equation	$pH = pK_a - \log[acid]/[base]$	

Further information

Further information 4.1 *The contribution of autoprotolysis to pH*

Some acids are so weak and undergo so little deprotonation that the autoprotolysis of water can contribute significantly to the pH. We must also take autoprotolysis into account when we find by using the procedures in *Example* 4.5 that the pH of a solution of a weak acid is greater than 6.

We begin the calculation by noting that, apart from water, there are four species in solution: HA, A^-, H_3O^+, and OH^-. Because there are four unknown quantities, we need four equations to solve the problem. Two of the equations are the expressions for K_a and K_w (eqns 4.20 and 4.24), written here in terms of molar concentrations:

$$K_a = \frac{[H_3O^+][A^-]}{[HA]} \qquad K_w = [H_3O^+][OH^-] \qquad (4.36)$$

A third equation takes **charge balance**, the requirement that the solution be electrically neutral, into account. That is, the sum of the concentrations of the cations must be equal to the sum of the concentrations of the anions. In our case, the charge balance equation is

$$[H_3O^+] = [OH^-] + [A^-] \qquad (4.37)$$

We also know that the total concentration of A groups in all forms in which they occur, which we denote as A, must be equal to the initial concentration of the weak acid. This condition, known as **material balance**, gives our final equation:

$$A = [HA] + [A^-] \qquad (4.38)$$

Now we are ready to proceed with a calculation of the hydronium ion concentration in the solution. First, we combine eqns 4.36 and 4.37 and write

$$[A^-] = [H_3O^+] - \frac{K_w}{[H_3O^+]} \qquad (4.39)$$

We continue by substituting this expression into eqn 4.39 and solving for [HA]:

$$[HA] = A - [H_3O^+] + \frac{K_w}{[H_3O^+]} \qquad (4.40)$$

On substituting the expressions for $[A^-]$ (eqn 4.39) and [HA] (eqn 4.40) into the first of eqn 4.36, we obtain

$$K_a = \frac{[H_3O^+]\left([H_3O^+] - \dfrac{K_w}{[H_3O^+]}\right)}{A - [H_3O^+] + \dfrac{K_w}{[H_3O^+]}} \qquad (4.41)$$

Rearrangement of this expression gives

$$[H_3O^+]^3 + K_a[H_3O^+]^2 - (K_w + K_a A)[H_3O^+] - K_a K_w = 0 \qquad (4.42)$$

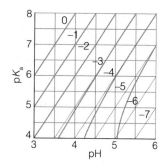

Fig. 4.18 The orange lines are the exact solutions of eqn 4.41 for a series of values of A (expressed as 10^{-x}, with $-x$ displayed). The dotted lines are the corresponding approximate solutions obtained from eqn 4.42b.

and we see that $[H_3O^+]$ is determined by solving this cubic equation, a task that is best accomplished with a calculator or mathematical software. Figure 4.18 summarizes the outcome.

There are several experimental conditions that allow us to simplify eqn 4.42. For example, when $K_aA \gg K_w$ and $A[H_3O^+] \gg K_w$ it becomes

$$[H_3O^+]^2 + K_a[H_3O^+] - K_aA = 0 \qquad (4.43a)$$

which can be solved for $[H_3O^+]$. If the extent of deprotonation is very small, we let $[H_3O^+] \ll A$ and write

$$K_a = \frac{[H_3O^+]^2}{A}, \text{ so } [H_3O^+] = (AK_a)^{1/2} \qquad (4.43b)$$

Equations 4.43 are similar to the expressions used in *Example* 4.5, where we set $[H_3O^+]$ equal to x.

Further information 4.2 *The pH of an amphiprotic salt solution*

The aim is to calculate the pH of a solution of a salt MHA, where HA^- is an amphiprotic anion (HCO_3^- is an example). We consider the following two equilibria:

$$H_2A(aq) + H_2O(l) \rightleftharpoons H_3O^+(aq) + HA^-(aq)$$

$$K_{a1} = \frac{[H_3O^+][HA^-]}{[H_2A]}$$

$$HA^-(aq) + H_2O(l) \rightleftharpoons H_3O^+(aq) + A^{2-}(aq)$$

$$K_{a2} = \frac{[H_3O^+][HA^{2-}]}{[HA^-]}$$

together with the autoprotolysis equilibrium, K_w. The total concentration of the group A is the sum of the concentrations of the species in which it occurs, which is equal to the concentration of the salt that was added. If we denote the concentration of added salt as S,

Mass conservation: $[H_2A] + [HA^-] + [A^{2-}] = S \qquad (4.44)$

The solution is electrically neutral overall, so

Charge balance: $[M^+] + [H_3O^+] = [OH^-] + [HA^-] + 2[A^{2-}] \qquad (4.45)$

with $[M^+] = S$. The difference of these two equations can be expressed as

$$[H_3O^+] = [OH^-] + [A^{2-}] - [H_2A]$$
$$= \frac{K_w}{[H_3O^+]} + \frac{[HA^-]K_{a2}}{[H_3O^+]} - \frac{[HA^-][H_3O^+]}{K_{a1}} \qquad (4.46)$$

Multiplication through by $[H_3O^+]K_{a1}$ turns this expression into

$$K_{a1}[H_3O^+]^2 = K_wK_{a1} + [HA^-]K_{a1}K_{a2} - [H_3O^+]^2[HA^-] \qquad (4.47)$$

The mass conservation expression can also be written in terms of $[H_3O^+]$:

$$S = \frac{[H_3O^+][HA^-]}{K_{a1}} + [HA^-] + \frac{[HA^-]K_{a2}}{[H_3O^+]}$$
$$= \left\{ 1 + \frac{[H_3O^+]}{K_{a1}} + \frac{K_{a2}}{[H_3O^+]} \right\}[HA^-] = H[HA^-]$$

When this expression is rearranged for $[HA^-]$ in terms of S and substituted into eqn 4.46 we find

$$[H_3O^+] = \frac{K_w}{[H_3O^+]} + \left(\frac{K_{a2}}{[H_3O^+]} - \frac{[H_3O^+]}{K_{a1}} \right)\frac{S}{H} \qquad (4.48)$$

To simplify this expression we multiply through by $[H_3O^+]H$:

$$H[H_3O^+]^2 = HK_w + K_{a2}S - \frac{[H_3O^+]^2S}{K_{a1}}$$

and then collect terms in $[H_3O^+]^2$:

$$\left(H + \frac{S}{K_{a1}} \right)[H_3O^+]^2 = HK_w + K_{a2}S \qquad (4.49)$$

Of course, H depends on $[H_3O^+]$, but we take care of that feature later (it will turn out that under the conditions to be specified, $H \approx 1$). The solution is

$$[H_3O^+] = \left(\frac{HK_w + K_{a2}S}{H + S/K_{a1}} \right)^{1/2} = \left(\frac{HK_wK_{a1} + K_{a1}K_{a2}S}{HK_{a1} + S} \right)^{1/2}$$
$$= \left(\frac{HK_wK_{a1}/S + K_{a1}K_{a2}}{HK_{a1}/S + 1} \right)^{1/2}$$
$$= (K_{a1}K_{a2})^{1/2}\left\{ \frac{1 + HK_w/SK_{a2}}{1 + HK_{a1}/S} \right\}^{1/2} \qquad (4.50)$$

On taking the negative logarithm, we obtain

$$pH = \tfrac{1}{2}(pK_{a1} + pK_{a2}) - \tfrac{1}{2}\log f(H) \qquad (4.51a)$$

where

$$f(H) = \frac{1 + HK_w/SK_{a2}}{1 + HK_{a1}/S} \qquad (4.51b)$$

Provided $HK_w/SK_{a2} \ll 1$ and $HK_{a1}/S \ll 1$, $f(H) \approx 1$, eqn 4.8a simplifies to

$$pH = \tfrac{1}{2}(pK_{a1} + pK_{a2}) \qquad (4.52)$$

which is the form quoted in the text.

The conditions for the validity of eqn 4.52 may be expressed more simply because if eqn 4.52 is a good approximation, we may use it in the form $[H_3O^+] = (K_{a1}K_{a2})^{1/2}$ to estimate the value of f by noting that

$$H = 1 + \frac{[H_3O^+]}{K_{a1}} + \frac{K_{a2}}{[H_3O^+]} \approx 1 + 2(K_{a2}/K_{a1})^{1/2} \qquad (4.53)$$

For small polyprotic species, $K_{a2} \ll K_{a1}$ (this is not always the case for polyprotic enzymes), so $H \approx 1$ and so

$$f(H) \approx \frac{1 + K_w/SK_{a2}}{1 + K_{a1}/S} \qquad (4.54)$$

Therefore, eqn 4.52 is valid provided $K_{a2} \ll K_{a1}$, $S \gg K_w/K_{a2}$, and $S \gg K_{a1}$. If these conditions are not fulfilled, then eqn 4.51 must be solved iteratively.

> **A brief illustration**
>
> Suppose we want to know the pH of 0.010 M KHS(aq). Because $K_{a1} = 1.3 \times 10^{-7}$ and $K_{a2} = 7.1 \times 10^{-15}$ ($pK_{a1} = 6.89$,

$pK_{a2} = 14.15$), we test whether we can use eqn 4.52 by forming $K_w/K_{a2} = 1.4$. This ratio is not small compared to $S = 0.010$, so eqn 4.52 cannot be used to obtain the final pH. However, it can be used as a first approximation, and we find pH $= 10.52$, which corresponds to $[H_3O^+] = 3.0 \times 10^{-11}$. This value corresponds to $f(H) = 142.318$, which gives the following improved estimate of the pH:

$$pH = \tfrac{1}{2}(6.89 + 14.15) - \tfrac{1}{2}\log 142.318 = 9.44$$

Next, we use this value to calculate an improved value of f, namely $f(H) = 142.650$, so

$$pH = \tfrac{1}{2}(6.89 + 14.15) - \tfrac{1}{2}\log 142.650 = 9.44$$

This value is unchanged from the previous estimate, so we accept it as the final value. Note that the estimate from eqn 4.51 is significantly different but that only one round of improvement is necessary in this case.

Discussion questions

4.1 Explain how the mixing of reactants and products affects the position of chemical equilibrium.

4.2 Explain how a reaction that is not spontaneous may be driven forward by coupling to a spontaneous reaction.

4.3 At blood temperature, $\Delta_r G^\oplus = -218$ kJ mol^{-1} and $\Delta_r H^\oplus = -120$ kJ mol^{-1} for the production of lactate ion during glycolysis. Provide a molecular interpretation for the observation that the reaction is more exergonic than it is exothermic.

4.4 Explain Le Chatelier's principle in terms of thermodynamic quantities.

4.5 Use the Boltzmann distribution to describe the molecular features that determine the magnitudes of equilibrium constants and their variation with temperature.

4.6 Describe the basis of buffer action.

4.7 State the limits to the generality of the following expressions: (a) pH $= \tfrac{1}{2}(pK_{a1} + pK_{a2})$, (b) pH $= pK_a - \log([\text{acid}]/[\text{base}])$, and (c) the van 't Hoff equation, written as $\ln K_2 = \ln K_1 + (\Delta_r H^\circ/R)(1/T_1 - 1/T_2)$.

Exercises

4.8 Write the expressions for the equilibrium constants for the following reactions, making the approximation of replacing activities by molar concentrations or partial pressures:

(a) $G6P(aq) + H_2O(l) \rightleftharpoons G(aq) + P_i(aq)$, where G6P is glucose-6-phosphate, G is glucose, and P_i is inorganic phosphate.

(b) $Gly(aq) + Ala(aq) \rightleftharpoons Gly–Ala(aq) + H_2O(l)$

(c) $Mg^{2+}(aq) + ATP^{4-}(aq) \rightleftharpoons MgATP^{2-}(aq)$

(d) $2\,CH_3COCOOH(aq) + 5\,O_2(g) \rightleftharpoons 6\,CO_2(g) + 4\,H_2O(l)$

4.9 The equilibrium constant for the reaction $A + B \rightleftharpoons 2\,C$ is reported as 3.4×10^4. What would it be for the reaction written as (a) $2\,C \rightleftharpoons A + B$, (b) $2\,A + 2\,B \rightleftharpoons 4\,C$, and (c) $\tfrac{1}{2}A + \tfrac{1}{2}B \rightleftharpoons C$?

4.10 The equilibrium constant for the hydrolysis of the dipeptide alanylglycine by a peptidase enzyme is $K = 8.1 \times 10^2$ at 310 K. Calculate the standard reaction Gibbs energy for the hydrolysis.

4.11 One enzyme-catalyzed reaction in a biochemical cycle has an equilibrium constant that is 10 times the equilibrium constant of a second reaction. If the standard Gibbs energy of the former reaction is -300 kJ mol^{-1}, what is the standard reaction Gibbs energy of the second reaction?

4.12 What is the value of the equilibrium constant of a reaction for which $\Delta_r G^\circ = 0$?

4.13 The standard reaction Gibbs energies (at pH = 7) for the hydrolysis of glucose-1-phosphate, glucose-6-phosphate, and glucose-3-phosphate are -21, -14, and -9.2 kJ mol^{-1}, respectively. Calculate the equilibrium constants for the hydrolyses at 37°C.

4.14 The standard Gibbs energy for the hydrolysis of ATP to ADP is -31 kJ mol^{-1}. What is the Gibbs energy of reaction in an environment at 37°C in which the ATP, ADP, and P_i concentrations are all (a) 1.0 mmol dm^{-3} and (b) 1.0 μmol dm^{-3}?

4.15 The distribution of Na^+ ions across a typical biological membrane is 10 mmol dm^{-3} inside the cell and 140 mmol dm^{-3} outside the cell. At equilibrium the concentrations are equal. What is the Gibbs energy difference across the membrane at 37°C? The difference in concentration must be sustained by coupling to reactions that have at least that difference of Gibbs energy.

4.16 For the hydrolysis of ATP at 37°C, $\Delta_r H^{\oplus} = -20$ kJ mol^{-1} and $\Delta_r S^{\oplus} = +34$ J K^{-1} mol^{-1}. Assuming that these quantities remain constant, estimate the temperature at which the equilibrium constant for the hydrolysis of ATP becomes greater than 1.

4.17 Calculate the standard biological Gibbs energy for the reaction

pyruvate$^-$(aq) + NADH(aq) + H^+(aq) → lactate$^-$(aq) + NAD^+(aq)

at 310 K given that $\Delta_r G^{\oplus} = -66.6$ kJ mol^{-1}. (NAD^+ is the oxidized form of nicotinamide dinucleotide.) This reaction occurs in muscle cells deprived of oxygen during strenuous exercise and can lead to cramping.

4.18 The standard biological reaction Gibbs energy for the removal of the phosphate group from adenosine monophosphate is -14 kJ mol^{-1} at 298 K. What is the value of the thermodynamic standard reaction Gibbs energy?

4.19 Estimate the values of the biological standard Gibbs energies of the following phosphate transfer reactions:

(a) GTP(aq) + ADP(aq) → GDP(aq) + ATP(aq)

(b) Glycerol(aq) + ATP(aq) → glycerol-1-phosphate + ADP(aq) + H^+(aq)

(c) 3-Phosphoglycerate(aq) + ATP(aq) → 1,3-bis(phospho) glycerate(aq) + ADP(aq)

4.20 Two polynucleotides with sequences $A_n U_n$ (where A and U denote adenine and uracil, respectively) interact through A–U base pairs, forming a double helix. When $n = 5$ and $n = 6$, the equilibrium constants for formation of the double helix are 5.0×10^3 and 2.0×10^5, respectively. (a) Suggest an explanation for the increase in the value of the equilibrium constant with n. (b) Calculate the contribution of a single A–U base pair to the Gibbs energy of formation of a double helix between $A_n U_n$ polypeptides.

4.21 Under biochemical standard conditions, aerobic respiration produces approximately 38 molecules of ATP per molecule of glucose that is completely oxidized. (a) What is the percentage efficiency of aerobic respiration under biochemical standard conditions? (b) The following conditions are more likely to be observed in a living cell: $p_{CO_2} = 53$ mbar, $p_{O_2} = 132$ mbar, [glucose] = 5.6×10^{-2} mol dm^{-3}, [ATP] = [ADP] = [P_i] = 1.0×10^{-4} mol dm^{-3}, pH = 7.4, $T = 310$ K. Assuming that activities can be replaced by the numerical values of molar concentrations, calculate the efficiency of aerobic respiration under these physiological conditions.

4.22 The second step in glycolysis is the isomerization of glucose-6-phosphate (G6P) to fructose-6-phosphate (F6P). Example 4.3 considered the equilibrium between F6P and G6P. Draw a graph to show how the reaction Gibbs energy varies with the fraction f of F6P in solution. Label the regions of the graph that correspond to the formation of F6P and G6P being spontaneous, respectively.

4.23 The saturation curves shown in Fig. 4.9 may also be modeled mathematically by the equation

$$\log \frac{s}{1-s} = v \log p - v \log K$$

where s is the saturation, p is the partial pressure of O_2, K is a constant (not the equilibrium constant for binding of one ligand), and v is the *Hill coefficient*, which varies from 1, for no cooperativity, to N for all-or-none binding of N ligands ($N = 4$ in Hb). The Hill coefficient for Mb is 1, and for Hb it is 2.8. (a) Determine the constant K for both Mb and Hb from the graph of fractional saturation (at $s = 0.5$) and then calculate the fractional saturation of Mb and Hb for the following values of p/kPa: 1.0, 1.5, 2.5, 4.0, 8.0. (b) Calculate the value of s at the same p values assuming v has the theoretical maximum value of 4.

4.24 Classify the following compounds as endergonic or exergonic: (a) glucose, (b) urea, (c) octane, and (d) ethanol.

4.25 Consider the combustion of sucrose:

$$C_{12}H_{22}O_{11}(aq) + 12\ O_2(g) \rightleftharpoons 12\ CO_2(g) + 11\ H_2O(l)$$

(a) Combine the standard reaction entropy with the standard reaction enthalpy and calculate the standard reaction Gibbs energy at 298 K. (b) In assessing metabolic processes, we are usually more interested in the work that may be performed for the consumption of a given mass of compound than the heat it can produce (which merely keeps the body warm). Recall from Chapter 2 that the change in Gibbs energy can be identified with the maximum nonexpansion work that can be extracted from a process. What is the maximum energy that can be extracted as (i) heat and (ii) nonexpansion work when 1.0 kg of sucrose is burned under standard conditions at 298 K?

4.26 Is it more energy effective to ingest sucrose or glucose? Calculate the nonexpansion work, the expansion work, and the total work that can be obtained from the combustion of 1.0 kg of glucose under standard conditions at 298 K when the product includes liquid water. Compare your answer with your results from *Exercise* 4.25b.

4.27 The oxidation of glucose in the mitochondria of energy-hungry brain cells leads to the formation of pyruvate ions, which are then decarboxylated to ethanal (acetaldehyde, CH_3CHO) in the course of the ultimate formation of carbon dioxide. (a) The standard Gibbs energies of formation of pyruvate ions in aqueous solution and gaseous ethanal are -474 and -133 kJ mol^{-1}, respectively. Calculate the Gibbs energy of the reaction in which pyruvate ions are converted to ethanal by the action of pyruvate decarboxylase with the release of carbon dioxide. (b) Ethanal is soluble in water. Would you expect the standard Gibbs energy of the enzyme-catalyzed decarboxylation of pyruvate ions to ethanal in solution to be larger or smaller than the value for the production of gaseous ethanal?

4.28 Show that if the logarithm of an equilibrium constant is plotted against the reciprocal of the temperature, then the standard reaction enthalpy may be determined.

4.29 The conversion of fumarate ion to malate ion is catalyzed by the enzyme fumarase:

fumarate^{2-}(aq) + H_2O(l) → malate^{2-}(aq)

Use the following data to determine the standard reaction enthalpy:

θ/°C	15	20	25	30	35	40	45	50
K	4.786	4.467	4.074	3.631	3.311	3.090	2.754	2.399

4.30 What is the standard enthalpy of a reaction for which the equilibrium constant is (a) doubled and (b) halved when the temperature is increased by 10 K at 298 K?

4.31 Numerous acidic species are found in living systems. Write the proton transfer equilibria for the following biochemically important acids in aqueous solution: (a) $H_2PO_4^-$ (dihydrogenphosphate ion), (b) lactic acid ($CH_3CHOHCOOH$), (c) glutamic acid

(HOOCCH$_2$CH$_2$CH(NH$_2$)COOH), (**d**) glycine (NH$_2$CH$_2$COOH), and (**e**) oxalic acid (HOOCCOOH).

4.32 For biological and medical applications we often need to consider proton transfer equilibria at body temperature (37°C). The value of K_w for water at body temperature is 2.5×10^{-14}. (**a**) What is the value of [H$_3$O$^+$] and the pH of neutral water at 37°C? (**b**) What is the molar concentration of OH$^-$ ions and the pOH of neutral water at 37°C?

4.33 Suppose that something had gone wrong in the Big Bang, and instead of ordinary hydrogen there was an abundance of deuterium in the Universe. There would be many subtle changes in equilibria, particularly the deuteron transfer equilibria of heavy atoms and bases. The K_w for D$_2$O, heavy water, at 25°C is 1.35×10^{-15}. (**a**) Write the chemical equation for the autoprotolysis (more precisely, autodeuterolysis) of D$_2$O. (**b**) Evaluate pK_w for D$_2$O at 25°C. (**c**) Calculate the molar concentrations of D$_3$O$^+$ and OD$^-$ in neutral heavy water at 25°C. (**d**) Evaluate the pD and pOD of neutral heavy water at 25°C. (**e**) Formulate the relation between pD, pOD, and pK_w(D$_2$O).

4.34 The molar concentration of H$_3$O$^+$ ions in the following solutions was measured at 25°C. Calculate the pH and pOH of the solutions: (**a**) 15 μmol dm^{-3} (a sample of rainwater), (**b**) 1.5 mmol dm^{-3}, (**c**) 5.1×10^{-14} mol dm^{-3}, and (**d**) 5.01×10^{-5} mol dm^{-3}.

4.35 Calculate the molar concentration of H$_3$O$^+$ ions and the pH of the following solutions: (**a**) 25.0 cm^3 of 0.144 M HCl(aq) was added to 25.0 cm^3 of 0.125 M NaOH(aq), (**b**) 25.0 cm^3 of 0.15 M HCl(aq) was added to 35.0 cm^3 of 0.15 M KOH(aq), and (**c**) 21.2 cm^3 of 0.22 M HNO$_3$(aq) was added to 10.0 cm^3 of 0.30 M NaOH(aq).

4.36 Determine whether aqueous solutions of the following salts have a pH equal to, greater than, or less than 7; if pH > 7 or pH < 7, write a chemical equation to justify your answer: (**a**) NH$_4$Br, (**b**) Na$_2$CO$_3$, (**c**) KF, (**d**) KBr.

4.37 (**a**) A sample of potassium acetate, KCH$_3$CO$_2$, of mass 8.4 g is used to prepare 250 cm^3 of solution. What is the pH of the solution? (**b**) What is the pH of a solution when 3.75 g of ammonium bromide, NH$_4$Br, is used to make 100 cm^3 of solution? (**c**) An aqueous solution of volume 1.0 dm^3 contains 10.0 g of potassium bromide. What is the percentage of Br$^-$ ions that are protonated?

4.38 There are many organic acids and bases in our cells, and their presence modifies the pH of the fluids inside them. It is useful to be able to assess the pH of solutions of acids and bases and to make inferences from measured values of the pH. A solution of equal concentrations of lactic acid and sodium lactate was found to have pH = 3.08. (**a**) What are the values of pK_a and K_a of lactic acid? (**b**) What would the pH be if the acid had twice the concentration of the salt?

4.39 Calculate the pH, pOH, and fraction of solute protonated or deprotonated in the following aqueous solutions: (**a**) 0.120 M CH$_3$CH(OH)COOH(aq) (lactic acid), (**b**) 1.4×10^{-4} M CH$_3$CH(OH)COOH(aq), (**c**) 0.15 M NH$_4$Cl(aq), (**d**) 0.15 M NaCH$_3$CO$_2$(aq), and (**e**) 0.112 M (CH$_3$)$_3$N(aq) (trimethylamine).

4.40 Show graphically the variation with pH of the composition of the following aqueous solutions: (**a**) 0.010 M glycine(aq) and (**b**) 0.010 M tyrosine(aq).

4.41 Calculate the pH of the following acid solutions at 25°C; ignore second deprotonations only when that approximation is justified: (**a**) 1.0×10^{-4} M H$_3$BO$_3$(aq) (boric acid acts as a monoprotic acid), (**b**) 0.015 M H$_3$PO$_4$(aq), and (**c**) 0.10 M H$_2$SO$_3$(aq).

4.42 The amino acid tyrosine has pK_a = 2.20 for deprotonation of its carboxylic acid group. What are the relative concentrations of tyrosine and its conjugate base at a pH of (**a**) 7, (**b**) 2.2, and (**c**) 1.5?

4.43 Appreciable concentrations of the potassium and calcium salts of oxalic acid, (COOH)$_2$, are found in many leafy green plants, such as rhubarb and spinach. (**a**) Calculate the molar concentrations of HOOCCO$_2^-$, (CO$_2$)$_2^{2-}$, H$_3$O$^+$, and OH$^-$ in 0.15 M (COOH)$_2$(aq). (**b**) Calculate the pH of a solution of 0.15 M potassium hydrogenoxalate.

4.44 In green sulfur bacteria, hydrogen sulfide, H$_2$S, is the agent that brings about the reduction of CO$_2$ to carbohydrates during photosynthesis. Calculate the molar concentrations of H$_2$S, HS$^-$, S^{2-}, H$_3$O$^+$, and OH$^-$ in 0.065 M H$_2$S(aq).

4.45 The *isoelectric point*, pI, of an amino acid is the pH at which the predominant species in solution is the zwitterionic form of the amino acid and only small but equal concentrations of positively and negatively charged forms of the amino acid are present. It follows that at the isoelectric point, the average charge on the amino acid is zero. Show that (**a**) pI = $\frac{1}{2}$(pK_{a1} + pK_{a2}) for amino acids with side chains that are neither acidic nor basic (such as glycine and alanine), (**b**) pI = $\frac{1}{2}$(pK_{a1} + pK_{a2}) for amino acids with acidic side chains (such as aspartic acid and glutamic acid), and (**c**) pI = $\frac{1}{2}$(pK_{a2} + pK_{a3}) for amino acids with basic side chains (such as lysine and histidine), where pK_{a1}, pK_{a2}, and pK_{a3} are given in Table 4.6. *Hint*: See *Case study* 4.4 and Section 4.10.

4.46 Predict the pH region in which each of the following buffers will be effective, assuming equal molar concentrations of the acid and its conjugate base: (**a**) sodium lactate and lactic acid, (**b**) sodium benzoate and benzoic acid, (**c**) potassium hydrogenphosphate and potassium phosphate, (**d**) potassium hydrogenphosphate and potassium dihydrogenphosphate, and (**e**) hydroxylamine and hydroxylammonium chloride.

4.47 From the information in Tables 4.4 and 4.5, select suitable buffers for (**a**) pH = 2.2 and (**b**) pH = 7.0.

4.48 The weak base colloquially known as Tris, and more precisely as tris(hydroxymethyl)aminomethane, has pK_a = 8.3 at 20°C and is commonly used to produce a buffer for biochemical applications. (**a**) At what pH would you expect Tris to act as a buffer in a solution that has equal molar concentrations of Tris and its conjugate acid? (**b**) What is the pH after the addition of 3.3 mmol NaOH to 100 cm^3 of a buffer solution with equal molar concentrations of Tris and its conjugate acid form? (**c**) What is the pH after the addition of 6.0 mmol HNO$_3$ to 100 cm^3 of a buffer solution with equal molar concentrations of Tris and its conjugate acid?

Projects

4.49 The denaturation of a biological macromolecule can be described by the equilibrium

macromolecule in native form $\rightleftharpoons$ macromolecule in denatured form

(a) Show that the fraction θ of denatured macromolecules is related to the equilibrium constant K_d for the denaturation process by

$$\theta = \frac{K_d}{1 + K_d}$$

(b) Write an expression for the temperature dependence of K_d in terms of the standard enthalpy and standard entropy of denaturation.

(c) At pH = 2, the standard enthalpy and entropy of denaturation of the enzyme chymotrypsin are +418 kJ mol^{-1} and +1.32 kJ K^{-1} mol^{-1}, respectively. Using these data and your results from parts (a) and (b), plot θ against T. Compare the shape of your plot with that of the plot shown in Fig. 3.16.

(d) The 'melting temperature' of a biological macromolecule is the temperature at which $\theta = \frac{1}{2}$. Use your results from part (c) to calculate the melting temperature of chymotrypsin at pH = 2.

(e) Calculate the standard Gibbs energy and the equilibrium constant for the denaturation of chymotrypsin at pH = 2.0 and $T = 310$ K (body temperature). Is the protein stable under these conditions?

4.50 The unfolding of a protein may be brought about by treatment with *denaturants*, substances such as guanidinium hydrochloride (GuHCl; the guanidinium ion is $(NH_2)_2C=NH_2^+$) that disrupt the intermolecular interactions responsible for the native three-dimensional conformation of a biological macromolecule. Data for a number of proteins denatured by urea or guanidinium hydrochloride suggest a linear relationship between the Gibbs energy of denaturation of a protein, ΔG_d, and the molar concentration of a denaturant [D]:

$$\Delta G_d^{\oplus} = \Delta G_{d,water}^{\oplus} - m[D]$$

where m is an empirical parameter that measures the sensitivity of unfolding to denaturant concentration and $\Delta G_{d,water}^{\oplus}$ is the Gibbs energy of denaturation of the protein in the absence of denaturant and is a measure of the thermal stability of the macromolecule.

(a) At 27°C and pH 6.5, the fraction θ of native chymotrypsin molecules varies with the concentration of GuHCl as follows:

θ	1.00	0.99	0.78	0.44	0.23	0.08	0.06	0.01
[GuHCl]/ (mol dm^{-3})	0.00	0.75	1.35	1.70	2.00	2.35	2.70	3.00

Calculate m and $\Delta G_{d,water}^{\oplus}$ for chymotrypsin under these experimental conditions.

(b) Using the same data and the expression for θ from Exercise 4.49, plot θ against [GnHCl]. Comment on the shape of the curve.

(c) To gain insight into your results from part (b), you will now derive an equation that relates θ to [D]. Begin by showing that $\Delta G_{d,water}^{\oplus} = m[D]_{1/2}$, where $[D]_{1/2}$ is the concentration of denaturant corresponding to $\theta = \frac{1}{2}$. Then write an expression for θ as a function of [D], $[D]_{1/2}$, m, and T. Finally, plot the expression using the values of $[D]_{1/2}$, m, and T from part (a). Is the shape of your plot consistent with your results from part (b)?

4.51 In *Case study* 4.5 we discussed the role of hemoglobin in regulating the pH of blood. Now we explore the mechanism of regulation in detail.

(a) If we denote the protonated and deprotonated forms of hemoglobin as HbH and Hb$^-$, respectively, then the proton transfer equilibria for deoxygenated and fully oxygenated hemoglobin can be written as:

$$HbH \rightleftharpoons Hb^- + H^+ \qquad pK_a = 8.18$$
$$HbHO_2 \rightleftharpoons HbO_2^- + H^+ \qquad pK_a = 6.62$$

where we take the view (for the sake of simplicity) that the protein contains only one acidic proton. **(i)** What fraction of deoxygenated hemoglobin is deprotonated at pH = 7.4, the value for normal blood? **(ii)** What fraction of oxygenated hemoglobin is deprotonated at pH = 7.4? **(iii)** Use your results from parts (a.i) and (a.ii) to show that deoxygenation of hemoglobin is accompanied by the uptake of protons by the protein.

(b) It follows from the discussion in *Case study* 4.4 and part (a) that the exchange of CO_2 for O_2 in tissue is accompanied by complex proton transfer equilibria: the release of CO_2 into blood produces hydronium ions that can be bound tightly to hemoglobin once it releases O_2. These processes prevent changes in the pH of blood. To treat the problem more quantitatively, let us calculate the amount of CO_2 that can be transported by blood without a change in pH from its normal value of 7.4. **(i)** Begin by calculating the amount of hydronium ion bound per mole of oxygenated hemoglobin molecules at pH = 7.4. **(ii)** Now calculate the amount of hydronium ion bound per mole of deoxygenated hemoglobin molecules at pH = 7.4. **(iii)** From your results for parts (b.i) and (b.ii), calculate the amount of hydronium ion that can be bound per mole of hemoglobin molecules as a result of the release of O_2 by the fully oxygenated protein at pH = 7.4. **(iv)** Finally, use the result from part (b.iii) to calculate the amount of CO_2 that can be released into the blood per mole of hemoglobin molecules at pH = 7.4.

Thermodynamics of ion and electron transport

5

Measurements such as the ones we describe in this chapter lead to collections of data that are very useful for discussing the characteristics of electrolyte solutions and the migration of ions across biological membranes. They are used to discuss the details of the propagation of signals in neurons and of the synthesis of ATP.

We shall also see that such apparently unrelated processes as oxidation of fuels, respiration, and photosynthesis are actually all closely related, for in each of them an electron, sometimes accompanied by a group of atoms, is transferred from one species to another. Indeed, together with the proton transfer typical of acid–base reactions, processes in which electrons are transferred, the so-called **redox reactions**, account for many of the reactions encountered in chemistry and biology.

Transport of ions across biological membranes

The cell membrane may be regarded as a barrier that slows down the transfer of material into or out of the cell. Here we focus on the transport of ions across biological membranes. We begin by developing some general ideas about solutions of electrolytes. Then we describe the thermodynamics of ion transport mediated by special membrane-spanning proteins. In Section 5.10 we shall see how electron transfer reactions during the later stages of aerobic metabolism of glucose couple to the movement of protons across biological membranes and contribute to the synthesis of ATP.

5.1 Ions in solution

To prepare for the discussion of biological redox reactions and the role of ions in physiological processes, we need to describe the factors that influence the activities of ions in aqueous solutions.

The most significant difference between the solution of an electrolyte and a non-electrolyte is that there are long-range Coulombic interactions between the ions in the former. As a result, electrolyte solutions exhibit nonideal behavior even at very low concentrations because the solute particles, the ions, do not move independently of one another. Some idea of the importance of ion–ion interactions is obtained by noting their average separations in solutions of different molar concentration c and, to appreciate the scale, the typical number of H_2O molecules that can fit between them:

c/mol dm^{-3}	0.001	0.01	0.1	1	10
Separation/nm	90	40	20	9	4
Number of H$_2$O molecules	30	14	6	3	1

(a) Activity coefficients

To take the interactions into account—which become very serious for concentrations of 0.01 mol dm^{-3} and more—we work with the activities of the charged solutes. We saw in Section 3.8 that the activity, a_J, is a kind of effective concentration and is related to concentrations by multiplication by an activity coefficient, γ_J. There are various ways of expressing concentration; in the first part of this chapter we use the molality, b_J, and write

$$a_J = \gamma_J b_J / b^{\ominus} \tag{5.1a}$$

with $b^{\ominus} = 1$ mol kg^{-1}. For notational simplicity, we often replace $b_J/b^{\ominus}$ by b_J, interpret b_J as the numerical value of the molality, and write

$$a_J = \gamma_J b_J \tag{5.1b}$$

Because the solution becomes more ideal as the molality approaches zero, we know that $\gamma_J \to 1$ as $b_J \to 0$. Once we know the activity of the species J, we can write its chemical potential by using

$$\mu_J = \mu_J^{\ominus} + RT \ln a_J \tag{5.2}$$

Mathematical tookit 5.1   *Exponentials and logarithms*

The **exponential function**, e^x (where e = 2.718 . . .) has the following important properties:

$$e^x \times e^y = e^{x+y}$$

$$\frac{e^x}{e^y} = e^{x-y}$$

$$(e^x)^a = e^{ax}$$

The **natural logarithm** of a number x is denoted as $\ln x$ and is defined as the power to which e must be raised for the result to be equal to x. It follows from the definition of logarithms that

$$\ln x + \ln y = \ln xy$$

$$\ln x - \ln y = \ln \frac{x}{y}$$

$$a \ln x = \ln x^a$$

Useful points about logarithms are summarized in the graph below:

- Logarithms increase only very slowly as x increases.
- The logarithm of 1 is 0: $\ln 1 = 0$.
- The logarithms of numbers less than 1 are negative.
- In elementary mathematics the logarithms of negative numbers are not defined.

The **common logarithm** of a number is the logarithm compiled with 10 in place of e; common logarithms are denoted $\log x$. Common logarithms follow the same rules of addition and subtraction as natural logarithms. Common and natural logarithms (log and ln, respectively) are related by

$$\ln x = \ln 10 \times \log x \approx 2.303 \log x$$

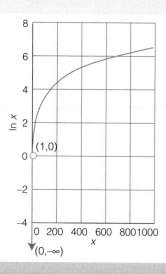

The thermodynamic properties of the solution—such as the equilibrium constants of reactions involving ions—can then be derived in the same way as for ideal solutions but with activities in place of concentrations. However, when we want to relate the results we derive, we need to know how to relate activities to concentrations. We ignored that problem when discussing acids and bases and simply assumed that all activity coefficients were 1. The cytoplasm and other fluids in organisms have ion concentrations that are far too high to behave ideally, so $\gamma = 1$ is a poor approximation; in this chapter, we see how to improve that approximation.

One problem that confronts us from the outset is that cations and anions always occur together in solution. As a result, there is no experimental procedure for distinguishing the deviations from ideal behavior due to the cations from those of the anions: we cannot measure the activity coefficients of cations and anions separately. The best we can do experimentally is to ascribe deviations from ideal behavior equally to each kind of ion and to talk in terms of a **mean activity coefficient**, $\gamma_\pm$. For a salt MX, such as NaCl, we show in the following *Justification* that the mean activity coefficient is related to the activity coefficients of the individual ions as follows:

$$\gamma_\pm = (\gamma_+\gamma_-)^{1/2}$$

| Mean activity coefficient for a salt MX | (5.3a) |

For a salt M_pX_q, the mean activity coefficient is related to the activity coefficients of the individual ions as follows:

$$\gamma_\pm = (\gamma_+^p \gamma_-^q)^{1/s} \quad s = p + q$$

| Mean activity coefficient for a salt M_pX_q | (5.3b) |

Justification 5.1 *Mean activity coefficients*

In this *Justification*, we use the relation $\ln xy = \ln x + \ln y$ several times (sometimes as $\ln x + \ln y = \ln xy$) and its implication (by setting $y = x$) that $\ln x^2 = 2\ln x$ (see *Mathematical toolkit* 5.1). For a salt MX that dissociates completely in solution, the molar Gibbs energy of the ions is

$$G_m = \mu_+ + \mu_-$$

where μ_+ and μ_- are the chemical potentials of the cations and anions, respectively. Each chemical potential can be expressed in terms of a molality b and an activity coefficient γ by using eqn 5.2 ($\mu = \mu^\ominus + RT\ln a$) and then eqn 5.1 ($a = \gamma b$) together with $\ln \gamma b = \ln \gamma + \ln b$, which gives

$$G_m = (\mu_+^\ominus + RT\ln \gamma_+ b_+) + (\mu_-^\ominus + RT\ln \gamma_- b_-)$$
$$= (\mu_+^\ominus + RT\ln \gamma_+ + RT\ln b_+) + (\mu_-^\ominus + RT\ln \gamma_- + RT\ln b_-)$$

We now use $\ln x + \ln y = \ln xy$ again to combine the two terms involving the activity coefficients as

$$RT\ln \gamma_+ + RT\ln \gamma_- = RT(\ln \gamma_+ + \ln \gamma_-) = RT\ln \gamma_+\gamma_-$$

and write

$$G_m = (\mu_+^\ominus + RT\ln b_+) + (\mu_-^\ominus + RT\ln b_-) + RT\ln \gamma_+\gamma_-$$

We now write the term inside the logarithm as $\gamma_\pm^2$ and use $\ln x^2 = 2 \ln x$ to obtain

$$
\begin{aligned}
G_m &= (\mu_+^\ominus + RT \ln b_+) + (\mu_-^\ominus + RT \ln b_-) + 2RT \ln \gamma_\pm \\
&= (\mu_+^\ominus + RT \ln b_+ + RT \ln \gamma_\pm) + (\mu_-^\ominus + RT \ln b_- + RT \ln \gamma_\pm) \\
&= (\mu_+^\ominus + RT \ln \gamma_\pm b_+) + (\mu_-^\ominus + RT \ln \gamma_\pm b_-)
\end{aligned}
$$

We see that, with the mean activity coefficient defined as in eqn 5.3a, the deviation from ideal behavior (as expressed by the activity coefficient) is now shared equally between the two types of ion. In exactly the same way, the Gibbs energy of a salt M_pX_q can be written

$$
G_m = p(\mu_+^\ominus + RT \ln \gamma_\pm b_+) + q(\mu_-^\ominus + RT \ln \gamma_\pm b_-)
$$

with the mean activity coefficient defined as in eqn 5.3b.[1]

A brief illustration

Suppose that we have devised a method for determining the activity coefficients of Na^+ and SO_4^{2-} ions in 0.010 m Na_2SO_4(aq) and found them to be 0.98 and 0.84, respectively. It follows from eqn 5.3b that the mean activity coefficient is

$$
\gamma_\pm = \{(0.98)^2 \times (0.84)\}^{1/3} = 0.93
$$

because $p = 2$, $q = 1$, and $s = 3$. From eqn 5.1b, the activities of the two ions are

$$
a_+ = \gamma_\pm b_+ = 0.93 \times (2 \times 0.010) = 0.019
$$
$$
a_- = \gamma_\pm b_- = 0.93 \times (0.010) = 0.0093
$$

Self-test 5.1   Write an expression for the mean activity coefficient of Mg^{2+} and PO_4^{3-} in an aqueous solution of $Mg_3(PO_4)_2$.

Answer: $\gamma_\pm = (\gamma_+^3 \gamma_-^2)^{1/5}$

(b) Debye–Hückel theory

The question still remains about how the mean activity coefficients can be estimated. A theory that accounts for their values in very dilute solutions was developed by Peter Debye and Erich Hückel in 1923. They supposed that each ion in solution is surrounded by an **ionic atmosphere** of counter-charge. This 'atmosphere' is actually the slight imbalance of charge arising from the competition between the stirring effect of thermal motion, which tends to keep all the ions distributed uniformly throughout the solution, and the Coulombic interaction between ions, which tends to attract counter-ions (ions of opposite charge) into each other's vicinity and repel ions of like charge (Fig. 5.1). As a result of this competition, there is a slight preponderance of cations near any anion, giving a positively charged ionic atmosphere around the anion, and a slight preponderance of anions near any cation, giving a negatively charged ionic atmosphere around the cation. Because each ion is in an atmosphere of opposite charge, its

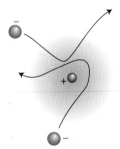

Fig. 5.1 The ionic atmosphere surrounding an ion consists of a slight excess of opposite charge as ions move through the vicinity of the central ion, with counter-ions lingering longer than ions of the same charge. The ionic atmosphere lowers the energy of the central ion.

---

[1] For the details of this general case, see our *Physical chemistry* (2010).

energy is lower than in a uniform, ideal solution, and therefore its chemical potential is lower than in an ideal solution. A lowering of the chemical potential of an ion below its ideal solution value is equivalent to the activity coefficient of the ion being less than 1 (because $\ln \gamma$ is negative when $\gamma < 1$). Debye and Hückel were able to derive an expression that is a limiting law in the sense that it becomes increasingly valid as the concentration of ions approaches zero. The **Debye–Hückel limiting law**[2] is

$$\log \gamma_{\pm} = -A|z_+ z_-|I^{1/2} \qquad \boxed{\text{Debye–Hückel limiting law}} \qquad (5.4)$$

(Note the common logarithm.) In this expression, A is a constant that for water at 25°C works out as 0.509. The z_j are the charge numbers of the ions (so $z_+ = +1$ for Na^+ and $z_- = -2$ for SO_4^{2-}); the vertical bars mean that we ignore the sign of the product. The quantity I is the dimensionless **ionic strength** of the solution, which is defined in terms of the molalities of the ions as

$$I = \tfrac{1}{2}(z_+^2 b_+ + z_-^2 b_-)/b^{\ominus} \qquad (5.5a)$$

When using this expression, we must include all the ions present in the solution, not just those of interest. For instance, if you are calculating the ionic strength of a solution of silver chloride and potassium nitrate, there are contributions to the ionic strength from all four types of ion. When more than two ions contribute to the ionic strength, we write

$$I = \tfrac{1}{2}\sum_i z_i^2 b_i/b^{\ominus} \qquad \boxed{\text{General expression for the ionic strength}} \qquad (5.5b)$$

where the symbol $\sum$ denotes a sum (in this case of all terms of the form $z_i^2 b_i$), z_i is the charge number of an ion i (positive for cations and negative for anions), and b_i is its molality.

(**A brief illustration**)

The sulfate ion, SO_4^{2-}, is an important source of sulfur used in the synthesis of the amino acids cysteine and methionine in plants and bacteria. To estimate the mean activity coefficient for the ions in 0.0010 m $Na_2SO_4(aq)$ at 25°C, we begin by evaluating the ionic strength of the solution from eqn 5.5:

$$I = \tfrac{1}{2}\{(+1)^2 \times (2 \times 0.0010) + (-2)^2 \times (0.0010)\} = 0.0030$$

Then we use the Debye–Hückel limiting law (eqn 5.4), with $A = 0.509$, to calculate $\log \gamma_{\pm}$:

$$\log \gamma_{\pm} = -0.509 \times |(+1)(-2)| \times (0.0030)^{1/2} = -2 \times 0.509 \times (0.0030)^{1/2}$$

(This expression evaluates to −0.056.) On taking the antilogarithm of $\log \gamma_{\pm}$ (by using $x = 10^{\log x}$), we conclude that $\gamma_{\pm} = 0.88$.

(**Self-test 5.2**) Estimate the mean activity coefficient of NaCl in a solution that is 0.020 m NaCl(aq) and 0.035 m $Ca(NO_3)_2(aq)$.

Answer: 0.661

---

[2] For a derivation of the Debye–Hückel limiting law, see our *Physical chemistry* (2010).

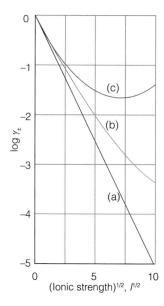

Fig. 5.2 The variation of the mean activity coefficient with ionic strength according to the extended Debye–Hückel theory. (a) The limiting law for a 1,1-electrolyte. (b) The extended law with $B = 0.5$. (c) The extended law, extended further by the addition of a term CI; in this case with $C = 0.02$. The last form of the law reproduces the observed behavior reasonably well.

As we have stressed, eqn 5.4 is a *limiting* law and is reliable only in very dilute solutions. For solutions more concentrated than about 10^{-3} M, it is better to use an empirical modification known as the **extended Debye–Hückel law**:

$$\log \gamma_{\pm} = \frac{A|z_+ z_-| I^{1/2}}{1 + BI^2} + CI \qquad \boxed{\text{Extended Debye–Hückel law}} \qquad (5.6)$$

with B and C empirically determined constants (Fig. 5.2).

5.2 Passive and active transport of ions across biological membranes

Nature has devised complex strategies for controlling the flow of ions across cell membranes, some of which are thermodynamic and others kinetic. Here we consider thermodynamic aspects of ion transport.

With a better understanding of how ions behave in solution, we are now prepared to discuss their behavior in cells. The concepts developed in the previous sections can be applied to the aqueous environment inside and outside cells. But we also need to know how ions traverse cell membranes and how such transport controls important biological processes, such as the synthesis of ATP and neuronal activity.

The thermodynamic tendency to transport a species A through a biological cell membrane is partially determined by an activity gradient across the membrane, which results in a difference in molar Gibbs energy between the inside and the outside of the cell

$$\Delta G_m = G_{m,in} - G_{m,out} = RT \ln a_{in}/a_{out} \qquad (5.7)$$

The equation implies that transport into the cell of either neutral or charged species is thermodynamically favorable if $a_{in} < a_{out}$ or, if we set the activity coefficients to 1, if $[A]_{in} < [A]_{out}$. An ion also needs to cross a membrane potential difference $\Delta\phi = \phi_{in} - \phi_{out}$ that arises from differences in Coulomb repulsions on each side of the bilayer. This potential difference is measured in volts (V, where $1\ \text{V} = 1\ \text{J C}^{-1}$). We show in the following *Justification* that the Gibbs energy of transfer of an ion of charge number z across a potential difference $\Delta\phi$ adds a term $zF\Delta\phi$ to eqn 5.7, where F is **Faraday's constant**, the magnitude of electric charge per mole of electrons:

$$F = eN_A = 96.485\ \text{kC mol}^{-1}$$

The final expression for G_m is then

$$\Delta G_m = RT \ln \frac{[A]_{in}}{[A]_{out}} + zF\Delta\phi \qquad \boxed{\begin{array}{l}\text{Gibbs energy of transfer of} \\ \text{an ion across a membrane} \\ \text{potential gradient}\end{array}} \qquad (5.8)$$

Justification 5.2   *The Gibbs energy of transfer of an ion across a membrane potential gradient*

The charge transferred per mole of ions of charge number z that cross a lipid bilayer is $N_A \times (ze)$, or zF, where $F = eN_A$. The work w' of transporting this charge is equal to the product of the charge and the potential difference $\Delta\phi$:

$$w' = zF \times \Delta\phi$$

Provided the work is done reversibly at constant temperature and pressure, we can equate this work to the molar Gibbs energy of transfer and write

$$\Delta G_m = zF\Delta\phi$$

Adding this term to eqn 5.7 gives eqn 5.8, the total Gibbs energy of transfer of an ion across both an activity and a membrane potential gradient.

Example 5.1   *Estimating a membrane potential*

Estimate the equilibrium membrane potential of a cell at 298 K by using the fact that the concentration of K^+ inside the cell is about 20 times that on the outside. Repeat the calculation, this time using the fact that the concentration of Na^+ outside the cell is about 10 times that on the inside.

Strategy Because the cell is at equilibrium, set $\Delta G_m = 0$ in eqn 5.8 and, after rearrangement, write

$$\Delta\phi = -\frac{RT}{zF}\ln\frac{[A]_{in}}{[A]_{out}}$$

where $z = +1$ for both K^+ and Na^+. Then calculate the equilibrium membrane potential from the given temperature and concentration ratios.

Solution When $[K^+]_{in}/[K^+]_{out} = 20$, we obtain

$$\Delta\phi = -\frac{(8.3145\ J\ K^{-1}\ mol^{-1}) \times (298\ K)}{9.648 \times 10^4\ C\ mol^{-1}}\ln 20 = -7.69 \times 10^{-2}\ V = -76.9\ mV$$

where we have used $1\ V = 1\ J\ C^{-1}$. The negative sign denotes that the inside has the lower potential. When $[Na^+]_{in}/[Na^+]_{out} = 0.10$, we obtain

$$\Delta\phi = -\frac{(8.3145\ J\ K^{-1}\ mol^{-1}) \times (298\ K)}{9.648 \times 10^4\ C\ mol^{-1}}\ln 0.10 = +5.91 \times 10^{-2}\ V = +59.1\ mV$$

and the positive sign denotes that the outside has the lower potential.

Self-test 5.3   Is the transport of Na^+ ions across a cell membrane spontaneous when $[Na^+]_{in}/[Na^+]_{out} = 0.10$ and $\Delta\phi = +50\ mV$?

Answer: Yes, because $\Delta G_m < 0$

Equation 5.8 implies that there is a tendency, called **passive transport**, for a species to move down concentration and membrane potential gradients. In **active transport**, a species moves against these gradients and the process is driven by its coupling to the exergonic hydrolysis of ATP. That is, when the sum of $RT\ln([A]_{in}/[A]_{out})$ and $zF\Delta\phi$ is positive, the overall Gibbs energy of transport can be made negative (and the process becomes spontaneous) by a large and negative Gibbs energy of ATP hydrolysis. It follows that the overall Gibbs energy of transport into a cell may be written as

$$\Delta G_m = RT\ln\frac{[A]_{in}}{[A]_{out}} + zF\Delta\phi + \Delta_r G^{ATP}$$

| Overall Gibbs energy of transfer of an ion across a biological membrane | (5.9) |

where $\Delta_r G^{ATP}$ is the Gibbs energy of hydrolysis of ATP at specific concentrations of ATP, ADP, P_i, and hydronium ion.

5.3 Ion channels and ion pumps

The mechanism of signal propagation along neurons in organisms is due to the migration of ions through membranes.

The transport of ions into or out of a cell needs to be mediated (that is, involve other species) because charged species do not partition well into the hydrophobic environment of the membrane. There are two mechanisms for ion transport: mediation by a carrier molecule or transport through a **channel former**, a protein that creates a hydrophilic pore through which the ion can pass. An example of a channel former is the polypeptide gramicidin A, which increases the membrane permeability to cations such as H^+, K^+, and Na^+.

Ion channels are proteins that permit the movement of specific ions down a membrane potential gradient. They are highly selective, so there is a channel protein for Ca^{2+}, another for Cl^-, and so on. In a *voltage-gated channel*, the opening of the gate is triggered by a membrane potential. In a *ligand-gated channel* ion transport is initiated by the binding of a molecule (called an 'effector molecule') to a specific receptor site on the channel.

Ions such as H^+, Na^+, K^+, and Ca^{2+} are often transported actively across membranes by membrane-spanning proteins called **ion pumps**. Ion pumps are molecular machines that work by adopting conformations that are permeable to one type of ion but not others, depending on the state of phosphorylation of the protein. Because protein phosphorylation requires dephosphorylation of ATP, the conformational change that opens or closes the pump is endergonic and requires the use of energy stored during metabolism. In Sections 5.10 and 8.8 we discuss the ion pump H^+-ATPase, which plays an important role in oxidative phosphorylation.

Case study 5.1 *Action potentials*

A striking example of the importance of ion channels is their role in the propagation of impulses by neurons, the fundamental units of the nervous system. Here we give a thermodynamic description of the process.

The cell membrane of a neuron is more permeable to K^+ ions than to either Na^+ or Cl^- ions. The key to the mechanism of action of a nerve cell is its use of Na^+ and K^+ channels to move ions across the membrane, modulating its potential. For example, the concentration of K^+ inside an inactive nerve cell is about 20 times that on the outside, whereas the concentration of Na^+ outside the cell is about 10 times that on the inside. The difference in concentrations of ions results in a transmembrane potential difference of about −62 mV. This potential difference is also called the **resting potential** of the cell membrane.

To estimate the resting potential, we need to understand that the cell is never at equilibrium, so the approach taken in *Example* 5.1 is not appropriate. Ions continually cross the membrane, which is more permeable to some ions than others. To take into account membrane permeability, we use the **Goldman equation** to calculate the resting potential:

$$\Delta\phi = \frac{RT}{F}\ln\frac{y}{y'} \qquad \boxed{\text{Goldman equation}} \quad (5.10)$$

with

$$y = \sum_i P_i[M_i^+]_{\text{out}} + \sum_j P_j[X_j^-]_{\text{in}}, \; y' = \sum_i P_i[M_i^+]_{\text{in}} + \sum_j P_j[X_j^-]_{\text{out}}$$

where P_i and P_j are the relative permeabilities, respectively, for the cation M_i^+ and the anion X_j^- and the sum is over all ions.

A brief illustration

For example, taking the permeabilities of the K^+, Na^+, and Cl^- ions as $P_{K^+} = 1.0$, $P_{Na^+} = 0.04$, and $P_{Cl^-} = 0.45$, respectively, the temperature as 298 K, and the concentrations as $[K^+]_{\text{in}} = 400$ mmol dm^{-3}, $[Na^+]_{\text{in}} = 50$ mmol dm^{-3}, $[Cl^-]_{\text{in}} = 50$ mmol dm^{-3}, $[K^+]_{\text{out}} = 20$ mmol dm^{-3}, $[Na^+]_{\text{out}} = 500$ mmol dm^{-3}, and $[Cl^-]_{\text{out}} = 560$ mmol dm^{-3}, we obtain

$$y = (1.0 \times 20) + (0.04 \times 500) + (0.45 \times 50) \text{ mmol dm}^{-3} = 62.5 \text{ mmol dm}^{-3}$$

$$y' = (1.0 \times 400) + (0.04 \times 50) + (0.45 \times 560) \text{ mmol dm}^{-3} = 654 \text{ mmol dm}^{-3}$$

$$\Delta\phi = \frac{(8.3145 \text{ J K}^{-1} \text{ mol}^{-1}) \times (298 \text{ K})}{9.648 \times 10^4 \text{ J mol}^{-1}} \times \ln\left(\frac{62.5}{645}\right)$$

$$= -6.0 \times 10^{-2} \text{ V} = -60 \text{ mV}$$

(The concentration units in the logarithm all cancel.) We see that the Goldman equation leads to an estimate that agrees well with the experimental value of −62 mV.

The transmembrane potential difference plays a particularly interesting role in the transmission of nerve impulses. On receiving an impulse, which is called an **action potential**, a site in the nerve cell membrane becomes transiently permeable to Na^+ and the transmembrane potential changes. To propagate along a nerve cell, the action potential must change the transmembrane potential by at least 20 mV to values that are less negative than −40 mV. Propagation occurs when an action potential at one site of the membrane triggers an action potential at an adjacent site, with sites behind the moving action potential relaxing back to the resting potential.

Redox reactions

We now embark on an investigation of the thermodynamics of redox reactions. Our ultimate goal is a description of electron transfer in plant photosynthesis and in the last stages of the oxidative breakdown of glucose. However, before we can understand these complex processes, we must examine a very much simpler system with a more controllable environment where precise measurements can be made. That is, we must consider electron transfer in an **electrochemical cell**, a device that consists of two electronic conductors (metal or graphite, for instance) dipping into an electrolyte (an ionic conductor), which may be a solution, a liquid, or a solid.

5.4 Half-reactions

A redox reaction, such as the breakdown of glucose by O_2 in biological cells, is the outcome of the loss of electrons, and perhaps atoms, from one species and their gain by another species; we need to be able to write chemical equations for redox reactions and the corresponding reaction quotients.

A brief comment

The oxidation number of a monatomic ion is equal to its charge number. An oxidation number is assigned to an element in a compound by supposing that it is present as an ion with a characteristic charge; for instance, oxygen is supposed—for this purpose—to be present as O^{2-} in most of its compounds, and hydrogen is supposed to be present as H^+.

It will be familiar from introductory chemistry that we identify the loss of electrons (oxidation) by noting whether or not an element has undergone an increase in oxidation number. We identify the gain of electrons (reduction) by noting whether or not an element has undergone a decrease in oxidation number. The requirement to break and form covalent bonds in some redox reactions, as in the conversion of H_2O to O_2 (during plant photosynthesis) or of N_2 to NH_3 (during nitrogen fixation by certain microorganisms) is one of the reasons why redox reactions often achieve equilibrium quite slowly, often much more slowly than acid–base proton transfer reactions.

Any redox reaction may be expressed as the difference of two reduction **half-reactions**. Two examples are

reduction of Cu^{2+}: $Cu^{2+}(aq) + 2\,e^- \rightarrow Cu(s)$

reduction of Zn^{2+}: $Zn^{2+}(aq) + 2\,e^- \rightarrow Zn(s)$

difference: $Cu^{2+}(aq) + Zn(s) \rightarrow Cu(s) + Zn^{2+}(aq)$ **(A)**

A half-reaction in which atom transfer accompanies electron transfer is

reduction of MnO_4^-:
$MnO_4^-(aq) + 8\,H^+(aq) + 5\,e^- \rightarrow Mn^{2+}(aq) + 4\,H_2O(l)$ **(B)**

where oxygen atoms are transferred from $MnO_4^-(aq)$ to $H_2O(l)$. In the discussion of redox reactions, the hydrogen ion is commonly denoted simply $H^+(aq)$ rather than treated as a hydronium ion, $H_3O^+(aq)$, as proton transfer is less of an issue and the chemical equations are simplified.

Half-reactions are *conceptual*. Redox reactions normally proceed by a much more complex mechanism in which the electron is never free. The electrons in these conceptual reactions are regarded as being 'in transit' and are not ascribed a state. The oxidized and reduced species in a half-reaction form a **redox couple**, denoted Ox/Red. Thus, the redox couples mentioned so far are Cu^{2+}/Cu, Zn^{2+}/Zn, and MnO_4^-, H^+/Mn^{2+}, H_2O. In general, we adopt the notation

couple: Ox/Red

half-reaction: $Ox + \nu\,e^- \rightarrow Red$

Example 5.2   *Expressing a reaction in terms of half-reactions*

Express the oxidation of nicotinamide adenine dinucleotide, which participates in aerobic metabolism, to NAD^+ (Atlas N4) by oxygen, when the latter is reduced to H_2O_2, in aqueous solution as the difference of two reduction half-reactions. The overall reaction is $NADH(aq) + O_2(g) + H^+(aq) \rightarrow NAD^+(aq) + H_2O_2(aq)$.

Strategy To express a reaction as the difference of two reduction half-reactions, identify one reactant species that undergoes reduction and its corresponding reduction product, then write the half-reaction for this process. To find the second half-reaction, subtract the first half-reaction from the overall reaction and rearrange the species so that all the stoichiometric coefficients are positive and the equation is written as a reduction.

Solution Oxygen is reduced to H_2O_2, so one half-reaction is

$$O_2(g) + 2\,H^+(aq) + 2\,e^- \rightarrow H_2O_2(aq)$$

Subtraction of this half-reaction from the overall equation gives

$$NADH(aq) - H^+(aq) - 2\,e^- \rightarrow NAD^+(aq)$$

Addition of $H^+(aq) + 2\,e^-$ to both sides gives

$$NADH(aq) \rightarrow NAD^+(aq) + H^+(aq) + 2\,e^-$$

This is an oxidation half-reaction. We reverse it to find the corresponding reduction half-reaction:

$$NAD^+(aq) + H^+(aq) + 2\,e^- \rightarrow NADH(aq)$$

Self-test 5.4 Express the formation of H_2O from H_2 and O_2 in acidic solution as the difference of two reduction half-reactions.

Answer: $4\,H^+(aq) + 4\,e^- \rightarrow 2\,H_2(g)$, $O_2(g) + 4\,H^+(aq) + 4\,e^- \rightarrow 2\,H_2O(l)$

We saw in Chapter 4 that a natural way to express the composition of a system is in terms of the reaction quotient Q. The quotient for a half-reaction is defined like the quotient for the overall reaction, but with the electrons ignored. Thus, for the half-reaction of the $NAD^+/NADH$ couple in Example 5.2 we would write

$$NAD^+(aq) + H^+(aq) + 2\,e^- \rightarrow NADH(aq) \qquad Q = \frac{a_{NADH}}{a_{NAD^+}a_{H^+}} \approx \frac{[NADH]}{[NAD^+][H^+]}$$

In elementary work, and provided the solution is very dilute, the activities are interpreted as the numerical values of the molar concentrations (see Table 3.3). The replacement of activities by molar concentrations is very hazardous for ionic solutions and intracellular fluids, as we have seen, so wherever possible we delay taking that final step.

Example 5.3 *Writing the reaction quotient for a half-reaction*

During the last stage of oxidative phosphorylation in mitochondria (Section 5.10), oxygen is reduced to water with the accompanying uptake of protons. Write the half-reaction and the reaction quotient for the reduction of oxygen to water in acidic solution.

Strategy Write the chemical equation for the half-reaction. Then express the reaction quotient in terms of the activities and the corresponding stoichiometric coefficients, with products in the numerator and reactants in the denominator. Pure (and nearly pure) solids and liquids do not appear in Q (because their activities are 1), nor does the electron. The activity of a gas is set equal to the numerical value of its partial pressure in bar (more formally: $a_J = p_J/p^\ominus$).

Solution The equation for the reduction of O_2 in acidic solution is

$$O_2(g) + 4\,H^+(aq) + 4\,e^- \rightarrow 2\,H_2O(l)$$

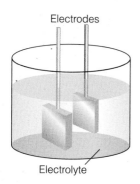

Electrodes

Electrolyte

Fig. 5.3 The arrangement for an electrochemical cell in which the two electrodes share a common electrolyte.

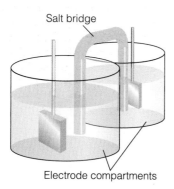

Salt bridge

Electrode compartments

Fig. 5.4 When the electrolytes in the electrode compartments of a cell are different, they need to be joined so that ions can travel from one compartment to another. One device for joining the two compartments is a salt bridge.

The reaction quotient for the half-reaction is therefore

$$Q = \frac{1}{p_{O_2} a_{H^+}^4}$$

Note the very strong dependence of Q on the hydrogen ion activity.

Self-test 5.5   Write the half-reaction and the reaction quotient for the reduction of chlorine gas to chloride ion.

Answer: $Cl_2(g) + 2\,e^- \rightarrow 2\,Cl^-(aq)$, $Q = a_{Cl^-}^2/p_{Cl_2}$

5.5 Reactions in electrochemical cells

The electron transfer processes that occur in respiration and photosynthesis can be modeled by electrochemical cells in which electrons are transferred between proteins.

In an electrochemical cell, the electronic conductor and its surrounding electrolyte is an **electrode**. The physical structure containing them is called an **electrode compartment**. The two electrodes may share the same compartment (Fig. 5.3). If the electrolytes are different, then the two compartments may be joined by a **salt bridge**, which is an electrolyte solution that completes the electrical circuit by permitting ions to move between the compartments (Fig. 5.4). Alternatively, the two solutions may be in direct physical contact (for example, through a porous membrane) and form a **liquid junction**. However, a liquid junction introduces complications to the interpretation of measurements, and we shall not consider it further.

(a) Galvanic and electrolytic cells

A **galvanic cell** is an electrochemical cell that produces electricity as a result of the spontaneous reaction occurring inside it.[3] An **electrolytic cell** is an electrochemical cell in which a nonspontaneous reaction is driven by an external source of direct current. The commercially available dry cells, mercury cells, nickel–cadmium ('nicad'), and lithium ion cells used to power electrical equipment are all galvanic cells and produce electricity as a result of the spontaneous chemical reaction between the substances built into them at manufacture. A **fuel cell** is a galvanic cell in which the reagents, such as hydrogen and oxygen or methane and oxygen, are supplied continuously from outside. Fuel cells are used on manned spacecraft, are beginning to be considered for use in automobiles, and gas supply companies hope that one day they may be used as a convenient, compact source of electricity in homes. Electric eels and electric catfish are biological versions of fuel cells in which the fuel is food and the cells are adaptations of muscle cells. It is hoped that biological versions of fuel cells will be developed to power the nanostructure-based entities that are being devised for inclusion in organisms.

In an electrochemical cell, the **anode** is where oxidation takes place; the **cathode** is where reduction takes place. As the reaction proceeds in a galvanic cell, the electrons released at the anode travel through the external circuit (Fig. 5.5). They re-enter the cell at the cathode, where they bring about reduction. This flow of current in the external circuit, from anode to cathode, corresponds to the cathode having a higher potential than the anode and arises from the tendency of

---

[3] The term *voltaic cell* is also used.

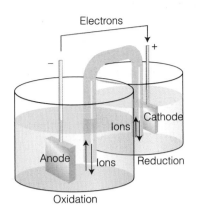

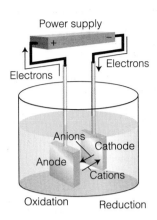

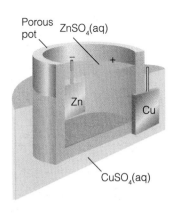

Fig. 5.5 The flow of electrons in the external circuit is from the anode of a galvanic cell, where they have been lost in the oxidation reaction, to the cathode, where they are used in the reduction reaction. Electrical neutrality is preserved in the electrolytes by the flow of cations and anions in opposite directions through the salt bridge.

Fig. 5.6 The flow of electrons and ions in an electrolytic cell. An external supply forces electrons into the cathode, where they are used to bring about a reduction, and withdraws them from the anode, which results in an oxidation reaction at that electrode. Cations migrate toward the negatively charged cathode and anions migrate toward the positively charged anode. An electrolytic cell usually consists of a single compartment, but a number of industrial versions have two compartments.

Fig. 5.7 A Daniell cell consists of copper in contact with copper(II) sulfate solution and zinc in contact with zinc sulfate solution; the two compartments are in contact through the porous pot that contains the zinc sulfate solution. The copper electrode is the cathode and the zinc electrode is the anode.

negatively charged electrons to travel to regions of higher potential. In an electrolytic cell, the anode is also the location of oxidation (by definition). Now, however, electrons must be withdrawn from the species in the anode compartment, so the anode must be connected to the positive terminal of an external supply. Similarly, electrons must pass from the cathode to the species undergoing reduction, so the cathode must be connected to the negative terminal of a supply (Fig. 5.6).

The simplest type of galvanic cell has a single electrolyte common to both electrodes (as in Fig. 5.3). In some cases it is necessary to immerse the electrodes in different electrolytes, as in the *Daniell cell* (Fig. 5.7), in which the redox couple at one electrode is Cu^{2+}/Cu and at the other is Zn^{2+}/Zn. In an **electrolyte concentration cell**, which would be constructed like the cell in Fig. 5.4, the electrode compartments are of identical composition except for the concentrations of the electrolytes. In an **electrode concentration cell** the electrodes themselves have different concentrations, either because they are gas electrodes operating at different pressures or because they are amalgams (solutions in mercury) with different concentrations.

In an electrochemical cell with two different electrolyte solutions in contact, as in the Daniell cell or an electrolyte concentration cell, the **liquid junction potential**, E_j, the potential difference across the interface of the two electrolytes, contributes to the overall potential difference generated by the cell. The contribution of the liquid junction to the potential can be decreased (to about 1 to 2 mV) by joining the electrolyte compartments through a salt bridge consisting of a saturated electrolyte solution (usually KCl) in agar jelly (as in Fig. 5.4). The reason for the success of the salt bridge is that the liquid junction potentials at either end are largely independent of the concentrations of the two more dilute solutions in the electrode compartments and so nearly cancel each other out.

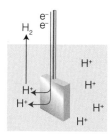

Fig. 5.8 The schematic structure of a hydrogen electrode, which is like other gas electrodes. Hydrogen is bubbled over a black (that is, finely divided) platinum surface that is in contact with a solution containing hydrogen ions. The platinum, as well as acting as a source or sink for electrons, speeds the electrode reaction because hydrogen attaches to (adsorbs on) the surface as atoms.

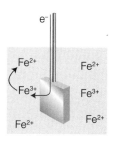

Fig. 5.9 The schematic structure of a redox electrode. The platinum metal acts as a source or sink for electrons required for the interconversion of (in this case) Fe^{2+} and Fe^{3+} ions in the surrounding solution.

(b) Varieties of electrodes

In a **gas electrode** (Fig. 5.8), a gas is in equilibrium with a solution of its ions in the presence of an inert metal. The inert metal, which is often platinum, acts as a source or sink of electrons but takes no other part in the reaction except perhaps to act as a catalyst. One important example is the *hydrogen electrode*, in which hydrogen is bubbled through an aqueous solution of hydrogen ions and the redox couple is H^+/H_2.

The term **redox electrode** is normally reserved for an electrode in which the couple consists of the same element in two nonzero oxidation states (Fig. 5.9). An example is an electrode in which the couple is Fe^{3+}/Fe^{2+}. In general, the reaction is

$$Ox + ve^- \rightarrow Red \qquad Q = \frac{a_{Red}}{a_{Ox}}$$

> **A brief illustration**
>
> For the electrode corresponding to the Fe^{3+}/Fe^{2+} couple the reduction half-reaction and reaction quotient are
>
> $$Fe^{3+}(aq) + e^- \rightarrow Fe^{2+}(aq) \qquad Q = \frac{a_{Fe^{2+}}}{a_{Fe^{3+}}}$$

(c) Electrochemical cell notation

It is useful to develop a short-hand to denote the processes taking place in an electrochemical cell. The notation scheme has a few rules:

- An interface between phases is denoted by a vertical bar, |.
- A double vertical line || denotes an interface for which the junction potential has been eliminated.
- A redox electrode is denoted M|Red,Ox, where M is an inert metal (typically platinum) making electrical contact with the solution.

It follows that the electrode corresponding to the Fe^{3+}/Fe^{2+} couple discussed in the previous *brief illustration* is denoted $Pt(s)|Fe^{2+}(aq),Fe^{3+}(aq)$. An electrochemical cell in which the left-hand electrode, in an arrangement like that in Fig. 5.4, is zinc in contact with aqueous zinc sulfate and the right-hand electrode is copper in contact with aqueous copper(II) sulfate is denoted

$$Zn(s)|ZnSO_4(aq)||CuSO_4(aq)|Cu(s)$$

> **A brief illustration**
>
> The hydrogen electrode is denoted $Pt(s)|H_2(g)|H^+(aq)$. In this electrode, the junctions are between the platinum and the gas and between the gas and the liquid containing its ions.

The current produced by a galvanic cell arises from the spontaneous reaction taking place inside it. The **cell reaction** is the reaction in the electrochemical cell written on the assumption that the right-hand electrode is the cathode and hence that reduction is taking place in the right-hand compartment. Later we see how to predict if the right-hand electrode is in fact the cathode; if it is, then the cell reaction is spontaneous as written. If the left-hand electrode turns out to be the cathode, then the reverse of the cell reaction is spontaneous.

To write the cell reaction corresponding to the electrochemical cell diagram, we first write the half-reactions at both electrodes as reductions and then subtract the equation for the left-hand electrode from the equation for the right-hand electrode. Thus, we saw in *Example* 5.2 that for the electrochemical cell used to study the reaction between NADH and O_2,

$$Pt(s)\,|\,NADH(aq),NAD^+(aq),H^+(aq)\,||\,H_2O_2(aq),H^+(aq)\,|\,O_2(g)\,|\,Pt(s)$$

the two reduction half-reactions are

right (R): $O_2(g) + 2\,H^+(aq) + 2\,e^- \rightarrow H_2O_2(aq)$

left (L): $NAD^+(aq) + H^+(aq) + 2\,e^- \rightarrow NADH(aq)$

The equation for the cell reaction is the difference:

overall (R − L): $NADH(aq) + O_2(g) + H^+(aq) \rightarrow NAD^+(aq) + H_2O_2(aq)$

In other cases, it may be necessary to match the numbers of electrons in the two half-reactions by multiplying one of the equations through by a numerical factor: there should be no spare electrons showing in the overall equation.

5.6 The Nernst equation

The concentrations of electroactive species in biological systems do not normally have their standard values, so we need to be able to relate the potential difference of a cell to the actual concentrations.

A galvanic cell does electrical work as the reaction drives electrons through an external circuit. The work done by a given transfer of electrons depends on the potential difference between the two electrodes. When the potential difference is large (for instance, 2 V), a given number of electrons traveling between the electrodes can do a lot of electrical work. When the potential difference is small (such as 2 mV), the same number of electrons can do only a little work. An electrochemical cell in which the reaction is at equilibrium can do no work and the potential difference between its electrodes is zero.

According to the discussion in Section 2.8, we know that the maximum non-expansion work that a system (in this context, the electrochemical cell performing electrical work, w') can do is given by the value of ΔG and in particular that

$$w'_{max} = \Delta G \qquad \boxed{\text{Maximum non-expansion work at constant temperature and pressure}} \qquad (5.11)$$

Therefore, by measuring the potential difference and converting it to the electrical work done by the reaction, we have a means of determining a thermodynamic quantity, the reaction Gibbs energy. Conversely, if we know ΔG for a reaction, then we have a route to the prediction of the potential difference between the electrodes of an electrochemical cell. However, to use eqn 5.11, we need to recall that maximum work is achieved only when a process occurs reversibly. In the present context, reversibility means that the electrochemical cell should be connected to an external source of potential difference that opposes and exactly matches the potential difference generated by the cell. Then an infinitesimal change of the external potential difference will allow the reaction to proceed in its spontaneous direction and an opposite infinitesimal change will drive the

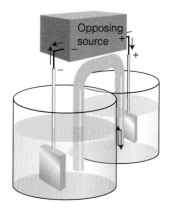

Fig. 5.10 The zero-current cell potential is measured by balancing the cell against an external potential that opposes the reaction in the cell. When there is no current flow, the external potential difference is equal to the zero-current cell potential.

reaction in its reverse direction.[4] The potential difference measured when an electrochemical cell is balanced against an external source of potential is called the **zero-current cell potential** (or simply the *cell potential*) and denoted E_{cell} (Fig. 5.10).[5] In practice, to determine the cell potential all we need do is to measure the potential difference with a voltmeter that draws negligible current.

As we show in the following *Justification*, the relation between the cell potential and the Gibbs energy of the electrochemical cell reaction is

$$-vFE_{cell} = \Delta_r G \qquad \boxed{\text{Relation between the cell potential and } \Delta_r G} \quad (5.12)$$

where F is Faraday's constant and v is the stoichiometric coefficient of the electrons in the matching half-reactions used to construct the overall reaction.

Justification 5.3   *The electrochemical cell potential*

Suppose the cell reaction can be broken down into half-reactions of the form $A + ve^- \rightarrow B$. Then, when the reaction takes place, vN_A electrons are transferred from the reducing agent to the oxidizing agent per mole of reaction events, so the charge transferred between the electrodes is $vN_A \times (-e)$, or $-vF$. Now we proceed as in *Justification* 5.2 and write the electrical work w' done when this charge travels from the anode to the cathode as the product of the charge and the potential difference E_{cell}:

$$w' = -vF \times E_{cell}$$

Provided the work is done reversibly at constant temperature and pressure, we can equate this electrical work to the reaction Gibbs energy and obtain eqn 5.12.

Equation 5.12 shows that the sign of E_{cell} is opposite to that of the reaction Gibbs energy, which we should recall is the slope of a graph of G plotted against the composition of the reaction mixture (Section 4.1). When the reaction is spontaneous in the forward direction, $\Delta_r G < 0$ and $E_{cell} > 0$. When $\Delta_r G > 0$, the reverse reaction is spontaneous and $E_{cell} < 0$. At equilibrium $\Delta_r G = 0$ and therefore $E_{cell} = 0$ too.

Equation 5.12 provides an electrical method for measuring a reaction Gibbs energy at any composition of the reaction mixture: we simply measure the zero-current cell potential and convert it to $\Delta_r G$. Conversely, if we know the value of $\Delta_r G$ at a particular composition, then we can predict the cell potential.

A brief illustration

Suppose $\Delta_r G \approx -1 \times 10^2 \, kJ \, mol^{-1}$ and $v = 1$; then

$$E_{cell} = \frac{-\Delta_r G}{vF} = \frac{-(-1 \times 10^5 \, J \, mol^{-1})}{1 \times (9.6485 \times 10^5 \, C \, mol^{-1})} = 1 \, V$$

Most electrochemical cells bought commercially are indeed rated at between 1 and 2 V.

---

[4] We saw in Chapter 1 that the criterion of thermodynamic reversibility is the reversal of a process by an infinitesimal change in the external conditions.

[5] This quantity was called the *electromotive force*, emf, of the electrochemical cell, but that name is deprecated by IUPAC because a potential difference is not a force.

Our next step is to see how E_{cell} varies with composition by combining eqn 5.12 and eqn 4.6, showing how the reaction Gibbs energy varies with composition:

$$\Delta_r G = \Delta_r G^\ominus + RT \ln Q$$

In this expression, $\Delta_r G^\ominus$ is the standard reaction Gibbs energy and Q is the reaction quotient for the cell reaction. When we substitute this relation into eqn 5.12 written as $E_{cell} = -\Delta_r G / \nu F$, we obtain the **Nernst equation**:

$$E_{cell} = E_{cell}^\ominus - \frac{RT}{\nu F} \ln Q \qquad \boxed{\text{Nernst equation}} \quad (5.13)$$

$E_{cell}^\ominus$ is the standard cell potential:

$$E_{cell}^\ominus = -\frac{\Delta_r G^\ominus}{\nu F} \qquad \boxed{\text{The standard cell potential}} \quad (5.14)$$

The standard cell potential is often interpreted as the cell potential when all the reactants and products are in their standard states (unit activity for all solutes, pure gases, and solids, a pressure of 1 bar). However, because such an electrochemical cell is not in general attainable, it is better to regard $E_{cell}^\ominus$ simply as the standard Gibbs energy of the reaction expressed as a potential. Note that if all the stoichiometric coefficients in the equation for a cell reaction are multiplied by a factor, then $\Delta_r G^\ominus$ is increased by the same factor, but so too is ν, so the standard cell potential is unchanged. Likewise, Q is raised to a power equal to the factor (so if the factor is 2, Q is replaced by Q^2), and because $\ln Q^2 = 2 \ln Q$, and likewise for other factors, the second term on the right-hand side of the Nernst equation is also unchanged. That is, E_{cell} is independent of how we write the balanced equation for the cell reaction.

A brief illustration

At 25.00°C,

$$\frac{RT}{F} = \frac{(8.31447 \text{ J K}^{-1}) \times (298.15 \text{ K})}{9.6485 \times 10^4 \text{ C mol}^{-1}} = 2.5693 \times 10^{-2} \text{ J C}^{-1}$$

Because $1 \text{ J} = 1 \text{ V C}$, $1 \text{ J C}^{-1} = 1 \text{ V}$, and $10^{-3} \text{ V} = 1 \text{ mV}$, we can write this result as

$$\frac{RT}{F} = 25.693 \text{ mV}$$

or approximately 25.7 mV. It follows from the Nernst equation that for a reaction in which $\nu = 1$, if Q is decreased by a factor of 10, then the potential of the electrochemical cell becomes more positive by $(25.7 \text{ mV}) \times \ln 10 = 59.2 \text{ mV}$. The reaction has a greater tendency to form products. If Q is increased by a factor of 10, then the cell potential falls by 59.2 mV and the reaction has a lower tendency to form products.

5.7 Standard potentials

To discuss the thermodynamics of biological processes, we need to be able to predict the standard reaction Gibbs energies of biological electron transfer reactions and their variation with pH.

Each electrode in a galvanic cell makes a characteristic contribution to the overall cell potential. Although it is not possible to measure the contribution of a single electrode, one electrode can be assigned a value zero and the others assigned relative values on that basis. The specially selected electrode is the **standard hydrogen electrode** (SHE):

$$\text{Pt(s)}|\text{H}_2\text{(g)}|\text{H}^+\text{(aq)} \qquad E^{\ominus} = 0 \text{ at all temperatures}$$

(a) Thermodynamic standard potentials

The (thermodynamic) **standard potential**, $E^{\ominus}(\text{Ox/Red})$, of a couple Ox/Red is measured by constructing an electrochemical cell in which the couple of interest forms the right-hand electrode and the standard hydrogen electrode is on the left. For example, the standard potential of the Ag^+/Ag couple is the standard potential of the cell

$$\text{Pt(s)}|\text{H}_2\text{(g)}|\text{H}^+\text{(aq)}||\text{Ag}^+\text{(aq)}|\text{Ag(s)}$$

and is $+0.80$ V. Table 5.1 lists a selection of standard potentials; a longer list will be found in the *Resource section*.

We saw in Section 4.2 that in biochemical work it is common to adopt the biological standard state (pH $= 7$, corresponding to neutral solution), rather than the thermodynamic standard state (pH $= 0$). To convert standard potentials to **biological standard potentials**, $E^{\oplus}$, we must first consider the variation of potential with pH. The two potentials differ when hydrogen ions are involved in the half-reaction, as in the fumaric acid/succinic acid couple fum/suc with fum $=$ HOOCCH=CHCOOH and suc $=$ HOOCCH$_2$CH$_2$COOH, which plays a role in the citric acid cycle (*Case study* 4.3):

$$\text{HOOCCH=CHCOOH(aq)} + 2\,\text{H}^+\text{(aq)} + 2\,\text{e}^- \rightarrow \text{HOOCCH}_2\text{CH}_2\text{COOH(aq)}$$

When hydrogen ions occur as reactants, as in this example, an increase in pH, corresponding to a decrease in hydrogen ion activity, favors the formation of reactants, so the fumaric acid has a lower thermodynamic tendency to become reduced. We expect, therefore, the potential of the fumaric/succinic acid couple to decrease as the pH is increased.

(b) Variation of potential with pH

To establish the quantitative variation of reduction potential with pH for a reaction as a first step in determining the effect of changing from pH $= 0$ to pH $= 7$ we use the Nernst equation. Thus, for fixed fumaric acid and succinic acid concentrations, the potential of the fumaric/succinic redox couple is

$$E = E^{\ominus} - \frac{RT}{2F}\ln\frac{a_{\text{suc}}}{a_{\text{fum}}a_{\text{H}^+}^2} = \overbrace{E^{\ominus} - \frac{RT}{2F}\ln\frac{a_{\text{suc}}}{a_{\text{fum}}}}^{E'} + \frac{RT}{T}\ln a_{\text{H}^+}$$

We then use a result from *Mathematical toolkit* 5.1 to write

$$\ln a_{\text{H}^+} = \ln 10 \times \log a_{\text{H}^+} = -\ln 10 \times \text{pH}$$

and obtain

$$E = E' - \frac{RT\ln 10}{F} \times \text{pH} \tag{5.15a}$$

A brief comment
Standard potentials are also called *standard electrode potentials* and *standard reduction potentials*. If in an older source of data you come across a 'standard oxidation potential,' reverse its sign and use it as a standard reduction potential.

Table 5.1 Standard potentials at 25°C

Reduction half-reaction			$E^{\ominus}/V$
Oxidizing agent		Reducing agent	
Strongly oxidizing			
F_2	$+2\,e^-$	$\rightarrow 2\,F^-$	+2.87
$S_2O_8^{2-}$	$+2\,e^-$	$\rightarrow 2\,SO_4^{2-}$	+2.05
Au^+	$+e^-$	$\rightarrow Au$	+1.69
Pb^{4+}	$+2\,e^-$	$\rightarrow Pb^{2+}$	+1.67
Ce^{4+}	$+e^-$	$\rightarrow Ce^{3+}$	+1.61
$MnO_4^- + 8\,H^+$	$+5\,e^-$	$\rightarrow Mn^{2+} + 4\,H_2O$	+1.51
Cl_2	$+2\,e^-$	$\rightarrow 2\,Cl^-$	+1.36
$Cr_2O_7^{2-} + 14\,H^+$	$+6\,e^-$	$\rightarrow 2\,Cr^{3+} + 7\,H_2O$	+1.33
$O_2 + 4\,H^+$	$+4\,e^-$	$\rightarrow 2\,H_2O$	+1.23,
			+0.81 at pH = 7
Br_2	$+2\,e^-$	$\rightarrow 2\,Br^-$	+1.09
Ag^+	$+e^-$	$\rightarrow Ag$	+0.80
Hg_2^{2+}	$+2\,e^-$	$\rightarrow 2\,Hg$	+0.79
Fe^{3+}	$+e^-$	$\rightarrow Fe^{2+}$	+0.77
I_2	$+e^-$	$\rightarrow 2\,I^-$	+0.54
$O_2 + 2\,H_2O$	$+4\,e^-$	$\rightarrow 4\,OH^-$	+0.40,
			0.81 at pH = 7
Cu^{2+}	$+2\,e^-$	$\rightarrow Cu$	+0.34
$AgCl$	$+e^-$	$\rightarrow Ag + Cl^-$	+0.22
$2\,H^+$	$+2\,e^-$	$\rightarrow H_2$	0, by definition
Fe^{3+}	$+3\,e^-$	$\rightarrow Fe$	−0.04
$O_2 + H_2O$	$+2\,e^-$	$\rightarrow HO_2^- + OH^-$	−0.08
Pb^{2+}	$+2\,e^-$	$\rightarrow Pb$	−0.13
Sn^{2+}	$+2\,e^-$	$\rightarrow Sn$	−0.14
Fe^{2+}	$+2\,e^-$	$\rightarrow Fe$	−0.44
Zn^{2+}	$+2\,e^-$	$\rightarrow Zn$	−0.76
$2\,H_2O$	$+2\,e^-$	$\rightarrow H_2 + 2\,OH^-$	−0.83,
			−0.42 at pH = 7
Al^{3+}	$+3\,e^-$	$\rightarrow Al$	−1.66
Mg^{2+}	$+2\,e^-$	$\rightarrow Mg$	−2.36
Na^+	$+e^-$	$\rightarrow Na$	−2.71
Ca^{2+}	$+2\,e^-$	$\rightarrow Ca$	−2.87
K^+	$+e^-$	$\rightarrow K$	−2.93
Li^+	$+e^-$	$\rightarrow Li$	−3.05
		Strongly reducing	

For a more extensive table, see the *Resource section*.

Note that this result is valid only for a half-reaction in which $v_e = 2$ and the stoichiometric coefficient of $H^+(aq)$ is 2 and appears as a reactant. In general:

$$E = E' - \frac{v_{H^+}RT\ln 10}{v_e F} \times pH$$

Variation of potential with pH   (5.15b)

> **A brief illustration**
>
> At 25°C and when $v_{H^+} = v_e$,
>
> $$E = E' - (59.2\text{ mV}) \times \text{pH}$$
>
> We see that an increase of 1 unit in pH decreases the potential by 59.2 mV, which is in agreement with the remark above, that the reduction of fumaric acid is discouraged by an increase in pH.

For a hydrogen electrode half-reaction, $\frac{1}{2} H_2(g) + e^- \rightarrow H^+(aq)$, with $E^{\ominus} = 0$, the same calculation gives

$$E = -\frac{RT \ln 10}{F} \times \text{pH}$$

This expression is the basis of a method for measuring the pK_a of an acid electrically. As we saw in Section 4.11, the pH of a solution containing equal amounts of the acid and its conjugate base is $\text{pH} = pK_a$. Therefore, by measuring the potential of the cell SHE||solution|HE, where SHE is a standard hydrogen electrode and HE is a hydrogen electrode dipping into the solution, we can determine the latter's pH and therefore the pK_a of the acid. It is in fact unwieldy to use an actual hydrogen electrode, and a far more convenient approach is developed later (*In the laboratory* 5.1).

(c) The biological standard potential

We can now use eqn 5.15 to convert standard potentials to biological standard potentials. If the hydrogen ions appear as reactants in the reduction half-reaction, then the potential is decreased below its standard value (for the fumaric/succinic couple, by 7×59.2 mV = 414 mV, or about 0.4 V). If the hydrogen ions appear as products, then the biological standard potential is higher than the thermodynamic standard potential. The precise change depends on the number of electrons and protons participating in the half-reaction, as expressed by eqn 5.15b and illustrated in *Example* 5.4. Biological standard potentials are important in the discussion of the electron transfer reactions of oxidative phosphorylation (Section 5.10). Table 5.2 is a partial list of biological standard potentials for redox couples that participate in important biochemical electron transfer reactions.

> **Example 5.4**   *Converting a standard potential to a biological standard value*
>
> Calculate the biological standard potential of the NAD/NADH couple at 25°C (*Example* 5.2) from its thermodynamic value. The reduction half-reaction is
>
> $$NAD^+(aq) + H^+(aq) + 2\,e^- \rightarrow NADH(aq) \qquad E^{\ominus} = -0.11\text{ V}$$
>
> *Strategy* Write the Nernst equation for the potential, and express the reaction quotient in terms of the activities of the species. All species except H^+ are in their standard states, so their activities are all equal to 1. The remaining task is to express the hydrogen ion activity in terms of the pH, exactly as was done in the text, and set pH = 7.
>
> *Solution* The Nernst equation for the half-reaction, with $v_e = 2$, is

Table 5.2 Biological standard potentials at 25°C

Reduction half-reaction			$E^{\oplus}/V$
Oxidizing agent		Reducing agent	
Strongly oxidizing			
$O_2 + 4\,H^+$	$+4\,e^-$	$\rightarrow 2\,H_2O$	+0.81
$Fe^{3+}\,(Cyt\,f)$	$+e^-$	$\rightarrow Fe^{2+}\,(Cyt\,f)$	+0.36
$O_2 + 2\,H_2O$	$+4\,e^-$	$\rightarrow 2\,H_2O_2$	+0.30
$Fe^{3+}\,(Cyt\,c)$	$+e^-$	$\rightarrow Fe^{2+}\,(Cyt\,c)$	+0.25
$Fe^{3+}\,(Cyt\,b)$	$+e^-$	$\rightarrow Fe^{2+}\,(Cyt\,b)$	+0.08
Dehydroascorbic acid $+ 2\,H^+$	$+2\,e^-$	$\rightarrow$ Ascorbic acid	+0.08
Coenzyme Q $+ 2\,H^+$	$+2\,e^-$	$\rightarrow$ Coenzyme QH_2	+0.04
Oxaloacetate$^{2-} + 2\,H^+$	$+2\,e^-$	$\rightarrow$ Malate^{2-}	−0.17
Pyruvate$^- + 2\,H^+$	$+2\,e^-$	$\rightarrow$ Lactate$^-$	−0.18
FAD $+ 2\,H^+$	$+2\,e^-$	$\rightarrow FADH_2$	−0.22
Glutathione (ox) $+ 2\,H^+$	$+2\,e^-$	$\rightarrow$ Glutathione (red)	−0.23
Lipoic acid (ox) $+ 2\,H^+$	$+2\,e^-$	$\rightarrow$ Lipoic acid (red)	−0.29
$NAD^+ + H^+$	$+2\,e^-$	$\rightarrow$ NADH	−0.32
$2\,H_2O$	$+2\,e^-$	$\rightarrow H_2 + 2\,OH^-$	−0.42
Ferredoxin (ox)	$+e^-$	$\rightarrow$ Ferredoxin (red)	−0.43
O_2	$+e^-$	$\rightarrow O_2^-$	−0.45
		Strongly reducing	

For a more extensive table, see the *Resource section*.

$$E = E^{\oplus} - \frac{RT}{2F}\ln\frac{\overbrace{a_{NADH}}^{1}}{a_{H^+}\underbrace{a_{NAD^+}}_{1}} = E^{\oplus} + \frac{RT}{2F}\ln a_{H^+}$$

We rearrange this expression to

$$E = E^{\oplus} + \frac{RT}{2F}\ln a_{H^+} = E^{\oplus} - \frac{RT\ln 10}{2F}\times pH = E^{\oplus} - (29.58\ \text{mV})\times pH$$

The biological standard potential (at pH = 7) is therefore

$$E^{\oplus} = (-0.11\ \text{V}) - (29.58\times 10^{-3}\ \text{V})\times 7 = -0.32\ \text{V}$$

A note on good practice
Whenever possible, avoid replacing activities by concentrations, especially when aiming to relate the electrode potential to pH, for the latter is defined in terms of the activity of hydrogen ions.

Self-test 5.6 Calculate the biological standard potential of the half-reaction $O_2(g) + 4\,H^+(aq) + 4\,e^- \rightarrow 2\,H_2O(l)$ at 25°C given its value +1.23 V under thermodynamic standard conditions.

Answer: +0.82 V

In the laboratory 5.1 *Ion-selective electrodes*

Special electrodes can be constructed to measure concentrations of ionic species, such as Na^+, K^+, Ca^{2+}, and hydronium ions, which are important in biochemical reactions. The potential of a hydrogen electrode is directly

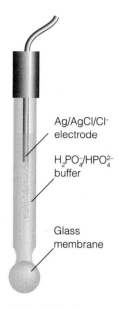

Ag/AgCl/Cl⁻
electrode

$H_2PO_4^-/HPO_4^{2-}$
buffer

Glass
membrane

Fig. 5.11 A glass electrode has a potential that varies with the hydrogen ion concentration in the medium in which it is immersed. It consists of a thin glass membrane containing an electrolyte and a silver chloride electrode, $Ag(s)|AgCl(s)|Cl^-(aq)$. The electrode is used in conjunction with a reference electrode, such as a calomel electrode, $Hg(l)|Hg_2Cl_2(s)|Cl^-(aq)$, that makes contact with the test solution through a salt bridge.

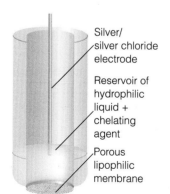

Silver/
silver chloride
electrode

Reservoir of
hydrophilic
liquid +
chelating
agent

Porous
lipophilic
membrane

Fig. 5.12 The structure of an ion-selective electrode. Ions bound to a compound, the chelating agent, in the hydrophilic liquid are able to migrate through the lipophilic membrane.

proportional to the pH of the solution. However, in practice, indirect methods are much more convenient to use than one based on the standard hydrogen electrode, and the hydrogen electrode is replaced by a *glass electrode* (Fig. 5.11). A glass electrode is an example of an **ion-selective electrode**, an electrode that generates a potential in response to the presence of a solution of specific ions.

The glass of a glass electrode is based on lithium silicate doped with heavy-metal oxides; it is filled with a phosphate buffer solution containing Cl^- ions. Conveniently, the electrode has $E_{cell} \approx 0$ when the external medium is at $pH = 7$. The electrode is calibrated using solutions of known pH (for example, one of the buffer solutions described in Section 4.11).

Self-test 5.7   What range should a voltmeter have (in volts) to display changes of pH from 1 to 14 at 25°C if it is arranged to give a reading of zero when $pH = 7$?

Answer: From −0.42 V to +0.35 V, a range of 0.77 V

Glass electrodes can be made responsive to Na^+, K^+, and NH_4^+ ions by using glasses doped with Al_2O_3 and B_2O_3. More sophisticated devices can extend the range of ions that can be detected in a test solution. For example, a porous hydrocarbon-attracting membrane can be attached to a small reservoir of a hydrophobic liquid, such as dioctylphenylphosphonate, that saturates it (Fig. 5.12). The liquid contains a compound, such as $(RO)_2PO_2^-$ with R a C_8 to C_{18} chain, which binds to the ion. The bound ions traverse the membrane and give rise to a transmembrane potential, which is detected by an electrode in the assembly. Electrodes of this construction can be designed to be sensitive to a variety of ionic species, including Ca^{2+} ions.

Applications of standard potentials

The measurement of the potential of an electrochemical cell is a convenient source of thermodynamic information on reactions. In practice the standard values (and the biological standard values) of these quantities are the ones normally determined.

5.8 The determination of thermodynamic functions

Calorimetry is not always practicable, especially for biochemically important reactions, but in some cases their thermodynamic properties can be measured electrochemically.

We have seen that the standard potential of an electrochemical cell is related to the standard reaction Gibbs energy by eqn 5.14 ($\Delta_r G^{\ominus} = -\nu F E_{cell}^{\ominus}$), therefore, by measuring the standard potential of a cell driven by the reaction of interest, we can obtain the standard reaction Gibbs energy. If we were interested in the biological standard state, then we would use the same expression but with the standard potential at $pH = 7$ ($\Delta_r G^{\oplus} = -\nu F E_{cell}^{\oplus}$). From the standard reaction Gibbs energy, the equilibrium constant, standard entropy, and standard enthalpy can be calculated.

(a) Calculation of the equilibrium constant

A special case of the Nernst equation has great importance in chemistry. Suppose the reaction has reached equilibrium; then $Q = K$, where K is the equilibrium constant of the cell reaction. However, because a chemical reaction at equilibrium cannot do work, it generates zero potential difference between the electrodes. Setting $Q = K$ and $E_{cell} = 0$ in the Nernst equation gives

$$\ln K = \frac{\nu F E^{\ominus}_{cell}}{RT}$$

The equilibrium constant in terms of the standard cell potential (5.16)

This very important equation—which is simply eqn 4.10 expressed electrochemically—lets us predict equilibrium constants from the standard potential of an electrochemical cell. Note that

- If $E^{\ominus}_{cell} > 0$, then $K > 1$ and at equilibrium the cell reaction lies in favor of products.
- If $E^{\ominus}_{cell} < 0$, then $K < 1$ and at equilibrium the cell reaction lies in favor of reactants.

A brief illustration

Because the standard potential of the Daniell cell is +1.10 V, the equilibrium constant for the cell reaction (reaction A) is

$$\ln K = \frac{2 \times (9.6485 \times 10^4 \, C \, mol^{-1}) \times (1.10 \, V)}{(8.3145 \, J \, K^{-1} \, mol^{-1}) \times (298.15 \, K)}$$

$$= \frac{2 \times 9.6485 \times 1.10 \times 10^4}{8.3145 \times 298.15}$$

(where we have used $1 \, C \, V = 1 \, J$ to cancel units) and therefore $K = 1.5 \times 10^{37}$. Hence, the displacement of copper by zinc goes virtually to completion in the sense that the ratio of concentrations of Zn^{2+} ions to Cu^{2+} ions at equilibrium is about 10^{37}. This value is far too large to be measured by classical analytical techniques, but its electrochemical measurement is straightforward. Note that a standard cell potential of +1 V corresponds to a very large equilibrium constant (and −1 V would correspond to a very small one).

It is also possible to use the data in Tables 5.1 and 5.2 to calculate the standard potential of an electrochemical cell formed from any pair of electrodes and, from the standard cell potential, the equilibrium constant of the cell reaction. To calculate the standard potential of an electrochemical cell, we take the difference of the standard potentials of the appropriate electrodes:

$$E^{\ominus}_{cell} = E^{\ominus}_{R} - E^{\ominus}_{L}$$

The standard cell potential from standard potentials (5.17a)

where $E^{\ominus}_{R}$ is the standard potential of the right-hand electrode and $E^{\ominus}_{L}$ is that of the left. The analogous expression for the biological standard state is

$$E_{cell}^{\oplus} = E_R^{\oplus} - E_L^{\oplus}$$

The biological standard cell potential from biological standard potentials   (5.17b)

When dealing with biological systems, the focus is not necessarily on reactions occurring at electrodes but on electron transfer processes in the cytosol or membranes of biological cells. We can still estimate the standard reaction Gibbs energy (and hence the equilibrium constant) of biological electron transfer reactions by using eqn 5.14 (written as $\Delta_r G^{\oplus} = -\nu F E_{cell}^{\oplus}$) if we express the chemical equation for the redox reaction as the difference of two reduction half-reactions with known standard potentials. We then find $E_{cell}^{\ominus}$ or $E_{cell}^{\oplus}$ from eqn 5.17 and use eqn 5.14 for the calculation of the standard reaction Gibbs energy or eqn 5.16 for the calculation of the equilibrium constant. The approach is illustrated in the following example.

Example 5.5   *Calculating the equilibrium constant of a biological electron transfer reaction*

The reduced and oxidized forms of riboflavin form a couple with $E^{\oplus} = -0.21$ V and the acetate/acetaldehyde couple has $E^{\oplus} = -0.60$ V under the same conditions. What is the equilibrium constant for the reduction of riboflavin (Rib) by acetaldehyde (ethanal) in neutral solution at 25°C? The reaction is

$$RibO(aq) + CH_3CHO(aq) \rightleftharpoons Rib(aq) + CH_3COOH(aq)$$

where RibO is the oxidized form of riboflavin and Rib is the reduced form.

Strategy The aim is to find the values of $E_{cell}^{\oplus}$ and ν corresponding to the reaction, for then we can use a modified form of eqn 5.16 to calculate the value of K in neutral solution from $E_{cell}^{\oplus}$. To do so, we express the equation as the difference of two reduction half-reactions. The stoichiometric number of the electron in these matching half-reactions is the value of ν we require. We then look up the biological standard potentials for the couples corresponding to the half-reactions and calculate their difference to find $E_{cell}^{\oplus}$.

Solution The two reduction half-reactions are

right: $RibO(aq) + 2 H^+(aq) + 2 e^- \rightarrow Rib(aq) + H_2O(l)$ $E^{\oplus} = -0.21$ V

left: $CH_3COOH(aq) + 2 H^+(aq) + 2 e^- \rightarrow CH_3CHO(aq) + H_2O(l)$
    $E^{\oplus} = -0.60$ V

and their difference is the redox reaction required. Note that $\nu = 2$. The corresponding standard cell potential is

$$E_{cell}^{\oplus} = (-0.21 \text{ V}) - (-0.60 \text{ V}) = +0.39 \text{ V}$$

It follows that

$$\ln K = \frac{2 F E_{cell}^{\oplus}}{RT} = \frac{2 \times (9.6485 \times 10^4 \text{ C mol}^{-1}) \times (0.39 \text{ V})}{(8.3145 \text{ J K}^{-1} \text{ mol}^{-1}) \times (298.15 \text{ K})}$$

$$= \frac{2 \times 9.6485 \times 0.39 \times 10^4}{8.3145 \times 298.15}$$

Therefore, because $K = e^{\ln K}$,

$$K = e^{(2 \times 9.6485 \times 0.39 \times 10^4)/(8.3145 \times 298.15)} = 1.5 \times 10^{13}$$

We conclude that riboflavin can be reduced by acetaldehyde in neutral solution. However, there may be mechanistic reasons—the energy required to break covalent bonds, for instance—that make the reduction too slow to be feasible in practice. Note that, because hydrogen ions do not appear in the chemical equation, the equilibrium constant is independent of pH.

Self-test 5.8 What is the equilibrium constant for the reduction of riboflavin with rubredoxin, a bacterial iron–sulfur protein, in the reaction

$$\text{riboflavin(ox)} + \text{rubredoxin(red)} \rightleftharpoons \text{riboflavin(red)} + \text{rubredoxin(ox)}$$

given that the biological standard potential of the rubredoxin couple is −0.06 V?

Answer: 8.5×10^{-6}; the reactants are favored

(b) Calculation of standard potentials

The relation between the standard cell potential and the standard reaction Gibbs energy is a convenient route for the calculation of the standard potential of a couple from two other standard potentials. We make use of the fact that G is a state function and that the Gibbs energy of an overall reaction is the sum of the Gibbs energies of the reactions into which it can be divided. In general, we cannot combine the E values directly because they depend on the value of ν, which may be different for the two couples.

Example 5.6 *Calculating a standard potential from two other standard potentials*

The superoxide ion (O_2^-) is an undesirable by-product of some enzyme-catalyzed reactions. It is metabolized by the enzyme superoxide dismutase (SOD) in a *disproportionation* (or *dismutation*), a reaction that both oxidizes and reduces a species. The reaction catalyzed by SOD is

$$2\,O_2^-(aq) + 2\,H^+(aq) \rightarrow H_2O_2(aq) + O_2(g)$$

where O_2^- is oxidized to O_2 and reduced to O_2^{2-} (in H_2O_2). Hydrogen peroxide, H_2O_2, is also produced by other biochemical reactions. It is a toxic substance that is metabolized by catalases and peroxidases. The disproportionation catalyzed by catalase is

$$2\,H_2O_2(aq) \rightarrow 2\,H_2O(l) + O_2(g)$$

Given the standard potentials $E^{\oplus}(O_2, O_2^-) = -0.45$ V and $E^{\oplus}(O_2, H_2O_2) = +0.30$ V, calculate $E^{\oplus}(O_2^-, H_2O_2)$, the biological standard potential for the SOD-catalyzed reduction of O_2^- to H_2O_2.

Strategy We need to convert the two $E^{\oplus}$ to $\Delta_r G^{\oplus}$ by using eqn 5.14 modified for the biological standard state, add them appropriately, and then convert the overall $\Delta_r G^{\oplus}$ so obtained to the required $E^{\oplus}$ by using eqn 5.14 again. Because the Fs cancel at the end of the calculation, carry them through.

Solution The electrode reactions are as follows:

(a) $O_2(g) + e^- \rightarrow O_2^-(aq)$

$E^{\oplus} = -0.45$ V $\Delta_r G^{\oplus}(a) = -F \times (-0.45\text{ V}) = (+0.45\text{ V}) \times F$

(b) $O_2(g) + 2\,H^+(aq) + 2\,e^- \rightarrow H_2O_2(aq)$

$E^{\oplus} = +0.30$ V $\Delta_r G^{\oplus}(b) = -2F \times (0.30\text{ V}) = (-0.60\text{ V}) \times F$

The required reaction is

(c) $O_2^-(aq) + 2 H^+(aq) + e^- \rightarrow H_2O_2(aq)$ $\Delta_r G^\oplus(c) = -FE^\oplus$

Because (c) = (b) − (a), it follows that

$$\Delta_r G^\oplus(c) = \Delta_r G^\oplus(b) - \Delta_r G^\oplus(a)$$

Therefore, from eqn 5.14,

$$FE^\oplus(c) = -\{(-0.60\ V)F - (+0.45\ V)F\}$$

The F_s cancel, and we are left with $E^\oplus(c) = +1.05\ V$.

Self-test 5.9 Given the standard potentials $E^\oplus(Fe^{3+},Fe) = -0.04\ V$ and $E^\oplus(Fe^{2+},Fe) = -0.44\ V$, calculate $E^\oplus(Fe^{3+},Fe^{2+})$.

Answer: +0.76 V

A note on good practice
Whenever combining standard potentials to obtain the standard potential of a third couple, always work via the Gibbs energies because they are additive, whereas in general standard potentials are not.

(c) Calculation of the standard reaction entropy and enthalpy

Once $\Delta_r G^\oplus$ has been measured, we can use thermodynamic relations to determine other properties. For instance, the entropy of the cell reaction can be obtained from the change in the potential with temperature:

$$\Delta_r S^\oplus = vF \frac{dE^\oplus_{cell}}{dT}$$

> The standard reaction entropy from the standard cell potential   (5.18)

Justification 5.4   *The reaction entropy from the electrochemical cell potential*

In Section 3.3 we used the fact that, at constant pressure, when the temperature changes by dT, the Gibbs energy changes by $dG = -SdT$. Because this equation applies to the reactants and the products, it follows that

$$d(\Delta_r G^\oplus) = -\Delta_r S^\oplus \times dT$$

Substitution of $\Delta_r G^\oplus = -vFE^\oplus_{cell}$ then gives

$$vF \times dE^\oplus_{cell} = \Delta_r S^\oplus \times dT$$

which rearranges into eqn 5.18.

A brief comment
Infinitesimally small quantities may be treated like any other quantity in algebraic manipulations. Thus, the expression $dy = adx$ may be rewritten as $dy/dx = a$, $dx/dy = 1/a$, and so on.

We see from eqn 5.18 that the standard cell potential increases with temperature if the standard reaction entropy is positive and that the slope of a plot of potential against temperature is proportional to the reaction entropy (Fig. 5.13). An implication is that if the cell reaction produces a lot of gas (corresponding to a positive reaction entropy), then its potential will increase with temperature. The opposite is true for a reaction that consumes gas.

Finally, we can combine the results obtained so far by using $G = H - TS$ in the form $H = G + TS$ to obtain the standard reaction enthalpy:

$$\Delta_r H^\oplus = \Delta_r G^\oplus + T\Delta_r S^\oplus$$

> The standard reaction enthalpy from the standard reaction Gibbs energy and entropy   (5.19)

with $\Delta_r G^\oplus$ determined from the cell potential and $\Delta_r S^\oplus$ from its temperature variation. Thus, we now have a noncalorimetric method of measuring a reaction enthalpy.

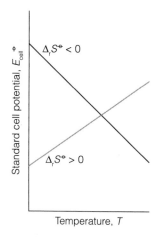

Fig. 5.13 The variation of the standard potential of a cell with temperature depends on the standard entropy of the cell reaction.

5.9 The electrochemical series

Some organic co-factors and metal centers in proteins act as electron transfer agents in a number of biological processes; we need to be able to predict which species is reduced or oxidized in a redox reaction.

We have seen that a cell reaction has $K > 1$ if $E_{cell}^{\oplus} > 0$ and that $E_{cell}^{\oplus} > 0$ corresponds to reduction at the right-hand electrode. We have also seen that $E_{cell}^{\oplus}$ may be written as the difference of the standard potentials of the redox couples in the right and left electrodes (eqn 5.17, $E_{cell}^{\oplus} = E_R^{\oplus} - E_L^{\oplus}$). A reaction corresponding to reduction at the right-hand electrode therefore has $K > 1$ if $E_L^{\oplus} < E_R^{\oplus}$, and we can conclude that

A couple with a low standard potential has a thermodynamic tendency to reduce a couple with a high standard potential.

More briefly: *low reduces high* and, equivalently, *high oxidizes low*. The same arguments apply to the biological standard values of the potentials.

A brief illustration

Consider the iron-containing protein ferredoxin, which participates in plant photosynthesis (Section 5.11), and cytochrome c, which participates in the last steps of respiration (Section 5.10). It follows from Table 5.2 that

$E^{\oplus}(\text{ferredoxin}_{ox},\text{ferredoxin}_{red}) = -0.43 \text{ V} < E^{\oplus}(\text{cyt } c_{ox},\text{Cyt } c_{red}) = +0.25 \text{ V}$

and ferredoxin has a thermodynamic tendency to reduce cytochrome c at $pH = 7$. Hence, $K > 1$ for the reaction

$\text{Cyt } c_{ox}(aq) + \text{ferredoxin}_{red}(aq) \rightleftharpoons \text{Cyt } c_{red}(aq) + \text{ferredoxin}_{ox}(aq)$

Self-test 5.10 Does NAD^+ have a thermodynamic tendency to oxidize the pyruvate ion at $pH = 7$?

Answer: No

Electron transfer in bioenergetics

Electron transfer between protein-bound cofactors or between proteins plays a role in a number of biological processes, such as the oxidative breakdown of foods, photosynthesis, nitrogen fixation, the reduction of atmospheric N_2 to NH_3 by certain microorganisms, and the mechanisms of action of oxidoreductases, which are enzymes that catalyze redox reactions. Here, we examine the redox reactions associated with photosynthesis and the aerobic oxidation of glucose. These processes are related by the reactions

$$C_6H_{12}O_6(s) + 6 \, O_2(g) \underset{\text{photosynthesis}}{\overset{\text{aerobic oxidation}}{\rightleftharpoons}} 6 \, CO_2(g) + 6 \, H_2O(l)$$

5.10 The respiratory chain

The centrally important processes of biochemistry include the electrochemical reactions between proteins in the mitochondrion of the cell, for they are responsible for delivering the electrons extracted from glucose to water.

The half-reactions for the oxidation of glucose and the reduction of O_2 are

$$C_6H_{12}O_6(s) + 6 H_2O(l) \rightarrow 6 CO_2(g) + 24 H^+(aq) + 24 e^-$$
$$6 O_2(g) + 24 H^+(aq) + 24 e^- \rightarrow 12 H_2O(l)$$

We see that the exergonic oxidation of one $C_6H_{12}O_6$ molecule requires the transfer of 24 electrons to six O_2 molecules. However, the electrons do not flow directly from glucose to O_2. In biological cells, glucose is oxidized to CO_2 by NAD^+ and FAD during glycolysis and the citric acid cycle (*Case study* 4.3):

$$C_6H_{12}O_6(s) + 10 NAD^+ + 2 FAD + 4 ADP + 4 P_i + 2 H_2O$$
$$\rightarrow 6 CO_2 + 10 NADH + 2 FADH_2 + 4 ATP + 6 H^+$$

In the **respiratory chain**, electrons from the powerful reducing agents NADH and $FADH_2$ pass through four membrane-bound protein complexes and two mobile electron carriers before reducing O_2 to H_2O. We shall see that the electron transfer reactions drive the synthesis of ATP at three of the membrane protein complexes.

(a) Electron transfer reactions

The respiratory chain begins in complex I (NADH-Q oxidoreductase), where NADH is oxidized by coenzyme Q (Q, Atlas M5) in a two-electron reaction:

$$H^+ + NADH + Q \xrightarrow{\text{complex I}} NAD^+ + QH_2$$
$$E_{cell}^{\ominus} = +0.42 \text{ V}, \Delta_r G^{\ominus} = -81 \text{ kJ mol}^{-1}$$

where the reduction of Q to Q^{2-} is accompanied by uptake of two H^+ ions to yield QH_2. Additional Q molecules are reduced by $FADH_2$ in complex II (succinate-Q reductase):

$$FADH_2 + Q \xrightarrow{\text{complex II}} FAD + QH_2$$
$$E_{cell}^{\ominus} = +0.32 \text{ V}, \Delta_r G^{\ominus} = -62 \text{ kJ mol}^{-1}$$

Reduced Q migrates to complex III (Q-cytochrome c oxidoreductase), which catalyzes the reduction of the protein cytochrome c (Cyt c). Cytochrome c contains the heme c group, the central iron ion of which can exist in oxidation states +3 and +2. The net reaction catalyzed by complex III is

$$QH_2 + 2 Fe^{3+}(Cyt\ c) \xrightarrow{\text{complex III}} Q + 2 Fe^{2+}(Cyt\ c) + 2 H^+$$
$$E_{cell}^{\ominus} = +0.15 \text{ V}, \Delta_r G^{\ominus} = -29 \text{ kJ mol}^{-1}$$

Reduced cytochrome c carries electrons from complex III to complex IV (cytochrome c oxidase), where O_2 is reduced to H_2O:

$$2 Fe^{2+}(Cyt\ c) + 2 H^+ + \tfrac{1}{2} O_2 \xrightarrow{\text{complex IV}} 2 Fe^{3+}(Cyt\ c) + H_2O$$
$$E_{cell}^{\ominus} = +0.56 \text{ V}, \Delta_r G^{\ominus} = -108 \text{ kJ mol}^{-1}$$

(b) Oxidative phosphorylation

The reactions that occur in complexes I, II, III, and IV are exergonic and together could drive the synthesis of ATP:

$$ADP + P_i + H^+ \rightarrow ATP \qquad \Delta_r G^{\ominus} = +31 \text{ kJ mol}^{-1}$$

We saw in *Case study* 4.2 that the phosphorylation of ADP to ATP can be coupled to the exergonic dephosphorylation of other molecules. Indeed, this is the mechanism by which ATP is synthesized during glycolysis and the citric acid cycle (*Case study* 4.3). However, the process of **oxidative phosphorylation** taking place in mitochondria operates by a different mechanism.

The structure of a mitochondrion is shown in Fig 5.14. The protein complexes associated with the electron transport chain span the inner membrane, and phosphorylation takes place in the intermembrane space. The Gibbs energy of the reactions in complexes I, III, and IV is first used to do the work of moving protons across the mitochondrial membrane. The complexes are oriented asymmetrically in the inner membrane so that the protons abstracted from one side of the membrane can be deposited on the other side. For example, the oxidation of NADH by Q in complex I is coupled to the transfer of four protons across the membrane. The coupling of electron transfer and proton pumping in complexes III and IV contribute further to a gradient of proton concentration across the membrane. Then the enzyme H^+-ATPase uses the energy stored in the proton gradient to phosphorylate ADP to ATP. Experiments show that 11 molecules of ATP are made for every three molecules of NADH and one molecule of $FADH_2$ that are oxidized by the respiratory chain. The ATP is then hydrolyzed on demand to perform useful biochemical work throughout the cell. Complex II does not contribute to oxidative phosphorylation because it does not have a proton pump.

The **chemiosmotic theory** proposed by Peter Mitchell explains how H^+-ATPases use the energy stored in a transmembrane proton gradient to synthesize ATP from ADP. It follows from eqn 5.8 that we can estimate the Gibbs energy available for phosphorylation by writing

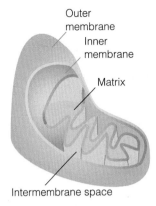

Fig. 5.14 The general structure of a mitochondrion.

$$\Delta G_m = RT \ln \frac{[H^+]_{in}}{[H^+]_{out}} + F\Delta\phi \qquad \begin{array}{l} \text{The Gibbs energy available for} \\ \text{phosphorylation according to} \\ \text{the chemiosmotic theory} \end{array} \qquad (5.20)$$

where $\Delta\phi = \phi_{in} - \phi_{out}$ is the membrane potential difference and we have used $z = +1$. After using $\ln[H^+] = (\ln 10)\log[H^+]$ and substituting $\Delta pH = pH_{in} - pH_{out} = -\log[H^+]_{in} + \log[H^+]_{out}$, it follows that

$$\Delta G_m = F\Delta\phi - (RT\ln 10)\Delta pH \qquad (5.21)$$

A brief illustration

In the mitochondrion, $\Delta pH \approx -1.4$ and $\Delta\phi \approx 0.14$ V, so it follows from eqn 5.20 that $\Delta G_m \approx +21.5$ kJ mol^{-1}. Because 31 kJ mol^{-1} is needed for phosphorylation (*Case study* 4.2), we conclude that at least 2 mol H^+ (and probably more) must flow through the membrane for the phosphorylation of 1 mol ADP.

5.11 Plant photosynthesis

We need to appreciate that the mechanism of formation of glucose from carbon dioxide and water in photosynthetic organisms is distinctly different from the mechanism of glucose breakdown.

In plant photosynthesis, solar energy drives the endergonic reduction of CO_2 to glucose, with concomitant oxidation of water to O_2 ($\Delta_r G^{\oplus} = +2880$ kJ mol^{-1}). The process takes place in the *chloroplast*, a special organelle of the plant cell. Electrons flow from reductant to oxidant via a series of electrochemical reactions that are coupled to the synthesis of ATP. First, the leaf absorbs solar energy and transfers it to membrane protein complexes known as photosystem I and photosystem II.[6] The absorption of energy from light decreases the reduction potential of special dimers of chlorophyll *a* molecules (Atlas R3) known as P700 (in photosystem I) and P680 (in photosystem II). In their high-energy or excited states, P680 and P700 initiate electron transfer reactions that culminate in the oxidation of water to O_2 and the reduction of $NADP^+$ (Atlas N5) to NADPH:

$$2\,NADP^+ + 2\,H_2O \xrightarrow{\text{light}} O_2 + 2\,NADPH + 2\,H^+$$

It is clear that energy from light is required to drive this reaction because, in the dark, $E_{cell}^{\oplus} = -1.135$ V and $\Delta_r G^{\oplus} = +438.0$ kJ mol^{-1}.

Working together, photosystem I and the enzyme ferredoxin:$NADP^+$ oxido-reductase catalyze the light-induced reduction of $NADP^+$ to NADPH. The electrons required for this process come initially from P700 in its excited state. The resulting P700 is then reduced by the mobile carrier plastocyanin (Pc), a protein in which the bound copper ion can exist in oxidation states +2 and +1. The net reaction is

$$NADP^+ + 2\,Cu^+(Pc) + H^+ \xrightarrow{\text{light, photosystem I}} NADPH + 2\,Cu^{2+}(Pc)$$

Oxidized plastocyanin accepts electrons from reduced plastoquinone (PQ). The process is catalyzed by the cytochrome $b_6 f$ complex, a membrane protein complex that resembles complex III of mitochondria:

$$PQH_2 + 2\,Cu^{2+}(Pc) \xrightarrow{\text{Cyt } b_6 f \text{ complex}} PQ + 2\,H^+ + 2\,Cu^+(Pc)$$

$$E_{cell}^{\oplus} = +0.370 \text{ V}, \Delta_r G^{\oplus} = -71.4 \text{ kJ mol}^{-1}$$

Plastoquinone is reduced by water in a process catalyzed by light and photosystem II. The electrons required for the reduction of plastoquinone come initially from P680 in its excited state. The resulting P680 is then reduced ultimately by water. The net reaction is

$$H_2O + PQ \xrightarrow{\text{light, photosystem II}} \tfrac{1}{2}O_2 + PQH_2$$

Electron transfer reactions are coupled to the movement of protons across membranes. **Photophosphorylation** uses the energy stored in the transmembrane proton gradient to phosphorylate ADP to ATP in H^+-ATPases (Fig. 5.15). We see that plant photosynthesis uses an abundant source of electrons (water) and of energy (the Sun) to drive the endergonic reduction of $NADP^+$, with concomitant synthesis of ATP. Experiments show that for each molecule of NADPH formed in the chloroplast of green plants, one molecule of ATP is synthesized.

The ATP and NADPH molecules formed by the light-induced electron transfer reactions of plant photosynthesis participate directly in the reduction of CO_2 to glucose in the chloroplast:

$$6\,CO_2 + 12\,NADPH + 12\,ATP + 12\,H^+$$

$$\rightarrow C_6H_{12}O_6 + 12\,NADP^+ + 12\,ADP + 12\,P_i + 6\,H_2O$$

In summary, electrochemical reactions mediated by membrane protein complexes harness energy in the form of ATP. Plant photosynthesis uses solar energy to transfer electrons from a poor reductant (water) to carbon dioxide. In the process, high-energy molecules (carbohydrates, such as glucose) are synthesized in the cell. Animals feed on the carbohydrates derived from photosynthesis.

---

[6] See Chapter 13 for details of the energy transfer process.

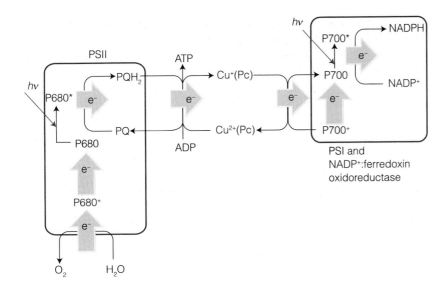

Fig. 5.15 In plant photosynthesis, light-induced electron transfer processes lead to the oxidation of water to O_2 and the reduction of $NADP^+$ to NADPH, with concomitant production of ATP. The energy stored in ATP and NADPH is used to reduce CO_2 to carbohydrate in a separate set of reactions. The scheme summarizes the general patterns of electron flow and does not show all the intermediate electron carriers in photosystems I and II, the cytochrome b_6f complex, and ferredoxin:$NADP^+$ oxidoreductase.

During aerobic metabolism, the O_2 released by photosynthesis as a waste product is used to oxidize carbohydrates to CO_2, driving biological processes such as biosynthesis, muscle contraction, cell division, and nerve conduction. Hence, the sustenance of life on Earth depends on a tightly regulated carbon–oxygen cycle that is driven by solar energy.

Checklist of key concepts

☐ 1. Deviations from ideal behavior in ionic solutions are ascribed to the interaction of an ion with its ionic atmosphere.

☐ 2. According to the Debye–Hückel limiting law, the mean activity coefficient of ions in a solution is related to the ionic strength of the solution.

☐ 3. The Gibbs energy of transfer of an ion across a cell membrane is determined by an activity gradient and a membrane potential difference that arises from differences in Coulomb repulsions on each side of the bilayer.

☐ 4. A galvanic cell is an electrochemical cell in which a spontaneous chemical reaction produces a potential difference.

☐ 5. An electrolytic cell is an electrochemical cell in which an external source of current is used to drive a non-spontaneous chemical reaction.

☐ 6. A redox reaction is expressed as the difference of two reduction half-reactions.

☐ 7. In an electrochemical cell, a cathode is the site of reduction; an anode is the site of oxidation.

☐ 8. The cell potential is the potential difference that the cell produces when operating reversibly.

☐ 9. The standard potential of a couple is the standard potential of a cell in which it forms the right-hand electrode and a hydrogen electrode is on the left.

☐ 10. Biological standard potentials are measured in neutral solution (pH = 7).

☐ 11. A couple with a low standard potential has a thermodynamic tendency (in the sense $K > 1$) to reduce a couple with a high standard potential.

☐ 12. In the respiratory chain, electrons from NADH and $FADH_2$ pass through four membrane-bound protein complexes and two mobile electron carriers before reducing O_2 to H_2O.

☐ 13. The chemiosmotic theory explains how H^+-ATPases use the energy stored in a transmembrane proton gradient to synthesize ATP from ADP.

☐ 14. Plant photosynthesis uses solar energy to transfer electrons from a poor reductant (water) to carbon dioxide.

Checklist of key equations

Property	Equation	Comment
Mean activity coefficient	$\gamma_\pm = (\gamma_+\gamma_-)^{1/2}$	MX salt
	$\gamma_\pm = (\gamma_+^p\gamma_-^q)^{1/s} \quad s = p + q$	M_pX_q salt
Debye–Hückel limiting law	$\log \gamma_\pm = -A\|z_+z_-\|I^{1/2}$	Valid as $I \to 0$
Ionic strength	$I = \frac{1}{2}\sum_i z_i^2 b_i/b^\circ$	Definition
Extended Debye–Hückel law	$\log \gamma_\pm = -\{A\|z_+z_-\|I^{1/2}/(1 + BI^2)\} + CI$	
Gibbs energy of transfer of an ion across a biological membrane	$\Delta G_m = RT\ln([A]_{in}/[A]_{out}) + zF\,\Delta\phi + \Delta_rG^{ATP}$	
Maximum non-expansion work	$w'_{max} = \Delta G$	Constant pressure and temperature
Relation between the cell potential and the reaction Gibbs energy	$-\nu FE_{cell} = \Delta_rG$	
Nernst equation	$E_{cell} = E_{cell}^\circ - (RT/\nu F)\ln Q$	
Standard cell potential	$E_{cell}^\circ = -\Delta_rG^\circ/\nu F$	
Equilibrium constant in terms of the standard cell potential	$\ln K = \nu FE_{cell}^\circ/RT$	
Standard reaction entropy from the standard cell potential	$\Delta_rS^\circ = \nu F(dE_{cell}^\circ/dT)$	
Standard reaction enthalpy	$\Delta_rH^\circ = \Delta_rG^\circ + T\Delta_rS^\circ$	
Gibbs energy available for phosphorylation	$\Delta G_m = RT\ln([H^+]_{in}/[H^+]_{out}) + F\Delta\phi$	Chemiosmotic theory

Discussion questions

5.1 Describe the general features of the Debye–Hückel theory of electrolyte solutions.

5.2 Describe the mechanism of proton conduction in water.

5.3 Distinguish between galvanic, electrolytic, and fuel cells.

5.4 Explain why some reactions that are not redox reactions may be used to generate an electric current.

5.5 Describe a method for the determination of the standard potential of an electrochemical cell.

5.6 Review the concepts in Chapters 1 through 5 and prepare a summary of the experimental and calculational methods that can be used to measure or estimate the Gibbs energies of phase transitions and chemical reactions.

5.7 Review the concepts in Chapters 1 through 5 and discuss how ATP is formed during the metabolism of glucose.

Exercises

5.8 Relate the ionic strengths of (a) KCl, (b) $FeCl_3$, and (c) $CuSO_4$ solutions to their molalities, b.

5.9 Calculate the ionic strength of a solution that is 0.10 mol kg^{-1} in KCl(aq) and 0.20 mol kg^{-1} in $CuSO_4$(aq).

5.10 Calculate the masses of (a) $Ca(NO_3)_2$ and, separately, (b) NaCl to add to a 0.150 mol kg^{-1} solution of KNO_3(aq) containing 500 g of solvent to raise its ionic strength to 0.250.

5.11 Express the mean activity coefficient of the ions in a solution of $CaCl_2$ in terms of the activity coefficients of the individual ions.

5.12 Estimate the mean ionic activity coefficient and activity of a solution that is 0.010 mol kg^{-1} $CaCl_2$(aq) and 0.030 mol kg^{-1} NaF(aq).

5.13 The mean activity coefficients of HBr in three dilute aqueous solutions at 25°C are 0.930 (at 5.0 mmol kg^{-1}), 0.907 (at 10.0 mmol kg^{-1}), and 0.879 (at 20.0 mmol kg^{-1}). Estimate the value of B in the extended Debye–Hückel law, with $C = 0$.

5.14 The addition of a small amount of a salt, such as $(NH_4)_2SO_4$, to a solution containing a charged protein increases the solubility of the protein in water. This observation is called the *salting-in effect*. However, the addition of large amounts of salt can decrease the solubility of the protein to such an extent that the protein precipitates from solution. This observation is called the *salting-out effect* and is used widely by biochemists to isolate and purify proteins. Consider the equilibrium $PX_\nu(s) \rightleftharpoons P^{\nu+}(aq) + \nu X^-(aq)$, where $P^{\nu+}$ is a polycationic protein of charge $+\nu$ and X^- is its counter-ion. Use Le Chatelier's principle and the physical principles behind the Debye–Hückel theory to provide a molecular interpretation for the salting-in and salting-out effects.

5.15 The overall reaction for the active transport of Na^+ and K^+ ions by the Na^+/K^+ pump is

$$3\, Na^+(aq,inside) + 2\, K^+(aq,outside) + ATP$$
$$\rightarrow ADP + P_i + 3\, Na^+(aq,outside) + 2\, K^+(aq,inside)$$

At 310 K, $\Delta_r G^\oplus$ for the hydrolysis of ATP is $-31.3\ kJ\ mol^{-1}$. Given that the [ATP]/[ADP] ratio is of the order of 100, is the hydrolysis of 1 mol ATP sufficient to provide the energy for the transport of Na^+ and K^+ according to the equation above? Take $[P_i] = 1.0\ mol\ dm^{-3}$.

5.16 Vision begins with the absorption of light by special cells in the retina. Ultimately, the energy is used to close ligand-gated ion channels, causing sizable changes in the transmembrane potential. The pulse of electric potential travels through the optical nerve and into the optical cortex, where it is interpreted as a signal and incorporated into the web of events we call visual perception (see Chapter 13). Taking the resting potential as $-30\ mV$, the temperature as 310 K, the permeabilities of the K^+ and Cl^- ions as $P_{K^+} = 1.0$ and $P_{Cl^-} = 0.45$, respectively, and the concentrations as $[K^+]_{in} = 100\ mmol\ dm^{-3}$, $[Na^+]_{in} = 10\ mmol\ dm^{-3}$, $[Cl^-]_{in} = 10\ mmol\ dm^{-3}$, $[K^+]_{out} = 5\ mmol\ dm^{-3}$, $[Na^+]_{out} = 140\ mmol\ dm^{-3}$, and $[Cl^-]_{out} = 100\ mmol\ dm^{-3}$, calculate the relative permeability (*Case study* 5.1) of the Na^+ ion.

5.17 Is the conversion of pyruvate ion to lactate ion

$$CH_3COCO_2^-(aq) + NADH(aq) + H^+(aq)$$
$$\rightarrow CH_3CH_2(OH)CO_2^-(aq) + NAD^+(aq)$$

a redox reaction?

5.18 Express the reaction in Exercise 5.17 as the difference of two half-reactions.

5.19 Express the reaction in which ethanol is converted to acetaldehyde (propanal) by NAD^+ in the presence of alcohol dehydrogenase as the difference of two half-reactions and write the corresponding reaction quotients for each half-reaction and the overall reaction.

5.20 Express the oxidation of cysteine, $HSCH_2CH(NH_2)COOH$, to cystine, $HOOCCH(NH_2)CH_2SSCH_2CH(NH_2)COOH$, as the difference of two half-reactions, one of which is $O_2(g) + 4\, H^+(aq) + 4\, e^- \rightarrow 2\, H_2O(l)$.

5.21 One of the steps in photosynthesis is the reduction of $NADP^+$ by ferredoxin (fd) in the presence of ferredoxin:NADP oxidoreductase: $2\, fd_{red}(aq) + NADP^+(aq) + 2\, H^+(aq) \rightarrow 2\, fd_{ox}(aq) + NADPH(aq)$. Express this reaction as the difference of two half-reactions. How many electrons are transferred in the reaction event?

5.22 From the biological standard half-cell potentials $E_{cell}^\oplus(O_2,H^+,H_2O) = +0.82\ V$ and $E_{cell}^\oplus(NAD,H^+,NADH) = -0.32\ V$, calculate the standard potential arising from the reaction in which NADH is oxidized to NAD^+ and the corresponding biological standard reaction Gibbs energy.

5.23 Cytochrome c oxidase receives electrons from reduced cytochrome c (cyt-c_{red}) and transmits them to molecular oxygen, with the formation of water. **(a)** Write a chemical equation for this process, which occurs in an acidic environment. **(b)** Estimate the values of $E_{cell}^\oplus$, $\Delta_r G^\oplus$, and K for the reaction at 25°C.

5.24 A fuel cell develops an electric potential from the chemical reaction between reagents supplied from an outside source. What is the cell potential of a cell fuelled by **(a)** hydrogen and oxygen, each at 1 bar and 298 K, and **(b)** the combustion of butane at 1.0 bar and 298 K?

5.25 Consider a hydrogen electrode in HBr(aq) at 25°C operating at 1.45 bar. Estimate the change in the electrode potential when the solution is changed from $5.0\ mmol\ dm^{-3}$ to $25.0\ mmol\ dm^{-3}$.

5.26 A hydrogen electrode can, in principle, be used to monitor changes in the molar concentrations of weak acids in biologically active solutions. Consider a hydrogen electrode in a solution of lactic acid as part of an overall galvanic cell at 25°C and 1 bar. Estimate the change in the electrode potential when the concentration of lactic acid in the solution is changed from $5.0\ mmol\ dm^{-3}$ to $25.0\ mmol\ dm^{-3}$.

5.27 Write the cell reactions and electrode half-reactions for the following cells:

(a) $Pt(s)|H_2(g, p_L)|HCl(aq)|H_2(g, p_R)|Pt(s)$

(b) $Pt(s)|Cl_2(g)|HCl(aq)||HBr(aq)|Br_2(l)|Pt(s)$

(c) $Pt(s)|NAD^+(aq), H^+(aq),NADH(aq)||oxaloacetate^{2-}(aq), H^+(aq),malate^{2-}(aq)|Pt(s)$

(d) $Fe(s)|Fe^{2+}(aq)||Mn^{2+}(aq), H^+(aq)|MnO_2(s)|Pt(s)$

5.28 Write the Nernst equations for the cells in the preceding exercise.

5.29 Devise cells to study the following biochemically important reactions. In each case state the value for ν to use in the Nernst equation.

(a) $CH_3CH_2OH(aq) + NAD^+(aq)$
$\rightarrow CH_3CHO(aq) + NADH(aq) + H^+(aq)$

(b) $ATP^{4-}(aq) + Mg^{2+}(aq) \rightarrow MgATP^{2-}(aq)$

(c) $2\ cyt$-$c(red, aq) + CH_3COCO_2^-(aq) + 2\ H^+(aq)$
$\rightarrow 2\ cyt$-$c(ox, aq) + CH_3CH(OH)CO_2^-(aq)$

5.30 Use the standard potentials of the electrodes to calculate the standard potentials of the cells devised in Exercise 5.29.

5.31 The permanganate ion is a common oxidizing agent. What is the standard potential of the $MnO_4^-,H^+/Mn^{2+}$ couple at **(a)** pH = 6.00 and **(b)** general pH?

5.32 State what you would expect to happen to the cell potential when the following changes are made to the corresponding cells in Exercise 5.27. Confirm your prediction by using the Nernst equation in each case. **(a)** The pressure of hydrogen in the left-hand compartment is increased. **(b)** The concentration of HCl is increased. **(c)** Acid is added to both compartments. **(d)** Acid is added to the right-hand compartment.

5.33 State what you would expect to happen to the cell potential when the following changes are made to the corresponding cells devised in Exercise 5.29. Confirm your prediction by using the Nernst equation in each case. **(a)** The pH of the solution is raised. **(b)** A solution of Epsom salts (magnesium sulfate) is added. **(c)** Sodium lactate is added to the solution.

5.34 **(a)** Calculate the standard potential of the cell $Hg(l)|Hg_2Cl_2(aq)||TlNO_3(aq)|Tl(s)$ at 25°C. **(b)** Calculate the cell potential when the molar concentration of the Hg^{2+} ion is $0.150\ mol\ dm^{-3}$ and that of the Tl^+ ion is $0.93\ mol\ dm^{-3}$.

5.35 Calculate the biological standard Gibbs energies of reactions of the following reactions and half-reactions:

(a) $2 \text{NADH(aq)} + O_2(g) + 2 H^+(aq) \rightarrow 2 \text{NAD}^+(aq) + 2 H_2O(l)$
$$E^{\oplus}_{cell} = +1.14 \text{ V}$$

(b) $\text{Malate}^{2-}(aq) + \text{NAD}^+(aq)$
$\rightarrow \text{oxaloacetate}^{2-}(aq) + \text{NADH(aq)} + H^+(aq) \quad E^{\oplus}_{cell} = -0.154 \text{ V}$

(c) $O_2(g) + 4 H^+(aq) + 4 e^- \rightarrow 2 H_2O(l)$
$$E^{\oplus}_{cell} = +0.81 \text{ V}$$

5.36 The silver–silver chloride electrode, $\text{Ag(s)}|\text{AgCl(s)}|\text{Cl}^-(aq)$, consists of metallic silver coated with a layer of silver chloride (which does not dissolve in water) in contact with a solution containing chloride ions. (a) Write the half-reaction for the silver–silver chloride half-electrode. (b) Estimate the potential of the cell $\text{Ag(s)}|\text{AgCl(s)}|\text{KCl(aq, 0.025 mol kg}^{-1})||\text{AgNO}_3(aq, 0.010 \text{ mol kg}^{-1})|\text{Ag(s)}$ at 25°C.

5.37 (a) Calculate the standard potential of the cell $\text{Pt(s)}|\text{cysteine(aq)}, \text{cystine(aq)}||H^+(aq)|O_2(g)|\text{Pt(s)}$ and the standard Gibbs energy and enthalpy of the cell reaction at 25°C. (b) Estimate the value of $\Delta_r G^{\ominus}$ at 35°C. Use $E^{\ominus} = -0.34 \text{ V}$ for the cystine/cysteine couple.

5.38 The biological standard potential of the couple pyruvic acid/lactic acid is -0.19 V. What is the thermodynamic standard potential of the couple? Pyruvic acid is $CH_3COCOOH$ and lactic acid is $CH_3CH(OH)COOH$.

5.39 Calculate the biological standard values of the potentials (the two potentials and the cell potential) for the system in Exercise 5.37 at 310 K.

5.40 (a) Does $FADH_2$ have a thermodynamic tendency to reduce coenzyme Q at pH = 7? (b) Does oxidized cytochrome b have a thermodynamic tendency to oxidize reduced cytochrome f at pH = 7?

5.41 Radicals, very reactive species containing one or more unpaired electrons, are among the by-products of metabolism. Evidence is accumulating that radicals are involved in the mechanism of aging and in the development of a number of conditions, ranging from cardiovascular disease to cancer. *Antioxidants* are substances that reduce radicals readily. Which of the following known antioxidants is the most efficient (from a thermodynamic point of view): ascorbic acid (vitamin C), reduced glutathione, reduced lipoic acid, or reduced coenzyme Q?

5.42 The biological standard potential of the redox couple pyruvic acid/lactic acid is -0.19 V and that of the fumaric acid/succinic acid couple is $+0.03$ V at 298 K. What is the equilibrium constant at pH = 7 for the reaction

pyruvic acid + succinic acid $\rightleftharpoons$ lactic acid + fumaric acid

5.43 Tabulated thermodynamic data can be used to predict the standard potential of a cell even if it cannot be measured directly. The presence of glyoxylate ion produced by the action of the enzyme glycolate oxidase on glycolate ion can be monitored by the following redox reaction:

$2 \text{cyt-}c(\text{ox,aq}) + \text{glycolate}^-(aq)$
$\rightleftharpoons 2 \text{cyt-}c(\text{red,aq}) + \text{glyoxylate}^-(aq) + 2 H^+(aq)$

The equilibrium constant for the reaction above is 2.14×10^{11} at pH = 7.0 and 298 K. (a) Calculate the biological standard potential of the corresponding galvanic cell and (b) the biological standard potential of the glyoxylate$^-$/glycolate$^-$ couple.

5.44 One ecologically important equilibrium is that between carbonate and hydrogencarbonate (bicarbonate) ions in natural water. (a) The standard Gibbs energies of formation of $CO_3^{2-}(aq)$ and $HCO_3^-(aq)$ are $-527.81 \text{ kJ mol}^{-1}$ and $-586.77 \text{ kJ mol}^{-1}$, respectively. What is the standard potential of the $HCO_3^-/CO_3^{2-},H_2$ couple? (b) Calculate the standard potential of a cell in which the cell reaction is $\text{Na}_2CO_3(aq) + H_2O(l) \rightarrow \text{NaHCO}_3(aq) + \text{NaOH(aq)}$. (c) Write the Nernst equation for the cell, (d) Predict and calculate the change in potential when the pH is changed to 7.0. (e) Calculate the value of pK_a for $HCO_3^-(aq)$.

5.45 The dichromate ion in acidic solution is a common oxidizing agent for organic compounds. Derive an expression for the potential of an electrode for which the half-reaction is the reduction of $Cr_2O_7^{2-}$ ions to Cr^{3+} ions in acidic solution.

5.46 The potential of the cell $\text{Pt(s)}|H_2(g)|\text{HCl(aq)}|\text{AgCl(s)}|\text{Ag(s)}$ is $+0.312$ V at 25°C. What is the pH of the electrolyte solution?

5.47 The standard potential of the AgCl/Ag,Cl^- couple fits the expression

$$E^{\ominus}/\text{V} = 0.23659 - 4.8564 \times 10^{-4}(\theta/°C) - 3.4205 \times 10^{-6}(\theta/°C)^2 + 5.869 \times 10^{-9}(\theta/°C)^3$$

Calculate the standard Gibbs energy and enthalpy of formation of $Cl^-(aq)$ and its entropy (relative to H^+) at 298 K.

5.48 If the mitochondrial electric potential between the matrix and the intermembrane space were 70 mV, as is common for other membranes, how much ATP could be synthesized from the transport of 4 mol H^+, assuming the pH difference remains the same?

5.49 Under certain stress conditions, such as viral infection or hypoxia, plants have been shown to have an intercellular pH increase of about 0.1 pH. Suppose this pH change also occurs in the mitochondrial intermembrane space. How much ATP can now be synthesized for the transport of 2 mol H^+, assuming no other changes occur?

5.50 In anaerobic bacteria, the source of carbon may be a molecule other than glucose and the final electron acceptor some molecule other than O_2. Could a bacterium evolve to use the ethanol/nitrate pair instead of the glucose/O_2 pair as a source of metabolic energy?

5.51 The following reaction occurs in the cytochrome $b_6 f$ complex, a component of the electron transport chain of plant photosynthesis:

cyt-b(red) + cyt-f(ox) $\rightleftharpoons$ cyt-b(ox) + cyt-f(red)

(a) Calculate the biological standard Gibbs energy of this reaction. (b) The Gibbs energy for hydrolysis of ATP under conditions found in the chloroplast is -50 kJ mol^{-1} and the synthesis of ATP by ATPase requires the transfer of four protons across the membrane. How many electrons must pass through the cytochrome $b_6 f$ complex to lead to the generation of a transmembrane proton gradient that is large enough to drive ATP synthesis in the chloroplast?

Project

5.52 The standard potentials of proteins are not commonly measured by the methods described in this chapter because proteins often lose their native structure and their function when they react on the surfaces of electrodes. In an alternative method, the oxidized protein is allowed to react with an appropriate electron donor in solution. The standard potential of the protein is then determined from the Nernst equation, the equilibrium concentrations of all species in solution, and the known standard potential of the electron donor. We shall illustrate this method with the protein cytochrome c.

(a) The one-electron reaction between cytochrome c, cyt-c, and 2,6-dichloroindophenol, D, can be written as

$$\text{cyt-}c_{\text{ox}} + D_{\text{red}} \rightleftharpoons \text{cyt-}c_{\text{red}} + D_{\text{ox}}$$

Consider $E^{\ominus}_{\text{cyt}}$ and $E^{\ominus}_{D}$ to be the standard potentials of cytochrome c and D, respectively. Show that, at equilibrium (eq), a plot of $\ln([D_{\text{ox}}]_{\text{eq}}/[D_{\text{red}}]_{\text{eq}})$ against $\ln([\text{cyt-}c_{\text{ox}}]_{\text{eq}}/[\text{cyt-}c_{\text{red}}]_{\text{eq}})$ is linear with a slope of 1 and

y-intercept $F(E^{\ominus}_{\text{cyt}} - E^{\ominus}_{D})/RT$, where equilibrium activities are replaced by the numerical values of equilibrium molar concentrations.

(b) The following data were obtained for the reaction between oxidized cytochrome c and reduced D at pH 6.5 buffer and 298 K. The ratios $[D_{\text{ox}}]_{\text{eq}}/[D_{\text{red}}]_{\text{eq}}$ and $[\text{cyt-}c_{\text{ox}}]_{\text{eq}}/[\text{cyt-}c_{\text{red}}]_{\text{eq}}$ were adjusted by adding known volumes of a solution of sodium ascorbate, a reducing agent, to a solution containing oxidized cytochrome c and reduced D. From the data and the standard potential of D of +0.237 V, determine the standard potential of cytochrome c at pH = 6.5 and 298 K.

$[D_{\text{ox}}]_{\text{eq}}/$ $[D_{\text{red}}]_{\text{eq}}$	0.002 79	0.008 43	0.0257	0.0497	0.0748	0.238	0.534
$[\text{cyt-}c_{\text{ox}}]_{\text{eq}}/$ $[\text{cyt-}c_{\text{red}}]_{\text{eq}}$	0.0106	0.0230	0.0894	0.197	0.335	0.809	1.39

PART 2 The kinetics of life processes

The branch of physical chemistry called chemical kinetics is concerned with the rates of chemical reactions. Chemical kinetics deals with how rapidly reactants are consumed and products formed, how reaction rates respond to changes in the conditions or the presence of a catalyst, and the identification of the steps by which a reaction takes place.

One reason for studying the rates of reactions is the practical importance of being able to predict how quickly a reaction mixture approaches equilibrium. The rate might depend on variables under our control, such as the temperature and the presence of a catalyst, and we might be able to optimize it by the appropriate choice of conditions. Another reason is that the study of reaction rates leads to an understanding of the mechanism of a reaction, its analysis into a sequence of elementary steps. For example, by analyzing the rates of biochemical reactions, we may discover how they take place in an organism and contribute to the activity of a cell. Enzyme kinetics, the study of the effect of enzymes on the rates of reactions, is also an important window on how these macromolecules work and is treated in Chapter 8 using the concepts developed in Chapters 6 and 7.

The rates of reactions

6

When dealing with physical and chemical changes, we need to cope with a wide variety of different rates. Even a process that appears to be slow may be the outcome of many faster steps. This is particularly true in the chemical reactions that underlie life. Some of the earlier steps in photosynthesis may take place in about 1–100 ps. The binding of a neurotransmitter can have an effect after about 1 s. Once a gene has been activated, a protein may emerge in about 100 s, but even that timescale incorporates many others, including the wriggling of a newly formed polypeptide chain into its working conformation, each step of which may take about 1 ps. On a grander view, some of the equations of chemical kinetics are applicable to the behavior of whole populations of organisms: such societies change on timescales of 10^7–10^9 s.

Reaction rates

The raw data from experiments to measure reaction rates are the concentrations or (for gases) partial pressures of reactants and products at a series of times after the reaction is initiated. Ideally, information on any intermediates should also be obtained, but often intermediates cannot be studied because their existence is too fleeting or their concentration too low. More information about the reaction can be extracted if data are obtained at a series of different temperatures.

The first step in the investigation of the rate and mechanism of a reaction is the determination of the overall stoichiometry of the reaction and the identification of any side reactions. The next step is to determine how the concentrations of the reactants and products change with time after the reaction has been initiated. Because the rates of chemical reactions are sensitive to temperature, the temperature of the reaction mixture must be held constant throughout the course of the reaction, for otherwise the observed rate would be a meaningless average of the rates for different temperatures. The next few sections look at these observations in more detail.

In the laboratory 6.1   *Experimental techniques*

Some of the more common methods for investigations of reaction kinetics are listed in Table 6.1.

(a) The determination of concentration

Spectrophotometry, the measurement of the absorption of light by a material, is used widely to monitor concentration. The technique is based on Beer's law (see Chapter 12), which states that the incident and transmitted intensities

Table 6.1 Kinetic techniques

Technique	Range of timescales/s
Flash photolysis	10^{-15}
Fluorescence decay[a]	10^{-10}–10^{-6}
Ultrasonic absorption	10^{-10}–10^{-4}
EPR[b]	10^{-9}–10^{-4}
Electric field jump[c]	10^{-7}–1
Temperature jump[c]	10^{-6}–1
Phosphorescence decay[a]	10^{-6}–10
NMR[b]	10^{-5}–1
Pressure jump[c]	$>10^{-5}$
Stopped flow	$>10^{-3}$

[a] Fluorescence and phosphorescence are modes of emission of radiation from a material; see Chapter 12.
[b] EPR is electron paramagnetic resonance (or electron spin resonance, ESR); NMR is nuclear magnetic resonance; see Chapter 13.
[c] These techniques are discussed in *In the laboratory* 7.1.

I and I_0, respectively, of light passing through a sample of length L are related to the molar concentration [J] of the absorbing species J by

$$I = I_0 10^{-\varepsilon[J]L} \qquad \text{Beer's law} \qquad (6.1a)$$

The molar absorption coefficient ε (epsilon) depends on the wavelength of the radiation. Once the value of ε has been measured (in a separate experiment) for an absorbing species taking part in a reaction, either as a reactant or a product, its concentration may be monitored by using eqn 6.1a in the form

$$[J] = \frac{1}{\varepsilon L} \log \frac{I_0}{I} \qquad (6.1b)$$

Note that log denotes a logarithm to the base 10.

Reactions that change the concentration of hydrogen ions can be studied by monitoring the pH of the solution with a glass electrode. Other methods of monitoring the composition include the detection of light emission, microscopy, mass spectrometry, gas chromatography, and magnetic resonance (both EPR and NMR; Chapter 13). Polarimetry and circular dichroism (Chapter 12), which monitor the optical activity of a reaction mixture, are occasionally applicable.

(b) Monitoring the time dependence

In a **real-time analysis**, the composition of a system is analyzed while the reaction is in progress by direct spectrophotometric observation of the reaction mixture. In the **flow method**, the reactants are mixed as they flow together in a chamber (Fig. 6.1). The reaction continues as the thoroughly mixed solutions flow through a capillary outlet tube at about 10 m s^{-1}, and different points along the tube correspond to different times after the start of the reaction. Spectrophotometric determination of the composition at different positions along the tube is equivalent to the determination of the composition of the reaction mixture at different times after mixing. This technique was originally developed in connection with the study of the rate at which oxygen combines with hemoglobin (*Case study* 4.1). Its disadvantage is that a large volume of reactant solution is necessary because the mixture must flow continuously through the apparatus. This disadvantage is particularly important for reactions that take place very rapidly, because the flow must be rapid if it is to spread the reaction over an appreciable length of tube.

The **stopped-flow technique** avoids this disadvantage (Fig. 6.2). The two solutions are mixed very rapidly (in less than 1 ms) by injecting them into a mixing chamber designed to ensure that the flow is turbulent and that complete mixing occurs very quickly. Behind the reaction chamber there is an observation cell fitted with a plunger that moves back as the liquids flood in, but that comes up against a stop after a certain volume has been admitted. The filling of that chamber corresponds to the sudden creation of an initial sample of the reaction mixture. The reaction then continues in the thoroughly mixed solution and is monitored spectrophotometrically. Because only a small, single charge of the reaction chamber is prepared, the technique is much more economical than the flow method. Modern techniques of monitoring composition spectrophotometrically can span repetitively a wavelength range of 300 nm at 1 ms intervals. The suitability of the stopped-flow technique to the study of small samples means that it is appropriate for biochemical reactions, and it has been widely used to study the kinetics of protein folding and unfolding. In a typical

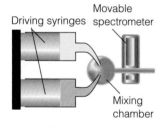

Movable spectrometer
Driving syringes

Mixing chamber

Fig. 6.1 The arrangement used in the flow technique for studying reaction rates. The reactants are squirted into the mixing chamber at a steady rate from the syringes or by using peristaltic pumps (pumps that squeeze the fluid through flexible tubes, like in our intestines). The location of the spectrometer (acting as a detector) corresponds to different times after initiation.

experiment, a sample of the protein with a high concentration of a chemical denaturant, such as urea or guanidinium hydrochloride, is mixed with a solution containing a much lower concentration of the same denaturant. On entering the mixing chamber, the denaturant is diluted and the protein re-folds. Unfolding is observed by mixing a sample of folded protein with a solution containing a high concentration of denaturant. These experiments probe conformational changes that occur on a millisecond timescale, such as the formation of contacts between helical segments in a large protein.

Very fast reactions can be studied by **flash photolysis**, in which the sample is exposed to a brief flash of light that initiates the reaction and then the contents of the reaction chamber are monitored spectrophotometrically. Biological processes that depend on the absorption of light, such as photosynthesis and vision, can be studied in this way. Lasers can be used to generate nanosecond flashes routinely, picosecond flashes quite readily, and flashes as brief as a few femtoseconds in special arrangements. Spectra are recorded at a series of times following the flash, using instrumentation described in Chapter 12.

In a **relaxation technique** the reaction mixture is initially at equilibrium but is then disturbed by a rapid change in conditions, such as a sudden increase in temperature. The equilibrium composition before the application of the perturbation becomes the initial state for the return of the system to its equilibrium composition at the new temperature, and the return to equilibrium—the 'relaxation' of the system—is monitored spectroscopically. Relaxation techniques are described in more detail in *In the laboratory* 7.1.

In contrast to real-time analysis, **quenching methods** are based on stopping, or quenching, the reaction after it has been allowed to proceed for a certain time and the composition is analyzed at leisure. In the **chemical quench flow method**, the reactants are mixed in much the same way as in the flow method, but the reaction is quenched by another reagent, such as a solution of acid or base, after the mixture has traveled along a fixed length of the outlet tube. Different reaction times can be selected by varying the flow rate along the outlet tube. An advantage of the chemical quench flow method over the stopped-flow method is that spectroscopic fingerprints are not needed in order to measure the concentration of reactants and products. Once the reaction has been quenched, the solution may be examined by rather 'slow' techniques, such as gel electrophoresis, mass spectrometry, and chromatography. In the **freeze quench method**, the reaction is quenched by cooling the mixture within milliseconds, and the concentrations of reactants, intermediates, and products are measured spectroscopically.

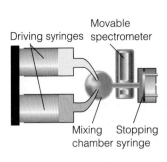

Fig. 6.2 In the stopped-flow technique the reagents are driven quickly into the mixing chamber and then the time dependence of the concentrations is monitored.

6.1 The definition of reaction rate

The concepts introduced here for the description of reaction rates are used whenever we explore such biological processes as enzymatic transformations, electron transfer reactions in metabolism, and the transport of molecules and ions across membranes.

The **average rate** of a reaction is defined in terms of the rate of change of the concentration of a designated species:

$$\text{average rate} = \frac{|\Delta[\text{J}]|}{\Delta t} \qquad \boxed{\text{Definition of average rate}} \quad (6.2a)$$

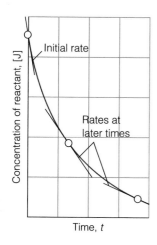

Fig. 6.3 The rate of a chemical reaction is the slope (without the sign) of the tangent to the curve showing the variation of concentration of a species with time. This graph is a plot of the concentration of a reactant, which is consumed as the reaction progresses. The rate of consumption decreases in the course of the reaction as the concentration of reactant decreases.

where $\Delta[J]$ is the change in the molar concentration of the species J that occurs during the time interval Δt. We have put the change in concentration between modulus signs (that is, the instruction to disregard the sign of the change) to ensure that all rates are positive: if J is a reactant, its concentration will decrease and $\Delta[J]$ itself is negative but $|\Delta[J]|$ is positive. With the concentration measured in moles per cubic decimeter (moles per liter) and the time in seconds, the average rate is reported in moles per cubic decimeter per second ($mol\ dm^{-3}\ s^{-1}$).

Because the rates at which reactants are consumed and products are formed typically change in the course of a reaction, it is necessary to consider the **instantaneous rate**, v, of the reaction, its rate at a specific instant. The instantaneous rate of consumption of a reactant is the slope of a graph of its molar concentration plotted against the time, with the slope evaluated as the tangent to the graph at the instant of interest (Fig. 6.3) and reported as a positive quantity. The instantaneous rate of formation of a product is also the slope of the tangent to the graph of its molar concentration plotted and also reported as a positive quantity. More formally:

$$\text{instantaneous rate} = \frac{|d[J]|}{dt} \qquad \boxed{\begin{array}{c}\text{Definition of}\\\text{instantaneous rate}\end{array}} \qquad (6.2b)$$

In general, the various reactants in a given reaction are consumed at different rates, and the various products are also formed at different rates. However, these rates are related by the stoichiometry of the reaction. For example, in the decomposition of urea, $(NH_2)_2CO$, in acidic solution

$$(NH_2)_2CO(aq) + 2\ H_2O(l) \rightarrow 2\ NH_4^+(aq) + CO_3^{2-}(aq)$$

provided any intermediates are not present in significant quantities, the rate of formation of NH_4^+ is twice the rate of disappearance of $(NH_2)_2CO$ because for 1 mol $(NH_2)_2CO$ consumed, 2 mol NH_4^+ is formed:

rate of formation of $NH_4^+ = 2 \times$ rate of consumption of $(NH_2)_2CO$

or, in terms of derivatives,

$$v = \frac{d[NH_4^+]}{dt} = -2\frac{d[(NH_2)_2CO]}{dt}$$

One consequence of this kind of relation is that we have to be careful to specify exactly what species we mean when we report a reaction rate.

(**Self-test 6.1**) The rate of formation of NH_3 in the reaction $N_2(g) + 3\ H_2(g) \rightarrow 2\ NH_3(g)$ was reported as 1.2 mmol dm^{-3} s^{-1} under a certain set of conditions. What is the rate of consumption of H_2?

Answer: 1.8 mmol dm^{-3} s^{-1}

The problem of having a variety of different rates for the same reaction is avoided by bringing the stoichiometric coefficients into the definition of the rate. Thus, for a reaction of the type

$$a\ A + b\ B \rightarrow c\ C + d\ D$$

we write the **unique reaction rate** as any of the four following quantities:

$$v = \frac{1}{d}\frac{d[D]}{dt} = \frac{1}{c}\frac{d[C]}{dt} = -\frac{1}{a}\frac{d[A]}{dt} = -\frac{1}{b}\frac{d[B]}{dt} \qquad \boxed{\begin{array}{c}\text{Definition of}\\\text{unique rate}\end{array}} \qquad (6.3)$$

Now there is a single rate for the reaction.

> **A brief illustration**
>
> For the decomposition of urea in the reaction specified above (with $a = 1, c = 2$, and $d = 1$ (the water is in excess and its abundance would not be measured)), the unique reaction rate would be calculated from any of the following three quantities:
>
> $$v = \frac{1}{2}\frac{d[NH_4^+]}{dt} = \frac{d[CO_3^{2-}]}{dt} = -\frac{d[(NH_2)_2CO]}{dt}$$

6.2 Rate laws and rate constants

> The observed dependence of rate on the composition of the reaction mixture is often exploited for the purpose of slowing down some processes and speeding up others; it is also a window on the underlying mechanism of the reaction.

An empirical observation of the greatest importance is that *the rate of reaction is often found to be proportional to the molar concentrations of the reactants raised to a simple power*. For example, it may be found that the rate is directly proportional to the concentrations of the reactants A and B, so

$$v = k_r[A][B] \tag{6.4}$$

The coefficient k_r, which is characteristic of the reaction being studied, is called the **rate constant**. The rate constant is independent of the concentrations of the species taking part in the reaction but depends on the temperature. An *experimentally determined* equation of this kind is called the 'rate law' of the reaction. More formally:

> A **rate law** is an equation that expresses the rate of reaction in terms of the molar concentrations (or partial pressures) of the species in the overall reaction (including, possibly, the products).

The units of k_r are always such as to convert the product of concentrations into a rate expressed as a change in concentration divided by time. For example, if the rate law is the one shown above, with concentrations expressed in moles per cubic decimeter ($mol\ dm^{-3}$), then the units of k_r will be cubic decimeters per mole per second ($dm^3\ mol^{-1}\ s^{-1}$) because

$$\underbrace{dm^3\ mol^{-1}\ s^{-1}}_{k_r} \times \underbrace{mol\ dm^{-3}}_{[A]} \times \underbrace{mol\ dm^{-3}}_{[B]} = \underbrace{mol\ dm^{-3}\ s^{-1}}_{v}$$

In gas-phase studies, such as those used to study reactions in planetary atmospheres, concentrations are commonly expressed in molecules per cubic centimeter ($molecules\ cm^{-3}$), so the rate constant for the reaction above would be expressed in $cm^3\ molecule^{-1}\ s^{-1}$. We can use the same approach to determine the units of the rate constant from rate laws of any form. For example, the rate constant for a reaction with a rate law of the form $k_r[A]$ is commonly expressed in s^{-1}.

> **Self-test 6.2** A reaction has a rate law of the form $k_r[A]^2[B]$. What are the units of the rate constant k_r if the reaction rate is measured in $mol\ dm^{-3}\ s^{-1}$?
>
> **Answer:** $dm^6\ mol^{-2}\ s^{-1}$

Once the rate law and the rate constant of the reaction have been determined, we can predict the rate of the reaction for any given composition of the reaction mixture. We shall also see that we can use the observed rate law to predict the concentrations of the reactants and products at any time after the start of the reaction. Furthermore, a rate law is an important guide to the **mechanism** of the reaction, the individual molecular steps by which it takes place, for any proposed mechanism must be consistent with the observed rate law.

6.3 Reaction order

Once a reaction has been classified according to its rate law, we can use the same expressions to predict the composition of the reaction mixture at any stage of the reaction: specifically, many enzyme-catalyzed reactions and biological electron transfer reactions are kinetically similar.

Many reactions can be classified on the basis of their **order**, the power to which the concentration of a species is raised in the rate law:

first order in A: $v = k_r[A]$ (6.5a)

first order in A, first order in B: $v = k_r[A][B]$ (6.5b)

second order in A: $v = k_r[A]^2$ (6.5c)

The **overall order** of a reaction is the sum of the orders of all the components. The rate laws in eqns 6.5b and 6.5c both correspond to reactions that are *second order* overall.

A brief illustration

The re-formation of a DNA double helix after the double helix has been separated into two strands by raising the temperature or the pH:

strand + complementary strand → double helix

is found to obey the rate law

$v = k_r[\text{strand}][\text{complementary strand}]$

This reaction is first order in each strand and second order overall. The reduction of nitrogen dioxide by carbon monoxide,

$NO_2(g) + CO(g) \rightarrow NO(g) + CO_2(g)$

is found to obey the rate law

$v = k_r[NO_2]^2$

which is second order in NO_2 and, because no other species occurs in the rate law, second order overall. The rate of the latter reaction is independent of the concentration of CO provided that some CO is present. It is therefore *zero order* in CO because a concentration raised to the power zero is 1 ($[CO]^0 = 1$, just as $x^0 = 1$ in algebra).

A reaction need not have an integral order, and many gas-phase reactions do not. For example, if a reaction is found to have the rate law

$$v = k_r[\text{A}]^{1/2}[\text{B}] \tag{6.6}$$

then it is *half order* in A, first order in B, and three-halves order overall.

If a rate law is not of the form $[\text{A}]^x[\text{B}]^y[\text{C}]^z \ldots$, then the reaction does not have an overall order. For example, a typical rate law for the action of an enzyme E on a substrate S is (see Chapter 8)

$$v = \frac{k_r[\text{E}][\text{S}]}{[\text{S}] + K_M} \tag{6.7a}$$

where K_M is a constant (not a rate constant). This rate law is first order in the enzyme but does not have a specific order with respect to the substrate.

Under certain circumstances a complicated rate law without an overall order may simplify into a law with a definite order. For example, if the substrate concentration in the enzyme-catalyzed reaction is so low that $[\text{S}] \ll K_M$, then eqn 6.7a simplifies to

$$v = \frac{k_r}{K_M}[\text{E}][\text{S}], \; [\text{S}] \ll K_M \tag{6.7b}$$

which is first order in S, first order in E, and second order overall with rate constant $k_r' = k_r/K_M$. On the other hand, when $[\text{S}] \gg K_M$, the rate law in eqn 6.7a becomes

$$v = k_r[\text{E}], \; [\text{S}] \gg K_M \tag{6.7c}$$

and the reaction is first order in E and zeroth order in S.

It is very important to note that *a rate law is established experimentally and cannot in general be inferred from the chemical equation for the reaction.* The reaction of an enzyme with a substrate, for example, has a very simple stoichiometry, but its rate law (eqn 6.7a) is moderately complicated. In some cases, however, the rate law does happen to reflect the reaction stoichiometry. This is the case with the re-naturation of DNA in the *brief illustration*.

6.4 The determination of the rate law

Because reaction order is such an important concept for the classification and investigation of biochemical reactions, we need to know how it is determined experimentally.

At the simplest level, a quick comparison of rates with two different concentrations of a reactant can indicate the order of the reaction. Thus, if the rate doubles when the concentration is doubled, we can infer that the reaction is first order in that reactant, and if it quadruples, then it is second order. However, to assess the data more fully, we need to be more systematic. There are two principal approaches. In one, we use rate measurements directly; in the other, we use concentration measurements, not rates. In this section we prepare the ground for the first approach and describe its implementation. The second approach needs more preparation and is described in Section 6.5.

(a) Isolation and pseudo-order reactions

The determination of a rate law is simplified by the **isolation method**, in which all the reactants except one are present in large excess. The dependence of the rate on each of the reactants can be found by isolating each of them in turn—focusing on a single species by having all the others present in large excess—and piecing together a picture of the overall rate law.

If a reactant B is in large excess it is a good approximation to take its concentration as constant throughout the reaction. Then, although the true rate law might be $v = k_r[A][B]^2$, we can approximate [B] by its initial value $[B]_0$ (from which it hardly changes in the course of the reaction) and write

$$v = k_r'[A] \text{ with } k_r' = k_r[B]_0^2 \qquad \boxed{\text{A pseudo-first-order reaction, B in excess}} \qquad (6.8a)$$

Because the true rate law has been forced into first-order form by assuming a constant B concentration, the effective rate law is classified as **pseudo-first order** and k_r' is called the **effective rate constant** for a given, fixed concentration of B. If, instead, the concentration of A were in large excess, and hence effectively constant, then the rate law $v = k_r[A][B]^2$ would simplify to

$$v = k_r''[B]^2 \text{ with } k_r'' = k_r[A]_0 \qquad \boxed{\text{A pseudo-second-order reaction, A in excess}} \qquad (6.8b)$$

This **pseudo-second-order rate law** is also much easier to analyze and identify than the complete law.

In a similar manner, a reaction may appear to be zeroth order. For instance, the oxidation of ethanol to acetaldehyde (ethanal) by NAD^+ in the liver in the presence of the enzyme liver alcohol dehydrogenase,

$$CH_3CH_2OH(aq) + NAD^+(aq) + H_2O(l) \rightarrow$$
$$CH_3CHO(aq) + NADH(aq) + H_3O^+(aq)$$

is zeroth order overall as the ethanol is in excess and the concentration of the NAD^+ is maintained at a constant level by normal metabolic processes. Many reactions in aqueous solution that are reported as first or second order are actually pseudo-first or pseudo-second order: the solvent water participates in the reaction, but it is in such large excess that its concentration remains constant.

(b) The method of initial rates

In the method of **initial rates**, which is often used in conjunction with the isolation method, the instantaneous rate is measured at the beginning of the reaction for several different initial concentrations of reactants. If the initial rate doubles when the initial concentration of the isolated reactant is doubled, then the reaction is first order in that reactant, and so on.

To use the data more fully we suppose that the rate law for a reaction with A isolated is $v = k_r'[A]^a$, then the initial rate of the reaction, v_0, is given by the initial concentration of A:

$$v_0 = k_r'[A]_0^a \qquad \boxed{\text{Initial rate of an ath-order reaction}} \qquad (6.9)$$

Taking logarithms[1] gives

$$\log v_0 = \log k_r' + a \log [A]_0 \qquad (6.10)$$

This equation has the form of the equation for a straight line:

$$y = \text{intercept} + \text{slope} \times x$$

with $y = \log v_0$ and $x = \log [A]_0$. It follows that, for a series of initial concentrations, a plot of the logarithms of the initial rates against the logarithms of the initial concentrations of A should be a straight line and that the slope of the graph will be

---

[1] For a review of logarithms, see *Mathematical toolkit* 5.1.

a, the order of the reaction with respect to the species A, and $\log k_r'$ is given by the intercept at $\log [A]_0 = 0$ (Fig. 6.4).

An important point to note is that the method of initial rates might not reveal the entire rate law, for in a complex reaction we may not be able to specify an order with respect to a reactant (see eqn 6.7a) or the products themselves might affect the rate.

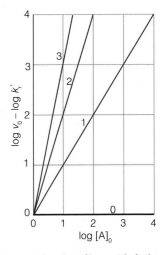

Fig. 6.4 The plot of $\log v_0$ (shifted by $\log k_r'$) against $\log [A]_0$ gives straight lines with slopes equal to the order of the reaction.

Example 6.1 *Using the method of initial rates*

The following data were obtained on the initial rate of binding of glucose to the enzyme hexokinase:

[glucose]$_0$/(mmol dm^{-3})		1.00	1.54	3.12	4.02
v_0/(mol dm^{-3} s^{-1})	(a)	5.0	7.6	15.5	20.0
	(b)	7.0	11.0	23.0	31.0
	(c)	21.0	34.0	70.0	96.0

The enzyme concentrations are (a) 1.34 mmol dm^{-3}, (b) 3.00 mmol dm^{-3}, and (c) 10.0 mmol dm^{-3}. Find the orders of reaction with respect to glucose and hexokinase and the rate constant.

Strategy We assume that the initial rate law has the form

$$v_0 = k_r[\text{glucose}]_0^a[\text{hexokinase}]_0^b$$

For constant [hexokinase]$_0$, the initial rate law has the form $v_0 = k_r'[\text{glucose}]_0^a$, with $k_r' = k_r[\text{hexokinase}]_0^b$, so

$$\log v_0 = \log k_r' + a \log [\text{glucose}]_0$$

We need to make a plot of $\log v_0$ against $\log [\text{glucose}]_0$ for a given [hexokinase]$_0$ and find the reaction order a from the slope and the value of k_r' from the intercept at $\log [\text{glucose}]_0 = 0$. Then, because

$$\log k_r' = \log k_r + b \log [\text{hexokinase}]_0$$

plot $\log k_r'$ against $\log [\text{hexokinase}]_0$ to find $\log k_r$ from the intercept and b from the slope.

Solution The data give the following points for the graph:

log ([glucose]$_0$/mol dm^{-3})		−3.00	−2.81	−2.51	−2.40
log (v_0/mol dm^{-3} s^{-1})	(a)	0.699	0.881	1.19	1.30
	(b)	0.844	1.04	1.36	1.49
	(c)	1.32	1.53	1.85	1.98

The graph of the data is shown in Fig. 6.5. The slopes of the lines are 1, so $a = 1$, and the effective rate constants k_r' are as follows:

[hexokinase]$_0$/(mol dm^{-3})	1.34×10^{-3}	3.00×10^{-3}	1.00×10^{-2}
log ([hexokinase]$_0$/mol dm^{-3})	−2.87	−2.52	−2.00
log (k_r'/dm^3 mol^{-1} s^{-1})	3.69	4.04	4.56

Figure 6.6 is the plot of $\log k_r'$ against $\log [\text{hexokinase}]_0$. The slope is 1, so $b = 1$. The intercept at $\log [\text{hexokinase}]_0 = 0$ is $\log k_r = 6.56$, so $k_r = 3.6 \times 10^6$ dm^3 mol^{-1} s^{-1}. The overall (initial) rate law is

$$v_0 = k_r[\text{glucose}]_0[\text{hexokinase}]_0$$

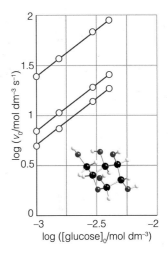

Fig. 6.5 The plots of the data in *Example* 6.1 for finding the order with respect to glucose.

A note on good practice
When taking the logarithm of a number of the form $x.xx \times 10^n$, there are *four* significant figures in the answer: the figure before the decimal point is simply the power of 10. Strictly, the logarithms are of the quantity divided by its units.

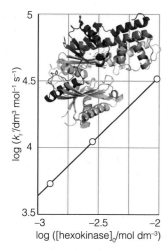

 Fig. 6.6 The plots of the data in *Example* 6.1 for finding the order with respect to hexokinase.

y-axis: $\log(k_r'/\text{dm}^3\,\text{mol}^{-1}\,\text{s}^{-1})$
x-axis: $\log([\text{hexokinase}]_0/\text{mol}\,\text{dm}^{-3})$

Self-test 6.3 The initial rate of a certain reaction depended on concentration of a substance J as follows:

$[J]_0/(\text{mmol dm}^{-3})$	5.0	10.2	17	30
$v_0/(10^{-7}\,\text{mol dm}^{-3}\,\text{s}^{-1})$	3.6	9.6	41	130

Find the order of the reaction with respect to J and the rate constant.

Answer: 2; $1.6 \times 10^{-2}\,\text{dm}^3\,\text{mol}^{-1}\,\text{s}^{-1}$

6.5 Integrated rate laws

The rate laws summarize useful information about the progress of a reaction and allow us to predict the composition of a reaction mixture at any time, including the concentrations of biochemically significant intermediates.

A rate law tells us the rate of the reaction at a given instant (when the reaction mixture has a particular composition). This is rather like being given the speed of a car at each point of its journey. For a car journey, we may want to know the distance that a car has traveled at a certain time given its varying speed. Similarly, for a chemical reaction, we may want to know the composition of the reaction mixture at a given time given the varying rate of the reaction. An **integrated rate law** is an expression that gives the concentration of a species as a function of the time.

Integrated rate laws have two principal uses. One is to predict the concentration of a species at any time after the start of the reaction. Another is to help find the rate constant and order of the reaction. Indeed, although we have introduced rate laws through a discussion of the determination of reaction rates, these rates are rarely measured directly because slopes are so difficult to determine accurately. Almost all experimental work in chemical kinetics deals with integrated rate laws; their great advantage being that they are expressed in terms of the experimental observables of concentration and time. Computers can be used to find numerical solutions of even the most complex rate laws. However, we now see that in a number of simple cases, solutions can be expressed as relatively simple functions and prove to be very useful.

(a) Zeroth-order reactions

For a chemical reaction and zeroth-order rate law of the form

$$A \rightarrow \text{products}, \quad v = -\frac{d[A]}{dt} = k_r \qquad \boxed{\text{Zeroth-order rate law}} \quad (6.11a)$$

the concentration of A falls linearly until all A has been consumed:

$$[A] = [A]_0 - k_r t \text{ for } k_r t \le [A]_0$$
$$[A] = 0 \text{ for } k_r t > [A]_0 \qquad \boxed{\substack{\text{Zeroth-order} \\ \text{integrated} \\ \text{rate law}}} \quad (6.11b)$$

(b) First-order reactions

For a chemical reaction with a first-order rate law of the form

$$A \rightarrow \text{products}, \quad v = -\frac{d[A]}{dt} = k_r[A] \qquad \boxed{\text{First-order rate law}} \quad (6.12a)$$

we show in the following *Justification* that the integrated rate law is

$$\ln \frac{[A]}{[A]_0} = -k_r t$$ First-order integrated rate law   (6.12b)

where $[A]_0$ is the initial concentration of A. Two alternative forms of this expression are

$$\ln [A] = \ln [A]_0 - k_r t$$ (6.12c)

$$[A] = [A]_0 e^{-k_r t}$$ (6.12d)

Equation 6.12d has the form of an **exponential decay** (Fig. 6.7). A common feature of all first-order reactions, therefore, is that *the concentration of the reactant decays exponentially with time.*

Justification 6.1   *First-order integrated rate laws*

A first-order rate equation has the form

$$-\frac{d[A]}{dt} = k_r [A]$$

and is an example of a 'first-order differential equation'. Because the terms $d[A]$ and dt may be manipulated like any algebraic quantity, we rearrange the differential equation into

$$\frac{d[A]}{[A]} = -k_r dt$$

and then integrate both sides. Integration from $t = 0$, when the concentration of A is $[A]_0$, to the time of interest, t, when the molar concentration of A is $[A]$, is written as

$$\int_{[A]_0}^{[A]} \frac{d[A]}{[A]} = -k_r \int_0^t dt$$

We now use the standard integral

$$\int \frac{dx}{x} = \ln x + \text{constant}$$

and obtain the expression

$$\ln [A] - \ln [A]_0 = -k_r t$$

which rearranges into eqn 6.12c.

Equation 6.12d lets us predict the concentration of A at any time after the start of the reaction. However, we can also use the result to confirm that a reaction is first order and to determine the rate constant: eqn 6.12c shows that if we plot $\ln [A]$ against t, then we will get a straight line if the reaction is first order (Fig. 6.8). If the experimental data do not give a straight line when plotted in this way, then the reaction is not first order. If the line is straight, then it follows from the same equation that its slope is $-k_r$, so we can also determine the rate constant from the graph.

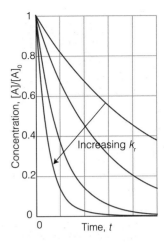

Fig. 6.7 The exponential decay of the reactant in a first-order reaction. The greater the rate constant, the more rapid is the decay.

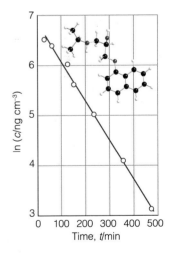

Fig. 6.8 The determination of the rate constant of a first-order reaction. A straight line is obtained when $\ln c$ is plotted against t; the slope is $-k_r$. The data are from *Case study* 6.1. The structure shown is that of propranolol, one of the first β-blockers.

Mathematical toolbox 6.1   *Differential equations*

An *ordinary differential equation* is a relation between derivatives of a function of one variable and the function itself, as in

$$a\frac{d^2y}{dx^2}+b\frac{dy}{dx}+cy+d=0$$

The coefficients a, b, etc., may be functions of x. The *order* of the equation is the order of the highest derivative that occurs in it, so eqn 6.12a is a first-order equation and the expression above is a second-order equation. 'Solving' a differential equation is the process of determining the function, in this case $y(x)$, that satisfies it.

In many cases it is found that various constants appear in the solution, such as $y(x)$ + constant. These constants are determined by imposing various *boundary conditions* on the solutions, values that the solution must have at specified points. A second-order differential equation requires two boundary conditions, a first-order equation requires one. For time-dependent solutions, the boundary condition is termed an *initial condition*, and is typically the value that the solution must have at $t=0$.

A useful indication of the rate of a first-order chemical reaction is the **half-life**, $t_{1/2}$, of a reactant, which is the time it takes for the concentration of the species to fall to half its initial value. We can find the half-life of a species A that decays in a first-order reaction (eqn 6.12a) by substituting $[A]=\frac{1}{2}[A]_0$ and $t=t_{1/2}$ into eqn 6.12b:

$$k_r t_{1/2}=-\ln\frac{\frac{1}{2}[A]_0}{[A]_0}=-\ln\frac{1}{2}=\ln 2$$

it follows that

$$t_{1/2}=\frac{\ln 2}{k_r}$$

Half-life of a first-order reaction   (6.13)

A brief illustration

Because the rate constant for the first-order denaturation of hemoglobin is $2.00\times 10^{-4}\text{ s}^{-1}$ at 60°C, the half-life of properly folded hemoglobin is 57.7 min. Hence, the concentration of folded hemoglobin falls to half its initial value in 57.7 min, and then to half that concentration again in a further 57.7 min, and so on (Fig. 6.9).

The main point to note about eqn 6.13 is that *for a first-order reaction, the half-life of a reactant is independent of its concentration*. It follows that if the concentration of A at some arbitrary stage of the reaction is $[A]$, then the concentration will fall to $\frac{1}{2}[A]$ after an interval of $(\ln 2)/k_r$ whatever the actual value of $[A]$ (Fig. 6.10).

A brief illustration

In acidic solution, the disaccharide sucrose (cane sugar, Atlas S5) is converted to a mixture of the monosaccharides glucose (Atlas S4), and fructose (Atlas S3) in a pseudo-first-order reaction. Under certain conditions of pH, the

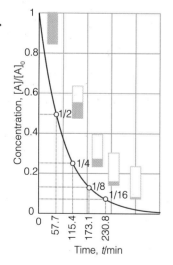

Fig. 6.9 The molar concentration of properly folded hemoglobin after a succession of half-lives.

half-life of sucrose is 28.4 min. To calculate how long it takes for the concentration of a sample to fall from 8.0 mmol dm^{-3} to 1.0 mmol dm^{-3}, we note that

$$\text{molar concentration/(mmol dm}^{-3}\text{): } 8.0 \xrightarrow{28.4\,\text{min}} 4.0 \xrightarrow{28.4\,\text{min}} 2.0 \xrightarrow{28.4\,\text{min}} 1.0$$

The total time required is 3×28.4 min $= 85.2$ min.

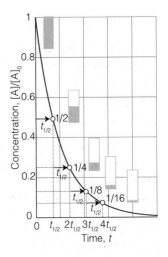

Self-test 6.4 The half-life of a substrate in a certain enzyme-catalyzed first-order reaction is 138 s. How long does it take for the concentration of substrate to fall from 1.28 mmol dm^{-3} to 0.040 mmol dm^{-3}?

Answer: 690 s

Fig. 6.10 In each successive period of duration $t_{1/2}$, the concentration of a reactant in a first-order reaction decays to half its value at the start of that period. After n such periods, the concentration is $(\frac{1}{2})^n$ of its initial concentration.

Another indication of the rate of a first-order reaction is the **time constant**, τ, the time required for the concentration of a reactant to fall to 1/e of its initial value. From eqn 6.12b it follows that

$$k_r\tau = -\ln\left(\frac{[A]_0/e}{[A]_0}\right) = -\ln\frac{1}{e} = 1$$

Hence, the time constant is the reciprocal of the rate constant:

$$\tau = \frac{1}{k_r} \qquad \boxed{\text{Time constant of a first-order decay}} \qquad (6.14)$$

The longer the time constant of a first-order reaction, the slower the decay and the longer the reactants survive.

(c) Second-order reactions

Now we need to see how concentration varies with time for a reaction with a second-order rate law of the form

$$A \rightarrow \text{products} \qquad v = -\frac{d[A]}{dt} = k_r[A]^2 \qquad \boxed{\text{Second-order rate law}} \qquad (6.15a)$$

As before, we suppose that the concentration of A at $t = 0$ is $[A]_0$ and, as shown in the following *Justification*, find that

$$\frac{1}{[A]_0} - \frac{1}{[A]} = -k_r t \qquad \boxed{\text{Second-order integrated rate law}} \qquad (6.15b)$$

Two alternative forms of eqn 6.15b are

$$\frac{1}{[A]} = \frac{1}{[A]_0} + k_r t \qquad (6.15c)$$

$$[A] = \frac{[A]_0}{1 + k_r t[A]_0} \qquad (6.15d)$$

Justification 6.2 *Second-order integrated rate laws I*

To solve the differential equation

$$-\frac{d[A]}{dt} = k_r[A]^2$$

we rearrange it into

$$-\frac{d[A]}{[A]^2} = k_r dt$$

and integrate it between $t = 0$, when the concentration of A is $[A]_0$, and the time of interest t, when the concentration of A is $[A]$:

$$\int_{[A]_0}^{[A]} \frac{d[A]}{[A]^2} = -k_r \int_0^t dt$$

The term on the right is $-k_r t$. We evaluate the integral on the left by using the standard form

$$\int \frac{dx}{x^2} = -\frac{1}{x} + \text{constant}$$

which implies that

$$\int_a^b \frac{dx}{x^2} = \left\{ -\frac{1}{x} + \text{constant} \right\}\Big|_b - \left\{ -\frac{1}{x} + \text{constant} \right\}\Big|_a = -\frac{1}{b} + \frac{1}{a}$$

and so obtain eqn 6.15b.

Equation 6.15b shows that to test for a second-order reaction, we should plot $1/[A]$ against t and expect a straight line. If the line is straight, then the reaction is second order in A and the slope of the line is equal to the rate constant (Fig. 6.11). Equation 6.15c enables us to predict the concentration of A at any time after the start of the reaction (Fig. 6.12). We see that the concentration of A approaches zero more slowly in a second-order reaction than in a first-order reaction with the same initial rate (Fig. 6.13).

It follows from eqn 6.15a by substituting $t = t_{1/2}$ and $[A] = \frac{1}{2}[A]_0$ that the half-life of a species A that is consumed in a second-order reaction is

$$t_{1/2} = \frac{1}{k_r[A]_0} \qquad \text{Half-life for a second-order reaction} \qquad (6.16)$$

Therefore, unlike a first-order reaction, the half-life of a substance in a second-order reaction varies with the initial concentration. A practical consequence of this dependence is that species that decay by second-order reactions (which includes some environmentally harmful substances) may persist in low concentrations for long periods because their half-lives are long when their concentrations are low.

Another type of second-order reaction is one that is first order in each of two reactants A and B:

$$\frac{d[A]}{dt} = -k_r[A][B] \qquad \text{Overall second-order rate law} \qquad (6.17a)$$

We have already seen that the rate of formation of DNA from two complementary strands can be modeled by this rate law. We cannot integrate eqn 6.17a until we know how the concentration of B is related to that of A. For example, if the

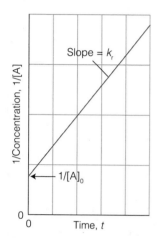

Fig. 6.11 The determination of the rate constant of a second-order reaction. A straight line is obtained when $1/[A]$ is plotted against t; the slope is k_r.

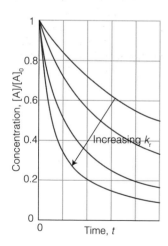

Fig. 6.12 The variation with time of the concentration of a reactant in a second-order reaction.

Mathematical toolkit 6.2   *Integration by the method of partial fractions*

To solve an integral of the form

$$I = \int \frac{1}{(a-x)(b-x)} dx$$

where a and b are constants, we use the method of partial fractions in which a fraction that is the product of terms (as in the denominator of this integrand) is written as a sum of fractions. To implement this procedure we write the integrand as

$$\frac{1}{(a-x)(b-x)} = \frac{1}{b-a}\left(\frac{1}{a-x} - \frac{1}{b-x}\right)$$

Then we integrate each term on the right by using the standard integral already given in *Justification 6.1*. It follows that

$$I = \frac{1}{b-a}\left[\int \frac{dx}{a-x} - \int \frac{dx}{b-x}\right]$$

$$= \frac{1}{b-a}\left(\ln\frac{1}{a-x} - \ln\frac{1}{b-x}\right) + \text{constant}$$

reaction is $A + B \rightarrow P$, where P denotes products and the initial concentrations are $[A]_0$ and $[B]_0$, then we show in the following *Justification* that at a time t after the start of the reaction, the concentrations satisfy the relation

$$\ln\left(\frac{[B]/[B]_0}{[A][A]_0}\right) = ([B]_0 - [A]_0)k_r t \qquad \boxed{\text{Integrated overall second-order rate law}} \qquad (6.17b)$$

Therefore, a plot of the expression on the left against t should be a straight line from which k_r can be obtained. Note that if $[A]_0 = [B]_0$, then the solutions are those already given in eqn 6.13b (but this solution cannot be found simply by setting $[A]_0 = [B]_0$ in eqn 6.17b.)

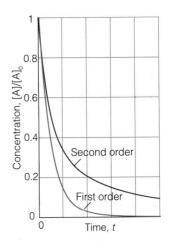

Fig. 6.13 Although the initial decay of a second-order reaction may be rapid, later the concentration approaches zero more slowly than in a first-order reaction with the same initial rate (compare Fig. 6.7).

Justification 6.3   *Second-order integrated rate laws II*

It follows from the reaction stoichiometry that when the concentration of A has fallen to $[A]_0 - x$, the concentration of B will have fallen to $[B]_0 - x$ because each A that disappears entails the disappearance of one B. It follows that

$$\frac{d[A]}{dt} = -k_r([A]_0 - x)([B]_0 - x)$$

Then, because $[A] = [A]_0 - x$ and $d[A]/dt = -dx/dt$, the rate law is

$$\frac{dx}{dt} = k_r([A]_0 - x)([B]_0 - x)$$

The initial condition is that $x = 0$ when $t = 0$; so the integration required is

$$\int_0^x \frac{dx}{([A]_0 - x)([B]_0 - x)} = k_r \int_0^t dt$$

The integral on the right is simply $k_r t$. The integral on the left is evaluated by using the method of partial fractions (see *Mathematical toolkit 6.2*):

$$\int_0^x \frac{dx}{([A]_0 - x)([B]_0 - x)} = \frac{1}{[B]_0 - [A]_0}\left\{\ln\left(\frac{[A]_0}{[A]_0 - x}\right) - \ln\left(\frac{[B]_0}{[B]_0 - x}\right)\right\}$$

The two logarithms can be combined as follows:

$$\ln\left(\frac{[A]_0}{[A]_0-x}\right)-\ln\left(\frac{[B]_0}{[B]_0-x}\right)=\ln[A]_0-\ln([A]_0-x)-\ln[B]_0+\ln([B]_0-x)$$

$$=\ln[A]_0-\ln[A]-\ln[B]_0+\ln[B]$$

$$=\{\ln[B]-\ln[B]_0\}-\{\ln[A]-\ln[A]_0\}$$

$$=\ln\left(\frac{[B]}{[B]_0}\right)-\ln\left(\frac{[A]}{[A]_0}\right)$$

$$=\ln\left(\frac{[B]/[B]_0}{[A]/[A]_0}\right)$$

where we have used $[A]=[A]_0-x$ and $[B]=[B]_0-x$. Combining all the results so far gives eqn 6.17b.

Equation 6.17b can be rearranged to give the concentration of either reactant. To do this, we substitute $[A]=[A]_0-x$ and $[B]=[B]_0-x$ into the equation written in the form

$$\frac{[B][A]_0}{[A][B]_0}=e^{([B]_0-[A]_0)k_rt}$$

and obtain

$$\frac{[A]_0([B]_0-x)}{[B]_0([A]_0-x)}=e^{([B]_0-[A]_0)k_rt}$$

This expression is then solved for x:

$$x=\frac{[A]_0[B]_0(e^{([B]_0-[A]_0)k_rt}-1)}{[B]_0e^{([B]_0-[A]_0)k_rt}-[A]_0}$$

At this point, we can form either [A] or [B]. For instance, from $[A]=[A]_0-x$ we find

$$[A]=\frac{[A]_0([B]_0-[A]_0)}{[B]_0e^{([B]_0-[A]_0)k_rt}-[A]_0}\tag{6.18}$$

The time dependence of [A] and [B] is illustrated in Fig. 6.14.

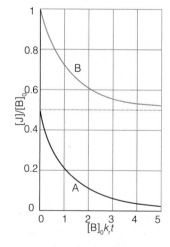

Fig. 6.14 The time dependence of the concentrations of the reactants in a reaction with the overall second-order rate law in eqn 6.17a. We have taken $[B]_0=2[A]_0$.

Case study 6.1 *Pharmacokinetics*

Pharmacokinetics is the study of the rates of absorption and elimination of drugs by organisms. In most cases, elimination is slower than absorption and is a more important determinant of availability of a drug for binding to its target. A drug can be eliminated by many mechanisms, such as metabolism in the liver, intestine, or kidney followed by excretion of breakdown products through urine or feces.

As an example of pharmacokinetic analysis, consider the elimination of β-adrenergic blocking agents ('β-blockers'), drugs used in the treatment of hypertension. After intravenous administration of a β-blocker, the blood plasma of a patient was analyzed for remaining drug, and the data are shown below, where c is the mass concentration of the drug measured at a time t after the injection.

t/min	30	60	120	150	240	360	480
$c/(\text{ng cm}^{-3})$	699	622	413	292	152	60	24

To see if the removal is a first-order process, we draw up the following table:

t/min	30	60	120	150	240	360	480
$\ln(c/(\text{ng cm}^{-3}))$	6.550	6.433	6.023	5.677	5.024	4.09	3.18

The graph of the data is shown in Fig. 6.8. The plot is straight, confirming a first-order process. Its least-squares best-fit slope is -7.6×10^{-3}, so $k_r = 7.6 \times 10^{-3} \text{ min}^{-1}$ and $t_{1/2} = 91$ min at 310 K, body temperature.

Most drugs are eliminated from the body by a first-order process. An essential aspect of drug development is the optimization of the half-life of elimination, which needs to be long enough to allow the drug to find and act on its target organ but not so long that harmful side effects become important.

The temperature dependence of reaction rates

The rates of most chemical reactions increase as the temperature is raised. Many organic reactions in solution lie somewhere in the range spanned by the hydrolysis of methyl ethanoate (for which the rate constant at 35°C is 1.8 times that at 25°C) and the hydrolysis of sucrose (for which the factor is 4.1). Reactions in the gas phase typically have rates that are only weakly sensitive to the temperature. Enzyme-catalyzed reactions may show a more complex temperature dependence because raising the temperature may provoke conformational changes and even denaturation and degradation, which lower the effectiveness of the enzyme. We saw in the discussion of the hydrophobic effect (Section 2.7) that *lowering* the temperature can also result in denaturation, so an enzyme may lose its effectiveness at low temperatures too.

6.6 The Arrhenius equation

The balance of reactions in organisms depends strongly on the temperature: that is one function of a fever, which modifies reaction rates in the infecting organism and hence destroys it. To discuss the effect quantitatively, we need to know the factors that make a reaction rate more or less sensitive to temperature.

As data on reaction rates were accumulated toward the end of the nineteenth century, the Swedish chemist Svante Arrhenius noted that almost all of them showed a similar dependence on temperature. In particular, he noted that a graph of $\ln k_r$, where k_r is the rate constant for the reaction, against $1/T$, where T is the (absolute) temperature at which k_r is measured, gives a straight line with a slope that is characteristic of the reaction (Fig. 6.15). The mathematical expression of this conclusion is that the rate constant varies with temperature as

$$\ln k_r = \text{intercept} + \text{slope} \times \frac{1}{T}$$

Empirical temperature dependence   (6.19a)

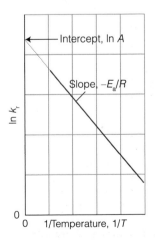

Fig. 6.15 The general form of an Arrhenius plot of $\ln k_r$ against $1/T$. The slope is equal to $-E_a/R$ and the intercept at $1/T = 0$ is equal to $\ln A$.

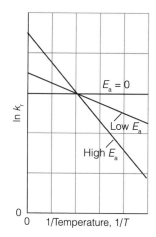

Fig. 6.16 These Arrhenius plots correspond to three different activation energies. Note the fact that the plot corresponding to the higher activation energy indicates that the rate of that reaction is more sensitive to temperature.

This expression is normally written as the **Arrhenius equation**:

$$\ln k_r = \ln A - \frac{E_a}{RT}$$

(Arrhenius equation)   (6.19b)

or alternatively as

$$k_r = Ae^{-E_a/RT}$$

(6.19c)

The parameter A (which has the same units as k_r) is called the **pre-exponential factor** and the parameter E_a (which is a molar energy and normally expressed as kilojoules per mole) is called the **activation energy**. Collectively, A and E_a are called the **Arrhenius parameters** of the reaction.

A practical point to note by comparing eqns 6.19a and 6.19b is that a high activation energy corresponds to a reaction rate that is very sensitive to temperature (the Arrhenius plot has a steep slope, Fig. 6.16). Conversely, a small activation energy indicates a reaction rate that varies only slightly with temperature (the slope is shallow). A reaction with zero activation energy (so $k_r = A$), such as for some radical recombination reactions in the gas phase, has a rate that is largely independent of temperature. Marked deviations from a straight line at high or low temperature may indicate that an enzyme has lost its effectiveness through denaturation.

Example 6.2   *Determining the Arrhenius parameters*

The rate constant of the acid hydrolysis of sucrose discussed in Section 6.6b varies with temperature as follows. Find the activation energy and the pre-exponential factor.

T/K	297	301	305	309	313
$k_r/(10^{-3}\,s^{-1})$	4.8	7.8	13	20	32

Strategy We plot $\ln k_r$ against $1/T$ and expect a straight line. The slope is $-E_a/R$ and the intercept of the extrapolation to $1/T = 0$ is $\ln A$. It is best to do a least-squares fit of the data to a straight line. Note that, as remarked in the text, A has the same units as k_r.

Solution The Arrhenius plot is shown in Fig. 6.17. The least-squares best fit of the line has slope -1.10×10^4 and intercept 31.7 (which is well off the graph), therefore

$$E_a = -R \times \text{slope}$$
$$= -(8.3145\,\text{J K}^{-1}\,\text{mol}^{-1}) \times (-1.10 \times 10^4\,\text{K}) = 91.5\,\text{kJ mol}^{-1}$$

and

$$A = e^{31.7}\,s^{-1} = 5.8 \times 10^{13}\,s^{-1}$$

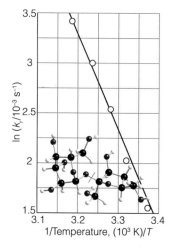

Fig. 6.17 The Arrhenius plot for the acid hydrolysis of sucrose, and the best (least-squares) straight line fitted to the data points. The data are from *Example* 6.2.

Self-test 6.5   Determine A and E_a from the following data:

T/K	300	350	400	450	500
$k_r/(\text{dm}^3\,\text{mol}^{-1}\,s^{-1})$	7.9×10^6	3.0×10^7	7.9×10^7	1.7×10^8	3.2×10^8

Answer: $8 \times 10^{10}\,\text{dm}^3\,\text{mol}^{-1}\,s^{-1}$, $23\,\text{kJ mol}^{-1}$

Once the activation energy of a reaction is known, it is a simple matter to predict the value of a rate constant $k_{r,2}$ at a temperature T_2 from its value $k_{r,1}$ at another temperature T_1 that lies within the functional range of the enzyme. To do so, we write

$$\ln k_{r,2} = \ln A - \frac{E_a}{RT_2}$$

and then subtract eqn 6.19b (with T identified as T_1 and k_r as $k_{r,1}$), so obtaining

$$\ln k_{r,2} - \ln k_{r,1} = -\frac{E_a}{RT_2} + \frac{E_a}{RT_1}$$

and therefore

$$\ln \frac{k_{r,2}}{k_{r,1}} = \frac{E_a}{R}\left(\frac{1}{T_1} - \frac{1}{T_2}\right) \qquad \boxed{\text{Temperature dependence of the rate constant}} \qquad (6.20)$$

A brief illustration

For a reaction with an activation energy of $50\,kJ\,mol^{-1}$, an increase in the temperature from 25°C to 37°C (body temperature) corresponds to

$$\ln \frac{k_{r,2}}{k_{r,1}} = \frac{50 \times 10^3\,J\,mol^{-1}}{8.3145\,J\,K^{-1}\,mol^{-1}}\left(\frac{1}{298\,K} - \frac{1}{310\,K}\right) = \frac{50 \times 10^3}{8.3145}\left(\frac{1}{298} - \frac{1}{310}\right)$$

By taking natural antilogarithms (that is, by forming e^x), $k_{r,2} = 2.18k_{r,1}$. This result corresponds to slightly more than a doubling of the rate constant.

Self-test 6.6   The activation energy of one of the reactions in the citric acid cycle (Case study 4.3) is $87\,kJ\,mol^{-1}$. What is the change in rate constant when the temperature falls from 37°C to 15°C?

Answer: $k_{r,2} = 0.076k_{r,1}$

6.7 Preliminary interpretation of the Arrhenius parameters

Once we know the molecular interpretation of the pre-exponential factor and the activation energy, we can identify the strategies that special biological macromolecules adopt to accelerate and regulate the rates of biochemical reactions.

The simplest interpretation of a chemical reaction is that it takes place when two molecules collide either in a gas or, more relevantly to biology, as they move through a solution. We shall refine this picture in Chapter 7, but it is adequate for our present purpose.

The pre-exponential factor A is a measure of the rate at which collisions occur, irrespective of their energy. More precisely, A is the constant of proportionality between the collision rate and the product of the molar concentrations of the reactants.

To interpret E_a, we consider how the energy of the reactant molecules A and B (specifically, their total molecular potential energy) changes in the course of a collision. As the reaction proceeds, A and B come into contact, distort, and begin to exchange or discard atoms. The potential energy rises to a maximum and the

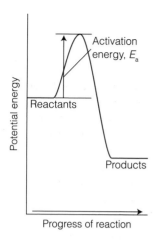

Fig. 6.18 A potential energy profile for an exothermic reaction. The graph depicts schematically the changing potential energy of two species that approach, collide, and then go on to form products. The activation energy is the height of the barrier above the potential energy of the reactants.

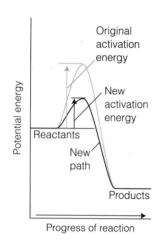

Fig. 6.19 A catalyst acts by providing a new reaction pathway between reactants and products, with a lower activation energy than the original pathway.

cluster of atoms that corresponds to the region close to the maximum is called the **activated complex** (Fig. 6.18). After the maximum, the potential energy falls as the atoms rearrange in the cluster and reaches a value characteristic of the products. The climax of the reaction is at the peak of the potential energy, which corresponds to the activation energy, E_a. Here two reactant molecules have come to such a degree of closeness and distortion that a small further distortion will send them in the direction of products. This crucial configuration is called the **transition state** of the reaction. Although some molecules entering the transition state might revert to reactants, if they pass through this configuration it is inevitable that products will emerge from the encounter.[2]

We can infer from the preceding discussion that to react when they meet, two reactant molecules must have sufficient kinetic energy to surmount the barrier and pass through the transition state. It follows that *the activation energy is the minimum relative kinetic energy that reactants must have in order to form products.* For example, in a gas phase reaction there are numerous collisions each second, but only a tiny proportion are sufficiently energetic to lead to reaction. Hence, the exponential factor in eqn 6.19c can be interpreted as the fraction of collisions that have enough kinetic energy to lead to reaction.

An enzyme, and a catalyst in general, lowers the activation energy of the reaction either by providing a different reaction pathway with a lower activation energy or by lowering the energy of the transition state to make it more accessible (Fig. 6.19). Enzymes are very specific and can have a dramatic effect on the reactions they control. For example, the enzyme catalase reduces the activation energy for the decomposition of hydrogen peroxide from 76 kJ mol^{-1} to 8 kJ mol^{-1}, corresponding to an acceleration of the reaction by a factor of 10^{12} at 298 K.

> **A brief illustration**
>
> The effect of catalase on the rate of decomposition of H_2O_2 can be assessed by evaluating the ratio of rate constants as follows:
>
> $$\frac{k_{r,catalyzed}}{k_{r,uncatalyzed}} = \frac{Ae^{-E_{a,catalyzed}/RT}}{Ae^{-E_{a,uncatalyzed}/RT}} = e^{-(E_{a,catalyzed} - E_{a,uncatalyzed})/RT}$$
>
> $$= e^{(68 \times 10^3 \text{ J mol}^{-1})/(8.3145 \text{ J K}^{-1} \text{ mol}^{-1}) \times (298 \text{ K})} = 8.3 \times 10^{11}$$

---

[2] The terms *activated complex* and *transition state* are often used as synonyms; however, we shall preserve a distinction.

Checklist of key concepts

☐ 1. The rates of chemical reactions are measured by using techniques that monitor the concentrations of species present in the reaction mixture (Table 6.1).

☐ 2. Techniques for the study of reactions include real-time and quenching procedures, flow and stopped-flow techniques, and flash photolysis.

☐ 3. The instantaneous rate of a reaction is the slope of the tangent to the graph of concentration against time (expressed as a positive quantity).

☐ 4. A rate law is an expression for the reaction rate in terms of the concentrations of the species that occur in the overall chemical reaction.

☐ 5. For a rate law of the form $rate = k_r[A]^a[B]^b \ldots$, the order with respect to A is a and the overall order is $a + b + \cdots$.

☐ 6. An integrated rate law is an expression for the rate of a reaction as a function of time.

☐ 7. The half-life $t_{1/2}$ of a reaction is the time it takes for the concentration of a species to fall to half its initial value.

☐ 8. The temperature dependence of the rate constant of a reaction typically follows the Arrhenius law.

☐ 9. The greater the activation energy, the more sensitive the rate constant is to the temperature.

☐ 10. The activation energy is the minimum relative kinetic energy that reactants must have in order to form products; the pre-exponential factor is a measure of the rate at which collisions occur irrespective of their energy.

Checklist of key equations

Property	Equation			Comment
Half-life	$t_{1/2} = (\ln 2)/k_r$			First-order reaction
	$t_{1/2} = 1/k_r[A]_0$			Second-order reaction
Arrhenius equation	$\ln k_r = \ln A - E_a/RT$			
Integrated rate laws	Zeroth order	$[A] = [A]_0 - k_r t$ for $k_r t \leq [A]_0$		$A \rightarrow P$ $v = k_r$
	First order	$[A] = [A]_0 e^{-k_r t}$		$A \rightarrow P$ $v = k_r[A]$
	Second order	$[A] = [A]_0/(1 + k_r t[A]_0)$		$A \rightarrow P$ $v = k_r[A]^2$
		$[A] = [A]_0([B]_0 - [A]_0)/f(t)$		$A + B \rightarrow P$ $v = k_r[A][B]$
		$f(t) = [B]_0 e^{([B]_0 - [A]_0)k_r t} - [A]_0$		

Discussion questions

6.1 Consult literature sources and list the observed timescales during which the following processes occur: proton transfer reactions, the initial event of vision, energy transfer in photosynthesis, the initial electron transfer events of photosynthesis, and the helix-to-coil transition in polypeptides.

6.2 Write a brief report on a recent research article in which at least one of the following techniques was used to study the kinetics of a biochemical reaction: stopped-flow techniques, flash photolysis, chemical quench-flow methods, or freeze-quench methods. Your report should be similar in content and extent to one of the *Case studies* found throughout this book.

6.3 Describe the main features, including advantages and disadvantages, of the following experimental methods for determining the rate law of a reaction: the isolation method, the method of initial rates, and fitting data to integrated rate law expressions.

6.4 Distinguish between zeroth-order, first-order, second-order, and pseudo-first-order reactions.

6.5 Define the terms in and limit the generality of the expression $\ln k_r = \ln A - E_a/RT$.

6.6 Provide molecular interpretations of the activation energy and the pre-exponential factor.

Exercises

6.7 The molar absorption coefficient of a substance dissolved in water is known to be 855 $dm^3 mol^{-1} cm^{-1}$ at 270 nm. To determine the rate of decomposition of this substance, a solution with a concentration of 3.25 $mmol dm^{-3}$ was prepared. Calculate the percentage reduction in intensity when light of that wavelength passes through 2.5 mm of this solution.

6.8 The molar absorption coefficient of cytochrome P450, an enzyme involved in the breakdown of harmful substances in the liver and small intestine, at 522 nm is 291 $dm^3 mol^{-1} cm^{-1}$. In a study of its enzymatic activity, a solution of cytochrome P450 was prepared in a cell of length 6.5 mm. When light of 522 nm passes through this cell, 39.8 per cent of the light is absorbed. What is the molar concentration of cytochrome P450 in the solution?

6.9 (a) The rate of formation of C in the reaction $2 A + B \rightarrow 3 C + 2 D$ is 2.2 $mol dm^{-3} s^{-1}$. State the rates of formation and consumption of A, B, and D. (b) The rate law for this reaction was reported as rate = $k_r[A][B][C]$ with the molar concentrations in $mol dm^{-3}$ and the time in seconds. What are the units of k_r?

6.10 If the rate laws are expressed with (a) concentrations in numbers of molecules per cubic meter (molecules m^{-3}) and (b) pressures in kilopascals, what are the units of the second-order and third-order rate constants?

6.11 The growth of microorganisms may be described in general terms as follows: (a) initially, cells do not grow appreciably; (b) after the initial period, cells grow rapidly with first-order kinetics; (c) after this period of growth, the number of cells reaches a maximum level and then begins to decrease. Sketch a plot of log(number of microorganisms) against t that reflects the kinetic behavior just described.

6.12 Laser flash photolysis is often used to measure the binding rate of CO to heme proteins, such as myoglobin (Mb), because CO dissociates from the bound state relatively easily on absorption of energy from an intense and short pulse of light. The reaction is usually run under pseudo-first-order conditions. For a reaction in which $[Mb]_0 = 10 mmol dm^{-3}$, $[CO] = 400 mmol dm^{-3}$, and the rate constant is $5.8 \times 10^5 dm^3 mol^{-1} s^{-1}$, plot a curve of [Mb] against time. The observed reaction is $Mb + CO \rightarrow MbCO$.

6.13 The oxidation of ethanol to acetaldehyde (ethanal) by NAD^+ in the liver in the presence of the enzyme liver alcohol dehydrogenase:

$$CH_3CH_2OH(aq) + NAD^+(aq) + H_2O(l) \rightarrow$$
$$CH_3CHO(aq) + NADH(aq) + H_3O^+(aq)$$

is effectively zeroth order overall as the ethanol is in excess and the concentration of the NAD^+ is maintained at a constant level by normal metabolic processes. Calculate the rate constant for the conversion of ethanol to ethanal in the liver if the concentration of ethanol in body fluid drops by 50 per cent from 1.5 $g dm^{-3}$, a level that results in lack of coordination and slurring of speech, in 49 min at body temperature. Express your answer in units of $g dm^{-3} h^{-1}$.

6.14 In a study of the alcohol dehydrogenase catalyzed oxidation of ethanol, the molar concentration of ethanol decreased in a first-order reaction from 220 $mmol dm^{-3}$ to 56.0 $mmol dm^{-3}$ in 1.22×10^4 s. What is the rate constant of the reaction?

6.15 The elimination of carbon dioxide from pyruvate ions by a decarboxylase enzyme was monitored by measuring the partial pressure of the gas as it was formed in a 250 cm^3 flask at 293. In one experiment, the partial pressure increased from 0 to 100 Pa in 522 s in a first-order reaction when the initial concentration of pyruvate ions in 100 cm^3 of solution was 3.23 $mmol dm^{-3}$. What is the rate constant of the reaction?

6.16 Carbonic anhydrase is a zinc-based enzyme that catalyzes the conversion of carbon dioxide to carbonic acid. In an experiment to study its effect, it was found that the molar concentration of carbon dioxide in solution decreased from 220 $mmol dm^{-3}$ to 56.0 $mmol dm^{-3}$ in 1.22×10^4 s. What is the rate constant of the first-order reaction?

6.17 The formation of NOCl from NO in the presence of a large excess of chlorine is pseudo-second order in NO. When the initial pressure of NO was 300 Pa, the partial pressure of NOCl increased from zero to 100 Pa in 522 s. What is the rate constant of the reaction?

6.18 The following data were obtained on the initial rate of isomerization of a compound S catalyzed by an enzyme E:

$[S]_0/(mmol dm^{-3})$		1.00	2.00	3.00	4.00
$v_0/(mmol dm^{-3} s^{-1})$	(a)	4.5	9.0	15.0	18.0
	(b)	14.8	25.0	45.0	59.7
	(c)	58.9	120.0	180.0	238.0

The enzyme concentrations are (a) 1.00 $mmol dm^{-3}$, (b) 3.00 $mol dm^{-3}$, and (c) 10.0 $mmol dm^{-3}$. Find the orders of reactions with respect to S and E, and the rate constant.

6.19 Sucrose is readily hydrolyzed to glucose and fructose in acidic solution. An experiment on the hydrolysis of sucrose in 0.50 M HCl(aq) produced the following data:

t/min	0	14	39	60	80
[sucrose]/($mol dm^{-3}$)	0.316	0.300	0.274	0.256	0.238
t/min	110	140	170	210	
[sucrose]/($mol dm^{-3}$)	0.211	0.190	0.170	0.146	

Determine the order of the reaction with respect to sucrose and the rate constant of the reaction.

6.20 Iodoacetamide and N-acetylcysteine react with 1:1 stoichiometry. The following data were collected at 298 K for the reaction of 1.00 $mmol dm^{-3}$ N-acetylcysteine with 1.00 $mmol dm^{-3}$ iodoacetamide:

t/s	10	20	40
[N-acetylcysteine]/($mmol dm^{-3}$)	0.770	0.580	0.410
t/s	60	100	150
[N-acetylcysteine]/($mmol dm^{-3}$)	0.315	0.210	0.155

(a) Explain why analysis of these data yield the overall order of the reaction and not the order with respect to N-acetylcysteine (or iodoacetamide). (b) Plot the data in an appropriate fashion to determine the overall order of the reaction. (c) From the graph, determine the rate constant.

6.21 The following data were collected at 298 K for the reaction of 1.00 $mmol dm^{-3}$ N-acetylcysteine with 2.00 $mmol dm^{-3}$

iodoacetamide under conditions that are different from those in Exercise 6.20:

t/s	5	10	25	35	50	60
[N-acetylcysteine]/(mmol dm⁻³)	0.74	0.58	0.33	0.21	0.12	0.09

(a) Use these data and your result from Exercise 6.20a to determine the order of the reaction with respect to each reactant. (b) Determine the rate constant.

6.22 The composition of a liquid phase reaction $2\,A \rightarrow B$ was followed spectrophotometrically with the following results:

t/min	0	10	20	30	40	∞
[B]/(mol dm⁻³)	0	0.089	0.153	0.200	0.230	0.312

Determine the order of the reaction with respect to sucrose and the rate constant of the reaction.

6.23 Establish the integrated form of a third-order rate law of the form $v = k_r[A]^3$. What would it be appropriate to plot to confirm that a reaction is third order?

6.24 Derive an integrated expression for a second-order rate law $v = k_r[A][B]$ for a reaction of stoichiometry $2\,A + 3\,B \rightarrow P$.

6.25 Derive the integrated form of a third-order rate law $v = k_r[A]^2[B]$ in which the stoichiometry is $2\,A + B \rightarrow P$ and the reactants are initially present in (a) their stoichiometric proportions and (b) with B present initially in twice the amount.

6.26 The half-life of pyruvic acid in the presence of an aminotransferase enzyme (which converts it to alanine) was found to be 221 s. How long will it take for the concentration of pyruvic acid to fall to $\frac{1}{64}$ of its initial value in this first-order reaction?

6.27 Radioactive decay of unstable atomic nuclei is a first-order process. The half-life for the (first-order) radioactive decay of ^{14}C is 5730 a (1 a is the SI unit 1 annum, for 1 year; the nuclide emits β particles, high-energy electrons, with an energy of 0.16 MeV). An archaeological sample contained wood that had only 69 per cent of the ^{14}C found in living trees. What is its age?

6.28 One of the hazards of nuclear explosions is the generation of ^{90}Sr and its subsequent incorporation in place of calcium in bones. This nuclide emits β particles of energy 0.55 MeV and has a half-life of 28.1 a (1 a is the SI unit 1 annum, for 1 year). Suppose 1.00 μg was absorbed by a newborn child. How much will remain after (a) 19 a and (b) 75 a if none is lost metabolically?

6.29 The estimated half-life for P–O bonds is 1.3×10^5 a (1 a is the SI unit 1 annum, for 1 year). Approximately 10^9 such bonds are present in a strand of DNA. How long (in terms of its half-life) would a single strand of DNA survive with no cleavage in the absence of repair enzymes?

6.30 To prepare a dog for surgery, about 30 mg (kg body mass)⁻¹ of phenobarbital must be administered intravenously. The anesthetic is metabolized with first-order kinetics and a half-life of 4.5 h. After about 2 h, the drug begins to lose its effect in a 15-kg dog. What mass

of phenobarbital must be re-injected to restore the original level of anesthetic in the dog?

6.31 The second-order rate constant for the reaction $CH_3COOC_2H_5(aq) + OH^-(aq) \rightarrow CH_3CO_2^-(aq) + CH_3CH_2OH(aq)$ is 0.11 dm³ mol⁻¹ s⁻¹. What is the concentration of ester after (a) 15 s and (b) 15 min when ethyl acetate is added to sodium hydroxide so that the initial concentrations are [NaOH] = 0.055 mol dm⁻³ and [CH₃COOC₂H₅] = 0.150 mol dm⁻³?

6.32 A reaction $2\,A \rightarrow P$ has a second-order rate law with $k_r = 1.24$ cm³ mol⁻¹ s⁻¹. Calculate the time required for the concentration of A to change from 0.260 mol dm⁻³ to 0.026 mol dm⁻³.

6.33 Show that the ratio $t_{1/2}/t_{3/4}$, where $t_{1/2}$ is the half-life and $t_{3/4}$ is the time for the concentration of A to decrease to $\frac{3}{4}$ of its initial value (implying that $t_{3/4} < t_{1/2}$), can be written as a function of n alone and can therefore be used as a rapid assessment of the order of a reaction.

6.34 (a) Show that, for a reaction that is n-order in A, $t_{1/2}$ is given by

$$t_{1/2} = \frac{2^{n-1} - 1}{(n-1)k_r[A]_0^{n-1}}$$

(b) Deduce an expression for the time it takes for the concentration of a substance to fall to one-third the initial value in an nth-order reaction.

6.35 A rate constant is 1.78×10^4 dm³ mol⁻¹ s⁻¹ at 19°C and 1.38×10^{-3} dm³ mol⁻¹ s⁻¹ at 37°C. Evaluate the Arrhenius parameters of the reaction.

6.36 The activation energy for the denaturation of the O_2-binding protein hemocyanin is 408 kJ mol⁻¹. At what temperature will the rate be 10 per cent greater than its rate at 25°C?

6.37 Which reaction responds more strongly to changes of temperature, one with an activation energy of 52 kJ mol⁻¹ or one with an activation energy of 25 kJ mol⁻¹?

6.38 The rate constant of a reaction increases by a factor of 1.23 when the temperature is increased from 20°C to 27°C. What is the activation energy of the reaction?

6.39 Make an appropriate Arrhenius plot of the following data for the binding of an inhibitor to the enzyme carbonic anhydrase and calculate the activation energy for the reaction.

T/K	289.0	293.5	298.1	303.2	308.0	313.5
k_r/(10⁶ dm³ mol⁻¹ s⁻¹)	1.04	1.34	1.53	1.89	2.29	2.84

6.40 Food rots about 40 times more rapidly at 25°C than when it is stored at 4°C. Estimate the overall activation energy for the processes responsible for its decomposition.

6.41 The enzyme urease catalyzes the reaction in which urea is hydrolyzed to ammonia and carbon dioxide. The half-life of urea in the pseudo-first-order reaction for a certain amount of urease doubles when the temperature is lowered from 20°C to 10°C and the equilibrium constant for binding of urea to the enzyme is largely unchanged. What is the activation energy of the reaction?

Project

6.42* Prebiotic reactions are reactions that might have occurred under the conditions prevalent on the Earth before the first living creatures emerged and can lead to analogs of molecules necessary for life as we now know it. To qualify, a reaction must proceed with a favorable rate and have a reasonable value for the equilibrium constant. An example of a prebiotic reaction is the formation of 5-hydroxymethyluracil (HMU) from uracil and formaldehyde (HCHO). Amino acid analogs can be formed from HMU under prebiotic conditions by reaction with various nucleophiles, such as H_2S, HCN, indole, and imidazole. For the synthesis of HMU at pH = 7, the temperature dependence of the rate constant is given by

* Adapted from an exercise provided by Charles Trapp, Carmen Giunta, and Marshall Cady.

$$\log k_r/(dm^3\, mol^{-1}\, s^{-1}) = 11.75 - 5488/(T/K)$$

and the temperature dependence of the equilibrium constant is given by

$$\log K = -1.36 + 1794/(T/K)$$

(a) Calculate the rate constants and equilibrium constants over a range of temperatures corresponding to possible prebiotic conditions, such as 0–50°C, and plot them against temperature.

(b) Calculate the activation energy and the standard reaction Gibbs energy and enthalpy at 25°C.

(c) Prebiotic conditions are not likely to be standard conditions. Speculate about how the actual values of the reaction Gibbs energy and enthalpy might differ from the standard values. Do you expect that the reaction would still be favorable?

Accounting for the rate laws

7

Even quite simple rate laws can give rise to complicated behavior. The observation that the heart maintains a steady pulse throughout a lifetime, but may break into fibrillation during a heart attack, is one sign of that complexity. On a less personal scale, intermediates come and go in the course of reactions, and all reactions approach equilibrium with the forward and reverse reactions proceeding at the same rate. However, the complexity of the behavior of reaction rates means that the study of reaction rates can give deep insight into the way that reactions actually take place. As we remarked in Chapter 6, rate laws are a window on to the mechanism, the sequence of elementary molecular events that leads from the reactants to the products, of the reactions they summarize. In this chapter, we see how to interpret an observed rate law in terms of a proposed mechanism in preparation for dealing with biological systems in Chapter 8.

Reaction mechanisms

So far, we have considered very simple rate laws, in which reactants are consumed or products formed. However, all reactions actually proceed toward a state of equilibrium in which the reverse reaction becomes increasingly important. Moreover, many reactions—particularly those in organisms—proceed to products through a series of intermediates. In organisms (and in chemical plants), one of these intermediates may be of crucial importance and the ultimate products may represent waste.

7.1 The approach to equilibrium

Many biochemical reactions take place in steps that reach equilibrium quickly, and to understand their role we need to understand the relation between their kinetic behavior and their equilibrium composition.

All forward reactions are accompanied by their reverse reactions. At the start of a reaction, when little or no product is present, the rate of the reverse reaction is negligible. However, as the concentration of products increases, the rate at which they decompose into reactants becomes greater. At equilibrium, the reverse rate matches the forward rate and the reactants and products are present in abundances given by the equilibrium constant for the reaction.

(a) The relation between equilibrium constants and rate constants

We can analyze the relation between rates and equilibrium by thinking of a very simple reaction of the form

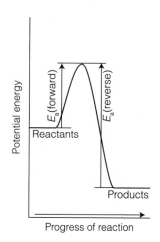

Fig. 7.1 The reaction profile for an exothermic reaction. The activation energy is greater for the reverse reaction than for the forward reaction, so the rate of the forward reaction increases less sharply with temperature. As a result, the equilibrium constant shifts in favor of the reactants as the temperature is raised.

Forward: A → B   Rate of formation of B = $k_r[A]$

Reverse: B → A   Rate of decomposition of B = $k_r'[B]$

For instance, we could envisage this scheme as the interconversion of coiled (A) and uncoiled (B) DNA molecules. The *net* rate of formation of B, the difference of its rates of formation and decomposition, is

$$\text{Net rate of formation of B} = \frac{d[B]}{dt} = k_r[A] - k_r'[B]$$

When the reaction has reached equilibrium, the concentrations of A and B are $[A]_{eq}$ and $[B]_{eq}$ and there is no net formation of either substance. It follows that $d[B]/dt = 0$ and hence that $k_r[A]_{eq} = k_r'[B]_{eq}$. Therefore, the equilibrium constant for the reaction is related to the forward and reverse rate constants by

$$K = \frac{[B]_{eq}}{[A]_{eq}} = \frac{k_r}{k_r'} \qquad \boxed{\text{The equilibrium constant in terms of rate constants}} \quad (7.1)$$

If the forward rate constant is much larger than the reverse rate constant, then $K \gg 1$. If the opposite is true, then $K \ll 1$. Equation 7.1 provides a crucial connection between the kinetics of a reaction and its equilibrium properties. It is also very useful in practice, for we may be able to measure the equilibrium constant and one of the rate constants and can then calculate the missing rate constant from eqn 7.1. Alternatively, we can use the relation to calculate the equilibrium constant from kinetic measurements. Equation 7.1 is not valid when the forward and reverse reactions have different orders; we introduce a more general version of eqn 7.1 in Section 7.2.

Equation 7.1 also gives us insight into the temperature dependence of equilibrium constants. First, we suppose that both the forward and reverse reactions show Arrhenius behavior (Section 6.6). As we see from Fig. 7.1, for an exothermic reaction the activation energy of the forward reaction is smaller than that of the reverse reaction. Therefore, the forward rate constant increases less sharply with

---

Mathematical toolkit 7.1   *Differential equations for kinetics*

We need two types of differential equations in this chapter. One has the form

$$\frac{dy}{dx} + ay = b$$

where a and b are constants. We encounter an equation of this form in *Justification* 7.1. The solution is

$$y = ce^{-ax} + b/a$$

where c is a constant. To verify this solution you should use the relation

$$\frac{d}{dx}e^{\pm ax} = \pm ae^{\pm ax}$$

The second type has the more complicated form

$$\frac{dy}{dx} + a(x)y = f(x)$$

This equation has the solution

$$y = e^{-F(x)} \int e^{F(x)} f(x)dx + ce^{-F(x)}$$

where c is a constant and

$$F(x) = \int a(x)dx$$

We encounter an equation of this type in Section 7.3(a), where a is a constant (so $F(x) = ax$) and where $f(x) = be^{-b'x}$ with b and b' constants. In that special case, the solution is

$$y = \frac{be^{-b'x}}{a - b'} + ce^{-ax}$$

temperature than the reverse reaction does (recall Fig. 6.18). Consequently, when we increase the temperature of a system at equilibrium, k'_r increases more steeply than k_r does, and the ratio k_r/k'_r, and therefore K, decreases. This is exactly the conclusion we drew from the van 't Hoff equation (eqn 4.14), which was based on thermodynamic arguments.

(b) The time dependence of the approach to equilibrium

Equation 7.1 tells us the ratio of concentrations after a long time has passed and the reaction has reached equilibrium. To find the concentrations at an intermediate stage, we need the integrated rate equation. If no B is present initially, we show in the following *Justification* that

$$[A] = \frac{(k'_r + k_r e^{-(k_r+k'_r)t})[A]_0}{k_r + k'_r} \tag{7.2a}$$

$$[B] = \frac{k_r(1 - e^{-(k_r+k'_r)t})[A]_0}{k_r + k'_r} \tag{7.2b}$$

where $[A]_0$ is the initial concentration of A.

Justification 7.1   *The approach to equilibrium*

The concentration of A is reduced by the forward reaction (at a rate $k_r[A]$), but it is increased by the reverse reaction (at a rate $k'_r[B]$). Therefore, the net rate of change is

$$\frac{d[A]}{dt} = -k_r[A] + k'_r[B]$$

If the initial concentration of A is $[A]_0$ and no B is present initially, then at all times $[A] + [B] = [A]_0$. Therefore,

$$\frac{d[A]}{dt} = -k_r[A] + k'_r([A]_0 - [A]) = -(k_r + k'_r)[A] + k'_r[A]_0$$

The solution of this differential equation is eqn 7.2a (it is an example of the first type of differential equation in *Mathematical toolkit* 7.1, with $a = k_r + k'_r$ and $b = k'_r[A]_0$). To obtain eqn 7.2b, we use eqn 7.2a and $[B] = [A]_0 - [A]$.

As we see in Fig. 7.2, the concentrations start from their initial values and approach their final equilibrium values as t approaches infinity. We find the latter by letting t approach infinity in eqn 7.2 and using $e^{-x} = 0$ at $x = \infty$:

$$[B]_{eq} = \frac{k_r[A]_0}{k_r + k'_r} \qquad [A]_{eq} = \frac{k'_r[A]_0}{k_r + k'_r} \tag{7.3}$$

As may be verified, the ratio of these two expressions is the equilibrium constant in eqn 7.1.

(**In the laboratory 7.1**)   *Relaxation techniques in biochemistry*

Because many biochemical reactions are fast, we need to know how to measure their rates. Rapid mixing and stopped-flow techniques (*In the laboratory* 6.1) are ideal for studying events on a millisecond timescale, such as some

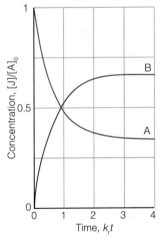

Fig. 7.2 The approach to equilibrium of a reaction that is first order in both directions. Here we have taken $k_r = 2k'_r$. Note how, at equilibrium, the ratio of concentrations is 2:1, corresponding to $K = 2$.

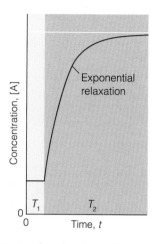

Fig. 7.3 The relaxation to the new equilibrium composition when a reaction initially at equilibrium at a temperature T_1 is subjected to a sudden change of temperature, which takes it to T_2.

enzyme-catalyzed reactions and the formation of contacts between helical segments in a large protein. However, many biochemical processes occur in less than 1 ms, a timescale not accessible by the stopped-flow technique. For example, the formation of a loop between helical or sheet segments in a protein may take as little as 1 μs, and the formation of tightly packed cores with significant tertiary structure occurs in 10–100 μs. Here we explore additional experimental techniques that extend the range of investigations in biochemical kinetics.

We noted in *In the laboratory* 6.1 that the term 'relaxation' denotes the return of a system to equilibrium. It is used in chemical kinetics to indicate that an externally applied influence has shifted the equilibrium position of a reaction, usually suddenly, and that the reaction is adjusting to the equilibrium composition characteristic of the new conditions (Fig. 7.3). We shall consider the response of reaction rates to a **temperature jump**, a sudden change in temperature. We know from Section 4.6 that the equilibrium composition of a reaction depends on the temperature (provided $\Delta_r H^{\ominus}$ is nonzero), so a change of temperature acts as a perturbation. Temperature jumps of between 5 K and 10 K can be achieved in about 1 μs with electrical discharges. The high energy output of pulsed lasers is sufficient to generate temperature jumps of between 10 K and 30 K within nanoseconds in aqueous samples. The laser-induced temperature-jump technique is very useful in studies of protein unfolding because a protein unfolds, or 'melts', at a characteristic temperature (*In the laboratory* 1.1 and Section 3.5). Proteins also lose their native structures at very low temperatures, a process known as **cold denaturation**, and re-fold when the temperature is increased but kept significantly below the melting temperature. Hence, a temperature-jump experiment can be configured to monitor either folding or unfolding of a polypeptide, depending on the initial and final temperatures of the sample.

When a sudden temperature increase is applied to a simple $A \rightleftharpoons B$ equilibrium that is first order in each direction, we show in the following *Justification* that the composition relaxes exponentially to the new equilibrium composition:

$$x = x_0 e^{-t/\tau} \qquad \frac{1}{\tau} = k_r + k_r' \qquad \boxed{\text{Relaxation time}} \quad (7.4)$$

where x is the departure from equilibrium at the new temperature, x_0 is the departure from equilibrium immediately after the temperature jump, and τ is the **relaxation time**, the time constant characteristic of the return to the new equilibrium composition.

Justification 7.2   *Relaxation to equilibrium*

When the temperature of a system at equilibrium is increased suddenly, the rate constants change from their earlier values to the new values k_r and k_r' characteristic of that temperature, but the concentrations of A and B remain for an instant at their old equilibrium values. As the system is no longer at equilibrium, it readjusts to the new equilibrium concentrations at a rate that depends on the new rate constants. We write the deviation of [A] from its new equilibrium value as x, so $[A] = [A]_{eq} + x$ and $[B] = [B]_{eq} - x$. The net rate of change of the concentration of A is

$$\frac{d[A]}{dt} = -k_r[A] + k_r'[B]$$

$$= -k_r([A]_{eq} + x) + k_r'([B]_{eq} - x)$$

$$= -(k_r + k_r')x$$

We have used the equilibrium relation $k_r[A]_{eq} = k_r'[B]_{eq}$ to cancel the two terms involving the equilibrium concentrations. From $[A] = [A]_{eq} + x$ it follows that $d[A]/dt = dx/dt$ and therefore

$$\frac{dx}{dt} = -(k_r + k_r')x$$

To solve this equation, we divide both sides by x and multiply by dt:

$$\frac{dx}{x} = -(k_r + k_r')dt$$

Now integrate both sides. When $t = 0$, $x = x_0$, its initial value, so the integrated equation has the form

$$\int_{x_0}^{x} \frac{dx}{x} = -(k_r + k_r')\int_0^t dt$$

The integral on the left is $\ln(x/x_0)$ (see *Mathematical toolkit* 3.1) and that on the right is t. The integrated equation is therefore

$$\ln \frac{x}{x_0} = -(k_r + k_r')t$$

When antilogarithms are taken of both sides, the result is eqn 7.4.

Equation 7.4 shows that the concentrations of A and B relax into the new equilibrium at a rate determined by the *sum* of the two new rate constants. Because the equilibrium constant under the new conditions is $K = k_r/k_r'$, its value may be combined with the relaxation time measurement to find the individual k_r and k_r':

$$k_r = \frac{K}{(1+K)\tau} \qquad k_r' = \frac{1}{(1+K)\tau} \tag{7.5}$$

The mathematical strategies described in *Justification* 7.2 can be used to write expressions for the relaxation time as a function of rate constants for more complex processes. In Exercise 7.13 you are invited to show that for the equilibrium $2A \rightleftharpoons A_2$, with second-order forward rate constant k_r and first-order reverse rate constant k_r', the relaxation time is

$$\tau = \frac{1}{k_r' + 4k_r[A]_{eq}} \tag{7.6}$$

7.2 Elementary reactions

To relate kinetic data on biochemical processes to a proposed reaction mechanism, we need to know how to write the rate law for each of the reaction steps.

Many reactions occur in a series of steps called **elementary reactions**, each of which involves only one or two molecules. We shall denote an elementary reaction by writing its chemical equation without displaying the physical state of the

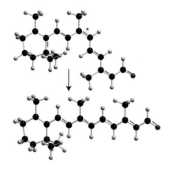

Fig. 7.4 In a unimolecular elementary reaction, an energetically excited species decomposes into products or undergoes a conformational change. Shown is an example of the latter process: the isomerization of energetically excited retinal (denoted with an asterisk). In the protein rhodopsin, bound retinal undergoes a similar isomerization when excited by light, initiating the cascade involved in vision.

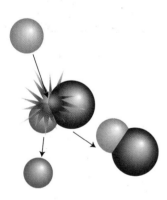

Fig. 7.5 In a bimolecular elementary reaction, two species are involved in the process.

species or using stoichiometric coefficients, as in a reaction that takes place in the upper atmosphere:

$$O + O_3 \rightarrow O_2 + O_2$$

This equation signifies that a specific O atom attacks a specific O_3 molecule to produce two O_2 molecules. Ordinary chemical equations summarize the overall stoichiometry of the reaction and do not imply any specific mechanism.

The **molecularity** of an elementary reaction is the number of molecules coming together to react. In a **unimolecular reaction** a single molecule shakes itself apart or its atoms into a new arrangement. An example is the isomerization of energetically excited retinal, a process that initiates the biochemical cascade involved in vision (Fig. 7.4). The radioactive decay of nuclei (for example, the emission of a β-particle from the nucleus of a tritium atom, which is used in mechanistic studies of biochemical reactions to follow the course of particular groups of atoms) is 'unimolecular' in the sense that a single nucleus shakes itself apart. In a **bimolecular reaction**, two molecules collide and exchange energy, atoms, or groups of atoms, or undergo some other kind of change (Fig. 7.5). The reaction between O and O_3, for instance, is bimolecular.

It is important to distinguish molecularity from order:

The order of a reaction is an empirical quantity and is obtained by inspection of the experimentally determined rate law.

The molecularity of a reaction refers to an individual elementary reaction that has been postulated as a step in a proposed mechanism.

Many substitution reactions in organic chemistry (for instance, S_N2 nucleophilic substitutions) are bimolecular and involve an activated complex that is formed from two reactant species. Many enzyme-catalyzed reactions can be regarded, to a good approximation, as bimolecular in the sense that they depend on the encounter of a substrate molecule and an enzyme molecule.

We can write down the rate law of an elementary reaction from its chemical equation. First, consider a unimolecular reaction. In a given interval, 10 times as many A molecules decay when there are initially 1000 A molecules as when there are only 100 A molecules present. Therefore the rate of decomposition of A is proportional to its concentration and we can conclude that *a unimolecular reaction is first order*:

$$A \rightarrow products \qquad \text{Rate of formation of products} = k_r[A] \qquad (7.7)$$

The rate of a bimolecular reaction is proportional to the rate at which the reactants meet, which in turn is proportional to both their concentrations. Therefore, the rate of the reaction is proportional to the product of the two concentrations and *an elementary bimolecular reaction is second order overall*. Two possibilities are:

$$A + B \rightarrow products \qquad \text{Rate of formation of products} = k_r[A][B] \qquad (7.8a)$$
$$A + A \rightarrow products \qquad \text{Rate of formation of products} = k_r[A]^2 \qquad (7.8b)$$

The relation between the rate constants and the equilibrium constant developed in Section 7.1 for *overall* reactions of the *same* order applies to *elementary* reactions of *any* order, because we can use the reaction stoichiometry to write down both the rate laws and the equilibrium constant. (Recall that in general although we can write down the equilibrium constant from the reaction stoichiometry, we cannot write down the rate law of an overall reaction simply by inspecting its stoichiometry.) However, we need to be careful about the units of concentration. For example, suppose a reaction takes place by the following elementary steps:

$A + A \rightarrow B$ Rate of formation of B $= k_r[A]^2$

$B \rightarrow A + A$ Rate of formation of A $= k_r'[B]$

Then the net rate of formation of B is

Net rate of formation of B $= k_r[A]^2 - k_r'[B]$

It follows that at equilibrium $k_r[A]_{eq}^2 = k_r'[B]_{eq}$. The (dimensionless) equilibrium constant for this pair of elementary steps is

$$K = \frac{[B]_{eq}/c^\ominus}{([A]_{eq}/c^\ominus)^2} = \frac{[B]_{eq}c^\ominus}{[A]_{eq}^2} = \frac{k_r c^\ominus}{k_r'} \tag{7.9}$$

(where $c^\ominus = 1$ mol dm^{-3}; see Section 3.8). We see that the criterion for $K \gg 1$ is now $k_r c^\ominus \gg k_r'$, and the units of each term match.

A brief illustration

The rates of the forward and reverse reactions for the dimerization of proflavin (**1**), an antibacterial agent that inhibits the biosynthesis of DNA by intercalating between adjacent base pairs, were found to be 8.1×10^8 dm^3 mol^{-1} s^{-1} (second order) and 2.0×10^6 s^{-1} (first order), respectively. The equilibrium constant for the dimerization is therefore

$$K = \frac{(8.1 \times 10^8 \, \text{dm}^3 \, \text{mol}^{-1} \, \text{s}^{-1}) \times (1 \, \text{mol dm}^{-3})}{2.0 \times 10^6 \, \text{s}^{-1}} = 4.0 \times 10^2$$

1 Proflavin

7.3 Consecutive reactions

In general, biological processes have complex mechanisms and we need to know how to build an overall rate law from the rate law of each step of a proposed mechanism.

A reactant commonly produces an **intermediate**, a species that does not appear in the overall chemical equation for the reaction but which has been invoked in the mechanism. Biochemical processes are often elaborate versions of this simple model. For instance, the restriction enzyme EcoRI catalyzes the cleavage of DNA at a specific sequence of nucleotides (at GAATTC, making the cut between G and A on both strands). The reaction sequence it brings about is

supercoiled DNA $\rightarrow$ open-circle DNA $\rightarrow$ linear DNA

We can discover the characteristics of this type of reaction by setting up the rate laws for the net rate of change of the concentration of each substance.

(a) The variation of concentration with time

To illustrate the kinds of considerations involved in dealing with a mechanism, let's suppose that a reaction takes place in two first-order steps, in one of which the intermediate I (the open-circle DNA, for instance) is formed from the reactant A (the supercoiled DNA) in a first-order reaction, and then I decays in a first-order reaction to form the product P (the linear DNA):

$$A \rightarrow I \qquad \text{Rate of formation of } I = k_a[A]$$
$$I \rightarrow P \qquad \text{Rate of formation of } P = k_b[I]$$

For simplicity, we are ignoring the reverse reactions, which is permissible if they are slow. The first of these rate laws implies that A decays with a first-order rate law and therefore that

$$[A] = [A]_0 e^{-k_a t} \tag{7.10}$$

The net rate of formation of I is the difference between its rate of formation and its rate of consumption, so we can write

$$\text{Net rate of formation of } I = \frac{d[I]}{dt} = k_a[A] - k_b[I] \tag{7.11}$$

with [A] given by eqn 7.10. This equation is more difficult to solve, but has the form of the second example in *Mathematical toolkit 7.1* and the solution is

$$[I] = \frac{k_a}{k_b - k_a} (e^{-k_a t} - e^{-k_b t})[A]_0 \tag{7.12}$$

Self-test 7.1 Show that eqn 7.12 follows from eqn 7.11 by (i) recasting eqn 7.11 so that it resembles the standard form in the second example of *Mathematical toolkit 7.1* and (ii) showing that the solution to the standard form can be rearranged into eqn 7.12 with the following substitutions: $x = t$, $f(x) = k_a[A]$ (with [A] given by eqn 7.10), $F(x) = k_b t$, $a = k_b$, $b = k_a[A]_0$, $b' = k_a$, and $c = k_a[A]_0/(k_a - k_b)$.

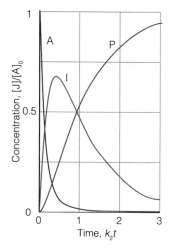

Fig. 7.6 The concentrations of the substances involved in a consecutive reaction of the form A → I → P, where I is an intermediate and P a product. We have used $k_1 = 5k_2$. Note how at each time the sum of the three concentrations is a constant.

Finally, because $[A] + [I] + [P] = [A]_0$ at all stages of the reaction, the concentration of P is

$$[P] = \left(1 + \frac{k_a e^{-k_b t} - k_b e^{-k_a t}}{k_b - k_a}\right)[A]_0 \tag{7.13}$$

These solutions are illustrated in Fig. 7.6. We see that the intermediate grows in concentration initially, then decays as A is exhausted. Meanwhile, the concentration of P rises smoothly to its final value. As we see in the following *Justification*, the intermediate reaches its maximum concentration at

$$t = \frac{1}{k_a - k_b} \ln \frac{k_a}{k_b} \qquad \boxed{\begin{array}{l}\text{Time of maximum} \\ \text{concentration}\end{array}} \tag{7.14}$$

A brief illustration

Consider a manufacturing process of a pharmaceutical in which $k_a = 0.120 \text{ h}^{-1}$ and $k_b = 0.012 \text{ h}^{-1}$. It follows that the intermediate is at a maximum at $t = 21 \text{ h}$ after the start of the process. This is the optimum time for a manufacturer trying to make the intermediate in a batch process to extract it.

Justification 7.3 *The time of maximum concentration*

To find the time corresponding to the maximum concentration of intermediate, we differentiate eqn 7.12 and look for the time at which $d[I]/dt = 0$. First we obtain

$$\frac{d[I]}{dt} = \frac{k_a}{k_b - k_a}(-k_a e^{-k_a t} + k_b e^{-k_b t})[A]_0 = 0$$

This equation is satisfied if

$$k_a e^{-k_a t} = k_b e^{-k_b t}$$

Because $e^{at}e^{bt} = e^{(a+b)t}$, this relation can be written as

$$\frac{k_a}{k_b} = e^{(k_a - k_b)t}$$

Taking logarithms of both sides leads to eqn 7.14.

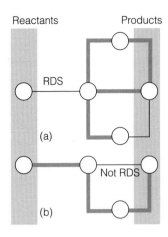

Fig. 7.7 The rate-determining step is the slowest step of a reaction *and* acts as a bottleneck. In this schematic diagram, fast reactions are represented by heavy lines (freeways) and slow reactions by thin lines (country roads). Circles represent substances. (a) The first step is rate determining. (b) Although the second step is the slowest, it is not rate determining because it does not act as a bottleneck (there is a faster route that circumvents it).

(b) The rate-determining step

We now suppose that the second step in the reaction we are considering is very fast, so that whenever an I molecule is formed, it decays rapidly into P. We can use the condition $k_b \gg k_a$ to write $e^{-k_b t} \ll e^{-k_a t}$ and $k_b - k_a \approx k_b$. Equation 7.13 then becomes

$$[P] \approx (1 - e^{-k_a t})[A]_0 \tag{7.15}$$

This equation shows that the formation of the final product P depends on only the *smaller* of the two rate constants, k_a. That is, the rate of formation of P depends on the rate at which I is formed, not on the rate at which I changes into P. For this reason, the step A → I is called the 'rate-determining step' of the reactions. Similar remarks apply to more complicated reactions mechanisms, and in general the **rate-determining step** is the slowest step in a mechanism on a pathway that controls the overall rate of the reaction. The rate-determining step is not just the slowest step: it must be slow *and* be a crucial gateway for the formation of products. If a faster reaction can also lead to products, then the slowest step is irrelevant because the slow reaction can then be sidestepped (Fig. 7.7). The rate-determining step is like a slow ferry crossing between two fast highways: the overall rate at which traffic can reach its destination is determined by the rate at which it can make the ferry crossing. If a bridge is built that avoids the ferry, the ferry remains the slowest step, but it is no longer rate-determining.

The rate law of a reaction that has a rate-determining step can often be written down almost by inspection. If the first step in a mechanism is rate determining, then the rate of the overall reaction is equal to the rate of the first step because all subsequent steps are so fast that once the first intermediate is formed, it results immediately in the formation of products. Figure 7.8 shows the reaction profile for a mechanism of this kind in which the slowest step is the one with the highest activation energy. Once over the initial barrier, the intermediates cascade into products.

However, we need to be alert to the possibility that a rate-determining step may also stem from the low concentration of a crucial reactant or catalyst and need not correspond to the step with the highest activation barrier. A rate-determining

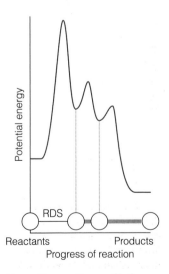

Fig. 7.8 The reaction profile for a mechanism in which the first step is rate determining.

step arising from the low activity of a crucial enzyme can sometimes be identified by determining whether or not the reactants and products for that step are in equilibrium: if the reaction is not at equilibrium, it suggests that the step may be slow enough to be rate determining.

Example 7.1   *Identifying a rate-determining step*

The following reaction is one of the early steps of glycolysis (*Case study* 4.3):

$$\text{F6P} + \text{ATP} \underset{}{\overset{\text{phosphofructokinase}}{\rightleftarrows}} \text{F16bP} + \text{ADP}$$

where F6P is fructose-6-phosphate and F16bP is fructose-1,6-bis(phosphate). The equilibrium constant for this step is 1.2×10^3. An analysis of the composition of heart tissue gave the following results:

	F16bP	F6P	ADP	ATP
Concentration/(mmol dm^{-3})	0.019	0.089	1.30	11.4

Might the phosphorylation of F6P be rate determining under these conditions?

Strategy Compare the value of the reaction quotient, Q (Section 4.2), with the equilibrium constant. If $Q \ll K$, the reaction step is far from equilibrium and it is so slow that it may be rate determining.

Solution From the data, the reaction quotient is

$$Q = \frac{[\text{F16bP}][\text{ADP}]}{[\text{F6P}][\text{ATP}]} = \frac{(1.9 \times 10^{-5}) \times (1.30 \times 10^{-3})}{(8.9 \times 10^{-5}) \times (1.14 \times 10^{-2})} = 0.024$$

Because $Q \ll K$, we conclude that the reaction step is not at equilibrium and so may be rate determining.

Self-test 7.2 Consider the reaction of *Example* 7.1. When the ratio [ADP]/[ATP] is equal to 0.10, what value should the ratio [F16bP]/[F6P] have for phosphorylation of F6P not to be a likely rate-determining step in glycolysis?

Answer: 1.2×10^4

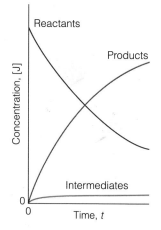

Fig. 7.9 The basis of the steady-state approximation. It is supposed that the concentrations of intermediates remain small and hardly change during most of the course of the reaction.

(c) The steady-state approximation

One feature of the calculation so far has probably not gone unnoticed: there is a considerable increase in mathematical complexity as soon as the reaction mechanism has more than a couple of steps. A reaction mechanism involving many steps is nearly always unsolvable analytically and alternative methods of solution are necessary. One approach is to integrate the rate laws numerically with a computer. An alternative approach, which continues to be widely used because it leads to convenient expressions and more readily digestible results, is to make an approximation.

The **steady-state approximation** assumes that after an initial **induction period**, an interval during which the concentrations of intermediates, I, rise from zero, and during the major part of the reaction, the rates of change of concentrations of all reaction intermediates are negligibly small (Fig. 7.9):

$$\frac{d[I]}{dt} \approx 0 \qquad \boxed{\text{Condition for the steady-state approximation}} \qquad (7.16)$$

This approximation greatly simplifies the discussion of reaction mechanisms. For example, when we apply the approximation to the consecutive first-order mechanism, we set $d[I]/dt = 0$ in eqn 7.11, which then becomes

$$k_a[A] - k_b[I] \approx 0$$

Then

$$[I] \approx (k_a/k_b)[A] \tag{7.17}$$

The product P is formed by unimolecular decay of I, so it follows that

$$\text{Rate of formation of P} = \frac{d[P]}{dt} = k_b[I] \approx k_a[A] \tag{7.18}$$

We see that P is formed by a first-order decay of A, with a rate constant k_a, the rate constant of the slower, rate-determining, step. We can write down the solution of this equation at once by substituting the solution for [A], eqn 7.10, and integrating:

$$[P] = k_a[A]_0 \int_0^t e^{-k_a t} dt = k_a[A]_0 \left(-\frac{1}{k_a} e^{-k_a t} + \frac{1}{k_a} \right) = [A]_0(1 - e^{-k_a t}) \tag{7.19}$$

A brief comment
For the integration we have used the standard result

$$\int e^{-kx} dx = -\frac{1}{k} e^{-kx} + \text{constant}$$

This equation is the same (approximate) result as before, eqn 7.15, but obtained more quickly.

(d) Pre-equilibria

From a simple sequence of consecutive reactions we now turn to a slightly more complicated mechanism proposed to account for the assembly of a DNA molecule from two polynucleotide chains, A and B. The first step in the mechanism is the formation of an intermediate that may be thought of as an unstable double helix:

A + B → unstable double helix

We must also allow for the reverse process:

unstable double helix → A + B

Competing with the latter process is the decay of the intermediate into a stable double helix:

unstable double helix → stable double helix

This sequence of steps is an example of the general mechanism

A + B → I	Rate of formation of I = $k_a[A][B]$	(7.20a)
I → A + B	Rate of decomposition of I = $k_a'[I]$	(7.20b)
I → P	Rate of formation of P = $k_b[I]$	(7.20c)

When the rates of formation of the intermediate and its decay back into reactants are much faster than its rate of formation of products, we are justified in assuming that A, B, and I are in equilibrium through the course of the reaction. This condition, called a **pre-equilibrium**, is possible when $k_a[A][B] \gg k_b[I]$ but not when $k_b[I] \gg k_a'[I]$. For the equilibrium between the intermediate and the reactants, we write (see Section 7.2)

$$K = \frac{[I]c^{\ominus}}{[A][B]} \qquad K = \frac{k_a c^{\ominus}}{k_a'} \tag{7.21}$$

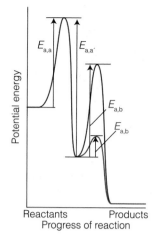

Fig. 7.10 For a reaction with a pre-equilibrium, there are three activation energies to take into account, two referring to the reversible steps of the pre-equilibrium and one for the final step. The relative magnitudes of the activation energies determine whether the overall activation energy is (a) positive or (b) negative.

In writing these equations, we are presuming that the rate of reaction of I to form P is too slow to affect the maintenance of the pre-equilibrium (see the example below). The rate of formation of P may now be written

$$\frac{d[P]}{dt} = k_b[I] = (k_b K/c^\ominus)[A][B] \tag{7.22}$$

This rate law has the form of a second-order rate law with a composite rate constant:

$$\frac{d[P]}{dt} = k_r[A][B] \qquad k_r = \frac{k_b K}{c^\ominus} = \frac{k_a k_b}{k_a'} \tag{7.23}$$

One feature to note is that although each of the rate constants in eqn 7.23 increases with temperature, that might not be true of k_r itself. Thus, if the rate constant k_a' increases more rapidly than the product $k_a k_b$ increases, then k_r will decrease with increasing temperature and the reaction will go more slowly as the temperature is raised. Mathematically, we would say that the composite reaction had a 'negative activation energy'. For example, suppose that each rate constant in eqn 7.23 exhibits an Arrhenius temperature dependence. It follows from the Arrhenius equation (eqn 6.19, $k_r = A e^{-E_a/RT}$) that

$$k_r = \frac{(A_a e^{-E_{a,a}/RT})(A_b e^{-E_{a,b}/RT})}{A_a' e^{-E_{a,a}'/RT}} = \frac{A_a A_b}{A_a'} \frac{e^{-E_{a,a}/RT} e^{-E_{a,b}/RT}}{e^{-E_{a,a}'/RT}}$$

$$= \frac{A_a A_b}{A_a'} e^{-(E_{a,a}+E_{a,b}-E_{a,a}')/RT}$$

where we have used the relations: $e^{x+y} = e^x e^y$ and $e^{x-y} = e^x/e^y$. The effective activation energy of the reaction is therefore

$$E_a = E_{a,a} + E_{a,b} - E_{a,a}' \tag{7.24}$$

This activation energy is positive if $E_{a,a} + E_{a,b} > E_{a,a}'$ (Fig. 7.10a) but negative if $E_{a,a}' > E_{a,a} + E_{a,b}$ (Fig. 7.10b). An important consequence of this discussion is that we have to be very cautious about making predictions about the effect of temperature on reactions that are the outcome of several steps. Enzyme-catalyzed reactions may also exhibit strongly non-Arrhenius temperature dependence if the enzyme denatures at high temperatures and ceases to function.

Case study 7.1 | *Mechanisms of protein folding and unfolding*

Much of the kinetic work on the mechanism of unfolding of a helix into a random coil has been conducted on small synthetic polypeptides rich in alanine, an amino acid known to stabilize helical structures. Experimental and theoretical results suggest that the mechanism of unfolding consists of at least two steps: a very fast step, in which amino acids at either end of a helical segment undergo transitions to coil regions, and a slower rate-determining step, which corresponds to the cooperative melting of the rest of the chain and loss of helical content. Using *h* and *c* to denote an amino acid residue belonging to a helical and coil region, respectively, the mechanism may be summarized as follows:

$hhhh \ldots \rightleftharpoons chhh$ very fast

$chhh \ldots \rightarrow cccc$ rate-determining step

The rate-determining step is thought to account for the relaxation time of 160 ns measured with a laser-induced temperature jump from 280 K to 300 K in an alanine-rich polypeptide containing 21 residues. It is thought that the limitation on the rate of the helix–coil transition in this peptide arises from an activation energy barrier of 1.7 kJ mol^{-1} associated with initial events of the form ...*hhhh*...→...*hhch*... in the middle of the chain. Therefore, initiation is not only thermodynamically unfavorable but also kinetically slow. Theoretical models also suggest that a *hhhh*...→ *chhh*... transition at either end of a helical segment has a significantly lower activation energy on account of the converting residue not being flanked by *h* regions.

The kinetics of unfolding have also been measured in naturally occurring proteins. In the engrailed homeodomain (En-HD) protein, which contains three short helical segments (Fig. 7.11), unfolding occurs with a half-life of about 630 μs at 298 K. It is difficult to interpret these results because we do not yet know how the amino acid sequence or interactions between helices in a folded protein affect the helix–coil relaxation time.

As remarked in the *Prologue*, a protein does not fold into its active conformation by sampling every possible three-dimensional arrangement of the chain, as the process would take far too long—up to 10^{21} years for a protein with 100 amino acids. Moreover, folding times have been measured in synthetic peptides and naturally occurring proteins and have been found to be very fast. For example, the En-HD protein folds with a half-life of 18 μs at 298 K. In fact, Nature's search for the active conformation of a large polypeptide appears to be highly streamlined, and the identification of specific mechanisms of protein folding is a major focus of current research in biochemistry. Although it is unlikely that a single model can describe the folding of every protein, progress has been made in the identification of some general mechanistic features.

Two models have received attention. In the **framework model**, regions with well-defined and stable secondary structure form independently and then coalesce to yield the correct tertiary structure. The En-HD protein and other proteins that are predominantly helical fold according to the framework model. In the **nucleation–condensation model**, rather loose and unstable helices and sheets are thought to form early in the folding process. However, the molecule can be stabilized by interactions that also give rise to some degree of tertiary structure. That is, formation of secondary structure is fostered by the formation of tertiary structure and vice versa. It is easy to imagine that some regions, called 'nuclei', of the loosely packed protein resemble the active conformation of the protein rather closely, whereas other regions do not. Far away from the nuclei, similarities to the active conformation are thought to be less prominent, but these regions eventually coalesce, or 'condense,' around nuclei to give the properly folded protein. Proteins containing mostly α-helices, mostly β-sheets, or a mixture of the two have been observed to fold in a manner consistent with the nucleation condensation model.

A key feature of the framework and nucleation–condensation models is the formation of secondary structure—which might or might not be coupled to the formation of tertiary structure—early in the folding process. It follows that a full description of the mechanism of protein folding also requires an understanding of the rules that stabilize molecular interactions in polypeptides. We consider these rules in Chapter 11.

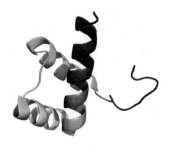

Fig. 7.11 The engrailed homeodomain (En-HD) protein contains three short helical segments.

7.4 Diffusion control

Most biochemical processes require that two or more molecules encounter each other as they travel through the aqueous environment of the cell, so one contribution to the overall rate of enzyme-catalyzed reactions that we need to analyze is the rate at which species diffuse through a solution.

The concept of the rate-determining step plays an important role for reactions in solution, where it leads to the distinction between 'diffusion control' and 'activation control'. To develop this point, let's suppose that a reaction between two solute molecules A and B occurs by the following mechanism. First, we assume that A and B drift into each other's vicinity by diffusion,[1] the process by which the molecules of different substances mingle with each other, and form an **encounter pair**, AB:

$$A + B \rightarrow AB \qquad \text{Rate of formation of } AB = k_d[A][B]$$

The subscript d reminds us that this process is diffusional. The encounter pair persists for some time as a result of the **cage effect**, the trapping of A and B near each other by their inability to escape rapidly through the surrounding solvent molecules. However, the encounter pair can break up when A and B have the opportunity to diffuse apart, and so we must allow for the following process:

$$AB \rightarrow A + B \qquad \text{Rate of loss of } AB = k_d'[AB]$$

We have supposed that this process is first order in AB. Competing with this process is the reaction between A and B while they exist as an encounter pair. This process depends on their ability to acquire sufficient energy to react. That energy might come from the jostling of the thermal motion of the solvent molecules. We shall assume that the reaction of the encounter pair is first order in AB, but if the solvent molecules are involved, it is more accurate to regard it as pseudo first order with the solvent molecules in great and constant excess. In any event, we can suppose that the reaction is

$$AB \rightarrow \text{products} \qquad \text{Rate of reactive loss of } AB = k_a[AB]$$

The subscript a on k reminds us that this process is activated in the sense that it depends on the acquisition by AB of at least a minimum energy.

Now we use the steady-state approximation to set up the rate law for the formation of products and deduce in the following *Justification* that the rate $v = d[P]/dt$ and rate constant k_r of formation of products are given by

$$v = k_r[A][B] \qquad k_r = \frac{k_a k_d}{k_a + k_d'} \qquad \boxed{\text{Rate constant in the presence of diffusion and activation}} \qquad (7.25)$$

Justification 7.4 *Rate in the presence of diffusion and activation*

The net rate of formation of AB is

$$\frac{d[AB]}{dt} = k_d[A][B] - k_d'[AB] - k_a[AB]$$

In a steady state, this rate is zero, so we can write

$$k_d[A][B] - k_d'[AB] - k_a[AB] = 0$$

[1] Diffusion is treated in more detail in Chapter 8.

which we can rearrange to find [AB]:

$$[AB] = \frac{k_d[A][B]}{k_a + k_d'}$$

The rate of formation of products (which is the same as the rate of reactive loss of AB) is therefore

$$v = k_a[AB] = \frac{k_a k_d[A][B]}{k_a + k_d'}$$

which is eqn 7.25.

Now we distinguish two limits. Suppose the rate of reaction is much faster than the rate at which the encounter pair breaks up. In this case, $k_a \gg k_d'$ and we can neglect k_d' in the denominator of the expression for k_r in eqn 7.25. The k_a in the numerator and denominator then cancel, and we are left with

$$v = k_d[A][B] \qquad \boxed{\text{Diffusion-controlled limit}} \qquad (7.26a)$$

In this **diffusion-controlled limit** the rate of the reaction is controlled by the rate at which the reactants diffuse together (as expressed by k_d), for the reaction once they have encountered is so fast that they will certainly go on to form products rather than diffuse apart before reacting. Alternatively, we may suppose that the rate at which the encounter pair accumulated enough energy to react is so low that it is highly likely that the pair will break up. In this case, we can set $k_a \ll k_d'$ in the expression for k_r and obtain

$$v = \frac{k_a k_d}{k_d'}[A][B] \qquad \boxed{\text{Activation-controlled limit}} \qquad (7.26b)$$

In this **activation-controlled limit** the reaction rate depends on the rate at which energy accumulates in the encounter pair (as expressed by k_a).

A lesson to learn from this analysis is that the concept of the rate-determining step is rather subtle. Thus, in the diffusion-controlled limit, the condition for the encounter rate to be rate determining is not that it is the slowest step, but that the reaction rate of the encounter pair is much greater than the rate at which the pair breaks up. In the activation-controlled limit, the condition for the rate of energy accumulation to be rate determining is likewise a competition between the rate of reaction of the pair and the rate at which it breaks up, and all three rate constants contribute to the overall rate. The best way to analyze competing rates is to do as we have done here: to set up the overall rate law and then to analyze how it simplifies as we allow particular elementary processes to dominate others.

We can go one stage further and end this part of the discussion on a more encouraging note. A detailed analysis of the rates of diffusion of molecules in liquids shows that the rate constant k_d is related to the viscosity, η, of the medium by[2]

$$k_d = \frac{8RT}{3\eta} \qquad \boxed{\begin{array}{l}\text{The diffusional rate constant in} \\ \text{terms of the viscosity of the medium}\end{array}} \qquad (7.27)$$

We see that the higher the viscosity, the smaller the diffusional rate constant, and therefore the slower the reaction of a diffusion-controlled reaction.

---

[2] See our *Physical chemistry* (2010).

> **A brief illustration**
>
> We shall see in Section 8.1 that the following simple mechanism can explain a variety of enzyme-catalyzed reactions:
>
> | | |
> |---|---|
> | $E + S \rightarrow ES$ | Rate of formation of $ES = k_d[E][S]$ |
> | $ES \rightarrow E + S$ | Rate of dissociation of $ES = k_d'[ES]$ |
> | $ES \rightarrow E + P$ | Rate of formation of $P = k_a[ES]$ |
>
> where E is the enzyme, S is the substrate (the substance processed by the enzyme), ES is an encounter pair between the enzyme and the substrate, and P is the product. When the reaction is controlled by diffusion of enzyme and substrate in solution, the rate is $v = k_d[E][S]$. In water at 25°C, $\eta = 8.9 \times 10^{-4}$ kg m^{-1} s^{-1} and it follows that $k_d = 7.4 \times 10^9$ dm^3 mol^{-1} s^{-1}. This value is a useful indication of the upper limit of the rate of an enzyme-catalyzed reaction. A number of enzymes operate near this diffusion limit.

7.5 Kinetic and thermodynamic control

Many biochemical processes never reach equilibrium, so we need to distinguish between thermodynamic and kinetic factors that control the relative concentrations of reaction products.

In some cases reactants can give rise to a variety of products. Suppose two products, P_1 and P_2, are produced by the following competing reactions:

$A + B \rightarrow P_1$	Rate of formation of $P_1 = k_1[A][B]$
$A + B \rightarrow P_2$	Rate of formation of $P_2 = k_2[A][B]$

The relative proportion in which the two products have been produced at a given stage of the reaction (before it has reached equilibrium) is given by the ratio of the two rates and therefore of the two rate constants:

$$\frac{[P_2]}{[P_1]} = \frac{k_2}{k_1} \qquad \boxed{\text{Kinetic control}} \quad (7.28)$$

This ratio represents the **kinetic control** over the proportions of products and is a common feature of biochemical reactions where an enzyme facilitates a specific pathway—one with a low activation energy—favoring the formation of a desired product. If a reaction is allowed to reach equilibrium, then the proportion of products is determined by thermodynamic rather than kinetic factors, and the ratio of concentrations is controlled by considerations of the standard Gibbs energies of all the reactants and products.

> **Self-test 7.3** Two products are formed in reactions in which there is kinetic control of the ratio of products. The activation energy for the reaction leading to product 1 is greater than that leading to product 2. Will the ratio of product concentrations $[P_1]/[P_2]$ increase or decrease if the temperature is raised?
>
> **Answer:** The ratio $[P_1]/[P_2]$ will increase

Reaction dynamics

We now identify and investigate the factors that control the value of the rate constant. In Section 6.6, we considered the Arrhenius equation

$$k_r = Ae^{-E_a/RT} \qquad \boxed{\text{Arrhenius equation}} \quad (7.29)$$

as a collection of empirical parameters, the activation energy, E_a, and the pre-exponential factor, A, that determine the temperature dependence of the rate constant. Here we describe two theories of **reaction dynamics**, the study of the details of molecular events that transform reactants into products, and thereby provide a richer interpretation of the Arrhenius parameters.

7.6 Collision theory

Reactions in the gas phase introduce a number of concepts relating to the rates of reaction without the complication of having to take into account the role of the solvent.

We can understand the origin of the Arrhenius parameters most simply by considering gas-phase bimolecular reactions, such as the $O + O_3$ reaction and others like it that occur in the upper atmosphere. In this **collision theory** of reaction rates it is supposed that reaction occurs only if two molecules collide with a certain minimum kinetic energy along their line of approach (Fig. 7.12). In collision theory, a reaction resembles the collision of two defective billiard balls: the balls bounce apart if they collide with only a small energy but might smash each other into fragments (products) if they collide with more than a certain minimum kinetic energy. This model of a reaction is a reasonable first approximation to the types of processes that take place in planetary atmospheres and govern their compositions and temperature profiles.

A **reaction profile** in collision theory is a graph showing the variation in potential energy as one reactant molecule approaches another and the products then separate (as in Fig. 7.1). On the left, the horizontal line represents the potential energy of the two reactant molecules that are far apart from each other. The potential energy rises from this value only when the separation of the molecules is so small that they are in contact, when it rises as bonds bend and start to break. The potential energy reaches a peak when the two molecules are highly distorted. Then it starts to decrease as new bonds are formed. At separations to the right of the maximum, the potential energy rapidly falls to a low value as the product molecules separate. For the reaction to be successful, the reactant molecules must approach with sufficient kinetic energy along their line of approach to carry them over the **activation barrier**, the peak in the reaction profile. As we shall see, we can identify the height of the activation barrier with the activation energy of the reaction.

With the reaction profile in mind, it is quite easy to establish that collision theory accounts for Arrhenius behavior. Thus, the **collision frequency**, the rate of collisions between species A and B, is proportional to both their concentrations: if the concentration of B is doubled, then the rate at which A molecules collide with B molecules is doubled, and if the concentration of A is doubled, then the rate at which B molecules collide with A molecules is also doubled. It follows that the collision frequency of A and B molecules is directly proportional to the concentrations of A and B, and we can write

collision frequency $\propto$ [A][B]

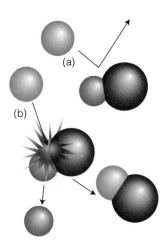

Fig. 7.12 In the collision theory of gas-phase chemical reactions, reaction occurs when two molecules collide, but only if the collision is sufficiently vigorous. (a) An insufficiently vigorous collision: the reactant molecules collide but bounce apart unchanged. (b) A sufficiently vigorous collision results in a reaction.

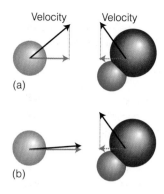

Velocity   Velocity

(a)

(b)

Fig. 7.13 The criterion for a successful collision is that the two reactant species should collide with a kinetic energy along their line of approach, which exceeds a certain minimum value E_a that is characteristic of the reaction. The two molecules might also have components of velocity (and an associated kinetic energy) in other directions (for example, the two molecules depicted here as (a) and (b) might be moving up the page as well as toward each other), but only the energy associated with their mutual approach can be used to overcome the activation energy.

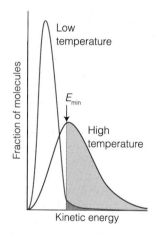

Fraction of molecules

Low temperature

E_{min}

High temperature

Kinetic energy

Fig. 7.14 According to the Maxwell distribution of speeds (*Further information 7.1*), as the temperature increases, so does the fraction of gas phase molecules with a speed that exceeds a minimum value. Because the kinetic energy is proportional to the square of the speed, it follows that more molecules can collide with a minimum kinetic energy $E_{min} = E_a$ (the activation energy) at higher temperatures.

Not all collisions occur with sufficient energy to result in reaction, therefore we need to multiply the collision frequency by a factor f that represents the fraction of collisions that occur with at least a kinetic energy E_a along the line of approach (Fig. 7.13), for only these collisions will lead to the formation of products. Molecules that approach with less than a kinetic energy E_a will behave like a ball that rolls toward the activation barrier, fails to surmount it, and rolls back. We saw in *Fundamentals* F.3 that only small fractions of molecules in the gas phase have very high speeds and that the fraction with very high speeds increases sharply as the temperature is raised. Because the kinetic energy increases as the square of the speed, we expect that, at higher temperatures, a larger fraction of molecules will have speed and kinetic energy that exceed the minimum values required for collisions that lead to formation of products (Fig. 7.14). The fraction of collisions that occur with at least a kinetic energy E_a can be calculated from the Boltzmann distribution and the result is[3]

$$f = e^{-E_a/RT} \tag{7.30}$$

This fraction increases with increasing temperature: at $T = 0$, $f = e^{-\infty} = 0$ and no collisions are successful; at $T = \infty$, $f = e^0 = 1$ and every collision has enough energy to result in reaction.

> **Self-test 7.4** What is the fraction of collisions that have sufficient energy for reaction if the activation energy is 50 kJ mol^{-1} and the temperature is (a) 25°C and (b) 500°C?
>
> **Answer:** (a) 1.7×10^{-9}; (b) 4.2×10^{-4}

At this stage we can conclude that the rate of reaction, which is proportional to the collision frequency multiplied by the fraction of successful collisions, is

$$v \propto [A][B]e^{-E_a/RT}$$

If we compare this expression with a second-order rate law,

$$v = k_r[A][B]$$

it follows that

$$k_r \propto e^{-E_a/RT}$$

This expression has exactly the Arrhenius form (eqn 6.19) if we identify the constant of proportionality with A. Collision theory therefore suggests the following interpretations:

- The *pre-exponential factor*, A, is the constant of proportionality between the concentrations of the reactants and the rate at which the reactant molecules collide.

- The *activation energy*, E_a, is the minimum kinetic energy along the line of approach required for a collision to result in reaction.

The value of A can be calculated from the kinetic model of gases (*Further information 7.1*), and the result is

$$A = \sigma\left(\frac{8kT}{\pi\mu}\right)^{1/2}N_A \qquad \mu = \frac{m_A m_B}{m_A + m_B} \qquad \boxed{\text{The pre-exponential factor according to collision theory}} \tag{7.31}$$

---

[3] For details of the calculation, see our *Physical chemistry* (2010).

where m_A and m_B are the masses of the molecules A and B and σ is the **collision cross section**, the target area presented by one molecule to another (Table 7.1). However, it is often found that the experimental value of A is smaller than that calculated from the kinetic model. One possible explanation is that not only must the molecules collide with sufficient kinetic energy, but they must also come together in a specific relative orientation (Fig. 7.15). It follows that the reaction rate is proportional to the probability that the encounter occurs in the correct relative orientation. The pre-exponential factor A should therefore include a **steric factor**, P, which usually lies between 0 (no relative orientations lead to reaction) and 1 (all relative orientations lead to reaction). Some reactions have $P > 1$ if specific molecular interactions during collisions (for example, Coulomb interactions between charged species) effectively extend the cross-section for the reactive encounter beyond the value expected from simple mechanical contact between reactants.

7.7 Transition state theory

The concepts we introduce here form the basis of a theory that explains the rates of biochemical reactions in fluid environments.

Although collision theory is a useful starting point for the discussion of reactions in the atmosphere, it has little relevance to the reactions that interest biologists the most: those taking place in the aqueous environment of a cell. A more sophisticated theory, 'transition state theory' (or *activated complex theory*), builds on collision theory but is applicable to a wider range of reaction environments and introduces a more sophisticated interpretation of the empirical Arrhenius parameters A and E_a.

(a) Formulation of the theory

In the **transition state theory** of reactions it is supposed that as two reactants approach, their potential energy rises and reaches a maximum, as illustrated by the reaction profile in Fig. 7.1. This maximum corresponds to the formation of an **activated complex**, a cluster of atoms that is poised to pass on to products or to collapse back into the reactants from which it was formed (Fig. 7.16). The concept of an activated complex is applicable to reactions in solutions as well as to the gas phase because we can think of the activated complex as perhaps involving any solvent molecules that may be present.

To describe the essential features of transition state theory, we follow the progress of a bimolecular reaction, possibly occurring in solution. Initially only the reactants A and B are present. As the reaction event proceeds, A and B come into contact, distort, and begin to exchange or discard atoms. The potential energy rises to a maximum, and the cluster of atoms that corresponds to the region close to the maximum is the activated complex. The potential energy falls as the atoms rearrange in the cluster and reaches a value characteristic of the products. The climax of the reaction is at the peak of the potential energy. Here two reactant molecules have come to such a degree of closeness and distortion that a small further distortion will send them in the direction of products. This crucial configuration is called the **transition state** of the reaction. Although some molecules entering the transition state might revert to reactants, if they pass through this configuration it is probable that products will emerge from the encounter.

The **reaction coordinate** is an indication of the stage reached in this process. On the left we have undistorted, widely separated reactants. On the right are the products. Somewhere in the middle is the stage of the reaction corresponding

Table 7.1 Collision cross-sections of atoms and molecules

Species	σ/nm^{2}*
Argon, Ar	0.36
Benzene, C_6H_6	0.88
Carbon dioxide, CO_2	0.52
Chlorine, Cl_2	0.93
Ethene, C_2H_4	0.64
Helium, He	0.21
Hydrogen, H_2	0.27
Methane, CH_4	0.46
Nitrogen, N_2	0.43
Oxygen, O_2	0.40
Sulfur dioxide, SO_2	0.58

*$1\ nm^2 = 10^{-18}\ m^2$.

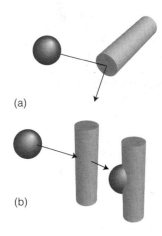

(a)

(b)

Fig. 7.15 Energy is not the only criterion of a successful reactive encounter, for relative orientation can also play a role. (a) In this collision, the reactants approach in an inappropriate relative orientation and no reaction occurs even though their energy is sufficient. (b) In this encounter, both the energy and the orientation are suitable for reaction.

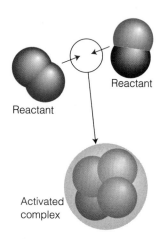

Reactant

Reactant

Activated complex

Fig. 7.16 In the transition state theory of chemical reactions, two reactants encounter each other (either in a gas-phase collision or as a result of diffusing together through a solvent) and, if they have sufficient energy, form an activated complex. The activated complex is depicted here by a relatively loose cluster of atoms that may undergo rearrangement into products. In an actual reaction, only some atoms—those at the actual reaction site—might be significantly loosened in the complex; the bonding of the others remaining almost unchanged. This would be the case for CH_3 groups attached to a carbon atom that was undergoing substitution.

to the formation of the activated complex. The principal goal of transition state theory is to write an expression for the rate constant by tracking the history of the activated complex from its formation by encounters between the reactants to its decay into product. Here we outline the steps involved in the calculation, with an eye toward gaining insight into the molecular events that affect the rate constant.

The activated complex $C^{\ddagger}$ is formed from the reactants A and B and it is supposed—without much justification—that there is an equilibrium between the concentrations of A, B, and $C^{\ddagger}$:

$$A + B \rightleftharpoons C^{\ddagger} \qquad K^{\ddagger} = \frac{[C^{\ddagger}]c^{\ominus}}{[A][B]}$$

At the transition state, motion along the reaction coordinate corresponds to some complicated collective vibration-like motion of all the atoms in the complex (and the motion of the solvent molecules if they are involved too). However, it is possible that not every motion along the reaction coordinate takes the complex through the transition state and to the product P. By taking into account the equilibrium between A, B, and $C^{\ddagger}$ and the rate of successful passage of $C^{\ddagger}$ through the transition state, it is possible to derive the **Eyring equation** for the rate constant k_r:[4]

$$k_r = \kappa \times \frac{kT}{h} \times \frac{K^{\ddagger}}{c^{\ominus}} \qquad \boxed{\text{Eyring equation}} \quad (7.32)$$

where $k = R/N_A = 1.381 \times 10^{-23}$ J K^{-1} is Boltzmann's constant and $h = 6.626 \times 10^{-34}$ J s is Planck's constant (*Fundamentals* F3). The factor κ (kappa) is the **transmission coefficient**, which takes into account the fact that the activated complex does not always pass through to the transition state. In the absence of information to the contrary, κ is assumed to be about 1. The term kT/h in eqn 7.32 has the dimensions of a frequency, as kT is an energy and division by Planck's constant turns an energy into a frequency (with kT in joules, kT/h has the units s^{-1}). It arises from a consideration of the motions of atoms that lead to the decay of $C^{\ddagger}$ into products, as specific bonds are broken and formed.

Calculation of the equilibrium constant $K^{\ddagger}$ is very difficult, except in certain simple model cases. For example, if we suppose that the reactants are two structureless atoms and that the activated complex is a weakly bound diatomic molecule of bond length R, then k_r turns out to be the same as for collision theory, provided we interpret the collision cross-section in eqn 7.31 as πR^2.

(b) Thermodynamic parameterization

It is more useful, especially for biological reactions in aqueous environments, to express the Eyring equation in terms of thermodynamic parameters and to discuss reactions in terms of their empirical values. Thus, we saw in Section 4.3 that an equilibrium constant may be expressed in terms of the standard reaction Gibbs energy ($-RT \ln K = \Delta_r G^{\ominus}$). We do the same here, and express $K^{\ddagger}$ in terms of the **activation Gibbs energy**, $\Delta^{\ddagger}G$:

$$\Delta^{\ddagger}G = -RT \ln K^{\ddagger} \qquad \text{and} \qquad K^{\ddagger} = e^{-\Delta^{\ddagger}G/RT}$$

Therefore, by writing

$$\Delta^{\ddagger}G = \Delta^{\ddagger}H - T\Delta^{\ddagger}S \qquad\qquad\qquad (7.33)$$

---

[4] In some expositions, you will see Boltzmann's constant denoted k_B to emphasize its significance.

where $\Delta^{\ddagger}H$ and $\Delta^{\ddagger}S$ are the **enthalpy of activation** and the **entropy of activation**, respectively, we conclude that (with $\kappa = 1$)

$$k_r = \frac{kT}{h}e^{-(\Delta^{\ddagger}H - T\Delta^{\ddagger}S)/RT} = \left(\frac{kT}{h}e^{\Delta^{\ddagger}S/R}\right)e^{-\Delta^{\ddagger}H/RT} \qquad \text{(7.34)}$$

> The Eyring equation in terms of thermodynamic parameters

This expression has the form of the Arrhenius expression, eqn 7.29, if we identify the $\Delta^{\ddagger}H$ with the activation energy and the term in parentheses with the pre-exponential factor.

The advantage of transition state theory over collision theory is that it is applicable to reactions in solution as well as in the gas phase. It also gives some clue to the calculation of the steric factor P, for the orientation requirements are carried in the entropy of activation. Thus, if there are strict orientation requirements (for example, in the approach of a substrate molecule to an enzyme), then the entropy of activation will be strongly negative (representing a decrease in disorder when the activated complex forms), and the pre-exponential factor will be small. In practice, it is occasionally possible to estimate the sign and magnitude of the entropy of activation and hence to estimate the rate constant. The general importance of transition state theory is that it shows that even a complex series of events—not only a collisional encounter in the gas phase—displays Arrhenius-like behavior and that the concept of activation energy is widely applicable.

Self-test 7.5 In a certain reaction in water, it is proposed that two ions of opposite charge come together to form an electrically neutral activated complex. Is the contribution of the solvent to the entropy of activation likely to be positive or negative?

> **Answer:** Positive, as H_2O is less organized around the neutral species

---

In the laboratory 7.2   *Time-resolved spectroscopy for kinetics*

The ability of lasers to produce pulses as brief as 1 fs (10^{-15} s) is particularly useful in chemistry when we want to monitor very fast processes in time. In **time-resolved spectroscopy**, a form of flash photolysis (*In the laboratory* 6.1), laser pulses are used to obtain the spectra of reactants, intermediates, products, and even activated complexes of reactions. Lasers that produce nanosecond pulses are generally suitable for the observation of reactions with rates controlled by the speed with which reactants can move through a fluid medium. However, pulses in the range 1 fs to 1 ps are needed to study energy transfer, molecular vibrations, and conversion from one mode of motion into another. The arrangement shown in Fig. 7.17 is often used to study ultrafast chemical reactions that can be initiated by light. An intense and short laser pulse, the *pump*, initiates the process, possibly by elevating a molecule A to a high-energy state, A*, which can either release the additional energy, perhaps as light, or react with another species B to yield a product C:

$$A \xrightarrow{\text{laser pulse}} A^* \qquad \text{(absorption of energy from light)}$$
$$A^* \rightarrow A \qquad \text{(energy release)}$$
$$A^* + B \rightarrow AB \rightarrow C \qquad \text{(reaction)}$$

where AB denotes either an intermediate or an activated complex.

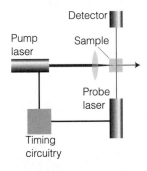

Fig. 7.17 A configuration used for time-resolved absorption spectroscopy, in which an intense laser pulse is used to generate a monochromatic pump pulse and a second laser pulse, the probe, arrives at the sample some time later to measure a spectroscopic feature of the reaction mixture.

The rates of appearance and disappearance of the various species are determined by observing time-dependent changes in the spectrum of the sample during the course of the reaction. This monitoring may be done by passing a second, weaker laser pulse, the *probe*, through the sample at different times after the laser pulse. For example, the wavelength of the probe pulse may be set at an absorption of an intermediate (to probe its formation and decay) or a product (to probe the rate of its formation).

Biological processes that are open to study by time-resolved spectroscopy include the energy-converting processes of photosynthesis (Section 5.11 and *Case study* 12.3) and the light-induced processes of vision (*Case study* 12.2). In other experiments, the laser-induced ejection of carbon monoxide from myoglobin and the attachment of O_2 to the exposed heme site have been studied to obtain rate constants for the two processes.

The technique may also be used to detect and study clusters of atoms that resemble activated complexes. In a typical experiment, energy from a femtosecond pump pulse is used to dissociate a molecule, and then a femtosecond probe pulse is fired at an interval after the pulse. The frequency of the probe pulse is set at an absorption of one of the free fragmentation products, so its absorption is a measure of the abundance of the dissociation product. For example, when ICN is dissociated by the first pulse, the emergence of CN can be monitored by watching the growth of the free CN absorption. In this way it has been found that the CN signal remains zero until the fragments have separated by about 600 pm, which takes about 205 fs. Time-resolved techniques have also been used to examine analogs of the activated complex involved in more complex reactions, such as the Diels–Alder reaction, nucleophilic substitution reactions, and pericyclic addition and cleavage reactions.

7.8 The kinetic salt effect

Many biochemical reactions in solution are between ions; to treat them, we need to combine transition state theory and the Debye–Hückel limiting law.

The thermodynamic version of transition state theory simplifies the discussion of reactions in solution, particularly those involving ions. For instance, the **kinetic salt effect** is the effect on the rate of a reaction of adding an inert salt to the reaction mixture. The physical origin of the effect is the difference in stabilization of the reactant ions and the activated complex by the ionic atmosphere (Section 5.1) formed around each of them by the added ions. Thus, in a reaction in which the activated complex forms in the pre-equilibrium

$$A^+ + B^- \rightleftharpoons C^{\ddagger}$$

both reactants are stabilized by their atmospheres, but the activated complex $C^{\ddagger}$ is not, less $C^{\ddagger}$ is present at the (presumed) equilibrium, so the rate of formation of products is decreased. On the other hand, if the reaction is between ions of like charge, as in

$$A^+ + B^+ \rightleftharpoons C^{\ddagger 2+}$$

the ionic atmosphere around the doubly charged activated complex has a greater effect than around each singly charged ion, the complex is stabilized more than either ion, so its abundance at equilibrium is increased and the rate of formation of products is increased too. We show in the following *Justification* that quantitative treatment of the problem leads to the result that

$$\log k_{r} = \log k_{r}^{o} + 2Az_{A}z_{B}I^{1/2} \qquad \boxed{\text{The kinetic salt effect}} \quad (7.35)$$

where k_{r}^{o} is the rate constant in the absence of added salt and $A = 0.509$ for water at 25°C. The charge numbers of A and B are z_{A} and z_{B}, so the charge number of the activated complex is $z_{A} + z_{B}$; the z_{J} are positive for cations and negative for anions. The quantity I is the ionic strength due to the added salt (Section 5.1), and for a 1:1 electrolyte (such as NaCl) is equal to the numerical value of the molality (that is, $I = b/b^{\oplus}$, with $b^{\oplus} = 1 \text{ mol kg}^{-1}$).

Justification 7.5   *The kinetic salt effect*

Consider a reaction with the mechanism $A + B \rightleftharpoons C^{\ddagger} \rightarrow$ products, where A and B are charged species, and the rate of formation of products is

$$v = k_{r}[A][B]$$

Our goal is to write an expression for the rate constant k_{r}. We begin by writing the thermodynamic equilibrium constant in terms of activities a and activity coefficients γ:

$$K = \frac{a_{C^{\ddagger}}}{a_{A}a_{B}} = K_{\gamma}\frac{[C^{\ddagger}]c^{\oplus}}{[A][B]}, \qquad \text{where} \qquad K_{\gamma} = \frac{\gamma_{C^{\ddagger}}}{\gamma_{A}\gamma_{B}}$$

It follows that

$$[A][B] = \frac{K_{\gamma}c^{\oplus}}{K}[C^{\ddagger}]$$

The combination of this expression with $v = k_{r}[A][B]$ gives

$$v = k_{r}\left(\frac{K_{\gamma}c^{\oplus}}{K}[C^{\ddagger}]\right)$$

If we let $k^{\ddagger}$ be the rate constant for formation of products from the activated complex $C^{\ddagger}$, then we may also write

$$v = k^{\ddagger}[C^{\ddagger}]$$

It follows that

$$k_{r} = \frac{k^{\ddagger}K}{K_{\gamma}c^{\oplus}}$$

If k_{r}^{o} is the rate constant when the activity coefficients are 1 (that is, $k_{r}^{o} = k^{\ddagger}K/c^{\oplus}$), we can write

$$k_{r} = \frac{k_{r}^{o}}{K_{\gamma}}$$

At low concentrations the activity coefficients can be expressed in terms of the ionic strength, I, of the solution by using the Debye–Hückel limiting law (eqn 5.4, $\log \gamma_{J} = -Az_{J}^{2}I^{1/2}$). Then

$$\log k_{r} = \log k_{r}^{o} - A\{z_{A}^{2} + z_{B}^{2} - (z_{A} + z_{B})^{2}\}I^{1/2} = \log k_{r}^{o} + 2Az_{A}z_{B}I^{1/2}$$

as in eqn 7.35.

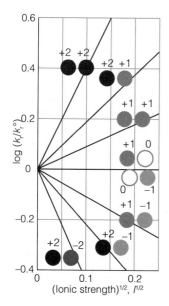

Fig. 7.18 An illustration of the kinetic salt effect. If the reactants have opposite charges, then the rate decreases as the ionic strength, I, is increased. However, if the charges of the reactant ions have the same sign, then the rate increases when a salt is added.

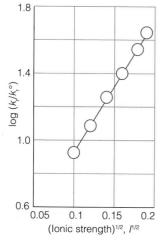

Fig. 7.19 The plot for the data in *Example 7.2*.

Equation 7.35 confirms that if the reactants have opposite charges (so $z_A z_B$ is negative), then the rate decreases as the ionic strength is increased (Fig. 7.18), just as the qualitative description suggested. However, if the charges of the reactant ions have the same sign (and $z_A z_B$ is positive), then the rate increases when a salt is added. Information of this kind is useful in unraveling the reaction mechanism of reactions in solution and identifying the nature of the activated complex.

Example 7.2   *Analyzing the kinetic salt effect*

The study of conditions that optimize the association of proteins in solution guides the design of protocols for the formation of large crystals that are amenable to analysis by the X-ray diffraction techniques discussed in Chapter 11. It is important to characterize protein dimerization because the process is considered to be the rate-determining step in the growth of crystals of many proteins. Consider the variation with ionic strength of the rate constant of dimerization in aqueous solution of a cationic protein P:

I	0.0100	0.0150	0.0200	0.0250	0.0300	0.0350
k_r/k_r^o	8.10	13.30	20.50	27.80	38.10	52.00

What can be deduced about the charge of P?

Strategy Although the dimer is not an activated complex in the same sense as in transition state theory, eqn 7.35 applies if we assume that the activated complex and the product (the dimer) are similar in the sense that two protein molecules associate to form the activated complex. Thus, the equilibrium constant for the dimerization is related to the rate constants for the formation of the dimer and its decomposition by $K = k_r/k_r'c^\ominus$ and $K = K_c K_\gamma$. Hence $k_r = K_c K_\gamma k_r' c^\ominus = K_\gamma k_r^o$. It then follows, as in *Justification 7.5*, that

$$\log(k_r/k_r^o) = 1.02z^2 I^{1/2}$$

Therefore, to infer the protein charge number z from the slope, $1.02z^2$, we need to plot $\log(k_r/k_r^o)$ against $I^{1/2}$.

Answer: We draw up the following table:

$I^{1/2}$	0.100	0.122	0.141	0.158	0.173	0.187
$\log(k_r/k_r^o)$	0.908	1.124	1.312	1.444	1.581	1.716

These points are plotted in Fig. 7.19. The slope of the straight line is 9.2, indicating that $z^2 = 9$. Because the protein is cationic, its charge number is +3.

Self-test 7.6   An ion of charge number +1 is known to be involved in the activated complex of a reaction. Deduce the charge number of the other ion from the following data:

I	0.0050	0.010	0.015	0.020	0.025	0.030
k_r/k_r^o	0.850	0.791	0.750	0.717	0.689	0.666

Answer: −1

Checklist of key concepts

1. In relaxation methods of kinetic analysis, the equilibrium position of a reaction is first shifted suddenly and then allowed to readjust to the equilibrium composition characteristic of the new conditions.

2. An elementary unimolecular reaction has first-order kinetics; an elementary bimolecular reaction has second-order kinetics.

3. The molecularity of an elementary reaction is the number of molecules coming together to react.

4. The rate-determining step is the slowest step in a reaction mechanism that controls the rate of the overall reaction.

5. In the steady-state approximation, it is assumed that the concentrations of all reaction intermediates remain constant and small throughout the reaction.

6. Provided a reaction has not reached equilibrium, the products of competing reactions are controlled by kinetics.

7. In collision theory, it is supposed that the rate is proportional to the collision frequency, a steric factor, and the fraction of collisions that occur with at least the kinetic energy E_a along their lines of centers.

8. In transition state theory, it is supposed that an activated complex is in equilibrium with the reactants and that the rate at which that complex forms products depends on the rate at which it passes through a transition state.

Checklist of key equations

Property	Equation	Comment
Equilibrium constant in terms of rate constants	$K = k_r/k_r'$	First order in each direction
Relaxation time	$x = x_0 e^{-t/\tau} \quad 1/\tau = k_r + k_r'$	Temperature jump experiment
Rate constant of a diffusion-controlled reaction	$k_r = k_d = 8RT/3\eta$	
Rate constant of an activation-controlled reaction	$k_r = k_a k_d/k_d'$	
Kinetic control	$[P_2]/[P_1] = k_2/k_1$	
Eyring equation	$k_r = \kappa(kT/h)(K^\ddagger/c^\ominus)$	
Thermodynamic parameterization	$k_r = (kT/h)e^{-\Delta^\ddagger G/RT} = \{(kT/h)e^{\Delta^\ddagger S/R}\}e^{-\Delta^\ddagger H/RT}$	Eyring equation with $\kappa = 1$
Kinetic salt effect	$\log k_r = \log k_r^\circ + 2Az_A z_B I^{1/2}$	Dilute solutions

Further information

Further information 7.1 *Collisions in the gas phase*

To gain insight into collisions in the gas phase, first we need to explore a model that can explain the properties of perfect gases. Then we consider the distribution of molecular speeds, a concept we previewed in *Fundamentals* F.2 and F.3. Finally, we define some of the parameters used in the collision theory of chemical reactions in the gas phase.

(a) The kinetic model of gases

The basis for our discussion is the **kinetic model of gases** (also called the 'kinetic molecular theory', KMT, of gases), which makes the following three assumptions:

1. A gas consists of molecules in ceaseless random motion (Fig. F.9).

2. The size of the molecules is negligible in the sense that their diameters are much smaller than the average distance traveled between collisions.

3. The molecules do not interact, except during collisions.

The assumption that the molecules do not interact unless they are in contact implies that the potential energy of the molecules (their energy due to their position) is independent of their separation and may be set equal to zero. The total energy of a sample of gas is therefore the sum of the kinetic energies (the energy due to motion) of all the molecules present in it. It follows that the faster the molecules travel (and hence the greater their kinetic energy), the greater the total energy of the gas.

The kinetic model accounts for the steady pressure exerted by a gas in terms of the collisions the molecules make with the walls of the container. Each collision gives rise to a brief force on the wall, but as billions of collisions take place every second, the walls experience a virtually constant force, and hence the gas exerts a steady pressure. On the basis of this model, the pressure exerted by a gas of molar mass M in a volume V is[5]

$$p = \frac{nMc^2}{3V} \qquad \boxed{\text{The pressure of a perfect gas according to the kinetic model}} \quad (7.36)$$

where c is the **root-mean-square speed** (r.m.s. speed) of the molecules. This quantity is defined as the square root of the mean value of the squares of the speeds, v, of the molecules. That is, for a sample consisting of N molecules with speeds v_1, $v_2, \ldots, v_N$, we square each speed, add the squares together, divide by the total number of molecules (to get the mean, denoted by $\langle \cdots \rangle$), and finally take the square root of the result:

$$c = \langle v^2 \rangle^{1/2} = \left(\frac{v_1^2 + v_2^2 + \cdots + v_N^2}{N} \right)^{1/2} \qquad \boxed{\text{Definition of the r.m.s. speed}} \quad (7.37)$$

The r.m.s. speed might at first encounter seem to be a rather peculiar measure of the mean speeds of the molecules, but its significance becomes clear when we make use of the fact that the kinetic energy of a molecule of mass m traveling at a speed v is $E_k = \frac{1}{2}mv^2$, which implies that the mean kinetic energy, $\langle E_k \rangle$, is the average of this quantity, or $\frac{1}{2}mc^2$. It follows that

$$c = \left(\frac{2\langle E_k \rangle}{m} \right)^{1/2} \qquad \boxed{\text{The r.m.s. speed in terms of the mean kinetic energy}} \quad (7.38)$$

Therefore, wherever c appears, we can think of it as a measure of the mean kinetic energy of the molecules of the gas.

We also note that eqn 7.36 resembles the perfect gas equation of state, $pV = nRT$, for we can rearrange it into

$$pV = \tfrac{1}{3}nMc^2$$

Equating the expression on the right to nRT gives

$$\tfrac{1}{3}nMc^2 = nRT$$

where the ns now cancel. The great usefulness of this expression is that we can rearrange it into a formula for the r.m.s. speed of the gas molecules at any temperature:

$$c = \left(\frac{3RT}{M} \right)^{1/2} \qquad \boxed{\text{The r.m.s. speed in terms of the temperature}} \quad (7.39)$$

A brief illustration

Substitution into eqn 7.39 of the molar mass of O_2 (32.0 g mol^{-1}) and a temperature corresponding to 25°C (that is, 298 K) gives an r.m.s. speed for these molecules of 482 m s^{-1}. The same calculation for nitrogen molecules gives 515 m s^{-1}.

---

[5]   For a derivation of eqn 7.36, see our *Physical chemistry* (2010).

The important conclusion to draw from eqn 7.39 is that *the r.m.s. speed of molecules in a gas is proportional to the square root of the temperature*. Because the mean speed is proportional to the r.m.s. speed, the same is true of the mean speed too. Therefore, doubling the temperature (on the Kelvin scale) increases the mean and the r.m.s. speed of molecules by a factor of $2^{1/2} = 1.414 \ldots$.

(b) The Maxwell distribution of speeds

The mathematical expression that tells us the fraction of molecules that have a particular speed at any instant is called the **distribution of molecular speeds**. The precise form of the distribution was worked out by James Clerk Maxwell towards the end of the nineteenth century, and his expression is known as the **Maxwell distribution of speeds**. According to Maxwell, the fraction f of molecules that have a speed in a narrow range between s and $s + \Delta s$ (for example, between 300 m s^{-1} and 310 m s^{-1}, corresponding to $s = 300$ m s^{-1} and $\Delta s = 10$ m s^{-1}) is

$$f = F(s)\Delta s \quad \text{with}$$

$$F(s) = 4\pi \left(\frac{M}{2\pi RT} \right)^{3/2} s^2 e^{-Ms^2/2RT} \qquad \boxed{\begin{array}{l}\text{Maxwell}\\\text{distribution}\\\text{of speeds}\end{array}} \quad (7.40)$$

Although eqn 7.40 looks complicated, its features can be picked out quite readily:

1. Because f is proportional to the range of speeds Δs, we see that the fraction in the range Δs increases in proportion to the width of the range. If at a given speed we double the range of interest (but still ensure that it is narrow), then the fraction of molecules in that range doubles too.

2. Equation 7.40 includes a decaying exponential function, the term $e^{-Ms^2/2RT}$. Its presence implies that the fraction of molecules with very high speeds will be very small because e^{-x^2} becomes very small when x^2 is large.

3. The factor $M/2RT$ multiplying s^2 in the exponent is large when the molar mass, M, is large, so the exponential factor goes most rapidly towards zero when M is large. That tells us that heavy molecules are unlikely to be found with very high speeds.

4. The opposite is true when the temperature, T, is high: then the factor $M/2RT$ in the exponent is small, so the exponential factor falls towards zero relatively slowly as s increases. This tells us that at high temperatures, a greater fraction of the molecules can be expected to have high speeds than at low temperatures.

5. A factor s^2 (the term before the e) multiplies the exponential. This factor goes to zero as s goes to zero, so the fraction of molecules with very low speeds will also be very small.

The remaining factors (the term in parentheses in eqn 7.40 and the 4π) simply ensure that when we add together the fractions over the entire range of speeds from zero to infinity, then we get 1. These features are summarized in Figs F.10 and F.11.

(c) Molecular collisions

The average distance that a molecule travels between collisions is called its **mean free path**, λ (lambda). The mean free path in a liquid is less than the diameter of the molecules because a molecule in a liquid meets a neighbor even if it moves only a fraction of a diameter. However, in gases, the mean free paths of molecules can be several hundred molecular diameters. If we think of a molecule as the size of a tennis ball, then the mean free path in a typical gas would be about the length of a tennis court.

The **collision frequency**, z, is the average rate of collisions made by one molecule. Specifically, z is the average number of collisions one molecule makes in a given time interval divided by the length of the interval. It follows that the inverse of the collision frequency, $1/z$, is the **time of flight**, the average time that a molecule spends in flight between two collisions (for instance, if there are 10 collisions per second, so the collision frequency is 10 s^{-1}, then the average time between collisions is $\frac{1}{10}$ of a second and the time of flight is $\frac{1}{10}$ s). As we shall see, the collision frequency in a typical gas is about 10^9 s^{-1} at 1 atm and room temperature, so the time of flight in a gas is typically 1 ns.

Because speed is distance traveled divided by the time taken for the journey, the r.m.s. speed c, which we can loosely think of as the average speed, is the average length of the flight of a molecule between collisions (that is, the mean free path, λ) divided by the time of flight ($1/z$). It follows that the mean free path and the collision frequency are related by

$$c = \frac{\text{mean free path}}{\text{time of flight}} = \frac{\lambda}{1/z} = \lambda z \qquad \boxed{\begin{array}{l}\text{The r.m.s. speed in} \\ \text{terms of the mean} \\ \text{free path and the} \\ \text{collision frequency}\end{array}} \qquad (7.41)$$

To find expressions for λ and z, we need a slightly more elaborate version of the kinetic model of gases. The basic kinetic model supposes that the molecules are effectively pointlike; however, to obtain collisions, we need to assume that two 'points' score a hit whenever they come within a certain range d of each other, where d can be thought of as the diameter of the molecules (Fig. 7.20). The **collision cross-section**, σ (sigma), the target area presented by one molecule to another, is therefore the area of a circle of radius d, so $\sigma = \pi d^2$. When this quantity is built into the kinetic model, we find that

$$\lambda = \frac{kT}{2^{1/2}\sigma p} \qquad z = \frac{2^{1/2}\sigma p}{kT}c \qquad \boxed{\begin{array}{l}\text{The mean free path and} \\ \text{the collision frequency} \\ \text{in terms of the collision} \\ \text{cross-section}\end{array}} \qquad (7.42)$$

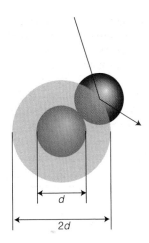

Fig. 7.20 To calculate features of a perfect gas that are related to collisions, a point is regarded as being surrounded by a sphere of diameter d. A molecule will hit another molecule if the center of the former lies within a circle of radius d. The collision cross-section is the target area, πd^2.

We can identify the following features:

1. Because $\lambda \propto 1/p$, we see that *the mean free path decreases as the pressure increases.* This decrease is a result of the increase in the number of molecules present in a given volume as the pressure is increased, so each molecule travels a shorter distance before it collides with a neighbor.

For example, the mean free path of an O_2 molecule decreases from 73 nm to 36 nm when the pressure is increased from 1.0 bar to 2.0 bar at 25°C.

2. Because $\lambda \propto 1/\sigma$, *the mean free path is shorter for molecules with large collision cross-sections.*

For instance, the collision cross-section of a benzene molecule (0.88 nm^2) is about four times greater than that of a helium atom (0.21 nm^2), and at the same pressure and temperature its mean free path is four times shorter.

3. Because $z \propto p$, *the collision frequency increases with the pressure of the gas.* This dependence follows from the fact that, provided the temperature is the same, the molecule take less time to travel to its neighbor in a denser, higher pressure gas.

For example, although the collision frequency for an O_2 molecule in oxygen gas at 298.15 K and 1.0 bar is $6.2 \times 10^9 \text{ s}^{-1}$, at 2.0 bar and the same temperature the collision frequency is doubled, to $1.2 \times 10^{10} \text{ s}^{-1}$.

Discussion questions

7.1 Sketch, without carrying out the calculation, the variation of concentration with time for the approach to equilibrium when both forward and reverse reactions are second order. How does your graph differ from that in Fig. 7.2?

7.2 Write a brief report on a recent research article in which at least one of the following techniques was used to study the kinetics of a biochemical reaction: stopped-flow techniques, time-resolved spectroscopy, chemical quench-flow methods, freeze-quench methods, temperature-jump methods. Your report should be similar in content and extent to one of the *Case studies* found throughout this text.

7.3 Assess the validity of the following statement: the rate-determining step is the slowest step in a reaction mechanism.

7.4 Distinguish between a pre-equilibrium approximation and a steady-state approximation.

7.5 Distinguish between a diffusion-controlled reaction and an activation-controlled reaction.

7.6 Distinguish between kinetic and thermodynamic control of a reaction. Suggest criteria for expecting one over the other.

7.7 Describe the formulation of the Eyring equation and interpret its form.

7.8 Is it possible for the activation energy of a reaction to be negative? Explain your conclusion and provide a molecular explanation.

7.9 Discuss the physical origin of the kinetic salt effect.

Exercises

7.10 The equilibrium constant for the attachment of a substrate to the active site of an enzyme was measured as 235. In a separate experiment, the rate constant for the second-order attachment was found to be $7.4 \times 10^7 \, \text{dm}^3 \, \text{mol}^{-1} \, \text{s}^{-1}$. What is the rate constant for the loss of the unreacted substrate from the active site?

7.11 Find the solutions of the same rate laws that led to eqn 7.2, but for some B present initially. Go on to confirm that the solutions you find reduce to those in eqn 7.2 when $[B]_0 = 0$.

7.12 The reaction $H_2O(l) \rightleftharpoons H^+(aq) + OH^-(aq)$ ($pK_w = 14.01$) relaxes to equilibrium with a relaxation time of 37 μs at 298 K and pH ≈ 7.
(a) Given that the forward reaction (with rate constant k_r) is first order and the reverse is second order overall (with rate constant k_r'), show that

$$\frac{1}{\tau} = k_r + k_r'([H^+]_{eq} + [OH^-]_{eq})$$

(b) Calculate the rate constants for the forward and reverse reactions.

7.13 A protein dimerizes according to the reaction $2A \rightleftharpoons A_2$ with forward rate constant k_r and reverse rate constant k_r'. Show that the relaxation time is

$$\tau = \frac{1}{k_r' + 4k_r[A]_{eq}}$$

7.14 Consider the dimerization of a protein, as in *Exercise* 7.13.
(a) Derive the following expression for the relaxation time in terms of the total concentration of protein, $[A]_{tot} = [A] + 2[A_2]$:

$$\frac{1}{\tau^2} = k_r'^2 + 8k_r k_r'[A]_{tot}$$

(b) Describe the computational procedures that lead to the determination of the rate constants k_r and k_r' from measurements of τ for different values of $[A]_{tot}$.

7.15 An understanding of the kinetics of formation of molecular complexes held together by hydrogen bonds gives insight into the formation of base pairs in nucleic acids. Use the data provided below

2 Pyridone

and the procedure you outlined in *Exercise* 7.14 to calculate the rate constants k_r and k_r' and the equilibrium constant K for formation of hydrogen-bonded dimers of 2-pyridone (**2**):

$[P]/(\text{mol dm}^{-3})$	0.500	0.352	0.251	0.151	0.101
τ/ns	2.3	2.7	3.3	4.0	5.3

7.16 Confirm (by differentiation) that the three expressions in eqns 7.10, 7.12, and 7.13 are correct solutions of the rate laws for consecutive first-order reactions.

7.17 Two radioactive nuclides decay by successive first-order processes:

$$X \xrightarrow{22.5 \, d} Y \xrightarrow{33.0 \, d} Z$$

(The times are half-lives in days.) Suppose that Y is an isotope that is required for medical applications. At what stage after X is first formed will Y be most abundant?

7.18 Use mathematical software or an electronic spreadsheet to examine the time dependence of [I] in the reaction mechanism $A \rightarrow I \rightarrow P$ (k_a, k_b) by plotting the expression in eqn 7.12. In the following calculations, use $[A]_0 = 1 \, \text{mol dm}^{-3}$ and a time range of 0 to 5 s. (a) Plot [I] against t for $k_a = 10 \, \text{s}^{-1}$ and $k_b = 1 \, \text{s}^{-1}$. (b) Increase the ratio k_b/k_a steadily by decreasing the value of k_a and examine the plot of [I] against t at each turn. What approximation about $d[I]/dt$ becomes increasingly valid?

7.19 The reaction $2 \, H_2O_2(aq) \rightarrow H_2O(l) + O_2(g)$ is catalyzed by Br^- ions. If the mechanism is as shown below give the predicted order of the reaction with respect to the various participants.

$H_2O_2(aq) + Br^-(aq) \rightarrow H_2O(l) + BrO^-(aq)$ (slow)

$BrO^-(aq) + H_2O_2(aq) \rightarrow H_2O(l) + O_2(g) + Br^-(aq)$ (fast)

7.20 The reaction mechanism

$A_2 \rightleftharpoons A + A$ (fast)

$A + B \rightarrow P$ (slow)

involves an intermediate A. Deduce the rate law for the formation of P.

7.21 Consider the following mechanism for formation of a double helix from its strands A and B:

$A + B \rightleftharpoons$ unstable helix   (fast)

unstable helix $\rightarrow$ stable double helix   (slow)

Derive the rate equation for the formation of the double helix and express the rate constant of the reaction in terms of the rate constants of the individual steps.

7.22 The following mechanism has been proposed for the decomposition of ozone in the atmosphere:

(1) $O_3 \rightarrow O_2 + O$ and its reverse (k_a, k_a')

(2) $O + O_3 \rightarrow O_2 + O_2$ (k_b; the reverse reaction is negligibly slow)

Use the steady-state approximation, with O treated as the intermediate, to find an expression for the rate of decomposition of O_3. Show that if step 2 is slow, then the rate is second order in O_3 and -1 order in O_2.

7.23 The condensation reaction of acetone, $(CH_3)_2CO$ (propanone), in aqueous solution is catalyzed by bases, B, which react reversibly with acetone to form the carbanion $C_3H_5O^-$. The carbanion then reacts with a molecule of acetone to give the product. A simplified version of the mechanism is

(1) $AH + B \rightarrow BH^+ + A^-$

(2) $A^- + BH^+ \rightarrow AH + B$

(3) $A^- + HA \rightarrow$ product

where AH stands for acetone and A^- its carbanion. Use the steady-state approximation to find the concentration of the carbanion and derive the rate equation for the formation of the product.

7.24 Consider the acid-catalyzed reaction

$HA + H^+ \rightleftharpoons HAH^+$ (fast)

$HAH^+ + B \rightarrow BH^+ + AH$ (slow)

Deduce the rate law and show that it can be made independent of the specific term $[H^+]$.

7.25 Models of population growth are analogous to chemical reaction rate equations. In the model due to Malthus (1798) the rate of change of the population N of the planet is assumed to be given by $dN/dt =$ births − deaths. The numbers of births and deaths are proportional to the population, with proportionality constants b and d. Obtain the integrated rate law. How well does it fit the (very approximate) data below on the population of the planet as a function of time?

Year	1750	1825	1922	1960	1974	1987	2000
$N/10^9$	0.5	1	2	3	4	5	6

7.26 The compound α-tocopherol, a form of vitamin E (Atlas M3), is a powerful antioxidant that may help to maintain the integrity of biological membranes. The light-induced reaction between duroquinone and the antioxidant in ethanol is bimolecular and

diffusion controlled. Estimate the rate constant for the reaction at 298 K, given that the viscosity of ethanol is 1.06×10^{-3} kg m^{-1} s^{-1}.

7.27 Collision theory demands knowing the fraction of molecular collisions having at least the kinetic energy E_a along the line of flight. What is this fraction when (**a**) $E_a = 10$ kJ mol^{-1} and (**b**) $E_a = 100$ kJ mol^{-1} at (**i**) 300 K and (**ii**) 1000 K?

7.28 Calculate the percentage increase in the fractions in *Exercise 7.27* when the temperature is raised by 10 K.

7.29 Calculate the ratio of rates of catalyzed to non-catalyzed reactions at 37°C given that the Gibbs energy of activation for a particular reaction is reduced from 100 kJ mol^{-1} to 10 kJ mol^{-1}.

7.30 Estimate the pre-exponential factor for the reaction between molecular hydrogen and ethene at 400°C. *Hint*: The steric factor is $P = 1.7 \times 10^{-6}$.

7.31 The mechanism of a composite reaction consists of a fast pre-equilibrium step with forward and reverse activation energies of 25 kJ mol^{-1} and 38 kJ mol^{-1}, respectively, followed by an elementary step of activation energy 10 kJ mol^{-1}. What is the activation energy of the composite reaction?

7.32 Rhodopsin is the protein in the retina that absorbs light, starting a cascade of chemical events that we call vision (see *Case study* 12.2 for additional information). Bovine rhodopsin undergoes a transition from one form (metarhodopsin I) to another form (metarhodopsin II) with a half-life of 600 μs at 37°C to 1 s at 0°C. On the other hand, studies of a frog retina show that the same transformation has a half-life that increases by only a factor of 6 over the same temperature range. Suggest an explanation and speculate on the survival advantages that this difference represents for the frog.

7.33 Estimate the activation Gibbs energy for the decomposition of urea in the reaction $(NH_2)_2CO(aq) + 2 H_2O(l) \rightarrow 2 NH_4^+(aq) + CO_3^{2-}(aq)$ for which the pseudo-first-order rate constant is 1.2×10^{-7} s^{-1} at 60°C and 4.6×10^{-7} s^{-1} at 70°C.

7.34 Calculate the entropy of activation of the reaction in *Exercise 7.33* at the two temperatures.

7.35 Calculate the Gibbs energy, enthalpy, and entropy of activation (at 300 K) for the binding of an inhibitor to the enzyme carbonic anhydrase using the following data:

T/K	289.0	293.5	298.1
$k_r/(10^6$ dm^3 mol^{-1} s$^{-1})$	1.04	1.34	1.53

T/K	303.2	308.0	313.5
$k_r/(10^6$ dm^3 mol^{-1} s$^{-1})$	1.89	2.29	2.84

7.36 The reaction $A^- + H^+ \rightarrow P$ has a rate constant given by the empirical expression $k_r = (8.72 \times 10^{12})e^{(6134\,K)/T}$ dm^3 mol^{-1} s^{-1}. Evaluate the energy and entropy of activation at 25°C.

7.37 The conversion of fumarate ion to malate ion is catalyzed by the enzyme fumarase:

fumarate^{2-}(aq) + H_2O(l) $\rightleftharpoons$ malate^{2-}(aq)

(**a**) Sketch the reaction profile for this reaction given that (**i**) the standard enthalpy of formation of the fumarate–fumarase complex from fumarate ion and enzyme is 17.6 kJ mol^{-1}, (**ii**) the enthalpy of activation of the forward reaction is 41.3 kJ mol^{-1}, (**iii**) the standard enthalpy of formation of the malate–fumarase complex from malate ion and enzyme is −5.0 kJ mol^{-1}, and (**iv**) the standard reaction enthalpy is −20.1 kJ mol^{-1}. (**b**) What is the enthalpy of activation of the reverse reaction?

7.38 The activation Gibbs energy is composed of two terms: the activation enthalpy and the activation entropy. Differences in the latter can lead to the activation Gibbs energy for a process having the same values despite species inhabiting environments that differ widely in temperature. Show how the data depicted in Fig. 7.21 support this remark. The data relate to the enthalpy and entropy of activation of myosin ATPase in different species of fish living in environments ranging from the Arctic to hot springs.

7.39 At 25°C, $k_r = 1.55$ dm^6 mol^{-2} min^{-1} at an ionic strength of 0.0241 for a reaction in which the rate-determining step involves the encounter of two singly charged cations. Use the Debye–Hückel limiting law to estimate the rate constant at zero ionic strength.

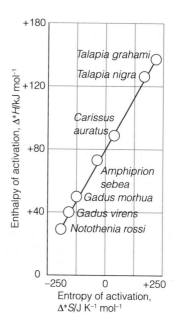

Fig. 7.21 (*right*) The correlation of the enthalpy and entropy of activation of the reaction catalyzed by myosin ATPase in a variety of fish species. (Data from I.A. Johnson and G. Goldspink, *Nature* **257**, 620 (1970), recalculated by H. Guttfreund, *Kinetics for the life sciences*. Cambridge University Press (1995).

Projects

7.40 The absorption and elimination of a drug in the body may be modeled with a mechanism consisting of two consecutive reactions:

$$\begin{array}{ccccc} A & \rightarrow & B & \rightarrow & C \\ \text{drug at site of} & & \text{drug dispersed} & & \text{eliminated} \\ \text{administration} & & \text{in blood} & & \text{drug} \end{array}$$

where the rate constants of absorption (A → B) and elimination are, respectively, k_a and k_b.

(a) Consider a case in which absorption is so fast that it may be regarded as instantaneous, and a dose of A at an initial concentration $[A]_0$ immediately leads to a drug concentration in blood of $[B]_0$. Also, assume that elimination follows first-order kinetics.

(i) Show that, after the administration of N equal doses separated by a time interval τ, the peak concentration of drug B in the blood, $[P]_N$, rises beyond the value of $[B]_0$ and eventually reaches a constant, maximum peak value given by

$$[P]_\infty = [B]_0(1 - e^{-k_b\tau})^{-1}$$

$[P]_N$ is the (peak) concentration of B immediately after administration of the Nth dose and $[P]_\infty$ is the value at very large N.

(ii) Write a mathematical expression for the residual concentration of B, $[R]_N$, which we define to be the concentration of drug B immediately before the administration of the $(N+1)$th dose. Note that $[R]_N$ is always smaller than $[P]_N$ because of drug elimination during the period τ between drug administrations. Show that $[P]_\infty - [R]_\infty = [B]_0$.

(b) Consider a drug for which $k_b = 0.0289$ h^{-1}.
(i) Calculate the τ value required to achieve $[P]_\infty/[B]_0 = 10$. Prepare a graph that plots both $[P]_N/[B]_0$ and $[R]_N/[B]_0$ against N.
(ii) How many doses must be administered to achieve a $[P]_N$ value that is 75 per cent of the maximum value? What time has passed during the administration of these doses?
(iii) What actions can be taken to reduce the variation $[P]_\infty - [R]_\infty$ while maintaining the same value of $[P]_\infty$?

(c) Now consider the administration of a single dose $[A]_0$ for which absorption follows first-order kinetics and elimination follows zeroth-order kinetics. Show that with the initial concentration $[B]_0 = 0$, the concentration of drug in the blood is given by

$$[B] = [A]_0(1 - e^{-k_a t}) - k_b t$$

Plot $[B]/[A]_0$ against t for the case $k_a = 10$ h^{-1}, $k_b = 4.0 \times 10^{-3}$ mmol dm^{-3} h^{-1}, and $[A]_0 = 0.1$ mmol dm^{-3}. Comment on the shape of the curve.

(d) Using the model from part (c), set $d[B]/dt = 0$ and show that the maximum value of $[B]$ occurs at the time

$$t_{max} = \frac{1}{k_a} \ln\left(\frac{k_a[A]_0}{k_b}\right)$$

Also, show that the maximum concentration of drug in blood is given by

$$[B]_{max} = [A]_0 - k_b/k_a - k_b t_{max}.$$

7.41 Consider a mechanism for the helix–coil transition in which nucleation occurs in the middle of the chain:

$$hhhh\ldots \rightleftharpoons hchh \qquad hchh\ldots \rightleftharpoons cccc$$

We saw in *Case study* 7.1 that this type of nucleation is relatively slow, so neither step may be rate determining.

(a) Set up the rate equations for this alternative mechanism.

(b) Apply the steady-state approximation and show that, under these circumstances, the mechanism is equivalent to $hhhh\ldots \rightleftharpoons cccc\ldots$

(c) Use your knowledge of experimental techniques and your results from parts (a) and (b) to support or refute the following statement: It is very difficult to obtain experimental evidence for intermediates in protein folding by performing simple rate measurements and one must resort to special flow, relaxation, or trapping techniques to detect intermediates directly.

Complex biochemical processes

Biochemical processes use a number of strategies to achieve kinetic control. Chief among them is the use of enzymes to accelerate and regulate the rates of chemical reactions that, although thermodynamically favorable under intracellular conditions, would be too slow to account for the observed rate of growth of organisms and the processes of life in general. With the constant development of powerful experimental techniques, biochemists are beginning to decipher the mechanisms of even the most complex biological processes, such as the transport of nutrients across cell membranes and the transfer of electrons between proteins during glucose metabolism and photosynthesis. In this chapter we describe these processes and develop the physical and chemical concepts that will be used throughout the remainder of the text.

Enzymes

Enzymes are homogeneous biological catalysts that work by lowering the activation energy of a reaction pathway or providing a new pathway with a low activation energy. Enzymes are special biological polymers that contain an **active site**, which is responsible for binding the **substrates**, the reactants, and processing them into products. As is true of any catalyst, the active site returns to its original state after the products are released. Many enzymes consist primarily of proteins, some featuring organic or inorganic cofactors in their active sites. However, certain ribonucleic acid (RNA) molecules can also be biological catalysts, forming **ribozymes**. A very important example of a ribozyme is the **ribosome**, a large assembly of proteins and catalytically active RNA molecules responsible for the synthesis of proteins in the cell.

The structure of the active site is specific to the reaction that it catalyzes, with groups in the substrate interacting with groups in the active site through intermolecular interactions, such as hydrogen bonding, electrostatic, or van der Waals interactions (see Chapter 11). Figure 8.1 shows two models that explain the binding of a substrate to the active site of an enzyme. In the **lock-and-key model**, the active site and substrate have complementary three-dimensional structures and dock perfectly without the need for major atomic rearrangements. Experimental evidence favors the **induced fit model**, in which binding of the substrate induces a conformational change in the active site. Only after the change does the substrate fit snugly in the active site.

Enzyme-catalyzed reactions are prone to inhibition by molecules that interfere with the formation of product. As we remarked in the *Prolog*, many drugs for the treatment of disease inhibit enzymes of infectious agents, such as bacteria and viruses. Here we focus on the kinetic analysis of enzyme inhibition, and in

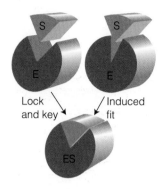

Fig. 8.1 Two models that explain the binding of a substrate to the active site of an enzyme. In the lock-and-key model, the active site and substrate have complementary three-dimensional structures and dock perfectly without the need for major atomic rearrangements. In the induced fit model, binding of the substrate induces a conformational change in the active site. The substrate fits well in the active site after the conformational change has taken place.

Chapters 10 and 11 we shall see how computational methods contribute to the design of efficient inhibitors and potent drugs.

8.1 The Michaelis–Menten mechanism of enzyme catalysis

Because enzyme-controlled reactions are so important in biochemistry, we need to build a model of their mechanism. The simplest approach proposed by Michaelis and Menten is our starting point.

Experimental studies of enzyme kinetics are typically conducted by monitoring the initial rate of product formation in a solution in which the enzyme is present at very low concentration. Indeed, enzymes are such efficient catalysts that significant accelerations may be observed even when their concentrations are more than three orders of magnitude smaller than those of their substrates.

The principal features of many enzyme-catalyzed reactions are as follows (Fig. 8.2):

1. For a given initial concentration of substrate, $[S]_0$, the initial rate of product formation is proportional to the total concentration of enzyme, $[E]_0$.

2. For a given $[E]_0$ and low values of $[S]_0$, the rate of product formation is proportional to $[S]_0$.

3. For a given $[E]_0$ and high values of $[S]_0$, the rate of product formation becomes independent of $[S]_0$, reaching a maximum value known as the **maximum velocity**, v_{max}.

The **Michaelis–Menten** mechanism accounts for these features.[1] According to this mechanism, an enzyme–substrate complex, ES, is formed in the first step and the substrate is released either unchanged or after modification to form products:

Michaelis–Menten mechanism:

$$E + S \rightarrow ES \qquad v = k_a[E][S]$$
$$ES \rightarrow E + S \qquad v = k_a'[ES]$$
$$ES \rightarrow P + E \qquad v = k_b[ES]$$

As we show in the following *Justification*, this mechanism implies that the rate of product formation is given by the **Michaelis–Menten equation**:

$$v = \frac{k_b[E]_0}{1 + K_M/[S]_0} \qquad \text{or} \qquad v = \frac{k_b[E]_0[S]_0}{[S]_0 + K_M} \qquad \boxed{\text{Michaelis–Menten equation}} \qquad (8.1)$$

where $K_M = (k_a' + k_b)/k_a$ is the **Michaelis constant**, characteristic of a given enzyme acting on a given substrate, $[E]_0$ is the molar concentration of enzyme added and $[S]_0$ is the molar concentration of the substrate.

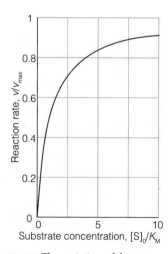

Fig. 8.2 The variation of the rate of an enzyme-catalyzed reaction with substrate concentration. The approach to a maximum rate, v_{max}, for large $[S]_0$ is explained by the Michaelis–Menten mechanism. (The constant K_M is explained shortly.)

Justification 8.1 *The Michaelis–Menten equation*

To derive eqn 8.1, we note that the rate of product formation ($v = k_b[ES]$) requires us to know [ES]. We can obtain the concentration of the enzyme–substrate complex by invoking the steady-state approximation (Section 7.3c) and writing

---

[1] Michaelis and Menten derived their rate law in 1913 in a more restrictive way, by assuming a rapid equilibrium. The approach we take is a generalization using the steady-state approximation made by Briggs and Haldane in 1925.

$$\frac{d[ES]}{dt} = k_a[E][S] - k'_a[ES] - k_b[ES] = 0$$

It follows that

$$[ES] = \left(\frac{k_a}{k'_a + k_b}\right)[E][S]$$

where [E] and [S] are the concentrations of *free* enzyme and substrate, respectively. Now we define the Michaelis constant as

$$K_M = \frac{k'_a + k_b}{k_a} = \frac{[E][S]}{[ES]}$$

and note that K_M has the units of a molar concentration. To express the rate law in terms of the concentrations of enzyme, we note that $[E]_0 = [E] + [ES]$, so

$$[E]_0 = [E] + \frac{[E][S]}{K_M} = \left(1 + \frac{[S]}{K_M}\right)[E]$$

Moreover, because the substrate is typically in large excess relative to the enzyme, the free substrate concentration is approximately equal to the initial substrate concentration and we can write $[S] \approx [S]_0$. It then follows that

$$[ES] = \frac{1}{K_M}\frac{[E]_0[S]_0}{1 + [S]_0/K_M} = \frac{[E]_0}{(1 + K_M)/[S]_0}$$

We obtain eqn 8.1 when we substitute this expression for [ES] into that for the rate of product formation.

Equation 8.1 shows that, in accord with experimental observations:

1. When $[S]_0 \ll K_M$, the rate is proportional to $[S]_0$:

$$v = \frac{k_b}{K_M}[S]_0[E]_0 \tag{8.2a}$$

2. When $[S]_0 \gg K_M$, the rate reaches its maximum value and is independent of $[S]_0$:

$$v = v_{max} = k_b[E]_0 \tag{8.2b}$$

We can rearrange eqn 8.1 into a form that is amenable to data analysis by linear regression. Substitution of the definition of v_{max} into eqn 8.2b gives

$$v = \frac{v_{max}}{1 + K_M/[S]_0}$$

Then, on taking reciprocals of both sides, we obtain

$$\frac{1}{v} = \frac{1}{v_{max}} + \left(\frac{K_M}{v_{max}}\right)\frac{1}{[S]_0} \qquad \boxed{\text{Lineweaver–Burk plot}} \tag{8.3}$$

A **Lineweaver–Burk plot** is a plot of $1/v$ against $1/[S]_0$ and, according to eqn 8.3, it should yield a straight line with slope of K_M/v_{max}, a y-intercept at $1/v_{max}$, and an x-intercept at $-1/K_M$ (Fig. 8.3). The value of k_b is then calculated from the y-intercept and eqn 8.2b. However, the plot cannot give the individual rate constants k_a and k'_a that appear in the expression for K_M. The stopped-flow technique

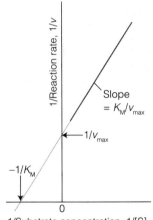

Fig. 8.3 A Lineweaver–Burk plot is used to analyze kinetic data on enzyme-catalyzed reactions. The reciprocal of the rate of formation of products ($1/v$) is plotted against the reciprocal of the substrate concentration ($1/[S]_0$). All the data points (which typically lie in the full region of the line) correspond to the same overall enzyme concentration, $[E]_0$. The intercept of the extrapolated (dotted) straight line with the horizontal axis is used to obtain the Michaelis constant, K_M. The intercept with the vertical axis is used to determine $v_{max} = k_b[E]_0$ and hence k_b. The slope may also be used, for it is equal to K_M/v_{max}.

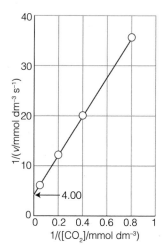

Fig. 8.4 The Lineweaver–Burk plot based on the data in *Example* 8.1.

A note on good practice
The slope and the intercept are unit-less: we have remarked previously that all graphs should be plotted as pure numbers by dividing the physical variables by their units (not just by ignoring the units!).

described in Section 6.1b gives the additional data needed because we can find the rate of formation of the enzyme–substrate complex by monitoring the concentration after mixing the enzyme and substrate. This procedure gives a value for k_a, and k_a' is then found by combining this result with the values of k_b and K_M.

Example 8.1 *Analyzing a Lineweaver–Burk plot*

The enzyme carbonic anhydrase (Atlas P2) catalyzes the hydration of CO_2 in red blood cells to give bicarbonate (hydrogencarbonate) ion:

$$CO_2(g) + H_2O(l) \rightarrow HCO_3^-(aq) + H^+(aq)$$

The following data were obtained for the reaction at pH $= 7.1$, 273.5 K, and an enzyme concentration of 2.3 nmol dm^{-3}:

$[CO_2]_0/(\text{mmol dm}^{-3})$	1.25	2.5	5	20
$v/(\text{mmol dm}^{-3}\text{ s}^{-1})$	2.78×10^{-2}	5.00×10^{-2}	8.33×10^{-2}	1.67×10^{-1}

Determine the maximum velocity and the Michaelis constant for the reaction.

Strategy We construct a Lineweaver–Burk plot by drawing up a table of $1/[S]_0$ and $1/v$. The intercept at $1/[S]_0 = 0$ is $1/v_{max}$ and the slope of the line through the points is K_M/v_{max}, so K_M is found from the slope divided by the intercept.

Solution We draw up the following table:

$1/([CO_2]_0/(\text{mmol dm}^{-3}))$	0.800	0.400	0.200	0.0500
$1/(v/(\text{mmol dm}^{-3}\text{ s}^{-1}))$	36.0	20.0	12.0	6.00

The graph is plotted in Fig. 8.4. A least-squares analysis gives an intercept at 4.00 and a slope of 40.0. It follows that

$$v_{max}/(\text{mmol dm}^{-3}\text{ s}^{-1}) = \frac{1}{\text{intercept}} = \frac{1}{4.00} = 0.250$$

and

$$K_M/(\text{mmol dm}^{-3}) = \frac{\text{slope}}{\text{intercept}} = \frac{40.0}{4.00} = 10.0$$

Self-test 8.1 The enzyme α-chymotrypsin (Atlas P3) is secreted in the pancreas of mammals and cleaves peptide bonds made between certain amino acids. Several solutions containing the small peptide N-glutaryl-L-phenylalanine-*p*-nitroanilide at different concentrations were prepared, and the same small amount of α-chymotrypsin was added to each one. The following data were obtained on the initial rates of the formation of product:

$[S]_0/(\text{mmol dm}^{-3})$	0.334	0.450	0.667	1.00	1.33	1.67
$v/(\text{mmol dm}^{-3}\text{ s}^{-1})$	0.152	0.201	0.269	0.417	0.505	0.667

Determine the maximum velocity and the Michaelis constant for the reaction.

Answer: 2.76 mmol dm^{-3} s^{-1}, 5.77 mmol dm^{-3}

Many enzyme-catalyzed reactions are consistent with a modified version of the Michaelis–Menten mechanism, in which the release of product from the ES complex is also reversible with the step

$$P \rightarrow ES \qquad v = k_b'[P]$$

added to the mechanism. In *Exercise* 8.10 you are invited to show that application of the steady-state approximation for [ES] then results in the following expression for the rate of the reaction:

$$v = \frac{(v_{max}/K_M)[S]_0 - (v_{max}'/K_M')[P]}{1 + [S]_0/K_M + [P]/K_M'} \tag{8.4a}$$

where

$$v_{max} = k_b[E]_0 \qquad v_{max}' = k_a'[E]_0 \tag{8.4b}$$

$$K_M = \frac{k_a' + k_b}{k_a} \qquad K_M' = \frac{k_a' + k_b}{k_b'} \tag{8.4c}$$

Equation 8.4a tells us that the reaction rate depends on the concentration of product. However, at the early stages of the reaction, when $[S] = [S]_0 \gg [P]$, terms containing [P] can be ignored and it is easy to show that eqn 8.4a reduces to eqn 8.1.

8.2 The analysis of complex mechanisms

The simple mechanism described in the previous section is only a starting point: to account for the full range of enzyme-controlled reactions, we need to consider more involved mechanisms.

Many enzymes can generate several intermediates as they process a substrate into one or more products. An example is the enzyme chymotrypsin, which we treat in detail in *Case study* 8.1. Other enzymes act on multiple substrates. An example is hexokinase, which catalyzes the reaction between ATP and glucose (the two substrates of the enzyme), the first step of glycolysis (Section 4.8). The same strategies developed in Section 8.1 can be used to deal with such complex reaction schemes, and we shall focus on reactions involving two substrates.

(a) Sequential reactions

In **sequential reactions** the active site binds all the substrates before processing them into products. The binding can be ordered:

Sequential reaction mechanism

$$E + S_1 \rightleftharpoons ES_1 \qquad K_{M1} = \frac{[E][S_1]}{[ES_1]}$$

$$ES_1 + S_2 \rightleftharpoons ES_1S_2 \qquad K_{M12} = \frac{[ES_1][S_2]}{[ES_1S_2]}$$

$$ES_1S_2 \rightarrow E + P \qquad v = k_b[ES_1S_2]$$

for the two substrates S_1 and S_2. (Note that the Michaelis constants are, apart from their units, the reciprocals of the equilibrium constants for each step; we are supposing that there are two fast pre-equilibrium steps.) Alternatively, substrate binding can be random and the following steps can also lead to formation of the ES_1S_2 complex:

Sequential mechanism with random attachment

$$E + S_2 \rightleftharpoons ES_2 \qquad K_{M2} = \frac{[E][S_2]}{[ES_2]}$$

$$ES_2 + S_1 \rightleftharpoons ES_1S_2 \qquad K_{M21} = \frac{[ES_2][S_1]}{[ES_1S_2]}$$

The resulting rate law, based on the relation

$$[E]_0 = [E] + [ES_1] + [ES_2] + [ES_1S_2]$$

for the total concentration of enzyme in its bound and unbound forms, is

$$v = \frac{v_{max}[S_1]_0[S_2]_0}{K_{M1}K_{M12} + K_{M12}[S_1]_0 + K_{M12}[S_2]_0 + [S_1]_0[S_2]_0} \tag{8.5a}$$

where $v_{max} = k_b[E]_0$ and we have supposed that both S_1 and S_2 are in such excess over the enzyme concentration that they are equal to their nominal concentrations.

This equation can be rearranged into a form more suitable for plotting by holding the concentration of one substrate (S_2, for instance) constant and writing first

$$v = \frac{v_{max}[S_1]_0}{K_{M1}K_{M12}/[S_2]_0 + K_{M12}[S_1]_0/[S_2]_0 + K_{M21} + [S_1]_0}$$

and then forming the reciprocal of both sides:

$$\frac{1}{v} = \frac{1 + K_{M12}/[S_2]_0}{v_{max}} + \frac{K_{M21} + K_{M1}K_{M12}/[S_2]_0}{v_{max}} \frac{1}{[S_1]_0} \qquad \boxed{\begin{array}{l}\text{Analysis of}\\\text{sequential}\\\text{reaction}\end{array}} \tag{8.5b}$$

It follows that a plot of $1/v$ against $1/[S_1]_0$ for constant $[S_2]_0$ is linear with

$$\text{slope} = \frac{K_{M21} + K_{M1}K_{M12}/[S_2]_0}{v_{max}}$$

$$y\text{-intercept} = \frac{1 + K_{M12}/[S_2]_0}{v_{max}} \tag{8.5c}$$

(b) Ping-pong reactions

In so-called **ping-pong reactions** products are released in a stepwise fashion. In a two-substrate reaction, the first substrate (S_1) binds to the enzyme E and a product (P_1) is released, leaving the enzyme chemically modified (denoted E*), perhaps by a fragment of the substrate. Then the second substrate (S_2) binds to the modified enzyme and is processed into a second product, P_2, returning the enzyme to its native form. The scheme can be summarized as follows:

Ping-pong mechanism

$$E + S_1 \rightleftharpoons ES_1 \qquad K_{M1} = \frac{[E][S_1]}{[ES_1]}$$

$$ES_1 \rightarrow E^* + P_1 \qquad v_1 = k_{b1}[ES_1]$$

$$E^* + S_2 \rightleftharpoons E^*S_2 \qquad K_{M2} = \frac{[E^*][S_2]}{[E^*S_2]}$$

$$E^*S_2 \rightarrow E + P_2 \qquad v_2 = k_{b2}[E^*S_2]$$

Enzymes with ping-pong mechanisms include various transferases, oxido-reductases, and proteases. The intermediate E^* in the action of the protease chymotrypsin (Atlas P3), for instance, is formed by modification of a serine residue in the active site.

If we suppose that ES_1 rapidly turns into E^*, we may identify $[E^*]$ with the value of $[ES_1]$ due to the first rapid equilibrium and write

$$[E]_0 = [E] + [E^*] + [E^*S_2] = [E] + [ES_1] + [E^*S_2]$$

$$= [E] + \frac{[E][S_1]}{K_{M1}} + \frac{[E^*][S_2]}{K_{M2}} = [E] + \frac{[E][S_1]}{K_{M1}} + \frac{[ES_1][S_2]}{K_{M2}}$$

$$= \left(1 + \frac{[S_1]}{K_{M1}} + \frac{[S_1][S_2]}{K_{M1}K_{M2}}\right)[E]$$

The rate of the production of P_2 is then given by

$$v_2 = k_{b2}[E^*S_2] = \frac{k_{b2}}{K_{M2}}[E^*][S_2] = \frac{k_{b2}}{K_{M2}}[ES_1][S_2] = \frac{k_{b2}}{K_{M1}K_{M2}}[E][S_1][S_2]$$

which, with the value of $[E]$ replaced by the expression derived above, becomes

$$v_2 = \frac{v_{2max}[S_1][S_2]}{K_{M2}[S_1] + K_{M1}[S_2] + [S_1][S_2]} \tag{8.6a}$$

with $v_{2max} = k_{b2}[E]_0$. Again, we can rearrange this equation to obtain

$$\frac{1}{v_2} = \frac{1 + K_{M2}/[S_2]}{v_{2max}} + \frac{K_{M1}}{v_{2max}} \cdot \frac{1}{[S_1]} \tag{8.6b}$$

Analyzing a ping-pong mechanism

It follows that a plot of $1/v_2$ against $1/[S_1]$ for constant $[S_2]$ is linear with

$$\text{slope} = \frac{K_{M1}}{v_{2max}} \qquad y\text{-intercept} = \frac{1 + K_{M2}/[S_2]}{v_{2max}} \tag{8.6c}$$

Equations 8.5 and 8.6 form the basis of a graphical method for distinguishing between sequential and ping-pong reactions. For sequential reactions, the slope of a plot of $1/v$ against $1/[S_1]$ depends on $[S_2]$, so a series of such plots for different values of [B] form a family of nonparallel lines (Fig. 8.5a). However, for ping-pong reactions the lines described by plots of $1/v_2$ against $1/[S_1]$ for different values of $[S_2]$ are parallel because the slopes are independent of $[S_2]$ (Fig. 8.5b).

8.3 The catalytic efficiency of enzymes

To discuss the effectiveness of enzymes, it is useful to have a quantitative measure of their kinetic efficiencies for the acceleration of biochemical reactions.

The **turnover frequency**, or **catalytic constant**, of an enzyme, k_{cat}, is the number of catalytic cycles (turnovers) performed by the active site in a given interval divided by the duration of the interval. This quantity has the same units as a first-order rate constant and, in terms of the Michaelis–Menten mechanism, is numerically equivalent to k_b, the rate constant for release of product from the enzyme–substrate complex. It follows from the identification of k_{cat} with k_b and from eqn 8.2b that

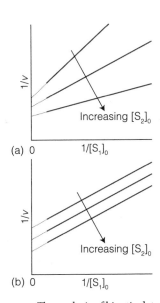

Fig. 8.5 The analysis of kinetic data for enzyme-catalyzed reactions involving two substrates. Plots of $1/v$ against $1/[S_1]_0$ for different values of $[S_2]_0$ can be used to distinguish between (a) a sequential reaction, which gives rise to a family of nonparallel lines, and (b) a 'ping-pong' reaction, which give rise to a family of parallel lines.

$$k_{cat} = k_b = \frac{v_{max}}{[E]_0}$$

Turnover frequency   (8.7)

The **catalytic efficiency**, η (eta), of an enzyme is the ratio k_{cat}/K_M. The higher the value of η, the more efficient is the enzyme. We can think of the catalytic activity as the effective rate constant of the enzymatic reaction. From $K_M = (k_a' + k_b)/k_a$ and eqn 8.7, it follows that

$$\eta = \frac{k_{cat}}{K_M} = \frac{k_a k_b}{k_a' + k_b}$$

Catalytic efficiency   (8.8)

The catalytic efficiency reaches its maximum value of k_a when $k_b \gg k_a'$. Because k_a is the rate constant for the formation of a complex from two species that are diffusing freely in solution, the maximum efficiency is related to the maximum rate of diffusion of E and S in solution (Section 7.5). This limit leads to rate constants of about 10^8–10^9 dm^3 mol^{-1} s^{-1} for molecules as large as enzymes at room temperature. The enzyme catalase has $\eta = 4.0 \times 10^8$ dm^3 mol^{-1} s^{-1} and is said to have attained 'catalytic perfection' in the sense that the rate of the reaction it catalyzes is essentially diffusion controlled: it acts as soon as a substrate makes contact.

Self-test 8.2   Calculate k_{cat} and the catalytic efficiency of carbonic anhydrase by using the data from *Example* 8.1.

Answer: $k_{cat} = 1.1 \times 10^5$ s^{-1}, $\eta = 1.1 \times 10^4$ dm^3 $mmol^{-1}$ s^{-1}

8.4 Enzyme inhibition

We now need to take the analysis a stage further to see how to accommodate reaction steps that prevent an enzyme from forming product.

An inhibitor, I, decreases the rate of product formation from the substrate by binding to the enzyme, to the ES complex, or to the enzyme and ES complex simultaneously. The most general kinetic scheme for enzyme inhibition is then

Reaction with inhibition

$E + S \rightarrow ES$ $\qquad v = k_a[E][S]$

$ES \rightarrow E + S$ $\qquad v = k_a'[ES]$

$ES \rightarrow E + P$ $\qquad v = k_b[ES]$

$EI \rightleftharpoons E + I$ $\qquad K_I = \dfrac{[E][I]}{[EI]}$

$ESI \rightleftharpoons ES + I$ $\qquad K_I' = \dfrac{[ES][I]}{[ESI]}$

The lower the values of K_I and K_I', the more efficient is the inhibition. As shown in the following *Justification*, the rate of reaction in the presence of an inhibitor is

$$v = \frac{v_{max}}{\alpha' + \alpha K_M/[S]_0}$$

Inhibited rate   (8.9a)

where

$$\alpha = 1 + \frac{[I]}{K_I} \qquad \alpha' = 1 + \frac{[I]}{K_I'}$$

(8.9b)

This equation is very similar to the Michaelis–Menten equation for the uninhibited enzyme (eqn 8.1) and is also amenable to analysis by a version of the Lineweaver–Burk plot:

$$\frac{1}{v} = \frac{\alpha'}{v_{max}} + \frac{\alpha K_M}{v_{max}} \frac{1}{[S]_0}$$

| Lineweaver–Burk plot with inhibition | (8.9c) |

Justification 8.2 *Enzyme inhibition*

By mass balance, the total concentration of enzyme is

$$[E]_0 = [E] + [EI] + [ES] + [ESI]$$

By using the definitions in eqn 8.9 and the two equilibrium constants it follows that

$$[E]_0 = \alpha[E] + \alpha'[ES]$$

Then, because $K_M = [E][S]/[ES]$ and $[S] \approx [S]_0$, we can write

$$[E]_0 = \frac{\alpha K_M[ES]}{[S]_0} + \alpha'[ES] = \left(\frac{\alpha K_M}{[S]_0} + \alpha' \right)[ES]$$

The expression for the rate of product formation is then

$$v = k_b[ES] = \frac{k_b[E]_0}{\alpha K_M/[S]_0 + \alpha'}$$

which, on rearrangement, gives eqn 8.9c.

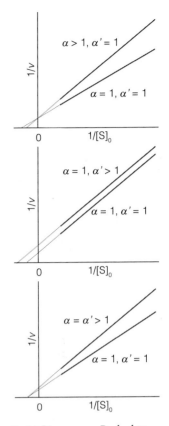

Fig. 8.6 Lineweaver–Burk plots characteristic of the three major modes of enzyme inhibition: (a) competitive inhibition, (b) uncompetitive inhibition, and (c) noncompetitive inhibition, showing the special case $\alpha = \alpha' > 1$.

There are three major modes of inhibition that give rise to distinctly different kinetic behavior (Fig. 8.6):

- **Competitive inhibition:** the inhibitor binds only to the active site of the enzyme and thereby inhibits the attachment of the substrate.

This condition corresponds to $\alpha > 1$ and $\alpha' = 1$ (because ESI does not form). The slope of the Lineweaver–Burk plot increases by a factor of α relative to the slope for data on the uninhibited enzyme ($\alpha = \alpha' = 1$). The y-intercept does not change as a result of competitive inhibition.

- **Uncompetitive inhibition:** the inhibitor binds to a site of the enzyme that is removed from the active site but only if the substrate is already present.

The inhibition occurs because ESI reduces the concentration of ES, the active type of complex. In this case $\alpha = 1$ (because EI does not form) and $\alpha' > 1$. The y-intercept of the Lineweaver–Burk plot increases by a factor of α' relative to the y-intercept for data on the uninhibited enzyme, but the slope does not change.

- **Non-competitive inhibition** (or *mixed inhibition*): the inhibitor binds to a site other than the active site, and its presence reduces the ability of the substrate to bind to the active site.

Inhibition occurs at both the E and ES sites. This condition corresponds to $\alpha > 1$ and $\alpha' > 1$. Both the slope and y-intercept of the Lineweaver–Burk plot increase on addition of the inhibitor. Figure 8.6c shows the special case of $K_I = K_I'$ and $\alpha = \alpha'$, which results in intersection of the lines at the x-axis.

A brief comment
Because several plots are
involved in the extraction of
information from the data,
it is sometimes difficult to keep
track of units. The following
lengthy Example shows in
detail how to keep track of
them in analyses of this kind
(and in general).

In all cases, the efficiency of the inhibitor may be obtained by determining K_M and v_{max} from a control experiment with uninhibited enzyme and then repeating the experiment with a known concentration of inhibitor. From the slope and y-intercept of the Lineweaver–Burk plot for the inhibited enzyme (eqn 8.9), the mode of inhibition, the values of α or α', and the values of K_I or K_I' can be obtained.

> **Example 8.2**   *Distinguishing between types of inhibition*
>
> Five solutions of a substrate, S, were prepared with the concentrations given in the first column below, and each one was divided into five equal volumes. The same concentration of enzyme was present in each one. An inhibitor, I, was then added in different concentrations to the samples, and the initial rate of formation of product was determined with the results given below. Does the inhibitor act competitively or noncompetitively? Determine K_I and K_M.

	$[I]/(\text{mmol dm}^{-3})$					
$[S]/(\text{mmol dm}^{-3})$	0	0.20	0.40	0.60	0.80	
0.050		0.033	0.026	0.021	0.018	0.016
0.10		0.055	0.045	0.038	0.033	0.029
0.20		0.083	0.071	0.062	0.055	0.050	$v/(\mu\text{mol dm}^{-3}\,\text{s}^{-1})$
0.40		0.111	0.100	0.091	0.084	0.077
0.60		0.126	0.116	0.108	0.101	0.094

Strategy We draw a series of Lineweaver–Burk plots for different inhibitor concentrations. If the plots resemble those in Fig. 8.6a, then the inhibition is competitive. On the other hand, if the plots resemble those in Fig. 8.6c, then the inhibition is noncompetitive. To find K_I, we need to determine the slope at each value of [I], which is equal to $\alpha K_M/v_{max}$, or $K_M/v_{max} + K_M[I]/K_I v_{max}$, then plot this slope against [I]: the intercept at [I] = 0 is the value of K_M/v_{max} and the slope is $K_M/K_I v_{max}$. To conform to the rule that all graphs should be plots of dimensionless quantities, we express eqn 8.9c as

$$\frac{1}{v/\mu\text{mol dm}^{-3}\,\text{s}^{-1}} = \frac{\alpha'}{v_{max}/\mu\text{mol dm}^{-3}\,\text{s}^{-1}} + \frac{\alpha K_M/\text{mmol dm}^{-3}}{v_{max}/\mu\text{mol dm}^{-3}\,\text{s}^{-1}}\frac{1}{[S]_0/\text{mmol dm}^{-3}}$$

and plot $1/(v/\mu\text{mol dm}^{-3}\,\text{s}^{-1})$ against $1/([S]_0/\text{mmol dm}^{-3})$, then the slope (which we shall call slope₁ for this first graph) is identified with

$$\text{slope}_1 = \frac{\alpha K_M/\text{mmol dm}^{-3}}{v_{max}/\mu\text{mol dm}^{-3}\,\text{s}^{-1}}, \quad \text{so} \quad \alpha = \text{slope}_1 \times \frac{v_{max}/\mu\text{mol dm}^{-3}\,\text{s}^{-1}}{K_M/\text{mmol dm}^{-3}}$$

and the y-intercept with

$$\text{intercept}_1 = \frac{\alpha'}{v_{max}/\mu\text{mol dm}^{-3}\,\text{s}^{-1}}, \quad \text{so} \quad v_{max} = \frac{\alpha'}{\text{intercept}_1}\mu\text{mol dm}^{-3}\,\text{s}^{-1}$$

For the determination of K_I and K_M we write eqn 8.9b, $\alpha = 1 + [I]/K_I$, which becomes

$$\text{slope}_1 \times \frac{v_{max}/\mu\text{mol dm}^{-3}\,\text{s}^{-1}}{K_M/\text{mmol dm}^{-3}} = 1 + \frac{[I]/\text{mmol dm}^{-3}}{K_I/\text{mmol dm}^{-3}}$$

and therefore

$$\text{slope}_1 = \frac{K_M/\text{mmol dm}^{-3}}{v_{max}/\mu\text{mol dm}^{-3}\text{ s}^{-1}} + \frac{K_M/\text{mmol dm}^{-3}}{v_{max}/\mu\text{mol dm}^{-3}\text{ s}^{-1}} \times \frac{[I]/\text{mmol dm}^{-3}}{K_I/\text{mmol dm}^{-3}}$$

$$= \frac{K_M/\text{mmol dm}^{-3}}{v_{max}/\mu\text{mol dm}^{-3}\text{ s}^{-1}} + \frac{K_M/K_I}{v_{max}/\mu\text{mol dm}^{-3}\text{ s}^{-1}} \times [I]/\text{mmol dm}^{-3}$$

We conclude that when in the second graph slope_1 is plotted against $[I]/\text{mmol dm}^{-3}$,

$$\text{slope}_2 = \frac{K_M/K_I}{v_{max}/\mu\text{mol dm}^{-3}\text{ s}^{-1}}, \quad \text{so} \quad K_M/K_I = \text{slope}_2 \times v_{max}/\mu\text{mol dm}^{-3}\text{ s}^{-1}$$

$$\text{intercept}_2 = \frac{K_M/\text{mmol dm}^{-3}}{v_{max}/\mu\text{mol dm}^{-3}\text{ s}^{-1}}, \quad \text{so}$$

$$\frac{K_M}{\text{mmol}}\text{dm}^{-3} = \text{intercept}_2 \times v_{max}/\mu\text{mol dm}^{-3}\text{ s}^{-1}$$

Solution First, we draw up a table of $1/[S]$ and $1/v$ for each value of $[I]$:

	$[I]/(\text{mmol dm}^{-3})$				
$1/([S]/(\text{mmol dm}^{-3}))$	0	0.20	0.40	0.60	0.80
20	30	38	48	56	62
10	18	22	26	30	34
5.0	12	14	16	18	20
2.5	9.01	10.0	11.0	11.9	13.0
1.7	7.94	8.62	9.26	9.90	10.6

$1/(v/(\mu\text{mol dm}^{-3}\text{ s}^{-1}))$

The five plots (one for each $[I]$) are given in Fig. 8.7. We see that they pass through the same intercept on the vertical axis, so the inhibition is competitive and we can set $\alpha' = 1$. The mean of the (least-squares) intercepts is $\text{intercept}_1 = 5.83$, so

$$v_{max} = \frac{1}{5.83}\mu\text{mol dm}^{-3}\text{ s}^{-1} = 0.17\ \mu\text{mol dm}^{-3}\text{ s}^{-1}$$

The (least-squares) slopes of the lines are as follows:

$[I]/\text{mmol dm}^{-3}$	0	0.20	0.40	0.60	0.80
slope_1	1.219	1.627	2.090	2.489	2.832

These values are plotted in Fig. 8.8. The intercept at $[I] = 0$ is $\text{intercept}_2 = 1.234$, so

$$K_M/\text{mmol dm}^{-3} = 1.234 \times 0.17 = 0.21$$

and therefore $K_M = 0.21\ \text{mmol dm}^{-3}$. The (least-squares) slope of the line is $\text{slope}_2 = 2.045$, so

$$K_M/K_I = 2.045 \times 0.17 = 0.34_8$$

It follows that

$$K_I = \frac{K_M}{2.045 \times 0.17} = \frac{0.21\ \text{mmol dm}^{-3}}{2.045 \times 0.17}$$

$$= 0.60\ \text{mmol dm}^{-3}$$

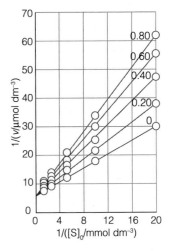

Fig. 8.7 Lineweaver–Burk plots for the data in *Example* 8.2. Each line corresponds to a different concentration of inhibitor.

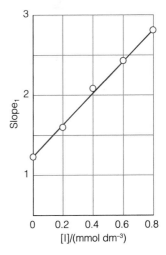

Fig. 8.8 Plot of the slopes of the plots in Fig. 8.7 against $[I]$ based on the data in *Example* 8.2.

> **Self-test 8.3** Repeat the question using the following data:
>
> $$[I]/(\text{mmol dm}^{-3})$$
>
> | $[S]/(\text{mmol dm}^{-3})$ | 0 | 0.20 | 0.40 | 0.60 | 0.80 | |
> |---|---|---|---|---|---|---|
> | 0.050 | | 0.020 | 0.015 | 0.012 | 0.0098 | 0.0084 |
> | 0.10 | | 0.035 | 0.026 | 0.021 | 0.017 | 0.015 |
> | 0.20 | | 0.056 | 0.042 | 0.033 | 0.028 | 0.024 |
> | 0.40 | | 0.080 | 0.059 | 0.047 | 0.039 | 0.034 |
> | 0.60 | | 0.093 | 0.069 | 0.055 | 0.046 | 0.039 |
>
> $v/(\mu\text{mol dm}^{-3}\,\text{s}^{-1})$
>
> **Answer:** Noncompetitive, $K_M = 0.30$ mmol dm^{-3}, $K_I = 0.57$ mmol dm^{-3}

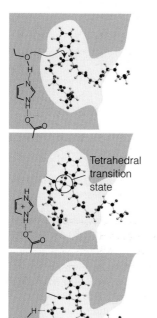

Fig. 8.9 The sequence of steps by which chymotrypsin cuts through the C–N bond of a peptide link and releases an amine.

Tetrahedral transition state

Escapes

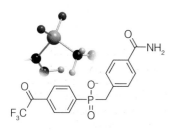

1 Phosphonate transition state analog

Case study 8.1 *The molecular basis of catalysis by hydrolytic enzymes*

One protein enzyme that has been studied in considerable detail is chymotrypsin (Atlas P3), which functions by hydrolyzing peptide bonds in polypeptides in the small intestine. The sequence of steps by which the enzyme carries out the first part of its task—to snip through the C–N bond of the peptide link—is shown in Fig. 8.9. The crucial point to notice is the formation of a tetrahedral transition state in the course of the reaction. The second sequence by which the carboxylic acid group is eliminated from the polypeptide is shown in Fig. 8.10. This step involves the attack by an H_2O molecule on the carboxyl group and the subsequent cleavage of the original C–O bond. Once again, the crucial point is the formation of a tetrahedral transition state. In each case, the catalytic activity of the enzyme can be traced to the structure of the active site, in this case featuring a *catalytic triad*, which enhances reactivity of the enzyme toward the substrate, and an *oxoanion hole*, which stabilizes the tetrahedral transition state.

The catalytic triad consists of the serine, histidine, and aspartic acid residues shown in Figs 8.9 and 8.10. There, proton transfer between the residues deprotonates serine's hydroxyl group, resulting in an alkoxide ion that is particularly reactive toward the carbonyl group of the polypeptide. In the oxoanion hole, NH groups from the peptide backbone of the enzyme are strategically placed to form hydrogen bonds with the negatively charged oxygen atom (formerly the carbonyl oxygen of the polypeptide substrate) of the tetrahedral transition state. By helping to accommodate a nascent negative charge, the oxoanion hole lowers the energy of the transition state and enhances the rate of hydrolysis.

The entities known as 'catalytic antibodies' combine the insight that studies on molecules such as chymotrypsin provide with an organism's natural defence system. In that way, they open routes to alternative enzymes for carrying out particular reactions. The key idea is that an organism generates a flood of antibodies when an antigen—a foreign body—is introduced. The organism maintains a wide range of latent antibodies, but they proliferate in the presence of the antigen. It follows that, if we can introduce an antigen that emulates the tetrahedral transition state typical of a peptide hydrolysis reaction, then an organism should produce a supply of antibodies that may be able to act as enzymes for that and related functions.

This procedure has been applied to the search for enzymes for the hydrolysis of esters. The compound used to mimic the tetrahedral transition state is a tetrahedral phosphonate (**1**). When the antibody stimulated to interact with this antigen is used to catalyze the hydrolysis of an ester, pronounced activity is indeed found, with $K_M = 1.9$ μmol dm^{-3} and an enhancement of rate over the uncatalyzed reaction by a factor of 10^3. The hope is that catalytic antibodies can be formed that catalyze reactions currently untouched by enzymes, such as those that target destruction of viruses and tumors.

Transport across biological membranes

At this stage we can begin to explore the molecular features that govern the rates of reactions. We saw in Chapter 5 that many cellular processes, such as the propagation of impulses in neurons and the synthesis of ATP by ATPases, are controlled by the transport of molecules and ions across biological membranes. **Passive transport** is the spontaneous movement of species down concentration and membrane potential gradients; **active transport** is nonspontaneous movement against these gradients and driven by the hydrolysis of ATP. Here we complement the thermodynamic treatment of Chapter 5 with a kinetic analysis that begins with a consideration of the laws governing the motion of molecules and ions in liquids and then describes modes of transport across cell membranes.

8.5 Molecular motion in liquids

Because the rate at which molecules move in solution may be a controlling factor of the maximum rate of a biochemical reaction in the intracellular medium, we need to understand the factors that limit molecular motion in a liquid.

A molecule in a liquid is surrounded by other molecules and can move only a fraction of a diameter in each step it takes, perhaps because its neighbors briefly move aside. Molecular motion in liquids is a series of short steps, with ever-changing directions, like people in an aimless, milling crowd.

The process of migration by means of a random jostling motion through a liquid is called **diffusion**. We can think of the motion of the molecule as a series of short jumps in random directions, a so-called **random walk** (Fig. 8.11). If there is an initial concentration gradient in the liquid (for instance, a solution may have a high concentration of solute in one region), then the rate at which the molecules spread out is proportional to the concentration gradient and we write

rate of diffusion ∝ concentration gradient

To express this relation mathematically, we introduce the **flux**, J, which is the number of particles passing through an imaginary window in a given time interval, divided by the area of the window and the duration of the interval:

$$J = \frac{\text{number of particles passing through window}}{\text{area of window} \times \text{time interval}} \qquad \boxed{\text{Definition of flux}} \qquad (8.10a)$$

The flux may also be expressed in terms of the amount (in moles) of molecules:

$$J = \frac{\text{amount of particles passing through window}}{\text{area of window} \times \text{time interval}} \qquad (8.10b)$$

To calculate the number or amount of molecules passing through a given window in a given time interval, we multiply the flux by the area of the window

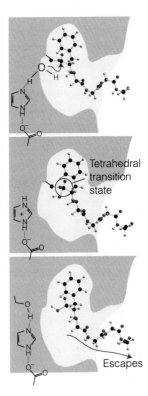

Tetrahedral transition state

Escapes

Fig. 8.10 The following sequence of steps by which chymotrypsin cuts through the C–O bond and releases a carboxylic acid.

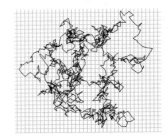

Fig. 8.11 One possible path of a random walk in three dimensions. In this general case, the step length is also a random variable. (Available at http://www.ki.inf.tu-dresden.de/~fritzke/research/TS/example1.html.)

Table 8.1 Diffusion coefficients in water, $D/(10^{-9}\,\mathrm{m^2\,s^{-1}})$

Water, H_2O*	2.26
Glycine, NH_2CH_2COOH*	1.055
Sucrose, $C_{12}H_{22}O_{11}$*	0.522
Lysozyme[†]	0.112
Serum albumin[†]	0.0594
Catalase[†]	0.0410
Fibrinogen[†]	0.0202
Bushy stunt virus[†]	0.0115

*Measured at 5°C.
[†]Measured at 20°C.

and the time interval. **Fick's first law** of diffusion (see *Further information* 8.1 for a derivation) then states:

$$J = -D\frac{dc}{dx}$$

Fick's first law (8.11)

where dc/dx is the gradient of either the number concentration (molecules m^{-3}, for instance) when the flux is in terms of numbers or the molar concentration (mol dm^{-3}, for instance) when the flux is in terms of amounts. The coefficient D, which has the dimensions of area divided by time (typically with units $m^2\,s^{-1}$), is called the **diffusion coefficient** (Table 8.1) and depends on the solute species, the solvent, and the temperature. For a given concentration gradient, large values of D correspond to rapid diffusion. The negative sign in eqn 8.11 simply means that if the concentration gradient is negative (down from left to right, Fig. 8.12), then the flux is positive (flowing from left to right).

A brief illustration

For sucrose in water at 25°C, $D = 5.22 \times 10^{-10}\,\mathrm{m^2\,s^{-1}}$. Suppose that in a region of an unstirred aqueous solution of sucrose the molar concentration gradient is $-0.10\,\mathrm{mol\,dm^{-3}\,cm^{-1}}$. Then, because $1\,\mathrm{dm} = 10^{-1}\,\mathrm{m}$ (so $1\,\mathrm{dm^{-3}} = 10^3\,\mathrm{m^{-3}}$) and $1\,\mathrm{cm} = 10^{-2}\,\mathrm{m}$ (so $1\,\mathrm{cm^{-1}} = 10^2\,\mathrm{m^{-1}}$), the flux arising from this gradient is

$$J = -(5.22 \times 10^{-10}\,\mathrm{m^2\,s^{-1}}) \times (-0.10\,\mathrm{mol\,dm^{-3}\,cm^{-1}})$$
$$= 5.22 \times 0.10 \times 10^{-10}\,\mathrm{m^2\,s^{-1}\,mol} \times (10^3\,\mathrm{m^{-3}}) \times (10^2\,\mathrm{m^{-1}})$$
$$= 5.2 \times 10^{-6}\,\mathrm{mol\,m^{-2}\,s^{-1}}$$

The amount of sucrose molecules passing through a 1.0-cm square window in 10 minutes is therefore

$$n = JA\Delta t = (5.2 \times 10^{-6}\,\mathrm{mol\,m^{-2}\,s^{-1}}) \times (1.0 \times 10^{-2}\,\mathrm{m})^2 \times (10 \times 60\,\mathrm{s})$$
$$= 3.1 \times 10^{-7}\,\mathrm{mol, or\ 0.31\ \mu mol}$$

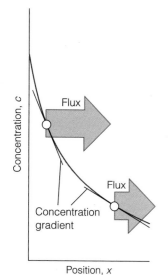

Fig. 8.12 The flux of solute particles is proportional to the concentration gradient. Here we see a solution in which the concentration falls from left to right. The gradient is negative (down from left to right) and the flux is positive (towards the right). The greatest flux is found where the gradient is steepest (at the left).

Diffusion coefficients are of the greatest importance for discussing the spread of pollutants in lakes and through the atmosphere. In both cases, the spread of pollutant may be assisted—and is normally greatly dominated—by bulk motion of the fluid as a whole (as when a wind blows in the atmosphere). This motion is called **convection**. Because diffusion is often a slow process, we speed up the spread of solute molecules by inducing convection by stirring a fluid or turning on an extractor fan.

One of the most important equations in the physical chemistry of fluids is the **diffusion equation**, which enables us to predict the rate at which the concentration of a solute changes in a nonuniform solution. In essence, the diffusion equation expresses the fact that wrinkles in the concentration tend to disperse. The formal statement of the diffusion equation, which is also known as **Fick's second law** of diffusion, is

$$\frac{\partial c}{\partial t} = D\frac{\partial^2 c}{\partial x^2}$$

Fick's second law (8.12)

Mathematical toolkit 8.1 *Partial derivatives*

When a function depends on more than one variable, such as the function $f(x,y)$, there are two first derivatives: one with respect to x with y held constant, and the other with respect to y with x held constant. These derivatives are referred to as 'partial derivatives' and denoted

$$\left(\frac{\partial f}{\partial x}\right)_y \quad \text{and} \quad \left(\frac{\partial f}{\partial y}\right)_x$$

respectively (note the 'curly d') and the variable held constant as a right-subscript (this may be omitted if there is no ambiguity). The first of these expressions is the slope of the function parallel to the x-axis and the second is the slope parallel to the y-axis (see the illustration). For instance, if $f = x^2y^3$, then

$$\left(\frac{\partial f}{\partial x}\right)_y = 2xy^3 \quad \text{and} \quad \left(\frac{\partial f}{\partial y}\right)_x = 3x^2y^2$$

Higher derivatives are defined analogously, as in

$$\left(\frac{\partial^2 f}{\partial x^2}\right)_y \quad \text{and} \quad \left(\frac{\partial^2 f}{\partial y^2}\right)_x$$

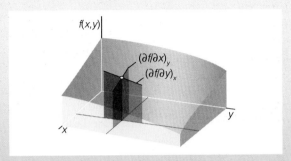

where $\partial c/\partial t$ is the rate of change of concentration in a region and $\partial^2 c/\partial x^2$ may be thought of as the curvature of the concentration in the region. Because the concentration c depends on both position and time, we have to express Fick's second law using 'partial derivative' notation (see *Mathematical toolkit 8.1*). The 'curvature' $\partial^2 c/\partial x^2$ is a measure of the wrinkliness of the concentration (see below). The derivation of this expression from Fick's first law is also given in *Further information* 8.1. The concentrations on the left and right of this equation may be either number concentrations or molar concentrations.

The diffusion equation tells us that a uniform concentration and a concentration with unvarying slope through the region (so $\partial^2 c/\partial x^2 = 0$ in each case) results in no net change in concentration in the region ($\partial c/\partial t = 0$) because the rate of influx through one wall of the region is equal to the rate of efflux through the opposite wall. Only if the slope of the concentration varies through a region— only if the concentration is wrinkled—is there a change in concentration. Where the curvature is positive (a dip, Fig. 8.13), the change in concentration is positive: the dip tends to fill. Where the curvature is negative (a heap), the change in concentration is negative: the heap tends to spread.

The diffusion coefficient increases with temperature because an increase in temperature enables a molecule to escape more easily from the attractive forces exerted by its neighbors. If we suppose that the rate of random motion follows an Arrhenius temperature dependence with an activation energy E_a, then the diffusion coefficient will follow the relation

$$D = D_0 e^{-E_a/RT} \qquad \boxed{\text{Temperature-dependence of } D} \qquad (8.13)$$

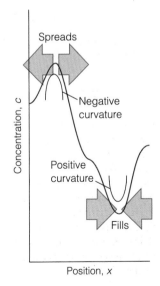

Fig. 8.13 Nature abhors a wrinkle. The diffusion equation tells us that peaks in a distribution (regions of negative curvature) spread and troughs (regions of positive curvature) fill in.

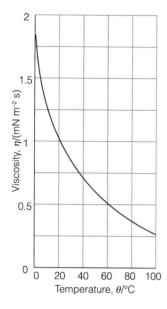

Fig. 8.14 The experimental temperature dependence of the viscosity of water. As the temperature is increased, more molecules are able to escape from the potential wells provided by their neighbors, so the liquid becomes more fluid.

The rate at which particles diffuse through a liquid is related to the viscosity, and we should expect a high diffusion coefficient to be found for fluids that have a low viscosity. That is, we can suspect that $D \propto 1/\eta$, where η (eta) is the **coefficient of viscosity**. In fact, the **Stokes–Einstein relation** states that

$$D = \frac{kT}{6\pi\eta a}$$ Stokes–Einstein relation   (8.14)

where a is the effective radius of the molecule. Figure 8.14 shows the observed temperature dependence of the viscosity of water.

Self-test 8.4   Estimate the activation energy for the diffusion of a solute molecule in water from the graph in Fig. 8.14 by using the viscosities at 40°C and 80°C. *Hint*: Use an equation such as eqn 8.13 to formulate an expression for the logarithm of the ratio of the two diffusion coefficients.

Answer: 19 kJ mol⁻¹

8.6 Molecular motion across membranes

A crucial aspect of biochemical change is the rate at which species are transported across a membrane, so we need to understand the kinetic factors that facilitate or impede transport.

Consider the passive transport of an uncharged species A across a lipid bilayer of thickness l. To simplify the problem, we assume that the concentration of A is always maintained at $[A] = [A]_0$ on one surface of the membrane and at $[A] = 0$ on the other surface, perhaps by a perfect balance between the rate of the process that produces A on one side and the rate of another process that consumes A completely on the other side. Then $\partial[A]/\partial t = 0$ because the two boundary conditions ensure that the interior of the membrane is maintained at a constant but not necessarily uniform concentration, and eqn 8.12 simplifies to

$$D\frac{d^2[A]}{dx^2} = 0$$ (8.15)

where D is the diffusion coefficient. (We can use d in place of the partial ∂ because this $[A]$ is independent of time, so x is the only variable.) We use the conditions $[A](0) = [A]_0$ and $[A](l) = 0$ to solve this differential equation and the result, which may be verified by differentiation, is

$$[A](x) = [A]_0\left(1 - \frac{x}{l}\right)$$ Concentration profile   (8.16)

which implies that $[A]$ decreases linearly inside the membrane. We now use Fick's first law to calculate the flux J of A through the membrane. From eqn 8.16, it follows that

$$\frac{d[A]}{dx} = -\frac{[A]_0}{l}$$

and from this result and eqn 8.11 obtain

$$J = D\frac{[A]_0}{l}$$ (8.17)

Before using this simple result we need to take into account the fact that the concentration of A on the surface of a membrane is not always equal to its concentration measured in the bulk solution, which we assume to be aqueous. This difference arises from the significant difference in the solubility of A in an aqueous environment and in the solution–membrane interface. One way to deal with this problem is to define a **partition coefficient** κ (kappa) as

$$\kappa = \frac{[A]_0}{[A]_s} \qquad \text{Definition of partition coefficient} \qquad (8.18)$$

where $[A]_s$ is the molar concentration of A in the bulk aqueous solution. It follows that

$$J = \kappa D \frac{[A]_s}{l} \qquad \text{Diffusion flux} \qquad (8.19)$$

We see, as intuition would suggest, that the flux is high when the concentration of A in the bulk solution is high and the membrane is thin.

In spite of the assumptions that led to its final form, eqn 8.19 describes adequately the passive transport of many nonelectrolytes through membranes of blood cells. In many cases, however, eqn 8.19 underestimates the flux, which suggests that the membrane is more permeable than expected. However, because the permeability increases only for certain species, we can infer that in these cases, transport is facilitated by carrier molecules. One example is the transporter protein that carries glucose into cells. But we issue a word of caution: there is little justification for supposing that D in the membrane is equal to its value in aqueous solution or that κ has any particular value, and the conclusion that facilitated transport is involved needs additional evidence before it can be accepted.

To treat facilitated transport we suppose that a characteristic of a carrier C is that it binds to the transported species A and that the dissociation of the AC complex is described by

$$AC \rightleftharpoons A + C \qquad K = \frac{[A][C]}{[AC]c^\ominus} \qquad (8.20a)$$

where we have used concentrations instead of activities (and $c^\ominus = 1$ mol dm^{-3}, the standard molar concentration). After writing $[C]_0 = [C] + [AC]$, where $[C]_0$ is the total concentration of carrier, it follows that

$$[AC] = \frac{[A][C]_0}{[A] + Kc^\ominus} \qquad (8.20b)$$

Then the flux through the membrane of the species AC is given by a version of eqn 8.19 as

$$J = \kappa_{AC} D_{AC} \frac{[AC]}{l} = \frac{\kappa_{AC} D_{AC}[C]_0}{l} \frac{[A]}{[A] + Kc^\ominus} = J_{max} \frac{[A]}{[A] + Kc^\ominus} \qquad \text{Mediated flux} \qquad (8.21)$$

where κ_{AC} and D_{AC} are the partition coefficient and diffusion coefficient of the species AC, respectively and $J_{max} = \kappa_{AC} D_{AC}[C]_0/l$. We see from Fig. 8.15 that:

- when $[A] \ll Kc^\ominus$, $J = J_{max}[A]$ and the flux is proportional to $[A]$
- when $[A] \gg Kc^\ominus$, $J = J_{max}$ and the flux has its maximum value.

This behavior is characteristic of mediated transport.

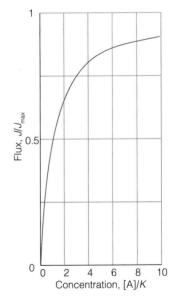

Fig. 8.15 The flux of the species AC through a membrane varies with the concentration of the species A. The behavior shown in the figure and explained in the text is characteristic of mediated transport of A, with C as a carrier molecule.

8.7 The mobility of ions

Ion transport through membranes is central to the operation of many biological processes, particularly signal transduction in neurons, and we need to be equipped to describe ion migration quantitatively.

An ion in solution responds to the presence of an electric field, migrates through the solution, and carries charge from one location to another. The study of the motion of ions down a potential gradient gives an indication of their size, the effect of solvation, and details of the type of motion they undergo.

When an ion is subjected to an electric field $\mathcal{E}$, it accelerates. However, the faster it travels through the solution, the greater the retarding force it experiences from the viscosity of the medium. As a result, as we show in the following *Justification*, the ion settles down into a limiting velocity called its **drift velocity**, s, which is proportional to the strength of the applied field:

$$s = u\mathcal{E}$$

Definition of mobility   (8.22)

The **mobility**, u, depends on the radius, a, and charge number, z, of the ion and the viscosity, η, of the solution:

$$u = \frac{ez}{6\pi\eta a}$$

Relation of mobility to size and viscosity   (8.23)

Justification 8.3   *The ionic mobility*

An ion of charge ze in an electric field $\mathcal{E}$ (typically in volts per meter, V m^{-1}) experiences a force of magnitude $ze\mathcal{E}$, which accelerates it. However, the ion experiences a frictional force due to its motion through the medium, and that retarding force increases the faster the ion travels. The viscous drag (the retarding force), F, on a spherical particle of radius a traveling at a speed s is given by Stokes' law:

$$F = 6\pi\eta as$$

When the particle has reached its drift speed, the accelerating and viscous retarding forces are equal, so we can write

$$ze\mathcal{E} = 6\pi\eta as$$

and solve this expression for s:

$$s = \frac{ez\mathcal{E}}{6\pi\eta a}$$

At this point we can compare this expression for the drift speed with eqn 8.22 and hence find the expression for mobility given in eqn 8.23.

Equation 8.23 tells us that the mobility of an ion is high if it is highly charged, is small, and is in a solution with low viscosity. These features appear to contradict the trends in Table 8.2, which lists the mobilities of a number of ions. For instance, the mobilities of the Group 1 cations *increase* down the group despite their increasing radii (Section 9.12). The explanation is that the radius to use in eqn 8.23 is the **hydrodynamic radius**, the *effective* radius for the migration of the ions taking into account the entire object that moves. When an ion migrates, it carries

Table 8.2 Ionic mobilities in water at 298 K, $u/(10^{-8}\,m^2\,s^{-1}\,V^{-1})$

Cations		Anions	
H^+ (H_3O^+)	36.23	OH^-	20.64
Li^+	4.01	F^-	5.74
Na^+	5.19	Cl^-	7.92
K^+	7.62	Br^-	8.09
Rb^+	8.06	I^-	7.96
Cs^+	8.00	CO_3^{2-}	7.46
Mg^{2+}	5.50	NO_3^-	7.41
Ca^{2+}	6.17	SO_4^{2-}	8.29
Sr^{2+}	6.16		
NH_4^+	7.62		
$[N(CH_3)_4]^+$	4.65		
$[N(CH_2CH_3)_4]^+$	3.38		

its hydrating water molecules with it, and as small ions are more extensively hydrated than large ions (because they give rise to a stronger electric field in their vicinity), ions of small radius actually have a large hydrodynamic radius. Thus, hydrodynamic radius decreases down Group 1 because the extent of hydration decreases with increasing ionic radius.

One significant deviation from this trend is the very high mobility of the proton in water. It is believed that this high mobility reflects an entirely different mechanism for conduction, the **Grotthus mechanism**, in which the proton on one H_2O molecule migrates to its neighbors, the proton on that H_2O molecule migrates to its neighbors, and so on along a chain (Fig. 8.16). The motion is therefore an *effective* motion of a proton, not the actual motion of a single proton.

| **In the laboratory 8.1** | *Electrophoresis* |

An important application of the preceding material is to the determination of the molar mass of biological macromolecules. **Electrophoresis** is the motion of a charged species, such as DNA and ionic forms of amino acids, in response to an electric field. Electrophoretic mobility is a result of a constant drift speed, so the mobility of a macromolecule in an electric field depends on its net charge, size (and hence molar mass), and shape.

Electrophoresis is a very valuable tool for the separation of biopolymers from complex mixtures, such as those resulting from fractionation of biological cells. We shall consider several strategies controlling the drift speeds of biomolecules in order to achieve separation of a mixture into its components.

In **gel electrophoresis**, migration takes place through a slab of a porous gel, a semi-rigid dispersion of a solid in a liquid. Because the molecules must pass through the pores in the gel, the larger the macromolecule, the less mobile it is in the electric field and, conversely, the smaller the macromolecule, the more swiftly it moves through the pores. In this way, gel electrophoresis allows for

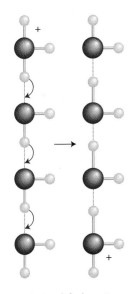

Fig. 8.16 A simplified version of the Grotthus mechanism of proton conduction through water. The proton entering the chain at the top is not the same as the proton leaving the chain at the bottom.

the separation of components of a mixture according to their molar masses. Two common gel materials for the study of proteins and nucleic acids are agarose and cross-linked polyacrylamide. Agarose has large pores and is better suited for the study of large macromolecules, such as DNA and enzyme complexes. Polyacrylamide gels with varying pore sizes can be made by changing the concentration of acrylamide in the polymerization solution. In general, smaller pores form as the concentration of acrylamide is increased, making possible the separation of relatively small macromolecules by **polyacrylamide gel electrophoresis** (PAGE).

The separation of very large pieces of DNA, such as chromosomes, by conventional gel electrophoresis is not effective, making the analysis of genomic material rather difficult. Double-stranded DNA molecules are thin enough to pass through gel pores, but long and flexible DNA coils can become trapped in the pores and the result is impaired mobility along the direction of the applied electric field. This problem can be avoided with **pulsed-field electrophoresis**, in which a brief burst of the electric field is applied first along one direction and then along a perpendicular direction. In response to the switching back and forth between field directions, the DNA coils writhe about and eventually pass through the gel pores. In this way, the mobility of the macromolecule can be related to its molar mass.

We have seen that charge also determines the drift speed. For example, proteins of the same size but different net charge travel along the slab at different speeds. One way to avoid this problem and to achieve separation by molar mass is to denature the proteins in a controlled way. Sodium dodecyl sulfate is an anionic detergent that is very useful in this respect: it denatures proteins, whatever their initial shapes, into rods by forming a complex with them. Moreover, most protein molecules bind a constant number of ions, so the net charge per protein is well regulated. Under these conditions, different proteins in a mixture may be separated according to size only. The molar mass of each constituent protein is estimated by comparing its mobility in its rod-like complex form with a standard sample of known molar mass. However, molar masses obtained by this method, often referred to as **SDS-PAGE** when polyacrylamide gels are used, are not as accurate as those obtained by the sophisticated techniques discussed in Chapter 11.

Another technique that deals with the effect of charge on drift speed takes advantage of the fact that the overall charge of proteins and other biopolymers depends on the pH of the medium. For instance, in acidic environments protons attach to basic groups and the net charge is positive; in basic media the net charge is negative as a result of proton loss. At the **isoelectric point**, the pH is such that there is no net charge on the biopolymer. Consequently, the drift speed of a biopolymer depends on the pH of the medium, with $s = 0$ at the isoelectric point (see Example 8.3 and Fig. 8.17). **Isoelectric focusing** is an electrophoresis method that exploits the dependence of drift speed on pH. In this technique, a mixture of proteins is dispersed in a medium with a pH gradient along the direction of an applied electric field. Each protein in the mixture will stop moving at a position in the gradient where the pH is equal to the isoelectric point. In this manner, the protein mixture can be separated into its components.

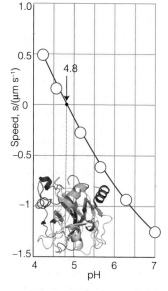

Fig. 8.17 The plot of the speed of a moving macromolecule against pH allows the isoelectric point to be detected as the pH at which the speed is zero. The data are from *Example* 8.3.

Example 8.3   *The isoelectric point of a protein*

The speed with which bovine serum albumin (BSA) moves through water under the influence of an electric field was monitored at several values of pH, and the data are listed below. What is the isoelectric point of the protein?

pH	4.20	4.56	5.20	5.65	6.30	7.00
Velocity/(μm s^{-1})	0.50	0.18	−0.25	−0.65	−0.90	−1.25

Strategy If we plot speed against pH, we can use interpolation to find the pH at which the speed is zero, which is the pH at which the molecule has zero net charge.

Solution The data are plotted in Fig. 8.17. The velocity passes through zero at pH = 4.8; hence pH = 4.8 is the isoelectric point.

Self-test 8.5   The following data were obtained for another protein:

pH	4.5	5.0	5.5	6.0
Velocity/(μm s^{-1})	−0.10	−0.20	−0.30	−0.35

Estimate the pH of the isoelectric point.

Answer: 4.1

The separation of complicated mixtures of macromolecules may be difficult by SDS-PAGE or isoelectric focusing alone. However, the two techniques can be combined in **two-dimensional (2D) electrophoresis**. In a typical experiment, a protein mixture is separated first by isoelectric focusing, yielding a pattern of bands in a gel slab such as the one shown in Fig. 8.18a. To improve the separation of closely spaced bands, the first slab is attached to a second slab and SDS-PAGE is performed with the electric field being applied in a direction that is perpendicular to the direction in which isoelectric focusing was performed. The macromolecules separate according to their molar masses along this second dimension of the experiment, and the result is that spots are spread widely over the surface of the slab, leading to enhanced separation of the mixture's components (Fig. 8.18b).

The techniques described so far give good separations, but the drift speeds attained by macromolecules in traditional electrophoresis methods are rather low; as a result, several hours are often necessary to achieve good separation of complex mixtures. According to eqn 8.22, one way to increase the drift speed is to increase the electric field strength. However, there are limits to this strategy because very large electric fields can heat the large surfaces of an electrophoresis apparatus unevenly, leading to a nonuniform distribution of electrophoretic mobilities and poor separation.

In **capillary electrophoresis**, the sample is dispersed in a medium (such as methylcellulose) and held in a thin glass or plastic tube with diameters ranging from 20 to 100 μm. The small size of the apparatus makes it easy to dissipate heat when large electric fields are applied. Excellent separations may be achieved in minutes rather than hours.

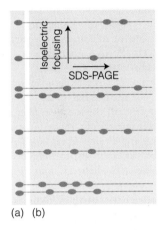

(a) (b)

Fig. 8.18 The experimental steps taken during separation of a mixture of biopolymers by two-dimensional electrophoresis. (a) Isoelectric focusing is performed on a thin gel slab, resulting in separation along the vertical direction of the illustration. (b) The first slab is attached to a second, larger slab and SDS-PAGE is performed with the electric field oriented in the horizontal direction of the illustration, resulting in further separation by molar mass. The dashed horizontal lines show how the bands in the two-dimensional gel correspond to the bands in the gel on which isoelectric focusing was performed.

8.8 Transport across ion channels and ion pumps

We now have enough background information about ion transport to consider the centrally important processes of ion transport mediated by ion channels and ion pumps, which are involved in the propagation of action potentials and the synthesis of ATP.

The thermodynamic treatment of ion transport in Chapter 5 does not explain the fact that ion channels and pumps discriminate between ions. For example, it is found experimentally that a K^+ ion channel is not permeable to Na^+ ions. We shall see that the key to the selectivity of an ion channel or pump lies in the mechanism of transport and, consequently, in the structure of the protein and the size of the ion.

The structures of a number of channel proteins have been obtained by the X-ray diffraction techniques that will be described in greater detail in Chapter 12. Information about the flow of ions across channels and pumps is supplied by the **patch clamp technique**. One of many possible experimental arrangements is shown in Fig. 8.19. With mild suction, a 'patch' of membrane from a whole cell or a small section of a broken cell can be attached tightly to the tip of a micropipette filled with an electrolyte solution and containing an electronic conductor, the *patch electrode*. A potential difference (the 'clamp') is applied between the patch electrode and an intracellular electronic conductor in contact with the cytosol of the cell. If the membrane is permeable to ions at the applied potential difference, a current flows through the completed circuit. Using narrow micropipette tips with diameters of less than 1 μm, ion currents of a few picoamperes (1 pA = 10^{-12} A) have been measured across sections of membranes containing only one ion channel protein.

(a) The potassium channel

A detailed picture of the mechanism of action of ion channels has emerged from analysis of patch clamp data and structural data. Here we focus on the K^+ ion

Fig. 8.19 A representation of the patch clamp technique for the measurement of ionic currents through membranes in intact cells. (a) A section of membrane containing an ion channel is in tight contact with the tip of a micropipette containing an electrolyte solution and the patch electrode. (b) A schematic representation of the cross-section of a membrane-spanning K^+ ion channel and (c) the protein. (d) The selectivity filter has a number of carbonyl groups that grip K^+ ions. As explained in the text, electrostatic repulsions between two bound K^+ ions encourage ionic movement through the selectivity filter and across the membrane.

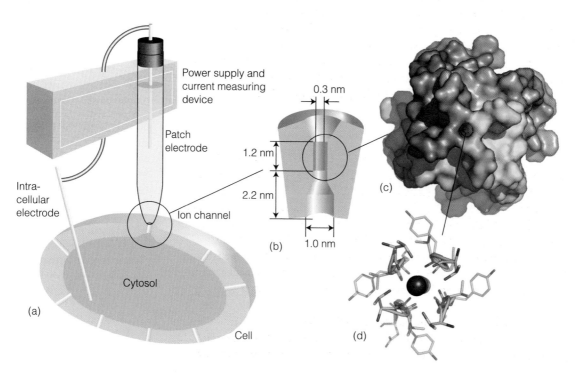

channel protein, which, like all other mediators of ion transport, spans the membrane bilayer. The pore through which ions move has a length of 3.4 nm and is divided into two regions: a wide region with a length of 2.2 nm and diameter of 1.0 nm, and a narrow region with a length of 1.2 nm and diameter of 0.3 nm. The narrow region is called the *selectivity filter* of the K^+ ion channel because it allows only K^+ ions to pass.

Filtering is a subtle process that depends on ionic size and the thermodynamic tendency of an ion to lose its hydrating water molecules. On entering the selectivity filter, the K^+ ion is stripped of its hydrating shell and is then gripped by carbonyl groups of the protein. Dehydration of the K^+ ion is endergonic ($\Delta_{dehyd}G^{\ominus} = +203$ kJ mol^{-1}) but is driven by the energy of interaction between the ion and the protein. The Na^+ ion, although smaller than the K^+ ion, does not pass through the selectivity filter of the K^+ ion channel because interactions with the protein are not sufficient to compensate for the high Gibbs energy of dehydration of Na^+ ($\Delta_{dehyd}G^{\ominus} = +301$ kJ mol^{-1}). More specifically, a dehydrated Na^+ ion is too small and cannot be held tightly by the protein carbonyl groups, which are positioned for ideal interactions with the larger K^+ ion. In its hydrated form, the Na^+ ion is too large (larger than a dehydrated K^+ ion), does not fit in the selectivity filter, and does not cross the membrane.

Although very selective, a K^+ ion channel can still let other ions pass through. For example, K^+ and Tl^+ ions have similar radii and Gibbs energies of dehydration, so Tl^+ can cross the membrane. As a result, Tl^+ is a neurotoxin because it replaces K^+ in many neuronal functions and suppresses them.

The efficiency of transfer of K^+ ions through the channel can also be explained by structural features of the protein. For efficient transport to occur, a K^+ ion must enter the protein but then must not be allowed to remain inside for very long, so that as one K^+ ion enters the channel from one side, another K^+ ion leaves from the opposite side. An ion is lured into the channel by water molecules about halfway through the length of the membrane. Consequently, the thermodynamic cost of moving an ion from an aqueous environment to the less hydrophilic interior of the protein is minimized. The ion is encouraged to leave the protein by electrostatic interactions in the selectivity filter, which can bind two K^+ ions simultaneously, usually with a bridging water molecule. Electrostatic repulsion prevents the ions from binding too tightly, minimizing the residence time of an ion in the selectivity filter and maximizing the transport rate.

(b) The proton pump

Now we turn our attention to a very important ion pump, the H^+-ATPase responsible for coupling of proton flow to synthesis of ATP from ADP and P_i (Chapter 4). Structural studies show that the channel through which the protons flow is linked in tandem to a unit composed of six protein molecules arranged in pairs of α and β subunits to form three interlocked $\alpha\beta$ segments (Fig. 8.20). The conformations of the three pairs may be loose, (L), tight (T), or open (O), and one of each type is present at each stage. A protein at the centre of the interlocked structure, the subunit shown as an arrow, rotates and induces structural changes that cycle each of the three segments between L, T, and O conformations.

At the start of a cycle, a T unit holds an ATP molecule. Then ADP and a P_i group migrate into the L site, and as it closes into T, the earlier T site opens into O and releases its ATP. The ADP and P_i in the T site meanwhile condense into ATP,

Fig. 8.20 The mechanism of action of H⁺-ATPase, a molecular motor that transports protons across the mitochondrial membrane and catalyzes either the formation or hydrolysis of ATP. The yellow shapes represent the species ADP, ATP, and P_i.

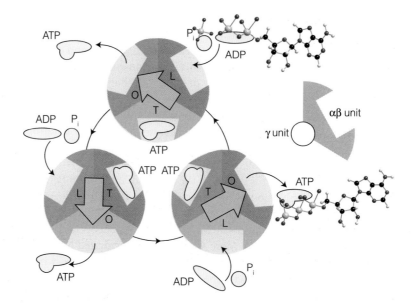

and the new L site is ready for the cycle to begin again. The proton flux drives the rotation of a γ subunit, and hence the conformational changes of the αβ segments, as well as providing the energy for the condensation reaction itself.

Several key aspects of this mechanism have been confirmed experimentally. For example, the rotation of a γ subunit has been portrayed directly by using single-molecule spectroscopy (*In the laboratory* 12.6).

Electron transfer in biological systems

We saw in *Case studies* 4.2 and 4.3 that exergonic electron transfer processes drive the synthesis of ATP in the mitochondrion during oxidative phosphorylation. Electron transfer between protein-bound co-factors or between proteins also plays a role in other biological processes, such as photosynthesis (Section 5.11 and *Case study* 12.3), nitrogen fixation, the reduction of atmospheric N_2 to NH_3 by certain microorganisms, and the mechanisms of action of oxidoreductases, which are enzymes that catalyze redox reactions.

We begin by examining the features of a theory that describes the factors governing the rates of electron transfer. Then we discuss the theory in the light of experimental results on a variety of systems, including protein complexes. We shall see that relatively simple expressions can be used to predict the rates of electron transfer between proteins with reasonable accuracy.

8.9 The rates of electron transfer processes

Electron transfer is of crucial importance in many biological reactions, and we need to see how to use the strategies we have developed to discuss them quantitatively.

Consider electron transfer from a donor species D to an acceptor species A in solution. The net reaction, the observed rate law, and the equilibrium constant are

$$D + A \rightarrow D^+ + A^- \quad v = k_{obs}[D][A] \quad K = \frac{[D^+][A^-]}{[D][A]} \qquad \text{Electron transfer} \quad (8.24)$$

The proposed mechanism is:

Electron transfer

$$D + A \rightleftharpoons DA \qquad k_a, k_a' \qquad K_{DA} = \frac{k_a c^{\ominus}}{k_a'} = \frac{[DA]c^{\ominus}}{[D][A]}$$

$$DA \rightarrow D^+A^- \qquad \nu_{et} = k_{et}[DA]$$

$$D^+A^- \rightarrow DA \qquad \nu_{ret} = k_{et}'[D^+A^-]$$

$$D^+A^- \rightarrow D^+ + A^- \qquad \nu_d = k_d[D^+A^-]$$

In the first step of the mechanism, D and A must diffuse through the solution and encounter to form a complex DA, in which the donor and acceptor are separated by a distance comparable to r, the distance between the edges of each species. Next, electron transfer occurs within the DA complex to yield D^+A^-. The D^+A^- complex has two possible fates. One is the regeneration of DA. The other is to break apart and for the ions to diffuse through the solution. We show in the following *Justification* that k_{obs} in eqn 8.24 is given by

$$\frac{1}{k_{obs}} = \frac{1}{k_a} + \frac{k_a'}{k_a k_{et}}\left(1 + \frac{k_{et}'}{k_d}\right) \qquad \boxed{\begin{array}{l}\text{Electron transfer}\\\text{rate constant}\end{array}} \qquad (8.25)$$

Justification 8.4 *The rate constant for electron transfer in solution*

We begin by equating the rate of the net reaction (eqn 8.24) to the rate of formation of separated ions, the reaction products:

$$\nu = k_{obs}[D][A] = k_d[D^+A^-]$$

Next, we apply the steady-state assumption to the intermediate D^+A^-:

$$\frac{d[D^+A^-]}{dt} = k_{et}[DA] - k_{et}'[D^+A^-] - k_d[D^+A^-] = 0$$

It follows that

$$[D^+A^-] = \frac{k_{et}}{k_{et}' + k_d}[DA]$$

However, DA is also an intermediate, so we apply the steady-state approximation again:

$$\frac{d[DA]}{dt} = k_a[D][A] - k_a'[DA] - k_{et}[DA] + k_{et}'[D^+A^-] = 0$$

Substitution of the initial expression for the steady-state concentration of D^+A^- into this expression for [DA] gives, after some algebra, a new expression for $[D^+A^-]$:

$$[D^+A^-] = \frac{k_a k_{et}}{k_a'k_{et}' + k_a'k_d + k_d k_{et}}[D][A]$$

When we multiply this expression by k_d, we see that the resulting equation has the form of the rate of electron transfer, $\nu = k_{obs}[D][A]$, with k_{obs} given by

$$k_{obs} = \frac{k_d k_a k_{et}}{k_a'k_{et}' + k_a'k_d + k_d k_{et}}$$

To obtain eqn 8.25, we divide the numerator and denominator on the right-hand side of this expression by $k_d k_{et}$ and solve for the reciprocal of k_{obs}.

To gain insight into eqn 8.25 and the factors that determine the rate of electron transfer reactions in solution, we assume that the main decay route for D^+A^- is dissociation of the complex into separated ions, or $k_d \gg k'_{et}$. It then follows that

$$\frac{1}{k_{obs}} \approx \frac{1}{k_a}\left(1 + \frac{k'_a}{k_{et}}\right)$$

(8.26)

There are two limits to consider:

- When $k_{et} \gg k'_a$, $k_{obs} \approx k_a$ and the rate of product formation is controlled by diffusion of D and A in solution, which fosters formation of the DA complex.
- When, $k_{et} \ll k'_a$, $k_{obs} \approx (k_a/k'_a)k_{et}$ and the process is controlled by the activation energy of electron transfer in the DA complex.

When the electron donor and acceptor are anchored at fixed distances within a single protein, the diffusion of D to A and their separation play no role and only k_{et} needs to be considered when calculating the rate of electron transfer. In terms of transition state theory (Section 7.7), we write

$$k_{et} = \kappa \frac{kT}{h} e^{-\Delta^\ddagger G/RT}$$

(8.27)

where κ is the transmission coefficient and $\Delta^\ddagger G$ is the Gibbs energy of activation. Cytochrome c oxidase is an example of a system where such intraprotein electron transfer is important. In that enzyme, bound copper ions and heme groups work together to reduce O_2 to water in the final step of respiration.

8.10 The theory of electron transfer processes

To gain insight into the rate constants for electron transfer, we need to know the factors that control their values and interpret them in terms of the specific arrangement of redox partners.

Our next task is to describe the **Marcus theory** of electron transfer, which gives clues about the factors that control the rate constant k_{et} for unimolecular electron transfer within the DA complex.[2] To do so, we examine the $\kappa(kT/h)$ term in eqn 8.27.

We saw in Chapter 7 that the transmission coefficient κ takes into account the fact that the activated complex does not always pass through to the transition state and the term kT/h arises from consideration of motions that lead to the decay of the activated complex into products. It follows that, in the case of an electron transfer process, $\kappa(kT/h)$ can be thought of as a measure of the probability that an electron will move from D to A in the transition state. The theory due to R.A. Marcus supposes that this probability decreases with increasing distance between D and A in the DA complex. More specifically, for given values of the temperature and $\Delta^\ddagger G$, the rate constant k_{et} varies with the edge-to-edge distance r as[3]

$$k_{et} \propto e^{-\beta r} \text{ (at constant } T \text{ and } \Delta^\ddagger G)$$

(8.28)

where β is a constant with a value that depends on the medium through which the electron must travel from donor to acceptor.

---

[2] The development of modern electron transfer theory began with independent work by R.A. Marcus, N.S. Hush, V.G. Levich, and R.R. Dogonadze between 1956 and 1959. Marcus received the Nobel Prize for chemistry in 1992 for his seminal contributions in this area.

[3] For a mathematical treatment of Marcus theory, see our *Physical chemistry* (2010).

In considering the factors that determine the value of the Gibbs energy of activation, Marcus noted that the DA complex and the medium surrounding it must rearrange spatially as charge is redistributed to form the ions D^+ and A^-. These molecular rearrangements include the relative reorientation of the D and A molecules in DA and the relative reorientation of the solvent molecules surrounding DA. The resulting expression for the Gibbs energy of activation is

$$\Delta^\ddagger G = \frac{(\Delta_r G^\ominus + \lambda)^2}{4\lambda} \qquad \boxed{\text{Marcus expression for the Gibbs energy of activation}} \qquad (8.29)$$

where $\Delta_r G^\ominus$ is the standard reaction Gibbs energy for the electron transfer process $DA \rightarrow D^+A^-$ and λ is the **reorganization energy**, the energy change associated with molecular rearrangements that must take place so that DA can take on the equilibrium geometry of D^+A^-. Equation 8.29 shows that when $\Delta_r G^\ominus = -\lambda$, corresponding to the cancelation of the reorganization energy term by the standard reaction Gibbs energy, then $\Delta^\ddagger G = 0$, with the implication that the reaction is not slowed down by an activation barrier.

Taken together, eqns 8.27 and 8.28 suggest that the expression for k_{et} has the form

$$k_{et} \propto e^{-\beta r} e^{-\Delta^\ddagger G / RT} \qquad \boxed{\text{Marcus expression for the rate constant of electron transfer}} \qquad (8.30)$$

where $\Delta^\ddagger G$ is given by eqn 8.29. In summary, Marcus theory predicts that k_{et} depends on

- The distance between the donor and acceptor, with electron transfer becoming more efficient as the distance between donor and acceptor decreases.

- The standard reaction Gibbs energy, $\Delta_r G^\ominus$, with electron transfer becoming more efficient as $\Delta_r G^\ominus$ becomes more negative. For example, kinetically efficient oxidation of D requires that its standard reduction potential be lower than the standard reduction potential of A.

- The reorganization energy, with electron transfer becoming more efficient as the reorganization energy is matched closely by the standard reaction Gibbs energy.

8.11 Experimental tests of the theory

Many of the key features of Marcus theory have been tested by experiments, showing in particular the predicted dependence of k_{et} on the standard reaction Gibbs energy and the edge-to-edge distance between electron donor and acceptor.

It is difficult to measure the distance dependence of k_{et} when the reactants are ions or molecules that are free to move in solution. In such cases, electron transfer occurs after a donor–acceptor complex forms and it is not possible to exert control over r, the edge-to-edge distance. The most meaningful experimental tests of the dependence of k_{et} on r are those in which the same donor and acceptor are positioned at a variety of distances, perhaps by covalent attachment to molecular linkers. Under these conditions, the term $e^{-\Delta^\ddagger G / RT}$ becomes a constant and, after taking the natural logarithm of eqn 8.30, we obtain

$$\ln k_{et} = -\beta r + \text{constant} \qquad (8.31)$$

which implies that a plot of $\ln k_{et}$ against r should be a straight line with slope $-\beta$. It is found experimentally that in a vacuum, $28 \text{ nm}^{-1} < \beta < 35 \text{ nm}^{-1}$, whereas

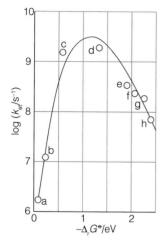

Fig. 8.21 Variation of log k_{et} with $-\Delta_r G^{\ominus}$ for a series of compounds with the structures given in (**2**). Kinetic measurements were conducted in 2-methyltetrahydrofuran at 296 K. The distance between the donor (the reduced biphenyl group) and the acceptor is constant for all compounds in the series because the molecular linker remains the same. Each acceptor has a characteristic standard potential, so it follows that the standard Gibbs energy for the electron transfer process is different for each compound in the series. The line is a fit to a version of eqn 8.32; the maximum of the parabola occurs at $-\Delta_r G^{\ominus} = \lambda = 1.2$ eV $= 1.2 \times 10^2$ kJ mol^{-1}. (Reproduced with permission from J.R. Miller, L.T. Calcaterra, and G.L. Closs, *J. Am. Chem. Soc.* **106**, 3047 (1984).)

$\beta \approx 9$ nm^{-1} when the intervening medium is a molecular link between donor and acceptor. Electron transfer between protein-bound cofactors can occur at distances of up to about 2.0 nm, a long distance on a molecular scale, corresponding to about 20 carbon atoms, with the protein providing an intervening medium between donor and acceptor.

There is, however, a great deal of controversy surrounding the interpretation of electron transfer data in proteins. Much of the available data may be interpreted with $\beta \approx 14$ nm^{-1}, a value that appears to be insensitive to the primary and secondary structures of the protein but does depend slightly on the density of atoms in the section of protein that separates donor from acceptor. More detailed work on the specific effect of secondary structure suggests that 12.5 nm$^{-1} < \beta < 16.0$ nm^{-1} when the intervening medium consists primarily of α helices and 9.0 nm$^{-1} < \beta < 11.5$ nm^{-1} when the medium is primarily β sheet. Yet another view suggests that the electron takes specific paths through covalent bonds and hydrogen bonds that exist in the protein for the purpose of optimizing the rate of electron transfer.

The dependence of k_{et} on the standard reaction Gibbs energy has been investigated in systems where the edge-to-edge distance and the reorganization energy are constant for a series of reactions. Then eqn 8.30 becomes

$$\ln k_{et} = \frac{RT}{4\lambda}\left(\frac{\Delta_r G^{\ominus}}{RT}\right)^2 - \frac{1}{2}\left(\frac{\Delta_r G^{\ominus}}{RT}\right) + \text{constant} \qquad (8.32)$$

and a plot of $\ln k_{et}$ (or log k_{et}) against $\Delta_r G^{\ominus}$ (or $-\Delta_r G^{\ominus}$) is predicted to be shaped like a downward parabola. Equation 8.32 implies that the rate constant increases as $\Delta_r G^{\ominus}$ decreases but only up to $-\Delta_r G^{\ominus} = \lambda$. Beyond that, the reaction enters the inverted region, in which the rate constant decreases as $\Delta_r G^{\ominus}$ becomes more negative. Figure 8.21 shows that the inverted region has been observed in compounds such as (**2**), in which the electron donor and acceptor are linked covalently to a molecular spacer of known and fixed size.

A = (a)   (b)   (c) A.

(d)   (e)   (f) R$_1$ = R$_2$ = H; (g) R$_1$ = H, R$_2$ = Cl; (h) R$_1$ = R$_2$ = Cl

2 An electron donor–acceptor complex

8.12 The Marcus cross-relation

Because electron transfer reactions are of such importance for metabolism and other biological processes, to discuss them quantitatively we need to be able to predict their rate constants: Marcus theory provides a way.

It follows from eqns 8.26 and 8.27 that the rate constant k_{obs} may be written as

$$k_{obs} = Z\, e^{-\Delta^{\ddagger}G/RT} \qquad (8.33)$$

where $Z = K_{DA}\kappa(kT/h)$. It is difficult to estimate k_{obs} because we often lack knowledge of β, λ, and κ. However, as we show in the following *Justification*, when $\lambda \gg |\Delta_r G^{\ominus}|$, k_{obs} may be estimated by a special case of the **Marcus cross-relation**:

$$k_{obs} = (k_{DD}k_{AA}K)^{1/2} \qquad \boxed{\text{Marcus cross-relation}} \quad (8.34)$$

where K is the equilibrium constant for the net electron transfer reaction (eqn 8.24) and k_{DD} and k_{AA} (in general, k_{ii}) are the experimental rate constants for the electron self-exchange processes (with the colors distinguishing one molecule from another):

$$D + D^+ \rightarrow D^+ + D \qquad k_{DD}$$
$$A^- + A \rightarrow A + A^- \qquad k_{AA}$$

Justification 8.5 *The Marcus cross-relation*

To derive the Marcus cross-relation, we use eqn 8.33 to write the rate constants for the self-exchange reactions as

$$k_{DD} = Z_{DD}\, e^{-\Delta^{\ddagger}G_{DD}/RT} \qquad k_{AA} = Z_{AA}\, e^{-\Delta^{\ddagger}G_{AA}/RT}$$

For the net reaction and the self-exchange reactions, the Gibbs energy of activation may be written from eqn 8.29 as

$$\Delta^{\ddagger}G = \frac{\Delta_r G^{\ominus 2}}{4\lambda} + \tfrac{1}{2}\Delta_r G^{\ominus} + \tfrac{1}{4}\lambda$$

For the self-exchange reactions $\Delta_r G_{DD}^{\ominus} = \Delta_r G_{AA}^{\ominus} = 0$ and hence $\Delta^{\ddagger}G_{DD} = \tfrac{1}{4}\lambda_{DD}$ and $\Delta^{\ddagger}G_{AA} = \tfrac{1}{4}\lambda_{AA}$. It follows that

$$k_{DD} = Z_{DD}\, e^{-\lambda_{DD}/4RT} \qquad k_{AA} = Z_{AA}\, e^{-\lambda_{AA}/4RT}$$

To make further progress, Marcus assumed that the reorganization energy of the net reaction is the arithmetic mean of the reorganization energies of the self-exchange reactions:

$$\lambda = \tfrac{1}{2}(\lambda_{DD} + \lambda_{AA})$$

Provided $\lambda \gg \Delta_r G^{\ominus}$ for the net reaction, the first term in $\Delta^{\ddagger}G$ may be neglected and the Gibbs energy of activation of the net reaction is

$$\Delta^{\ddagger}G = \tfrac{1}{2}\Delta_r G^{\ominus} + \tfrac{1}{8}\lambda_{DD} + \tfrac{1}{8}\lambda_{AA}$$

Therefore, the rate constant for the net reaction is

$$k_{obs} = Z e^{-\Delta_r G^{\ominus}/2RT} e^{-\lambda_{DD}/8RT} e^{-\lambda_{AA}/8RT}$$

We can use eqn 4.10 ($\ln K = -\Delta_r G^{\ominus}/RT$) in the form $K = e^{-\Delta_r G^{\ominus}/RT}$ to write

$$k_{obs} = (k_{DD}k_{AA}K)^{1/2}f$$

where

$$f = \frac{Z}{(Z_{AA}Z_{DD})^{1/2}}$$

In practice, the factor f is usually set to 1 and we obtain eqn 8.34.

The rate constants estimated by eqn 8.34 agree fairly well with experimental rate constants for electron transfer between proteins, as we see in the following example.

Example 8.4   *Using the Marcus cross-relation*

The following data were obtained for cytochrome c and cytochrome c_{551}, two proteins in which heme-bound iron ions shuttle between the oxidation states Fe(II) and Fe(III):

	$k_{ii}/(dm^3\,mol^{-1}\,s^{-1})$	$E^\ominus/V$
cytochrome c	1.5×10^2	+0.260
cytochrome c_{551}	4.6×10^7	+0.286

Estimate the rate constant k_{obs} for the process

cytochrome c_{551}(red) + cytochrome c(ox)
$\rightarrow$ cytochrome c_{551}(ox) + cytochrome c(red)

Then compare the estimated value with the observed value of $6.7 \times 10^4\,dm^3\,mol^{-1}\,s^{-1}$.

Strategy We use the standard potentials and eqns 5.16 ($\ln K = vFE^\ominus_{cell}/RT$) and 5.17a ($E^\ominus_{cell} = E^\ominus_R - E^\ominus_L$) to calculate the equilibrium constant K. Then we use eqn 8.34, the calculated value of K, and the self-exchange rate constants k_{ii} to calculate the rate constant k_{obs}.

Solution The two reduction half-reactions are

Right: cytochrome c(ox) + e^- $\rightarrow$ cytochrome c(red) $E^\ominus_R = +0.260$ V

Left: cytochrome c_{551}(ox) + e^- $\rightarrow$ cytochrome c_{551}(red) $E^\ominus_L = +0.286$ V

The difference is

$$E^\ominus_{cell} = (0.260\,V) - (0.286\,V) = -0.026\,V$$

It then follows from eqn 5.16 with $v = 1$ and $RT/F = 25.69$ mV that

$$\ln K = -\frac{0.026V}{25.69 \times 10^{-3}\,V} = -\frac{2.6}{2.569}$$

Therefore, $K = 0.36$. From eqn 8.34 and the self-exchange rate constants, we calculate

$$k_{obs} = \{(1.5 \times 10^2\,dm^3\,mol^{-1}\,s^{-1}) \times (4.6 \times 10^7\,dm^3\,mol^{-1}\,s^{-1}) \times 0.36\}^{1/2}$$
$$= 5.0 \times 10^4\,dm^3\,mol^{-1}\,s^{-1}$$

The calculated and observed values differ by only 25 per cent, indicating that the Marcus relation can lead to reasonable estimates of rate constants for electron transfer.

Self-test 8.6   Estimate k_{obs} for the reduction by cytochrome c of plastocyanin, a protein containing a copper ion that shuttles between the +2 and +1 oxidation states and for which $k_{AA} = 6.6 \times 10^2\,dm^3\,mol^{-1}\,s^{-1}$ and $E^\ominus_{cell} = +0.350$ V.

Answer: $1.8 \times 10^3\,dm^3\,mol^{-1}\,s^{-1}$

Checklist of key concepts

☐ 1. Catalysts are substances that accelerate reactions but undergo no net chemical change.

☐ 2. A homogeneous catalyst is a catalyst in the same phase as the reaction mixture.

☐ 3. Enzymes are homogeneous, biological catalysts.

☐ 4. The Michaelis–Menten mechanism of enzyme kinetics accounts for the dependence of rate on the concentration of the substrate.

☐ 5. A Lineweaver–Burk plot is used to determine the parameters that occur in the Michaelis–Menten mechanism.

☐ 6. In sequential reactions, the active site binds all the substrates before processing them into products. In 'ping-pong' reactions, products are released in a stepwise fashion.

☐ 7. In competitive inhibition of an enzyme, the inhibitor binds only to the active site of the enzyme and thereby inhibits the attachment of the substrate.

☐ 8. In uncompetitive inhibition, the inhibitor binds to a site of the enzyme that is removed from the active site but only if the substrate is already present.

☐ 9. In noncompetitive inhibition, the inhibitor binds to a site other than the active site, and its presence reduces the ability of the substrate to bind to the active site.

☐ 10. Fick's first law of diffusion states that the flux of molecules is proportional to the concentration gradient.

☐ 11. Fick's second law of diffusion (the diffusion equation) states that the rate of change of concentration in a region is proportional to the curvature of the concentration in the region.

☐ 12. Diffusion is an activated process.

☐ 13. The flux of molecules through biological membranes is often mediated by carrier molecules.

☐ 14. Protons migrate by the Grotthus mechanism, Fig. 8.16.

☐ 15. Electrophoresis is the motion of a charged macromolecule, such as DNA, in response to an electric field. Important techniques are gel electrophoresis, isoelectric focusing, pulsed-field electrophoresis, two-dimensional electrophoresis, and capillary electrophoresis.

☐ 16. According to the Marcus theory, the rate constant of electron transfer in a donor–acceptor complex depends on the distance between electron donor and acceptor, the standard reaction Gibbs energy, and the reorganization energy, λ.

Checklist of key equations

Property	Equation	Comment
Michaelis–Menten rate law	$v = v_{max}[S]_0/([S]_0 + K_M)$	$v_{max} = k_b[E]_0$; assumes S is in excess
Lineweaver–Burk plot	$1/v = 1/v_{max} + (K_M/v_{max})(1/[S]_0)$	Based on Michaelis–Menten mechanism
Fick's first law	$J = -Ddc/dx$	
Fick's second law	$\partial c/\partial t = D\partial^2 c/\partial x^2$	Also known as the diffusion equation
Temperature-dependence of D	$D = D_0 e^{-E_a/RT}$	
Mobility of an ion	$u = ez/6\pi\eta a$	Assumes the validity of Stokes' law
Marcus expression	$k_{et} \propto e^{-\beta r}e^{-\Delta^\ddagger G/RT}$	
	with	
	$\Delta^\ddagger G = (\Delta_r G^\circ + \lambda)^2/4\lambda$	
Marcus cross-relation	$k_{obs} = (k_{DD}k_{AA}K)^{1/2}$	

Further information

Further information 8.1 *Fick's laws of diffusion*

1. Fick's first law of diffusion

Consider the arrangement in Fig. 8.22. In an interval Δt the number of molecules passing through the window of area A from the left is proportional to the number in the slab of thickness l and area A, and therefore volume lA, just to the left of the window where the average (number) concentration is $c(x - \frac{1}{2}l)$, and to the length of the interval Δt:

$$\text{number coming from left} \propto c(x - \tfrac{1}{2}l)lA\Delta t$$

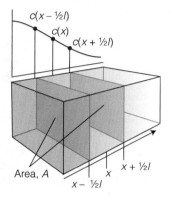

Fig. 8.22 The calculation of the rate of diffusion considers the net flux of molecules through a plane of area A as a result of arrivals from on average a distance $\frac{1}{2}l$ in each direction.

Likewise, the number coming from the right in the same interval is

$$\text{number coming from right} \propto c(x + \tfrac{1}{2}l)lA\Delta t$$

The net flux is therefore proportional to the difference in these numbers divided by the area and the time interval:

$$J \propto \frac{c(x - \frac{1}{2}l)lA\Delta t - c(x + \frac{1}{2}l)lA\Delta t}{A\Delta t} = \{c(x - \tfrac{1}{2}l) - c(x + \tfrac{1}{2}l)\}l$$

We now express the two concentrations in terms of the concentration at the window itself, $c(x)$, as follows:

$$c(x + \tfrac{1}{2}l) = c(x) + \tfrac{1}{2}l \times \frac{dc}{dx}$$

$$c(x - \tfrac{1}{2}l) = c(x) - \tfrac{1}{2}l \times \frac{dc}{dx}$$

From which it follows that

$$J \propto \left\{ \left(c(x) - \tfrac{1}{2}l \frac{dc}{dx} \right) - \left(c(x) + \tfrac{1}{2}l \frac{dc}{dx} \right) \right\} l$$

$$\propto -l^2 \frac{dc}{dx}$$

On writing the constant of proportionality as D (and absorbing l^2 into it), we obtain eqn 8.11.

2. Fick's second law

Consider the arrangement in Fig. 8.23. The number of solute particles passing through the window of area A located at x in an infinitesimal interval dt is $J(x)A dt$, where $J(x)$ is the flux at the location x. The number of particles passing out of the region through a window of area A at $x + dx$ is $J(x + dx)A dt$, where $J(x + dx)$ is the flux at the location of this window. The flux in and the flux out will be different if the concentration gradients are different at the two windows. The net change in the number of solute particles in the region between the two windows is

$$\text{net change in number} = J(x)A dt - J(x + dx)A dt$$
$$= \{J(x) - J(x + dx)\}A dt$$

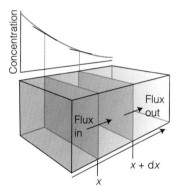

Fig. 8.23 To calculate the change in concentration in the region between the two walls, we need to consider the net effect of the influx of particles from the left and their efflux toward the right. Only if the slope of the concentrations is different at the two walls will there be a net change.

Now we express the flux at $x + dx$ in terms of the flux at x and the gradient of the flux, dJ/dx:

$$J(x + dx) = J(x) + \frac{dJ}{dx} \times dx$$

It follows that

$$\text{net change in number} = -\frac{dJ}{dx} \times dx\, A dt$$

The change in concentration inside the region between the two windows is the net change in number divided by the volume of the region (which is $A dx$), and the net rate of change is obtained by dividing that change in concentration by the time interval dt. Therefore, on dividing by both $A dx$ and dt, we obtain rate of change of concentration $\left(=\dfrac{dc}{dt}\right) = -\dfrac{dJ}{dx}$

Finally, we express the flux by using Fick's first law (at this point we need to acknowledge that c depends on both x and t and therefore use partial differential notation):

$$\frac{\partial c}{\partial t} = -\frac{\partial}{\partial x}\left(-D\frac{\partial c}{\partial x}\right) = D\frac{\partial^2 c}{\partial x^2}$$

which is eqn 8.12.

Discussion questions

8.1 Discuss the features and limitations of the Michaelis–Menten mechanism of enzyme action.

8.2 Prepare a report on the application of the experimental strategies described in Chapters 6 and 7 to the study of enzyme-catalyzed reactions. Devote some attention to the following topics: (a) the determination of reaction rates over a long time scale, (b) the determination of the rate constants and equilibrium constant of binding of substrate to an enzyme, and (c) the characterization of intermediates in a catalytic cycle. Your report should be similar in content and extent to one of the *Case studies* found throughout this text.

8.3 A plot of the rate of an enzyme-catalyzed reaction against temperature has a maximum, in an apparent deviation from the behavior predicted by the Arrhenius relation (eqn 6.19). Provide a molecular interpretation for this effect.

8.4 Describe graphical procedures for distinguishing between (a) sequential and ping-pong enzyme-catalyzed reactions and (b) competitive, uncompetitive, and noncompetitive inhibition of an enzyme.

8.5 Some enzymes are inhibited by high concentrations of their own products. (a) Sketch a plot of reaction rate against concentration of substrate for an enzyme that is prone to product inhibition. (b) How does product inhibition of hexokinase, the enzyme that phosphorylates glucose in the first step of glycolysis, provide a mechanism for regulation of glycolysis in the cell? *Hint*: Review *Case study* 4.3.

8.6 Provide a molecular interpretation for the observation that mediated transport through biological membranes leads to a maximum flux J_{max} when the concentration of the transported species becomes very large.

8.7 Discuss the mechanism of proton conduction in liquid water. For a more detailed account of the modern version of this mechanism, consult our *Quanta, matter, and change* (2009).

8.8 Discuss how the following factors determine the rate of electron transfer in biological systems: (a) the distance between electron donor and acceptor, and (b) the reorganization energy of redox active species and the surrounding medium.

Exercises

8.9 As remarked in the text, Michaelis and Menten derived their rate law by assuming a rapid pre-equilibrium of E, S, and ES. Derive the rate law in this manner, and identify the conditions under which it becomes the same as that based on the steady-state approximation (eqn 8.1).

8.10 Equation 8.4a gives the expression for the rate of formation of product by a modified version of the Michaelis–Menten mechanism in which the second step is also reversible. Derive the expression and find its limiting behavior for large and small concentrations of substrate.

8.11 For many enzymes, such as chymotrypsin (*Case study* 8.1), the mechanism of action involves the formation of two intermediates:

$E + S \rightarrow ES$ $v = k_a[E][S]$

$ES \rightarrow E + S$ $v = k_a'[ES]$

$ES \rightarrow ES'$ $v = k_b[ES]$

$ES' \rightarrow E + P$ $v = k_c[ES']$

Show that the rate of formation of product has the same form as that shown in eqn 8.1, written as:

$$v = \frac{v_{max}}{1 + K_M/[S]_0}$$

but with v_{max} and K_M given by

$$v_{max} = \frac{k_b k_c [E]_0}{k_b + k_c} \qquad \text{and} \qquad K_M = \frac{k_c(k_a' + k_b)}{k_a(k_b + k_c)}$$

8.12 The enzyme-catalyzed conversion of a substrate at 25°C has a Michaelis constant of 0.045 mol dm⁻³. The rate of the reaction is 1.15 mmol dm⁻³ s⁻¹ when the substrate concentration is 0.110 mol dm⁻³. What is the maximum velocity of this reaction?

8.13 Find the condition for which the reaction rate of an enzyme-catalyzed reaction that follows Michaelis–Menten kinetics is half its maximum value.

8.14 Isocitrate lyase catalyzes the following reaction:

isocitrate ion $\rightarrow$ glyoxylate ion + succinate ion

The rate, v, of the reaction was measured when various concentrations of isocitrate ion were present, and the following results were obtained at 25°C:

[isocitrate]/(μmol dm⁻³)	31.8	46.4	59.3	118.5	222.2
v/(pmol dm⁻³ s⁻¹)	70.0	97.2	116.7	159.2	194.5

Determine the Michaelis constant and the maximum velocity of the reaction.

8.15 The following results were obtained for the action of an ATPase on ATP at 20°C, when the concentration of the ATPase was 20 nmol dm^{-3}:

[ATP]/(μmol dm^{-3})	0.60	0.80	1.4	2.0	3.0
v/(μmol dm^{-3} s^{-1})	0.81	0.97	1.30	1.47	1.69

Determine the Michaelis constant, the maximum velocity of the reaction, the turnover number, and the catalytic efficiency of the enzyme.

8.16 Enzyme-catalyzed reactions are sometimes analyzed by use of the *Eadie–Hofstee plot*, in which $v/[S]_0$ is plotted against v. (a) Using the simple Michaelis–Menten mechanism, derive a relation between $v/[S]_0$ and v. (b) Discuss how the values of K_M and v_{max} are obtained from analysis of the Eadie–Hofstee plot. (c) Determine the Michaelis constant and the maximum velocity of the reaction from Exercise 8.14 by using an Eadie–Hofstee plot to analyze the data.

8.17 Enzyme-catalyzed reactions are sometimes analyzed by use of the *Hanes plot*, in which $[S]_0/v$ is plotted against $[S]_0$. (a) Using the simple Michaelis–Menten mechanism, derive a relation between $[S]_0/v$ and $[S]_0$. (b) Discuss how the values of K_M and v_{max} are obtained from analysis of the Hanes plot. (c) Determine the Michaelis constant and the maximum velocity of the reaction from *Exercise* 8.14 by using a Hanes plot to analyze the data.

8.18 An *allosteric enzyme* shows catalytic activity that changes on noncovalent binding of small molecules called *effectors*. For example, consider a protein enzyme consisting of several identical subunits and several active sites. In one mode of allosteric behavior, the substrate acts as effector, so that binding of a substrate molecule to one of the subunits either increases or decreases the catalytic efficiency of the other active sites. Consequently, reactions catalyzed by allosteric enzymes show significant deviations from Michaelis–Menten behaviour. (a) Sketch a plot of reaction rate against substrate concentration for a multi-subunit allosteric enzyme, assuming that the catalytic efficiency changes in such a way that the enzyme with all its active sites occupied is more efficient than the enzyme with one fewer bound substrate molecule, and so on. Compare your sketch with Fig. 8.2, which illustrates Michaelis–Menten behavior. (b) Your plot from part (a) should have a sigmoidal shape (S shape) that is typical for allosteric enzymes. The mechanism of the reaction can be written as

$$E + nS \rightleftharpoons ES_n \rightarrow E + nP$$

and the reaction rate v is given by

$$v = \frac{v_{max}}{1 + K'/[S]_0^n}$$

where K' is a collection of rate constants analogous to the Michaelis constant and n is the *interaction coefficient*, which may be taken as the number of active sites that interact to give allosteric behavior. Plot v/v_{max} against $[S]_0$ for a fixed value of K' of your choosing and several values of n. Confirm that the expression for v does predict sigmoidal kinetics and provide a molecular interpretation for the effect of n on the shape of the curve.

8.19 (a) Show that the expression for the rate of a reaction catalyzed by an allosteric enzyme of the type discussed in Exercise 8.18 may be rewritten as

$$\log \frac{v}{v_{max} - v} = n \log [S]_0 - \log K'$$

(b) Use the preceding expression and the following data to determine the interaction coefficient for an enzyme-catalyzed reaction showing sigmoidal kinetics:

[S]$_0$/(10^{-5} mol dm^{-3})	0.10	0.40	0.50
v/(μmol dm^{-3} s^{-1})	0.0040	0.25	0.46

[S]$_0$/(10^{-5} mol dm^{-3})	0.60	0.80	1.0
v/(μmol dm^{-3} s^{-1})	0.75	1.42	2.08

[S]$_0$/(10^{-5} mol dm^{-3})	1.5	2.0	3.0
v/(μmol dm^{-3} s^{-1})	3.22	3.70	4.02

For substrate concentrations ranging between 0.10 mmol dm^{-3} and 10 mmol dm^{-3}, the reaction rate remained constant at 4.17 μmol dm^{-3} s^{-1}.

8.20 A simple method for the determination of the interaction coefficient n for an enzyme-catalyzed reaction involves the calculation of the ratio $[S]_{90}/[S]_{10}$, where $[S]_{90}$ and $[S]_{10}$ are the concentrations of substrate for which the reaction rates are $0.90v_{max}$ and $0.10v_{max}$, respectively. (a) Show that $[S]_{90}/[S]_{10} = 81$ for an enzyme-catalyzed reaction that follows Michaelis–Menten kinetics. (b) Show that $[S]_{90}/[S]_{10} = (81)^{1/n}$ for an enzyme-catalyzed reaction that follows sigmoidal kinetics, where n is the interaction coefficient defined in *Exercise* 8.19. (c) Use the data from Exercise 8.19 to estimate the value of n.

8.21 Yeast alcohol dehydrogenase catalyzes the oxidation of ethanol by NAD$^+$ according to the reaction

$$CH_3CH_2OH(aq) + NAD^+(aq) \rightarrow CH_3CHO(aq) + NADH(aq) + H^+(aq)$$

The following results were obtained for the reaction:

[CH$_3$CH$_2$OH]$_0$/(10^{-2} mol dm^{-3})		1.0	2.0	4.0	20.0
v/(mol s^{-1} (kg protein)$^{-1}$)	(a)	0.30	0.44	0.57	0.76
v/(mol s^{-1} (kg protein)$^{-1}$)	(b)	0.51	0.75	0.99	1.31
v/(mol s^{-1} (kg protein)$^{-1}$)	(c)	0.89	1.32	1.72	2.29
v/(mol s^{-1} (kg protein)$^{-1}$)	(d)	1.43	2.11	2.76	3.67

where the concentrations of NAD$^+$ are (a) 0.050 mmol dm^{-3}, (b) 0.10 mmol dm^{-3}, (c) 0.25 mmol dm^{-3}, and (d) 1.0 mmol dm^{-3}. Is the reaction sequential or ping-pong? Determine v_{max} and the appropriate K constants for the reaction.

8.22 One of the key events in the transmission of chemical messages in the brain is the hydrolysis of the neurotransmitter acetylcholine by the enzyme acetylcholinesterase. The kinetic parameters for this reaction are $k_{cat} = 1.4 \times 10^4$ s^{-1} and $K_M = 9.0 \times 10^{-5}$ mol dm^{-3}. Is acetylcholinesterase catalytically perfect?

8.23 The enzyme carboxypeptidase catalyses the hydrolysis of polypeptides, and here we consider its inhibition. The following results were obtained when the rate of the enzymolysis of carbobenzoxy-glycyl-D-phenylalanine (CBGP) was monitored without inhibitor:

[CBGP]$_0$/(10^{-2} mol dm^{-3})	1.25	3.84	5.81	7.13
Relative reaction rate	0.398	0.669	0.859	1.000

(All rates in this exercise were measured with the same concentration of enzyme and are relative to the rate measured when [CBGP]$_0$ = 0.0713 mol dm^{-3} in the absence of inhibitor.) When 2.0 mmol dm^{-3} phenylbutyrate ion was added to a solution containing the enzyme and substrate, the following results were obtained:

[CBGP]$_0$/(10^{-2} mol dm^{-3})	1.25	2.50	4.00	5.50
Relative reaction rate	0.172	0.301	0.344	0.548

In a separate experiment, the effect of 50 mmol dm^{-3} benzoate ion was monitored and the results were

[CBGP]$_0$/(10^{-2} mol dm^{-3})	1.75	2.50	5.00	10.00
Relative reaction rate	0.183	0.201	0.231	0.246

Determine the mode of inhibition of carboxypeptidase by the phenylbutyrate ion and benzoate ion.

8.24 Consider an enzyme-catalyzed reaction that follows Michaelis–Menten kinetics with $K_M = 3.0$ mmol dm^{-3}. What concentration of a competitive inhibitor characterized by $K_I = 20$ μmol dm^{-3} will reduce the rate of formation of product by 50 per cent when the substrate concentration is held at 0.10 mmol dm^{-3}?

8.25 Some enzymes are inhibited by high concentrations of their own substrates. (a) Show that when substrate inhibition is important, the reaction rate v is given by

$$v = \frac{v_{max}}{1 + K_M/[S]_0 + [S]_0/K_I}$$

where K_I is the equilibrium constant for dissociation of the inhibited enzyme–substrate complex. (b) What effect does substrate inhibition have on a plot of $1/v$ against $1/[S]_0$?

8.26 What is (a) the flux of nutrient molecules down a concentration gradient of 0.10 mol dm^{-3} m^{-1}, (b) the amount of molecules (in moles) passing through an area of 5.0 mm^2 in 1.0 min? Take for the diffusion coefficient the value for sucrose in water (5.22 × 10^{-10} m^2 s^{-1}).

8.27 How long does it take a sucrose molecule in water at 25°C to diffuse (a) 1 mm, (b) 1 cm, and (c) 1 m from its starting point?

8.28 The mobility of species through fluids is of the greatest importance for nutritional processes. (a) Estimate the diffusion coefficient for a molecule that steps 150 pm each 1.8 ps. (b) What would be the diffusion coefficient if the molecule traveled only half as far on each step?

8.29 The diffusion coefficient of a particular kind of t-RNA molecule is $D = 1.0 \times 10^{-11}$ m^2 s^{-1} in the medium of a cell interior at 37°C. How long does it take molecules produced in the cell nucleus to reach the walls of the cell at a distance 1.0 μm, corresponding to the radius of the cell?

8.30 The diffusion coefficients for a lipid in a plasma membrane and in a lipid bilayer are 1.0×10^{-10} m^2 s^{-1} and 1.0×10^{-9} m^2 s^{-1}, respectively. How long will it take the lipid to diffuse 10 nm in a plasma membrane and a lipid bilayer?

8.31 Diffusion coefficients of proteins are often used as a measure of molar mass. For a spherical protein, $D \propto M^{-1/2}$. Considering only one-dimensional diffusion, compare the length of time it would take ribonuclease ($M = 13.683$ kg mol^{-1}) to diffuse 10 nm to the length of time it would take the enzyme catalase ($M = 250$ kg mol^{-1}) to diffuse the same distance.

8.32 Is diffusion important in lakes? How long would it take a small pollutant molecule about the size of H_2O to diffuse across a lake of width 100 m?

8.33 Pollutants spread through the environment by convection (winds and currents) and by diffusion. How many steps must a molecule take to be 1000 step lengths away from its origin if it undergoes a one-dimensional random walk?

8.34 The viscosity of water at 20°C is 1.0019×10^{-3} kg m^{-1} s^{-1} and at 30°C it is 7.982×10^{-4} kg m^{-1} s^{-1}. What is the activation energy for the motion of water molecules?

8.35 The mobility of a Na$^+$ ion in aqueous solution is 5.19×10^{-8} m^2 s^{-1} V^{-1} at 25°C. The potential difference between two electrodes placed in the solution is 12.0 V. If the electrodes are 1.00 cm apart, what is the drift speed of the ion? Use $\eta = 8.91 \times 10^{-4}$ kg m^{-1} s^{-1}.

8.36 It is possible to estimate the isoelectric point of a protein from its primary sequence. (a) A molecule of calf thymus histone contains one aspartic acid, one glutamic acid, 11 lysine, 15 arginine, and two histidine residues. Will the protein bear a net charge at pH = 7? If so, will the net charge be positive or negative? Is the isoelectric point of the protein less than, equal to, or greater than 7? *Hint*: See Exercise 4.45. (b) Each molecule of egg albumin has 51 acidic residues (aspartic and glutamic acid), 15 arginine, 20 lysine, and seven histidine residues. Is the isoelectric point of the protein less than, equal to, or greater than 7? (c) Can a mixture of calf thymus histone and egg albumin be separated by gel electrophoresis with the isoelectric focusing method?

8.37 We saw in Section 8.8 that to pass through a channel, the ion must first lose its hydrating water molecules. To explore the motion of hydrated Na$^+$ ions, we need to know that the diffusion coefficient D of an ion is related to its mobility u by the *Einstein relation*:

$$D = \frac{uRT}{zF}$$

where z is the charge number of the ion and F is Faraday's constant. (a) Estimate the diffusion coefficient and the effective hydrodynamic radius a of the Na$^+$ ion in water at 25°C. For water, $\eta = 8.91 \times 10^{-4}$ kg m^{-1} s^{-1}. (b) Estimate the approximate number of water molecules that are dragged along by the cations. Ionic radii are given in Table 9.3.

8.38 For a pair of electron donor and acceptor, $k_{et} = 2.02 \times 10^5$ s^{-1} for $\Delta_r G^\ominus = -0.665$ eV. The standard reaction Gibbs energy changes to $\Delta_r G^\ominus = -0.975$ eV when a substituent is added to the electron acceptor and the rate constant for electron transfer changes to $k_{et} = 3.33 \times 10^6$ s^{-1}. Assuming that the distance between donor and acceptor is the same in both experiments, estimate the value of the reorganization energy.

8.39 For a pair of electron donor and acceptor, $k_{et} = 2.02 \times 10^5$ s^{-1} when $r = 1.11$ nm and $k_{et} = 2.8 \times 10^4$ s^{-1} when $r = 1.23$ nm. (a) Assuming that $\Delta_r G^\ominus$ and λ are the same in both experiments, estimate the value of β. (b) Estimate the value of k_{et} when $r = 1.48$ nm.

8.40 Azurin is a protein containing a copper ion that shuttles between the +2 and +1 oxidation states, and cytochrome c is a protein in which a heme-bound iron ion shuttles between the +3 and +2 oxidation states. The rate constant for electron transfer from reduced azurin to oxidized cytochrome c is 1.6×10^3 dm^3 mol^{-1} s^{-1}. Estimate the electron self-exchange rate constant for azurin from the following data:

	$k_{ii}/(\text{dm}^3 \text{ mol}^{-1} \text{ s}^{-1})$	$E_{cell}^\ominus/\text{V}$
Cytochrome c	1.5×10^2	0.260
Azurin	?	0.304

Projects

8.41 Autocatalysis is the catalysis of a reaction by the products. For example, for a reaction A $\rightarrow$ P it can be found that the rate law is $v = k[A][P]$ and the reaction rate is proportional to the concentration of P. The reaction gets started because there are usually other reaction routes for the formation of some P initially, which then takes part in the autocatalytic reaction proper. Many biological and biochemical processes involve autocatalytic steps, and here we explore one case: the spread of infectious diseases.

(a) Integrate the rate equation for an autocatalytic reaction of the form A $\rightarrow$ P, with rate law $v = k[A][P]$, and show that

$$\frac{[P]}{[P]_0} = (1 + b)\frac{e^{at}}{1 + be^{at}}$$

where $a = ([A]_0 + [P]_0)k$ and $b = [P]_0/[A]_0$. *Hint:* Starting with the expression $v = -d[A]/dt = k[A][P]$, write $[A] = [A]_0 - x$, $[P] = [P]_0 + x$ and then write the expression for the rate of change of either species in terms of x. To integrate the resulting expression, the following relation will be useful:

$$\frac{1}{([A]_0 - x)([P]_0 + x)} = \frac{1}{[A]_0 + [P]_0}\left(\frac{1}{[A]_0 - x} + \frac{1}{[P]_0 + x}\right)$$

(b) Plot $[P]/[P]_0$ against at for several values of b. Discuss the effect of autocatalysis on the shape of a plot of $[P]/[P]_0$ against t by comparing your results with those for a first-order process, in which $[P]/[P]_0 = 1 - e^{-kt}$.

(c) Show that for the autocatalytic process discussed in parts (a) and (b), the reaction rate reaches a maximum at $t_{max} = -(1/a)\ln b$.

(d) In the so-called SIR model of the spread and decline of infectious diseases, the population is divided into three classes: the susceptibles, S, who can catch the disease, the infectives, I, who have the disease and can transmit it, and the removed class, R, who have either had the disease and recovered, are dead, are immune, or are isolated. The model mechanism for this process implies the following rate laws:

$$\frac{dS}{dt} = -rSI \qquad \frac{dI}{dt} = rSI - aI \qquad \frac{dR}{dt} = aI$$

(i) What are the autocatalytic steps of this mechanism?

(ii) Find the conditions on the ratio a/r that decide whether the disease will spread (an epidemic) or die out.

(iii) Show that a constant population is built into this system, namely that $S + I + R = N$, meaning that the timescales of births, deaths by other causes, and migration are assumed large compared to that of the spread of the disease.

8.42 In general, the catalytic efficiency of an enzyme depends on the pH of the medium in which it operates. One way to account for this behavior is to propose that the enzyme and the enzyme–substrate complex are active only in specific protonation states. This proposition can be summarized by the following mechanism:

$$EH + S \rightleftharpoons ESH \qquad k_a, k_a'$$

$$ESH \rightarrow E + P \qquad k_b$$

$$EH \rightleftharpoons E^- + H^+ \qquad K_{E,a} = \frac{[E^-][H^+]}{[EH]}$$

$$EH_2^+ \rightleftharpoons EH + H^+ \qquad K_{E,b} = \frac{[EH][H^+]}{[EH_2^+]}$$

$$ESH \rightleftharpoons ES^- + H^+ \qquad K_{ES,a} = \frac{[ES^-][H^+]}{[ESH]}$$

$$ESH_2^+ \rightleftharpoons ESH + H^+ \qquad K_{ES,b} = \frac{[ESH][H^+]}{[ESH_2^+]}$$

in which only the EH and ESH forms are active.

(a) For the mechanism above, show that

$$v = \frac{v_{max}'}{1 + K_M'/[S]_0}$$

with

$$v_{max}' = \frac{v_{max}}{1 + \dfrac{[H^+]}{K_{ES,b}} + \dfrac{K_{ES,a}}{[H^+]}}$$

$$K_M' = K_M \frac{1 + \dfrac{[H^+]}{K_{E,b}} + \dfrac{K_{E,a}}{[H^+]}}{1 + \dfrac{[H^+]}{K_{ES,b}} + \dfrac{K_{ES,a}}{[H^+]}}$$

where v_{max} and K_M correspond to the form EH of the enzyme.

(b) For pH values ranging from 0 to 14, plot v_{max}' against pH for a hypothetical reaction for which $v_{max} = 1.0\ \mu\text{mol dm}^{-3}\ \text{s}^{-1}$, $K_{ES,b} = 1.0\ \mu\text{mol dm}^{-3}$, and $K_{ES,a} = 10\ \text{nmol dm}^{-3}$. Is there a pH at which v_{max} reaches a maximum value? If so, determine the pH.

(c) Redraw the plot in part (b) by using the same value of v_{max}, but $K_{ES,b} = 0.10\ \text{mmol dm}^{-3}$ and $K_{ES,a} = 0.10\ \text{nmol dm}^{-3}$. Account for any differences between this plot and the plot from part (b).

8.43 Studies of biochemical reactions initiated by the absorption of light have contributed significantly to our understanding of the kinetics of electron transfer processes. The experimental arrangement is that for time-resolved spectroscopy (*In the laboratory* 7.2) and relies on the observation that many substances become more efficient electron donors on absorbing energy from a light source, such as a laser. With judicious choice of electron acceptor, it is possible to set up an experimental system in which electron transfer will not occur in the dark (when only a poor electron donor is present) but will proceed after application of a laser pulse (when a better electron donor is generated). Nature makes use of this strategy to initiate the chain of electron transfer events that leads ultimately to the phosphorylation of ATP in photosynthetic organisms.

(a) An elegant way to study electron transfer in proteins consists of attaching an electroactive species to the protein's surface and then measuring k_{et} between the attached species and an electroactive protein cofactor. J.W. Winkler and H.B. Gray, *Chem. Rev.* **92**, 369 (1992), summarize data for cytochrome *c* modified by replacement of the heme iron by a Zn^{2+} ion, resulting in a zinc–porphyrin (ZnP) moiety in the interior of the protein, and by attachment of a ruthenium ion complex to a surface histidine amino acid. The edge-to-edge distance between the electroactive species was thus fixed at 1.23 nm. A variety of ruthenium ion complexes with different standard reduction potentials were used. For each ruthenium-modified protein, either $Ru^{2+} \rightarrow ZnP^+$ or $ZnP^* \rightarrow Ru^{3+}$, in which the zinc-porphyrin is excited by a laser pulse, was monitored.

This arrangement leads to different standard reaction Gibbs energies because the redox couples ZnP$^+$/ZnP and ZnP$^+$/ZnP* have different standard potentials, with the electronically excited porphyrin being a more powerful reductant. Use the following data to estimate the reorganization energy for this system:

$\Delta_r G^\ominus$/eV	0.665	0.705	0.745	0.975	1.015	5.50
k_{et}/(10^6 s^{-1})	0.657	1.52	1.52	8.99	5.76	10.1

(b) The photosynthetic reaction center of the purple photosynthetic bacterium *Rhodopseudomonas viridis* is a protein complex containing a number of bound co-factors that participate in electron transfer reactions. The table below shows data compiled by Moser et al., *Nature* **355**, 796 (1992), on the rate constants for electron transfer between different co-factors and their edge-to-edge distances. (BChl, bacteriochlorophyll; BChl$_2$, bacteriochlorophyll dimer, functionally distinct from BChl; BPh, bacteriopheophytin; Q$_A$ and Q$_B$, quinone molecules bound to two distinct sites; cyt c_{559}, a cytochrome bound to the reaction center complex.) Are these data in agreement with the behavior predicted by eqn 8.31? If so, evaluate the value of β.

Reaction	BChl$^-\rightarrow$BPh	BPh$^-\rightarrow$Chl$_2^+$	BPh$^-\rightarrow$Q$_A$	cyt $c_{559}\rightarrow$Chl$_2^+$
r/nm	0.48	0.95	0.96	1.23
k_{et}/s^{-1}	1.58×10^{12}	3.98×10^9	1.00×10^9	1.58×10^8

Reaction	Q$_A^-\rightarrow$Q$_B$	Q$_A^-\rightarrow$BChl$_2^+$
r/nm	1.35	2.24
k_{et}/s^{-1}	3.98×10^7	63.1

PART 3 Biomolecular structure

We now begin our study of *structural biology*, the description of the molecular features that determine the structures of and the relationships between structure and function in biological macromolecules. In the following chapters, we shall see how concepts of physical chemistry can be used to establish some of the known 'rules' for the assembly of complex structures, such as proteins, nucleic acids, and biological membranes. However, not all the rules are known, so structural biology is a very active area of research that brings together biologists, chemists, physicists, and mathematicians.

Microscopic systems and quantization

9

The first goal of our study of biological molecules and assemblies is to gain a firm understanding of their ultimate structural components, atoms. To make progress, we need to become familiar with the principal concepts of quantum mechanics, the most fundamental description of matter that we currently possess and the only way to account for the structures of atoms. Such knowledge is applied to rational drug design (see the *Prolog*) when computational chemists use quantum mechanical concepts to predict the structures and reactivities of drug molecules. Quantum mechanical phenomena also form the basis for virtually all the modes of spectroscopy and microscopy that are now so central to investigations of composition and structure in both chemistry and biology. Present-day techniques for studying biochemical reactions have progressed to the point where the information is so detailed that quantum mechanics has to be used in its interpretation.

Atomic structure—the arrangement of electrons in atoms—is an essential part of chemistry and biology because it is the basis for the description of molecular structure and molecular interactions. Indeed, without intimate knowledge of the physical and chemical properties of elements, it is impossible to understand the molecular basis of biochemical processes, such as protein folding, the formation of cell membranes, and the storage and transmission of information by DNA.

Principles of quantum theory

The role—indeed, the existence—of quantum mechanics was appreciated only during the twentieth century. Until then it was thought that the motion of atomic and subatomic particles could be expressed in terms of the laws of classical mechanics introduced in the seventeenth century by Isaac Newton (see *Fundamentals* F.3), for these laws were very successful at explaining the motion of planets and everyday objects such as pendulums and projectiles. Classical physics is based on three 'obvious' assumptions:

1. A particle travels in a **trajectory**, a path with a precise position and momentum at each instant.

2. Any type of motion can be excited to a state of arbitrary energy.

3. Waves and particles are distinct concepts.

These assumptions agree with everyday experience. For example, a pendulum swings with a precise oscillating motion and can be made to oscillate with any energy simply by pulling it back to an arbitrary angle and then letting it swing freely. Classical mechanics lets us predict the angle of the pendulum and the speed at which it is swinging at any instant.

Towards the end of the nineteenth century, experimental evidence accumulated showing that classical mechanics failed to explain all the experimental evidence on very small particles, such as individual atoms, nuclei, and electrons. It took until 1926 to identify the appropriate concepts and equations for describing them. We now know that classical mechanics is in fact only an *approximate* description of the motion of particles and the approximation is invalid when it is applied to molecules, atoms, and electrons.

9.1 The emergence of the quantum theory

The structure of biological matter cannot be understood without understanding the nature of electrons. Moreover, because many of the experimental tools available to biochemists are based on interactions between light and matter, we also need to understand the nature of light. We shall see, in fact, that matter and light have a lot in common.

Quantum theory emerged from a series of observations made during the late nineteenth century, from which two important conclusions were drawn. The first conclusion, which countered what had been supposed for two centuries, is that energy can be transferred between systems only in discrete amounts. The second conclusion is that light and particles have properties in common: electromagnetic radiation (light), which had long been considered to be a wave, in fact behaves like a stream of particles, and electrons, which since their discovery in 1897 had been supposed to be particles, but in fact behave like waves. In this section we review the evidence that led to these conclusions, and establish the properties that a valid system of mechanics must accommodate.

(a) Atomic and molecular spectra

A **spectrum** is a display of the frequencies or wavelengths (which are related by $\lambda = c/v$; see *Fundamentals* F.3) of electromagnetic radiation that are absorbed or emitted by an atom or molecule. Figure 9.1 shows a typical atomic emission spectrum and Fig. 9.2 shows a typical molecular absorption spectrum. The obvious feature of both is that *radiation is absorbed or emitted at a series of discrete frequencies*. The emission or absorption of light at discrete frequencies can be understood if we suppose that

- the energy of the atoms or molecules is confined to discrete values, for then energy can be discarded or absorbed only in packets as the atom or molecule jumps between its allowed states (Fig. 9.3)
- the frequency of the radiation is related to the energy difference between the initial and final states.

These assumptions are brought together in the **Bohr frequency condition**, which relates the frequency v (nu) of radiation to the difference in energy ΔE between two states of an atom or molecule:

$$\Delta E = hv \qquad \text{Bohr frequency relation} \qquad (9.1)$$

where h is the constant of proportionality. The additional evidence that we describe below confirms this simple relation and gives the value $h = 6.626 \times 10^{-34}$ J s. This constant is now known as **Planck's constant**, for it arose in a context that had been suggested by the German physicist Max Planck.

At this point we can conclude that one feature of nature that any system of mechanics must accommodate is that the internal modes of atoms and molecules

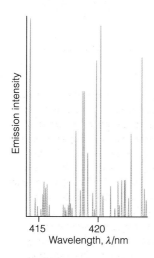

Fig. 9.1 A region of the spectrum of radiation emitted by excited iron atoms consists of radiation at a series of discrete wavelengths (or frequencies).

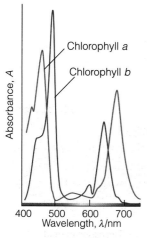

Fig. 9.2 When a molecule changes its state, it does so by absorbing radiation at definite frequencies. This spectrum of chlorophyll (Atlas R3) suggests that the molecule (and molecules in general) can possess only certain energies, not a continuously variable energy.

can possess only certain energies; that is, these modes are **quantized**. The limitation of energies to discrete values is called the **quantization of energy**.

(b) Wave–particle duality

In *Fundamentals* F.3 we saw that classical physics describes light as electromagnetic radiation, an oscillating electromagnetic field that spreads as a harmonic wave through empty space, the vacuum, at a constant speed c. A new view of electromagnetic radiation began to emerge in 1900 when the German physicist Max Planck discovered that the energy of an electromagnetic oscillator is limited to discrete values and cannot be varied arbitrarily. This proposal is quite contrary to the viewpoint of classical physics, in which all possible energies are allowed. In particular, Planck found that the permitted energies of an electromagnetic oscillator of frequency v are integer multiples of hv:

$$E = nhv \qquad n = 0, 1, 2, \ldots$$

Quantization of energy in electromagnetic oscillators (9.2)

where h is Planck's constant. This conclusion inspired Albert Einstein to conceive of radiation as consisting of a stream of particles, each particle having an energy hv. When there is only one such particle present, the energy of the radiation is hv, when there are two particles of that frequency, their total energy is $2hv$, and so on. These particles of electromagnetic radiation are now called **photons**. According to the photon picture of radiation, an intense beam of monochromatic (single-frequency) radiation consists of a dense stream of identical photons; a weak beam of radiation of the same frequency consists of a relatively small number of the same type of photons.

Evidence that confirms the view that radiation can be interpreted as a stream of particles comes from the **photoelectric effect**, the ejection of electrons from metals when they are exposed to ultraviolet radiation (Fig. 9.4). Experiments show that no electrons are ejected, regardless of the intensity of the radiation, unless the frequency exceeds a threshold value characteristic of the metal. On the other hand, even at low light intensities, electrons are ejected immediately if the frequency is above the threshold value. These observations strongly suggest an interpretation of the photoelectric effect in which an electron is ejected in a collision with a particle-like projectile, the photon, provided the projectile carries enough energy to expel the electron from the metal. When the photon collides with an electron, it gives up all its energy, so we should expect electrons to appear as soon as the collisions begin, provided each photon carries sufficient energy. That is, through the principle of conservation of energy, the photon energy should be equal to the sum of the kinetic energy of the electron and the **work function** Φ (uppercase phi) of the metal, the energy required to remove the electron from the metal (Fig. 9.5).

The photoelectric effect is strong evidence for the existence of photons and shows that light has certain properties of particles, a view that is contrary to the classical wave theory of light. A crucial experiment performed by the American physicists Clinton Davisson and Lester Germer in 1925 challenged another classical idea by showing that matter is wavelike: they observed the diffraction of electrons by a crystal (Fig. 9.6). **Diffraction** is the interference between waves caused by an object in their path and results in a series of bright and dark fringes where the waves are detected (Fig. 9.7). It is a typical characteristic of waves.

The Davisson–Germer experiment, which has since been repeated with other particles (including molecular hydrogen), shows clearly that 'particles' have

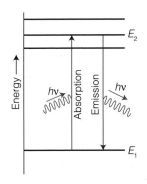

Fig. 9.3 Spectral features can be accounted for if we assume that a molecule emits (or absorbs) a photon as it changes between discrete energy levels. High-frequency radiation is emitted (or absorbed) when the two states involved in the transition are widely separated in energy; low-frequency radiation is emitted when the two states are close in energy. In absorption or emission, the change in the energy of the molecule, ΔE, is equal to hv, where v is the frequency of the radiation.

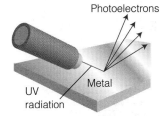

Fig. 9.4 The experimental arrangement to demonstrate the photoelectric effect. A beam of ultraviolet radiation is used to irradiate a patch of the surface of a metal, and electrons are ejected from the surface if the frequency of the radiation is above a threshold value that depends on the metal.

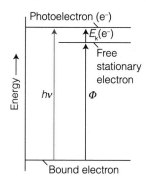

Fig. 9.5 In the photoelectric effect, an incoming photon brings a definite quantity of energy, $h\nu$. It collides with an electron close to the surface of the metal target and transfers its energy to it. The difference between the work function, Φ, and the energy $h\nu$ appears as the kinetic energy of the photoelectron, the electron ejected by the photon.

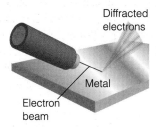

Fig. 9.6 In the Davisson–Germer experiment, a beam of electrons was directed on a single crystal of nickel, and the scattered electrons showed a variation in intensity with angle that corresponded to the pattern that would be expected if the electrons had a wave character and were diffracted by the layers of atoms in the solid.

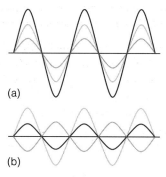

Fig. 9.7 When two waves (drawn as blue and orange lines) are in the same region of space they interfere (with the resulting wave drawn as a red line). Depending on the relative positions of peaks and troughs, they may interfere (a) constructively, to given an enhanced amplitude), or (b) destructively, to give a smaller amplitude.

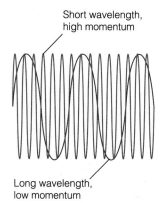

Fig. 9.8 According to the de Broglie relation, a particle with low momentum has a long wavelength, whereas a particle with high momentum has a short wavelength. A high momentum can result either from a high mass or from a high velocity (because $p = mv$). Macroscopic objects have such large masses that, even if they are traveling very slowly, their wavelengths are undetectably short.

wavelike properties. We have also seen that 'waves' have particle-like properties. Thus we are brought to the heart of modern physics. When examined on an atomic scale, the concepts of particle and wave melt together, particles taking on the characteristics of waves and waves the characteristics of particles. This joint wave–particle character of matter and radiation is called **wave–particle duality**. You should keep this extraordinary, perplexing, and at the time revolutionary idea in mind whenever you are thinking about matter and radiation at an atomic scale.

As these concepts emerged there was an understandable confusion—which continues to this day—about how to combine both aspects of matter into a single description. Some progress was made by Louis de Broglie when, in 1924, he suggested that any particle traveling with a linear momentum, p, should have (in some sense) a wavelength λ given by the **de Broglie relation**:

$$\lambda = \frac{h}{p} \qquad \boxed{\text{de Broglie relation}} \quad (9.3)$$

The wave corresponding to this wavelength, what de Broglie called a 'matter wave', has the mathematical form $\sin(2\pi x/\lambda)$. The de Broglie relation implies that the wavelength of a 'matter wave' should decrease as the particle's speed increases (Fig. 9.8). The relation also implies that, for a given speed, heavy particles should be associated with waves of shorter wavelengths than those of lighter particles. Equation 9.3 was confirmed by the Davisson–Germer experiment, for the wavelength it predicts for the electrons they used in their experiment agrees with the details of the diffraction pattern they observed. We shall build on the relation, and understand it more, in the next section.

| Example 9.1 | *Estimating the de Broglie wavelength of electrons* |

The wave character of the electron is the key to imaging small samples by electron microscopy (see *In the laboratory* 9.1). Consider an electron microscope in which electrons are accelerated from rest through a potential difference of 15.0 kV. Calculate the wavelength of the electrons.

Strategy To use the de Broglie relation, we need to establish a relation between the kinetic energy E_k and the linear momentum p. With $p = mv$ and $E_k = \frac{1}{2}mv^2$, it follows that $E_k = \frac{1}{2}m(p/m)^2 = p^2/2m$, and therefore $p = (2mE_k)^{1/2}$. The kinetic energy acquired by an electron accelerated from rest by falling through a potential difference V is eV, where $e = 1.602 \times 10^{-19}$ C is the magnitude of its charge, so we can write $E_k = eV$ and, after using $m_e = 9.109 \times 10^{-31}$ kg for the mass of the electron, $p = (2m_e eV)^{1/2}$.

Solution By using $p = (2m_e eV)^{1/2}$ in de Broglie's relation (eqn 9.3), we obtain

$$\lambda = \frac{h}{(2m_e eV)^{1/2}}$$

At this stage, all we need do is to substitute the data and use the relations $1\,C\,V = 1\,J$ and $1\,J = 1\,kg\,m^2\,s^{-2}$:

$$\lambda = \frac{6.626 \times 10^{-34}\,J\,s}{\{2 \times (9.109 \times 10^{-31}\,kg) \times (1.602 \times 10^{-19}\,C) \times (1.50 \times 10^4\,V)\}^{1/2}}$$

$$= 1.00 \times 10^{-11}\,m = 10.0\,pm$$

Self-test 9.1 Calculate the wavelength of an electron accelerated from rest in an electric potential difference of 1.0 MV ($1\,MV = 10^6\,V$).

Answer: 1.2 pm

In the laboratory 9.1 *Electron microscopy*

The basic approach of illuminating a small area of a sample and collecting light with a microscope has been used for many years to image small specimens. However, the **resolution** of a microscope, the minimum distance between two objects that leads to two distinct images, is in the order of the wavelength of light being used. Therefore, conventional microscopes employing visible light have resolutions in the micrometer range and cannot resolve features on a scale of nanometers.

There is great interest in the development of new experimental probes of very small specimens that cannot be studied by traditional light microscopy. For example, our understanding of biochemical processes, such as enzymatic catalysis, protein folding, and the insertion of DNA into the cell's nucleus, will be enhanced if it becomes possible to image individual biopolymers—with dimensions much smaller than visible wavelengths—at work. The concept of wave–particle duality is directly relevant to biology because the observation that electrons can be diffracted led to the development of important techniques for the determination of the structures of biologically active matter. One technique that is often used to image nanometer-sized objects is **electron microscopy**, in which a beam of electrons with a well-defined de Broglie wavelength replaces the lamp found in traditional light microscopes. Instead of glass or quartz lenses, magnetic fields are used to focus the beam. In **transmission electron microscopy** (TEM), the electron beam passes through the specimen and the image is collected on a screen. In **scanning electron microscopy** (SEM), electrons scattered back from a small irradiated area of the sample are detected and the electrical signal is sent to a video screen. An image of the surface is then obtained by scanning the electron beam across the sample.

As in traditional light microscopy, the resolution of the microscope is governed by the wavelength (in this case, the de Broglie wavelength of the electrons in the beam) and the ability to focus the beam. Electron wavelengths in typical electron microscopes can be as short as 10 pm, but it is not possible to focus electrons well with magnetic lenses so, in the end, typical resolutions of TEM and SEM instruments are about 2 nm and 50 nm, respectively. It follows that electron microscopes cannot resolve individual atoms (which have diameters of about 0.2 nm). Furthermore, only certain samples can be observed under certain conditions. The measurements must be conducted under high vacuum. For TEM observations, the samples must be very thin cross-sections of a specimen and SEM observations must be made on dry samples.

Bombardment with high-energy electrons can damage biological samples by excessive heating, ionization, and formation of radicals. These effects can lead to denaturation or more severe chemical transformation of biological molecules, such as the breaking of bonds and formation of new bonds not found in native structures. To minimize such damage, it has become common to cool samples to temperatures as low as 77 K or 4 K (by immersion in liquid N_2 or liquid He, respectively) prior to and during examination with the microscope. This technique is known as **electron cryomicroscopy**.[1]

A consequence of these stringent experimental requirements is that electron microscopy cannot be used to study living cells. In spite of these limitations, the technique is very useful in studies of the internal structure of cells (Fig. 9.9).

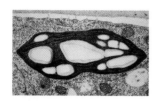

Fig. 9.9 A TEM image of a cross-section of a plant cell showing chloroplasts, organelles responsible for the reactions of photosynthesis (Chapter 12). Chloroplasts are typically 5 μm long. (Dr Jeremy Burgess/Science Photo Library.)

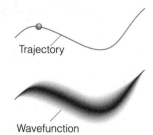

Trajectory

Wavefunction

Fig. 9.10 According to classical mechanics, a particle can have a well-defined trajectory, with a precisely specified position and momentum at each instant (as represented by the precise path in the diagram). According to quantum mechanics, a particle cannot have a precise trajectory; instead, there is only a probability that it may be found at a specific location at any instant. The wavefunction that determines its probability distribution is a kind of blurred version of the trajectory. Here, the wavefunction is represented by areas of shading: the darker the area, the greater the probability of finding the particle there.

9.2 The Schrödinger equation

The surprising consequences of wave–particle duality led not only to powerful techniques in microscopy and medical diagnostics but also to new views of the mechanisms of biochemical reactions, particularly those involving the transfer of electrons and protons. To understand these applications, it is essential to know how electrons behave under the influence of various forces.

We take the de Broglie relation as our starting point for the formulation of a new mechanics and abandon the classical concept of particles moving along trajectories. From now on, we adopt the quantum mechanical view that *a particle is spread through space like a wave*. Like for a wave in water, where the water accumulates in some places but is low in others, there are regions where the particle is more likely to be found than others. To describe this distribution, we introduce the concept of **wavefunction**, ψ (psi), in place of the trajectory, and then set up a scheme for calculating and interpreting ψ. A 'wavefunction' is the modern term for de Broglie's 'matter wave'. To a very crude first approximation, we can visualize a wavefunction as a blurred version of a trajectory (Fig. 9.10); however, we shall refine this picture in the following sections.

---

[1] The prefix 'cryo' originates from *kryos*, the Greek word for cold or frost.

(a) The formulation of the equation

In 1926, the Austrian physicist Erwin Schrödinger proposed an equation for calculating wavefunctions. The **Schrödinger equation** for a single particle of mass m moving with energy E in one dimension is

$$-\frac{\hbar^2}{2m}\frac{\mathrm{d}^2\psi}{\mathrm{d}x^2} + V\psi = E\psi$$

Schrödinger equation   (9.4a)

You will often see eqn 9.4a written in the very compact form

$$\hat{H}\psi = E\psi$$

Compact form of the Schrödinger equation   (9.4b)

where $\hat{H}\psi$ stands for everything on the left of eqn 9.4a. The quantity $\hat{H}$ is called the **hamiltonian** of the system after the mathematician William Hamilton, who had formulated a version of classical mechanics that used the concept. It is written with a caret (ˆ) to signify that it is an 'operator', something that acts in a particular way on ψ rather than just multiplying it (as E multiplies ψ in $E\psi$). You should be aware that much of quantum theory is formulated in terms of various operators, but we shall encounter them only very rarely in this text.[2]

Technically, the Schrödinger equation is a second-order differential equation. In it, V, which may depend on the position x of the particle, is the potential energy; $\hbar$ (which is read h-bar) is a convenient modification of Planck's constant:

$$\hbar = \frac{h}{2\pi} = 1.054 \times 10^{-34}\,\mathrm{J\,s}$$

We provide a justification of the form of the equation in *Further information 9.1*. The rare cases where we need to see the explicit forms of its solution will involve very simple functions. For example (and to become familiar with the form of wavefunctions in three simple cases, but not putting in various constants):

1. The wavefunction for a freely moving particle is $\sin x$ (exactly as for de Broglie's matter wave, $\sin(2\pi x/\lambda)$).

2. The wavefunction for the lowest energy state of a particle free to oscillate to and fro near a point is e^{-x^2}, where x is the displacement from the point (see Section 9.6),

3. The wavefunction for an electron in the lowest energy state of a hydrogen atom is e^{-r}, where r is the distance from the nucleus (see Section 9.8).

As can be seen, none of these wavefunctions is particularly complicated mathematically.

One feature of the solution of any given Schrödinger equation, a feature common to all differential equations, is that an infinite number of possible solutions are allowed mathematically. For instance, if $\sin x$ is a solution of the equation, then so too is $a \sin bx$, where a and b are arbitrary constants, with each solution corresponding to a particular value of E. However, it turns out that only some of these solutions are acceptable physically when the motion of a particle is constrained somehow (as in the case of an electron moving under the influence of the electric field of a proton in a hydrogen atom). In such instances, an acceptable solution must satisfy certain constraints called **boundary conditions**, which we describe shortly (Fig. 9.11). Suddenly, we are at the heart of quantum mechanics:

[2] See, for instance, our *Physical chemistry* (2010).

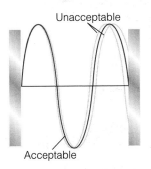

Unacceptable

Acceptable

Fig. 9.11 Although an infinite number of solutions of the Schrödinger equation exist, not all of them are physically acceptable. Acceptable wavefunctions have to satisfy certain boundary conditions, which vary from system to system. In the example shown here, where the particle is confined between two impenetrable walls, the only acceptable wavefunctions are those that fit between the walls (like the vibrations of a stretched string). Because each wavefunction corresponds to a characteristic energy and the boundary conditions rule out many solutions, only certain energies are permissible.

the fact that only some solutions of the Schrödinger equation are acceptable, together with the fact that each solution corresponds to a characteristic value of E, implies that only certain values of the energy are acceptable. That is, when the Schrödinger equation is solved subject to the boundary conditions that the solutions must satisfy, we find that the energy of the system is quantized. Planck and his immediate successors had to postulate the quantization of energy for each system they considered: now we see that quantization is an automatic feature of a single equation, the Schrödinger equation, which is applicable to all systems. Later in this chapter and the next we shall see exactly which energies are allowed in a variety of systems, the most important of which (for chemistry) is an atom.

(b) The interpretation of the wavefunction

Before going any further, it will be helpful to understand the physical significance of a wavefunction. The interpretation most widely used is based on a suggestion made by the German physicist Max Born. He made use of an analogy with the wave theory of light, in which the square of the amplitude of an electromagnetic wave is interpreted as its intensity and therefore (in quantum terms) as the number of photons present. The **Born interpretation** asserts:

> The probability of finding a particle in a small region of space of volume δV is proportional to $\psi^2 \delta V$, where ψ is the value of the wavefunction in the region.

In other words, ψ^2 is a **probability density**. As for other kinds of density, such as mass density (ordinary 'density'), we get the probability itself by multiplying the probability density by the volume of the region of interest.

The Born interpretation implies that wherever ψ^2 is large ('high probability density'), there is a high probability of finding the particle. Wherever ψ^2 is small ('low probability density'), there is only a small chance of finding the particle. The density of shading in Fig. 9.12 represents this **probabilistic interpretation**, an interpretation that accepts that we can make predictions only about the probability of finding a particle somewhere. This interpretation is in contrast to classical physics, which claims to be able to predict precisely that a particle will be at a given point on its path at a given instant.

A note on good practice
The symbol δ (see below, right) indicates a small (and, in the limit, infinitesimal) change in a parameter, as in x changing to $x + \delta x$. The symbol Δ indicates a finite (measurable) difference between two quantities, as in $\Delta X = X_{\text{final}} - X_{\text{initial}}$.

A brief comment
We are supposing throughout that ψ is a real function (that is, one that does not depend on $i = (-1)^{1/2}$). In general, y is complex (has both real and imaginary components); in such cases ψ^2 is replaced by $\psi^*\psi$, where ψ^* is the complex conjugate of ψ. We do not consider complex functions in this text.[3]

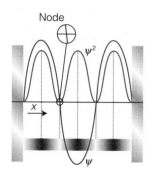

Fig. 9.12 A wavefunction ψ does not have a direct physical interpretation. However, its square (its square modulus if it is complex), ψ^2, tells us the probability of finding a particle at each point. The probability density implied by the wavefunction shown here is depicted by the density of shading in the band at the bottom of the figure.

> | Example 9.2 | *Interpreting a wavefunction* |

The wavefunction of an electron in the lowest energy state of a hydrogen atom is proportional to e^{-r/a_0}, with $a_0 = 52.9$ pm and r the distance from the nucleus (Fig. 9.13). Calculate the relative probabilities of finding the electron inside a small volume located at (a) $r = 0$ (that is, at the nucleus) and (b) $r = a_0$ away from the nucleus.

Strategy The probability is proportional to $\psi^2 \delta V$ evaluated at the specified location, with $\psi \propto e^{-r/a_0}$ and $\psi^2 \propto e^{-2r/a_0}$. The volume of interest is so small (even on the scale of the atom) that we can ignore the variation of ψ within it and write

probability $\propto \psi^2 \delta V$

with ψ evaluated at the point in question.

---

[3] For the role, properties, and interpretation of complex wavefunctions, see our *Physical chemistry* (2010).

Solution (a) When $r = 0$, $\psi^2 \propto 1.0$ (because $e^0 = 1$) and the probability of finding the electron at the nucleus is proportional to $1.0 \times \delta V$. (b) At a distance $r = a_0$ in an arbitrary direction, $\psi^2 \propto e^{-2}$, so the probability of being found there is proportional to $e^{-2} \times \delta V = 0.14 \times \delta V$. Therefore, the ratio of probabilities is $1.0/0.14 = 7.1$. It is more probable (by a factor of 7.1) that the electron will be found at the nucleus than in the same tiny volume located at a distance a_0 from the nucleus.

Self-test 9.2 The wavefunction for the lowest energy state in the ion He^+ is proportional to e^{-2r/a_0}. Calculate the ratio of probabilities as in *Example* 9.2, by comparing the cases for which $r = 0$ and $r = a_0$. Any comment?

> **Answer:** The ratio of probabilities is 55; a more compact wavefunction on account of the higher nuclear charge.

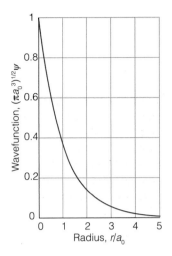

Fig. 9.13 The wavefunction for an electron in the ground state of a hydrogen atom is an exponentially decaying function of the form e^{-r/a_0}, where $a_0 = 52.9$ pm is the Bohr radius.

9.3 The uncertainty principle

Given that electrons behave like waves, we need to be able to reconcile the predictions of quantum mechanics with the existence of objects, such as biological cells and the organelles within them.

We have seen that, according to the de Broglie relation, a wave of constant wavelength, the wavefunction $\sin(2\pi x/\lambda)$, corresponds to a particle with a definite linear momentum $p = h/\lambda$. However, a wave does not have a definite location at a single point in space, so we cannot speak of the precise position of the particle if it has a definite momentum. Indeed, because a sine wave spreads throughout the whole of space, we cannot say anything about the location of the particle: because the wave spreads everywhere, the particle may be found anywhere in the whole of space. This statement is one half of the **uncertainty principle**, proposed by Werner Heisenberg in 1927, in one of the most celebrated results of quantum mechanics:

> It is impossible to specify simultaneously, with arbitrary precision, both the momentum and the position of a particle.

Before discussing the principle, we must establish the other half: that if we know the position of a particle exactly, then we can say nothing about its momentum. If the particle is at a definite location, then its wavefunction must be nonzero there and zero everywhere else (Fig. 9.14). We can simulate such a wavefunction by forming a **superposition** of many wavefunctions; that is, by adding together the amplitudes of a large number of sine functions (Fig. 9.15). This procedure is successful because the amplitudes of the waves add together at one location to give a nonzero total amplitude but cancel everywhere else. In other words, we can create a sharply localized wavefunction by adding together wavefunctions corresponding to many different wavelengths, and therefore, by the de Broglie relation, of many different linear momenta.

The superposition of a few sine functions gives a broad, ill-defined wavefunction. As the number of functions used to form the superposition increases, the wavefunction becomes sharper because of the more complete interference between the positive and negative regions of the components. When an infinite number of components are used, the wavefunction is a sharp, infinitely narrow spike like that in Fig. 9.14, which corresponds to perfect localization of the

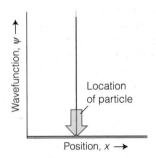

Fig. 9.14 The wavefunction for a particle with a well-defined position is a sharply spiked function that has zero amplitude everywhere except at the particle's position.

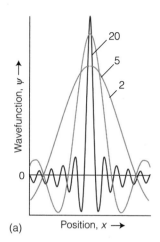

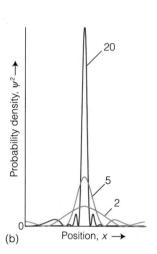

Fig. 9.15 The wavefunction for a particle with an ill-defined location can be regarded as the sum (superposition) of several wavefunctions of different wavelength that interfere constructively in one place but destructively elsewhere. As more waves are used in the superposition, the location becomes more precise at the expense of uncertainty in the particle's momentum. An infinite number of waves are needed to construct the wavefunction of a perfectly localized particle. The numbers against each curve are the number of sine waves used in the superposition. (a) The wavefunctions; (b) the corresponding probability densities.

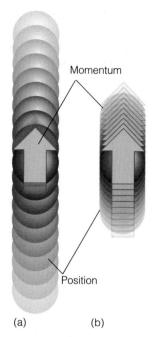

Fig. 9.16 A representation of the content of the uncertainty principle. The range of locations of a particle is shown by the circles and the range of momenta by the arrows. In (a), the position is quite uncertain, and the range of momenta is small. In (b), the location is much better defined, and now the momentum of the particle is quite uncertain.

A brief comment
Strictly, the uncertainty in momentum is the root mean square (r.m.s.) deviation of the momentum from its mean value, $\Delta p = (\langle p^2 \rangle - \langle p \rangle^2)^{1/2}$, where the angle brackets denote mean values. Likewise, the uncertainty in position is the r.m.s. deviation in the mean value of position, $\Delta x = (\langle x^2 \rangle - \langle x \rangle^2)^{1/2}$.

particle. Now the particle is perfectly localized, but at the expense of discarding all information about its momentum.

The exact, quantitative version of the position–momentum uncertainty relation is

$$\Delta p \Delta x \geq \tfrac{1}{2}\hbar \qquad \text{Position–momentum uncertainty relation (in one dimension)} \qquad (9.5)$$

The quantity Δp is the 'uncertainty' in the linear momentum and Δx is the uncertainty in position (which is proportional to the width of the peak in Fig. 9.15). Equation 9.5 expresses quantitatively the fact that the more closely the location of a particle is specified (the smaller the value of Δx), then the greater the uncertainty in its momentum (the larger the value of Δp) parallel to that coordinate and vice versa (Fig. 9.16).

The uncertainty principle applies to location and momentum along the same axis. It is silent on location on one axis and momentum along a perpendicular axis, such as location along the x-axis and momentum parallel to the y-axis.

Example 9.3   *Using the uncertainty principle*

To gain some appreciation of the biological importance—or lack of it—of the uncertainty principle, estimate the minimum uncertainty in the position of

each of the following, given that their speeds are known to within 1.0 μm s⁻¹:
(a) an electron in a hydrogen atom and (b) a mobile *E. coli* cell of mass 1.0 pg
that can swim in a liquid or glide over surfaces by flexing tail-like structures,
known as flagella. Comment on the importance of including quantum mechanical effects in the description of the motion of the electron and the cell.

Strategy We can estimate Δp from $m\Delta v$, where Δv is the uncertainty in the
speed v; then we use eqn 9.5 to estimate the minimum uncertainty in position,
Δx, where x is the direction in which the projectile is traveling.

Solution From $\Delta p \Delta x \geq \frac{1}{2}\hbar$, the uncertainty in position is

(a) for the electron, with mass 9.109×10^{-31} kg:

$$\Delta x \geq \frac{\hbar}{2\Delta p} = \frac{1.054 \times 10^{-34}\,\text{J s}}{2 \times (9.109 \times 10^{-31}\,\text{kg}) \times (1.0 \times 10^{-6}\,\text{m s}^{-1})} = 58\,\text{m}$$

(b) for the *E. coli* cell (using 1 kg = 10^3 g):

$$\Delta x \geq \frac{\hbar}{2\Delta p} = \frac{1.054 \times 10^{-34}\,\text{J s}}{2 \times (1.0 \times 10^{-15}\,\text{kg}) \times (1.0 \times 10^{-6}\,\text{m s}^{-1})} = 5.3 \times 10^{-14}\,\text{m}$$

For the electron, the uncertainty in position is far larger than the diameter of
the atom, which is about 100 pm. Therefore, the concept of a trajectory—the
simultaneous possession of a precise position and momentum—is untenable.
However, the degree of uncertainty is completely negligible for all practical
purposes in the case of the bacterium. Indeed, the position of the cell can be
known to within 0.05 per cent of the diameter of a hydrogen atom. It follows
that the uncertainty principle plays no direct role in cell biology. However, it
plays a major role in the description of the motion of electrons around nuclei
in atoms and molecules and, as we shall see soon, the transfer of electrons
between molecules and proteins during metabolism.

(**Self-test 9.3**) Estimate the minimum uncertainty in the speed of an electron
that can move along the carbon skeleton of a conjugated polyene (such as
β-carotene) of length 2.0 nm.

Answer: 29 km s⁻¹

The uncertainty principle epitomizes the difference between classical and
quantum mechanics. Classical mechanics supposed, falsely as we now know,
that the position and momentum of a particle can be specified simultaneously
with arbitrary precision. However, quantum mechanics shows that position and
momentum are **complementary**, that is, not simultaneously specifiable. Quantum
mechanics requires us to make a choice: we can specify position at the expense of
momentum or momentum at the expense of position.

Applications of quantum theory

We shall now illustrate some of the concepts that have been introduced and
gain some familiarity with the implications and interpretation of quantum
mechanics, including applications to biochemistry. We shall encounter many

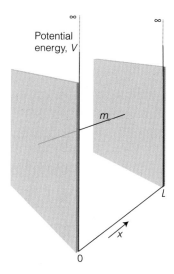

Fig. 9.17 A particle in a one-dimensional region with impenetrable walls at either end. Its potential energy is zero between $x = 0$ and $x = L$ and rises abruptly to infinity as soon as the particle touches either wall.

other illustrations in the following chapters, for quantum mechanics pervades the whole of chemistry. Just to set the scene, here we describe three basic types of motion: translation (motion in a straight line, like a beam of electrons in the electron microscope), rotation, and vibration.

9.4 Translation

The three primitive types of motion—translation, rotation, and vibration—occur throughout science, and we need to be familiar with their quantum mechanical description before we can understand the motion of electrons in atoms and molecules.

In this section we shall see how quantization of energy arises when a particle is confined between two walls. When the potential energy of the particle within the walls is not infinite, the solutions of the Schrödinger equation reveal surprising features, especially the ability of particles to tunnel into and through regions where classical physics would forbid them to be found.

(a) Motion in one dimension

Let's consider the translational motion of a 'particle in a box', a particle of mass m that can travel in a straight line in one dimension (along the x-axis) but is confined between two walls separated by a distance L. The potential energy of the particle is zero inside the box but rises abruptly to infinity at the walls (Fig. 9.17). The particle might be an electron free to move along the linear arrangement of conjugated double bonds in a linear polyene, such as β-carotene (Atlas E1), the molecule responsible for the orange color of carrots and pumpkins.

The boundary conditions for this system are the requirement that each acceptable wavefunction of the particle must fit inside the box exactly, like the vibrations of a violin string (as in Fig. 9.11). It follows that the wavelength, λ, of the permitted wavefunctions must be one of the values

$$\lambda = 2L, L, \tfrac{2}{3}L, \ldots \quad \text{or} \quad \lambda = \frac{2L}{n}, \text{ with } n = 1, 2, 3, \ldots \qquad (9.6)$$

A brief comment
More precisely, the boundary conditions stem from the requirement that the wavefunction is continuous everywhere: because the wavefunction is zero outside the box, it must therefore be zero at its edges, at $x = 0$ and at $x = L$.

Each wavefunction is a sine wave with one of these wavelengths; therefore, because a sine wave of wavelength λ has the form $\sin(2\pi x/\lambda)$, the permitted wavefunctions are

$$\psi_n = N \sin \frac{n\pi x}{L} \qquad n = 1, 2, \ldots$$

Wavefunctions for a particle in a one-dimensional box   (9.7)

As shown in the following *Justification*, the **normalization constant**, N, a constant that ensures that the total probability of finding the particle anywhere is 1, is equal to $(2/L)^{1/2}$.

Justification 9.1   *The normalization constant*

To calculate the constant N, we recall that the wavefunction ψ must have a form that is consistent with the interpretation of the quantity $\psi(x)^2 dx$ as the probability of finding the particle in the infinitesimal region of length dx at the point x given that its wavefunction has the value $\psi(x)$ at that point. Therefore, the total probability of finding the particle between $x = 0$ and $x = L$ is the sum (integral) of all the probabilities of its being in each infinitesimal region.

That total probability is 1 (the particle is certainly in the range somewhere), so we know that

$$\int_0^L \psi^2 dx = 1$$

Substitution of eqn 9.7 turns this expression into

$$N^2 \int_0^L \sin^2 \frac{n\pi x}{L} dx = 1$$

Our task is to solve this equation for N. Because

$$\int \sin^2 ax\, dx = \tfrac{1}{2}x - \frac{\sin 2ax}{4a} + \text{constant}$$

and $\sin b\pi = 0$ ($b = 0, 1, 2, \ldots$), it follows that, because the sine term is zero at $x = 0$ and $x = L$,

$$\int_0^L \sin^2 \frac{n\pi x}{L} dx = \tfrac{1}{2}L$$

Therefore,

$$N^2 \times \tfrac{1}{2}L = 1$$

and hence $N = (2/L)^{1/2}$. Note that, in this case but not in general, the same normalization factor applies to all the wavefunctions regardless of the value of n.

It is a simple matter to find the permitted energy levels because the only contribution to the energy is the kinetic energy of the particle: the potential energy is zero everywhere inside the box, and the particle is never outside the box. First, we note that it follows from the de Broglie relation, eqn 9.3, that the only acceptable values of the linear momentum are

$$p = \frac{h}{\lambda} = \frac{nh}{2L} \qquad n = 1, 2, \ldots \qquad (9.8)$$

Then, because the kinetic energy of a particle of momentum p and mass m is $E = p^2/2m$, it follows that the permitted energies of the particle are

$$E_n = \frac{n^2 h^2}{8mL^2} \qquad n = 1, 2, \ldots \qquad \boxed{\begin{array}{l}\text{Quantized energies}\\\text{of a particle in a}\\\text{one-dimensional box}\end{array}} \qquad (9.9)$$

As we see in eqns 9.7 and 9.9, the wavefunctions and energies of a particle in a box are labeled with the number n. A **quantum number**, of which n is an example, is an integer (in certain cases, as we shall see later, a half-integer) that labels the state of the system. As well as acting as a label, a quantum number specifies certain physical properties of the system: in the present example, n specifies the energy of the particle through eqn 9.9.

The permitted energies of the particle are shown in Fig. 9.18 together with the shapes of the wavefunctions for $n = 1$ to 6. All the wavefunctions except the one of

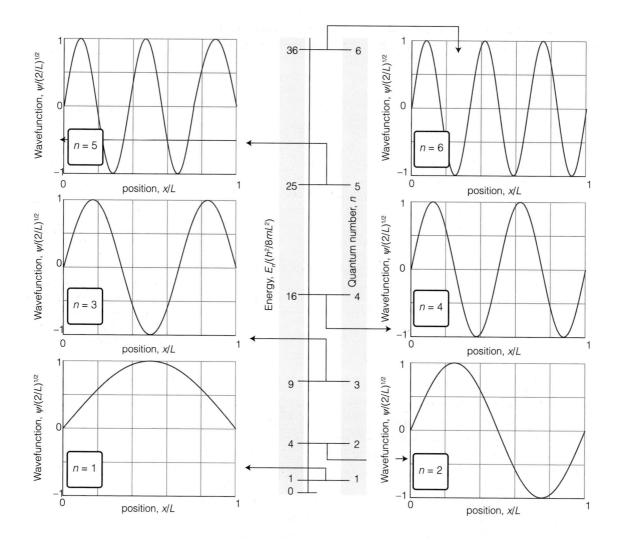

Fig. 9.18 The allowed energy levels and the corresponding (sine wave) wavefunctions for a particle in a box. Note that the energy levels increase as n^2, and so their spacing increases as n increases. Each wavefunction is a standing wave, and successive functions possess one more half-wave and a correspondingly shorter wavelength.

lowest energy ($n = 1$) possess points called **nodes** where the function passes through zero. Passing *through* zero is an essential part of the definition: just becoming zero is not sufficient. The points at the edges of the box where $\psi = 0$ are not nodes because the wavefunction does not pass through zero there.

The number of nodes in the wavefunctions shown in Fig. 9.18 increases from 0 (for $n = 1$) to 5 (for $n = 6$) and is $n - 1$ for a particle in a box in general. It is a general feature of quantum mechanics that the wavefunction corresponding to the state of lowest energy has no nodes, and as the number of nodes in the wavefunctions increases, the energy increases too.

The solutions of a particle in a box introduce another important general feature of quantum mechanics. Because the quantum number n cannot be zero (for this system), the lowest energy that the particle may possess is not zero, as would be allowed by classical mechanics, but $h^2/8mL^2$ (the energy when $n = 1$). This lowest, irremovable energy is called the **zero-point energy**. The existence of a zero-point energy is consistent with the uncertainty principle. If a particle is confined to a finite region, its location is not completely indefinite; consequently its momentum cannot be specified precisely as zero, and therefore its kinetic energy cannot be precisely zero either. The zero-point energy is not a special, mysterious kind of energy. It is simply the last remnant of energy that a particle cannot give up.

For a particle in a box it can be interpreted as the energy arising from a ceaseless fluctuating motion of the particle between the two confining walls of the box.

The energy difference between adjacent levels is

$$\Delta E = E_{n+1} - E_n = (n+1)^2 \frac{h^2}{8mL^2} - n^2 \frac{h^2}{8mL^2} = (2n+1)\frac{h^2}{8mL^2} \qquad (9.10)$$

This expression shows that the difference decreases as the length L of the box increases and that it becomes zero when the walls are infinitely far apart (Fig. 9.19). Atoms and molecules free to move in laboratory-sized vessels may therefore be treated as though their translational energy is not quantized, because L is so large. The expression also shows that the separation decreases as the mass of the particle increases. Particles of macroscopic mass (like balls and planets and even minute specks of dust) behave as though their translational motion is unquantized. Both these conclusions are true in general:

1. The greater the size of the system, the less important are the effects of quantization.

2. The greater the mass of the particle, the less important are the effects of quantization.

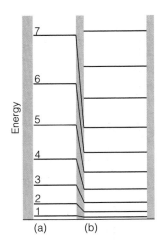

Fig. 9.19 (a) A narrow box has widely spaced energy levels; (b) a wide box has closely spaced energy levels. (In each case, the separations depend on the mass of the particle too.)

Case study 9.1   *The electronic structure of β-carotene*

Some linear polyenes, of which β-carotene is an example, are important biological co-factors that participate in processes as diverse as the absorption of solar energy in photosynthesis (Chapter 12) and protection against harmful biological oxidations. β-Carotene is a linear polyene in which 21 bonds, 10 single and 11 double, alternate along a chain of 22 carbon atoms. We already know from introductory chemistry that this bonding pattern results in *conjugation*, the sharing of π electrons among all the carbon atoms in the chain.[4] Therefore, the particle in a one-dimensional box may be used as a simple model for the discussion of the distribution of π electrons in conjugated polyenes. If we take each C–C bond length to be about 140 pm, the length L of the molecular box in β-carotene is

$$L = 21 \times (1.40 \times 10^{-10}\,\text{m}) = 2.94 \times 10^{-9}\,\text{m}$$

For reasons that will become clear in Sections 9.9 and 10.4, we assume that only one electron per carbon atom is allowed to move freely within the box and that, in the lowest energy state (called the *ground state*) of the molecule, each level is occupied by two electrons. Therefore, the levels up to $n = 11$ are occupied. From eqn 9.10 it follows that the separation in energy between the ground state and the state in which one electron is promoted from the $n = 11$ level to the $n = 12$ level is

$$\Delta E = E_{12} - E_{11} = (2 \times 11 + 1)\frac{(6.626 \times 10^{-34}\,\text{J s})^2}{8 \times (9.109 \times 10^{-31}\,\text{kg}) \times (2.94 \times 10^{-9}\,\text{m})^2}$$
$$= 1.60 \times 10^{-19}\,\text{J}$$

We can relate this energy difference to the properties of the light that can bring about the transition. From the Bohr frequency condition (eqn 9.1), this energy separation corresponds to a frequency of

---

[4] The quantum mechanical basis for conjugation is discussed in Chapter 10.

$$v \approx \frac{\Delta E}{h} = \frac{1.60 \times 10^{-19}\,\text{J}}{6.626 \times 10^{-34}\,\text{J s}} = 2.41 \times 10^{14}\,\text{Hz}$$

(we have used $1\,\text{s}^{-1} = 1\,\text{Hz}$) and a wavelength ($\lambda = c/v$) of 1240 nm; the experimental value is 497 nm.

This model of β-carotene is primitive and the agreement with experiment not very good, but the fact that the calculated and experimental values are of the same order of magnitude is encouraging as it suggests that the model is not ludicrously wrong. Moreover, the model gives us some insight into the origins of quantized energy levels in conjugated systems and predicts, for example, that the separation between adjacent energy levels decreases as the number of carbon atoms in the conjugated chain increases. In other words, the wavelength of the light absorbed by conjugated polyenes increases as the chain length increases. We shall develop better models in Chapter 10.

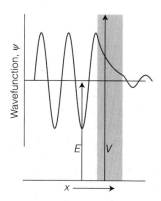

Fig. 9.20 A particle incident on a barrier from the left has an oscillating wavefunction, but inside the barrier there are no oscillations (for $E < V$). If the barrier is not too thick, the wavefunction is nonzero at its opposite face, and so oscillation begins again there.

(b) Tunneling

We now need to consider the case in which the potential energy of a particle does not rise to infinity when it is in the walls of the container and $E < V$. If the walls are thin (so that the potential energy falls to zero again after a finite distance, as for a biological membrane) and the particle is very light (as for an electron or a proton), the wavefunction oscillates inside the box (eqn 9.7), varies smoothly inside the region representing the wall, and oscillates again on the other side of the wall outside the box (Fig. 9.20). Hence, the particle might be found on the outside of a container even though according to classical mechanics it has insufficient energy to escape. Such leakage by penetration through classically forbidden zones is called **tunneling**. Tunneling is a consequence of the wave character of matter. So, just as radio waves pass through walls and X-rays penetrate soft tissue, so can 'matter waves' tunnel through thin walls.

The Schrödinger equation can be used to determine the probability of tunneling, the **transmission probability**, T, of a particle incident on a finite barrier. When the barrier is high (in the sense that $V/E \gg 1$) and wide (in the sense that the wavefunction loses much of its amplitude inside the barrier), we may write[5]

$$T \approx 16\varepsilon(1-\varepsilon)e^{-2\kappa L} \qquad \kappa = \frac{\{2m(V-E)\}^{1/2}}{\hbar}$$

Transmission probability for a high and wide one-dimensional barrier   (9.11)

where $\varepsilon = E/V$ and L is the thickness of the barrier. The transmission probability decreases exponentially with L and with $m^{1/2}$. It follows that particles of low mass are more able to tunnel through barriers than heavy ones (Fig. 9.21). Hence, tunneling is very important for electrons, moderately important for protons, and negligible for most other heavier particles.

The very rapid equilibration of proton transfer reactions (Chapter 4) is also a manifestation of the ability of protons to tunnel through barriers and transfer quickly from an acid to a base. Tunneling of protons between acidic and basic groups is also an important feature of the mechanism of some enzyme-catalyzed reactions. The process may be visualized as a proton passing *through* an activation barrier rather than having to acquire enough energy to travel over it (Fig. 9.22). Quantum mechanical tunneling can be the dominant process in reactions

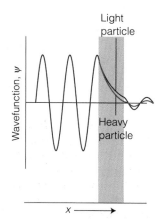

Fig. 9.21 The wavefunction of a heavy particle decays more rapidly inside a barrier than that of a light particle. Consequently, a light particle has a greater probability of tunneling through the barrier.

---

[5] For details of the calculation, see our *Physical chemistry* (2010).

involving hydrogen atom or proton transfer when the temperature is so low that very few reactant molecules can overcome the activation energy barrier. One indication that a proton transfer is taking place by tunneling is that an Arrhenius plot (Section 6.6) deviates from a straight line at low temperatures and the rate is higher than would be expected by extrapolation from room temperature.

Equation 9.11 implies that the rates of electron transfer processes should decrease exponentially with distance between the electron donor and acceptor. This prediction is supported by the experimental evidence that we discussed in Section 8.11, where we showed that, when the temperature and Gibbs energy of activation are held constant, the rate constant k_{et} of electron transfer is proportional to $e^{-\beta r}$, where r is the edge-to-edge distance between electron donor and acceptor and β is a constant with a value that depends on the medium through which the electron must travel from donor to acceptor. It follows that tunneling is an essential mechanistic feature of the electron transfer processes between proteins, such as those associated with oxidative phosphorylation.

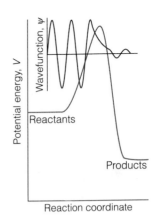

Fig. 9.22 A proton can tunnel through the activation energy barrier that separates reactants from products, so the effective height of the barrier is reduced and the rate of the proton transfer reaction increases. The effect is represented by drawing the wavefunction of the proton near the barrier. Proton tunneling is important only at low temperatures, when most of the reactants are trapped on the left of the barrier.

| In the laboratory 9.2 | *Scanning probe microscopy* |

Like electron microscopy, **scanning probe microscopy** (SPM) also opens a window into the world of nanometer-sized specimens and, in some cases, provides details at the atomic level. One version of SPM is **scanning tunneling microscopy** (STM), in which a platinum–rhodium or tungsten needle is scanned across the surface of a conducting solid. When the tip of the needle is brought very close to the surface, electrons tunnel across the intervening space (Fig. 9.23).

In the *constant-current mode* of operation, the stylus moves up and down corresponding to the form of the surface, and the topography of the surface, including any adsorbates, can be mapped on an atomic scale. The vertical motion of the stylus is achieved by fixing it to a piezoelectric cylinder, which contracts or expands according to the potential difference it experiences. In the *constant-z mode*, the vertical position of the stylus is held constant and the current is monitored. Because the tunneling probability is very sensitive to the size of the gap (remember the exponential dependence of T on L), the microscope can detect tiny, atom-scale variations in the height of the surface (Fig. 9.24). It is difficult to observe individual atoms in large molecules, such as biopolymers. However, Fig. 9.25 shows that STM can reveal some details of the double helical structure of a DNA molecule on a surface.

In **atomic force microscopy** (AFM), a sharpened tip attached to a cantilever is scanned across the surface. The force exerted by the surface and any molecules attached to it pushes or pulls on the tip and deflects the cantilever (Fig. 9.26). The deflection is monitored by using a laser beam. Because no current needs to pass between the sample and the probe, the technique can be applied to nonconducting surfaces and to liquid samples.

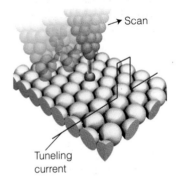

Fig. 9.23 A scanning tunneling microscope makes use of the current of electrons that tunnel between the surface and the tip of the stylus. That current is very sensitive to the height of the tip above the surface.

Two modes of operation of AFM are common. In *contact mode*, or *constant-force mode*, the force between the tip and surface is held constant and the tip makes contact with the surface. This mode of operation can damage fragile samples on the surface. In *noncontact*, or *tapping*, *mode*, the tip bounces up and down with a specified frequency and never quite touches the surface. The amplitude of the tip's oscillation changes when it passes over a species adsorbed on the surface.

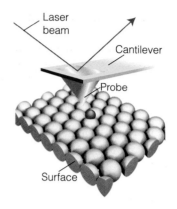

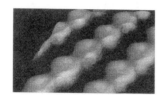

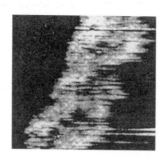

Fig. 9.24 An STM image of cesium atoms on a gallium arsenide surface.

Fig. 9.25 Image of a DNA molecule obtained by scanning tunneling microscopy, showing some features that are consistent with the double helical structure discussed in *Fundamentals* and Chapter 11. (Courtesy of J. Baldeschwieler, CIT.)

Fig. 9.26 In atomic force microscopy, a laser beam is used to monitor the tiny changes in position of a probe as it is attracted to or repelled by atoms on a surface.

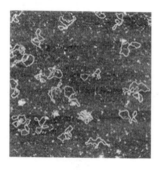

Fig. 9.27 An atomic force microscopy image of bacterial DNA plasmids on a mica surface. (Courtesy of Veeco Instruments.)

Figure 9.27 demonstrates the power of AFM, which shows bacterial DNA plasmids on a solid surface. The technique also can visualize in real time processes occurring on the surface, such as the enzymatic degradation of DNA, and conformational changes in proteins. The tip may also be used to cleave biopolymers, achieving mechanically on a surface what enzymes do in solution or in organisms.

(c) Motion in two dimensions

Now that we have described motion in one dimension, it is a simple matter to step into higher dimensions. The arrangement we consider is like a particle confined to a rectangular box of side L_X in the x-direction and L_Y in the y-direction (Fig. 9.28). The wavefunction varies across the floor of the box, so it is a function of the variables x and y, written as $\psi(x,y)$. We show in *Further information 9.2* that, according to the **separation of variables procedure**, the wavefunction can be expressed as a product of wavefunctions for each direction

$$\psi(x,y) = X(x)Y(y) \tag{9.12}$$

with each wavefunction satisfying a Schrödinger equation like that in eqn 9.4. The solutions are

$$\psi_{n_X,n_Y}(x,y) = X_{n_X}(x)Y_{n_Y}(y)$$

$$= \left(\frac{4}{L_X L_Y}\right)^{1/2} \sin\left(\frac{n_X \pi x}{L_X}\right) \sin\left(\frac{n_Y \pi y}{L_Y}\right) \qquad \boxed{\text{Wavefunctions of a particle in a two-dimensional box}} \tag{9.13a}$$

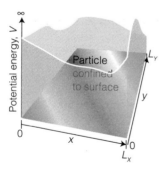

Fig. 9.28 A two-dimensional square well. The particle is confined to a rectangular plane bounded by impenetrable walls. As soon as the particle touches a wall, its potential energy rises to infinity.

Figure 9.29 shows some examples of these wavefunctions. The energies are

$$E_{n_X,n_Y} = E_{n_X} + E_{n_Y} = \frac{n_X^2 h^2}{8mL_X^2} + \frac{n_Y^2 h^2}{8mL_Y^2}$$

$$= \left(\frac{n_X^2}{L_X^2} + \frac{n_Y^2}{L_Y^2}\right)\frac{h^2}{8m} \qquad \boxed{\text{Energies of a particle in a two-dimensional box}} \tag{9.13b}$$

There are two quantum numbers, n_X and n_Y, each allowed the values 1, 2, ... independently.

An especially interesting case arises when the region is a square, with $L_X = L_Y = L$. The allowed energies are then

$$E_{n_X,n_Y} = (n_X^2 + n_Y^2)\frac{h^2}{8mL^2} \qquad (9.14)$$

This result shows that two different wavefunctions may correspond to the same energy. For example, the wavefunctions with $n_X = 1$, $n_Y = 2$ and $n_X = 2$, $n_Y = 1$ are different

$$\psi_{1,2}(x,y) = \frac{2}{L}\sin\left(\frac{\pi x}{L}\right)\sin\left(\frac{2\pi y}{L}\right)$$

$$\psi_{2,1}(x,y) = \frac{2}{L}\sin\left(\frac{2\pi x}{L}\right)\sin\left(\frac{\pi y}{L}\right) \qquad (9.15)$$

but both have the energy $5h^2/8mL^2$. Different states with the same energy are said to be **degenerate**. Degeneracy occurs commonly in atoms, and is a feature that underlies the structure of the periodic table.

The separation of variables procedure is very important because it tells us that energies of independent systems are additive and that their wavefunctions are products of simpler component wavefunctions. We shall encounter it several times in later chapters.

9.5 **Rotation**

Rotational motion is the starting point for our discussion of the atom, in which electrons are free to circulate around a nucleus.

To describe rotational motion we need to focus on the **angular momentum, J**, a vector with a length proportional to the rate of circulation and a direction that indicates the axis of rotation (Fig. 9.30). The magnitude of the angular momentum of a particle that is traveling on a circular path of radius r is defined as

$$J = pr \qquad \boxed{\text{Magnitude of the angular momentum}\atop\text{of a particle moving on a circular path}} \qquad (9.16)$$

where p is the magnitude of its linear momentum ($p = mv$) at any instant. A particle that is traveling at high speed in a circle has a higher angular momentum than a particle of the same mass traveling more slowly. An object with a high angular momentum (such as a flywheel) requires a strong braking force (more precisely, a strong torque) to bring it to a standstill.

(a) A particle on a ring

Consider a particle of mass m moving in a horizontal circular path of radius r. The energy of the particle is entirely kinetic because the potential energy is constant and can be set equal to zero everywhere. We can therefore write $E = p^2/2m$. By using eqn 9.16, we can express this energy in terms of the angular momentum as

$$E = \frac{J_z^2}{2mr^2} \qquad \boxed{\text{Kinetic energy of a particle}\atop\text{moving on a circular path}} \qquad (9.17)$$

where J_z is the angular momentum for rotation around the z-axis (the axis perpendicular to the plane). The quantity mr^2 is the **moment of inertia** of the particle

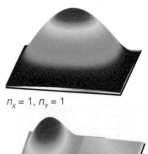

$n_X = 1$, $n_Y = 1$

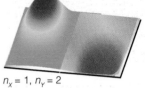

$n_X = 1$, $n_Y = 2$

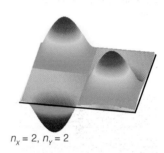

$n_X = 2$, $n_Y = 2$

Fig. 9.29 Three wavefunctions of a particle confined to a rectangular surface.

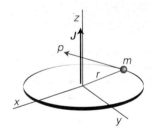

Fig. 9.30 The angular momentum of a particle of mass m on a circular path of radius r in the xy-plane is represented by a vector J perpendicular to the plane and of magnitude pr.

Mathematical toolkit 9.1   *Vectors*

A vector quantity has both magnitude and direction. The vector v shown in the figure has components on the x-, y-, and z-axes with magnitudes v_x, v_y, and v_z, respectively. The direction of each of the components is denoted with a plus sign or minus sign. For example, if $v_x = -1.0$, the x-component of the vector v has a magnitude of 1.0 and points in the $-x$ direction. The magnitude of the vector is denoted v or $|v|$ and is given by

$$v = (v_x^2 + v_y^2 + v_z^2)^{1/2}$$

Operations involving vectors are not as straightforward as those involving numbers. We describe the operations we need for this text in *Mathematical toolkit 11.1*.

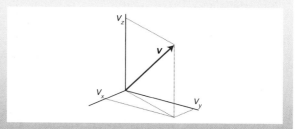

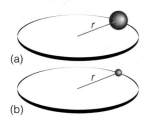

(a)

(b)

Fig. 9.31 A particle traveling on a circular path has a moment of inertia I that is given by mr^2. (a) This heavy particle has a large moment of inertia about the central point; (b) this light particle is traveling on a path of the same radius, but it has a smaller moment of inertia. The moment of inertia plays a role in circular motion that is the analog of the mass for linear motion: a particle with a high moment of inertia is difficult to accelerate into a given state of rotation and requires a strong braking force to stop its rotation.

about the z-axis and denoted I: a heavy particle in a path of large radius has a large moment of inertia (Fig. 9.31). It follows that the energy of the particle is

$$E = \frac{J_z^2}{2I}$$

> Kinetic energy of a particle on a ring in terms of the moment of inertia   (9.18)

Now we use the de Broglie relation to see that the energy of rotation is quantized. To do so, we express the angular momentum in terms of the wavelength of the particle:

$$J_z = pr = \frac{hr}{\lambda}$$

> The angular momentum in terms of the de Broglie wavelength   (9.19)

Suppose for the moment that λ can take an arbitrary value. In that case, the amplitude of the wavefunction depends on the angle ϕ as shown in Fig. 9.32. When the angle increases beyond 2π (that is, 360°), the wavefunction continues to change. For an arbitrary wavelength it gives rise to a different value at each point and the interference between the waves on successive circuits cancels the wave on its previous circuit. Thus, this arbitrarily selected wave cannot survive in the system. An acceptable solution is obtained only if the wavefunction reproduces itself on successive circuits: $\psi(\phi + 2\pi) = \psi(\phi)$. We say that the wavefunction must satisfy **cyclic boundary conditions**. It follows that acceptable wavefunctions have wavelengths that are given by the expression

$$\lambda = \frac{2\pi r}{n} \qquad n = 0, 1, \ldots \tag{9.20}$$

where the value $n = 0$, which gives an infinite wavelength, corresponds to a uniform amplitude. It follows that the permitted energies are

$$E_n = \frac{(hr/\lambda)^2}{2I} = \frac{(nh/2\pi)^2}{2I} = \frac{n^2 \hbar^2}{2I} \tag{9.21}$$

with $n = 0, \pm 1, \pm 2, \ldots$.

It is conventional in the discussion of rotational motion to denote the quantum number by m_l in place of n. Therefore, the final expression for the energy levels is

$$E_{m_l} = \frac{m_l^2 \hbar^2}{2I} \qquad m_l = 0, \pm 1, \ldots$$

> Quantized energies of a particle on a ring   (9.22)

These energy levels are drawn in Fig. 9.33. The occurrence of m_l^2 in the expression for the energy means that two states of motion, such as those with $m_l = +1$

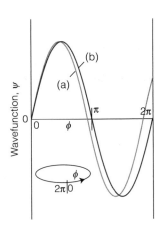

Fig. 9.32 Two solutions of the Schrödinger equation for a particle on a ring. The circumference has been opened out into a straight line; the points at $\phi = 0$ and 2π are identical. The solution labeled (a) is unacceptable because it has different values after each circuit and so interferes destructively with itself. The solution labeled (b) is acceptable because it reproduces itself on successive circuits.

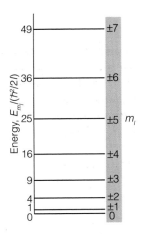

Fig. 9.33 The energy levels of a particle that can move on a circular path. Classical physics allowed the particle to travel with any energy; quantum mechanics, however, allows only discrete energies. Each energy level, other than the one with $m_l = 0$, is doubly degenerate because the particle may rotate either clockwise or counterclockwise with the same energy.

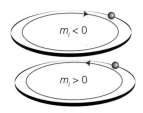

Fig. 9.34 The significance of the sign of m_l. When $m_l < 0$, the particle travels in a counterclockwise direction as viewed from below; when $m_l > 0$, the motion is clockwise.

and $m_l = -1$, both correspond to the same energy. This degeneracy arises from the fact that the direction of rotation, represented by positive and negative values of m_l, does not affect the energy of the particle. All the states with $|m_l| > 0$ are doubly degenerate because two states correspond to the same energy for each value of $|m_l|$. The state with $m_l = 0$, the lowest energy state of the particle, is **nondegenerate**, meaning that only one state has a particular energy (in this case, zero).

An important additional conclusion is that the *angular momentum of a particle is quantized*. We can use the relation between angular momentum and linear momentum (angular momentum $J = pr$), and between linear momentum and the allowed wavelengths of the particle ($\lambda = 2\pi r/m_l$), to conclude that the angular momentum of a particle around the z-axis is confined to the values

$$J_z = pr = \frac{hr}{\lambda} = \frac{hr}{2\pi r/m_l} = m_l \times \frac{h}{2\pi} \tag{9.23}$$

That is, the angular momentum of the particle around the axis is confined to the values

$$J_z = m_l \hbar \qquad \boxed{\text{z-component of the angular momentum of a particle on a ring}} \tag{9.24}$$

with $m_l = 0, \pm 1, \pm 2, \dots$. Positive values of m_l correspond to clockwise rotation (as seen from below) and negative values correspond to counterclockwise rotation (Fig. 9.34). The quantized motion can be thought of in terms of the rotation of a bicycle wheel that can rotate only with a discrete series of angular momenta, so that as the wheel is accelerated, the angular momentum jerks from the values 0 (when the wheel is stationary) to $\hbar, 2\hbar, \dots$ but can have no intermediate value.

A final point concerning the rotational motion of a particle is that it does not have a zero-point energy: m_l may take the value 0, so E may be zero. This conclusion is also consistent with the uncertainty principle. Although the particle is certainly between the angles 0 and 360° on the ring, that range is equivalent to not knowing anything about where it is on the ring. Consequently, the angular momentum may be specified exactly, and a value of zero is possible. When the angular momentum is zero precisely, the energy of the particle is also zero precisely.

Case study 9.2   *The electronic structure of phenylalanine*

Just as the particle in a box gives us some understanding of the distribution and energies of π electrons in linear conjugated systems, the particle on a ring is a useful model for the distribution of π electrons around a cyclic conjugated system.

Consider the π electrons of the phenyl group of the amino acid phenylalanine (Atlas A14). We may treat the group as a circular ring of radius 140 pm, with six electrons in the conjugated system moving along the perimeter of the ring. As in *Case study* 9.1, we assume that only one electron per carbon atom is allowed to move freely around the ring and that in the ground state of the molecule each level is occupied by two electrons. Therefore, only the $m_l = 0, +1$, and -1 levels are occupied (with the last two states being degenerate). From eqn 9.22, the energy separation between the $m_l = \pm 1$ and the $m_l = \pm 2$ levels is

$$\Delta E = E_{\pm 2} - E_{\pm 1} = (4 - 1)\frac{(1.054 \times 10^{-34}\,\mathrm{J\,s})^2}{2 \times (9.109 \times 10^{-31}\,\mathrm{kg}) \times (1.40 \times 10^{-10}\,\mathrm{m})^2}$$

$$= 9.33 \times 10^{-19}\,\mathrm{J}$$

This energy separation corresponds to an absorption frequency of 1409 THz and a wavelength of 213 nm; the experimental value for a transition of this kind is 260 nm.

Even though the model is primitive, it gives insight into the origin of the quantized π-electron energy levels in cyclic conjugated systems, such as the aromatic side chains of phenylalanine, tryptophan, and tyrosine, the purine and pyrimidine bases in nucleic acids, the heme group, and the chlorophylls.

(b) A particle on a sphere

We now consider a particle of mass m free to move around a central point at a constant radius r. That is, it is free to travel anywhere on the surface of a sphere of radius r. To calculate the energy of the particle, we let—as we did for motion on a ring—the potential energy be zero wherever it is free to travel. Furthermore, when we take into account the requirement that the wavefunction should match as a path is traced over the poles as well as around the equator of the sphere surrounding the central point, we define two cyclic boundary conditions (Fig. 9.35). Solution of the Schrödinger equation leads to the following expression for the permitted energies of the particle:

$$E = l(l+1)\frac{\hbar^2}{2I} \qquad l = 0, 1, 2, \ldots \qquad \boxed{\text{Quantized energies of a particle on a sphere}} \quad (9.25)$$

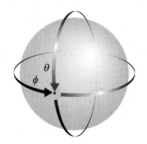

Fig. 9.35 The wavefunction of a particle on the surface of a sphere must satisfy two cyclic boundary conditions. The wavefunction must reproduce itself after the angles ϕ and θ are swept by 360° (or 2π radians). This requirement leads to two quantum numbers for its state of angular momentum.

As before, the energy of the rotating particle is related classically to its angular momentum J by $E = J^2/2I$. Therefore, by comparing $E = J^2/2I$ with eqn 9.25, we can deduce that because the energy is quantized, the magnitude of the angular momentum is also confined to the values

$$J = \{l(l+1)\}^{1/2}\hbar \qquad l = 0, 1, 2\ldots \qquad \boxed{\text{Magnitude of the angular momentum of a particle on a sphere}} \quad (9.26)$$

where l is the **orbital angular momentum quantum number**. For motion in three dimensions, the vector J has components J_x, J_y, and J_z along the x-, y-, and z-axes, respectively (Fig. 9.36). We have already seen (in the context of rotation in a plane) that the angular momentum about the z-axis is quantized and that it has the values $J_z = m_l \hbar$. However, it is a consequence of there being two cyclic boundary conditions that the values of m_l are restricted, so the z-component of the angular momentum is given by

$$J_z = m_l \hbar \qquad m_l = l, l-1, \ldots, -l \qquad \boxed{\text{Magnitude of the z-component of the angular momentum of a particle on a sphere}} \quad (9.27)$$

and m_l is now called the **magnetic quantum number**. We note that for a given value of l there are $2l + 1$ permitted values of m_l. Therefore, because the energy is independent of m_l (because m_l does not appear in the expression for the energy, eqn 9.25) a level with quantum number l is $(2l+1)$-fold degenerate.

9.6 Vibration

The atoms in a molecule vibrate about their equilibrium positions, and the following description of molecular vibrations sets the stage for a discussion of vibrational spectroscopy (Chapter 12), an important experimental technique for the structural characterization of biological molecules.

The simplest model that describes molecular vibrations is the **harmonic oscillator**, in which a particle is restrained by a spring that obeys **Hooke's law** of force, that the restoring force is proportional to the displacement, x:

$$\text{restoring force} = -k_f x \qquad \boxed{\text{Hooke's law}} \quad (9.28a)$$

The constant of proportionality k_f is called the **force constant**: a stiff spring has a high force constant and a weak spring has a low force constant. We show in the following *Justification* that the potential energy of a particle subjected to this force increases as the square of the displacement, and specifically

$$V(x) = \tfrac{1}{2}k_f x^2 \qquad \boxed{\text{Potential energy of a harmonic oscillator}} \quad (9.28b)$$

The variation of V with x is shown in Fig. 9.37: it has the shape of a parabola (a curve of the form $y = ax^2$), and we say that a particle undergoing harmonic motion has a 'parabolic potential energy'.

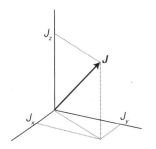

Fig. 9.36 For motion in three dimensions, the angular momentum vector J has components J_x, J_y, and J_z on the x-, y-, and z-axes, respectively.

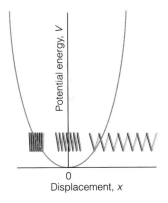

Fig. 9.37 The parabolic potential energy characteristic of a harmonic oscillator. Positive displacements correspond to extension of the spring; negative displacements correspond to compression of the spring.

Justification 9.2 *Potential energy of a harmonic oscillator*

Force is the negative slope of the potential energy: $F = -dV/dx$. Because the infinitesimal quantities may be treated as any other quantity in algebraic manipulations, we rearrange the expression into $dV = -Fdx$ and then integrate both sides from $x = 0$, where the potential energy is $V(0)$, to x, where the potential energy is $V(x)$:

$$V(x) - V(0) = -\int_0^x F\,dx$$

Now substitute $F = -k_f x$:

$$V(x) - V(0) = -\int_0^x (-k_f x)\,dx = k_f \int_0^x x\,dx = \tfrac{1}{2}k_f x^2$$

We are free to choose $V(0) = 0$, which then gives eqn 9.28b.

Unlike the earlier cases we considered, the potential energy varies with position, so we have to use $V(x)$ in the Schrödinger equation and solve it using the techniques for solving differential equations. Then we have to select the solutions that satisfy the boundary conditions, which in this case means that they must fit into the parabola representing the potential energy. More precisely, the wavefunctions must all go to zero for large displacements from $x = 0$: they do not have to go abruptly to zero at the edges of the parabola.

The solutions of the Schrödinger equation for a harmonic oscillator are quite hard to find, but once found, they turn out to be very simple. For instance, the energies of the solutions that satisfy the boundary conditions are

$$E_v = (v + \tfrac{1}{2})hv \quad v = 0, 1, 2 \dots \quad v = \frac{1}{2\pi}\left(\frac{k_f}{m}\right)^{1/2} \qquad \boxed{\begin{array}{l}\text{Quantized energies}\\\text{of a harmonic}\\\text{oscillator}\end{array}} \quad (9.29)$$

where m is the mass of the particle and v is the **vibrational quantum number**.[6] These energies form a uniform ladder of values separated by hv (Fig. 9.38). The separation is large for stiff springs and low masses.

Figure 9.39 shows the shapes of the first few wavefunctions of a harmonic oscillator. The ground-state wavefunction (corresponding to $v = 0$ and having the zero-point energy $\tfrac{1}{2}hv$) is a bell-shaped curve, a curve of the form e^{-x^2} (a Gaussian function; see *Mathematical toolkit F.2*), with no nodes. This shape shows that the particle is most likely to be found at $x = 0$ (zero displacement) but may be found at greater displacements with decreasing probability. The first excited wavefunction has a node at $x = 0$ and peaks on either side. Therefore, in this state, the particle will be found most probably with the 'spring' stretched or compressed to the same amount. In all the states of a harmonic oscillator the wavefunctions extend beyond the limits of motion of a classical oscillator (Fig. 9.40), but the extent decreases as v increases. This penetration into classically forbidden regions is another example of quantum mechanical tunneling, in this case tunneling into rather than through a barrier.

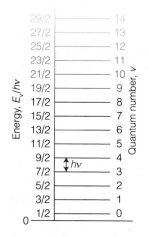

Fig. 9.38 The array of energy levels of a harmonic oscillator. The separation depends on the mass and the force constant. Note the zero-point energy.

1 The peptide link, –CONH–

Case study 9.3   *The vibration of the N–H bond of the peptide link*

Atoms vibrate relative to one another in molecules with the bond acting like a spring. Therefore, eqn 9.29 describes the allowed vibrational energy levels of molecules. Here we consider the vibration of the N–H bond of the peptide link (**1**), making the approximation that the relatively heavy C, N, and O atoms

[6] Be very careful to distinguish the quantum number v (italic vee) from the frequency v (Greek nu).

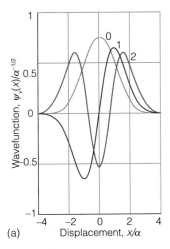

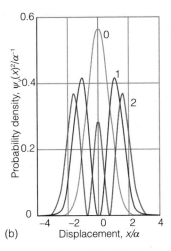

Fig. 9.39 (a) The wavefunctions and (b) the probability densities of the first three states of a harmonic oscillator. Note how the probability of finding the oscillator at large displacements increases as the state of excitation increases. The wavefunctions and displacements are expressed in terms of the parameter $\alpha = (\hbar^2/mk_f)^{1/4}$.

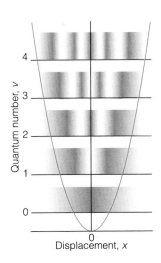

Fig. 9.40 A schematic illustration of the probability density for finding a harmonic oscillator at a given displacement. Classically, the oscillator cannot be found at displacements at which its total energy is less than its potential energy (because the kinetic energy cannot be negative). A quantum oscillator, however, can tunnel into regions that are classically forbidden.

form a stationary anchor for the very light H atom. That is, only the H atom moves, vibrating as a simple harmonic oscillator.

Because the force constant for an N–H bond can be set equal to 700 N m^{-1} and the mass of the ^{1}H atom is $m_H = 1.67 \times 10^{-27}$ kg, we write

$$\nu = \frac{1}{2\pi}\left(\frac{k_f}{m}\right)^{1/2} = \frac{1}{2\pi}\left(\frac{700\,\text{N m}^{-1}}{1.67 \times 10^{-27}\,\text{kg}}\right)^{1/2} = 1.03 \times 10^{14}\,\text{Hz}$$

or 103 THz. Therefore, we expect that radiation with a frequency of 103 THz, in the infrared range of the spectrum, induces a spectroscopic transition between $\upsilon = 0$ and the $\upsilon = 1$ levels of the oscillator. We shall see in Chapter 12 that the concepts just described represent the starting point for the interpretation of vibrational (infrared) spectroscopy, an important technique for the characterization of biopolymers both in solution and inside biological cells.

A note on good practice
To calculate the vibrational frequency precisely, we need to specify the nuclide. Also, the mass to use is the actual atomic mass in kilograms, not the element's molar mass. In Section 12.3 we explain how to take into account the motion of both atoms in a bond by introducing the 'effective mass' of an oscillator.

Hydrogenic atoms

Quantum theory provides the foundation for the description of atomic structure. A **hydrogenic atom** is a one-electron atom or ion of general atomic number Z. Hydrogenic atoms include H, He$^+$, Li^{2+}, C^{5+}, and even U^{91+}. A **many-electron atom** is an atom or ion that has more than one electron. Many-electron atoms include all neutral atoms other than H. For instance, helium, with its two electrons, is a many-electron atom in this sense. Hydrogenic atoms, and H in particular, are important because the Schrödinger equation can be solved for them and their structures can be discussed exactly. Furthermore, the concepts learned from a study of hydrogenic atoms can be used to describe the structures of many-electron atoms and of molecules too.

Much of the material in the remainder of this chapter is a review of introductory chemistry. However, we provide some detail not commonly covered in introductory chemistry, with the goal of showing how core concepts of quantum mechanics can be applied to atoms. The material also sets the stage for the discussion of molecules in Chapter 10.

9.7 The permitted energy levels of hydrogenic atoms

Hydrogenic atoms provide the starting point for the discussion of many-electron atoms and hence of the properties of all atoms and their abilities to form bonds and hence aggregate into molecules.

The quantum mechanical description of the structure of a hydrogenic atom is based on Rutherford's **nuclear model**, in which the atom is pictured as consisting of an electron outside a central nucleus of charge $+Ze$, where Z is the atomic number. To derive the details of the structure of this type of atom, we have to set up and solve the Schrödinger equation in which the potential energy, V, is the Coulombic potential energy (*Fundamentals* F.3 and eqn F.13) for the interaction between the nucleus of charge $Q_1 = +Ze$ and the electron of charge $Q_2 = -e$:

$$V = -\frac{Ze^2}{4\pi\varepsilon_0 r} \tag{9.30}$$

where $\varepsilon_0 = 8.854 \times 10^{-12}\,\mathrm{C^2\,J^{-1}\,m^{-1}}$ is the vacuum permittivity. We also need to identify the appropriate boundary conditions that the wavefunctions must satisfy in order to be acceptable. For a hydrogenic atom, these conditions are that the wavefunction must not become infinite anywhere and that it must repeat itself (just like the particle on a sphere) as we circle the nucleus either over the poles or around the equator.

With a lot of work, the Schrödinger equation with this potential energy and these boundary conditions can be solved, and we shall summarize the results. As usual, the need to satisfy boundary conditions leads to the conclusion that the electron can have only certain energies. Schrödinger himself found that for a hydrogenic atom of atomic number Z with a nucleus of mass m_N, the allowed energy levels are given by the expression

$$E_n = -hc\mathcal{R}\,\frac{Z^2}{n^2} \qquad hc\mathcal{R} = \frac{\mu e^4}{32\pi^2\varepsilon_0^2\hbar^2} \qquad \mu = \frac{m_e m_N}{m_e + m_N} \qquad \boxed{\text{Energy levels of a hydrogenic atom}} \tag{9.31}$$

and $n = 1, 2, \dots$. The quantity $\mathcal{R}$, the *Rydberg constant*, has the dimensions of a wavenumber and is commonly reported in units of reciprocal centimeters ($\mathrm{cm^{-1}}$). The quantity μ is the **reduced mass**. For all except the most precise considerations, the mass of the nucleus is so much bigger than the mass of the electron that the latter may be neglected in the denominator of μ, and then $\mu \approx m_e$.

Let's unpack the significance of eqn 9.31:

1. The quantum number n is called the **principal quantum number**. It gives the energy of the electron in the atom by substituting its value into eqn 9.31.

The resulting energy levels are depicted in Fig. 9.41. Note how they are widely separated at low values of n but then converge as n increases. At low values of n the electron is confined close to the nucleus by the pull between opposite charges and the energy levels are widely spaced like those of a particle in a narrow box. At high values of n, when the electron has such a high energy that it can travel out

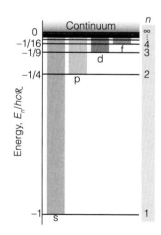

Fig. 9.41 The energy levels of the hydrogen atom. The energies are relative to a proton and an infinitely distant, stationary electron.

to large distances, the energy levels are close together, like those of a particle in a large box.

2. All the energies are negative, which signifies that an electron in an atom has a lower energy than when it is free.

The zero of energy (which occurs at $n = \infty$) corresponds to the infinitely widely separated (so that the Coulomb potential energy is zero) and stationary (so that the kinetic energy is zero) electron and nucleus. The state of lowest, most negative energy, the ground state of the atom, is the one with $n = 1$ (the lowest permitted value of n and hence the most negative value of the energy). The energy of this state is

$$E_1 = -hc\mathcal{R}Z^2$$

The negative sign means that the ground state lies at $hc\mathcal{R}Z^2$ below the energy of the infinitely separated stationary electron and nucleus.

The minimum energy needed to remove an electron completely from an atom is called the **ionization energy**, I. For a hydrogen atom, the ionization energy is the energy required to raise the electron from the ground state with energy $E_1 = -hc\mathcal{R}$ to the state corresponding to complete removal of the electron (the state with $n = \infty$ and zero energy). Therefore, the energy that must be supplied is (using $\mu \approx m_e$)

$$I_H = \frac{m_e e^4}{32\pi^2\varepsilon_0^2\hbar^2} = 2.179 \times 10^{-18}\,\text{J}$$

or 2.179 aJ (1 aJ $= 10^{-18}$ J). This energy corresponds to 13.60 eV and (after multiplication by N_A, Avogadro's constant) to 1312 kJ mol^{-1}.

3. The energy of a given level, and therefore the separation of neighboring levels, is proportional to Z^2.

This dependence on Z^2 stems from two effects. First, an electron at a given distance from a nucleus of charge $+Ze$ has a potential energy that is Z times larger than that of an electron at the same distance from a proton (for which $Z = 1$). However, the electron is drawn into the vicinity of the nucleus by the greater nuclear charge, so it is more likely to be found closer to the nucleus of charge Z than the proton. This effect is also proportional to Z, so overall the energy of an electron can be expected to be proportional to the square of Z, one factor of Z representing the Z times greater strength of the nuclear field and the second factor of Z representing the fact that the electron is Z times more likely to be found closer to the nucleus.

Self-test 9.4 Predict the ionization energy of He$^+$ given that the ionization energy of H is 13.60 eV. *Hint*: Decide how the energy of the ground state varies with Z.

Answer: $I_{\text{He}^+} = 4I_H = 54.40$ eV

9.8 Atomic orbitals

The properties of elements and the formation of chemical bonds are consequences of the shapes and energies of the wavefunctions that describe the distribution of electrons in atoms. We need information about the shapes of these wavefunctions to understand why compounds of carbon adopt the conformations that are responsible for the unique biological functions of such molecules as proteins, nucleic acids, and lipids.

The wavefunction of the electron in a hydrogenic atom is called an **atomic orbital**. The name is intended to express something less definite than the 'orbit' of classical mechanics. An electron that is described by a particular wavefunction is said to 'occupy' that orbital. So, in the ground state of the atom, the electron occupies the orbital of lowest energy (that with $n = 1$).

(a) Shells and subshells

We have remarked that there are three boundary conditions on the orbitals: that the wavefunctions must not become infinite, that they must match as they encircle the equator, and that they must match as they encircle the poles. Each boundary condition gives rise to a quantum number, so each orbital is specified by three quantum numbers that act as a kind of 'address' of the electron in the atom. We can suspect that the values allowed to the three quantum numbers are linked because, for instance, to get the right shape on a polar journey, we also have to note how the wavefunction changes shape as it wraps around the equator.

The quantum numbers are:

- The **principal quantum number** n, which determines the energy of the orbital through eqn 9.31 and has values

$$n = 1, 2, \ldots \text{(without limit)}$$ Principal quantum number

- The **orbital angular momentum quantum number** l,[7] which is restricted to the values

$$l = 0, 1, 2, \ldots, n - 1$$ Orbital angular momentum quantum number

For a given value of n, there are n allowed values of l: all the values are positive (for example, if $n = 3$, then l may be 0, 1, or 2).

- The **magnetic quantum number**, m_l, which is confined to the values

$$m_l = l, l - 1, l - 2, \ldots, -l$$ Magnetic quantum number

For a given value of l, there are $2l + 1$ values of m_l (for example, when $l = 3$, m_l may have any of the seven values $+3, +2, +1, 0, -1, -2, -3$).

A note on good practice
Always give the sign of m_l, even when it is positive. So, write $m_l = +1$, not $m_l = 1$.

It follows from the restrictions on the values of the quantum numbers that there is only one orbital with $n = 1$, because when $n = 1$ the only value that l can have is 0, and that in turn implies that m_l can have only the value 0. Likewise, there are four orbitals with $n = 2$, because l can take the values 0 and 1, and in the latter case m_l can have the three values $+1$, 0, and -1. In general, there are n^2 orbitals with a given value of n.

Because the energy of a hydrogenic atom depends only on the principal quantum number n, *orbitals of the same value of n but different values of l and m_l have the same energy*. It follows that all orbitals with the same value of n are degenerate. But be careful: this statement applies only to hydrogenic atoms. A second point is that the average distance of an electron from the nucleus of a hydrogenic atom of atomic number Z increases as n increases. As Z increases, the average distance is reduced because the increasing nuclear charge draws the electron closer in.

---

[7] This quantum number is also called by its older name, the *azimuthal quantum number*.

The degeneracy of all orbitals with the same value of n (remember that there are n^2 of them) and their similar mean radii is the basis of saying that they all belong to the same **shell** of the atom. It is common to refer to successive shells by letters:

$$n \quad 1 \quad 2 \quad 3 \quad 4\ldots$$
$$ \quad K \quad L \quad M \quad N\ldots$$

Thus, all four orbitals of the shell with $n = 2$ form the L shell of the atom.

Orbitals with the same value of n but different values of l belong to different **subshells** of a given shell. These subshells are denoted by the letters s, p, ... using the following correspondence:

$$l \quad 0 \quad 1 \quad 2 \quad 3\ldots$$
$$ \quad s \quad p \quad d \quad f\ldots$$

For the shell with $n = 1$, there is only one subshell, the one with $l = 0$. For the shell with $n = 2$ (which allows $l = 0, 1$), there are two subshells, namely the 2s subshell (with $l = 0$) and the 2p subshell (with $l = 1$). The general pattern of the first three shells and their subshells is shown in Fig. 9.42. In a hydrogenic atom, all the subshells of a given shell correspond to the same energy (because, as we have seen, the energy depends on n and not on l).

We have seen that if the orbital angular momentum quantum number is l, then m_l can take the $2l + 1$ values $m_l = 0, \pm 1, \ldots, \pm l$. Therefore, each subshell contains $2l + 1$ individual orbitals (corresponding to the $2l + 1$ values of ml for each value of l). It follows that in any given subshell, the number of orbitals is

$$s \quad p \quad d \quad f\ldots$$
$$1 \quad 3 \quad 5 \quad 7\ldots$$

An orbital with $l = 0$ (and necessarily $m_l = 0$) is called an **s orbital**. A p subshell ($l = 1$) consists of three **p orbitals** (corresponding to $m_l = +1, 0, -1$). An electron that occupies an s orbital is called an **s electron**. Similarly, we can speak of p, d, ... electrons according to the orbitals they occupy.

Self-test 9.5 How many orbitals are there in a shell with $n = 5$ and what is their designation?

Answer: 25; one s, three p, five d, seven f, nine g

(b) The shapes of s orbitals

We saw in Section 9.4c that in certain cases a wavefunction can be separated into factors that depend on different coordinates and that the Schrödinger equation separates into simpler versions for each variable. Application of this separation of variables procedure to the hydrogen atom leads to a Schrödinger equation that separates into one equation for the electron moving around the nucleus (the analog of the particle on a sphere) and an equation for the radial dependence. The wavefunction is written as

$$\psi_{n,l,m_l}(r,\theta,\phi) = Y_{l,m_l}(\theta,\phi)R_{n,l}(r) \qquad \boxed{\text{Wavefunctions of hydrogenic atoms}} \quad (9.32)$$

The factor $R(r)$ is a function of the distance r from the nucleus and is known as the **radial wavefunction**. Its form depends on the values of n and l but is

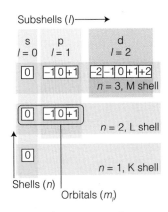

Fig. 9.42 The structures of atoms are described in terms of shells of electrons that are labeled by the principal quantum number n and a series of n subshells of these shells, with each subshell of a shell being labeled by the quantum number l. Each subshell consists of $2l + 1$ orbitals.

independent of m_l: that is, all orbitals of the same subshell of a given shell have the same radial wavefunction. In other words, all p orbitals of a shell have the same radial wavefunction, all d orbitals of a shell likewise (but different from that of the p orbitals), and so on. The other factor, $Y(\theta,\phi)$, is called the **angular wavefunction**; it is independent of the distance from the nucleus but varies with the angles θ and ϕ. This factor depends on the quantum numbers l and m_l. Therefore, regardless of the value of n, orbitals with the same value of l and m_l have the same angular wavefunction. In other words, for a given value of m_l, a d orbital has the same angular shape regardless of the shell to which it belongs.

The mathematical form of a 1s orbital (the wavefunction with $n = 1$, $l = 0$, and $m_l = 0$) for a hydrogen atom is

$$\psi = \frac{1}{(4\pi)^{1/2}}\left(\frac{4}{a_0^3}\right)^{1/2} e^{-r/a_0} = \frac{1}{(\pi a_0^3)^{1/2}} e^{-r/a_0}$$

$$a_0 = \frac{4\pi\varepsilon_0 \hbar^2}{m_e e^2}$$

Wavefunction of a 1s electron in a hydrogen atom   (9.33)

In this case the angular wavefunction, $Y_{0,0} = 1/(4\pi)^{1/2}$, is a constant, independent of the angles θ and ϕ. You should recall that in Section 9.2 we anticipated that a wavefunction for an electron in the ground state of a hydrogen atom has a wavefunction proportional to e^{-r}: eqn 9.33 is its precise form. The constant a_0 is called the **Bohr radius** (because it occurred in the equations based on an early model of the structure of the hydrogen atom proposed by the Danish physicist Niels Bohr) and has the value 52.92 pm.

The amplitude of a 1s orbital depends only on the radius, r, of the point of interest and is independent of angle (the latitude and longitude of the point). Therefore, the orbital has the same amplitude at all points at the same distance from the nucleus regardless of direction. Because, according to the Born interpretation (Section 9.2b), the probability density of the electron is proportional to the square of the wavefunction, we now know that the electron will be found with the same probability in any direction (for a given distance from the nucleus). We summarize this angular independence by saying that a 1s orbital is **spherically symmetrical**. Because the same factor Y occurs in all orbitals with $l = 0$, all s orbitals have the same spherical symmetry (but different radial dependences).

The wavefunction in eqn 9.33 decays exponentially toward zero from a maximum value at the nucleus (Fig. 9.43). It follows that *the most probable point at which the electron will be found is at the nucleus itself.* A method of depicting the probability of finding the electron at each point in space is to represent ψ^2 by the density of shading in a diagram (Fig. 9.44). A simpler procedure is to show only the **boundary surface**, the shape that captures about 90 per cent of the electron probability. For the 1s orbital, the boundary surface is a sphere centered on the nucleus (Fig. 9.45).

We often need to know the total probability that an electron will be found in the range r to $r + \delta r$ from a nucleus regardless of its angular position (Fig. 9.46). We can calculate this probability by combining the wavefunction in eqn 9.33 with the Born interpretation and find that for s orbitals, the answer can be expressed as

$$\text{probability} = P(r)\delta r \text{ with } P(r) = 4\pi r^2 \psi^2$$

Radial distribution function of an s orbital   (9.34)

The function P is called the **radial distribution function**.

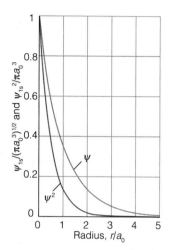

Fig. 9.43 The radial dependence of the wavefunction of a 1s orbital ($n = 1$, $l = 0$) and the corresponding probability density. The quantity a_0 is the Bohr radius (52.92 pm).

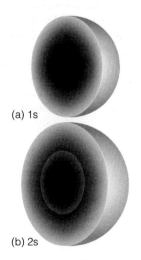

(a) 1s

(b) 2s

Fig. 9.44 Representations of the first two hydrogenic s orbitals, (a) 1s and (b) 2s, in terms of the electron densities (as represented by the density of shading).

Justification 9.3 *The radial distribution function*

Consider two spherical shells centered on the nucleus, one of radius r and the other of radius $r + \delta r$. The probability of finding the electron at a radius r regardless of its direction is equal to the probability of finding it between these two spherical surfaces. The volume of the region of space between the surfaces is equal to the surface area of the inner shell, $4\pi r^2$, multiplied by the thickness, δr, of the region and is therefore $4\pi r^2 \delta r$. According to the Born interpretation, the probability of finding an electron inside a small volume of magnitude δV is given, for a normalized wavefunction, by the value of $\psi^2 \delta V$. Therefore, interpreting V as the volume of the shell, we obtain

$$\text{probability} = \psi^2 \times (4\pi r^2 \delta r)$$

as in eqn 9.34. The result we have derived is for any s orbital. For orbitals that depend on angle, the more general form is $P(r) = r^2 R(r)^2$, where $R(r)$ is the radial wavefunction.

> **Self-test 9.6** Calculate the probability that an electron in a 1s orbital will be found between a shell of radius a_0 and a shell of radius 1.0 pm greater. *Hint*: Use $r = a_0$ in the expression for the probability density and $\delta r = 1.0$ pm in eqn 9.34.
>
> **Answer:** 0.010

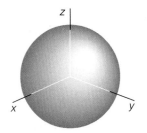

Fig. 9.45 The boundary surface of an s orbital within which there is a high probability of finding the electron.

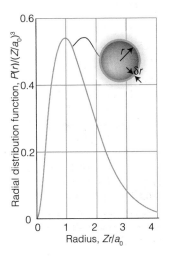

Fig. 9.46 The radial distribution function gives the probability that the electron will be found anywhere in a shell of radius r and thickness δr regardless of angle. The graph shows the output from an imaginary shell-like detector of variable radius and fixed thickness δr.

The radial distribution function tells us the total probability of finding an electron at a distance r from the nucleus regardless of its direction. Because r^2 increases from 0 as r increases but ψ^2 decreases toward 0 exponentially, P starts at 0, goes through a maximum, and declines to 0 again. The location of the maximum marks the most probable *radius* (not point) at which the electron will be found. For a 1s orbital of hydrogen, the maximum occurs at a_0, the Bohr radius. An analogy that might help to fix the significance of the radial distribution function for an electron is the corresponding distribution for the population of the Earth regarded as a perfect sphere. The radial distribution function is zero at the center of the Earth and for the next 6400 km (to the surface of the planet), when it peaks sharply and then rapidly decays again to zero. It remains virtually zero for all radii more than about 10 km above the surface. Almost all the population will be found very close to $r = 6400$ km, and it is not relevant that people are dispersed non-uniformly over a very wide range of latitudes and longitudes. The small probabilities of finding people above and below 6400 km anywhere in the world corresponds to the population that happens to be down mines or living in places as high as Denver or Tibet at the time.

A 2s orbital (an orbital with $n = 2$, $l = 0$, and $m_l = 0$) is also spherical, so its boundary surface is a sphere. Because a 2s orbital spreads farther out from the nucleus than a 1s orbital—because the electron it describes has more energy to climb away from the nucleus—its boundary surface is a sphere of larger radius. The orbital also differs from a 1s orbital in its radial dependence (Fig. 9.47), for although the wavefunction has a nonzero value at the nucleus (like all s orbitals), it passes through zero before commencing its exponential decay toward zero at large distances. We summarize the fact that the wavefunction passes through zero everywhere at a certain radius by saying that the orbital has a **radial node**. A 3s

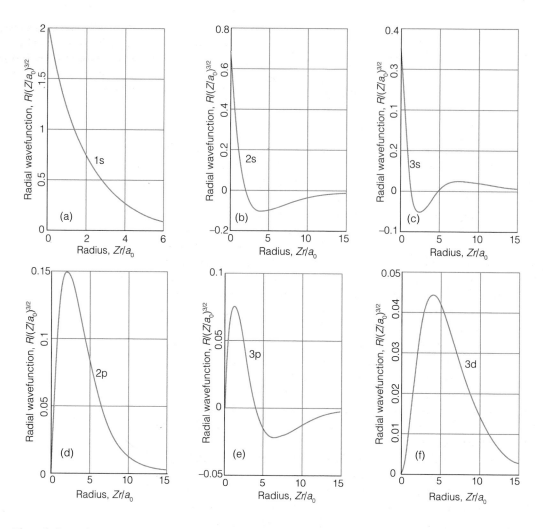

Fig. 9.47 The radial wavefunctions of the hydrogenic 1s, 2s, 3s, 2p, 3p, and 3d orbitals. Note that the s orbitals have a nonzero and finite value at the nucleus. The vertical scales are different in each case.

orbital has two radial nodes; a 4s orbital has three radial nodes. In general, an *ns* orbital has $n - 1$ radial nodes.

(c) The shapes of p orbitals

Now we turn our attention to the p orbitals (orbitals with $l = 1$), which have a double-lobed appearance like that shown in Fig. 9.48. The two lobes are separated by a **nodal plane** that cuts through the nucleus. There is zero probability density for an electron on this plane. Here, for instance, is the explicit form of the $2p_z$ orbital:

$$\psi = \left(\frac{3}{4\pi}\right)^{1/2} \cos\theta \times \tfrac{1}{2}\left(\frac{1}{6a_0^3}\right)^{1/2} \frac{r}{a_0} e^{-r/2a_0}$$

$$= \left(\frac{1}{32\pi a_0^5}\right)^{1/2} r\cos\theta\, e^{-r/2a_0}$$

Wavefunction associated with a $2p_z$ orbital   (9.35)

A brief comment
The radial wavefunction is zero at $r = 0$, but because r does not take negative values that is not a radial node: the wavefunction does not pass *through* zero there. A 2p orbital has an angular node, not a radial node.

Note that because ψ is proportional to r, it is zero at the nucleus, so there is zero probability of finding the electron in a small volume centered on the nucleus. The orbital is also zero everywhere on the plane with $\cos\theta = 0$, corresponding to $\theta = 90°$. The p_x and p_y orbitals are similar but have nodal planes perpendicular to the *x*- and *y*-axes, respectively.

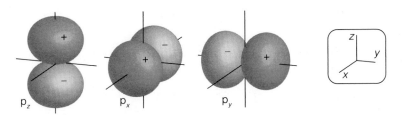

Fig. 9.48 The boundary surfaces of p orbitals. A nodal plane passes through the nucleus and separates the two lobes of each orbital.

The exclusion of the electron from the region of the nucleus is a common feature of all atomic orbitals except s orbitals. To understand its origin, we need to recall from Section 9.5 that the value of the quantum number l tells us the magnitude of the angular momentum of the electron around the nucleus (eqn 9.26, $J = \{l(l+1)\}^{1/2}\hbar$). For an s orbital, the orbital angular momentum is zero (because $l = 0$), and in classical terms the electron does not circulate around the nucleus. Because $l = 1$ for a p orbital, the magnitude of the angular momentum of a p electron is $2^{1/2}\hbar$. As a result, a p electron is flung away from the nucleus by the centrifugal force arising from its motion, but an s electron is not. The same centrifugal effect appears in all orbitals with angular momentum (those for which $l > 0$), such as d orbitals and f orbitals, and all such orbitals have nodal planes that cut through the nucleus.

Each p subshell consists of three orbitals ($m_l = +1, 0, -1$). The three orbitals are normally represented by their boundary surfaces, as depicted in Fig. 9.48. The p_x orbital has a symmetrical double-lobed shape directed along the x-axis, and similarly the p_y and p_z orbitals are directed along the y- and z-axes, respectively. As n increases, the p orbitals become bigger (for the same reason as s orbitals) and have $n - 2$ radial nodes. However, their boundary surfaces retain the double-lobed shape shown in the illustration.

We can now explain the physical significance of the quantum number m_l. It indicates the component of the electron's orbital angular momentum around an arbitrary axis passing through the nucleus. Positive values of m_l correspond to clockwise motion seen from below and negative values correspond to counterclockwise motion. The larger the value of $|m_l|$, the higher is the angular momentum around the arbitrary axis. Specifically:

component of angular momentum = $m_l\hbar$

An s electron (an electron described by an s orbital) has $m_l = 0$ and has no angular momentum about any axis. A p electron can circulate clockwise about an axis as seen from below ($m_l = +1$). Of its total angular momentum of $2^{1/2}\hbar = 1.414\hbar$, an amount $\hbar$ is due to motion around the selected axis (the rest is due to motion around the other two axes). A p electron can also circulate counterclockwise as seen from below ($m_l = -1$) or not at all ($m_l = 0$) about that selected axis.

Except for orbitals with $m_l = 0$, there is not a one-to-one correspondence between the value of m_l and the orbitals shown in the illustrations: we cannot say, for instance, that a p_x orbital has $m_l = +1$. For technical reasons, the orbitals we draw are combinations of orbitals with equal but opposite values of m_l (p_x, for instance, is a combination of the orbitals with $m_l = +1$ and -1).

(d) The shapes of d orbitals

When $n = 3$, l can be 0, 1, or 2. As a result, this shell consists of one 3s orbital, three 3p orbitals, and five 3d orbitals, corresponding to five different values of the magnetic quantum number ($m_l = +2, +1, 0, -1, -2$) for the value $l = 2$ of the orbital angular momentum quantum number. That is, an electron in the d

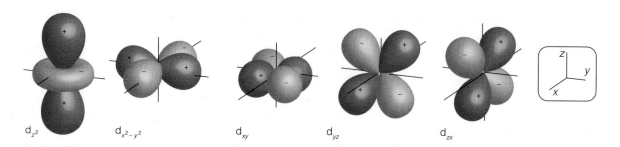

d_{z^2} $d_{x^2-y^2}$ d_{xy} d_{yz} d_{zx}

Fig. 9.49 The boundary surfaces of d orbitals. Two nodal planes in each orbital intersect at the nucleus and separate the four lobes of each orbital.

subshell can circulate with five different amounts of angular momentum about an arbitrary axis $(+2\hbar, +\hbar, 0, -\hbar, -2\hbar)$. As for the p orbitals, d orbitals with opposite values of m_l (and hence opposite senses of motion around an arbitrary axis) may be combined in pairs to give orbitals designated as d_{xy}, d_{yz}, d_{zx}, $d_{x^2-y^2}$, and d_{z^2} and having the shapes shown in Fig. 9.49.

The structures of many-electron atoms

The Schrödinger equation for a many-electron atom is highly complicated because all the electrons interact with one another. Even for a He atom, with its two electrons, no mathematical expression for the orbitals and energies can be given and we are forced to make approximations. Modern computational techniques, however, are able to refine the approximations we are about to make and permit highly accurate numerical calculations of energies and wavefunctions.

The periodic recurrence of analogous ground state electron configurations as the atomic number increases accounts for the periodic variation in the properties of atoms. Here we concentrate on two aspects of atomic periodicity—atomic radius and ionization energy—and see how they can help to explain the different biological roles played by different elements.

9.9 The orbital approximation and the Pauli exclusion principle

Here we begin to develop the rules by which electrons occupy orbitals of different energies and shapes. We shall see that our study of hydrogenic atoms was a crucial step toward our goal of 'building' many-electron atoms and associating atomic structure with biological function.

In the **orbital approximation** we suppose that a reasonable first approximation to the exact wavefunction is obtained by letting each electron occupy (that is, have a wavefunction corresponding to) its 'own' orbital and writing

$$\psi = \psi(1)\psi(2)\ldots \qquad \boxed{\text{Orbital approximation}} \quad (9.36)$$

where $\psi(1)$ is the wavefunction of electron 1, $\psi(2)$ that of electron 2, and so on. We can think of the individual orbitals as resembling the hydrogenic orbitals. For example, consider a model of the helium atom in which both electrons occupy the same 1s orbital, so the wavefunction for each electron is $\psi = (8/\pi a_0^3)^{1/2}e^{-2r/a_0}$ (because $Z = 2$). If electron 1 is at a radius r_1 and electron 2 is at a radius r_2 (and at any angle), then the overall wavefunction for the two-electron atom is

$$\psi = \psi(1)\psi(2) = \left(\frac{8}{\pi a_0^3}\right)^{1/2} e^{-2r_1/a_0} \times \left(\frac{8}{\pi a_0^3}\right)^{1/2} e^{-2r_2/a_0} = \left(\frac{8}{\pi a_0^3}\right)e^{-2(r_1+r_2)/a_0}$$

This description is only approximate because it neglects repulsions between electrons and does not take into account the fact that the nuclear charge is modified by the presence of all the other electrons in the atom.

The orbital approximation allows us to express the electronic structure of an atom by reporting its **configuration**, the list of occupied orbitals (usually, but not necessarily, in its ground state). For example, because the ground state of a hydrogen atom consists of a single electron in a 1s orbital, we report its configuration as $1s^1$ (read 'one s one'). A helium atom has two electrons. We can imagine forming the atom by adding the electrons in succession to the orbitals of the bare nucleus (of charge $+2e$). The first electron occupies a hydrogenic 1s orbital, but because $Z = 2$, the orbital is more compact than in H itself. The second electron joins the first in the same 1s orbital, and so the electron configuration of the ground state of He is $1s^2$ (read 'one s two').

To continue our description, we need to introduce the concept of **spin**, an *intrinsic* angular momentum that every electron possesses and that cannot be changed or eliminated (just like its mass or its charge). The name 'spin' is evocative of a ball spinning on its axis, and this classical interpretation can be used to help to visualize the motion. However, spin is a purely quantum mechanical phenomenon and has no classical counterpart, so the analogy must be used with care.

We shall make use of two properties of electron spin:

1. Electron spin is described by a **spin quantum number**, s (the analog of l for orbital angular momentum), with s fixed at the single (positive) value of $\frac{1}{2}$ for all electrons at all times.

2. The spin can be clockwise or counterclockwise; these two states are distinguished by the **spin magnetic quantum number**, m_s, which can take the values $+\frac{1}{2}$ or $-\frac{1}{2}$ but no other values (Fig. 9.50). An electron with $m_s = +\frac{1}{2}$ is called an **α electron** and commonly denoted α or $\uparrow$; an electron with $m_s = -\frac{1}{2}$ is called a **β electron** and denoted β or $\downarrow$.

When an atom contains more than one electron, we need to consider the interactions between the electron spin states. Consider lithium ($Z = 3$), which has three electrons. Two of its electrons occupy a 1s orbital drawn even more closely than in He around the more highly charged nucleus. The third electron, however, does not join the first two in the 1s orbital because a $1s^3$ configuration is forbidden by a fundamental feature of nature summarized by the Austrian physicist Wolfgang Pauli in the **Pauli exclusion principle**:

No more than two electrons may occupy any given orbital, and if two electrons do occupy one orbital, then their spins must be paired.

Electrons with **paired spins**, denoted $\uparrow\downarrow$, have zero net spin angular momentum because the spin angular momentum of one electron is canceled by the spin of the other. In *Further information 9.3* we see that the exclusion principle is a consequence of an even deeper statement about wavefunctions.

Lithium's third electron cannot enter the 1s orbital because that orbital is already full: we say that the K shell is **complete** and that the two electrons form a **closed shell**. Because a similar closed shell occurs in the He atom, we denote it [He]. The third electron is excluded from the K shell ($n = 1$) and must occupy the next available orbital, which is one with $n = 2$ and hence belonging to the L shell. However, we now have to decide whether the next available orbital is the 2s orbital or a 2p orbital and therefore whether the lowest energy configuration of the atom is $[\text{He}]2s^1$ or $[\text{He}]2p^1$.

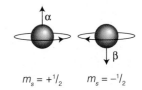

$m_s = +\frac{1}{2}$ $m_s = -\frac{1}{2}$

Fig. 9.50 A classical representation of the two allowed spin states of an electron. The magnitude of the spin angular momentum is $(3^{1/2}/2)\hbar$ in each case, but the directions of spin are opposite.

A note on good practice
The quantum number s should not be confused with or used in place of m_s. The spin quantum number s has a single, positive value ($\frac{1}{2}$; there is no need to write a + sign). Use m_s to denote the orientation of the spin ($m_s = +\frac{1}{2}$ or $-\frac{1}{2}$), and always include the + sign in $m_s = +\frac{1}{2}$.

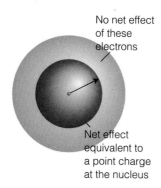

No net effect of these electrons

Net effect equivalent to a point charge at the nucleus

Fig. 9.51 An electron at a distance *r* from the nucleus experiences a Coulombic repulsion from all the electrons within a sphere of radius *r* that is equivalent to a point negative charge located on the nucleus. The effect of the point charge is to reduce the apparent nuclear charge of the nucleus from Ze to $Z_{eff}e$.

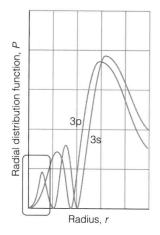

Radial distribution function, *P*

3p

3s

Radius, *r*

Fig. 9.52 An electron in an s orbital (here a 3s orbital) is more likely to be found close to the nucleus than an electron in a p orbital of the same shell. Hence it experiences less shielding and is more tightly bound.

9.10 **Penetration and shielding**

Penetration and shielding account for the general form of the periodic table and the physical and chemical properties of the elements. The two effects underlie all the varied properties of the elements and hence their contributions to biological systems.

An electron in a many-electron atom experiences a Coulombic repulsion from all the other electrons present. When the electron is at a distance *r* from the nucleus, the repulsion it experiences from the other electrons can be modeled by a point negative charge located on the nucleus and having a magnitude equal to the charge of the electrons within a sphere of radius *r* (Fig. 9.51). The effect of the point negative charge is to lower the full charge of the nucleus from Ze to $Z_{eff}e$, the **effective nuclear charge**.[8] To express the fact that an electron experiences a nuclear charge that has been modified by the other electrons present, we say that the electron experiences a **shielded nuclear charge**. The electrons do not actually 'block' the full Coulombic attraction of the nucleus: the effective charge is simply a way of expressing the net outcome of the nuclear attraction and the electronic repulsions in terms of a single equivalent charge at the center of the atom.

The effective nuclear charges experienced by s and p electrons are different because the electrons have different wavefunctions and therefore different distributions around the nucleus (Fig. 9.52). An s electron has a greater **penetration** through inner shells than a p electron of the same shell in the sense that an s electron is more likely to be found close to the nucleus than a p electron of the same shell (a p orbital, remember, is proportional to *r* and hence has zero probability density at the nucleus). As a result of this greater penetration, an s electron experiences less shielding than a p electron of the same shell and therefore experiences a larger Z_{eff}. Consequently, by the combined effects of penetration and shielding, an s electron is more tightly bound than a p electron of the same shell. Similarly, a d electron (which has a wavefunction proportional to r^2) penetrates less than a p electron of the same shell, and it therefore experiences more shielding and an even smaller Z_{eff}.

As a consequence of penetration and shielding, the energies of orbitals in the same shell of a many-electron atom lie in the order s < p < d < f. The individual orbitals of a given subshell (such as the three p orbitals of the p subshell) remain degenerate because they all have the same radial characteristics and so experience the same effective nuclear charge.

We can now complete the Li story. Because the shell with $n = 2$ has two non-degenerate subshells, with the 2s orbital lower in energy than the three 2p orbitals, the third electron occupies the 2s orbital. This arrangement results in the ground state configuration $1s^2 2s^1$, or $[He]2s^1$. It follows that we can think of the structure of the atom as consisting of a central nucleus surrounded by a complete helium-like shell of two 1s electrons and around that a more diffuse 2s electron. The electrons in the outermost shell of an atom in its ground state are called the **valence electrons** because they are largely responsible for the chemical bonds that the atom forms (and, as we shall see, the extent to which an atom can form bonds is called its 'valence'). Thus, the valence electron in Li is a 2s electron, and lithium's other two electrons belong to its core, where they take little part in bond formation.

---

[8] Commonly, Z_{eff} itself is referred to as the 'effective nuclear charge,' although strictly that quantity is $Z_{eff}e$.

9.11 The building-up principle

The exclusion principle and the consequences of shielding are our keys to understanding the structures of complex atoms and ions, chemical periodicity, and molecular structure.

The extension of the procedure used for H, He, and Li to other atoms is called the **building-up principle**.[9] The building-up principle specifies an order of occupation of atomic orbitals that in most cases reproduces the experimentally determined ground state configurations of atoms and ions.

(a) Neutral atoms

We imagine the bare nucleus of atomic number Z and then feed into the available orbitals Z electrons one after the other. The first two rules of the building-up principle are:

1. The order of occupation of orbitals is

   1s 2s 2p 3s 3p 4s 3d 4p 5s 4d 5p 6s 5d 4f 6p . . .

2. According to the Pauli exclusion principle, each orbital may accommodate up to two electrons.

The order of occupation is approximately the order of energies of the individual orbitals because in general the lower the energy of the orbital, the lower the total energy of the atom as a whole when that orbital is occupied. An s subshell is complete as soon as two electrons are present in it. Each of the three p orbitals of a shell can accommodate two electrons, so a p subshell is complete as soon as six electrons are present in it. A d subshell, which consists of five orbitals, can accommodate up to 10 electrons.

As an example, consider a carbon atom. Because $Z = 6$ for carbon, there are six electrons to accommodate. Two enter and fill the 1s orbital, two enter and fill the 2s orbital, leaving two electrons to occupy the orbitals of the 2p subshell. Hence its ground configuration is $1s^2 2s^2 2p^2$, or more succinctly $[He]2s^2 2p^2$, with $[He]$ the helium-like $1s^2$ core. On electrostatic grounds, we can expect the last two electrons to occupy different 2p orbitals, for they will then be farther apart on average and repel each other less than if they were in the same orbital. Thus, one electron can be thought of as occupying the $2p_x$ orbital and the other the $2p_y$ orbital, and the lowest energy configuration of the atom is $[He]2s^2 2p_x^1 2p_y^1$. The same rule applies whenever degenerate orbitals of a subshell are available for occupation. Therefore, another rule of the building-up principle is:

3. Electrons occupy different orbitals of a given subshell before doubly occupying any one of them.

It follows that a nitrogen atom ($Z = 7$) has the configuration $[He]2s^2 2p_x^1 2p_y^1 2p_z^1$. Only when we get to oxygen ($Z = 8$) is a 2p orbital doubly occupied, giving the configuration $[He]2s^2 2p_x^2 2p_y^1 2p_z^1$.

An additional point arises when electrons occupy degenerate orbitals (such as the three 2p orbitals) singly, as they do in C, N, and O, for there is then no requirement that their spins should be paired. We need to know whether the lowest energy is achieved when the electron spins are the same (both $\uparrow$, for instance,

---

[9] The building-up principle is still widely called the *Aufbau principle*, from the German word for 'building up'.

denoted ↑↑, if there are two electrons in question, as in C) or when they are paired (↑↓). This question is resolved by **Hund's rule**:

4. In its ground state, an atom adopts a configuration with the greatest number of unpaired electrons.

The explanation of Hund's rule is complicated, but it reflects the quantum mechanical property of **spin correlation**, that electrons in different orbitals with parallel spins have a quantum mechanical tendency to stay well apart (a tendency that has nothing to do with their charge: even two 'uncharged electrons' would behave in the same way). Their mutual avoidance allows the atom to shrink slightly, so the electron–nucleus interaction is improved when the spins are parallel. We can now conclude that in the ground state of a C atom, the two 2p electrons have the same spin, that all three 2p electrons in an N atom have the same spin, and that the two electrons that singly occupy different 2p orbitals in an O atom have the same spin (the two in the $2p_x$ orbital are necessarily paired).

Neon, with $Z = 10$, has the configuration $[He]2s^22p^6$, which completes the L shell. This closed-shell configuration is denoted [Ne] and acts as a core for subsequent elements. The next electron must enter the 3s orbital and begin a new shell, and so an Na atom, with $Z = 11$, has the configuration $[Ne]3s^1$. Like lithium with the configuration $[He]2s^1$, sodium has a single s electron outside a complete core.

Self-test 9.7) Predict the ground state electron configuration of sulfur.

Answer: $[Ne]3s^23p_x^23p_y^13p_z^1$

This analysis has brought us to the origin of chemical periodicity. The L shell is completed by eight electrons, and so the element with $Z = 3$ (Li) should have similar properties to the element with $Z = 11$ (Na). Likewise, Be ($Z = 4$) should be similar to Mg ($Z = 12$), and so on up to the noble gases He ($Z = 2$), Ne ($Z = 10$), and Ar ($Z = 18$).

Argon has complete 3s and 3p subshells, and as the 3d orbitals are high in energy, the atom effectively has a closed-shell configuration. Indeed, the 4s orbitals are so lowered in energy by their ability to penetrate close to the nucleus that the next electron (for potassium) occupies a 4s orbital rather than a 3d orbital and the K atom resembles an Na atom. The same is true of a Ca atom, which has the configuration $[Ar]4s^2$, resembling that of its congener Mg, which is $[Ne]3s^2$.

Ten electrons can be accommodated in the five 3d orbitals, which accounts for the electron configurations of scandium to zinc. The building-up principle has less clear-cut predictions about the ground-state configurations of these elements, and a simple analysis no longer works. Calculations show that for these atoms the energies of the 3d orbitals are always lower than the energy of the 4s orbital. However, experiments show that Sc has the configuration $[Ar]3d^14s^2$ instead of $[Ar]3d^3$ or $[Ar]3d^24s^1$. To understand this observation, we have to consider the nature of electron–electron repulsions in 3d and 4s orbitals. The most probable distance of a 3d electron from the nucleus is less than that for a 4s electron, so two 3d electrons repel each other more strongly than two 4s electrons. As a result, Sc has the configuration $[Ar]3d^14s^2$ rather than the two alternatives, for then the strong electron–electron repulsions in the 3d orbitals are minimized. The total energy of the atom is least despite the cost of allowing electrons to populate the high-energy 4s orbital (Fig. 9.53). The effect just described is generally true for Sc through Zn, so the electron configurations of these atoms are of the form $[Ar]3d^n4s^2$, where $n = 1$ for Sc and $n = 10$ for Zn. Experiments show that there are two notable exceptions: Cr, with electron configuration $[Ar]3d^54s^1$, and Cu, with electron configuration $[Ar]3d^{10}4s^1$.

At gallium, the energy of the 3d orbitals has fallen so far below those of the 4s and 4p orbitals that they (the full 3d orbitals) can be largely ignored, and the building-up principle can be used in the same way as in preceding periods. Now the 4s and 4p subshells constitute the valence shell, and the period terminates with krypton. Because 18 electrons have intervened since argon, this period is the first **long period** of the periodic table. The existence of the **d block** (the 'transition metals') reflects the stepwise occupation of the 3d orbitals, and the subtle shades of energy differences along this series give rise to the rich complexity of inorganic (and bioinorganic) d-metal chemistry (*Case study* 9.4 and Section 10.8). A similar intrusion of the f orbitals in Periods 6 and 7 accounts for the existence of the **f block** of the periodic table (the lanthanoids and actinoids; still commonly the lanthanides and actinides).

(b) Cations and anions

The configurations of cations of elements in the s, p, and d blocks of the periodic table are derived by removing electrons from the ground state configuration of the neutral atom in a specific order. First, we remove any valence p electrons, then the valence s electrons, and then as many d electrons as are necessary to achieve the stated charge. We consider a few examples below.

Calcium, an essential constituent of bone and a key player in a number of biochemical processes (such as muscle contraction, cell division, blood clotting, and the conduction of nerve impulses), is taken up by and functions in the cell as the Ca^{2+} ion. Because the configuration of Ca is $[Ar]4s^2$, the Ca^{2+} cation has the same configuration, $[Ar]$, as the argon atom.

Iron, copper, and manganese can shuttle between different cationic forms and participate in electron transfer reactions that form the core of bioenergetics. For instance, because the configuration of Fe is $[Ar]3d^64s^2$, the Fe^{2+} and Fe^{3+} cations have the configurations $[Ar]3d^6$ and $[Ar]3d^5$, respectively. These are the oxidation states adopted by the iron ions bound to the protein cytochrome *c* as it transfers electrons between complexes II and IV in the mitochondrial electron transport chain (Section 5.10).

The configurations of anions are derived by continuing the building-up procedure and adding electrons to the neutral atom until the configuration of the next noble gas has been reached. It is the chloride ion, and not elemental chlorine, that works together with Na^+ and K^+ ions to establish membrane potentials (Section 5.3) and to maintain osmotic pressure (Section 3.10) and charge balance in the cell. The configuration of a Cl^- ion is achieved by adding an electron to $[Ne]3s^23p^5$, giving the configuration of Ar.

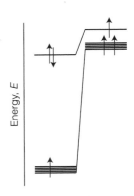

Fig. 9.53 Strong electron–electron repulsions in the 3d orbitals are minimized in the ground state of a scandium atom if the atom has the configuration $[Ar]3d^14s^2$ (shown on the left) instead of $[Ar]3d^24s^1$ (shown on the right). The total energy of the atom is lower when it has the configuration $[Ar]3d^14s^2$ despite the cost of populating the high-energy 4s orbital.

Self-test 9.8 Predict the electron configurations of (a) a Cu^{2+} ion and (b) an O^{2-} ion.

Answer: (a) $[Ar]3d^9$, (b) $[He]2s^22p^6$

9.12 Three important atomic properties

The fitness of an element for a biological role is a consequence of electronic structure. We now need to understand how electronic structure affects atomic and ionic radii, and the thermodynamic ability of an atom to release or acquire electrons to form ions or chemical bonds.

We now explore three important atomic properties: the atomic (and ionic) radius, the ionization energy, and the electron affinity. These properties are of great significance in chemistry and biology, for they are controls on the number and types of chemical bonds the atom can form. Indeed, we can use these properties to reveal an important reason for the unique role of carbon in biology.

(a) Atomic and ionic radii

The **atomic radius** of an element is half the distance between the centers of neighboring atoms in a solid (such as Cu) or, for nonmetals, in a homonuclear molecule (such as H_2 or S_8). If there is one single attribute of an element that determines its chemical properties (either directly, or indirectly through the variation of other properties), then it is atomic radius.

In general, atomic radii decrease from left to right across a period and increase down each group (Table 9.1 and Fig. 9.54). The decrease across a period can be traced to the increase in nuclear charge, which draws the electrons in closer to the nucleus. The increase in nuclear charge is partly canceled by the increase in the number of electrons, but because electrons are spread over a region of space, one electron does not fully shield one nuclear charge, so the increase in nuclear charge dominates. The increase in atomic radius down a group (despite the increase in nuclear charge) is explained by the fact that the valence shells of successive periods correspond to higher principal quantum numbers. That is, successive periods correspond to the start and then completion of successive (and more distant) shells of the atom that surround each other like the successive layers of an onion. The need to occupy a more distant shell leads to a larger atom despite the increased nuclear charge.

A modification of the increase down a group is encountered in Period 6, for the radii of the atoms late in the d block and in the following regions of the p block are not as large as would be expected by simple extrapolation down the group. The reason can be traced to the fact that in Period 6 the f orbitals are in the process of being occupied to form the 14 lanthanoids, cerium (Ce) to lutetium (Lu). An f electron is a very inefficient shielder of nuclear charge (for reasons connected

Table 9.1 Atomic radii of main-group elements, r/pm

Li	Be	B	C	N	O	F
157	112	88	77	74	66	64
Na	Mg	Al	Si	P	S	Cl
191	160	143	118	110	104	99
K	Ca	Ga	Ge	As	Se	Br
235	197	153	122	121	117	114
Rb	Sr	In	Sn	Sb	Te	I
250	215	167	158	141	137	133
Cs	Ba	Tl	Pb	Bi	Po	
272	224	171	175	182	167	

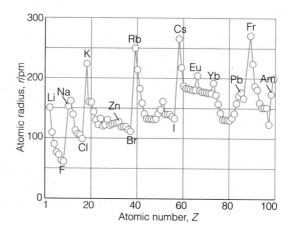

Fig. 9.54 The variation of atomic radius through the periodic table. Note the contraction of radius following the lanthanoids in Period 6 (following lutetium, $Z = 71$).

Table 9.2 Ionic radii of selected main-group elements*

Ion	Main biochemical function	r/pm
Mg^{2+}	Binds to ATP, constituent of chlorophyll, control of protein folding and muscle contraction	72
Ca^{2+}	Component of bone and teeth, control of protein folding, hormonal action, blood clotting, and cell division	100
Na^+		102
K^+	Control of osmotic pressure, charge balance, and membrane potentials	138
Cl^-		167

*The values are for ions surrounded by six counter-ions in a crystal.

with its radial extension), and as the atomic number increases from Ce to Lu, there is a considerable contraction in radius. By the time the d block resumes (at hafnium, Hf), the poorly shielded but considerably increased nuclear charge has drawn in the surrounding electrons, and the atoms are compact. They are so compact that the metals in this region of the periodic table (iridium to lead) are very dense. The reduction in radius below that expected by extrapolation from preceding periods is called the **lanthanide contraction**.

The **ionic radius** of an element is its share of the distance between neighboring ions in an ionic solid (**2**). That is, the distance between the centers of a neighboring cation and anion is the sum of the two ionic radii. Table 9.2 lists the radii of some ions that play important roles in biochemical processes.

When an atom loses one or more valence electrons to form a cation, the remaining atomic core is generally much smaller than the parent atom. Therefore, a cation is often smaller than its parent atom. For example, the atomic radius of Na, with the configuration [Ne]$3s^1$, is 191 pm, but the ionic radius of Na^+, with the configuration [Ne], is only 102 pm. Like atomic radii, cationic radii increase down each group because electrons are occupying shells with higher principal quantum numbers.

An anion is larger than its parent atom because the electrons added to the valence shell repel one another. Without a compensating increase in the nuclear charge, which would draw the electrons closer to the nucleus and each other, the ion expands. The variation in anionic radii shows the same trend as that for atoms and cations, with the smallest anions at the upper right of the periodic table, close to fluorine.

Atoms and ions with the same number of electrons are called **isoelectronic**. For example, Ca^{2+}, K^+, and Cl^- have the configuration [Ar] and are isoelectronic. However, their radii differ because they have different nuclear charges. The Ca^{2+} ion has the largest nuclear charge, so it has the strongest attraction for the electrons and the smallest radius. The Cl^- ion has the lowest nuclear charge of the three isoelectronic ions and, as a result, the largest radius.

Cation Anion

$r_{cation} + r_{anion}$

2

(b) Ionization energy

The minimum energy necessary to remove an electron from a many-electron atom is its **first ionization energy**, I_1. The **second ionization energy**, I_2, is the minimum energy needed to remove a second electron (from the singly charged cation):

Fig. 9.55 The periodic variation of the first ionization energies of the elements.

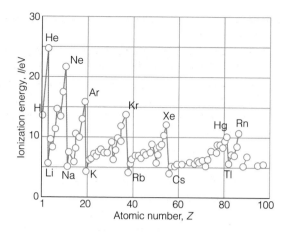

Table 9.3 First ionization energies of main-group elements, I/eV^*

H							He
13.60							24.59
Li	**Be**	**B**	**C**	**N**	**O**	**F**	**Ne**
5.32	9.32	8.30	11.26	14.53	13.62	17.42	21.56
Na	**Mg**	**Al**	**Si**	**P**	**S**	**Cl**	**Ar**
5.14	7.65	5.98	8.15	10.49	10.36	12.97	15.76
K	**Ca**	**Ga**	**Ge**	**As**	**Se**	**Br**	**Kr**
4.34	6.11	6.00	7.90	9.81	9.75	11.81	14.00
Rb	**Sr**	**In**	**Sn**	**Sb**	**Te**	**I**	**Xe**
4.18	5.70	5.79	7.34	8.64	9.01	11.81	14.00
Cs	**Ba**	**Tl**	**Pb**	**Bi**	**Po**	**At**	**Rn**
3.89	5.21	6.11	7.42	7.29	8.42	9.64	10.78

$^*1 \text{ eV} = 96.485 \text{ kJ mol}^{-1}$.

A note on good practice
The physical state of the electron is given because ionization (and electron attachment; see below) is an actual process, unlike in electrochemistry, where the half-reaction, such as $E(s) \rightarrow E^+(aq) + e^-$, is hypothetical and the electron is stateless.

$$E(g) \rightarrow E^+(g) + e^-(g) \qquad I_1 = E(E^+) - E(E)$$
$$E^+(g) \rightarrow E^{2+}(g) + e^-(g) \qquad I_2 = E(E^{2+}) - E(E^+) \qquad (9.37)$$

The variation of the first ionization energy through the periodic table is shown in Fig. 9.55, and some numerical values are given in Table 9.3. The ionization energy of an element plays a central role in determining the ability of its atoms to participate in bond formation (for bond formation, as we shall see in Chapter 10, is a consequence of the relocation of electrons from one atom to another). After atomic radius, it is the most important property for determining an element's chemical characteristics.

Lithium has a low first ionization energy: its outermost electron is well shielded from the weakly charged nucleus by the core ($Z_{eff} = 1.3$ compared with $Z = 3$) and it is easily removed. Beryllium has a higher nuclear charge than Li, and its outermost electron (one of the two 2s electrons) is more difficult to remove: its ionization energy is larger. The ionization energy decreases between Be and B because in the latter the outermost electron occupies a 2p orbital and is less strongly bound than if it had been a 2s electron. The ionization energy increases between B and C because the latter's outermost electron is also 2p and the nuclear charge has increased. Nitrogen has a still higher ionization energy because of the further increase in nuclear charge.

There is now a kink in the curve because the ionization energy of O is lower than would be expected by simple extrapolation. At O a 2p orbital must become doubly occupied, and the electron–electron repulsions are increased above what would be expected by simple extrapolation along the row. (The kink is less pronounced in the next row, between P and S, because their orbitals are more diffuse.) The values for O, F, and Ne fall roughly on the same line, the increase of their ionization energies reflecting the increasing attraction of the nucleus for the outermost electrons.

The outermost electron in Na is 3s. It is far from the nucleus, and the latter's charge is shielded by the compact, complete neon-like core. As a result, the ionization energy of Na is substantially lower than that of Ne. The periodic cycle starts again along this row, and the variation of the ionization energy can be traced to similar reasons.

(c) Electron affinity

The **electron affinity**, E_{ea}, is the difference in energy between a neutral atom and its anion. It is the energy *released* in the process

$$E(g) + e^-(g) \rightarrow E^-(g) \qquad E_{ea} = E(E) - E(E^-) \qquad (9.38)$$

The electron affinity is positive if the anion has a lower energy than the neutral atom.

Electron affinities (Table 9.4) vary much less systematically through the periodic table than ionization energies. Broadly speaking, however, the highest electron affinities are found close to F. In the halogens, the incoming electron enters the valence shell and experiences a strong attraction from the nucleus. The electron affinities of the noble gases are negative—which means that the anion has a higher energy than the neutral atom—because the incoming electron occupies an orbital outside the closed valence shell. It is then far from the nucleus and repelled by the electrons of the closed shells. The first electron affinity of O is

A note on good practice
We use the convention that $E_{ea} > 0$ signifies a 'positive affinity' for the added electron. Distinguish the electron affinity from the electron-gain enthalpy, which is negative for such an exothermic process (that is, has the opposite sign to the electron affinity, and differs very slightly in value).

Table 9.4 Electron affinities of main-group elements, E_{ea}/eV^*

H							He
+0.75							<0[†]
Li	Be	B	C	N	O	F	Ne
+0.62	−0.19	+0.28	+1.26	−0.07	+1.46	+3.40	−0.30[†]
Na	Mg	Al	Si	P	S	Cl	Ar
+0.55	−0.22	+0.46	+1.38	+0.46	+2.08	+3.62	−0.36[†]
K	Ca	Ga	Ge	As	Se	Br	Kr
+0.50	−1.99	+0.3	+1.20	+0.81	+2.02	+3.37	−0.47[†]
Rb	Sr	In	Sn	Sb	Te	I	Xe
+0.49	+1.51	+0.3	+1.20	+1.50	+1.97	+3.06	−0.42[†]
Cs	Ba	Tl	Pb	Bi	Po	At	Rn
+0.47	−0.48	+0.2	+0.36	+0.95	+1.90	+2.80	−0.42[†]

*1 eV = 96.485 kJ mol⁻¹.
[†]Calculated.

positive for the same reason as for the halogens. However, the second electron affinity (for the formation of O^{2-} from O^-) is strongly negative because although the incoming electron enters the valence shell, it experiences a strong repulsion from the net negative charge of the O^- ion.

Further analysis of ionization energies and electron affinities can begin to tell us why carbon is an essential building block of complex biological structures. Among the elements in Period 2, C has intermediate values of the ionization energy and electron affinity, so it can share electrons (that is, form covalent bonds) with many other elements, such as H, N, O, S, and, more importantly, other C atoms. As a consequence, such networks as long carbon–carbon chains (as in lipids) and chains of peptide links can form readily. Because the ionization energy and electron affinity of C are neither too high nor too low, the bonds in these covalent networks are neither too strong nor too weak. As a result, biological molecules are sufficiently stable to form viable organisms but are still susceptible to dissociation (essential to catabolism) and rearrangement (essential to anabolism). In Chapter 10 we shall develop additional concepts that will complete this aspect of carbon's biological role.

Case study 9.4   *The biological role of Zn^{2+}*

Here we explore how the ionic radius and charge can work together to impart unique chemical properties to an ion, leading to unique biochemical function. Consider the Zn^{2+} ion, which is found in the active sites of many enzymes. An example is carbonic anhydrase (Atlas P2), which catalyzes the hydration of CO_2 in red blood cells to give bicarbonate (hydrogencarbonate) ion:

$$CO_2 + H_2O \rightarrow HCO_3^- + H^+$$

To understand the catalytic role played by the Zn^{2+} ion, we need to know that a 'Lewis acid' is an electron-poor species that forms a complex with a 'Lewis base', an electron-rich species. Metal cations are good Lewis acids, and molecules with lone pairs of electrons, such as H_2O, are good Lewis bases.

The Lewis acidity of a metal cation increases with its effective nuclear charge, Z_{eff} (defined here as the charge experienced by a Lewis base on the 'surface' of the cation), and decreases with the ionic radius, r_{ion}. Among the M^{2+} d-metal ions found in the active sites of enzymes, Cu^{2+} and Zn^{2+} are the best Lewis acids because they have the largest Z_{eff}/r_{ion} ratios. Thermodynamically, organisms make use of the Cu^{2+}/Cu^+ redox couple for electron transport processes (Chapters 5 and 8) and, generally, the Cu^{2+} ion does not act as a Lewis acid in biochemical processes. On the other hand, the Zn^{2+} ion is not used in biological redox reactions but is a ubiquitous biological Lewis acid.

To illustrate the consequences of the Lewis acidity of the Zn^{2+} ion, we consider the mechanism of the hydration of CO_2 by carbonic anhydrase (Fig. 9.56). In the first two steps, a Lewis acid–base complex forms between the protein-bound Zn^{2+} ion and a water molecule, which is then deprotonated. The Zn^{2+} ion has a large Z_{eff}/r_{ion} ratio and gives rise to a strong electric field in its vicinity, so it stabilizes the negative charge on the bound OH^- ion, thus effectively lowering the pK_w of water from 14 to about 7. Thermodynamically, the Zn^{2+} ion facilitates the generation of a strong nucleophile, the OH^- ion, which can attack CO_2 more effectively than H_2O. In the next steps, CO_2 binds to the active site and then reacts with the bound OH^- ion, forming a hydrogencarbonate ion. Release of the bicarbonate ion poises the enzyme for another catalytic cycle.

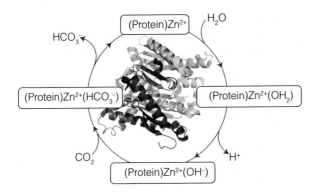

Fig. 9.56 The mechanism of the hydration of CO_2 by carbonic anhydrase. In the first two steps, a Lewis acid–base complex forms between the protein-bound Zn^{2+} ion and a water molecule, which is then deprotonated. In the next steps, CO_2 binds to the active site and then reacts with the bound OH^- ion, forming a bicarbonate ion. Release of the bicarbonate ion poises the enzyme for another catalytic cycle.

Checklist of key concepts

1. Atoms and molecules have discrete energy levels, which are revealed by their absorption or emission spectra.

2. Planck proposed that electromagnetic oscillators of frequency v could acquire or discard energy in quanta of magnitude hv.

3. The photoelectric effect is the ejection of electrons when radiation of greater than a threshold frequency is incident on a metal. The photon energy is equal to the sum of the kinetic energy of the electron and the work function Φ of the metal, the energy required to remove the electron from the metal.

4. The wavelike character of electrons was demonstrated by the Davisson–Germer diffraction experiment.

5. The joint wave–particle character of matter and radiation is called wave–particle duality.

6. A wavefunction, ψ, contains all the dynamical information about a system and is found by solving the appropriate Schrödinger equation subject to the constraints on the solutions known as boundary conditions.

7. According to the Born interpretation, the probability of finding a particle in a small region of space of volume δV is proportional to $\psi^2 \delta V$, where ψ is the value of the wavefunction in the region.

8. According to the Heisenberg uncertainty principle, it is impossible to specify simultaneously, with arbitrary precision, both the momentum and the position of a particle.

9. Because wavefunctions do not, in general, decay abruptly to zero, particles may tunnel into and through classically forbidden regions.

10. A particle undergoes harmonic motion if it is subjected to a Hooke's-law restoring force (a force proportional to the displacement).

11. Hydrogenic atoms are atoms with a single electron.

12. The wavefunctions of hydrogenic atoms are labeled with three quantum numbers: the principal quantum number $n = 1, 2, \ldots$, the orbital angular momentum quantum number $l = 0, 1, \ldots, n - 1$, and the magnetic quantum number $m_l = l, l - 1, \ldots, -l$.

13. s Orbitals are spherically symmetrical and have nonzero amplitude at the nucleus.

14. The p and d orbitals of a shell are shown in Figs. 9.48 and 9.49, respectively.

15. A radial distribution function, $P(r)$, is the probability density for finding an electron between r and $r + \delta r$ regardless of orientation.

16. An electron possesses an intrinsic angular momentum, its spin, which is described by the quantum numbers $s = \frac{1}{2}$ and $m_s = \pm\frac{1}{2}$.

17. In the orbital approximation, each electron in a many-electron atom is supposed to occupy its own orbital.

18. The Pauli exclusion principle states that no more than two electrons may occupy any given orbital and if two electrons do occupy one orbital, then their spins must be paired.

19. In a many-electron atom, the orbitals of a given shell lie in the order s < p < d < f as a result of the effects of penetration and shielding.

20. Atomic radii typically decrease from left to right across a period and increase down a group.

21. Ionization energies typically increase from left to right across a period and decrease down a group.

22. Electron affinities are highest toward the top right of the periodic table (near fluorine).

Checklist of key equations

Property	Equation	Comment
Bohr frequency relation	$\Delta E = h\nu$	
de Broglie relation	$\lambda = h/p$	
Schrödinger equation	$-(\hbar^2/2m)(d^2\psi/dx^2) + V\psi = E\psi$	Motion in one dimension
Heisenberg uncertainty relation	$\Delta p \Delta x \geq \frac{1}{2}\hbar$	Motion in one dimension
Particle in a box:		Motion in one dimension
energy levels	$E_n = n^2h^2/8mL^2$	$n = 1, 2, \ldots$
zero-point energy	$E_1 = h^2/8mL^2$	
wavefunctions	$\psi_n(x) = (2/L)^{1/2}\sin(n\pi x/L)$	
Transmission probability	$T \approx 16\varepsilon(1-\varepsilon)e^{-2\kappa L}$, $\kappa = \{2m(V-E)\}^{1/2}/\hbar$	Motion in one dimension; $V/E \gg 1$
Particle on a ring:		
energy levels	$E_{ml} = m_l^2\hbar^2/2I, I = mr^2$	$m_l = 0, \pm 1, \pm 2, \ldots$
z-component of the angular momentum	$J_z = m_l\hbar$	
Particle on a sphere		
energy levels	$E = l(l+1)(\hbar^2/2I)$	$l = 0, 1, 2, \ldots$
angular momentum	$J = \{l(l+1)\}^{1/2}\hbar$	
z-component of the angular momentum	$J_z = m_l\hbar$	$m_l = l, l-1, \ldots, -l$
Harmonic oscillator:		
potential energy	$V(x) = \frac{1}{2}k_f x^2$	Motion in one dimension
energy levels	$E_v = (v + \frac{1}{2})h\nu, \nu = (1/2\pi)(k_f/m)^{1/2}$	$v = 0, 1, 2, \ldots$
Hydrogenic atoms:		
energy levels	$E_n = -hc\mathcal{R}(Z^2/n^2)$, $hc\mathcal{R} = \mu e^4/(32\pi^2\varepsilon_0^2\hbar^2)$, $\mu = m_N m_N/(m_e + m_N)$	$n = 1, 2, \ldots$
wavefunctions	$\psi_{n,l,m_l}(r,\theta,\phi) = Y_{l,m_l}(\theta,\phi)R_{n,l}(r)$	$l = 0, 1, 2, \ldots, n-1$ $m_l = l, l-1, \ldots, -l$
Radial distribution function	$P(r) = 4\pi r^2\psi^2$	s orbitals

Further information

Further information 9.1 *A justification of the Schrödinger equation*

The form of the Schrödinger equation can be justified to a certain extent by showing that it implies the de Broglie relation for a freely moving particle. Free motion means motion in a region where the potential energy is zero ($V = 0$ everywhere). Then

$$\hat{H} = -\frac{\hbar^2}{2m}\frac{d^2}{dx^2}$$

and eqn 9.4 simplifies to

$$-\frac{\hbar^2}{2m}\frac{d^2\psi}{dx^2} = E\psi$$

A solution is

$$\psi = \sin kx \qquad k = (2mE)^{1/2}/\hbar$$

The function $\sin kx$ is a wave of wavelength $\lambda = 2\pi/k$, as we can see by comparing $\sin kx$ with $\sin(2\pi x/\lambda)$, the standard form of a harmonic wave with wavelength λ. To verify that $\sin kx$ is indeed a solution, we insert $\psi = \sin kx$ into both sides of the differential equation and use

$$\frac{d}{dx}\sin kx = k\cos kx \qquad \frac{d}{dx}\cos kx = -k\sin kx$$

$$\frac{d^2}{dx^2}\sin(kx) = -k^2\sin(kx)$$

Thus

$$-\frac{\hbar^2}{2m}\frac{d^2\psi}{dx^2} = -\frac{\hbar^2}{2m}\frac{d^2\sin(kx)}{dx^2} = -\frac{\hbar^2}{2m}(-k^2\sin(kx)) = \frac{k^2\hbar^2}{2m}\psi$$

According to the Schrödinger equation, the final term of this expression is equal to $E\psi$, so it follows that $E = k^2\hbar^2/2m$ and $k = (2mE)^{1/2}/\hbar$.

Next, we note that the energy of the particle is entirely kinetic (because $V = 0$ everywhere), so the total energy of the particle is just its kinetic energy:

$$E = E_k = p^2/2m$$

Because E is related to k by $E = k^2\hbar^2/2m$, it follows from a comparison of the two equations that $p = k\hbar$. Therefore, the linear momentum is related to the wavelength of the wavefunction by

$$p = \frac{2\pi}{\lambda} \times \frac{h}{2\pi} = \frac{h}{\lambda}$$

which is the de Broglie relation. We see, in the case of a freely moving particle, that the Schrödinger equation has led to an experimentally verified conclusion.

Further information 9.2 *The separation of variables procedure*

We illustrate the separation of variables procedure with motion in a rectangular box as example. The Schrödinger equation for the problem described in Section 9.4c is

$$-\frac{\hbar^2}{2m}\frac{\partial^2\psi(x,y)}{\partial x^2} - \frac{\hbar^2}{2m}\frac{\partial^2\psi(x,y)}{\partial y^2} = E\psi(x,y)$$

where we have noted that for a function of two variables the derivatives to be calculated are partial derivatives (see *Mathematical toolkit* 8.1). For simplicity, we can write this expression as

$$\hat{H}_X\psi(x,y) + \hat{H}_Y\psi(x,y) = E\psi(x,y)$$

where $\hat{H}_X$ operates only on functions of x and $\hat{H}_Y$ operates only on functions of y. To see if $\psi(x,y) = X(x)Y(y)$ is indeed a solution, we substitute this product on both sides of the last equation,

$$\hat{H}_X X(x)Y(y) + \hat{H}_Y X(x)Y(y) = EX(x)Y(y)$$

and note that $\hat{H}_X$ acts only on $X(x)$, with $Y(y)$ being treated as a constant, and $\hat{H}_Y$ acts only on $Y(y)$, with $X(x)$ being treated as a constant. Therefore, this equation becomes

$$Y(y)\hat{H}_X X(x) + X(x)\hat{H}_Y Y(y) = EX(x)Y(y)$$

When we divide both sides by $X(x)Y(y)$, we obtain

$$\frac{1}{X(x)}\hat{H}_X X(x) + \frac{1}{Y(y)}\hat{H}_Y Y(y) = E$$

Now we come to the crucial part of the argument. The first term on the left depends only on x and the second term depends only on y. Therefore, if x changes, only the first term can change. But its sum with the unchanging second term is the constant E. Therefore, the first term cannot in fact change when x changes. That is, the first term is equal to a constant, which we write as E_X. The same argument applies to the second term when y is changed, so it too is equal to a constant, which we write as E_Y. That is, we have shown that

$$\frac{1}{X(x)}\hat{H}_X X(x) = E_X \qquad \frac{1}{Y(y)}\hat{H}_Y Y(y) = E_Y$$

with $E_X + E_Y = E$. These two equations are easily turned into

$$\hat{H}_X X(x) = E_X X(x) \qquad \hat{H}_Y Y(y) = E_Y Y(y)$$

which we should recognize as the Schrödinger equation for one-dimensional motion, one along the x-axis and the other along the y-axis. Thus, the variables have been separated, and because the boundary conditions are essentially the same for each axis (the only difference being the actual values of the lengths L_X and L_Y), the individual wavefunctions are essentially the same as those already found for the one-dimensional case.

Further information 9.3 *The Pauli principle*

Some elementary particles have $s = 1$ and therefore have a higher intrinsic angular momentum than an electron. For our purposes the most important **spin-1 particle** is the photon. It is a very deep feature of nature that the fundamental particles from which matter is built have half-integral spin (such as electrons and quarks, all of which have $s = \frac{1}{2}$). The particles that transmit forces between these particles, so binding them together into entities such as nuclei, atoms, and planets, all have integral spin (such as $s = 1$ for the photon, which transmits the electromagnetic interaction between charged particles). Fundamental particles with half-integral spin are called **fermions**; those with integral spin are called bosons. Matter therefore consists of fermions bound together by **bosons**.

The Pauli exclusion principle is a special case of a general statement called the *Pauli principle*:

> When the labels of any two identical fermions are exchanged, the total wavefunction changes sign. When the labels of any two identical bosons are exchanged, the total wavefunction retains the same sign.

The Pauli *exclusion* principle applies only to fermions. 'Total wavefunction' means the entire wavefunction, including the spin of the particles.

Consider the wavefunction for two electrons $\psi(1,2)$. The Pauli principle implies that it is a fact of nature that the wavefunction must change sign if we interchange the labels 1 and 2 wherever they occur in the function: $\psi(2,1) = -\psi(1,2)$. Suppose the two electrons in an atom occupy an orbital ψ; then in the orbital approximation the overall wavefunction is $\psi(1)\psi(2)$. To apply the Pauli principle, we must deal with the total wavefunction, the wavefunction including spin. There are several possibilities for two spins: the state $\alpha(1)\alpha(2)$ corresponds to parallel spins, whereas (for technical reasons related to the cancelation of each spin's angular momentum by the other) the combination $\alpha(1)\beta(2) - \beta(1)\alpha(2)$ corresponds to paired spins. The total wavefunction of the system is one of the following:

Parallel spins: $\psi(1)\psi(2)\alpha(1)\alpha(2)$

Paired spins: $\psi(1)\psi(2)\{\alpha(1)\beta(2) - \beta(1)\alpha(2)\}$

The Pauli principle, however, asserts that for a wavefunction to be acceptable (for electrons), it must change sign when the electrons are exchanged. In each case, exchanging the labels 1 and 2 converts the factor $\psi(1)\psi(2)$ into $\psi(2)\psi(1)$, which is the same because the order of multiplying the functions does not change the value of the product. The same is true of $\alpha(1)\alpha(2)$. Therefore, the first combination is not allowed because it does not change sign. The second combination, however, changes to

$$\psi(2)\psi(1)\{\alpha(2)\beta(1) - \beta(2)\alpha(1)\}$$
$$= -\psi(1)\psi(2)\{\alpha(1)\beta(2) - \beta(1)\alpha(2)\}$$

This combination does change sign (it is 'antisymmetric') and is therefore acceptable.

Now we see that the only possible state of two electrons in the same orbital allowed by the Pauli principle is the one that has paired spins. This is the content of the Pauli exclusion principle. The exclusion principle is irrelevant when the orbitals occupied by the electrons are different and both electrons may then have (but need not have) the same spin state. Nevertheless, even then the overall wavefunction must still be antisymmetric overall and must still satisfy the Pauli principle itself.

Discussion questions

9.1 Summarize the evidence that led to the introduction of quantum theory.

9.2 Consult texts or online sources to establish the size range for the following particles: a plant cell, an animal cell, a bacterium, a ribosome, a protein (such as chymotrypsin), a small molecule (such as an amino acid), and an atom. Choose among light microscopy (which uses visible light as a probe), electron microscopy, AFM, and STM as suitable techniques for the study of the size and general shape (but not the internal structure) of these particles.

9.3 Discuss the physical origin of the quantization of energy of a particle confined to moving inside a one-dimensional box or on a ring.

9.4 Define, justify, and provide examples of zero-point energy.

9.5 Discuss the physical origins of quantum mechanical tunneling. Why is tunneling more likely to contribute to the mechanisms of electron transfer and proton transfer processes than to mechanisms of group transfer reactions, such as $AB + C \rightarrow A + BC$ (where A, B, and C are large molecular groups)?

9.6 Explain how the technique of separation of variables is used to simplify the discussion of multi-dimensional problems. When can it not be used?

9.7 List and describe the significance of the quantum numbers needed to specify the internal state of a hydrogenic atom.

9.8 Explain the significance of (a) a boundary surface and (b) the radial distribution function for hydrogenic orbitals.

9.9 Describe the orbital approximation for the wavefunction of a many-electron atom. What are the limitations of the approximation?

9.10 The d metals iron, copper, and manganese form cations with different oxidation states. For this reason they are found in many oxidoreductases and in several proteins of oxidative phosphorylation and photosynthesis (Section 5.10). Explain why many d metals form cations with different oxidation states.

Exercises

9.11 Calculate the size of the quantum involved in the excitation of (a) an electronic motion of frequency 1.0×10^{15} Hz, (b) a molecular vibration of period 20 fs, and (c) a pendulum of period 0.50 s. Express the results in joules and in kilojoules per mole.

9.12 Calculate the average power output of a photodetector that collects 8.0×10^7 photons in 3.8 ms from monochromatic light of wavelength (a) 470 nm, the wavelength produced by some commercially available light-emitting diodes (LED), and (b) 780 nm, a wavelength produced by lasers that are commonly used in compact disc (CD) players. *Hint*: The total energy emitted by a source or collected by a detector in a given interval is its power multiplied by the time interval of interest (1 J = 1 W s).

9.13 Calculate the de Broglie wavelength of (a) a mass of 1.0 g traveling at 1.0 m s^{-1}, (b) the same, traveling at 1.00×10^5 km s^{-1}, (c) an He atom traveling at 1000 m s^{-1} (a typical speed at room temperature), (d) yourself traveling at 8 km h^{-1}, and (e) yourself at rest.

9.14 Calculate the linear momentum per photon, energy per photon, and the energy per mole of photons for radiation of wavelength

(a) 600 nm (red), (b) 550 nm (yellow), (c) 400 nm (violet), (d) 200 nm (ultraviolet), (e) 150 pm (X-ray), and (f) 1.0 cm (microwave).

9.15 Electron microscopes can obtain images with several hundred-fold higher resolution than optical microscopes because of the short wavelength obtainable from a beam of electrons. For electrons moving at speeds close to c, the speed of light, the expression for the de Broglie wavelength (eqn 9.3) needs to be corrected for relativistic effects:

$$\lambda = \frac{h}{\left\{2m_e eV\left(1 + \dfrac{eV}{2m_e c^2}\right)\right\}^{1/2}}$$

where c is the speed of light in a vacuum and V is the potential difference through which the electrons are accelerated. (a) Calculate the de Broglie wavelength of electrons accelerated through 50 kV. (b) Is the relativistic correction important?

9.16 Suppose that you designed a spacecraft to work by photon pressure. The sail was a completely absorbing fabric of area 1.0 km^2 and you directed a red laser beam of wavelength 650 nm onto it at a rate of N_A photons per second from a base on the Moon. What are **(a)** the force and **(b)** the pressure exerted by the radiation on the sail? **(c)** Suppose the mass of the spacecraft was 1.0 kg. Given that, after a period of acceleration from standstill, speed = (force/mass) × time, how long would it take for the craft to accelerate to a speed of 1.0 m s^{-1}?

9.17 The speed of a certain proton is 350 km s^{-1}. If the uncertainty in its momentum is 0.0100 per cent, what uncertainty in its location must be tolerated?

9.18 An electron is confined to a linear region with a length of the same order as the diameter of an atom (about 100 pm). Calculate the minimum uncertainties in its position and speed.

9.19 Calculate the probability that an electron will be found **(a)** between $x = 0.1$ and 0.2 nm, and **(b)** between 4.9 and 5.2 nm in a box of length $L = 10$ nm when its wavefunction is $\psi = (2/L)^{1/2} \sin(2\pi x/L)$. *Hint*: Treat the wavefunction as a constant in the small region of interest and interpret δV as δx.

9.20 Repeat *Exercise* 9.19, but allow for the variation of the wavefunction in the region of interest. What are the percentage errors in the procedure used in *Exercise* 9.19? *Hint*: You will need to integrate $\psi^2 dx$ between the limits of interest. The indefinite integral you require is given in *Justification* 9.1.

9.21 What is the probability of finding a particle of mass m in **(a)** the left-hand one-third, **(b)** the central one-third, and **(c)** the right-hand one-third of a box of length L when it is in the state with $n = 1$?

9.22 A certain wavefunction is zero everywhere except between $x = 0$ and $x = L$, where it has the constant value A. Normalize the wavefunction.

9.23 The conjugated system of retinal consists of 11 carbon atoms and one oxygen atom. In the ground state of retinal, each level up to $n = 6$ is occupied by two electrons. Assuming an average internuclear distance of 140 pm, calculate **(a)** the separation in energy between the ground state and the first excited state in which one electron occupies the state with $n = 7$ and **(b)** the frequency of the radiation required to produce a transition between these two states.

9.24 Many biological electron transfer reactions, such as those associated with biological energy conversion, may be visualized as arising from electron tunneling between protein-bound cofactors, such as cytochromes, quinones, flavins, and chlorophylls. This tunneling occurs over distances that are often greater than 1.0 nm, with sections of protein separating electron donor from acceptor. For a specific combination of electron donor and acceptor, the rate of electron tunneling is proportional to the transmission probability, with $\kappa \approx 7$ nm^{-1} (eqn 9.11). By what factor does the rate of electron tunneling between two co-factors increase as the distance between them changes from 2.0 nm to 1.0 nm?

9.25 The rate, v, at which electrons tunnel through a potential barrier of height 2 eV, like that in a scanning tunneling microscope, and thickness d can be expressed as $v = Ae^{-d/l}$, with $A = 5 \times 10^{14}$ s^{-1} and $l = 70$ pm. **(a)** Calculate the rate at which electrons tunnel across a barrier of width 750 pm. **(b)** By what factor is the current reduced when the probe is moved away by a further 100 pm?

9.26 The particle in a two-dimensional well is a useful model for the motion of electrons around the indole ring (**3**), the conjugated cycle found in the side chain of tryptophan. We may regard indole as a rectangle with sides of length 280 pm and 450 pm, with 10 electrons in the conjugated π system. As in *Case study* 9.1, we assume that in the

ground state of the molecule each quantized level is occupied by two electrons. **(a)** Calculate the energy of an electron in the highest occupied level. **(b)** Calculate the frequency of radiation that can induce a transition between the highest occupied and lowest unoccupied levels.

3 Indole

9.27 The particle on a ring is a useful model for the motion of electrons around the porphine ring (**4**), the conjugated macrocycle that forms the structural basis of the heme group and the chlorophylls. We may treat the group as a circular ring of radius 440 pm, with 20 electrons in the conjugated system moving along the perimeter of the ring. As in Exercise 9.26, assume that in the ground state of the molecule quantized each level is occupied by two electrons. **(a)** Calculate the energy and angular momentum of an electron in the highest occupied level. **(b)** Calculate the frequency of radiation that can induce a transition between the highest occupied and lowest unoccupied levels.

4 Porphine (free base form)

9.28 Use mathematical software or an electronic spreadsheet to plot the wavefunctions $\psi_{1,1}$, $\psi_{1,2}$, $\psi_{2,1}$, and $\psi_{2,2}$, and the corresponding probability densities, for a particle in a square well.

9.29 **(a)** Use the separation of variables procedure to write expressions for the wavefunctions and energies of a particle trapped in a three-dimensional box with sides L_X, L_Y, and L_Z. **(b)** Using results from part (a), write expressions for the wavefunctions and energies of a particle in a cubic box with sides L. Investigate the existence of degeneracy in this system.

9.30 The HI molecule may be treated as a stationary I atom around which an H atom moves. Assuming that the H atom circulates in a plane at a distance of 161 pm from the I atom, calculate **(a)** the moment of inertia of the molecule and **(b)** the greatest wavelength of the radiation that can excite the molecule into rotation.

9.31 Consider again the HI molecule as you did in *Exercise* 9.30. Assuming that the H atom oscillates toward and away from the I atom and that the force constant of the HI bond is 314 N m^{-1}, calculate **(a)** the vibrational frequency of the molecule and **(b)** the wavelength required to excite the molecule into vibration. **(c)** Assuming that the force constant of the bond does not change upon isotopic substitution, by what factor will the vibrational frequency of HI change when H is replaced by deuterium?

9.32 The ground state wavefunction of a harmonic oscillator is proportional to $e^{-ax^2/2}$, where a depends on the mass and force constant. **(a)** Normalize this wavefunction. **(b)** At what displacement is the oscillator most likely to be found in its ground state? *Hint:* For part (a), you will need the integral $\int_{-\infty}^{+\infty} e^{-ax^2} dx = (\pi/a)^{1/2}$. For part (b), recall that the maximum (or minimum) of a function $f(x)$ occurs at the value of x for which $df/dx = 0$.

9.33 The solutions of the Schrödinger equation for a harmonic oscillator also apply to diatomic molecules. The only complication is that both atoms joined by the bond move, so the 'mass' of the oscillator has to be interpreted carefully. Detailed calculation shows that for two atoms of masses m_A and m_B joined by a bond of force constant k_f, the energy levels are given by eqn 9.29, but the vibrational frequency is

$$v = \frac{1}{2\pi}\left(\frac{k_f}{\mu}\right)^{1/2} \qquad \mu = \frac{m_A m_B}{m_A + m_B}$$

and μ is called the *effective mass* of the molecule. Consider the vibration of carbon monoxide, a poison that prevents the transport and storage of O_2 (see *Exercise* 9.48). The bond in a $^{12}C^{16}O$ molecule has a force constant of 1860 N m^{-1}. (a) Calculate the vibrational frequency, v, of the molecule. (b) In infrared spectroscopy it is common to convert the vibrational frequency of a molecule to its vibrational wavenumber, $\tilde{v}$, given by $\tilde{v} = v/c$. What is the vibrational wavenumber of a $^{12}C^{16}O$ molecule? (c) Assuming that isotopic substitution does not affect the force constant of the C≡O bond, calculate the vibrational wavenumbers of the following molecules: $^{12}C^{16}O$, $^{13}C^{16}O$, $^{12}C^{18}O$, $^{13}C^{18}O$.

9.34 Predict the ionization energy of Li^{2+} given that the ionization energy of He^+ is 54.40 eV.

9.35 How many orbitals are present in the N shell of an atom?

9.36 Consider the ground state of the H atom. (a) At what radius does the probability of finding an electron in a small volume located at a point fall to 25 per cent of its maximum value? (b) At what radius does the radial distribution function have 25 per cent of its maximum value? (c) What is the most probable distance of an electron from the nucleus? *Hint*: Look for a maximum in the radial distribution function.

9.37 What is the probability of finding an electron anywhere in one lobe of a p orbital given that it occupies the orbital?

9.38 The (normalized) wavefunction for a 2s orbital in a hydrogen atom is

$$\psi = \left(\frac{1}{32\pi a_0^3}\right)^{1/2}\left(2 - \frac{r}{a_0}\right)e^{-r/2a_0}$$

where a_0 is the Bohr radius. (a) Calculate the probability of finding an electron that is described by this wavefunction in a volume of 1.0 pm^3 (i) centered on the nucleus, (ii) at the Bohr radius, and (iii) at twice the Bohr radius. (b) Construct an expression for the radial distribution

function of a hydrogenic 2s electron and plot the function against r. What is the most probable radius at which the electron will be found? (c) For a more accurate determination of the most probable radius at which an electron will be found in an H2s orbital, differentiate the radial distribution function to find where it is a maximum.

9.39 Locate the radial nodes in (a) the 3s orbital and (b) the 4s orbital of an H atom.

9.40 The wavefunction of one of the d orbitals is proportional to $\sin\theta\cos\theta$. At what angles does it have nodal planes?

9.41 What is the orbital angular momentum (as multiples of $\hbar$) of an electron in the orbitals (a) 1s, (b) 3s, (c) 3d, (d) 2p, and (e) 3p? Give the numbers of angular and radial nodes in each case.

9.42 How many electrons can occupy subshells with the following values of l: (a) 0, (b) 3, (c) 5?

9.43 If we lived in a four-dimensional world, there would be one s orbital, four p orbitals, and nine d orbitals in their respective subshells. (a) Suggest what form the periodic table might take for the first 24 elements. (b) Which elements (using their current names) would be noble gases? (c) On what element would life be likely to be based?

9.44 The central iron ion of cytochrome c changes between the +2 and +3 oxidation states as the protein shuttles electrons between complex III and complex IV of the respiratory chain (Section 5.10). Which do you expect to be larger: Fe^{2+} or Fe^{3+}? Why?

9.45 Thallium, a neurotoxin, is currently the heaviest member of Group 13 of the periodic table and is most often found in the +1 oxidation state. Aluminum, which causes anemia and dementia, is also a member of the group, but its chemical properties are dominated by the +3 oxidation state. Examine this issue by plotting the first, second, and third ionization energies for the Group 13 elements against atomic number. Explain the trends you observe. *Hints*: The third ionization energy, I_3, is the minimum energy needed to remove an electron from the doubly charged cation: $E^{2+}(g) \rightarrow E^{3+}(g) + e^-(g)$, $I_3 = E(E^{3+}) - E(E^{2+})$. For data, see the links to databases of atomic properties provided in the text's website.

9.46 How is the ionization energy of an anion related to the electron affinity of the parent atom?

9.47 To perform many of their biological functions, the Lewis acids Mg^{2+} and Ca^{2+} must be bound to Lewis bases, such as nucleotides (with ATP^{4-} as an example) or the side chains of amino acids in proteins. The equilibrium constant for the association of a doubly charged cation M^{2+} to a Lewis base increases in the order: $Ba^{2+} < Sr^{2+} < Ca^{2+} < Mg^{2+}$. Provide a molecular interpretation for this trend, which does not depend on the nature of the Lewis base. *Hint*: Consider the effect of ionic radius.

Projects

9.48 Here we see how infrared spectroscopy can be used to study the binding of diatomic molecules to heme proteins. We focus on carbon monoxide, which is poisonous because it binds strongly to the Fe^{2+} ion of the heme group of hemoglobin and myoglobin and interferes with the transport and storage of O_2 (*Case study* 4.1).

(a) Estimate the vibrational frequency and wavenumber of CO bound to myoglobin by using the data in *Exercise* 9.33 and by making the following assumptions: the atom that binds to the heme group is immobilized, the protein is infinitely more massive than either the C or O atom, the C atom binds to the Fe^{2+} ion, and binding of CO to the protein does not alter the force constant of the C≡O bond.

(b) Of the four assumptions made in part (a), the last two are questionable. Suppose that the first two assumptions are still reasonable and that you have at your disposal a supply of myoglobin, a suitable buffer in which to suspend the protein, $^{12}C^{16}O$, $^{13}C^{16}O$, $^{12}C^{18}O$, $^{13}C^{18}O$, and an infrared spectrometer, an instrument used for the determination of vibrational frequencies. Describe a set of experiments that: **(i)** proves which atom, C or O, binds to the heme group of myoglobin and **(ii)** allows for the determination of the force constant of the C≡O bond for myoglobin-bound carbon monoxide.

9.49 The postulation of a plausible reaction mechanism requires careful analysis of many experiments designed to determine the fate of atoms during the formation of products. Observation of the *kinetic isotope effect*, a decrease in the rate of a chemical reaction on replacement of one atom in a reactant by a heavier isotope, facilitates the identification of bond-breaking events in the rate-determining step. A *primary kinetic isotope effect* is observed when the rate-determining step requires the scission of a bond involving the isotope. A *secondary kinetic isotope effect* is the reduction in reaction rate even though the bond involving the isotope is not broken to form product. In both cases, the effect arises from the change in activation energy that accompanies the replacement of an atom by a heavier isotope on account of changes in the zero-point vibrational energies. We now explore the primary kinetic isotope effect in some detail.

Consider a reaction, such as the rearrangements catalyzed by vitamin B_{12}, in which a C–H bond is cleaved. If scission of this bond is the rate-determining step, then the reaction coordinate corresponds to the stretching of the C–H bond and the potential energy profile is shown in Fig. 9.57. On deuteration, the dominant change is the reduction of the zero-point energy of the bond (because the deuterium atom is heavier). The whole reaction profile is not lowered, however, because the relevant vibration in the activated complex has a very low force constant, so there is little zero-point energy associated with the reaction coordinate in either form of the activated complex.

(a) Assume that the change in the activation energy arises only from the change in zero-point energy of the stretching vibration and show that

$$E_a(C-D) - E_a(C-H) = \frac{1}{2}N_A hc\tilde{v}(C-H)\left\{1 - \left(\frac{\mu_{C-H}}{\mu_{C-D}}\right)^{1/2}\right\}$$

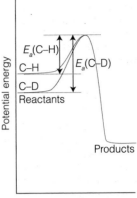

Fig. 9.57 Changes in the reaction profile when a C–H bond undergoing cleavage is deuterated. In this illustration, the C–H and C–D bonds are modeled as simple harmonic oscillators. The only significant change is in the zero-point energy of the reactants, which is lower for C–D than for C–H. As a result, the activation energy is greater for C–D cleavage than for C–H cleavage.

where $\tilde{v}$ is the relevant vibrational wavenumber and μ is the relevant effective mass (*Exercise* 9.33).

(b) Now consider the effect of deuteration on the rate constant, k_r, of the reaction. **(i)** Starting with the Arrhenius equation (eqn 6.19) and assuming that the pre-exponential factor does not change on deuteration, show that the rate constants for the two species should be in the ratio

$$\frac{k_r(C-D)}{k_r(C-H)} = e^{-\lambda} \quad \text{with} \quad \lambda = \frac{hc\tilde{v}(C-H)}{2kT}\left\{1 - \left(\frac{\mu_{C-H}}{\mu_{C-D}}\right)^{1/2}\right\}$$

(ii) Does $k_r(C-D)/k_r(C-H)$ increase or decrease with decreasing temperature?

(c) From infrared spectroscopy, the fundamental vibrational wavenumber for stretching of a C–H bond is about 3000 cm^{-1}. Predict the value of the ratio $k_r(C-D)/k_r(C-H)$ at 298 K.

(d) In some cases (including several enzyme-catalyzed reactions), substitution of deuterium for hydrogen results in values of $k_r(C-D)/k_r(C-H)$ that are too low to be accounted for by the model described above. Explain this effect.

10

The chemical bond

The **chemical bond**, a link between atoms, is central to all aspects of chemistry and biochemistry. The theory of the origin of the numbers, strengths, and three-dimensional arrangements of chemical bonds between atoms is called **valence theory**. Valence theory is an attempt to explain the properties of molecules ranging from the smallest to the largest. For instance, it explains why N_2 is so inert that it acts as a diluent for the aggressive oxidizing power of atmospheric oxygen. At the other end of the scale, valence theory deals with the structural origins of the function of protein molecules and the molecular biology of DNA.

Certain ideas of valence theory will be familiar from introductory chemistry. We know that chemical bonds may be classified on the basis of the degree of redistribution of electron density among interacting atomic nuclei:

- An **ionic bond** is formed by the transfer of electrons from one atom to another and the consequent attraction between the ions so formed.
- A **covalent bond** is formed when two atoms share a pair of electrons.

The character of a covalent bond, the main focus of this chapter, was identified by G.N. Lewis in 1916, before quantum mechanics was fully developed. Lewis's original theory was unable to account for the shapes adopted by molecules. The most elementary (but qualitatively quite successful) explanation of the shapes adopted by molecules is the **valence-shell electron pair repulsion model** (VSEPR model). In this model, which should be familiar from introductory chemistry courses, the shape of a molecule is ascribed to the repulsions between electron pairs in the valence shell. The purpose of this chapter is to extend these elementary arguments and to indicate some of the contributions that quantum theory has made to understanding why atoms form bonds and molecules adopt characteristic shapes.

There are two major approaches to the calculation of molecular structure, **valence bond theory** (VB theory) and **molecular orbital theory** (MO theory). Almost all modern computational work makes use of MO theory, and we concentrate on that theory in this chapter. Valence bond theory, however, has left its imprint on the language of chemistry, and it is important to know the significance of terms that chemists use every day. The structure of this chapter is therefore as follows. First, we present VB theory and the terms it introduces. Next, we present in more detail the basic ideas of MO theory. Finally, we see how computational techniques based on MO theory pervade all current discussions of molecular structure, including the prediction of the physiological properties of therapeutic agents.

Both theories of molecular structure adopt the **Born–Oppenheimer approximation** in which it is supposed that the nuclei, being so much heavier than an electron, move relatively slowly and may be treated as stationary while the electrons move around them. We can therefore think of the nuclei as being fixed at arbitrary locations and then solve the Schrödinger equation for the electrons alone. The approximation is quite good for molecules in their electronic ground states, for calculations suggest that (in classical terms) the nuclei in H_2 move through only about 1 pm while the electron speeds through 1000 pm.

By invoking the Born–Oppenheimer approximation, we can select an internuclear separation in a diatomic molecule and solve the Schrödinger equation for the electrons for that nuclear separation. Then we can choose a different separation and repeat the calculation, and so on. In this way we can explore how the energy of the molecule varies with bond length and obtain a **molecular potential energy** curve, a graph showing how the energy of the molecule depends on the internuclear separation (Fig. 10.1). The graph is called a *potential energy* curve because the nuclei are stationary and contribute no kinetic energy. Once the curve has been calculated, we can identify the **equilibrium bond length**, R_e, the internuclear separation at the minimum of the curve, and D_e, the depth of the minimum below the energy of the infinitely widely separated atoms. In Chapter 12 we shall also see that the narrowness of the potential well is an indication of the stiffness of the bond. Similar considerations apply to polyatomic molecules, where bond angles may be varied as well as bond lengths.

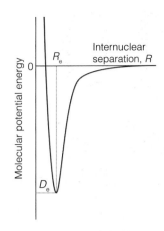

Fig. 10.1 A molecular potential energy curve. The equilibrium bond length R_e corresponds to the energy minimum D_e.

Valence bond theory

In VB theory, a bond is regarded as forming when an electron in an atomic orbital on one atom pairs its spin with that of an electron in an atomic orbital on another atom. To understand why this pairing leads to bonding, we have to examine the wavefunction for the two electrons that form the bond.

10.1 Diatomic molecules

There are many diatomic molecules of biological importance, including O_2 (the source of oxidizing power for catabolism), N_2 (the ultimate source of nitrogen for the synthesis of a host of biomolecules, including proteins and nucleic acids), and NO (a versatile carrier of biochemical messages). We need to know how bonding in these molecules determines their physical and chemical properties and hence their biological function.

We begin by considering the simplest possible chemical bond, the one in molecular hydrogen, H–H, and then see how the concepts it introduces can be extended to other diatomic molecules.

(a) Formulation of the VB wavefunction

When two ground-state H atoms are far apart, we can be confident that electron 1 is in the 1s orbital of atom A, which we denote $\psi_A(1)$, and that electron 2 is in the 1s orbital of atom B, which we denote $\psi_B(2)$. It is a general rule in quantum mechanics that the wavefunction for several noninteracting particles is the product of the wavefunctions for each particle, so we can write $\psi(1,2) = \psi_A(1)\psi_B(2)$. When the two atoms are at their bonding distance, it may still be true that electron 1 is on A and electron 2 is on B. However, an equally likely arrangement is for electron 1 to escape from A and be found on B and for electron 2 to be on A. In this case the wavefunction is $\psi(1,2) = \psi_A(2)\psi_B(1)$. Whenever two outcomes are equally likely, the rules of quantum mechanics tell us to add together the two corresponding wavefunctions. Therefore, the (unnormalized) wavefunction for the two electrons in a hydrogen molecule is

$$\psi_{H-H}(1,2) = \psi_A(1)\psi_B(2) + \psi_A(2)\psi_B(1) \qquad \boxed{\text{A valence-bond wavefunction}} \quad (10.1)$$

This expression is the VB wavefunction for the bond in molecular hydrogen. For technical reasons related to the Pauli exclusion principle (see the following *Justification*), this wavefunction can exist only if the two electrons it describes have opposite spins. Bonds do not form *because* electrons tend to pair their spins: bonds are *allowed* to form when the electrons pair their spins.

Justification 10.1 *The role of spin pairing in VB theory*

The spatial wavefunction in eqn 10.1 does not change sign when the labels 1 and 2 are interchanged:

$$\psi_{H-H}(2,1) = \psi_A(2)\psi_B(1) + \psi_A(1)\psi_B(2)$$
$$= \psi_A(1)\psi_B(2) + \psi_A(2)\psi_B(1)$$
$$= \psi_{H-H}(1,2)$$

According to the Pauli principle (Further information 9.3), the *overall* wavefunction of the molecule (the wavefunction including spin) must change sign when we interchange the labels 1 and 2. Therefore, we must multiply $\psi_{A-B}(2,1)$ by an antisymmetric spin function of the form shown in Further information 9.3. There is only one choice:

$$\psi_{A-B}(1,2) = \{\psi_A(1)\psi_B(2) + \psi_A(2)\psi_B(1)\} \times \{\alpha(1)\beta(2) - \beta(1)\alpha(2)\}$$

For this combination, $\psi_{A-B}(2,1) = -\psi_{A-B}(1,2)$ as required. Because the spin state $\alpha(1)\beta(2) - \beta(1)\alpha(2)$ corresponds to paired electron spins, we conclude that the two electron spins in the bond must be paired in order for the bond to form.

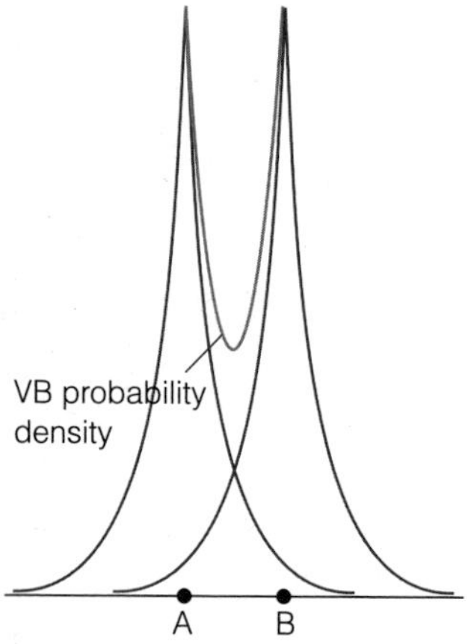

Fig. 10.2 The electron density in H_2 according to the valence-bond model of the chemical bond and the electron densities corresponding to the contributing atomic orbitals. The nuclei are denoted by large dots on the horizontal line. Note the accumulation of electron density in the internuclear region.

(b) The energy of interaction

Why, though, does the VB wavefunction result in bonding? As can be seen from Fig. 10.2, as the two atoms approach each other, there is an accumulation of electron density between the two nuclei where the two atomic orbitals overlap and their amplitudes add together. The electrons that have accumulated between the nuclei attract them and the potential energy is lowered. However, this decrease in energy is counteracted by an increase in energy from the Coulombic repulsion between the two positively charged nuclei. At intermediate internuclear separations the attraction dominates the internuclear repulsion, but at very short distances the repulsion dominates the attraction and the total energy rises above that of the widely separated atoms. Qualitatively at least, we see that this description leads to a molecular potential energy curve like that depicted in Fig. 10.1 and hence accounts for the existence of a bond.

To test the model quantitatively we calculate the energy of a molecule for a series of internuclear separations by substituting the VB wavefunction into the Schrödinger equation for the molecule and calculate the corresponding values of the energy. When this energy is plotted against R, we do indeed get a curve very much like that shown in Fig. 10.1, although numerically the agreement between the calculated and experimental bond length and depth of the well is not very good.

(c) σ and π bonds

Because the wavefunction in eqn 10.1 is built from two H1s orbitals we can expect the overall distribution of the electrons in the molecule to be sausage shaped

(Fig. 10.3). A VB wavefunction with cylindrical symmetry around the internuclear axis is called a **σ bond**. It is so called because, when viewed along the bond, it resembles a pair of electrons in an s orbital (and σ, sigma, is the Greek equivalent of s). All VB wavefunctions are constructed in a similar way, by using the atomic orbitals available on the participating atoms. In general, therefore, the (unnormalized) VB wavefunction for an A–B bond has the form given in eqn 10.1 with the two contributing wavefunctions the atomic orbitals that are being used to form the bond (for instance, the 2p orbitals of carbon atoms).

We can use a similar description for molecules built from atoms that contribute more than one electron to the bonding. For example, to construct the VB description of N_2, we consider the valence-electron configuration of each atom, which is $2s^2 2p_x^1 2p_y^1 2p_z^1$. It is conventional to take the z-axis to be the internuclear axis, so we can imagine each atom as having a $2p_z$ orbital pointing toward a $2p_z$ orbital on the other atom, with the $2p_x$ and $2p_y$ orbitals perpendicular to the axis (Fig. 10.4). Each of these p orbitals is occupied by one electron, so we can think of bonds as being formed by the merging of matching orbitals on neighbouring atoms and the pairing of the electrons that occupy them. We get a cylindrically symmetric σ bond from the merging of the two $2p_z$ orbitals and the pairing of the electrons they contain.

The remaining N2p orbitals cannot merge to give σ bonds because they do not have cylindrical symmetry around the internuclear axis. Instead, the $2p_x$ orbitals merge and the two electrons pair to form a **π bond**, so called because, viewed along the internuclear axis, it resembles a pair of electrons in a p orbital (and π is the Greek equivalent of p). Similarly, the $2p_y$ orbitals merge and their electrons pair to form another π bond. In general, a π bond arises from the merging of two p orbitals that approach side by side and the pairing of the electrons that they contain. It follows that the overall bonding pattern in N_2 is a σ bond plus two π bonds (Fig. 10.5), which is consistent with the Lewis structure :N≡N: in which the atoms are linked by a triple bond.

> **Self-test 10.1** Describe the VB ground state of an O_2 molecule.
>
> **Answer:** One σ($O2p_z, O2p_z$) bond and one π($O2p_x, O2p_x$) bond.
> See Case study 10.1 for an important comment.

10.2 Polyatomic molecules

To understand the role of molecules in the processes of life, including self-assembly, metabolism, and self-replication, we need to extend the discussion to include the electronic structures and shapes of polyatomic molecules, ranging in size from H_2O to DNA.

The ideas we have introduced so far are easily extended to polyatomic molecules. Each σ bond in a polyatomic molecule is formed by the merging of orbitals with cylindrical symmetry about the internuclear axis and the pairing of the spins of the electrons they contain. Likewise, each π bond (if there is one) is formed by pairing electrons that occupy atomic orbitals of the appropriate symmetry. The description of the electronic structure of H_2O will make this clear, but also bring to light a deficiency of the theory.

The valence electron configuration of an O atom is $2s^2 2p_x^2 2p_y^1 2p_z^1$. The two unpaired electrons in the O2p orbitals can each pair with an electron in a H1s orbital, and each combination results in the formation of a σ bond (each bond

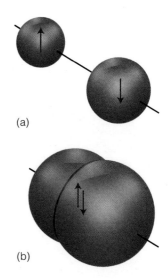

Fig. 10.3 In the valence bond theory, a σ bond is formed when two electrons in orbitals on neighboring atoms, as in (a), pair and the orbitals merge to form a cylindrical electron cloud, as in (b).

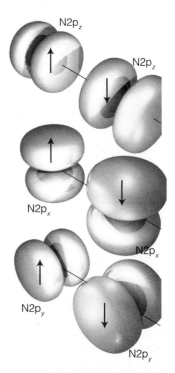

Fig. 10.4 The bonds in N_2 are built by allowing the electrons in the N2p orbitals to pair. However, only one orbital on each atom can form a σ bond: the orbitals perpendicular to the axis form π bonds.

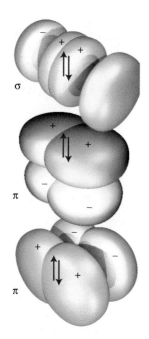

Fig. 10.5 The electrons in the 2p orbitals of two neighboring N atoms merge to form σ and π bonds. The electrons in the N2p$_z$ orbitals pair to form a bond of cylindrical symmetry. Electrons in the N2p orbitals that lie perpendicular to the axis also pair to form two π bonds.

has cylindrical symmetry about the respective O–H internuclear axis, Fig. 10.6). Because the 2p$_y$ and 2p$_z$ orbitals lie at 90° to each other, the two σ bonds they form also lie at 90° to each other. We predict, therefore, that H$_2$O should be an angular ('bent') molecule, which it is. However, the model predicts a bond angle of 90°, whereas the actual bond angle is 104°. Clearly, the VB model needs to be improved.

A further major deficiency becomes apparent as soon as we apply these arguments to carbon. The ground state valence configuration of a carbon atom is $2s^2 2p_x^1 2p_y^1$, which suggests that it should be capable of forming only two bonds, not the four bonds that are so characteristic of this element.

Self-test 10.2 Give a VB description of NH$_3$, and predict the bond angle of the molecule on the basis of this description. The experimental bond angle is 107°.

Answer: Three σ(N2p,H1s) bonds; 90°.

(a) Promotion

Two modifications solve both deficiencies of VB theory. They are both based on the fact that it might be appropriate to invest energy initially in order to achieve a greater overall lowering of energy by allowing bond angles to change and stronger and perhaps more bonds to form.

Suppose we imagine that a valence electron is **promoted** from a full atomic orbital to an empty atomic orbital. In carbon, with ground state configuration $2s^2 2p_x^1 2p_y^1$, for example, the promotion of a 2s electron to a 2p orbital leads to the configuration $2s^1 2p_x^1 2p_y^1 2p_z^1$, with four unpaired electrons in separate orbitals. These electrons may pair with four electrons in orbitals provided by four other atoms (such as four H1s orbitals if the molecule is CH$_4$), and as a result the atom can form four σ bonds. Promotion is worthwhile if the energy it requires can be more than recovered in the greater strength or greater number of bonds that can be formed. We should not think of the atom as making an initial transition to an excited state: promotion is just a way of analyzing the electron rearrangement that takes place as bonds form and achieve the lowest possible energy.

We can now see why tetravalent carbon is so common. The promotion energy of carbon is small because the promoted electron leaves a doubly occupied 2s orbital and enters a vacant 2p orbital, hence significantly relieving the electron–electron repulsion it experiences in the former. Furthermore, the energy required for promotion is more than recovered by the atom's ability to form four bonds in place of the two bonds of the unpromoted atom.

(b) Hybridization

Promotion, however, appears to imply the presence of three σ bonds of one type (in CH$_4$, from the merging of H1s and C2p orbitals) and a fourth σ bond of a distinctly different type (formed from the merging of H1s and C2s). It is well known, however, that all four bonds in methane are exactly equivalent in terms of both their chemical and their physical properties (their lengths, strengths, and stiffness).

This problem is overcome in VB theory by drawing on another feature of quantum mechanics that allows the same electron distribution to be described in different ways. In this case, we can describe the electron distribution in the promoted atom either as arising from four electrons in one s and three p orbitals or as arising from four electrons in four different mixtures of these orbitals. Mixtures (more formally, linear combinations) of atomic orbitals on the same atom are called **hybrid orbitals**. We can picture them by thinking of the four original atomic orbitals, which are waves centered on a nucleus, as being like ripples spreading from a single point on the surface of a lake. These waves interfere destructively (where their amplitudes cancel) or constructively (where their amplitudes add) in different regions and give rise to four new shapes. The specific linear combinations that give rise to four equivalent hybrid orbitals are

$$h_1 = s + p_x + p_y + p_z \qquad h_2 = s - p_x - p_y + p_z$$
$$h_3 = s - p_x + p_y - p_z \qquad h_4 = s + p_x - p_y - p_z \qquad \boxed{sp^3 \text{ hybrid orbitals}} \quad (10.2a)$$

As a result of the constructive and destructive interference between the positive and negative regions of the component orbitals, each hybrid orbital has a large lobe pointing toward one corner of a regular tetrahedron (Fig. 10.7). Because each hybrid is built from one s orbital and three p orbitals, it is called an **sp^3 hybrid orbital**.

We can now see how the VB description of CH_4 leads to a tetrahedral molecule containing four equivalent C–H bonds. It is energetically favorable (in the end, after bonding has been taken into account) for the C atom to undergo promotion. The promoted configuration has a distribution of electrons that is equivalent to one electron occupying each of four tetrahedral hybrid orbitals. Each hybrid orbital of the promoted atom contains a single unpaired electron; an H1s electron can pair with each one, giving rise to a σ bond pointing in a tetrahedral direction. Because each sp^3 hybrid orbital has the same composition, all four σ bonds are identical apart from their orientation in space (Fig. 10.8).

Hybridization is also used in the VB description of alkenes. Consider ethene (ethylene), which is not only an important industrial gas but also a hormone associated with the ripening of fruit. An ethene molecule is planar, with HCH and HCC bond angles close to 120°. To reproduce this σ-bonding structure, we think of each C atom as being promoted to a $2s^1 2p_x^1 2p_y^1 2p_z^1$ configuration. However, instead of using all four orbitals to form hybrids, we form **sp^2 hybrid orbitals** by allowing the s orbital and two of the p orbitals to interfere. As shown in Fig. 10.9a, the three hybrid orbitals

$$h_1 = s + 2^{1/2} p_y$$
$$h_2 = s + \left(\tfrac{3}{2}\right)^{1/2} p_x - \left(\tfrac{1}{2}\right)^{1/2} p_y$$
$$h_3 = s - \left(\tfrac{3}{2}\right)^{1/2} p_x - \left(\tfrac{1}{2}\right)^{1/2} p_y \qquad \boxed{sp^2 \text{ hybrid orbitals}} \quad (10.2b)$$

lie in a plane and point toward the corners of an equilateral triangle. The third 2p orbital ($2p_z$) is not included in the hybridization, and its axis is perpendicular to the plane in which the hybrids lie (Fig. 10.9b). The coefficients $2^{1/2}$, etc., in the hybrids have been chosen to give the correct directional properties of the hybrids.

The sp^2-hybridized C atoms each form three σ bonds with either the h_1 hybrid of the other C atom or with the H1s orbitals. The σ framework therefore consists of bonds at 120° to each other. Moreover, provided the two CH_2 groups lie in the same plane, the two electrons in the unhybridized $C2p_z$ orbitals can pair and form a π bond (Fig. 10.10). The formation of this π bond locks the framework into the planar arrangement, for any rotation of one CH_2 group relative to the other leads to a weakening of the π bond (and consequently an increase in energy of the molecule).

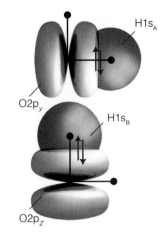

Fig. 10.6 The bonding in an H_2O molecule can be pictured in terms of the pairing of an electron belonging to one H atom with an electron in an O2p orbital; the other bond is formed likewise, but using a perpendicular O2p orbital. The predicted bond angle is 90°, which is in poor agreement with the experimental bond angle (104°).

Fig. 10.7 The 2s and three 2p orbitals of a carbon atom hybridize, and the resulting hybrid orbitals point toward the corners of a regular tetrahedron.

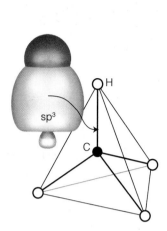

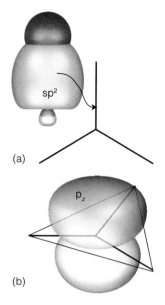

(a)

(b)

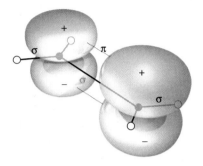

Fig. 10.8 The valence bond description of the structure of CH_4. Each σ bond is formed by the pairing of an electron in an H1s orbital with an electron in one of the hybrid orbitals shown in Fig. 10.7. The resulting molecule is regular tetrahedral.

Fig. 10.9 (a) Trigonal planar hybridization is obtained when an s and two p orbitals are hybridized. The three lobes lie in a plane and make an angle of 120° to each other. (b) The remaining p orbital in the valence shell of an sp^2-hybridized atom lies perpendicular to the plane of the three hybrids.

Fig. 10.10 The valence bond description of the structure of a carbon–carbon double bond, as in ethene. The electrons in the two sp^2 hybrids that point toward each other pair and form a σ bond. Electrons in the two p orbitals that are perpendicular to the plane of the hybrids pair and form a π bond. The electrons in the remaining hybrid orbitals are used to form bonds to other atoms (in ethene itself, to H atoms).

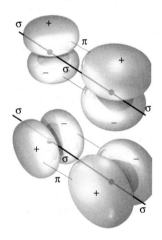

Fig. 10.11 The electronic structure of ethyne (acetylene). The electrons in the two sp hybrids on each atom pair to form σ bonds either with the other C atom or with an H atom. The remaining two unhybridized 2p orbitals on each atom are perpendicular to the axis: the electrons in corresponding orbitals on each atom pair to form two π bonds. The overall electron distribution is cylindrical.

A similar description applies to a linear ethyne (acetylene) molecule, H–C≡C–H. Now the carbon atoms are **sp hybridized**, and the σ bonds are built from hybrid atomic orbitals of the form

$$h_1 = s + p_z \qquad h_2 = s - p_z$$

sp hybrid orbitals (10.2c)

The two hybrids lie along the z-axis. The electrons in them pair either with an electron in the corresponding hybrid orbital on the other C atom or with an electron in an H1s orbital. Electrons in the two remaining p orbitals on each atom, which are perpendicular to the molecular axis, pair to form two perpendicular π bonds (as in Fig. 10.11).

It is possible to form hybrid orbitals with intermediate proportions of atomic orbitals. For example, as more p-orbital character is included in an sp-hybridization scheme, the hybridization changes toward sp^2 and the angle between the hybrids changes from 180° for pure sp hybridization to 120° for pure sp^2 hybridization. If the proportion of p character continues to be increased (by reducing the proportion of s orbital), then the hybrids eventually become pure p orbitals at an angle of 90° to each other (Fig. 10.12). Hybridization schemes involving d orbitals (Table 10.1) are often invoked to account for (or at least be consistent with) other molecular geometries but are not commonly invoked in biology. Regardless of the types of orbitals used, an important point is that:

The hybridization of N atomic orbitals always results in the formation of N hybrid orbitals.

Table 10.1 Hybrid orbitals

Number	Shape	Hybridization*
2	Linear	sp
3	Trigonal planar	sp^2
4	Tetrahedral	sp^3
5	Trigonal bipyramidal	sp^3d
6	Octahedral	sp^3d^2

*Other combinations are possible.

Hybridization accounts for—or at least is consistent with—the structure of H_2O, with its bond angle of 104°. Each O–H σ bond is formed from an O atom hybrid orbital with a composition that lies between pure p (which would lead to a bond angle of 90°) and pure sp^2 (which would lead to a bond angle of 120°). The actual bond angle and hybridization adopted are found by calculating the energy of the molecule as the bond angle is varied and looking for the angle at which the energy is a minimum.

Example 10.1 *Bonding in the peptide group*

Use VB theory to describe the CO, CN, and NH bonds of the peptide group based on the structure shown in (**1**).

Strategy To calculate the number of hybrid orbitals, we note that each orbital can hold either one or two electrons. If it contains one electron, the orbital is ready to make a σ bond with an orbital on another atom. If it contains a pair of electrons, then it does not participate in bonding but acts as a lone pair. It follows that the number of hybrid orbitals on an atom is equal to the sum of the number of σ bonds to the atom and the number of lone pairs on the atom. Unhybridized p orbitals can participate in π bonds, as described in Section 10.4. As noted in Section 10.2, a double bond consists of a σ and a π bond.

Solution The O atom is sp^2 hybridized because it has two lone pairs and makes a σ bond with the C atom. The C atom is sp^2 hybridized because it makes three σ bonds: one with the O atom, one with the $C_{\alpha 1}$ atom, and one with the N atom. The N atom is sp^3 hybridized because it has one lone pair and makes three σ bonds: one with the H atom, one with the C atom, and one with the $C_{\alpha 2}$ atom.

We can infer that the CO group has a σ bond between Csp^2 and Osp^2 hybrid orbitals and a π bond between unhybridized $C2p_z$ and $O2p_z$ orbitals (where again we have taken the z-axis to be perpendicular to the plane containing the hybrid orbitals). The CN group has a σ bond between Csp^2 and Nsp^3 hybrid orbitals. Finally, the NH group has a σ bond between a Nsp^3 hybrid orbital and a H1s atomic orbital. This pattern of hybridizations is summarized in Fig. 10.13; but read on!

Self-test 10.3 Estimate the values of the $C_{\alpha 1}CN$ and $CNC_{\alpha 2}$ bond angles for the structure shown in (**1**).

Answer: 120°, <109°

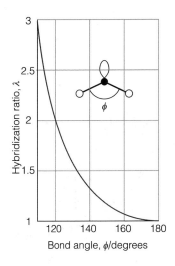

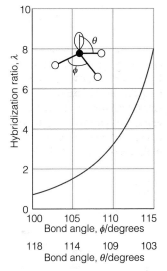

Fig. 10.12 The variation of hybridization with bond angle in (a) angular and (b) trigonal pyramidal molecules. The vertical axis gives the ratio of p to s character, so high values indicate mostly p character.

1 The peptide group

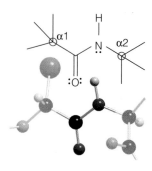

Fig. 10.13 The pattern of bonding in the peptide group.

$$
\begin{array}{c}
\ddot{\text{O}}: \\
\parallel \\
C_{\alpha1}\!-\!C\!-\!\underset{\underset{H}{|}}{\overset{\cdot\cdot}{N}}\!-\!C_{\alpha2}
\end{array}
$$

2

↕

$$
\begin{array}{c}
:\!\ddot{\text{O}}:^{-} \\
| \\
C_{\alpha1}\!-\!C\!=\!\underset{\underset{H}{|}}{\overset{+}{N}}\!-\!C_{\alpha2}
\end{array}
$$

3

4

(c) Resonance

The VB theory as presented so far fails to account for some experimental observations. For example, data (from X-ray diffraction, Section 11.4) on the peptide group show that all six of the atoms shown in (**1**) lie in the same plane. This geometry is not consistent with the sp^3 hybridization of the N atom, which implies a tetrahedral arrangement of bonded and non-bonded electron pairs and hence a non-planar arrangement of the C, N, H, and $C_{\alpha2}$ atoms. We need to refine the theory further.

What has been omitted so far? We have supposed that the peptide group has a structure that matches the Lewis structure in (**2**). But suppose instead we had assumed that the Lewis structure is (**3**), with equal and opposite charges on the O and N atoms. There is now a double bond between the CO and CN groups and both C and N are sp^2 hybridized and hence planar. If this were the structure, then the O, C, $C_{\alpha1}$, N, H, and $C_{\alpha2}$ atoms would lie in a single plane, as is observed.

When two Lewis structures have a similar energy, the true wavefunction is a linear combination of them both and in this case we would write

$$\psi = a\psi_1 + b\psi_2 \qquad \boxed{\text{A resonance hybrid}} \quad (10.3)$$

where ψ_1 is the wavefunction for structure (**2**), ψ_2 is that for structure (**3**), and a and b are numerical coefficients that are determined by minimizing the energy. According to quantum mechanics, we interpret the value of a^2 as the probability that the peptide group has structure (**2**) and the value of b^2 as the probability that it has structure (**3**), with $a^2 + b^2 = 1$. We say that the true wavefunction is a **resonance hybrid** of the contributing structures. The superposition of contributing structures is called **resonance**. Resonance is not a flickering between the contributing states: it is a blending of their characteristics, much as a mule is a blend of a horse and a donkey.

Resonance has two main effects: it distributes multiple-bond character over the molecule and it lowers the overall energy. The most famous example is that of benzene, where the two Kekulé structures (**4**), having the same energy, contribute equally to the resonance hybrid. Resonance between structures of the same energy results in the greatest lowering of energy, and accounts (in VB terms) for the considerable chemical stability of the phenyl group wherever it occurs in a molecule. Resonance also distributes double-bond character over the ring so that all the CC links are equivalent.

(d) The language of valence bonding

It might be helpful at this point to summarize the concepts that VB theory has introduced into chemistry and which still survive even though MO theory is the dominant computational mode:

1. *The names of bond types*: σ and π bonds are formed by spin pairing of electrons on adjacent atoms.

2. *Promotion*: valence electrons may be promoted to empty orbitals if overall that results in a lowering of energy.

3. *Hybridization*: atomic orbitals may be hybridized to match the observed geometry of a molecule.

4. *Resonance*: the superposition of individual structures. Resonance distributes multiple-bond character over the molecule and lowers the overall energy.

Molecular orbital theory

In MO theory, electrons are treated as spreading throughout the entire molecule: every electron contributes to the strength of every bond. As we have remarked, this theory has been more fully developed than VB theory and provides the language that is widely used in modern discussions of bonding in organic and inorganic molecules and d-metal complexes. It is also the basis for the calculation of spectroscopic properties, the modeling of molecular interactions (such as those between therapeutic agents and receptor sites in the cell), and the prediction of the outcome of chemical reactions.

To introduce the theory, we follow the same strategy as in Chapter 9, where the one-electron hydrogen atom was taken as the fundamental species for discussing atomic structure and then developed into a description of many-electron atoms. In this section we use the simplest molecule of all, the one-electron hydrogen molecule-ion, H_2^+, to introduce the essential features of bonding and then use H_2^+ as a guide to the structures of more complex systems. The hydrogen molecule-ion has no direct importance to biology, but is of crucial importance for establishing the concepts of MO theory.

10.3 Linear combinations of atomic orbitals

To formulate orbitals that spread around a molecule as small as O_2 or as large as DNA, we need to develop a mathematical procedure for combining atomic orbitals.

A **molecular orbital** is a one-electron wavefunction for an electron that spreads throughout the molecule. The mathematical forms of such orbitals are highly complicated, even for such a simple species as H_2^+, and they are unknown in general. All modern work builds approximations to the true molecular orbital by building them from the atomic orbitals of the atoms present in the molecule.

First, we use once again (as in VB theory) the general principle that if there are several possible outcomes of an observation, then we add together the wavefunctions that represent each outcome. In H_2^+, there are two possible outcomes of locating the electron: it may be found either in an atomic orbital centered on A, ψ_A, or in an orbital centered on B, ψ_B. Therefore, we write

$$\psi = c_A\psi_A + c_B\psi_B \qquad \boxed{\text{An LCAO}} \quad (10.4a)$$

where c_A and c_B are numerical coefficients. This wavefunction is called a **linear combination of atomic orbitals** (LCAO). The squares of the coefficients tell us the relative proportions of the atomic orbitals contributing to the molecular orbital. In a homonuclear diatomic molecule, an electron can be found with equal probability in orbital A or orbital B, so the *squares* of the coefficients must be equal, which implies that $c_B = \pm c_A$. The two possible wavefunctions are therefore

$$\psi = \psi_A \pm \psi_B \qquad \boxed{\begin{array}{l}\text{LCAOs for a homonuclear} \\ \text{diatomic molecule}\end{array}} \quad (10.4b)$$

where, for simplicity, and to focus on the structure of molecular orbitals rather than their numerical details, we are ignoring the overall normalization factor.

(a) Bonding orbitals

First, we consider the LCAO with the plus sign, $\psi = \psi_A + \psi_B$, as this molecular orbital will turn out to have the lower energy of the two. The form of this orbital is

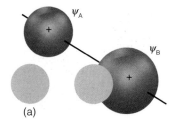

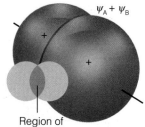

(a)

(b)

Region of constructive interference

Fig. 10.14 The formation of a bonding molecular orbital (a σ orbital). (a) Two H1s orbitals come together. (b) The atomic orbitals overlap, interfere constructively, and give rise to an enhanced amplitude in the internuclear region. The resulting orbital has cylindrical symmetry about the internuclear axis. When it is occupied by two paired electrons, to give the configuration σ^2, we have a σ bond.

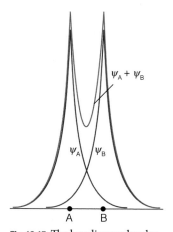

Fig. 10.15 The bonding molecular orbital wavefunction along the internuclear axis. Note that there is an enhancement of amplitude between the nuclei, so there is an increased probability of finding the bonding electrons in that region.

shown in Fig. 10.14. It is called a **σ orbital** because it resembles an s orbital when viewed along the axis. Because (as we shall see) it is the σ orbital of lowest energy, it is labeled 1σ. An electron that occupies a σ orbital is called a **σ electron**. In the ground state of the H_2^+ ion, there is a single 1σ electron, so we report the ground state configuration of H_2^+ as $1\sigma^1$.

By examining the LCAO-MO in eqn 10.4, we can identify the origin of the lowering of energy that is responsible for the formation of the bond. The two atomic orbitals are like waves centered on adjacent nuclei. In the internuclear region, the amplitudes interfere constructively and the wavefunction has an enhanced amplitude (Fig. 10.15). Because the amplitude is increased, there is an increased probability of finding the electron between the two nuclei, where it is in a good position to interact strongly with both of them. Hence the energy of the molecule is lower than that of the separate atoms, where each electron can interact strongly with only one nucleus. In elementary MO theory, the bonding effect of an electron that occupies a molecular orbital is ascribed to its accumulation in the internuclear region as a result of the constructive interference of the contributing atomic orbitals.

A 1σ orbital is an example of a **bonding orbital**, a molecular orbital that, if occupied, contributes to the strength of a bond between two atoms. As in VB theory, we can substitute the wavefunction in eqn 10.4 into the Schrödinger equation for the molecule-ion with the nuclei at a fixed separation R and solve the equation for the energy. The molecular potential energy curve obtained by plotting the energy against R is very similar to the one drawn in Fig. 10.1. The energy of the molecule falls as R is decreased from large values because the electron is increasingly likely to be found in the internuclear region as the two atomic orbitals interfere more effectively. However, at small separations, there is too little space between the nuclei for significant accumulation of electron density there. In addition, the nucleus–nucleus repulsion becomes large. As a result, after an initial decrease, at small internuclear separations the potential energy curve passes through a minimum and then rises sharply to high values. Calculations on H_2^+ give the equilibrium bond length as 130 pm and the bond dissociation energy as 171 kJ mol^{-1}; the experimental values are 106 pm and 250 kJ mol^{-1}, so this simple LCAO-MO description of the molecule, while inaccurate, is not absurdly wrong.

(b) Antibonding orbitals

Now consider the alternative LCAO, the one with a minus sign, $\psi = \psi_A - \psi_B$. Because this wavefunction is also cylindrically symmetrical around the internuclear axis, it is also a σ orbital and is denoted 1σ* (Fig. 10.16). When substituted into the Schrödinger equation, we find that it has a higher energy than the 1σ orbital and, indeed, it has a higher energy than either of the two atomic orbitals.

Self-test 10.4 Show that the molecular orbital written above is zero on a plane cutting through the internuclear axis at its midpoint. Take each atomic orbital to be of the form e^{-r/a_0}, with r_A measured from nucleus A and r_B measured from nucleus B.

Answer: The atomic orbitals cancel for values equidistant from the two nuclei

We can trace the origin of the high energy of $1\sigma^*$ to the existence of a **nodal plane**, a plane on which the wavefunction passes through zero. This plane lies halfway between the nuclei and cuts through the internuclear axis. The two atomic orbitals cancel on this plane as a result of their destructive interference because they have opposite signs. In drawings like that in Figs 10.14 and 10.16, we represent overlap of orbitals with the same sign (as in the formation of 1σ) by shading of the same color; the overlap of orbitals of opposite sign (as in the formation of $1\sigma^*$) is represented by one orbital of one color and another orbital of a different color.

The $1\sigma^*$ orbital is an example of an **antibonding orbital**, an orbital that, if occupied, decreases the strength of a bond between two atoms. The antibonding character of the $1\sigma^*$ orbital is partly a result of the exclusion of the electron from the internuclear region and its relocation outside the bonding region, where it helps to pull the nuclei apart rather than pulling them together (Fig. 10.17). An antibonding orbital is often slightly more strongly antibonding than the corresponding bonding orbital is bonding. This is partly because, although the 'gluing' effect of a bonding electron and the 'anti-gluing' effect of an antibonding electron are similar, the nuclei repel each other in both cases, and this repulsion pushes both levels up in energy.

(c) Inversion symmetry

A final point about notation is important for the discussion of electronic transitions (Chapter 12). For homonuclear diatomic molecules, it is helpful to identify the **inversion symmetry** of a molecular orbital, the behavior of the wavefunction when it is inverted through the center (more formally, the *center of inversion*) of the molecule. Thus, if we consider any point of the 1σ orbital and then project it through the center of the molecule and out an equal distance on the other side, we arrive at an identical value of the wavefunction (Fig. 10.18). This so-called **gerade symmetry** (from the German word for 'even') is denoted by a subscript g, as in $1\sigma_g$. On the other hand, the same procedure applied to the antibonding $1\sigma^*$ orbital results in the same size but opposite sign of the wavefunction. This **ungerade symmetry** ('odd symmetry') is denoted by a subscript u, as in $1\sigma_u$. This inversion symmetry classification is not applicable to heteronuclear diatomic molecules (such as CO) because they do not have a center of inversion.

We shall use the g,u notation because it is helpful when discussing the electronic spectra of molecules and when labeling orbitals in many-electron species. However, to keep track of the bonding or antibonding character of an orbital, when we judge it appropriate we shall attach a * to the orbital label. The g,u classification is fundamental as it is based on symmetry; the * designation is just an aid to interpretation.

10.4 Homonuclear diatomic molecules

To make MO theory relevant to biological systems, we need to describe procedures for describing molecules that are more complex than H_2^+.

In Chapter 9 we used the hydrogenic atomic orbitals and the building-up principle to deduce the ground electronic configurations of many-electron atoms. Here we use the same procedure for many-electron diatomic molecules (such as H_2 with two electrons and even Br_2 with 70), but using the H_2^+ molecular orbitals as a basis.

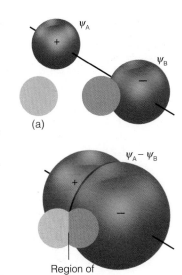

Fig. 10.16 The formation of an antibonding molecular orbital (a σ^* orbital). (a) Two H1s orbitals come together. (b) The atomic orbitals overlap with opposite signs (as depicted by different colors), interfere destructively, and give rise to a decreased amplitude in the internuclear region. There is a nodal plane exactly halfway between the nuclei, on which any electrons that occupy the orbital will not be found.

$dE = (\text{peak anti bond } E) - (\text{peak bond } E)$

$]dE$

(a) Criteria for the formation of molecular orbitals

To use MO theory to build biological molecules we need to know why some atomic orbitals combine whereas some do not. When building molecular orbitals, *we need to consider linear combinations only of atomic orbitals of the same symmetry with respect to the internuclear axis.* Because an s orbital has cylindrical symmetry around the internuclear axis, but a p_x orbital (with x perpendicular to the bond) does not, the two atomic orbitals cannot contribute to the same molecular orbital. The reason for this distinction based on symmetry can be understood by considering the interference between an s orbital and a p_x orbital (Fig. 10.19): although there is constructive interference between the two orbitals on one side of the axis, there is an exactly compensating amount of destructive interference on the other side of the axis, and the net bonding or antibonding effect is zero.

The extent to which two orbitals overlap is measured by the **overlap integral**, S:

$$S = \int \psi_A \psi_B \, d\tau$$

vague   *measure of wave interference*   The overlap integral   (10.5)

where the integration is over all space. If the atomic orbital ψ_A on A is small wherever the orbital ψ_B on B is large or vice versa, then the product of their amplitudes is everywhere small and the integral—the sum of these products—is small (Fig. 10.20a). If ψ_A and ψ_B are simultaneously large in some region of space, then S may be large (Fig. 10.20b). If the two atomic orbitals are identical (for example, 1s orbitals on the same nucleus), $S = 1$. The overlap integral between two H1s orbitals separated by a distance R turns out to be

$$S = \left\{ 1 + \frac{R}{a_0} + \frac{1}{3}\left(\frac{R}{a_0}\right)^2 \right\} e^{-R/a_0} \qquad (10.6)$$

H —R— H

where a_0 is the Bohr radius. This function is plotted in Fig. 10.21: notice how the exponential factor ensures that S approaches zero for large separations. Typical values for orbitals with $n = 2$ are in the range 0.2 to 0.3.

Now consider the arrangement in Fig. 10.20c in which an s orbital overlaps a p_x orbital of a different atom. At some point the product $\psi_A \psi_B$ may be large. However, there is a point where $\psi_A \psi_B$ has exactly the same magnitude but an opposite sign. When the overlap integral is evaluated, these two contributions are added together and cancel out. For every point in the upper half of the diagram, there is a point in the lower half that cancels it, so $S = 0$. Therefore, there is no net overlap between the s and p orbitals in this arrangement, and no contribution to bonding.

Now consider the $2p_x$ and $2p_y$ orbitals of each atom, which are perpendicular to the internuclear axis and may overlap side by side. This overlap may be constructive or destructive and results in a bonding and an antibonding **π orbital**, which we label 1π and 1π*, respectively. The notation π is the analog of p in atoms, for when viewed along the axis of the molecule, a π orbital looks like a p orbital (Fig. 10.22). The two $2p_x$ orbitals overlap to give a bonding and an antibonding π orbital, as do the two $2p_y$ orbitals. The two bonding combinations have the same energy; likewise, the two antibonding combinations have the same energy. Hence, each π energy level is doubly degenerate and consists of two distinct orbitals. Two electrons in a π orbital constitute a **π bond**: such a bond resembles a π bond of VB theory, but the details of th e electron distribution are slightly different.

The inversion-symmetry classification also applies to π orbitals. As we see from Fig. 10.23, a bonding π orbital changes sign on inversion and is therefore classified

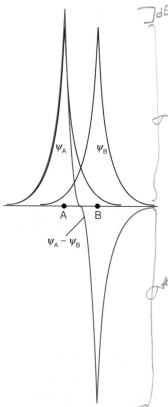

ψ_A ψ_B

A   B

$\psi_A - \psi_B$

Fig. 10.17 The antibonding molecular orbital wavefunction along the internuclear axis. Note that there is a decrease in amplitude between the nuclei, so there is a decreased probability of finding the bonding electrons in that region.

A brief comment

In quantum mechanics, it is conventional to use $d\tau$ (where τ is tau) to represent an infinitesimal volume. In cartesian coordinates, $d\tau = dxdydz$. In spherical coordinates, $d\tau = r^2 dr \sin\theta \, d\theta d\phi$.

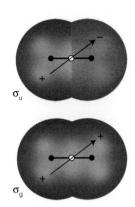

Fig. 10.18 The inversion (gerade/ungerade) character of σ bonding and antibonding orbitals.

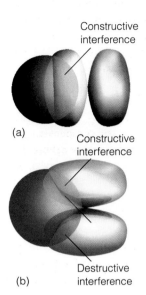

Fig. 10.19 Overlapping s and p orbitals. (a) End-on overlap leads to nonzero overlap and to the formation of an axially symmetric σ orbital. (b) Broadside overlap leads to no net accumulation or reduction of electron density and does not contribute to bonding.

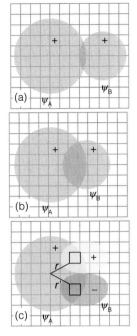

Fig. 10.20 A schematic representation of the contributions to the overlap integral. (a) $S \ll 1$ because the orbitals are far apart and their product is always small. (b) S is large (but less than 1) because the product $\psi_A \psi_B$ is large over a substantial region. (c) $S = 0$ because the positive region of overlap is exactly canceled by the negative region.

as u. On the other hand, the antibonding π* orbital does not change sign and is therefore g. The bonding and antibonding combinations can therefore be denoted $1\pi_u$ and $1\pi_g$ (or $1\pi_g^*$ when we want to emphasize its antibonding character).

We now have the criteria for selecting atomic orbitals from which molecular orbitals are to be built:

1. Use all available valence orbitals from both atoms (in polyatomic molecules, from all the atoms).

2. Classify the atomic orbitals as having σ and π symmetry with respect to the internuclear axis, and build σ and π orbitals from all atomic orbitals of a given symmetry.

3. From N_σ atomic orbitals of σ symmetry, N_σ σ orbitals can be built with progressively higher energy from strongly bonding to strongly antibonding.

4. From N_π atomic orbitals of π symmetry, N_π π orbitals can be built with progressively higher energy from strongly bonding to strongly antibonding. The π orbitals occur in doubly degenerate pairs.

As a general rule, the energy of each type of orbital (σ or π) increases with the number of internuclear nodes. The lowest-energy orbital of a given species has no internuclear nodes, and the highest-energy orbital has a nodal plane between each pair of adjacent atoms (Fig. 10.24).

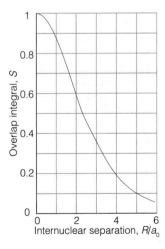

Fig. 10.21 The variation of the overlap integral with internuclear distance for two H1s orbitals.

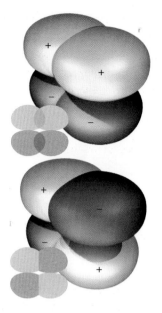

Fig. 10.22 (a) The interference leading to the formation of a π bonding orbital and (b) the corresponding antibonding orbital.

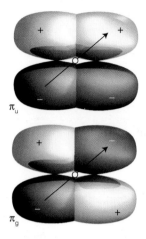

π_u

π_g

Fig. 10.23 The gerade/ungerade character of π bonding and antibonding orbitals.

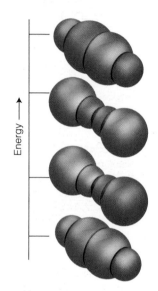

Energy ⟶

Fig. 10.24 A schematic representation of the four molecular orbitals that can be formed from four s orbitals in a chain of four atoms. The lowest-energy combination (the bottom diagram) is formed from atomic orbitals with the same sign, and there are no internuclear nodes. The next higher orbital has one node (at the center of the molecule). The next-higher orbital has two internuclear nodes, and the uppermost, highest energy orbital, has three internuclear nodes, one between each neighboring pair of atoms, and is fully antibonding. The sizes of the spheres reflect the contributions of each atom to the molecular orbital; the different colors represent different signs.

Example 10.2 *Assessing the contribution of d orbitals*

In Section 10.8, we shall see the need to include d orbitals in the description of bonding between d-metal ions, such as Fe^{2+}, and proteins, such as hemoglobin. To get a sense of how molecular orbitals can be built from d orbitals, show how they can contribute to the formation of σ and π orbitals in diatomic molecules.

Strategy We need to assess the symmetry of d orbitals with respect to the internuclear z-axis: orbitals of the same symmetry can contribute to a given molecular orbital.

Solution A d_{z^2} orbital has cylindrical symmetry around z and so can contribute to σ orbitals. The d_{zx} and d_{yz} orbitals have π symmetry with respect to the axis (Fig. 10.25), so they can contribute to π orbitals.

Self-test 10.5 Sketch the 'δ orbitals' (orbitals that resemble four-lobed d orbitals when viewed along the internuclear axis) that may be formed by the remaining two d orbitals (and which contribute to bonding in some d-metal cluster compounds). Give their inversion-symmetry classification.

Answer: see Fig. 10.25: bonding are g, antibonding are u

Once we have constructed the molecular orbitals, we build up the ground-state electron configuration as follows:

1. Accommodate the valence electrons supplied by the atoms so as to achieve the lowest overall energy subject to the constraint of the Pauli exclusion

principle, that no more than two electrons may occupy a single orbital (and then must be paired).

2. If more than one molecular orbital of the same energy is available, add the electrons to each individual orbital before doubly occupying any one orbital (because that minimizes electron–electron repulsions).

3. Take note of Hund's rule (Section 9.11), that if electrons occupy different degenerate orbitals, then they do so with parallel spins.

The following sections show how these rules are used in practice.

Self-test 10.6 How many molecular orbitals can be built from the valence shell orbitals in O_2?

Answer: 8

(b) The hydrogen molecule

The first step in the discussion of H_2, the simplest many-electron diatomic molecule, is to build the molecular orbitals. Because each H atom of H_2 contributes a 1s orbital (as in H_2^+), we can form the $1\sigma_g$ and $1\sigma_u^*$ bonding and antibonding orbitals from them, as we have seen already. At the equilibrium internuclear separation these orbitals will have the energies represented by the horizontal lines in Fig. 10.26.

There are two electrons to accommodate (one from each atom). Both can enter the $1\sigma_g$ orbital by pairing their spins (Fig. 10.27). The ground state configuration is therefore $1\sigma_g^2$, and the atoms are joined by a bond consisting of an electron pair in a bonding σ orbital. These two electrons bind the two nuclei together more strongly and closely than the single electron in H_2^+, and the bond length is reduced from 106 pm to 74 pm. A pair of electrons in a σ orbital is called a **σ bond** and is very similar to the σ bond of VB theory. The two differ in certain details of the electron distribution between the two atoms joined by the bond, but both have an accumulation of density between the nuclei.

We can conclude that *the importance of an electron pair in bonding stems from the fact that two is the maximum number of electrons that can enter a bonding molecular orbital*. Electrons do not 'want' to pair: they pair because, as we show in the following brief *Justification*, the Pauli exclusion principle implies that:

- only if electrons pair their spins can they both occupy a bonding orbital
- no more than two electrons can occupy any given orbital.

Justification 10.2 *Electron pairing in MO theory*

The spatial wavefunction for two electrons in a bonding molecular orbital ψ such as the bonding orbital in eqn 10.4b (with the plus sign) is $\psi(1)\psi(2)$. This two-electron wavefunction is obviously symmetric under interchange of the electron labels. To satisfy the Pauli principle, it must be multiplied by the antisymmetric spin state $\alpha(1)\beta(2) - \beta(1)\alpha(2)$ to give the overall antisymmetric state

$$\psi(1,2) = \psi(1)\,\psi(2) \times \{\alpha(1)\beta(2) - \beta(1)\alpha(2)\}$$

Because $\alpha(1)\beta(2) - \beta(1)\alpha(2)$ corresponds to paired electron spins, we see that two electrons can occupy the same molecular orbital (in this case, the bonding orbital) only if their spins are paired.

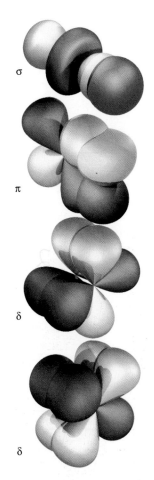

Fig. 10.25 The types of molecular orbital to which d orbitals can contribute. The σ and π combinations can be formed with s, p, and d orbitals of the appropriate symmetry, but the δ orbitals can be formed only by the d orbitals of the two atoms.

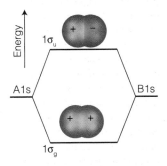

Fig. 10.26 A molecular orbital energy level diagram for orbitals constructed from (1s,1s) overlap, the separation of the levels corresponding to the equilibrium bond length.

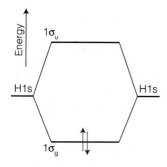

Fig. 10.27 The ground electronic configuration of H_2 is obtained by accommodating the two electrons in the lowest available orbital (the bonding orbital).

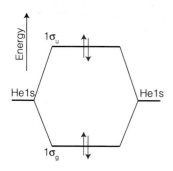

Fig. 10.28 The ground electronic configuration of the four-electron molecule He_2 has two bonding electrons and two antibonding electrons. It has a higher energy than the separated atoms, and so He_2 is unstable relative to two He atoms.

A similar argument shows why helium is a monatomic gas. Consider a hypothetical He_2 molecule. Each He atom contributes a 1s orbital to the linear combination used to form the molecular orbitals, and so we can construct $1\sigma_g$ and $1\sigma_u^*$ molecular orbitals. They differ in detail from those in H_2 because the He1s orbitals are more compact than H1s orbitals, but the general shape of the molecular orbitals is the same, and for qualitative discussions we can use the same molecular orbital energy level diagram as for H_2. Because each atom provides two electrons, there are four electrons to accommodate. Two can enter the $1\sigma_g$ orbital, but then it is full (by the Pauli exclusion principle). The next two electrons must enter the antibonding $1\sigma_u^*$ orbital (Fig. 10.28). The ground electronic configuration of He_2 is therefore $1\sigma_g^2 1\sigma_u^{*2}$. Because an antibonding orbital is slightly more antibonding than a bonding orbital is bonding, the He_2 molecule has a higher energy than the separated atoms and is unstable. Hence, two ground-state He atoms do not form bonds to each other, and helium is a monatomic gas.

(c) Many-electron homonuclear diatomic molecules

We shall now see how the concepts we have introduced apply to other homonuclear diatomic molecules, such as N_2 and O_2, and diatomic ions such as O_2^{2-}. In line with the building-up procedure, we first consider the molecular orbitals that can be formed from the valence orbitals and do not (at this stage) worry about how many electrons are available.

In an element of Period 2 (Li to Ne), the valence orbitals are 2s and 2p. Suppose first that we consider these two types of orbital separately. Then the 2s orbitals on each atom overlap to form bonding and antibonding combinations that we denote $1\sigma_g$ and $1\sigma_u^*$, respectively. Likewise, the two $2p_z$ orbitals (by convention, the internuclear axis is the z-axis) have cylindrical symmetry around the internuclear axis. They may therefore participate in σ-orbital formation to give the bonding and antibonding combinations $2\sigma_g$ and $2\sigma_u^*$, respectively (Fig. 10.29). The two $2p_x$ orbitals overlap to give a bonding and an antibonding π orbital, as do the two $2p_y$ orbitals. The resulting energy levels of the orbitals are shown in the molecular orbital energy level diagram in Fig. 10.30.

There is a minor complication: although the p_x and p_y orbitals have different symmetry from the p_z orbitals (in the sense of forming π and σ orbitals, respectively), the p_z orbital has the same symmetry as the s orbital (in the sense that both can contribute to σ orbitals). When the 2s and 2p orbitals differ considerably in energy, as they do on the right of Period 2, they can be treated separately as we have described. However, when their energies are similar, all four orbitals (the 2s orbitals on each atom and their $2p_z$ orbitals) all contribute to the formation of σ orbitals, and each orbital has the form $\psi = c_1 \psi_{A2s} + c_2 \psi_{B2s} + c_3 \psi_{A2p_z} + c_4 \psi_{B2p_z}$. To find the four coefficients and the energies of the molecular orbitals we need to solve the Schrödinger equation. However, in practice, the energies of the two lowest-energy combinations of this kind are similar to the energies of the $1\sigma_g$ and $1\sigma_u$ orbitals formed solely from 2s orbitals. Similarly, the energies of the two highest-energy combinations are very similar to the energies of the $2\sigma_g$ and $2\sigma_u$ combinations of $2p_z$ orbitals. Because the changes are not great, we can continue to think of $1\sigma_g$ and $1\sigma_u$ as being one bonding and antibonding pair and of $2\sigma_g$ and $2\sigma_u$ as being another pair.

The relative order of the σ and π orbitals in a molecule cannot be predicted without detailed calculation and varies with the energy separation between the 2s and 2p orbitals of the atoms. In molecules built from atoms in which the 2s and 2p orbitals are widely separated in energy (on the right of Period 2, specifically for

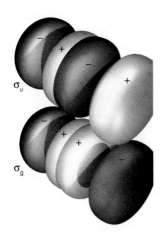

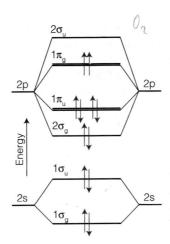

O_2

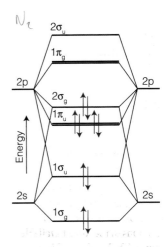

N_2

Fig. 10.29 (a) The interference leading to the formation of a σ bonding orbital and (b) the corresponding antibonding orbital when two p orbitals overlap along an internuclear axis.

Fig. 10.30 A typical molecular orbital energy level diagram for Period 2 homonuclear diatomic molecules. The valence atomic orbitals are drawn in the columns on the left and the right; the molecular orbitals are shown in the middle. Note that the π orbitals form doubly degenerate pairs. The sloping lines joining the molecular orbitals to the atomic orbitals show the principal composition of the molecular orbitals. This diagram is suitable for O_2 and F_2; the configuration of O_2 is shown.

Fig. 10.31 A typical molecular orbital energy level diagram for Period 2 homonuclear diatomic molecules up to and including N_2.

O and F) and can be treated separately, the order shown in Fig. 10.30 applies. When the 2s and 2p atomic orbitals have similar energies (on the left of Period 2, as far as N) and must be treated collectively, the order of molecular orbitals is more like that in Fig. 10.31. The change in order can be seen in Fig. 10.32, which shows the calculated energy levels for the Period 2 homonuclear diatomic molecules. In summary:

- Figure 10.30 is appropriate for O_2 and F_2.
- Figure 10.31 is appropriate for the preceding homonuclear diatomic molecules of the period.

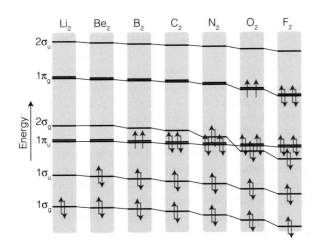

Fig. 10.32 The variation of the orbital energies of Period 2 homonuclear diatomic molecules. Only the valence shell orbitals are shown.

With the orbital energy level scheme established, we can predict the ground-state configuration of a molecule by using the building-up principle. For N_2, for instance, which has 10 valence electrons; we use Fig. 10.31. The first two electrons pair, enter, and fill the $1\sigma_g$ orbital. The next two electrons enter and fill the $1\sigma_u$ orbital. Six electrons remain. There are two $1\pi_u$ orbitals, so four electrons can be accommodated in them. The two remaining electrons enter the $2\sigma_g$ orbital. The ground state configuration of N_2 is therefore $1\sigma_g^2 1\sigma_u^2 1\pi_u^4 2\sigma_g^2$. This configuration is also depicted in Fig. 10.31.

(d) Bond order

The strength of a bond in a molecule is the net outcome of the bonding and anti-bonding effects of the electrons in the orbitals. The **bond order**, b, in a diatomic molecule is defined as

$$b = \tfrac{1}{2}(n - n^*)$$ Definition of bond order   (10.7)

where n is the number of electrons in bonding orbitals and n^* is the number of electrons in antibonding orbitals (as judged by the presence of a nodal plane between the two atoms due to destructive interference of the orbitals). Each electron pair in a bonding orbital increases the bond order by 1 and each pair in an antibonding orbital decreases it by 1. In N_2, $1\sigma_g$, $2\sigma_g$, and $1\pi_u$ are bonding orbitals, and $n = 2 + 2 + 4 = 8$; however, $1\sigma_u$ is antibonding, so $n^* = 2$ and the bond order of N_2 is $b = \tfrac{1}{2}(8 - 2) = 3$. This value is consistent with the Lewis structure :N≡N:, in which there is a triple bond between the two atoms.

The bond order is a useful parameter for discussing the characteristics of bonds because it correlates with bond length, in the sense that the greater the bond order between atoms of a given pair of atoms, then the shorter the bond. The bond order also correlates with bond strength, in the sense that the greater the bond order, then the greater the strength. The high bond order of N_2 is consistent with its high dissociation energy (942 kJ mol^{-1}).

Self-test 10.7 Write the ground-state electronic configuration and deduce the bond order of F_2 and Ne_2. Which of these elements is expected to exist as a monatomic species under normal conditions?

Answer: F_2: $1\sigma_g^2 1\sigma_u^2 2\sigma_g^2 1\pi_u^4 1\pi_g^4$, $b = 1$; Ne_2: $1\sigma_g^2 1\sigma_u^2 2\sigma_g^2 1\pi_u^4 1\pi_g^4 2\sigma_u^2$,
$b = 0$ (neon is a monatomic species)

Self-test 10.8 Which can be expected to have the higher dissociation energy, F_2 or F_2^+?

Answer: F_2^+

Case study 10.1 *The biochemical reactivity of O_2 and N_2*

Dinitrogen, N_2, the major component of the air we breathe, is so stable (on account of the triple bond connecting the atoms) and unreactive that *nitrogen fixation*, the reduction of atmospheric N_2 to NH_3, is among the most thermodynamically demanding of biochemical reactions, in the sense that it requires a great deal of energy derived from metabolism. So taxing is the process that only certain bacteria and archaea are capable of carrying it out, making nitrogen available first to plants and other microorganisms in the form of ammonia. Only after incorporation into amino acids by plants does nitrogen adopt a chemical form that, when consumed, can be used by animals in the synthesis of proteins and other nitrogen-containing molecules.

Figure 10.30 is the appropriate molecular orbital energy level diagram for O_2. There are 12 valence electrons to accommodate: the first 10 electrons recreate the N_2 configuration (with a reversal of the order of the $2\sigma_g$ and $1\pi_u$ orbitals) and the remaining two electrons must occupy the $1\pi_g$ orbitals. The configuration is therefore $1\sigma_g^2 1\sigma_u^2 2\sigma_g^2 1\pi_u^4 1\pi_g^2$ (as depicted in Fig. 10.30). Because $1\sigma_g$, $2\sigma_g$, and $1\pi_u$ are regarded as bonding and $1\sigma_u$ and $1\pi_g$ as antibonding, the bond order is $b = \frac{1}{2}(8-4) = 2$, a value that is consistent with the classical view that O_2 has a double bond.

According to the building-up principle, the two $1\pi_g$ electrons in O_2 occupy different orbitals. One enters the $1\pi_{g,x}$ orbital formed by overlap of the $2p_x$ orbitals. The other enters its degenerate partner, the $1\pi_{g,y}$ orbital formed from overlap of the $2p_y$ orbitals. Because the two electrons occupy different orbitals, by Hund's rule they will have parallel spins ($\uparrow\uparrow$), and an O_2 molecule is sometimes said to be a **biradical**, a radical containing two unpaired electrons. However, the term must be used with caution because in a true biradical the two electron spins have random relative orientations; O_2 is not a true biradical because the two spins are locked into a parallel arrangement.

A striking prediction of MO theory is that because the O_2 molecule has two unpaired spins, it is a **paramagnetic** substance, a substance that is drawn into a magnetic field. Most substances (those with paired electron spins) are **diamagnetic** and are pushed out of a magnetic field. That O_2 is in fact a paramagnetic gas is a striking confirmation of the superiority of the molecular orbital description of the molecule over the Lewis and VB descriptions (which require all the electrons to be paired; recall Self-test 10.1). The property of paramagnetism is utilized to monitor the oxygen content of incubators by measuring the magnetism of the gases they contain.

The reactivity of O_2, while important for biological energy conversion, also poses serious physiological problems. During the course of metabolism, some electrons escape from complexes I, II, and III of the respiratory chain (Chapter 5) and reduce O_2 to superoxide ion, O_2^-. From Fig. 10.30, the ground-state electronic configuration of O_2^- is expected to be $1\sigma_g^2 1\sigma_u^2 2\sigma_g^2 1\pi_u^4 1\pi_g^3$, so the ion is a radical with a bond order $b = 1.5$. We predict that the superoxide ion is a reactive species that must be scavenged to prevent damage to cellular components. The enzyme superoxide dismutase protects cells by catalyzing the disproportionation (or dismutation) of O_2^- into O_2 and H_2O_2:

$$2\,O_2^- + 2\,H^+ \rightarrow H_2O_2 + O_2$$

However, H_2O_2 (hydrogen peroxide), formed by the reaction above and by leakage of electrons out of the respiratory chain, is a powerful oxidizing agent and also harmful to cells. It is metabolized further by catalases and peroxidases. A catalase catalyzes the reaction

$$2\,H_2O_2 \rightarrow 2\,H_2O + O_2$$

and a peroxidase reduces hydrogen peroxide to water by oxidizing an organic molecule. For example, the enzyme glutathione peroxidase catalyzes the oxidation of glutathione (Atlas M4):

$$2 \text{ glutathione}_{red} + H_2O_2 \rightarrow 2 \text{ glutathione}_{ox} + 2 H_2O$$

There is growing evidence for the involvement of the damage caused by *reactive oxygen species* (ROS), such as O_2^-, H_2O_2, and OH (the hydroxyl radical), in the mechanism of aging and in the development of cardiovascular disease, cancer, stroke, inflammatory disease, and other conditions. For this reason, much effort has been expended on studies of the biochemistry of *antioxidants*, substances that can either deactivate ROS directly (as glutathione does) or halt the progress of cellular damage through reactions with radicals formed by processes initiated by ROS. Important examples of antioxidants are vitamin C (ascorbic acid, Atlas M1), vitamin E (α-tocopherol, Atlas M3), and uric acid (Atlas M2).

10.5 Heteronuclear diatomic molecules

We need to understand how electronic structure affects the reactivity of molecules such as NO (a biochemical messenger).

The characteristic feature of heteronuclear diatomic molecules that will be familiar from introductory chemistry is that the electron distribution is not symmetrical between the atoms because it is energetically favorable for a bonding electron pair to be found closer to one atom rather than the other. This imbalance results in a **polar bond**, which is a covalent bond in which the electron pair is shared unequally by the two atoms.

(a) Polarity and electronegativity

The imbalance of charge distribution is commonly expressed in terms of the **electronegativity**, χ (chi), the power of an element to draw electrons to itself when it is part of a compound. Linus Pauling formulated a numerical scale of electronegativity based on considerations of bond dissociation energies, $E(A-B)$:

$$|\chi_A - \chi_B| = 0.102 \times (\Delta E/\text{kJ mol}^{-1})^{1/2} \qquad \text{Pauling electronegativity scale} \qquad (10.8a)$$

with

$$\Delta E = E(A-B) - \tfrac{1}{2}\{E(A-A) + E(B-B)\} \qquad (10.8b)$$

Table 10.2 lists values for the main-group elements. Robert Mulliken proposed an alternative definition in terms of the ionization energy, I, and the electron affinity, E_{ea}, of the element expressed in electronvolts:

$$\chi = \tfrac{1}{2}(I + E_{ea}) \qquad \text{Mulliken electronegativity scale} \qquad (10.8c)$$

This relation is plausible because an atom that has a high electronegativity is likely to be one that has a high ionization energy (so that it is unlikely to lose electrons to another atom in the molecule) and a high electron affinity (so that it is energetically favorable for an electron to move toward it). The Mulliken electronegativities are broadly in line with the Pauling electronegativities. Electronegativities show a periodicity, and the elements with the highest electronegativities are those close to fluorine in the periodic table (with the exception of the noble gases).

Table 10.2 Electronegativities of the main-group elements*

H						
2.1						
Li	Be	B	C	N	O	F
1.01	1.5	2.0	2.5	3.0	3.5	4.0
Na	Mg	Al	Si	P	S	Cl
0.9	1.2	1.5	1.8	2.1	2.5	3.0
K	Ca	Ga	Ge	As	Se	Br
0.8	1.0	1.6	1.8	2.0	2.4	2.8
Rd	Sr	In	Sn	Sb	Te	I
8.0	1.0	1.7	1.8	1.9	2.1	2.5
Cs	Ba	Tl	Pb	Bi	Po	
0.7	0.9	1.8	1.8	1.9	2.0	

*Pauling values.

The location of the bonding electron pair close to one atom in a heteronuclear molecule results in that atom having a net negative charge, which is called a **partial negative charge** and denoted δ−. There is a compensating **partial positive charge**, δ+, on the other atom. In a typical heteronuclear diatomic molecule, the more electronegative element has the partial negative charge and the more electropositive element has the partial positive charge.

Self-test 10.9) Predict the (weak) polarity of a C–H bond.

Answer: $^{δ-}C–H^{δ+}$

(b) Molecular orbitals in heteronuclear species

Molecular orbital theory takes heteronuclear diatomic molecules and their polar bonds in its stride. Each molecular orbital has the form

$$\psi = c_A\psi_A + c_B\psi_B \qquad \boxed{\text{A general LCAO}} \quad (10.9)$$

If $c_B^2 > c_A^2$, then the electrons spend more time on B than on A and the bond is polar in the sense $^{δ+}A–B^{δ-}$. A nonpolar bond, a covalent bond in which the electron pair is shared equally between the two atoms and there are zero partial charges on each atom, has $c_A^2 = c_B^2$. A pure ionic bond, in which one atom has obtained virtually sole possession of the electron pair (as in Cs^+F^-, to a first approximation), has one coefficient zero (so that A^+B^- would have $c_A^2 = 0$ and $c_B^2 = 1$).

A general feature of molecular orbitals between dissimilar atoms is that the atomic orbital with the lower energy (that belonging to the more electronegative atom) makes the larger contribution to the lowest-energy molecular orbital. The opposite is true of the highest (most antibonding) orbital, for which the principal contribution comes from the atomic orbital with higher energy (the less electronegative atom):

	Bonding orbitals	**Antibonding orbitals**
For $\chi_A > \chi_B$:	$c_A^2 > c_B^2$	$c_B^2 > c_A^2$

Figure 10.33 shows a schematic representation of this point.

These features of polar bonds can be illustrated by considering the N–H bond in the peptide group (**1**). The electronegativity of N is greater than that of H, so we expect a polar bond with the charge distribution $^{δ-}N–H^{δ+}$. For the purposes of illustrating concepts and expressing this polarity in terms of molecular orbitals, we treat the NH fragment in isolation, disregarding its interactions with other atoms in the group. The general form of the molecular orbitals of the NH fragment is $\psi = c_H\psi_H + c_N\psi_N$, where ψ_H is an H1s orbital and ψ_N is an $N2p_z$ orbital. Because the ionization energy of a hydrogen atom is 13.6 eV, we know that the energy of the H1s orbital is −13.6 eV. As usual, the zero of energy is the infinitely separated electron and proton (Fig. 10.34). Similarly, from the ionization energy of nitrogen, which is 14.5 eV, we know that the energy of the $N2p_z$ orbital is −14.5 eV, about 0.9 eV lower than the H1s orbital. It follows that the bonding σ orbital in NH is mainly $N2p_z$ and the antibonding σ orbital is mainly H1s orbital in character. The two electrons in the bonding orbital are most likely to be found in the $N2p_z$ orbital, so there is a partial negative charge on the N atom and a partial positive charge on the H atom.

A systematic way of finding the coefficients in the linear combinations is to solve the Schrödinger equation and to look for the values of the coefficients that

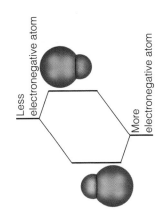

Fig. 10.33 A schematic representation of the relative contributions of atoms of different electronegativities to bonding and antibonding molecular orbitals. In the bonding orbital, the more electronegative atom makes the greater contribution (represented by the larger sphere), and the electrons of the bond are more likely to be found on that atom. The opposite is true of an antibonding orbital. A part of the reason why an antibonding orbital is of high energy is that the electrons that occupy it are likely to be found on the more electropositive atom.

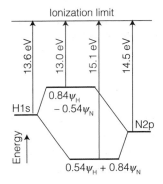

Fig. 10.34 The atomic orbital energy levels of H and N atoms and the molecular orbitals they form. The bonding orbital has predominantly N atom character and the antibonding orbital has predominantly H atom character. Energies are in electronvolts.

result in the lowest energy. For NH, the lowest energy is obtained for the orbital[1] $\psi = 0.54\psi_H + 0.84\psi_N$. We see that indeed the N2p$_z$ orbital does make the greater contribution to the bonding σ orbital.

Self-test 10.10 What percentage of its time does a electron in the NH fragment spend in a N2p$_z$ orbital?

Answer: 71 per cent $(= (0.84)^2 \times 100\%)$

Case study 10.2   *The biochemistry of NO*

Nitric oxide (nitrogen monoxide, NO) is a small molecule that diffuses quickly between cells, carrying chemical messages that help initiate a variety of processes, such as regulation of blood pressure, inhibition of platelet aggregation, and defense against inflammation and attacks to the immune system. Initially there was much opposition to the suggestion that such a small reactive molecule could be biologically relevant, but in due course the proposal was recognized by the award of the Nobel Prize for Medicine in 1998 (to R.F. Furchgott, L.J. Ignarro, and F. Murad). The molecule is synthesized from the amino acid arginine in a series of reactions catalyzed by nitric oxide synthase and requiring O_2 and NADPH.

To gain insight into the biochemistry of NO, we need to consider its electronic structure. Figure 10.35 shows the bonding scheme in NO and illustrates a number of points we have made about heteronuclear diatomic molecules. The ground configuration is $1\sigma^2 2\sigma^2 3\sigma^2 1\pi^4 2\pi^1$. (The g,u designation is not applicable because the molecule is heteronuclear, and we are numbering each species of orbital in sequence or increasing energy.) The 3σ and 1π orbitals are predominantly of O character because that is the more electronegative element. The **highest occupied molecular orbital** (HOMO) is 2π, contains one electron, and has more N character than O character. It follows that NO is a radical with an unpaired electron that can be regarded as localized more on the N atom than on the O atom. The **lowest unoccupied molecular orbital** (LUMO) is 4σ, which is also localized predominantly on N.

Because NO is a radical, we expect it to be reactive. Its half-life is estimated at approximately 1–5 s, so it needs to be synthesized often in the cell. As we saw in *Case study* 10.1, there is a biochemical price to be paid for the reactivity of biological radicals. Like O_2, NO participates in some reactions that are not beneficial to the cell. Indeed, the radicals O_2^- and NO combine to form the peroxynitrite ion (5):

$$NO + O_2^- \rightarrow ONOO^-$$

The peroxynitrite ion is a reactive oxygen species that damages proteins, DNA, and lipids, possibly leading to heart disease, amyotrophic lateral sclerosis (Lou Gehrig's disease), Alzheimer's disease, and multiple sclerosis. We note that the structure of the ion is consistent with the bonding scheme of Fig. 10.35: because the unpaired electron in NO is slightly more localized on the N atom, we expect that atom to form a bond with an O atom from the O_2^- ion.

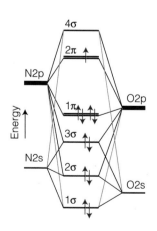

5 ONOO⁻

Fig. 10.35 The molecular orbital energy level diagram for NO.

---

[1] The values of the coefficients are best found by using software of the kind described in Section 10.9. For the purpose of this illustration, we are ignoring overlap between the atomic orbitals.

10.6 The structures of polyatomic molecules

Polyatomic molecules are the building blocks of living organisms, and to understand their electronic structures we need to use MO theory; by doing so, we shall come to understand the unique role of carbon.

The bonds in polyatomic molecules are built in the same way as in diatomic molecules, the only differences being that we use more atomic orbitals to construct the molecular orbitals and these molecular orbitals spread over the entire molecule, not just the adjacent atoms of the bond. In general, a molecular orbital is a linear combination of all the atomic orbitals of all the atoms in the molecule.

In H_2O, for instance, the atomic orbitals are the two H1s orbitals, the O2s orbital, and the three O2p orbitals (if we consider only the valence shell). From these six atomic orbitals we can construct six molecular orbitals that differ in energy. The lowest-energy, most strongly bonding orbital has the least number of nodes between adjacent atoms. The highest-energy, most strongly antibonding orbital has the greatest numbers of nodes between neighboring atoms (Fig. 10.36). According to MO theory, the bonding influence of a single electron pair is distributed over all the atoms, and each electron pair (the maximum number of electrons that can occupy any single molecular orbital) helps to bind all the atoms together.

In the LCAO approximation, each molecular orbital is modeled as a linear combination of atomic orbitals of matching symmetry, with atomic orbitals contributed by all the atoms in the molecule. Thus, a typical molecular orbital in H_2O constructed from H1s orbitals (denoted ψ_A and ψ_B) and O2s and O2p$_y$ and O2p$_z$ orbitals will have the composition

$$\psi = c_1\psi_A + c_2\psi_B + c_3\psi_{O2s} + c_4\psi_{O2p_y} + c_5\psi_{O2p_z} \qquad (10.10)$$

The O2p$_x$ orbital (with x perpendicular to the molecular frame) does not contribute because it has the wrong symmetry to overlap with the H1s orbitals. Because five atomic orbitals are being used to form the LCAO, there are five molecular orbitals of this kind: the lowest-energy (most bonding) orbital will have no internuclear nodes and the highest-energy (most antibonding) orbital will have a node between each pair of neighboring nuclei.

10.7 Hückel theory

Many biological systems, such as those responsible for photosynthesis, vision, and the colors of vegetation, consist of molecules with conjugated π-electron systems. We need a simple way to construct their molecular orbitals and assess their energies.

An important example of the application of MO theory is to the orbitals that may be formed from the p orbitals perpendicular to a molecular plane, such as that of the phenyl ring of the amino acid phenylalanine. A computational scheme was proposed by Erich Hückel and provides a simple way of establishing the molecular orbitals of π-electron systems, especially hydrocarbons such as ethene, benzene, and their derivatives. A common procedure is to treat the σ-bonding framework using the language of VB theory, and to treat the π-electron system separately by MO theory. We use that approach here.

(a) Ethene

Each carbon atom in ethene, $CH_2=CH_2$, is regarded as sp^2 hybridized and forming C–C and C–H σ-bonds at 120° to each other by spin-pairing and either (Csp^2,Csp^2)- or (Csp^2,H1s)-orbital overlap (note the VB language). The unhybridized C2p$_z$

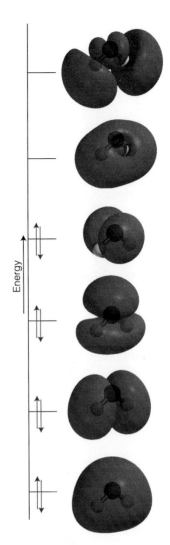

Energy

Fig. 10.36 Schematic form of the molecular orbitals of H_2O and their energies.

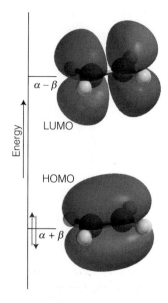

Energy

$\alpha - \beta$

LUMO

HOMO

$\alpha + \beta$

Fig. 10.37 The bonding and antibonding π molecular orbitals of ethene and their energies.

orbitals perpendicular to the σ-framework (ψ_A and ψ_B), each of which is occupied by a single electron, are then used to construct molecular orbitals (Fig. 10.37):

$$\psi = c_A \psi_A + c_B \psi_B \qquad (10.11)$$

We show in the following *Justification* that to find the energies and coefficients of the two molecular orbitals that can be formed from these two atomic orbitals we need to solve the following simultaneous equations:

$$(H_{AA} - ES_{AA})c_A + (H_{AB} - ES_{AB})c_B = 0$$
$$(H_{BA} - ES_{BA})c_A + (H_{BB} - ES_{BB})c_B = 0$$

Secular equations for ethene $\qquad (10.12)$

In the context of MO theory, these simultaneous equations are called **secular equations**. The H_{JK} are expressions that include various contributions to the energy, including the repulsion between electrons and their attractions to the nuclei; the S_{JK} are the overlap integrals between orbitals on atoms J and K.

Justification 10.3   *The secular equations*

We begin by substituting eqn 10.11 into the Schrödinger equation written in the form $\hat{H}\psi = E\psi$:

$$c_A \hat{H}\psi_A + c_B \hat{H}\psi_B = c_A E\psi_A + c_B E\psi_B$$

Then we multiply both sides by ψ_A

$$c_A \psi_A \hat{H}\psi_A + c_B \psi_A \hat{H}\psi_B = c_A \psi_A E\psi_A + c_B \psi_A E\psi_B$$

and integrate over all space (with the term $d\tau$ denoting an infinitesimal volume element in three dimensions; in cartesian coordinates, $d\tau = dxdydz$):

$$c_A \int \psi_A \hat{H}\psi_A d\tau + c_B \int \psi_A \hat{H}\psi_B d\tau = c_A E \int \psi_A \psi_A d\tau + c_B E \int \psi_A \psi_B d\tau$$

E is a constant, so we have been able to take it outside the integral. Now we write

$$H_{AA} = \int \psi_A \hat{H}\psi_A d\tau \quad H_{AB} = \int \psi_A \hat{H}\psi_B d\tau \quad S_{AA} = \int \psi_A \psi_A d\tau \quad S_{AB} = \int \psi_A \psi_B d\tau$$

The preceding equation then becomes

$$c_A H_{AA} + c_B H_{AB} = c_A E S_{AA} + c_B E S_{AB}$$

which is easy to rearrange into the first of eqn 10.12. If instead of multiplying through by ψ_A we multiply by ψ_B, we obtain the second of eqn 10.12.

To simplify the solution of the secular equations Hückel introduced the following drastic approximations:

- All H_{JJ} are set equal to a single quantity α called the **Coulomb integral**.

- All H_{JK} are set equal to zero unless atoms J and K are adjacent, when it is set equal to a single quantity β (a negative quantity) called the **resonance integral**.

- All S_{JJ} are set equal to 1 and all S_{JK} are set equal to 0 whether or not J and K are adjacent.

Mathematical toolbox 10.1 *Simultaneous equations and determinants*

Two simultaneous equations of the form

$$ax + by = 0$$
$$cx + dy = 0$$

have solutions only if the 'determinant' of the coefficients is equal to zero. In this case we write

$$\begin{vmatrix} a & b \\ c & d \end{vmatrix} = 0$$

where the term on the left is the determinant and has the following meaning:

$$\begin{vmatrix} a & b \\ c & d \end{vmatrix} = ad - bc$$

Three simultaneous equations of the form

$$ax + by + cz = 0$$

$$dx + ey + fz = 0$$
$$gx + hy + iz = 0$$

have a solution only if

$$\begin{vmatrix} a & b & c \\ d & e & f \\ g & h & i \end{vmatrix} = 0$$

This 3×3 determinant expands as follows:

$$\begin{vmatrix} a & b & c \\ d & e & f \\ g & h & i \end{vmatrix} = a \begin{vmatrix} e & f \\ h & i \end{vmatrix} - b \begin{vmatrix} d & f \\ g & i \end{vmatrix} + c \begin{vmatrix} d & e \\ g & h \end{vmatrix}$$

Note the alternation in signs for successive columns. The 2×2 determinants then expand like the one above.

With these 'Hückel approximations' the secular equations become

$$(\alpha - E)c_A + \beta c_B = 0$$
$$\beta c_A + (\alpha - E)c_B = 0$$

Hückel approximation for ethene · (10.13a)

As set out in Mathematical toolbox 10.1, these two simultaneous equations have a solution only if the **secular determinant** vanishes:

$$\begin{vmatrix} \alpha - E & \beta \\ \beta & \alpha - E \end{vmatrix} = (\alpha - E)^2 - \beta^2 = 0$$

Hückel secular determinant for ethene · (10.13b)

This condition is satisfied if

$$E = \alpha \pm \beta$$

Hückel energies for ethene · (10.13c)

When each value is substituted into eqn 10.13a, we find:

For $E = \alpha + \beta$ $c_A = c_B$, so $\psi = c_A(\psi_A + \psi_B)$

For $E = \alpha - \beta$ $c_A = -c_B$, so $\psi = c_A(\psi_A - \psi_B)$

(Remember that $\beta < 0$, so $E = \alpha + \beta$ is the lower energy of the two.) These energies and orbitals are represented in Fig. 10.37: they will be recognized as the bonding and antibonding combinations of the $C2p_z$ atomic orbitals. The value of the one unknown, c_A, is found by ensuring that each orbital is normalized, but we do not need its explicit value.

Because there are two electrons to be accommodated, both enter the lower energy orbital and contribute $2\alpha + 2\beta$ to the energy of the molecule. We can also infer that the energy needed to excite a π electron to the antibonding combination is $2|\beta|$. A typical value of β in hydrocarbons is about -2.4 eV, or -230 kJ mol^{-1}.

Self-test 10.11   Write down the Hückel secular determinant for butadiene.

Answer:
$$\begin{vmatrix} \alpha - E & \beta & 0 & 0 \\ \beta & \alpha - E & \beta & 0 \\ 0 & \beta & \alpha - E & \beta \\ 0 & 0 & \beta & \alpha - E \end{vmatrix}$$

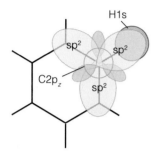

Fig. 10.38 The orbitals used to construct the molecular orbitals of benzene.

(b) Benzene

Exactly the same procedure can be used for benzene, C_6H_6. Each C atom is regarded as sp^2 hybridized (note the VB language again) and forms a planar hexagonal framework of σ bonds (Fig. 10.38). There is an unhybridized $C2p_z$ orbital on each atom perpendicular to the ring from which we form molecular orbitals. From these six atomic orbitals we construct six molecular orbitals of the form

$$\psi = c_A\psi_A + c_B\psi_B + c_C\psi_C + c_D\psi_D + c_E\psi_E + c_F\psi_F \qquad (10.14)$$

Then we set up the six simultaneous equations for the coefficients and the corresponding 6×6 secular determinant and apply the Hückel approximations. Its form resembles that for cyclobutadiene in Exercise 10.33, but with six rows and six columns. Full-frontal attack on it to determine the six values of E is rather tedious, especially as there are procedures that make use of symmetry that greatly simplifies the solution. As should be verified, the energies and the corresponding (unnormalized) molecular orbitals obtained are as follows (Fig. 10.39):

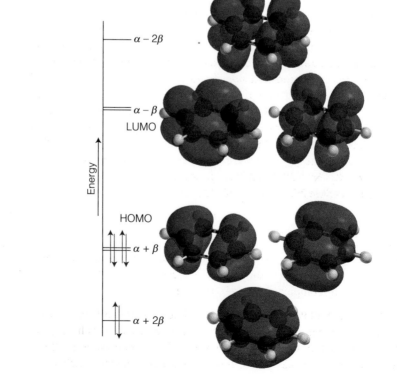

Fig. 10.39 The π orbitals of benzene and their energies. The lowest-energy orbital is fully bonding between neighboring atoms, but the uppermost orbital is fully antibonding. The two pairs of doubly degenerate molecular orbitals have an intermediate number of internuclear nodes. As usual, different colors represent different signs of the wavefunction.

Energy	Orbital
Highest (most antibonding)	
$\alpha - 2\beta$	$\psi = \psi_A - \psi_B + \psi_C - \psi_D + \psi_E - \psi_F$
$\alpha - \beta$ *degen.*	$\psi = 2^{1/2}\psi_A - \psi_B - \psi_C + 2^{1/2}\psi_D - \psi_E - \psi_F$
$\alpha - \beta$	$\psi = \psi_B - \psi_C + \psi_E - \psi_F$
$\alpha + \beta$ *degen.*	$\psi = 2^{1/2}\psi_A + \psi_B + \psi_C - 2^{1/2}\psi_D - \psi_E - \psi_F$
$\alpha + \beta$	$\psi = \psi_B + \psi_C - \psi_E - \psi_F$
$\alpha + 2\beta$	$\psi = \psi_A + \psi_B + \psi_C + \psi_D + \psi_E + \psi_F$
Lowest (most bonding)	

Note that the lowest-energy, most bonding orbital has no internuclear nodes. It is strongly bonding because the constructive interference between neighboring p orbitals results in a good accumulation of electron density between the nuclei (but slightly off the internuclear axis, as in the π bonds of diatomic molecules). In the most antibonding orbital the alternation of signs in the linear combination results in destructive interference between neighbors, and the molecular orbital has a nodal plane between each pair of neighbors, as shown in the illustration. The four intermediate orbitals form two doubly degenerate pairs, one net bonding and the other net antibonding.

There are six electrons to be accommodated (one is supplied by each C atom), and they occupy the lowest three orbitals in Fig. 10.39. The resulting electron distribution is like a double donut. It is an important feature of the configuration that the only molecular orbitals occupied have a net bonding character, for this is one contribution to the stability (in the sense of low energy) of the benzene molecule. It may be helpful to note the similarity between the molecular orbital energy level diagram for benzene and that for N_2 (see Fig. 10.31): the strong bonding, and hence the stability, of benzene and of the phenyl ring in aromatic amino acids is an echo of the strong bonding in the nitrogen molecule.

A feature of the molecular orbital description of benzene is that each molecular orbital spreads either all around or partially around the C_6 ring. That is, π bonding is **delocalized**, and each electron pair helps to bind together several or all of the C atoms. The delocalization of bonding influence is a primary feature of MO theory that we shall use time and again when discussing conjugated systems, such as those found in selected amino acid side chains (phenylalanine, tyrosine, histidine, and tryptophan), the purine and pyrimidine bases in nucleic acids, the heme group, and the pigments involved in photosynthesis and vision. The stabilization of the benzene molecule due to delocalization can be expressed quantitatively. If the six π electrons occupied three localized ethene-like orbitals, then their energy would be $3 \times (2\alpha + 2\beta) = 6\alpha + 6\beta$. However, their energy in benzene is $2(\alpha + 2\beta) + 4(\alpha + \beta) = 6\alpha + 8\beta$. The **delocalization energy**, the difference of these two energies, is therefore 2β, or about -460 kJ mol^{-1}.

Case study 10.3 *The unique role of carbon in biochemistry*

Now we can take stock of our knowledge of chemical bonding and continue the discussion in Section 9.12 of the properties of carbon that make it uniquely suitable for building complex biological structures.

Among the elements of Period 2, carbon has an intermediate electronegativity, so it can form covalent bonds with many other elements, such as hydrogen, nitrogen, oxygen, sulfur, and, more importantly, other carbon atoms. Furthermore, because it has four valence electrons, carbon atoms can form chains and rings containing single, double, or triple C–C bonds. Such a variety of bonding options leads to the intricate molecular architectures of proteins, nucleic acids, and cell membranes.

Bonds need to be sufficiently strong to maintain the structure of the cell yet need to be susceptible to dissociation and rearrangement during chemical reactions. To get a sense of the uniqueness of the C–C bond, consider the energetics of the N–N and Si–Si bonds. The comparison is useful because nitrogen and silicon are neighbors of carbon in the periodic table and are abundant elements on Earth. The atomic radius of silicon is greater than that of carbon, so we expect an Si–Si bond to be longer than a C–C bond and the orbital overlap to be weaker. The atomic radius of nitrogen is smaller than that of carbon, but the length and energy of an N–N bond, such as that in hydrazine ($H_2N–NH_2$), are influenced by the fact that sp^3 hybridization leaves lone pairs on the nitrogen atoms. These lone pairs repel each other, making an N–N bond weaker than a C–C bond. A C–C bond is sufficiently strong that it can be used as a motif for the formation of robust cellular components. Weaker bonds, such as C–N and C–O, are more reactive, breaking during catabolism and re-forming during anabolism.

10.8 d-Metal complexes

Ions of the d metals participate in biological electron transfer (Chapter 8), the binding and transport of O_2, and the mechanisms of action of many enzymes. To understand the biochemical function of d metal atoms, we need to develop a theory for the formation of bonds between them and biological molecules.

In Chapter 9 we saw that the d-metal ions typically have an incomplete shell of d electrons. These electrons play a special role in d-metal complexes, giving rise to their biochemical activity, their colors, and their magnetic properties. There are two approaches: one, **crystal field theory**, is a simple approach that accounts for the general structures of complexes; the other, **ligand field theory**, is an adaptation of MO theory and is much more powerful.

(a) Crystal field theory

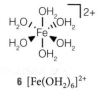

6 $[Fe(OH_2)_6]^{2+}$

In an octahedral d-metal complex six identical ions or molecules, the *ligands*, are at the vertices of a regular octahedron, with the metal atom at its center. An example of this arrangement is the complex $[Fe(OH_2)_6]^{2+}$ (**6**), in which the Fe^{2+} ion, a good Lewis acid, is surrounded by six H_2O molecules, which are good Lewis bases. In crystal field theory, each ligand is regarded as a point negative charge that repels the d electrons of the central ion.

The energy of the entire system decreases when the six ligands approach the central metal cation on account of the favorable Coulomb interactions between its positive charge and the lone electron pairs of the ligands. However, because the point charges representing the ligands repel the d electrons present on the atom, there is also a relatively small modification of that overall decrease in energy. Figure 10.40 shows that the five d orbitals of the central metal ion fall into two groups: $d_{x^2-y^2}$ and d_{z^2} point directly toward the ligand positions, whereas d_{xy}, d_{yz}, and d_{zx} point between them. According to crystal-field theory, an electron occupying an orbital of the former group has a less favorable potential energy than when

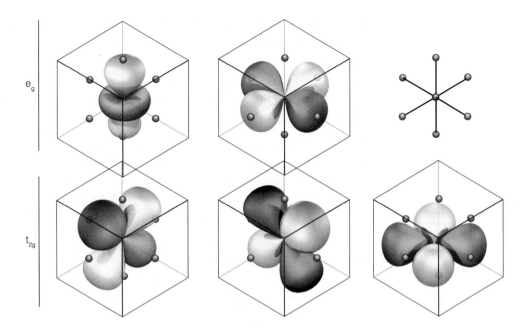

it occupies any of the three orbitals of the other group, and so the d orbitals split into two sets (**7**): a triply degenerate set comprising the d_{xy}, d_{yz}, and d_{zx} orbitals and labeled t_{2g} and a doubly degenerate set comprising the $d_{x^2-y^2}$ and d_{z^2} orbitals and labeled e_g. (The notation is derived from group theory, the mathematical theory of symmetry.) The energy difference between the two sets of orbitals is called the **crystal-field splitting** and denoted Δ_O. The splitting is about 10 per cent of the total energy of interaction of the ligands with the central metal ion.

If we know the number of electrons supplied by the central ion, then we can use the building-up principle to arrive at its electronic configuration by letting the electrons occupy the d orbitals so as to achieve the lowest possible energy bearing in mind, as usual, the Pauli exclusion principle. If the ion has one d electron, as in the case of Ti^{3+}, the configuration of the complex is t_{2g}^1. For two and three d electrons, the configurations are, respectively, t_{2g}^2 (as in V^{3+}) and t_{2g}^3 (as in Cr^{3+}). According to Hund's rule, these electrons can have parallel spins (Fig. 10.41).

A decision now has to be made: the fourth d electron (as in Mn^{3+}) can occupy either the half-filled t_{2g} set of orbitals or the empty e_g orbitals. The advantage of the former arrangement is that the t_{2g} orbitals lie lower in energy than the e_g orbitals; the disadvantage is the significant electron–electron repulsions in a doubly filled orbital. The disadvantage of the second arrangement, which gives the configuration $t_{2g}^3 e_g^1$, is the necessity of occupying a high-energy orbital, but the advantage is less electron–electron repulsion. This advantage is more important than might be expected because all four electrons may have parallel spins in $t_{2g}^3 e_g^1$ and Hund's rule indicates that parallel spins are energetically favorable.

Which configuration, $t_{2g}^3 e_g^1$ or t_{2g}^4, actually occurs depends on a variety of factors, an important one being the magnitude of the crystal-field splitting. If Δ_O is large, the t_{2g}^4 configuration, with its spin-paired arrangement, is favored. Such a molecule is called a **low-spin complex** (Fig. 10.42a). If Δ_O is small, the advantage of minimizing electron–electron repulsion outweighs the disadvantage of occupying a high-energy orbital and the $t_{2g}^3 e_g^1$ configuration is expected, giving rise to a **high-spin complex** with the maximum number of unpaired electrons (Fig. 10.42b).

Fig. 10.40 The classification of d orbitals in an octahedral environment.

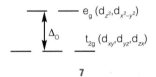

7

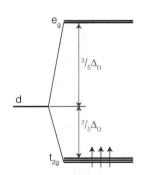

Fig. 10.41 The occupation of energy levels in a d^3 octahedral complex.

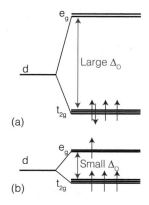

(a)

(b)

Fig. 10.42 The energy separation Δ_O controls the electronic configuration of an octahedral d-metal complex, as shown here for a metal with four d electrons. (a) If Δ_O is large, a low-spin complex results with a t_{2g}^4 configuration. (b) If Δ_O is small, a high-spin complex is favored with a $t_{2g}^3 e_g^1$ configuration.

Example 10.3 *Low- and high-spin complexes of Fe(II) in hemoglobin*

We saw in *Case study* 4.1 that O_2 binds to and is transported through the body by the protein hemoglobin (Atlas P7), which contains the heme group (Atlas P6), a complex of the Fe^{2+} ion. Deoxygenated heme is a high-spin complex that makes a transition to a low-spin complex on binding O_2 as a ligand of the Fe^{2+} ion. Predict the number of unpaired electrons in deoxygenated and oxygenated heme.

Strategy Determine the electronic configuration of the Fe^{2+} ion according to the rules described in Section 9.11. Then apply the building-up principle to the two sets of d orbitals, allowing the maximum number of unpaired electrons to be the dominant factor in high-spin complexes, but not in low-spin complexes.

Solution The ground-state electron configuration of an Fe atom is $[Ar]3d^6 4s^2$, so the configuration of an Fe^{2+} ion is $[Ar]3d^6$. In deoxygenated heme, a high-spin complex, Δ_O is small, so the first five electrons enter the t_{2g} and e_g orbitals with parallel spins. The sixth electron occupies the t_{2g} orbital and must pair. The configuration is, therefore, $t_{2g}^4 e_g^2$ and there are four unpaired electrons. In oxygenated heme, Δ_O is large and all six electrons occupy the t_{2g} orbitals. To do so, they must have paired spins. The configuration is t_{2g}^6 and there are no unpaired electrons.

Self-test 10.12 Cobalt is present in vitamin B_{12}. Predict the number of unpaired electron spins in high-spin and low-spin complexes of a Co^{2+} ion.

Answer: 3 and 1, respectively

(b) Ligand-field theory: σ bonding

Crystal-field theory has a major deficiency: it attempts to ascribe the bonding of the complex to Coulombic interactions between d electrons localized on a central metal ion and electron pairs localized in orbitals confined to the ligands. However, we know from our discussion of MO theory that molecular orbitals spread over both metal atoms and ligands. Ligand-field theory develops this point of view in terms of molecular orbitals. It proceeds in three steps:

- Identify combinations of the ligand orbitals that have symmetries that match the symmetries of the d orbitals of the central metal ion.

- Form molecular orbitals by allowing overlap between these combinations and d orbitals of the same symmetry.

- Use the building-up principle in the same way as in crystal-field theory.

We shall represent (only for purposes of visualization) the ligand orbitals by six spheres, each occupied by two electrons (for concreteness, think of the spheres as representing the lone pairs of NH_3 molecules). From these six atomic orbitals we construct the six combinations spreading over the six ligands shown in Fig. 10.43. We see that two of the six combinations have a shape that matches the two e_g orbitals of the central ion, and four have the wrong shape for any net overlap with either the e_g or t_{2g} metal orbitals. As a result, only e_g molecular orbitals can be formed between the d orbitals and the ligands, and as a result there are two e_g bonding molecular orbitals and two e_g^* antibonding molecular orbitals. The three

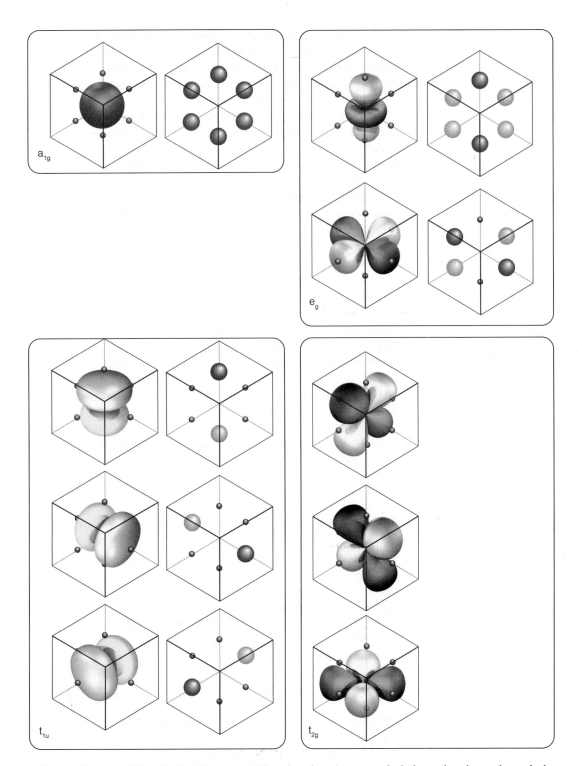

Fig. 10.43 The combinations of ligand orbitals (represented here by spheres) in an octahedral complex, shown alongside the atomic orbitals of the metal. Only the ligand orbitals labeled e_g have the right shape to give nonzero overlap with e_g orbitals of the metal. The metal's t_{2g} orbitals do not combine with the ligand orbitals.

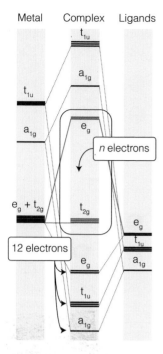

Metal   Complex   Ligands

Fig. 10.44 The molecular orbital energy level diagram for an octahedral complex. The 12 electrons provided by the six ligands fill the lowest six orbitals, which are all bonding orbitals. The n d electrons provided by the central metal atom or ion are accommodated in the orbitals inside the box.

metal t_{2g} orbitals are classified as non-bonding, in the sense that they do not interact to form bonding and antibonding combinations. The four remaining ligand orbitals (labeled, once again using notation that comes from group theory, as a_{1g} and t_{1u} in Fig. 10.43) have the appropriate symmetry to overlap with metal s and p orbitals, respectively, and form bonding and antibonding combinations with them. The energies of the full array of molecular orbitals are shown in Fig. 10.44.

According to the building-up principle, we need to accommodate the appropriate number of electrons into the molecular orbitals of the complex. Each ligand provides two electrons, and the d^n central ion provides n electrons, so we must accommodate $12 + n$ electrons. Of these electrons, four will occupy the two e_g bonding molecular orbitals, eight will occupy the t_{1u} and a_{1g} bonding orbitals, and the remaining n electrons need to be distributed among the metal-centered t_{2g} nonbonding orbitals and the e_g^* antibonding molecular orbitals. We see that there are similarities between the ligand-field and crystal-field formalisms because the n electrons contributed to the complex by the metal atom enter five orbitals split into a set of three and a set of two orbitals. The difference between the theories lies both in the source of the energy separation Δ_O and in the spread of the e_g^* orbitals onto the ligands; the occurrence of low-spin and high-spin complexes is accounted for in terms of the energy splittings that result from the formation of bonding and antibonding molecular orbitals and not just in terms of metal–ligand Coulombic interactions.

(c) Ligand-field theory: π bonding

So far, we have considered only ligand orbitals that point directly at the metal ion orbitals, forming σ molecular orbitals. Ligand-field theory also takes into account the effects of ligand orbitals that participate in the formation of π molecular orbitals with metal ion orbitals. Figure 10.45 shows that a π orbital on the ligand perpendicular to the axis of the metal–ligand bond can overlap with one of the t_{2g} orbitals. The resulting bonding combination lies below the energy of the original nonbonding t_{2g} orbitals and the antibonding combination lies above them.

Interactions between metal ion orbitals and ligand π orbitals can either decrease or increase Δ_O. To see how this is so, consider the bonding schemes in Fig. 10.45. If a ligand π orbital and a nonbonding t_{2g} orbital of the metal ion have similar energies, when they interact a likely outcome is shown in Fig. 10.45a. If the ligand π orbital supplies two electrons and the t_{2g} orbital supplies one, then the result is a decrease in Δ_O. On the other hand, if the metal t_{2g} and ligand π* orbitals have

Fig. 10.45 The effect of π bonding on the magnitude of Δ_O. (a) In this case, the antibonding π* orbital of the ligand is too high in energy to take part in bonding or it is absent; the interaction with the (full) π orbital of the ligand decreases Δ_O. (b) In this case, the antibonding π* orbital of a ligand matches the metal orbital in energy; the interaction with the (empty) π* orbital of the ligand increases Δ_O.

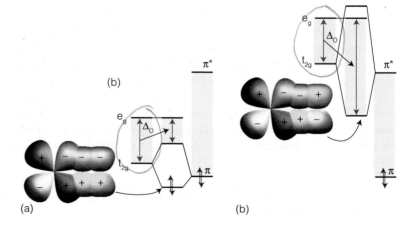

(a)   (b)

similar energies, then they may interact in the manner shown in Fig. 10.45b, If the π^* orbital is empty and t_{2g} orbital supplies one, then the result is an increase in Δ_O.

Case study 10.4   *Ligand-field theory and the binding of O_2 to hemoglobin*

Ligand-field theory provides excellent descriptions of the interactions between metal ions and ligands in metalloproteins. In this *Case study*, we apply the theory to an important biological process: the binding of O_2 to hemoglobin.

Nature makes unconscious use of ligand-field effects to pump and store oxygen throughout our bodies. Here we concentrate on hemoglobin (Hb, Atlas P7), the protein used to transport oxygen through our bodies, and myoglobin (Mb), the protein used to store oxygen in muscle tissue and to release it on demand (see also *Case study* 4.1). Hemoglobin is a tetramer of four myoglobin-like subunits, and each subunit, as in myoglobin, binds a single heme group, an almost flat ring-like structure with an iron atom at its center (Atlas P6). The oxygenated form of hemoglobin is called the *relaxed state* (R state) and the deoxygenated form is called the *tense state* (T state).

The heme group binds oxygen when the Fe atom is present as iron(II) (Fig. 10.46). The Fe–O_2 complex is held together by a σ bond between an empty Fe(II) e_g orbital and the full σ orbital of O_2 and a π bond between filled t_{2g} orbitals on Fe(II) and the half-full π^* orbitals of O_2. The bound O_2 molecule adopts a bent orientation with respect to the Fe atom, partly because that orientation maximizes interactions between orbitals, but also because it is consistent with the spatial constraints imposed by the arrangement of peptide residues in the pocket of the protein containing the heme group.

Another important change that occurs when the Fe atom is oxygenated is the transition from an Fe(II) high-spin d^6 configuration to an Fe(II) low-spin d^6 configuration. This change accompanies the increase in the number of ligands of the Fe ion from five to six. In the deoxygenated form, the fifth location is taken up by the N atom of a histidine residue (His); in the oxygenated form that link remains, but the O_2 molecule binds on the other side of the ring (as shown in Fig. 10.46b). The change from high spin to low spin results in a slightly smaller atom. As a result, instead of lying 60 pm above the plane of the heme ring, the Fe atom can fall back almost into the plane of the ring, and in the oxygenated form it lies only 20 pm above the plane. As it falls back, it pulls the histidine residue with it.

In *Case study* 4.1 we discussed the thermodynamic view of the binding of O_2 to hemoglobin. Now we can merge the thermodynamic and molecular views into a single model.[2] When one of the subunits binds the first O_2 molecule, the heme group and its ligands reorganize as described above with the further consequence that one pair of subunits rotates through 15° relative to the other pair and becomes offset by 80 pm. This realignment of two of the subunits relative to the other two disrupts an ionic His$^+$···Asp$^-$ interaction that helps to stabilize the deoxygenated form, and as a result the partially oxygenated hemoglobin molecule is more capable of taking up the next O_2 than the fully

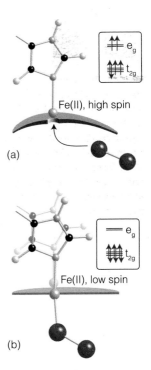

(a)

Fe(II), high spin

(b)

Fe(II), low spin

Fig. 10.46 The change in molecular geometry that takes place when an O_2 molecule attaches to an Fe atom in a hemoglobin molecule. (a) The deoxygenated heme group, with the Fe(II) ion in its low-spin configuration. (b) The oxygenated heme group, with the Fe(II) ion in its high-spin configuration. Note how the histidine residue is pulled into a different location by the motion of the iron atom.

---

[2] J. Monod, J. Wyman, and J.-P. Changeux and later D. Koshland proposed the essential features of the model, which has been refined by structural studies with diffraction and spectroscopic techniques (discussed in Chapters 11 and 12, respectively).

deoxygenated form was. In thermodynamic terms, the equilibrium constant for binding of the second O_2 molecule is greater than the equilibrium constant for binding of the first O_2 molecule. As each of the four subunits become oxygenated, the binding of O_2 to the remaining deoxygenated subunits becomes successively more favorable thermodynamically. In other words, in hemoglobin there is a *cooperative* uptake of O_2 molecules. The cooperative binding of O_2 by hemoglobin is an example of an *allosteric effect*, in which an adjustment of the conformation of a molecule when one substrate binds affects the ease with which a subsequent substrate molecule binds.

Oxygenated hemoglobin also unloads O_2 cooperatively when conditions demand it. The result of cooperativity is that hemoglobin can release its O_2 under conditions when myoglobin cannot, which is an ideal arrangement for a transport protein rather than a storage protein (see *Case study* 4.1).

Computational biochemistry

Computational chemistry is now a standard part of chemical research. One major application is in pharmaceutical chemistry, where the likely pharmacological activity of a molecule can be assessed computationally from its shape and electron density distribution before expensive clinical trials are started. Commercial software is now widely available for calculating the electronic structures of molecules and displaying the results graphically. All such calculations work within the Born–Oppenheimer approximation and express the molecular orbitals as linear combinations of atomic orbitals.

10.9 Computational techniques

Elaborate computational methods make reasonably accurate predictions of molecular properties, including their conformation, spectroscopic properties, and reactivity. Although these techniques tax computational resources heavily, they can be used in studies of moderately sized biological molecules.

There are two principal approaches to solving the Schrödinger equation for many-electron polyatomic molecules. In the **semi-empirical methods**, certain expressions that occur in the Schrödinger equation are set equal to parameters that have been chosen to lead to the best fit to experimental quantities, such as enthalpies of formation. Semi-empirical methods are applicable to a wide range of molecules with a virtually limitless number of atoms and are widely popular. In the more fundamental ***ab initio*** method, an attempt is made to calculate structures from first principles, using only the atomic numbers of the atoms present. Such an approach is intrinsically more reliable than a semi-empirical procedure.

Both types of procedure typically adopt a **self-consistent field** (SCF) procedure, in which an initial guess about the composition of the LCAO is successively refined until the solution remains unchanged in a cycle of calculation. For example, the potential energy of an electron at a point in the molecule depends on the locations of the nuclei and all the other electrons. Initially, we do not know the locations of those electrons (more specifically, we do not know the detailed form of the wavefunctions that describe their locations, the molecular orbitals they occupy). First, then, we guess the form of those wavefunctions—we guess the values of the coefficients in the LCAO used to build the molecular orbitals—and solve the Schrödinger equation for the electron of interest on the basis of that guess. Now we have a first approximation to the molecular orbital of our electron (a reasonable estimate of the coefficients for its LCAO) and we repeat the procedure for all the other molecular orbitals in the molecule. At this stage, we have a new set of molecular orbitals, which in general will have coefficients that differ from our first guess, and we also have an estimate of the energy of the molecule. We use that refined set of molecular orbitals to repeat the calculation and calculate a new energy. In general, the coefficients in the LCAOs and the energy will differ from the new starting point. However, there comes a stage when repetition of the calculation leaves the coefficients and energy unchanged. The orbitals are now said to be self-consistent, and we accept them as a description of the molecule.

(a) Semi-empirical methods

The Hückel method is a very primitive example of a semi-empirical method in which various integrals are set equal to either α or β and treated as empirical parameters; overlap integrals are ignored. The removal of the restriction of the Hückel method to planar hydrocarbon systems was achieved with the introduction of the **extended Hückel theory** (EHT) in about 1963. In heteroatomic non-planar systems (such as d-metal complexes) the separation of orbitals into π and σ is no longer appropriate and each type of atom has a different value of H_{JJ} (which in Hückel theory is set equal to α for all atoms). In this approximation, the overlap integrals are not set equal to zero but are calculated explicitly. Furthermore, the H_{JK}, which in Hückel theory are set equal to β, in EHT are made proportional to the overlap integral between the orbitals J and K.

Further approximations of the Hückel method were removed with the introduction of the **complete neglect of differential overlap** (CNDO) method, which is a slightly more sophisticated method for dealing with the terms H_{JK} that appear in the secular equations for the coefficients. The introduction of CNDO opened the door to an avalanche of similar but improved methods and their accompanying acronyms, such as **intermediate neglect of differential overlap** (INDO), **modified neglect of differential overlap** (MNDO), and the **Austin Model 1** (AM1, version 2 of MINDO). Software for all these procedures is now readily available, and reasonably sophisticated calculations can be run even on handheld computers.

(b) Density functional theory

A semi-empirical technique that has gained considerable ground in recent years to become one of the most widely used techniques for the calculation of molecular structure is **density functional theory** (DFT). Its advantages include less demanding computational effort, less computer time, and—in some cases (particularly d-metal complexes)—better agreement with experimental values than is obtained from other procedures.

The central focus of DFT is the electron density, ρ (rho), rather than the wavefunction ψ. When the Schrödinger equation is expressed in terms of ρ, it becomes a set of equations called the **Kohn–Sham equations**. As for the Schrödinger equation itself, this equation is solved iteratively and self-consistently. First, we guess the electron density. For this step it is common to use a superposition of atomic electron densities. Next, the Kohn–Sham equations are solved to obtain an initial set of orbitals. This set of orbitals is used to obtain a better

A brief comment

In mathematics, when an entire function $f(x)$ is associated with a single number, F, as when an entire wavefunction is associated with the energy of the state, the number is said to be a *functional* of the function and written $F[f]$. Thus, the energy E is a functional of the wavefunction and we could denote it $E[\psi]$ to denote the functional dependence of the energy on the entire wavefunction. In DFT, the energy is regarded as a functional of the electron density, and written $E[\rho]$.

approximation to the electron density, and the process is repeated until the density and the energy are constant to within some tolerance.

(c) *Ab initio* methods

The *ab initio* methods also simplify the calculations, but they do so by setting up the problem in a different manner, avoiding the need to estimate parameters by appeal to experimental data. In these methods, sophisticated techniques are used to solve the Schrödinger equation numerically. The difficulty with this procedure, however, is the enormous time it takes to carry out the detailed calculation. That time can be reduced by replacing the hydrogenic atomic orbitals used to form the LCAO by a **Gaussian-type orbital** (GTO) in which the exponential function e^{-r} characteristic of actual orbitals is replaced by a sum of Gaussian functions of the form e^{-r^2} (recall the relative shapes of exponential and Gaussian functions shown in *Mathematical Toolkit* F.2).

10.10 **Graphical output**

One of the most significant developments in computational chemistry and its application to biology has been the introduction of graphical representations of molecular geometries, molecular orbitals, and electron densities.

The raw output of a molecular structure calculation is a list of the coefficients of the atomic orbitals in each molecular orbital and the energies of these orbitals. The graphical representation of a molecular orbital uses stylized shapes to represent the basis set and then scales their size to indicate the value of the coefficient in the LCAO. Different signs of the wavefunctions are represented by different colors (as we saw in Figs 10.36, 10.37, and 10.39).

Once the coefficients are known, we can build up a representation of the electron density in the molecule by noting which orbitals are occupied and then forming the squares of those orbitals. The total electron density at any point is then the sum of the squares of the wavefunctions evaluated at that point. The outcome is commonly represented by an **isodensity surface**, a surface of constant total electron density (Fig. 10.47). There are several styles of representing an isodensity surface: as a solid form, as a transparent form with a ball-and-stick representation of the molecule within, or as a mesh. A related representation is a **solvent-accessible surface**, which is generated by plotting the location of the center of a sphere (representing a solvent molecule) that is imagined to roll across the exposed surfaces of the atoms.

Fig. 10.47 The isodensity surface of benzene.

One of the most important aspects of a molecule other than its geometrical shape is the distribution of electric potential over its surface. A common procedure begins with calculation of the potential energy of a 'probe' charge at each point on an isodensity surface and interpreting its energy as an interaction with an electric potential at that point. The result is an **electrostatic potential surface** (an 'elpot surface') in which net positive potential is shown in one color and net negative potential is shown in another, with intermediate gradations of color (Fig. 10.48).

10.11 **The prediction of molecular properties**

The results of quantum mechanical calculations are only approximate, with deviations from experimental values increasing with the size of the molecule. Therefore, one goal of computational biochemistry is to gain an insight into the trends in properties of biological molecules, without necessarily striving for ultimate accuracy.

Computation is now used to explore far more than the electronic structures of molecules. We already saw in Section 1.12 that computational techniques can be used to estimate the enthalpies of formation of conformational isomers and the effect of solvent on the enthalpy of formation.

(a) Electrochemical properties

Molecular orbital calculations may also be used to predict trends in electrochemical properties, such as standard potentials (Chapter 5). Several experimental and computational studies of aromatic hydrocarbons indicate that decreasing the energy of the LUMO enhances the ability of a molecule to accept an electron into the LUMO, with an attendant increase in the value of the molecule's standard potential. The effect is also observed in quinones and flavins, co-factors involved in biological electron transfer reactions. For example, stepwise substitution of the hydrogen atoms in *p*-benzoquinone by methyl groups ($-CH_3$) results in a systematic increase in the energy of the LUMO and a decrease in the standard potential for formation of the semiquinone radical:

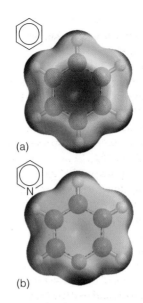

(a)

(b)

Fig. 10.48 The electrostatic potential surfaces of (a) benzene and (b) pyridine. Note the accumulation of electron density on the N atom of pyridine at the expense of the other atoms.

$$\text{O} \xrightarrow{e^-} \text{O}^{\bullet} \xrightarrow[2\,H^+]{e^-} \text{OH}$$

Semiquinone radical

The standard potentials of naturally occurring quinones are also modified by the presence of different substituents, a strategy that imparts specific functions to specific quinones. For example, the substituents in coenzyme Q are largely responsible for poising its standard potential so that the molecule can function as an electron shuttle between specific electroactive proteins in the respiratory chain (Section 5.10).

(b) Spectroscopic properties

We remarked in Chapter 9 that a molecule can absorb or emit a photon of energy hc/λ, resulting in a transition between two quantized molecular energy levels. The transition of lowest energy (and longest wavelength) occurs between the HOMO and LUMO. We can use calculations based on semi-empirical, *ab initio*, and DFT methods to correlate the HOMO–LUMO energy gap with the wavelength of absorption.

For example, consider the linear polyenes shown in Table 10.3: ethene (C_2H_4), butadiene (C_4H_6), hexatriene (C_6H_8), and octatetraene (C_8H_{10}), all of which absorb in the ultraviolet region of the spectrum. The table also shows that, as expected, the wavelength of the lowest-energy electronic transition decreases as the energy separation between the HOMO and LUMO increases. We also see that the smallest HOMO–LUMO gap and longest transition wavelength correspond to octatetraene, the longest polyene in the group. It follows that the wavelength of the transition increases with increasing number of conjugated double bonds in linear polyenes. Extrapolation of the trend suggests that a sufficiently long linear polyene should absorb light in the visible region of the electromagnetic spectrum. This is indeed the case for β-carotene (Atlas E1), which absorbs light with $\lambda \approx 450$ nm. The ability of β-carotene to absorb visible light is part of the strategy employed by plants to harvest solar energy for use in photosynthesis (Chapter 12).

Table 10.3 Summary of *ab initio* calculations and spectroscopic data for four linear polyenes

	$\Delta E_{\text{HOMO-LUMO}}/\text{eV}^{*}$	$\lambda_{\text{transition}}/\text{nm}$
	18.1	163
	14.5	217
	12.7	252
	11.6	304

$1\,\text{eV} = 1.602 \times 10^{-19}\,\text{J}.$

(c) Chemical reactivity

There are several ways in which molecular orbital calculations lend insight into reactivity. For example, electrostatic potential surfaces may be used to identify an electron-poor region of a molecule that is susceptible to association with or chemical attack by an electron-rich region of another molecule. Such considerations are important for assessing the pharmacological activity of potential drugs (Section 11.17(c)).

An attractive feature of computational chemistry is its ability to model species that may be too unstable or short-lived to be studied experimentally. For this reason, quantum mechanical methods are often used to study the transition state, with an eye toward describing factors that stabilize it and increase the reaction rate. Systems as complex as enzymes are amenable to study by computational methods.

Checklist of key concepts

☐ 1. An ionic bond is formed by transfer of electrons from one atom to another and the attraction between the ions. A covalent bond is formed when two atoms share a pair of electrons.

☐ 2. In the Born–Oppenheimer approximation, nuclei are treated as stationary while electrons move around them.

☐ 3. In valence bond theory (VB theory), a bond is regarded as forming when an electron in an atomic orbital on one atom pairs its spin with that of an electron in an atomic orbital on another atom.

☐ 4. A valence-bond wavefunction with cylindrical symmetry around the internuclear axis is a σ bond. A π bond arises from the merging of two p orbitals that approach side by side and the pairing of electrons that they contain.

☐ 5. Hybrid orbitals are mixtures of atomic orbitals on the same atom. In VB theory, hybridization is invoked to be consistent with molecular geometries.

☐ 6. Resonance is the superposition of the wavefunctions representing different electron distributions in the same nuclear framework.

☐ 7. In molecular orbital theory (MO theory), electrons are treated as spreading throughout the entire molecule.

☐ 8. A bonding orbital is a molecular orbital that, if occupied, contributes to the strength of a bond between two atoms. An antibonding orbital is a molecular orbital that, if occupied, decreases the strength of a bond between two atoms.

☐ 9. The building-up principle suggests procedures for constructing the electron configuration of molecules on the basis of their molecular orbital energy level diagram.

10. When constructing molecular orbitals, we need to consider only combinations of atomic orbitals of similar energies and of the same symmetry around the internuclear axis.

11. The electronegativity of an element is the power of its atoms to draw electrons to itself when it is part of a compound.

12. In a bond between dissimilar atoms, the atomic orbital belonging to the more electronegative atom makes the larger contribution to the molecular orbital with the lowest energy. For the molecular orbital with the highest energy, the principal contribution comes from the atomic orbital belonging to the less electronegative atom.

13. Hückel theory is a simple treatment of the molecular orbitals of π-electron systems. In hydrocarbons the technique consists of forming linear combinations of unhybridized C2p orbitals.

14. In crystal-field theory, bonding in d-metal complexes arises from Coulomb interactions between electrons from the central metal ion and electrons from the ligands. In an octahedral complex, the degenerate d atomic orbitals of the metal are split into two sets of orbitals separated by an energy Δ_O: a triply degenerate set comprising the d_{xy}, d_{yz}, and d_{zx} orbitals and labeled t_{2g} and a doubly degenerate set comprising the $d_{x^2-y^2}$ and d_{z^2} orbitals and labeled e_g.

15. In a high-spin complex, the t_{2g} and e_g orbitals are filled in such a way as to maximize the number of unpaired d electrons. In a low-spin complex, the number of unpaired electrons is minimized.

16. Ligand-field theory is an adaptation of MO theory for complexes of the d metals.

17. In the self-consistent field procedure, an initial guess about the composition of the molecular orbitals is successively refined until the solution remains unchanged in a cycle of calculations.

18. In semi-empirical methods for the determination of electronic structure, the Schrödinger equation is written in terms of parameters chosen to agree with selected experimental quantities.

Checklist of key equations

Property	Equation	Comment
A VB wavefunction	$\psi(1,2) = \psi_A(1)\psi_B(2) + \psi_A(2)\psi_B(1)$	ψ_A and ψ_B are atomic orbitals on different atoms
Resonance hybrid	$\psi = a\psi_1 + b\psi_2$	ψ_1 and ψ_2 are wavefunctions for molecules with different electron distributions and the same nuclear locations
Molecular orbital	$\psi = c_A\psi_A + c_B\psi_B$	An LCAO
Overlap integral	$S = \int \psi_A\psi_B \, d\tau$	
Bond order	$b = \frac{1}{2}(n - n^\star)$	

Discussion questions

10.1 Compare the approximations built into valence bond theory and molecular orbital theory.

10.2 Discuss the steps involved in the construction of sp^3, sp^2, and sp hybrid orbitals.

10.3 Distinguish between the Pauling and Mulliken electronegativity scales.

10.4 Use molecular orbital theory to discuss the biochemical reactivity of O_2, N_2, and NO.

10.5 Identify and justify the approximations used in the Hückel theory of conjugated hydrocarbons.

10.6 Using information found in this and the previous chapter, discuss the unique role that carbon plays in biochemistry.

10.7 In the laboratory, the Fe^{2+} ion in the heme group of hemoglobin can be removed and replaced by a Zn^{2+} ion. Discuss whether this modified protein is likely to bind O_2 efficiently.

10.8 Distinguish between semi-empirical, *ab initio*, and density functional theory methods of electronic structure determination.

Exercises

10.9 Write down the valence bond wavefunction for a nitrogen molecule.

10.10 Calculate the molar energy of repulsion between two hydrogen nuclei at the separation in H_2 (74.1 pm). The result is the energy that must be overcome by the attraction from the electrons that form the bond.

10.11 Give the valence bond description of SO_2 and SO_3 molecules.

10.12 Write the Lewis structure for the peroxynitrite ion, $ONOO^-$. Label each atom with its state of hybridization and specify the composition of each of the different types of bond.

10.13 The structure of the visual pigment retinal is shown in (8). Label each atom with its state of hybridization and specify the composition of each of the different types of bond.

8 11-*cis*-retinal

10.14 Show that $S = \int h_1 h_2 d\tau = 0$, where $h_1 = s + p_x + p_y + p_z$ and $h_2 = s - p_x - p_y + p_z$ are hybrid orbitals. *Hint*: Each atomic orbital is individually normalized to 1. Also, note that $S = \int sp \, d\tau = 0$, and that p orbitals with perpendicular orientations have zero overlap.

10.15 Show that the sp^2 hybrid orbital $(s + 2^{1/2}p)/3^{1/2}$ is normalized to 1 if the s and p orbitals are each normalized to 1.

10.16 Find another sp^2 hybrid orbital that has zero overlap with the hybrid orbital in the preceding problem.

10.17 Benzene is commonly regarded as a resonance hybrid of the two Kekulé structures shown in (4), but other structures can also contribute. Draw three other structures in which there are only covalent π bonds (allowing for bonding between some non-adjacent C atoms), and two structures in which there is one ionic bond. Why may these structures be ignored in simple descriptions of the molecule?

10.18 Before doing the calculation below, sketch how the overlap between a 1s orbital and a 2p orbital can be expected to depend on their separation. The overlap integral between a 1s orbital and a 2p orbital on nuclei separated by a distance R is $S = (R/a_0)\{1 + (R/a_0) + \frac{1}{3}(R/a_0)^2\}e^{-R/a_0}$. Plot this function, and find the separation for which the overlap is a maximum.

10.19 Suppose that a molecular orbital has the form $N(0.145A + 0.844B)$. Find a linear combination of the orbitals A and B that has zero overlap with this combination.

10.20 Show, if overlap is ignored, (**a**) that any molecular orbital expressed as a linear combination of two atomic orbitals may be written in the form $\psi = \psi_A \cos\theta + \psi_B \sin\theta$, where θ is a parameter that varies between 0 and $\frac{1}{2}\pi$, and (**b**) that if ψ_A and ψ_B are orthogonal and

normalized to 1, then ψ is also normalized to 1. (**c**) To what values of θ do the bonding and antibonding orbitals in a homonuclear diatomic molecule correspond?

10.21 Draw diagrams to show the various orientations in which a p orbital and a d orbital on adjacent atoms may form bonding and antibonding molecular orbitals.

10.22 How many molecular orbitals can be constructed from a diatomic molecule in which s, p, d, and f orbitals are all important in bonding?

10.23 Give the ground state electron configurations of (**a**) H_2^-, (**b**) N_2, and (**c**) O_2.

10.24 Three biologically important diatomic species, either because they promote or inhibit life, are (**a**) CO, (**b**) NO, and (**c**) CN^-. The first binds to hemoglobin, the second is a chemical messenger, and the third interrupts the respiratory electron transfer chain. Their biochemical action is a reflection of their orbital structure. Deduce their ground-state electron configurations.

10.25 Some chemical reactions proceed by the initial loss or transfer of an electron to a diatomic species. Which of the molecules N_2, NO, O_2, C_2, F_2, and CN would you expect to be stabilized by (**a**) the addition of an electron to form AB^- and (**b**) the removal of an electron to form AB^+?

10.26 Give the (g,u) parities of the wavefunctions for the first four levels of a particle in a box.

10.27 (**a**) Give the parities of the wavefunctions for the first four levels of a harmonic oscillator. (**b**) How may the parity be expressed in terms of the quantum number v?

10.28 State the parities of the six π orbitals of benzene (see Fig. 10.39).

10.29 Two important diatomic molecules for the welfare of humanity are NO and N_2: the former is both a pollutant and a chemical messenger, and the latter is the ultimate source of the nitrogen of proteins and other biomolecules. Use the electron configurations of NO and N_2 to predict which is likely to have the greater bond dissociation energy and the shorter bond length.

10.30 Arrange the species O_2^+, O_2, O_2^-, and O_2^{2-} in order of increasing bond length.

10.31 Construct the molecular orbital energy level diagrams of (**a**) ethene (ethylene) and (**b**) ethyne (acetylene) on the basis that the molecules are formed from the appropriately hybridized CH_2 or CH fragments.

10.32 Many of the colors of vegetation are due to electronic transitions in conjugated π-electron systems. In the *free-electron molecular orbital* (FEMO) theory, the electrons in a conjugated molecule are treated as independent particles in a box of length L. Sketch the form of the two occupied orbitals in butadiene predicted by this model and predict the minimum excitation energy of the molecule. The tetraene $CH_2{=}CHCH{=}CHCH{=}CHCH{=}CH_2$ can be treated as a box of length 8R, where $R = 140$ pm (as in this case, an extra half bond length is often added at each end of the box). Calculate the minimum excitation energy of the molecule and sketch the HOMO and LUMO.

10.33 Write down the Hückel secular determinant for *cyclo*-butadiene.

10.34 Solve the secular determinant for the allyl radical, $CH_2=CHCH_2\cdot$. *Hints:* (a) regard the unpaired electron on the $-CH_2\cdot$ fragment to be in a $C2p_z$ orbital, so that the electron can delocalize within the π system of the molecule. (b) See *Mathematical toolkit 10.1*.

10.35 It is important to understand the origins of stabilization of linear conjugated molecules because they play important biological roles in plants and animals (see *Case study 9.1*). According to Hückel theory, the energies of the bonding π molecular orbitals of butadiene, $CH_2=CH_2-CH_2=CH_2$, are $E = \alpha + 1.62\beta$ and $\alpha + 0.62\beta$. The energies of the antibonding π^* molecular orbitals are $E = \alpha - 1.62\beta$ and $\alpha - 0.62\beta$. The total π-*electron binding energy*, E_π, is the sum of the energies of each π electron. Recalling that there are four electrons to accommodate in the π molecular orbitals, calculate the π-electron binding energies of ethene (see Section 10.7) and butadiene. Is the energy of the butadiene molecule lower or higher than the sum of two individual π bonds?

10.36 Cyclic conjugated systems occur widely in biological macromolecules. Examples include the phenyl group of phenylalanine and a host of heterocyclic molecules, such as the purine and pyrimidine bases found in nucleic acids. In general, the delocalization energy of a conjugated system is

$$E_{deloc} = E_\pi - N_{db}(2\alpha + 2\beta)$$

where N_{db} is the number of double bonds, each contributing an energy $2\alpha + 2\beta$ in the absence of conjugation. The most notable example of delocalization conferring extra stability is benzene and the aromatic molecules based on its structure. Predict the electronic configuration and delocalization energy of **(a)** the benzene anion and **(b)** the benzene cation.

10.37 The FEMO theory (*Problem* 10.32) of conjugated molecules is rather crude and better results are obtained with simple Hückel theory. **(a)** For a linear conjugated polyene with each of N_C carbon atoms contributing an electron in a 2p orbital, the energies E_k of the resulting π molecular orbitals are given by

$$E_k = \alpha + 2\beta \cos\frac{k\pi}{N_C + 1} \qquad k = 1, 2, 3, \ldots, N_C$$

Use this expression to determine a reasonable empirical estimate of the parameter β for the series consisting of ethene, butadiene, hexatriene, and octatetraene given that light-induced absorptions from the HOMO to the LUMO occur at 61 500, 46 080, 39 750, and 32 900 cm^{-1}, respectively. **(b)** Calculate the delocalization energy of octatetraene (see *Exercise* 10.36). **(c)** In the context of this Hückel model, the π molecular orbitals are written as linear combinations of the carbon 2p orbitals. The coefficient of the jth atomic orbital in the kth molecular orbital is given by

$$c_{kj} = \left(\frac{2}{N_C + 1}\right)^{1/2} \sin\frac{jk\pi}{N_C + 1} \qquad j = 1, 2, 3, \ldots, N_C$$

Determine the values of the coefficients of each of the six 2p orbitals in each of the six π molecular orbitals of hexatriene. Match each set of coefficients (that is, each molecular orbital) with a value of the energy

calculated with the expression given in part (a) of the molecular orbital. Comment on trends that relate the energy of a molecular orbital with its 'shape', which can be inferred from the magnitudes and signs of the coefficients in the linear combination that describes the molecular orbital.

10.38 For monocyclic conjugated polyenes (such as cyclobutadiene and benzene) with each of N_C carbon atoms contributing an electron in a 2p orbital, simple Hückel theory gives the following expression for the energies E_k of the resulting π molecular orbitals:

$$E_k = \alpha + 2\beta \cos\frac{2k\pi}{N_C} \qquad k = 0, \pm 1, \pm 2, \ldots, \pm N_C/2 \qquad \text{(even } N\text{)}$$
$$k = 0, \pm 1, \pm 2, \ldots, \pm(N_C - 1)/2 \quad \text{(odd } N\text{)}$$

(a) Calculate the energies of the π molecular orbitals of benzene and cyclooctatetraene. Comment on the presence or absence of degenerate energy levels. **(b)** Calculate and compare the delocalization energies of benzene (using the expression above) and hexatriene (see *Exercise* 10.36). What do you conclude from your results? **(c)** Calculate and compare the delocalization energies of cyclooctaene and octatetraene. Are your conclusions for this pair of molecules the same as for the pair of molecules investigated in part (b)?

10.39 Experimentally, it is found that the value of Δ_O varies with the chemical nature of the ligand according to the spectrochemical series: $S^{2-} < Cl^- < OH^- \approx RCO_2^- < H_2O \approx RS^- < NH_3 \approx$ imidazole (the side chain of histidine) $< CN^- < CO$. **(a)** Draw an energy level diagram like those in Fig. 10.41 showing the configuration of the d electrons on the metal ion in $[Fe(OH_2)_6]^{3+}$ and $[Fe(CN)_6]^{3-}$. **(b)** Predict the number of unpaired electrons in each complex.

10.40 The terms *low spin* and *high spin* apply only to complexes of d-metal ions having certain numbers of d electrons. Put differently, certain d-metal ions can have only one electron configuration and a distinction between low- and high-spin complexes is not possible. For what number of d electrons are both high- and low-spin octahedral complexes possible?

10.41 Figures 10.41 and 10.42 show the result of an octahedral arrangement of ligands around a d-metal ion. In a tetrahedral complex, the $d_{x^2-y^2}$ and d_{z^2} orbitals form a degenerate pair that is separated in energy from the degenerate d_{xy}, d_{yz}, and d_{zx} orbitals by Δ_T. In a square-planar complex with the ligand orbitals in the xy plane, the metal d orbitals increase in energy as follows: $d_{xz} = d_{yz} < d_{z^2} < d_{xy} < d_{x^2-y^2}$. In a nickel-containing enzyme, the metal was shown to be in the +2 oxidation state and to have no unpaired electrons. What is the most probable geometry of the Ni^{2+} site?

10.42 Ligands that interact with d metals as shown in Fig. 10.43 are called σ-*donor ligands*. When π bonding is important, π-*acceptor* and π-*donor ligands* behave as shown in Fig. 10.45. If a ligand generates a weak ligand field around a d-metal ion, the result will be a small value of Δ_O and a high-spin complex. Conversely, a strong ligand field leads to a large value of Δ_O and a low-spin complex. **(a)** Justify the following statement: Cl^- is a weak-field ligand because it is a π acceptor and CO is a strong-field ligand because it is a π donor. **(b)** Show that O_2 is a π-acceptor ligand. **(c)** Using the information from parts (a) and (b) and from *Case studies* 4.1 and 10.4, propose a detailed mechanism for CO poisoning.

Projects

10.43 In Section 10.2c we used VB theory to account for the planarity of the peptide link (**1**). Now we develop a molecular orbital theory treatment that provides a richer description of the factors that stabilize the planar conformation of the peptide link.

(a) Taking a hint from VB theory, we can suspect that delocalization of the π bond between the oxygen, carbon, and nitrogen atoms can be modeled by making LCAO-MOs from 2p orbitals perpendicular to the plane define by the atoms. The three combinations have the form

$$\psi_1 = a\chi_O + b\chi_C + c\chi_N \qquad \psi_2 = d\chi_O - e\chi_N \qquad \psi_3 = f\chi_O - g\chi_C + h\chi_N$$

where the coefficients a to h are all positive. Sketch the orbitals ψ_1, ψ_2, and ψ_3 and characterize them as bonding, non-bonding, or antibonding molecular orbitals.

(b) Show that this treatment is consistent only with a planar conformation of the peptide link.

(c) Draw a diagram showing the relative energies of these molecular orbitals and determine the occupancy of the orbitals. *Hint*: Convince yourself that there are four electrons to be distributed among the molecular orbitals.

(d) Now consider a nonplanar conformation of the peptide link, in which the O2p and C2p orbitals are perpendicular to the plane defined by the O, C, and N atoms, but the N2p orbital lies on that plane. The LCAO-MOs are given by

$$\psi_4 = a\chi_O + b\chi_C \qquad \psi_5 = e\chi_N \qquad \psi_6 = f\chi_O - g\chi_C$$

Just as before, sketch these molecular orbitals and characterize them as bonding, nonbonding, or antibonding. Also, draw an energy level diagram and determine the occupancy of the orbitals.

(e) Why is this arrangement of atomic orbitals consistent with a nonplanar conformation for the peptide link?

(f) Does the bonding MO associated with the planar conformation have the same energy as the bonding MO associated with the nonplanar conformation? If not, which bonding MO is lower in energy? Repeat the analysis for the nonbonding and antibonding molecular orbitals.

(g) Use your results from parts (a)–(f) to construct arguments that support the planar model for the peptide link.

The following projects require the use of molecular modeling software.

10.44 Here we explore further the application of molecular orbital calculations to the prediction of spectroscopic properties of conjugated molecules.

(a) Using data from Table 10.3, plot the HOMO-LUMO energy separations against the experimental frequencies for π-to-π* ultraviolet absorptions for ethene, butadiene, hexatriene, and octatetraene. Then use mathematical software to find the polynomial equation that best fits the data.

(b) Using molecular modeling software and the computational method recommended by your instructor (extended Hückel, semiempirical, *ab initio*, or DFT methods), calculate the energy separation between the HOMO and LUMO of decapentaene.

(c) Use your polynomial fit from part (a) to estimate the frequency of the π-to-π* absorption of decapentaene from the calculated HOMO-LUMO energy separation.

(d) Discuss why the calibration procedure of part (a) is necessary.

(e) Electronic excitation of a molecule may weaken or strengthen some bonds because bonding and antibonding characteristics differ between the HOMO and the LUMO. For example, a carbon–carbon bond in a linear polyene may have bonding character in the HOMO and antibonding character in the LUMO. Therefore, promotion of an electron from the HOMO to the LUMO weakens this carbon–carbon bond in the excited electronic state relative to the ground electronic state. (i) Use molecular modeling software to display the HOMO and LUMO of each molecule discussed in this project. (ii) Discuss in detail any changes in bond order that accompany the π-to-π* ultraviolet absorptions in these molecules.

10.45 Molecular orbital calculations may be used to predict trends in the standard potentials of conjugated molecules, such as the quinones and flavins, that are involved in biological electron transfer reactions (Chapter 5). It is commonly assumed that decreasing the energy of the LUMO enhances the ability of a molecule to accept an electron into the LUMO, with an attendant increase in the value of the molecule's standard potential. Furthermore, a number of studies indicate that there is a linear correlation between the LUMO energy and the reduction potential of aromatic hydrocarbons.

(a) The biological standard potentials for the one-electron reduction of methyl-substituted *p*-benzoquinones (**9**) to their respective semiquinone radical anions are

R_2	R_3	R_5	R_6	$E_{cell}^{\ominus}/V$
H	H	H	H	0.078
CH_3	H	CH_3	H	0.023
CH_3	H	CH_3	H	−0.067
CH_3	CH_3	CH_3	H	−0.065
CH_3	CH_3	CH_3	CH_3	−0.260

9

Using molecular modeling software and the computational method recommended by your instructor (extended Hückel, semi-empirical, *ab initio*, or DFT methods), calculate E_{LUMO}, the energy of the LUMO of each substituted *p*-benzoquinone, and plot E_{LUMO} against $E_{cell}^{\ominus}$. Do your calculations support a linear relation between E_{LUMO} and $E_{cell}^{\ominus}$?

(b) The 1,4-benzoquinone for which $R_2 = R_3 = CH_3$ and $R_5 = R_6 = OCH_3$ is a suitable model of coenzyme Q, a component of the respiratory electron transport chain (Section 5.10). Determine E_{LUMO} of this quinone and then use your results from part (a) to estimate its biological standard potential.

(c) The *p*-benzoquinone for which $R_2 = R_3 = R_5 = CH_3$ and $R_6 = H$ is a suitable model of plastoquinone, a component of the photosynthetic electron transport chain (Section 5.11). Determine E_{LUMO} of this quinone and then use your results from part (a) to estimate its standard potential. Is plastoquinone expected to be a better or worse oxidizing agent than coenzyme Q?

(d) Based on your predictions and on basic concepts of biological electron transport (Sections 5.10 and 5.11), suggest a reason why coenzyme Q is used in respiration and plastoquinone is used in photosynthesis.

Macromolecules and self-assembly

11

Biological cells are complex devices with outer shells built largely from lipids, sterols, and, in some organisms, complex carbohydrates. Inside the cells are information storage and retrieval systems—the chromosomes—and molecular machines—enzymes, ion channels and pumps, and so on—made from small molecules and macromolecules, such as proteins, nucleic acids, and polysaccharides. The construction of functional structures in the cell proceeds largely through **self-assembly**, the spontaneous formation of complex aggregates of molecules or macromolecules held together by a variety of molecular interactions of the kind described later in the chapter. We have already encountered a few examples of self-assembly, such as the formation of biological membranes from lipids and of a DNA double helix from two polynucleotide chains (*Fundamentals* F.1). In this chapter, we describe several techniques for the determination of the size and shape of biological macromolecules and aggregates, and then explore the interactions responsible for the shapes so found. These interactions contribute to a whole hierarchy of structure, from 'no structure' in fluids all the way up to the elaborate and functionally important structures of proteins and nucleic acids. We also describe computer-aided methods for building three-dimensional models of macromolecules in which the molecular interactions that promote self-assembly are optimized.

Determination of size and shape

In this section we explore important methods used in modern biochemical research to determine the molar mass and structure of very large molecules. The most powerful of these techniques are based on the diffraction of X-rays from crystalline samples and reveal the position of almost every heavy atom (that is, every atom other than hydrogen) even in very large molecules.

11.1 Ultracentrifugation

Because molar mass is so important for the identification of a molecule and the determination of its structure, we need to discuss sophisticated and accurate methods for its determination.

In a gravitational field, heavy particles settle toward the foot of a column of solution by the process called **sedimentation**. The rate of sedimentation depends on the strength of the field and on the masses and shapes of the particles. Spherical molecules (and compact molecules in general) sediment faster than rodlike or extended molecules. For example, DNA helices sediment much faster when they are denatured to a random coil, so sedimentation rates can be used to study

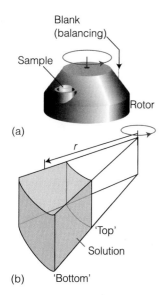

Fig. 11.1 (a) An ultracentrifuge head. The sample on one side is balanced by a blank diametrically opposite. (b) Detail of the sample cavity: the 'top' surface is the inner surface, and the centrifugal force causes sedimentation toward the outer surface; a particle at a radius r experiences a force of magnitude $m_{eff}r\omega^2$.

denaturation. When the sample is at equilibrium, the particles are dispersed over a range of heights because the gravitational field competes with the stirring effect of thermal motion. The spread of heights depends on the masses of the molecules, so the equilibrium distribution is another way to determine molar mass.

Sedimentation is normally very slow, but it can be accelerated by **ultracentrifugation**, a technique that replaces the gravitational field with a centrifugal field. The effect can be achieved in an ultracentrifuge, which is essentially a cylinder that can be rotated at high speed about its axis with a sample in a cell near its periphery (Fig. 11.1). Modern ultracentrifuges can produce accelerations equivalent to about 10^5 that of gravity ('10^5 g'). Initially the sample is uniform, but the 'top' (innermost) boundary of the solute moves outward as sedimentation proceeds.

(a) The sedimentation rate

Solute particles in a spinning rotor adopt a constant speed away from the rotational axis because the outward, centrifugal force is balanced by a retarding, frictional force. The **sedimentation constant**, S, a measure of the rate at which a particle migrates in the centrifugal field, is defined as

$$S = \frac{s}{r\omega^2}$$

Sedimentation constant (11.1)

where s is the speed of sedimentation, r is the distance of the sample from the rotational axis, and ω is the angular velocity of the rotor (in radians per second). For biological macromolecules, typical values of S are of the order of 10^{-13} s and depend on the shape and size of the particle, the temperature, and the viscosity of the solution. A common unit for S is the 'svedberg', denoted Sv and defined as 1 Sv = 10^{-13} s. For example, the sedimentation constant of the protein bovine serum albumin is 5.02 Sv in water at 25°C. We show in the following *Justification* that the molar mass of a macromolecule is related to its sedimentation constant, S, and diffusion constant, D, by the relation

$$M = \frac{SRT}{bD}$$

Relation between the molar mass and the sedimentation constant (11.2)

where $b = 1 - \rho v_s$ is a correction factor that takes into account the buoyancy of the solution, with ρ the mass density of the solvent (typically in grams per cubic centimeter) and v_s the specific volume of the solute (typically in cubic centimeters per gram).

Justification 11.1 *The sedimentation constant*

A solute particle of mass m has an effective mass $m_{eff} = bm$ in the solution. The solute particles at a distance r from the axis of a rotor spinning at an angular velocity ω experience a centrifugal force of magnitude $m_{eff}r\omega^2$. The acceleration outward is countered by a frictional force proportional to the speed, s, of the particles through the medium. This force is written fs, where f is the *frictional coefficient*. The particles therefore adopt a *drift speed*, a constant speed through the medium, which is found by equating the two forces $m_{eff}r\omega^2$ and fs. The forces are equal when

$$s = \frac{m_{eff}r\omega^2}{f} = \frac{bmr\omega^2}{f}$$

Next, we draw on the **Stokes–Einstein relation**, $f = kT/D$, between the frictional coefficient, f, and the diffusion coefficient, D, to write

$$s = \frac{bmr\omega^2 D}{kT} = \frac{bMr\omega^2 D}{RT}$$

where we have used the relation $m = M/N_A$ between the molecular mass and the molar mass M and the relation $R = kN_A$ between Boltzmann's constant and the gas constant. Use of eqn 11.1 and rearrangement of this expression gives eqn 11.2.

The diffusion coefficient is related to the rate at which molecules migrate down a concentration gradient (it is treated in detail in Section 8.5) and can be measured by observing the rate at which a concentration boundary moves or the rate at which a more concentrated solution diffuses into a less concentrated one. The diffusion coefficient can also be measured by using laser light-scattering methods (Section 11.3). It follows that we can find the molar mass by combining measurements of sedimentation and diffusion rates (to obtain S and D, respectively).

Self-test 11.1 Determine the molar mass of human hemoglobin, given that it has a sedimentation constant of 4.48 Sv and a diffusion coefficient of 6.9×10^{-11} m^2 s^{-1} in a solution with $\rho v_s = 0.748$ at 293 K.

Answer: 63 kg mol^{-1}

(b) Sedimentation equilibrium

It is sometimes more convenient to measure the equilibrium distribution of molecules than the rate at which they sediment. At equilibrium, when the tendency of the solute to settle is balanced by the spreading effect of thermal motion, the molar mass can be obtained from the ratio of concentrations c_2/c_1 of the macromolecules at two different radii r_2 and r_1, respectively, in a centrifuge operating at angular frequency ω:

$$M = \frac{2RT}{(r_2^2 - r_1^2)b\omega^2} \ln \frac{c_2}{c_1}$$

Molar mass from sedimentation data $\quad$ (11.3a)

where R is the gas constant. The centrifuge is run more slowly in this technique than in the sedimentation rate method to avoid having all the solute pressed in a thin film against the bottom of the cell. At these slower speeds, several days may be needed for equilibrium to be reached.

Example 11.1 *The molar mass of a protein from ultracentrifugation experiments*

The data from an equilibrium ultracentrifugation experiment performed at 300 K on an aqueous solution of a protein show that a graph of $\ln c$ against $(r/\text{cm})^2$ is a straight line with a slope of 0.729. The rotational rate of the centrifuge was 50 000 rotations per minute and $b = 0.70$. Calculate the molar mass of the protein.

Strategy We need to reinterpret eqn 11.3 in terms of the slope of a plot of $\ln c$ against r^2. To do so, we apply the relation $\ln(x/y) = \ln x - \ln y$ to eqn 11.3 and obtain, after minor rearrangement,

$$M = \frac{2RT}{b\omega^2} \times \frac{\ln c_2 - \ln c_1}{r_2^2 - r_1^2}$$

If a plot of $\ln c$ against r^2 is linear, then the ratio $(\ln c_2 - \ln c_1)/(r_2^2 - r_1^2)$ has the form of the slope of the line. In practice, $\ln(c/\text{g cm}^{-3})$ is plotted against $(r/\text{cm})^2$ to give a dimensionless slope. It follows that

$$M = \frac{2RT}{b\omega^2} \times (\text{slope} \times \text{cm}^{-2}) \tag{11.3b}$$

and we can use the data provided to calculate the molar mass M. Each full revolution of the rotor corresponds to an angular change of 2π radians, so to obtain the angular frequency ω, we multiply the rotation rate in cycles per second by 2π.

Solution The angular frequency is

$$\omega = 2\pi \times (50\,000\ \text{min}^{-1}) \times \frac{1\ \text{min}}{60\ \text{s}} = \frac{2\pi \times 50\,000}{60}\ \text{s}^{-1}$$

It follows from eqn 11.3 with $1\ \text{cm}^{-2} = 10^4\ \text{m}^{-2}$ and the slope 0.729 that the molar mass is

$$M = \frac{2 \times (8.3145\ \text{J K}^{-1}\ \text{mol}^{-1}) \times (300\ \text{K}) \times (0.729 \times 10^4\ \text{m}^{-2})}{(1 - 0.70) \times \left(\dfrac{2\pi \times 50\,000}{60}\ \text{s}^{-1}\right)^2}$$

where we have used $1\ \text{J} = 1\ \text{kg m}^2\ \text{s}^{-2}$. The molar mass is therefore $4.4\ \text{kg mol}^{-1}$.

A note on good practice
A *molar* mass (the mass per mole of molecules) of $4.4\ \text{kg mol}^{-1}$ corresponds to a *molecular* mass (the mass of one molecule) of 4.4 kDa. Be careful to use the unit dalton (Da) to denote molecular, not molar, mass.

> **Self-test 11.2** The data from a sedimentation equilibrium experiment performed at 293 K on a macromolecular solute in aqueous solution show that a graph of $\ln(c/\text{g cm}^{-3})$ against $(r/\text{cm})^2$ is a straight line with a slope of 0.821. The rotation rate of the centrifuge was 4500 Hz ($1\ \text{Hz} = 1\ \text{s}^{-1}$) and $\rho v_s = 0.40$. Calculate the molar mass of the solute.
>
> **Answer:** $8.3\ \text{kg mol}^{-1}$, corresponding to a molecular mass of 8.3 kDa

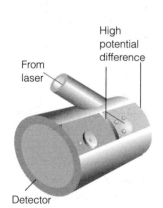

Fig. 11.2 Diagram of a matrix-assisted laser desorption/ionization time-of-flight (MALDI-TOF) mass spectrometer. A laser beam ejects macromolecules and ions from the solid matrix. The ionized macromolecules are accelerated by an electrical potential difference over a distance d and then travel through a drift region of length l. Ions with the smallest mass-to-charge ratio (m/z) reach the detector first.

11.2 Mass spectrometry

The most precise technique for the determination of molar mass is mass spectrometry, and we need to know how to adapt traditional techniques developed for small molecules to the study of biological macromolecules.

In mass spectrometry, the sample is first ionized in the gas phase and then the mass-to-charge number ratios, m/z, of all the resulting ions are measured. Macromolecules present a challenge because it is difficult to produce gaseous ions of large species without fragmentation. However, two new techniques have emerged that circumvent this problem: **matrix-assisted laser desorption/ionization** (MALDI) and **electrospray ionization**. We shall discuss **MALDI–TOF mass spectrometry**, so-called because the MALDI technique is coupled to a time-of-flight (TOF) ion detector.

Figure 11.2 shows a schematic view of a MALDI–TOF mass spectrometer. The macromolecule is first embedded in a solid matrix that often consists of an organic acid such as 2,5-dihydroxybenzoic acid, nicotinic acid, or an α-cyanocarboxylic acid.

This sample is then irradiated with a laser pulse. The pulse of electromagnetic energy ejects matrix ions, cations, and neutral macromolecules, thus creating a dense gas plume above the sample surface. The macromolecule is ionized by collisions and complexation with H^+ cations.

In the TOF spectrometer, the ions are accelerated over a short distance d by an electrical field of strength $\mathcal{E}$ and then travel through a drift region of length l. We show in the following *Justification* that the **time of flight**, t, required for an ion of mass m and charge number z to reach the detector at the end of the drift region is

$$t = l\left(\frac{m}{2ze\mathcal{E}d}\right)^2$$

Time of flight in a TOF spectrometer   (11.4)

where e is the fundamental charge. Because d, l, and $\mathcal{E}$ are fixed for a given experiment, the time of flight of the ion is a direct measure of its m/z ratio, which is given by

$$m/z = 2ze\mathcal{E}d\left(\frac{t}{l}\right)^2$$

The mass-to-charge number ratio in a TOF spectrometer   (11.5)

A note on good (in this case, common) practice Strictly, the units of m/z are kilograms; however, it is conventional to interpret m as the ratio of the molecular mass to the atomic mass constant m_u, in which case 'm/z' (strictly m/zm_u) is dimensionless.

Justification 11.2   *The time of flight of an ion in a mass spectrometer*

Consider an ion of charge ze and mass m that is accelerated from rest by an electric field of strength $\mathcal{E}$ applied over a distance d. The kinetic energy, E_k, of the ion is

$$E_k = \tfrac{1}{2}mv^2 = ze\mathcal{E}d$$

where v is the speed of the ion. The drift region, l, and the time of flight, t, in the mass spectrometer are both sufficiently short that we can ignore acceleration and write $v = l/t$. Then substitution into the expression for E_k gives

$$\tfrac{1}{2}m\left(\frac{l}{t}\right)^2 = ze\mathcal{E}d$$

Rearrangement of this equation gives eqn 11.5.

Figure 11.3 shows the MALDI-TOF mass spectrum of bovine albumin. The MALDI technique produces unfragmented molecular ions of varying charges, with the singly charged ion often giving rise to the most prominent feature in the spectrum. The spectrum of a mixture of biopolymers consists of multiple peaks arising from molecules with different molar masses. The intensity of each peak is proportional to the abundance of each biopolymer in the sample.

(Self-test 11.3) A MALDI-TOF mass spectrum consists of two intense features at $m/z = 9912$ and 4554. Does the sample contain one or two distinct biopolymers? Explain your answer.

Answer: Two distinct biopolymers because the feature at lower m/z probably does not arise from the unfragmented +2 cation of the species that gives rise to the feature at higher m/z.

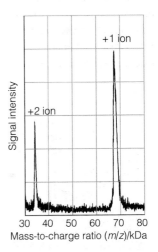

Signal intensity

+1 ion

+2 ion

30 40 50 60 70 80
Mass-to-charge ratio (m/z)/kDa

Fig. 11.3 The MALDI-TOF mass spectrum of bovine albumin, a protein with molar mass 66.43 kg mol⁻¹. During the MALDI process, the protein takes up one or two H^+ ions, making molecular ions of charge +1 and +2, respectively. Because the protein does not fragment, the +2 ion gives rise to a peak in the spectrum at a m/z value that is one-half the value for the peak associated with the +1 ion. (Adapted from B.S. Larsen and C.N. McEwen in *Mass spectrometry of biological materials*, Marcel Dekker, New York (1998).)

Fig. 11.4 Rayleigh scattering from a sample of point-like particles. The intensity of scattered light depends on the angle θ between the incident and scattered beams. The treatment developed in the text corresponds to an experimental arrangement in which the plane of polarization of the laser beam (the dark blue plane in the inset) is perpendicular to the plane defined by the incident ray and the line from the sample to the detector (the light blue plane in the inset).

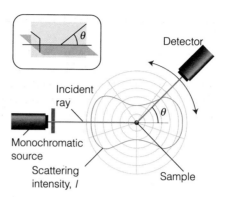

11.3 **Laser light scattering**

The analysis of the intensity of laser light scattered by a solution of a biological macromolecule yields information about its size and shape.

Scattering of light by particles with diameters much smaller than the wavelength of the incident radiation is called **Rayleigh scattering**. In the Rayleigh regime, the intensity of scattered light is proportional to the molar mass of the particle and to λ^{-4}, so shorter-wavelength radiation is scattered more intensely than longer wavelengths. For example, the blue of the sky arises from the more intense scattering of the blue component of white sunlight by the molecules of the atmosphere.

(a) Rayleigh scattering

Consider the experimental arrangement shown in Fig. 11.4 for the measurement of light scattering from solutions of macromolecules. Typically, the sample is irradiated with monochromatic light from a laser. The intensity of scattered light is then measured as a function of the angle θ that the direction of the laser beam makes with the direction of the detector from the sample at a distance r. Under these conditions, the intensity, $I(\theta)$, of light scattered is written as the **Rayleigh ratio**:

$$R(\theta) = \frac{I(\theta)}{I_0} \times r^2 \qquad \boxed{\begin{array}{l}\text{Definition of the}\\\text{Rayleigh ratio}\end{array}} \quad (11.6)$$

where I_0 is the intensity of the incident laser radiation.

A detailed examination of the scattering shows that the Rayleigh ratio depends on the mass concentration, c_M (units: kg m^{-3}), of the macromolecule and its molar mass M as:

$$R(\theta) = KP(\theta)c_M M \qquad \boxed{\begin{array}{l}\text{The relation of Rayleigh}\\\text{ratio to molar mass}\end{array}} \quad (11.7)$$

where the constant K depends on the refractive index of the solution, the incident wavelength, and the distance between the detector and the sample, which is held constant during the experiment. The quantity $P(\theta)$ is the **structure factor**, which is related to the size of the molecule. When the molecule is much smaller than the wavelength of the light, $P(\theta) \approx 1$. However, when the size of the molecule is about one-tenth the wavelength of the incident radiation, it is possible to show that

$$P(\theta) \approx 1 - \frac{16\pi^2 R_g^2 \sin^2 \frac{1}{2}\theta}{3\lambda^2} \qquad \boxed{\begin{array}{l}\text{Structure factor for}\\\text{small molecules}\end{array}} \quad (11.8)$$

A brief comment
The factor r^2 occurs in the definition of the Rayleigh ratio because the light wave spreads out over a sphere of radius r and surface area $4\pi r^2$, so any sample of the radiation has its intensity $I(\theta)$ decreased by a factor proportional to r^2. Therefore, the quantity $I(\theta) \times r^2$, and not simply $I(\theta)$, should be compared to I_0 in forming the Rayleigh ratio. We also note that the definition of the Rayleigh ratio given here applies only to the experimental conditions in Fig. 11.4.

where R_g is the radius of gyration of the macromolecule, a measure of its size (Section 11.12).

Equation 11.7 applies only to ideal solutions. In practice, even relatively dilute solutions of macromolecules can deviate considerably from ideality, as we saw in *In the laboratory* 3.1. Being so large, macromolecules displace a large quantity of solvent instead of replacing individual solvent molecules with negligible disturbance. To take deviations from ideality into account, it is common to rewrite eqn 11.7 as $Kc_M/R(\theta) = 1/P(\theta)M$ and to extend it to

$$\frac{Kc_M}{R(\theta)} = \frac{1}{P(\theta)M} + Bc_M \tag{11.9}$$

where B is an empirical constant analogous to the osmotic virial coefficient (*In the laboratory* 3.1) and indicative of the effect of excluded volume.

The preceding discussion shows that structural properties, such as the size and molar mass of a macromolecule, can be obtained from measurements of light scattering by a sample at several angles θ relative to the direction of propagation on an incident beam. In modern instruments, lasers are used as the radiation sources.

Example 11.2 *Determining the molar mass and size of a protein by laser light scattering*

The following data for an aqueous solution of a protein with $c_M = 2.00$ kg m^{-3} were obtained at 20°C with laser light at $\lambda = 532$ nm.

$\theta/°$	15.0	45.0	70.0	85.0	90.0
$R(\theta)/m^2$	23.8	22.9	21.6	20.7	20.4

In a separate experiment, it was determined that $K = 2.40 \times 10^{-2}$ mol m^5 kg^{-2}. From this information, calculate R_g and M for the protein. Assume that B is negligibly small and that the protein is small enough that eqn 11.7 holds.

Strategy Substituting the result of eqn 11.8 into eqn 11.7 we obtain, after some rearrangement:

$$\frac{1}{R(\theta)} = \frac{1}{Kc_M M} + \left(\frac{16\pi^2 R_g^2}{3\lambda^2}\right)\left(\frac{1}{R(\theta)}\sin^2 \tfrac{1}{2}\theta\right)$$

Hence, a plot of $1/R(\theta)$ against $\{1/R(\theta)\}\sin^2 \tfrac{1}{2}\theta$ should be a straight line with slope $16\pi^2 R_g^2/3\lambda^2$ and y-intercept $1/Kc_M M$. As usual, the plot should be of dimensionless quantities, so we actually plot $1/(R(\theta)/m^2)$ against $\{1/(R(\theta)/m^2)\} \times \sin^2 \tfrac{1}{2}\theta$, in which case the dimensionless slope is equal to the dimensionless quantity $16\pi^2 R_g^2/3\lambda^2$ and the dimensionless intercept is equal to $1/Kc_M M$.

Solution We construct a table of values of $1/R(\theta)$ and $\{1/R(\theta)\}\sin^2 \tfrac{1}{2}\theta$ and plot the data (Fig. 11.5).

$10^2/(R(\theta)/m^2)$	4.20	4.37	4.63	4.83	4.90
$10^3 \times (\sin^2 \tfrac{1}{2}\theta)/(R(\theta)/m^2)$	0.716	6.40	15.2	22.0	24.5

The best straight line through the data has a slope of 0.295 and a y-intercept of $1/(R(\theta)/m^2) = 4.18 \times 10^{-2}$. From these values, we calculate

$$R_g = \left(\frac{3\lambda^2 \times \text{slope}}{16\pi^2}\right)^{1/2} = \left(\frac{3 \times (532 \text{ nm})^2 \times 0.295}{16\pi^2}\right)^{1/2} = 39.8 \text{ nm}$$

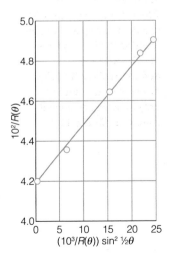

Fig. 11.5 Plot of the data for *Example* 11.2.

$$M = \frac{1 \text{ m}^2}{(2.40 \times 10^{-2} \text{ mol m}^5 \text{ kg}^{-2}) \times (2.00 \text{ kg m}^{-3}) \times (4.18 \times 10^{-2})}$$

$$= 4.98 \times 10^2 \text{ kg mol}^{-1}$$

We conclude that the radius of gyration is 39.8 nm and the molar mass is 498 kg mol^{-1}.

Self-test 11.4 The following data for an aqueous solution of a macromolecule were obtained at 20°C with plane-polarized light at $\lambda = 546$ nm.

$\theta/°$	26.0	36.9	66.4	90.0	113.6
$R(\theta)/\text{m}^2$	19.7	18.8	17.1	16.0	14.4

In separate experiments, it was determined that $K = 6.42 \times 10^{-5}$ mol m^5 kg^{-2}. From this information, and using $c_M = 311$ kg m^{-3}, calculate the R_g and M of the macromolecule. State any assumptions you must make to solve this problem.

Answer: $R_g = 46.9$ nm and $M = 987$ kg mol^{-1}

(b) Dynamic light scattering

A special laser scattering technique, **dynamic light scattering**, can be used to investigate the diffusion of macromolecules in solution. Consider two molecules being irradiated by a laser beam. Suppose that at a time t the scattered waves from these particles interfere constructively at the detector, leading to a large signal. However, as the molecules move through the solution, the scattered waves may interfere destructively at another time t' and result in no signal. When this behavior is extended to a very large number of molecules in solution, it results in fluctuations in light intensity that depend on the diffusion coefficient, D. Hence, analysis of the fluctuations gives the diffusion coefficient and molecular size in cases where the molecular shape is known.

Light scattering is a convenient method for the characterization of biological systems from proteins to viruses. Unlike mass spectrometry, laser light-scattering measurements may be performed in nearly intact samples; often the only preparation required is filtration of the sample.

11.4 X-ray crystallography

The success of modern biochemistry in explaining such processes as DNA replication, protein biosynthesis, and enzyme catalysis is a direct result of developments in preparatory, instrumental, and computational procedures that have led to the determination of large numbers of structures of biological macromolecules by techniques based on X-ray diffraction.

Because much of our knowledge of the three-dimensional structures of biological macromolecules comes from studies of crystals of proteins and nucleic acids, we need to study the arrangements adopted by molecules when they stack together to form a crystalline solid. One of the most important techniques for the determination of the structures of crystals is **X-ray diffraction**. In its most sophisticated version, known as **X-ray crystallography**, X-ray diffraction provides detailed information about the location of all the atoms in molecules as complicated as biological macromolecules.

(a) Diffraction

A characteristic property of waves is that they **interfere** with one another, which means that they give a greater amplitude where their displacements add and a smaller amplitude where their displacements subtract (Section 9.1). Because the intensity of electromagnetic radiation is proportional to the square of the amplitude of the waves, the regions of constructive and destructive interference show up as regions of enhanced and diminished intensities. The phenomenon of **diffraction** is the interference caused by an object in the path of waves, and the pattern of varying intensity that results is called the **diffraction pattern** (Fig. 11.6). Diffraction occurs when the dimensions of the diffracting object are comparable to the wavelength of the radiation. Sound waves, with wavelengths of the order of 1 m, are diffracted by macroscopic objects. Light waves, with wavelengths of the order of 500 nm, are diffracted by narrow slits. X-rays have wavelengths comparable to bond lengths in molecules and the spacing of atoms in crystals (about 100 pm), so they are diffracted by them. By analyzing the diffraction pattern, it is possible to draw up a detailed picture of the location of atoms.

The short-wavelength electromagnetic radiation we call X-rays is produced by bombarding a metal with high-energy electrons. The electrons decelerate as they plunge into the metal and generate radiation with a continuous range of wavelengths. This radiation is called **bremsstrahlung**.[1] Superimposed on the continuum are a few high-intensity, sharp peaks. These peaks arise from the interaction of the incoming electrons with the electrons in the inner shells of the atoms. A collision expels an electron (Fig. 11.7), and an electron of higher energy drops into the vacancy, emitting the excess energy as an X-ray photon. An example of the process is the expulsion of an electron from the K shell (the shell with $n = 1$) of a copper atom, followed by the transition of an outer electron into the vacancy. If an electron from the L shell undergoes the transition, then the energy so released gives rise to copper's K_α radiation of wavelength 154 pm.

In 1912, the German physicist Max von Laue suggested that X-rays might be diffracted when passed through a crystal, for the wavelengths of X-rays are comparable to the separation of atoms. Laue's suggestion was confirmed almost immediately by Walter Friedrich and Paul Knipping, and then developed by William and Laurence Bragg (father and son), who later jointly received the Nobel Prize. It has grown since then into a technique of extraordinary power.

(b) Crystal systems

X-ray diffraction is applied to crystalline arrays of molecules, so we need to know how to describe the arrangement of molecules in a crystal. The pattern that atoms, ions, or molecules adopt in a crystal is expressed in terms of an array of points making up the **lattice** that identify the locations of the individual species (Fig. 11.8). A **unit cell** of a crystal is the small three-dimensional figure obtained by joining typically eight of these points, which may be used to construct the entire crystal lattice by purely translational displacements, much as a wall may be constructed from bricks (Fig. 11.9). An infinite number of different unit cells can describe the same structure, but it is conventional to choose the cell with sides that have the shortest lengths and are most nearly perpendicular to one another.

Unit cells are classified into one of seven **crystal systems** according to the symmetry they possess under rotations about different axes. The *cubic system*, for example, has four threefold axes (Fig. 11.10). A threefold axis is an axis of

---

[1] *Bremse* is German for 'brake', *Strahlung* for 'radiation'.

Fig. 11.6 The X-ray diffraction pattern obtained from a fiber of B-DNA. The black dots are the reflections, the points of maximum constructive interference, that are used to determine the structure of the molecule (see *Case study* 11.1). (Adapted from an illustration that appears in J.P. Glusker and K.N. Trueblood, *Crystal structure analysis: A primer*. Oxford University Press (1972).)

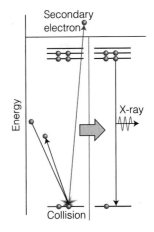

Fig. 11.7 The formation of X-rays. When a metal is subjected to a high-energy electron beam, an electron in an inner shell of an atom is ejected. When an electron falls into the vacated orbital from an orbital of much higher energy, the excess energy is released as an X-ray photon.

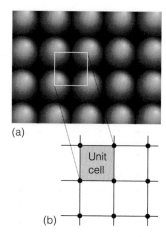

(a)

(b)

Fig. 11.8 (a) A crystal consists of a uniform array of atoms, molecules, or ions, as represented by these spheres. In many cases, the components of the crystal are far from spherical, but this diagram illustrates the general idea. (b) The location of each atom, molecule, or ion can be represented by a single point; here (for convenience only), the locations are denoted by a point at the center of the sphere. The unit cell, which is shown boxed, is the smallest block from which the entire array of points can be constructed without rotating or otherwise modifying the block.

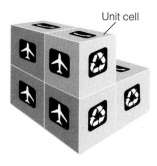

Unit cell

Fig. 11.9 A unit cell, here shown in three dimensions, is like a brick used to construct a wall. Once again, only pure translations are allowed in the construction of the crystal. (Some bonding patterns for actual walls use rotations of bricks, so for these patterns a single brick is not a unit cell.)

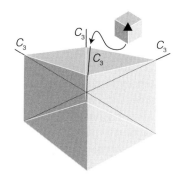

Fig. 11.10 A unit cell belonging to the cubic system has four threefold axes (denoted C_3) arranged tetrahedrally.

Table 11.1 The essential symmetries of the seven crystal systems

System	Essential symmetries
Triclinic	None
Monoclinic	One twofold axis
Orthorhombic	Three perpendicular twofold axes
Rhombohedral	One threefold axis
Tetragonal	One fourfold axis
Hexagonal	One sixfold axis
Cubic	Four threefold axes in a tetrahedral arrangement

a rotation that restores the unit cell to the same appearance three times during a complete revolution, after rotations through 120°, 240°, and 360°. The four axes make the tetrahedral angle to each other. The *monoclinic system* has one twofold axis (Fig. 11.11). A twofold axis is an axis of a rotation that leaves the cell apparently unchanged twice during a complete revolution, after rotations through 180° and 360°. The **essential symmetries**, the properties that must be present for the unit cell to belong to a particular system, are listed in Table 11.1.

A unit cell may have lattice points other than at its corners, so each crystal system can occur in a number of different varieties. For example, in some cases points may occur on the faces and in the body of the cell without destroying the cell's essential symmetry. These various possibilities give rise to 14 distinct types of unit cell, called **Bravais lattices**. Three examples are shown in Fig. 11.12.

(c) Crystal planes

To specify a unit cell fully, we need to know not only its symmetry but its size, such as the lengths of its sides. There is a useful relation between the spacing of the planes passing through the lattice points, which (as we shall see) we can measure, and the lengths we need to know. Because two-dimensional arrays of points are easier to visualize than three-dimensional arrays, we shall introduce the concepts we need by referring to two dimensions initially and then extend the conclusions to three dimensions.

Consider the two-dimensional rectangular lattice formed from a rectangular unit cell of sides a and b (Fig. 11.13). We can distinguish the four sets of planes shown in the illustration by the distances at which they intersect the axes. One way of labeling the planes would therefore be to denote each set by the smallest intersection distances. For example, we could denote the four sets in the illustration as $(1a,1b)$, $(3a,2b)$, $(-1a,1b)$, and $(\infty a,1b)$. If, however, we agreed always to quote distances along the axes as multiples of the lengths of the unit cell, then we could omit the a and b and label the planes more simply as $(1,1)$, $(3,2)$, $(-1,1)$, and $(\infty,1)$.

Now suppose that the array in Fig. 11.13 is the top view of a three-dimensional rectangular lattice in which the unit cell has a length c in the z direction. All four

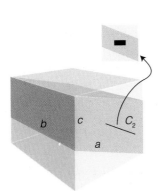

Fig. 11.11 A unit cell belonging to the monoclinic system has one twofold (denoted C_2) axis (along b).

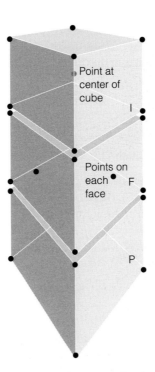

Fig. 11.12 The cubic unit cells. The letter P denotes a primitive unit cell, I a body-centered unit cell, and F a face-centered unit cell.

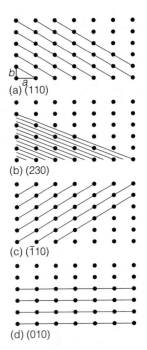

Fig. 11.13 Some of the planes that can be drawn through the points of the space lattice and their corresponding Miller indices (hkl).

sets of planes intersect the z-axis at infinity, so the full labels of the sets of planes of lattice points are $(1,1,\infty)$, $(3,2,\infty)$, $(-1,1,\infty)$, and $(\infty,1,\infty)$.

The presence of infinity in the labels is inconvenient. We can eliminate it by taking the reciprocals of the numbers in the labels; this step also turns out to have further advantages, as we shall see. The resulting **Miller indices**, (hkl), are the reciprocals of the numbers in the parentheses with fractions cleared. For example, the $(1,1,\infty)$ planes in Fig. 11.13 are the (110) planes in the Miller notation. Similarly, the $(3,2,\infty)$ planes become first $(\frac{1}{3},\frac{1}{2},0)$ when reciprocals are formed and then (2,3,0) when fractions are cleared by multiplication through by 6, so they are referred to as the (230) planes. We write negative indices with a bar over the number: Fig. 11.13c shows the $(\bar{1}10)$ planes. Figure 11.14 shows some planes in three dimensions, including an example of a lattice with axes that are not mutually perpendicular.

(**Self-test 11.5**) A representative member of a set of planes in a crystal intersects the axes at $3a$, $2b$, and $2c$. What are the Miller indices of the planes?

Answer: (233)

It is helpful to keep in mind the fact, as illustrated in Fig. 11.13, that the smaller the value of h in the Miller index (hkl), the more nearly parallel the plane is to the a axis. The same is true of k and the b axis and l and the c axis. When $h = 0$, the planes intersect the a axis at infinity, so the ($0kl$) planes are parallel to the a axis. Similarly, the ($h0l$) planes are parallel to b and the ($hk0$) planes are parallel to c.

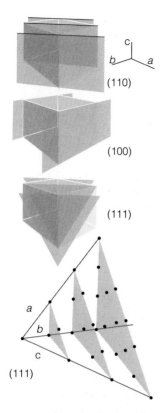

(110)

(100)

(111)

(111)

Fig. 11.14 Some representative planes in three dimensions and their Miller indices. Note that a 0 indicates that a plane is parallel to the corresponding axis. The indexing may also be used for unit cells with nonorthogonal axes.

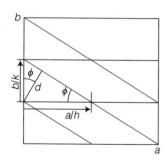

Fig. 11.15 The geometrical construction used to relate the separation of planes to the dimensions of a rectangular unit cell.

The Miller indices are very useful for calculating the separation of planes. For instance, they can be used to derive the following very simple expression for the separation, d, of the (hkl) planes in a rectangular lattice:

$$\frac{1}{d^2} = \frac{h^2}{a^2} + \frac{k^2}{b^2} + \frac{l^2}{c^2}$$

Separation of planes   (11.10)

Justification 11.3   *The separation of lattice planes*

Consider the $(hk0)$ planes of a rectangular lattice with sides of lengths a and b (Fig. 11.15). We can write the following trigonometric expressions for the angle ϕ shown in the illustration:

$$\sin \phi = \frac{d}{(a/h)} = \frac{hd}{a} \qquad \cos \phi = \frac{d}{(b/k)} = \frac{kd}{b}$$

Then, because $\sin^2 \phi + \cos^2 \phi = 1$, we obtain

$$\frac{h^2 d^2}{a^2} + \frac{k^2 d^2}{b^2} = 1$$

which we can rearrange into

$$\frac{1}{d^2} = \frac{h^2}{a^2} + \frac{k^2}{b^2}$$

Now consider an orthorhombic unit cell, a unit cell with perpendicular faces but different lengths of their edges (Fig. 11.16). In three dimensions, the expression above generalizes to eqn 11.10.

Example 11.3   *Using the Miller indices*

Calculate the separation of (a) the (123) planes and (b) the (246) planes of an orthorhombic cell with $a = 0.84$ nm, $b = 0.96$ nm, and $c = 0.77$ nm.

Strategy For the first part, we simply substitute the information into eqn 11.10. For the second part, instead of repeating the calculation, we should examine how d in eqn 11.10 changes when all three Miller indices are multiplied by 2 (or by a more general factor, n).

Solution Substituting the data into eqn 11.10 gives

$$\frac{1}{d^2} = \frac{1^2}{(0.84 \text{ nm})^2} + \frac{2^2}{(0.96 \text{ nm})^2} + \frac{3^2}{(0.77 \text{ nm})^2} = \frac{21}{\text{nm}^2}$$

It follows that $d = 0.22$ nm. When the indices are all increased by a factor of 2, the separation becomes

$$\frac{1}{d^2} = \frac{(2 \times 1)^2}{(0.84 \text{ nm})^2} + \frac{(2 \times 2)^2}{(0.96 \text{ nm})^2} + \frac{(2 \times 3)^2}{(0.77 \text{ nm})^2} = 4 \times \frac{21}{\text{nm}^2}$$

So, for these planes $d = 0.11$ nm. In general, increasing the indices uniformly by a factor n decreases the separation of the planes by n.

Self-test 11.6 Calculate the separation of the (133) and (399) planes in the same lattice.

Answer: 0.19 nm, 0.065 nm

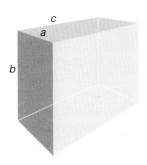

Fig. 11.16 An orthorhombic unit cell with sides of lengths a, b, and c.

(d) Bragg's law

The earliest approach to the analysis of X-ray diffraction patterns treated a plane of atoms as a semitransparent mirror and modeled the crystal as stacks of reflecting planes of separation d (Fig. 11.17). The model makes it easy to calculate the angle the crystal must make to the incoming beam of X-rays for constructive interference to occur. It has also given rise to the name **reflection** to denote an intense spot arising from constructive interference.

The path-length difference of the two rays shown in the illustration is

$$AB + BC = 2d \sin\theta$$

where the angle θ is often expressed as the **glancing angle** 2θ. When the path-length difference is equal to one wavelength ($AB + BC = \lambda$), the reflected waves interfere constructively. It follows that a reflection should be observed when the glancing angle satisfies **Bragg's law**:

$$\lambda = 2d \sin\theta \qquad \boxed{\text{Bragg's law}} \qquad (11.11a)$$

The primary use of Bragg's law is to determine the spacing between the layers of atoms, for once the angle θ corresponding to a reflection has been determined, d may readily be calculated. Equation 11.11a is sometimes written

$$n\lambda = 2d \sin\theta \qquad \boxed{\text{Alternative version of Bragg's law}} \qquad (11.11b)$$

with $n = 1, 2, \ldots$ denoting the *order* of the reflection, but the modern tendency is to incorporate n into the definition of d, as illustrated in Example 11.4.

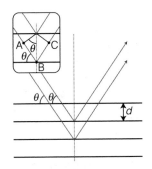

Fig. 11.17 The derivation of Bragg's law treats each lattice plane as reflecting the incident radiation. The path lengths differ by $AB + BC$, which depends on the angle θ. Constructive interference (a 'reflection') occurs when $AB + BC$ is equal to an integral number of wavelengths.

Example 11.4 *Using Bragg's law*

A reflection from the (111) planes of a cubic crystal was observed at a glancing angle of 11.2° when Cu K_α X-rays of wavelength 154 pm were used. What is the length of the side of the unit cell?

Strategy We can find the separation, d, of the lattice planes from eqn 11.11 and the data. Then we find the length of the side of the unit cell by using eqn 11.10. Because the unit cell is cubic, $a = b = c$, so eqn 11.10 simplifies to

$$\frac{1}{d^2} = \frac{h^2 + k^2 + l^2}{a^2}$$

which rearranges to

$$a = d \times (h^2 + k^2 + l^2)^{1/2}$$

Solution According to Bragg's law, the separation of the (111) planes responsible for the diffraction is

$$d = \frac{\lambda}{2 \sin\theta} = \frac{154\ \text{pm}}{2 \sin 11.2°}$$

It then follows that with $h = k = l = 1$,

$$a = \frac{154\ \text{pm}}{2 \sin 11.2°} \times 3^{1/2} = 687\ \text{pm}$$

Self-test 11.7 Calculate the angle at which the same lattice will give a reflection from the (123) planes.

Answer: 24.8°

(e) Fourier synthesis

Bragg's law is a very primitive approach to the interpretation of X-ray diffraction data. The huge amount of data obtained from a modern diffractometer has a much richer content than the separation of lattice planes, for in principle it contains information about the locations of individual atoms and the distribution of electron density throughout a unit cell.

To derive the structure of the crystal from the intensities, I_{hkl}, we need to convert them to the *amplitude* of the wave responsible for the signal. For simplicity we shall focus on a one-dimensional crystal (a line of atoms) and write the diffraction intensities as I_h. Because the intensity of electromagnetic radiation is given by the square of the amplitude, we need to form the **structure factors** $F_h = I_h^{1/2}$. (Note that this 'structure factor' is entirely distinct from the structure factor of Rayleigh scattering, Section 11.3.) Here is the first difficulty: we do not know the sign to take. For instance, if $I_h = 4$, then F_h can be either +2 or −2. This ambiguity is the **phase problem** of X-ray diffraction. However, once we have the structure factors, we can calculate the electron density $\rho(x)$ by forming the following sum:

A brief comment
Formally, a Fourier synthesis is a reconstruction of a repetitive function as a superposition of sine or cosine waves. Long-wavelength waves account for the general features of the structure, and the details are gradually filled in by incorporating shorter-wavelength waves.

$$\rho(x) = \frac{1}{V}\left\{F_0 + 2\sum_{h=1}^{\infty} F_h \cos(2h\pi x)\right\}$$

Fourier synthesis   (11.12)

where V is the volume of the unit cell. This expression is called a **Fourier synthesis** of the electron density: we show how it is used in the following *Example*. The point to note is that low values of the index h give the major features of the structure (they correspond to long-wavelength cosine terms), whereas the high values give the fine detail (short-wavelength cosine terms). Clearly, if we do not know the sign of F_h, we do not know whether the corresponding term in the sum is positive or negative and we get different electron densities, and hence crystal structures, for different choices of sign.

Example 11.5 *Calculating an electron density by Fourier synthesis*

The determination of the three-dimensional structure of molecules is a key step in the rational design of therapeutic agents that bind specifically to receptor sites on proteins and nucleic acids (*Case study* 11.2). Consider the (*h*00) planes of a crystal of an organic molecule regarded as a candidate for a drug. In an X-ray analysis the structure factors were found as follows:

h	0	1	2	3	4	5	6	7	8	9
F_h	16	−10	2	−1	7	−10	8	−3	2	−3

h	10	11	12	13	14	15
F_h	6	−5	3	−2	2	−3

Construct a plot of the electron density projected on to the *x*-axis of the unit cell.

Strategy Evaluate the sum in eqn 11.12 (stopping at $h = 15$) for points $0 \le x \le 1$:

$$V\rho(x) = 16 + 2\sum_{h=1}^{15} F_h \cos(2h\pi x)$$

The task is made easier by using an electronic spreadsheet, which also can generate a plot of the results.

Solution After introducing the data, eqn 11.12 takes the form

$$V\rho(x) = 16 - 20\cos(2\pi x) + 4\cos(4\pi x) - \cdots - 6\cos(30\pi x)$$

This function is shown in Fig. 11.18(a), and the locations of several types of atom are easy to identify as peaks in the electron density. The more terms there are included, the more accurate the density plot. Terms corresponding to high values of h (short-wavelength cosine terms in the sum) account for the finer details of the electron density; low values of h account for the broad features.

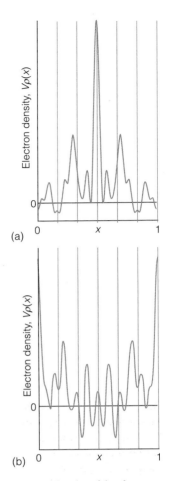

(a)

(b)

Fig. 11.18 The plot of the electron density calculated in (a) *Example* 11.5 and (b) *Self-test* 11.8.

Self-test 11.8 Use an electronic spreadsheet to experiment with different structure factors (including changes in signs as well as amplitudes). For example, use the same values of F_h as above, but with positive signs for all values of h.

Answer: Fig. 11.18(b)

The phase problem can be overcome to some extent by the method of **isomorphous replacement**, in which heavy atoms are introduced into the crystal. The technique relies on the fact that the scattering of X-rays is caused by the oscillations an incoming electromagnetic wave generates in the electrons of atoms, and heavy atoms with their large numbers of electrons give rise to stronger scattering than light atoms. Therefore, heavy atoms dominate the diffraction pattern and greatly simplify its interpretation. The phase problem can also be resolved by judging whether the calculated structure is chemically plausible, whether the electron density is positive throughout, and by using more refined mathematical techniques, with the help of powerful computers.

Because biopolymers contain a great many atoms, overcoming the phase problem requires repeated rounds of isomorphous replacement and computer-aided refinement, a process that can take several years to complete. As suggested by eqn 11.12 and *Example* 11.5, the more values of I_{hkl} that are collected, the richer the detail of the structure: analyzing few intensities leads to a fuzzy, low-resolution structure, whereas collecting more reflections results in a sharper, high-resolution structure. In practice, it is not the abundance of data, but rather the quality of the crystal—as determined by how perfectly ordered the molecules are packed in the solid—that limits the resolution of a structure. With current crystallization techniques, the best resolution of protein structures is approximately 200 pm, implying that two atoms cannot be located unambiguously if they are separated by less than this distance, which is greater than the average length of a carbon–carbon single bond (154 pm). In spite of this limitation, the identity and location of every atom in a biopolymer can be obtained by combining X-ray diffraction and sequencing data.

In the laboratory 11.1 *The crystallization of biopolymers*

The first and often very demanding step in the structural analysis of biological macromolecules by X-ray diffraction methods is to form crystals in which the large molecules lie in orderly ranks. A technique that works well for charged proteins consists of adding large amounts of a salt, such as $(NH_4)_2SO_4$, to a buffer solution containing the biopolymer. The increase in the ionic strength

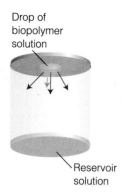

Drop of biopolymer solution

Reservoir solution

Fig. 11.19 In a common implementation of the vapor diffusion method of biopolymer crystallization, a single drop of biopolymer solution hangs above a reservoir solution that is very concentrated in a nonvolatile solute. Solvent evaporates from the more dilute drop until the vapor pressure of water in the closed container reaches a constant equilibrium value. In the course of evaporation (denoted by the downward arrows), the biopolymer solution becomes more concentrated and, at some point, crystals may form.

of the solution decreases the solubility of the protein to such an extent that the protein precipitates, sometimes as crystals that are amenable to analysis by X-ray diffraction (see *Exercise* 5.14 for an explanation of this effect).

Other common strategies for inducing crystallization involve the gradual removal of solvent from a biopolymer solution, either by *dialysis* (Section 3.10) or *vapor diffusion*. In one implementation of the vapor diffusion method, a single drop of biopolymer solution hangs above an aqueous solution (the reservoir), as shown in Fig. 11.19. If the reservoir solution is more concentrated in a nonvolatile solute (for example, a salt) than is the biopolymer solution, then solvent will evaporate slowly from the drop until the vapor pressure of water in the closed container reaches a constant, equilibrium value. At the same time, the concentration of biopolymer in the drop increases gradually until crystals begin to form.

Special techniques are used to crystallize hydrophobic proteins, such as those spanning the bilayer of a cell membrane. In such cases, surfactant molecules, which, like phospholipids, contain polar head groups and hydrophobic tails, are used to encase the protein molecules and make them soluble in aqueous buffer solutions. Dialysis or vapor diffusion may then be used to induce crystallization.

In the laboratory 11.2   *Data acquisition in X-ray crystallography*

After suitable crystals are obtained, X-ray diffraction data are collected and analyzed. Laue's original method consisted of passing a beam of X-rays of a wide range of wavelengths into a single crystal and recording the diffraction pattern photographically. The idea behind the approach was that a crystal might not be suitably oriented to act as a diffraction grating for a single wavelength, but whatever its orientation Bragg's law would be satisfied for at least one of the wavelengths when a range of wavelengths is present in the beam.

An alternative technique was developed by Peter Debye and Paul Scherrer and independently by Albert Hull. They used monochromatic (single-frequency) X-rays and a powdered sample. When the sample is a powder, we can be sure that some of the randomly distributed crystallites will be orientated so as to satisfy Bragg's law. For example, some of them will be orientated so that their (111) planes, of spacing *d*, give rise to a reflection at a particular angle, and others will be orientated so that their (230) planes give rise to a reflection at a different angle. Each set of (*hkl*) planes gives rise to reflections at a different angle. In the modern version of the technique, which uses a **powder diffractometer**, the sample is spread on a flat plate and the diffraction pattern is monitored electronically. The major application is for qualitative analysis because the diffraction pattern is a kind of fingerprint and may be recognizable (Fig. 11.20). The technique is also used for the characterization of substances that cannot be crystallized or the initial determination of the dimensions and symmetries of unit cells.

Modern X-ray crystallography, which utilizes an **X-ray diffractometer** (Fig. 11.21), is now a highly sophisticated technique. By far the most detailed information comes from developments of the techniques pioneered by the Braggs, in which a single crystal is employed as the diffracting object and a monochromatic beam of X-rays is used to generate the diffraction pattern. The single crystal (which may be only a fraction of a millimeter in length) is rotated relative to the beam, and the diffraction pattern is monitored and recorded electronically for each crystal orientation. The primary data are therefore a set of intensities arising from the Miller planes (hkl), with each set of planes giving a reflection of intensity I_{hkl}.

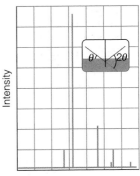

(a) NaCl

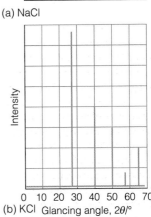

0 10 20 30 40 50 60 70
(b) KCl Glancing angle, 2θ/°

Fig. 11.20 A typical X-ray powder diffraction pattern that can be used to identify the material and determine the size of its unit cell: (a) NaCl, (b) KCl.

| Case study 11.1 | *The structure of DNA from X-ray diffraction studies* |

Bragg's law helps us understand the features of one of the most seminal X-ray images of all time, the characteristic X-shaped pattern obtained by Rosalind Franklin and Maurice Wilkins from strands of DNA, and used by James Watson and Francis Crick in their construction of the double-helix model of DNA (Fig. 11.22). To interpret this image by using Bragg's law, we have to be aware that it was obtained by using a fiber consisting of many DNA molecules oriented with their axes parallel to the axis of the fiber, with X-rays incident from a perpendicular direction. All the molecules in the fiber are parallel (or nearly so) but are randomly distributed in the perpendicular directions; as a result, the diffraction pattern exhibits the periodic structure parallel to the fiber axis superimposed on a general background of scattering from the distribution of molecules in the perpendicular directions.

There are two principal features in Fig. 11.22: the strong 'meridional' scattering upward and downward by the fiber and the X-shaped distribution at smaller scattering angles. Because scattering through large angles occurs for closely spaced features (from $\lambda = 2d \sin \theta$, if d is small then θ has to be large to preserve the equality), we can infer that the meridional scattering arises from closely spaced components and that the inner X-shaped pattern arises from features with a longer periodicity. Because the meridional pattern occurs at a distance of about 10 times that of the innermost spots of the X pattern, the large-scale structure is about 10 times bigger than the small-scale structure. From the geometry of the instrument, the wavelength of the radiation, and Bragg's law, we can infer that the periodicity of the small-scale feature is 340 pm, whereas that of the large-scale feature is 3400 pm (that is, 3.4 nm).

To see that the cross is characteristic of a helix, look at Fig. 11.22. Each turn of the helix defines two planes, one orientated at an angle α to the horizontal and the other at $-\alpha$. As a result, to a first approximation, a helix can be thought of as consisting of an array of planes at an angle α together with an array of planes at an angle $-\alpha$ with a separation within each set determined by the pitch of the helix. Thus, a DNA molecule is like two arrays of planes, each set corresponding to those treated in the derivation of the Bragg law, with a perpendicular separation $d = p \cos \alpha$, where p is the pitch of the helix, each canted at the angles $\pm\alpha$ to the horizontal. The diffraction spots from one set of planes therefore occur at an angle α to the vertical, giving one leg of the X, and those of the other set occur at an angle $-\alpha$, giving rise to the other leg of the X. The experimental arrangement has up–down symmetry, so the diffraction pattern repeats to produce the lower half of the X. The sequence of spots outward along a leg corresponds to first-, second-, ... order diffraction ($n = 1, 2, ...$ in eqn 11.11b). Therefore from the X-ray pattern, we see at once that the molecule is helical and we can measure the angle directly and find $\alpha = 40°$. Finally, with the angle

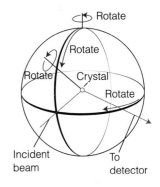

Fig. 11.21 The geometry of a four-circle diffractometer. The settings of the orientations of the components are controlled by computer; each reflection is monitored in turn, and their intensities are recorded.

Fig. 11.22 The origin of the X pattern characteristic of diffraction by a helix. (a) A helix can be thought of as consisting of an array of planes at an angle α together with an array of planes at an angle $-\alpha$. (b) The diffraction spots from one set of planes appear at an angle α to the vertical, giving one leg of the X, and those of the other set appear at an angle $-\alpha$, giving rise to the other leg of the X. The lower half of the X appears because the helix has up–down symmetry in this arrangement. (c) The sequence of spots outward along a leg of the X corresponds to first-, second-, ... order diffraction ($n = 1, 2, \ldots$).

Fig. 11.23 The effect of the internal structure of the helix on the X-ray diffraction pattern. (a) The residues of the macromolecule are represented by points. (b) Parallel planes passing through the residues are perpendicular to the axis of the molecule. (c) The planes give rise to strong diffraction with an angle that allows us to determine the layer spacing h from $\lambda = 2h \sin \theta$.

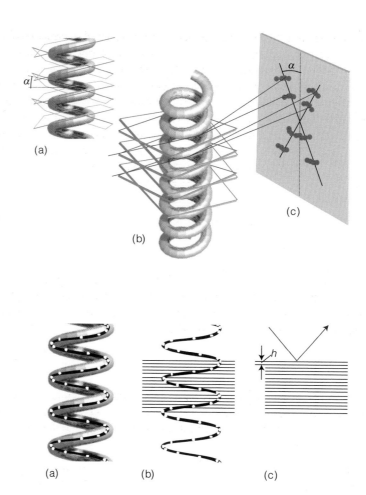

(a)

(b)

(c)

α and the pitch p determined, we can determine the radius r of the helix from $\tan \alpha = p/4r$, from which it follows that $r = (3.4\text{ nm})/(4 \tan 40°) = 1.0\text{ nm}$.

To derive the relation between the helix and the cross-like pattern, we have ignored the detailed structure of the helix, the fact that it is a periodic array of nucleotide bases, not a smooth wire. In Fig. 11.23 we represent the bases by points and see that there is an additional periodicity of separation h, forming planes that are perpendicular to the axis of the molecule (and the fiber). These planes give rise to the strong meridional diffraction with an angle that allows us to determine the layer spacing from Bragg's law in the form $\lambda = 2d \sin \theta$ as $h = 340\text{ pm}$.

The control of shape

The conformation of a biological molecule that has been determined by one of the techniques described so far is of crucial importance for its function, and we need to understand the forces that bring about the shape we observe. Interactions between molecules include the attractive and repulsive interactions between the partial electric charges of polar molecules and of polar functional groups in macromolecules and the repulsive interactions that prevent the complete collapse of matter to densities as high as those characteristic of atomic nuclei. The repulsive interactions arise from the exclusion of electrons from regions of space where the orbitals of closed-shell species overlap. One class of interaction, those proportional to the inverse sixth power of the separation, are termed **van der Waals interactions**. However, these are not the only interactions, and in the following paragraphs we describe the principal nonbonding interactions that occur between molecules and between different parts of the same molecule. All these interactions are much weaker—in some cases by several orders of magnitude—than those responsible for the formation of chemical bonds.

11.5 Interactions between partial charges

The Coulomb interaction between charges is our starting point for the discussion of the assembly of biological structures.

Atoms in molecules in general have partial charges. Table 11.2 gives the partial charges typically found on the atoms in peptides. If these charges were separated by a vacuum, they would attract or repel each other in accordance with Coulomb's law (*Fundamentals* F.3), and we would write

$$V = \frac{Q_1 Q_2}{4\pi\varepsilon_0 r} \qquad \boxed{\text{Coulomb's law (vacuum)}} \qquad (11.13a)$$

where Q_1 and Q_2 are the partial charges and r is their separation. However, we should take into account the possibility that other parts of the molecule, or other molecules, lie between the charges and decrease the strength of the interaction. We therefore write

$$V = \frac{Q_1 Q_2}{4\pi\varepsilon r} \qquad \boxed{\text{Coulomb's law (in any medium)}} \qquad (11.13b)$$

where ε is the permittivity of the medium lying between the charges. The permittivity is usually expressed as a multiple of the vacuum permittivity by writing $\varepsilon = \varepsilon_r \varepsilon_0$, where ε_r is the **relative permittivity** (formerly known as the *dielectric constant*). The effect of the medium can be very large: for water $\varepsilon_r = 78$, so the potential energy of two charges separated by bulk water is reduced by nearly two orders of magnitude compared to the value it would have if the charges were separated by a vacuum (Fig. 11.24). The problem is made worse in calculations on polypeptides and nucleic acids by the fact that two partial charges may have water and a biopolymer chain lying between them. Various models have been proposed to take this awkward effect into account, the simplest being to set $\varepsilon_r = 3.5$ and to hope for the best.

Table 11.2 Partial charges in polypeptides

Atom	Partial charge/e
C(=O)	+0.45
C(–CO)	+0.06
H(–C)	+0.02
H(–N)	+0.18
H(–O)	+0.42
N	−0.36
O	−0.38

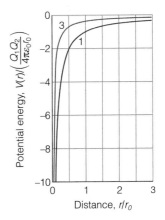

Fig. 11.24 The Coulomb potential energy of two charges Q_1 and Q_2 and its dependence on their separation. The two curves correspond to different relative permittivities ($\varepsilon_r = 1$ for a vacuum, 3 for a fluid).

A brief illustration

The energy of interaction between a partial charge of −0.36 (that is, $Q_1 = -0.36e$) on the N atom of a peptide link and the partial charge of +0.45 ($Q_2 = +0.45e$) on the carbonyl C atom at a distance of 3.0 nm on the assumption that the medium between them is a vacuum is

$$V = \frac{(-0.36e) \times (0.45e)}{4\pi\varepsilon_0 \times (3.0\ \text{nm})} = \frac{-0.36 \times 0.45 \times (1.602 \times 10^{-19}\ \text{C})^2}{4\pi \times (8.854 \times 10^{-12}\ \text{J}^{-1}\ \text{C}^2\ \text{m}^{-1}) \times (3.0 \times 10^{-9}\ \text{m})}$$

$$= -1.2 \times 10^{-20}\ \text{J}$$

This energy (after multiplication by Avogadro's constant) corresponds to $-7.5\,kJ\,mol^{-1}$. However, if the medium has a 'typical' relative permittivity of 3.5, then the interaction energy is reduced to $-2.1\,kJ\,mol^{-1}$. For bulk water as the medium, with the H_2O molecules able to rotate in response to a field, the energy of interaction would be reduced by a factor of 78, to only $-0.96\,kJ\,mol^{-1}$.

11.6 Electric dipole moments

Many physical and chemical properties are related to the distribution of partial charges in a molecule or group (such as the peptide group), and here we start to identify them.

At its simplest, an **electric dipole** consists of two charges Q and $-Q$ separated by a distance l. The product Ql is called the **electric dipole moment**, μ. We represent dipole moments by an arrow with a length proportional to μ and pointing from the negative charge to the positive charge (**1**).[2] Because a dipole moment is the product of a charge (in coulombs, C) and a length (in meters, m), the SI unit of dipole moment is the coulomb meter (C m). However, it is often much more convenient to report a dipole moment in the non-SI unit **debye**, D, where $1\,D = 3.335 \times 10^{-30}$ C m, because experimental values for molecules are then close to 1 D (Table 11.3).[3] The dipole moment of two charges e and $-e$ separated by 100 pm is 1.6×10^{-29} C m, corresponding to 4.8 D. Dipole moments of small molecules are typically smaller than that, at about 1 D.

A **polar molecule** is a molecule with a permanent electric dipole moment arising from the partial charges on its atoms (Section 10.5). A **nonpolar molecule** is a molecule that has no permanent electric dipole moment. All heteronuclear diatomic molecules are polar because the difference in electronegativities of their two atoms results in nonzero partial charges. Typical dipole moments are 1.08 D for HCl and 0.42 D for HI (Table 11.3). A very approximate relation between the dipole moment and the difference in Pauling electronegativities (Table 10.2) of the two atoms, $\Delta\chi$, is

$$\mu/D \approx \Delta\chi$$

Relation between dipole moment and electronegativity   (11.14)

Q μ $-Q$

1

Table 11.3 Dipole moments and mean polarizability volumes

	μ/D	$\alpha'/(10^{-30}\,m^3)$
Ar	0	1.66
CCl_4	0	10.3
C_6H_6	0	10.4
H_2	0	0.819
H_2O	1.85	1.48
NH_3	1.47	2.22
HCl	1.08	2.63
HBr	0.80	3.61
HI	0.42	5.45

A brief illustration

The electronegativities of hydrogen and bromine are 2.1 and 2.8, respectively. The difference is 0.7, so we predict an electric dipole moment of about 0.7 D for HBr. The experimental value is 0.80 D.

Because it attracts the electrons more strongly, the more electronegative atom is usually the negative end of the dipole. However, there are exceptions, particularly when antibonding orbitals are occupied. Thus, the dipole moment of NO is very small (0.07 D), but the negative end of the dipole is on the N atom even

---

[2] Be careful with this convention: for historical reasons the opposite convention is still widely adopted.

[3] The unit is named after Peter Debye, the Dutch pioneer of the study of dipole moments of molecules.

Mathematical toolbox 11.1   *Addition and subtraction of vectors*

Consider two vectors v_1 and v_2 making an angle θ shown below

The first step in the addition of v_2 to v_1 consists of joining the tail of v_2 to the head of v_1:

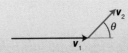

In the second step, we draw a vector v_{res}, the **resultant vector**, originating from the tail of v_1 to the head of v_2:

The subtraction of vectors follows the same principles outlined above for addition by noting that subtraction of v_2 from v_1 amounts to addition of $-v_2$ to v_1:

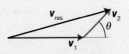

though the O atom is more electronegative. This apparent paradox is resolved as soon as we realize that antibonding orbitals are occupied in NO (see Fig. 10.35) and because electrons in antibonding orbitals tend to be found closer to the less electronegative atom, they contribute a negative partial charge to that atom. If this contribution is larger than the opposite contribution from the electrons in bonding orbitals, then the net effect will be a small negative partial charge on the *less* electronegative atom.

Molecular symmetry is of the greatest importance in deciding whether a polyatomic molecule is polar or not. Indeed, molecular symmetry is more important than the question of whether or not the atoms in the molecule belong to the same element. Homonuclear polyatomic molecules may be polar if they have low symmetry and the atoms are in inequivalent positions. For instance, the angular molecule ozone, O_3 (**2**), is homonuclear; however, it is polar because the central O atom is different from the outer two (it is bonded to two atoms, they are bonded only to one). Moreover, the dipole moments associated with each bond make an angle to each other and do not cancel (see *Mathematical toolkit* 11.1). Heteronuclear polyatomic molecules may be nonpolar if they have high symmetry, because individual bond dipoles may then cancel. The heteronuclear linear triatomic molecule CO_2, for example, is nonpolar because, although there are partial charges on all three atoms, the dipole moment associated with the OC bond points in the opposite direction to the dipole moment associated with the CO bond, and the two cancel (**3**).

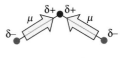

2 Ozone, O_3

3 Carbon dioxide, CO_2

Self-test 11.9 Ozone, carbon dioxide, water, and methane are all components of the Earth's atmosphere that absorb heat emanating from the surface of the planet, thus maintaining temperatures consistent with the proliferation of life. Predict whether methane and water molecules are polar or nonpolar.

Answer: An H_2O molecule is angular and polar; a CH_4 molecule is tetrahedral and nonpolar

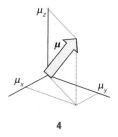

4

A useful approach to the calculation of dipole moments is to take into account the locations and magnitudes of the partial charges on all the atoms. These partial charges are included in the output of many molecular structure software packages. Indeed, the programs calculate the dipole moments of the molecules by noting that an electric dipole moment is actually a vector, $\boldsymbol{\mu}$, with three components, μ_x, μ_y, and μ_z (**4**). The direction of $\boldsymbol{\mu}$ shows the orientation of the dipole in the molecule, and the length of the vector is the magnitude, μ, of the dipole moment. In common with all vectors (*Mathematical toolkit* 9.1), the magnitude is related to the three components by

$$\mu = (\mu_x^2 + \mu_y^2 + \mu_z^2)^{1/2}$$

> Magnitude of the dipole moment vector   (11.15a)

To calculate μ, we need to calculate the three components and then substitute them into this expression. To calculate the x-component, for instance, we need to know the magnitude of the partial charge on each atom and the atom's x-coordinate relative to a point in the molecule and form the sum

$$\mu_x = \sum_J Q_J x_J$$

> Calculation of a component of the dipole moment vector   (11.15b)

Here Q_J is the partial charge of atom J, x_J is the x-coordinate of atom J, and the sum is over all the atoms in the molecule. Similar expressions are used for the y- and z-components. For an electrically neutral molecule, the origin of the coordinates is arbitrary, so it is best chosen to simplify the measurements.

Example 11.6 *Calculating the dipole moment of the peptide group*

Estimate the electric dipole moment of the peptide group (**5**) by using the partial charges and the locations of the atoms shown in pm.

Strategy We use eqn 11.15b to calculate each of the components of the dipole moment. Then we use eqn 11.15a to assemble the three components into the magnitude of the dipole moment. Note that the partial charges are multiples of the fundamental charge $e = 1.602 \times 10^{-19}$ C.

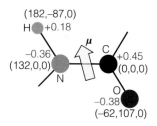

5

Solution The expression for μ_x is

$$\mu_x = (-0.36e) \times (132 \text{ pm}) + (0.45e) \times (0 \text{ pm})$$
$$+ (0.18e) \times (182 \text{ pm}) + (-0.38e) \times (-62.0 \text{ pm})$$
$$= 8.8e \text{ pm} = 8.8 \times (1.602 \times 10^{-19} \text{ C}) \times (10^{-2} \text{ m}) = 1.4 \times 10^{-30} \text{ C m}$$

corresponding to $\mu_x = +0.42$ D. The expression for μ_y is

$$\mu_y = (-0.36e) \times (0 \text{ pm}) + (0.45e) \times (0 \text{ pm}) + (0.18e) \times (-87 \text{ pm})$$
$$+ (-0.38e) \times (107 \text{ pm})$$
$$= -56e \text{ pm} = -9.0 \times 10^{-30} \text{ C m}$$

It follows that $\mu_y = -2.7$ D. Therefore, because $\mu_z = 0$,

$$\mu = \{(0.42 \text{ D})^2 + (-2.7 \text{ D})^2\}^{1/2} = 2.7 \text{ D}$$

We can find the orientation of the dipole moment by arranging an arrow of length 2.7 units of length to have x-, y-, and z-components of 0.42, −2.7, and 0 units; the orientation is superimposed on (**5**).

Self-test 11.10 Calculate the electric dipole moment of formaldehyde, using the information in (**6**).

Answer: 3.2 D

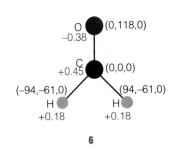

6

11.7 **Interactions between dipoles**

When molecules or groups are widely separated, it is simpler to express their interaction in terms of the dipole moments rather than with each partial charge. We need to know how to handle these interactions because they are important for the assembly of biological macromolecules.

The potential energy of a dipole μ_1 in the presence of a charge Q_2 is calculated by taking into account the interaction of the charge with the two partial charges of the dipole, one resulting in a repulsion and the other an attraction. The result for the arrangement shown in (**7**) is

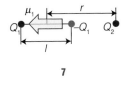

7

$$V = -\frac{Q_2\mu_1}{4\pi\varepsilon_0 r^2}$$

Charge–dipole interaction energy (as in **7**)   (11.16a)

Justification 11.4   *The interaction of a charge with a dipole*

When the charge and dipole are collinear, as in (**7**), the potential energy is

$$V = \frac{Q_1 Q_2}{4\pi\varepsilon_0(r+\frac{1}{2}l)} - \frac{Q_1 Q_2}{4\pi\varepsilon_0(r-\frac{1}{2}l)}$$

$$= \frac{Q_1 Q_2}{4\pi\varepsilon_0 r\left(1+\dfrac{l}{2r}\right)} - \frac{Q_1 Q_2}{4\pi\varepsilon_0 r\left(1-\dfrac{l}{2r}\right)}$$

$$= \frac{Q_1 Q_2}{4\pi\varepsilon_0 r}\left(\frac{1}{1+\dfrac{l}{2r}} - \frac{1}{1-\dfrac{l}{2r}}\right)$$

Next, we suppose that the separation of charges in the dipole is much smaller than the distance of the charge Q_2 in the sense that $l/2r \ll 1$. Then we can use (see *Mathematical toolkit 3.2*)

$$\frac{1}{1+x} \approx 1-x \qquad \frac{1}{1-x} \approx 1+x$$

to write

$$V \approx \frac{Q_1 Q_2}{4\pi\varepsilon_0 r}\left\{\left(1-\frac{l}{2r}\right)-\left(1+\frac{l}{2r}\right)\right\} = -\frac{Q_1 Q_2 l}{4\pi\varepsilon_0 r^2}$$

Now we recognize that $Q_1 l = \mu_1$, the dipole moment of molecule 1, and obtain eqn 11.16a.

A similar calculation for the more general orientation shown in (**8**) gives

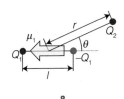

8

$$V = \frac{\mu_1 Q_2 \cos\theta}{4\pi\varepsilon_0 r^2}$$

Charge–dipole interaction energy (as in **8**)   (11.16b)

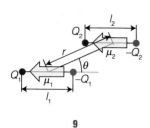

9

If Q_2 is positive, the energy is lowest when $\theta = 0$ (and $\cos\theta = 1$) because then the partial negative charge of the dipole lies closer than the partial positive charge to the point charge and the attraction outweighs the repulsion. This interaction energy decreases more rapidly with distance than that between two point charges (as $1/r^2$ rather than $1/r$) because, from the viewpoint of the single charge, the partial charges of the point dipole seem to merge and cancel as the distance r increases.

We can calculate the interaction energy between two dipoles μ_1 and μ_2 in the orientation shown in (9) in a similar way, by taking into account all four charges of the two dipoles. The outcome is[4]

$$V = \frac{\mu_1\mu_2(1 - 3\cos^2\theta)}{4\pi\varepsilon_0 r^3}$$

Dipole–dipole interaction energy (as in **9**)   (11.17)

This potential energy decreases even more rapidly than in eqn 11.16 (as $1/r^3$) because the charges of *both* dipoles seem to merge as the separation of the dipoles increases. The angular factor takes into account how the like or opposite charges come closer to one another as the relative orientation of the dipoles is changed. The energy is lowest when $\theta = 0$ or $180°$ (when $1 - 3\cos^2\theta = -2$) because opposite partial charges then lie closer together than like partial charges.

A brief illustration

We can use eqn 11.17 to calculate the molar potential energy of the dipolar interaction between two peptide groups. Supposing that the groups are separated by 3.0 nm in different regions of a polypeptide chain with $\theta = 180°$, we take $\mu_1 = \mu_2 = 2.7$ D, corresponding to 9.1×10^{-30} C m, and find

$$V = \frac{(9.0 \times 10^{-30}\ \text{C m})^2 \times (-2)}{4\pi \times (8.854 \times 10^{-12}\ \text{J}^{-1}\ \text{C}^2\ \text{m}^{-1}) \times (3.0 \times 10^{-9}\ \text{m})^3}$$

$$= \frac{(9.0 \times 10^{-30})^2 \times (-2)}{4\pi \times (8.854 \times 10^{-12}) \times (3.0 \times 10^{-9})^3}\ \frac{\text{C}^2\ \text{m}^2}{\text{J}^{-1}\ \text{C}^2\ \text{m}^{-1}\ \text{m}^3}$$

$$= -5.4 \times 10^{-23}\ \text{J}$$

where we have used 1 V C = 1 J. This value corresponds to -32 J mol^{-1}. If the medium lying between the two dipoles has a relative permittivity of 3.5, then the interaction energy will be reduced by this factor, to -9.3 J mol^{-1}. Note, however, that this energy is considerably less than that between two partial charges at the same separation (see the *brief illustration* at the end of Section 11.5).

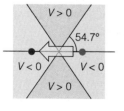

10

Equation 11.17 shows that the potential energy is negative (attractive) in some orientations when $\theta < 54.7°$ (the angle at which $1 - 3\cos^2\theta = 0$, $\cos\theta = (\frac{1}{3})^{1/2}$) because opposite charges are closer than like charges. It is positive (repulsive) when $\theta > 54.7°$ because then like charges are closer than unlike charges. The potential energy is zero on the lines at $54.7°$ and $180° - 54.7° = 125.3°$ because at those angles the two attractions and the two repulsions cancel (**10**).

The average potential energy of interaction between polar molecules that are freely rotating in a fluid (a gas or liquid) is zero because the attractions and

---

[4] For a derivation of eqn 11.17, see our *Physical chemistry* (2010).

repulsions cancel. However, because the potential energy of a dipole near another dipole depends on their relative orientations, the molecules exert forces on each other and therefore do not in fact rotate completely freely, even in a gas. As a result, the lower energy orientations are marginally favored, so there is a nonzero interaction between rotating polar molecules (Fig. 11.25). The detailed calculation of the average interaction energy is quite complicated, but the final answer is very simple:

$$V = -\frac{2\mu_1^2\mu_2^2}{3(4\pi\varepsilon_0)^2 kTr^6}$$

Average dipole–dipole interaction energy (freely rotating dipoles)   (11.18)

The important features of this expression are the dependence of the average interaction energy on the inverse sixth power of the separation (which identifies it as a van der Waals interaction) and its inverse dependence on the temperature. The temperature dependence reflects the way that the greater thermal motion overcomes the mutual orientating effects of the dipoles at higher temperatures. Equation 11.18 is applicable when both molecules are free to rotate or when one is fixed and only the other is free to rotate, as for a small polar molecule near a macromolecule.

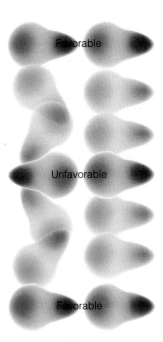

Fig. 11.25 A dipole–dipole interaction. When a pair of molecules can adopt all relative orientations with equal probability, the favorable orientations and the unfavorable ones cancel, and the average interaction is zero. In an actual fluid, the favorable interactions slightly predominate.

A brief illustration

Suppose a water molecule ($\mu = 1.85$ D) can rotate freely at 1.0 nm from a peptide group ($\mu = 2.7$ D): the energy of their interaction at 25°C (298 K) is

$$V = -\frac{2 \times (1.85 \times 3.336 \times 10^{-30}\,\text{C m})^2 \times (2.7 \times 3.336 \times 10^{-30}\,\text{C m})^2}{3(4\pi \times (8.854 \times 10^{-12}\,\text{J}^{-1}\,\text{C}^2\,\text{m}^{-1})^2 \times 1.381 \times 10^{-23}\,\text{J K}^{-1} \times 298\,\text{K} \times (1.0 \times 10^{-9}\,\text{m})^6}$$

$$= -4.0 \times 10^{-23}\,\frac{\text{C}^4\,\text{m}^4}{\text{J}^{-2}\,\text{C}^4\,\text{m}^{-2}\,\text{J K}^{-1}\,\text{K m}^6}$$

$$= -4.0 \times 10^{-23}\,\text{J}$$

This interaction energy corresponds (after multiplication by Avogadro's constant) to -24 J mol^{-1}. When the temperature is raised to body temperature, 37°C (310 K), the H_2O molecule rotates more vigorously and the average interaction is reduced to -23 J mol^{-1}.

A note on good practice
Note how the units are included in the calculation and cancel to give the result in joules. It is far better to include the units at each stage of the calculation and treat them as algebraic quantities that can be multiplied and canceled than to guess the units at the end of the calculation.

11.8 Induced dipole moments

The structures and properties of biological assemblies also emerge from interactions that involve nonpolar species, such as nonpolar groups on the peptide residues of a protein.

A nonpolar molecule may acquire a temporary **induced dipole moment**, μ^*, as a result of the influence of an electric field generated by a nearby ion or polar molecule. The field distorts the electron distribution of the molecule and gives rise to an electric dipole in it. The molecule is said to be **polarizable**. The magnitude of the induced dipole moment is proportional to the strength of the electric field, $\mathcal{E}$, and we write

$$\mu^* = \alpha\mathcal{E}$$

Polarizability   (11.19)

The proportionality constant α is the **polarizability** of the molecule. The larger the polarizability of the molecule, the greater is the distortion caused by a given strength of electric field. If the molecule has few electrons, they are tightly controlled by the nuclear charges and the polarizability of the molecule is low. If the molecule contains large atoms with electrons some distance from the nucleus, the nuclear control is less and the polarizability of the molecule is greater. The polarizability also depends on the orientation of the molecule with respect to the field unless the molecule is tetrahedral (such as CCl_4), octahedral (such as SF_6), or icosahedral (such as C_{60}). Atoms, tetrahedral, octahedral, and icosahedral molecules have isotropic (orientation-independent) polarizabilities; all other molecules have anisotropic (orientation-dependent) polarizabilities.

The polarizabilities reported in Table 11.3 are given as **polarizability volumes, α'**:

$$\alpha' = \frac{\alpha}{4\pi\varepsilon_0}$$

$\boxed{\text{Polarizability volume}}$ (11.20)

The polarizability volume has the dimensions of volume (hence its name) and is comparable in magnitude to the volume of the molecule.

Self-test 11.11 What strength of electric field is required to induce an electric dipole moment of 1.0 µD in a molecule of polarizability volume 2.6×10^{-30} m^3 (like CO_2)?

Answer: 11 kV m^{-1}

(a) Dipole–induced-dipole interactions

A polar molecule with dipole moment μ_1 can induce a dipole moment in a polarizable molecule (which may itself be either polar or nonpolar) because the partial charges of the polar molecule give rise to an electric field that distorts the second molecule. That induced dipole interacts with the permanent dipole of the first molecule, and the two are attracted together (Fig. 11.26). The formula for the **dipole–induced-dipole interaction energy** is

$$V = -\frac{\mu_1^2 \alpha_2'}{4\pi\varepsilon_0 r^6}$$

$\boxed{\begin{array}{l}\text{Dipole–induced-dipole}\\ \text{interaction energy}\end{array}}$ (11.21)

where α_2 is the polarizability of molecule 2. The negative sign shows that the interaction is attractive. For a molecule with $\mu = 1$ D (such as HCl) near a molecule of polarizability volume $\alpha' = 1.0 \times 10^{-29}$ m^3 (such as benzene, Table 11.3), the average interaction energy is about -0.8 kJ mol^{-1} when the separation is 0.3 nm.

(b) Dispersion interactions

Despite the absence of partial charges, we know that uncharged, nonpolar species can interact because they form condensed phases, such as benzene, liquid hydrogen, and liquid xenon. The **dispersion interaction**, or **London interaction**, between nonpolar species arises from the transient dipoles that they possess as a result of fluctuations in the instantaneous positions of their electrons (Fig. 11.27). Suppose, for instance, that the electrons in one molecule flicker into an arrangement that results in partial positive and negative charges and thus gives it an instantaneous dipole moment μ_1. While it exists, this dipole can polarize the other molecule and induce in it an instantaneous dipole moment μ_2. The two dipoles attract each other and the potential energy of the pair is lowered. Although the first molecule will go on to change the size and direction of its dipole (perhaps

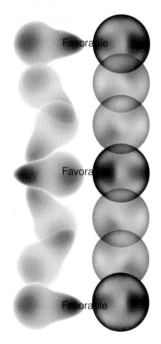

Fig. 11.26 A dipole–induced-dipole interaction. The induced dipole follows the changing orientation of the permanent dipole.

within 10^{-16} s), the second will follow it; that is, the two dipoles are *correlated* in direction like two meshing gears, with a positive partial charge on one molecule appearing close to a negative partial charge on the other molecule and vice versa. Because of this correlation of the relative positions of the partial charges, and their resulting attractive interaction, the attraction between the two instantaneous dipoles does not average to zero. Instead, it gives rise to a net attractive interaction. Polar molecules interact by a dispersion interaction as well as by dipole–dipole interactions.

The strength of the dispersion interaction depends on the polarizability of the first molecule because the magnitude of the instantaneous dipole moment μ_1 depends on the looseness of the control that the nuclear charge has over the outer electrons. If that control is loose, the electron distribution can undergo relatively large fluctuations. Moreover, if the control is loose, then the electron distribution can also respond strongly to applied electric fields and hence have a high polarizability. It follows that a high polarizability is a sign of large fluctuations in local charge density. The strength also depends on the polarizability of the second molecule, for that polarizability determines how readily a dipole can be induced in molecule 2 by molecule 1. We therefore expect that $V \propto \alpha_1 \alpha_2$. The actual calculation of the dispersion interaction is quite involved, but a reasonable approximation to the interaction energy is the **London formula**:

$$V = -\frac{3}{2} \times \frac{\alpha_1' \alpha_2'}{r^6} \times \frac{I_1 I_2}{I_1 + I_2} \qquad \boxed{\text{London formula}} \quad (11.22)$$

where I_1 and I_2 are the ionization energies of the two molecules.

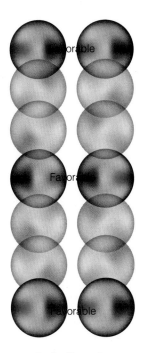

Fig. 11.27 In the dispersion interaction, an instantaneous dipole on one molecule induces a dipole on another molecule, and the two dipoles then interact to lower the energy. The directions of the two instantaneous dipoles are correlated and, although they occur in different orientations at different instants, the interaction does not average to zero.

$\boxed{\text{A brief illustration}}$

If two phenylalanine residues are separated by 3.0 nm in a polypeptide, the dispersion interaction between their phenyl groups is calculated from eqn 11.22 by setting $\alpha_1' = \alpha_2'$ and $I_1 = I_2 = I$:

$$V = -\frac{3}{4} \times \frac{\alpha_1'^2}{r^6} \times I$$

We treat the phenyl groups as benzene rings of polarizability volume 1.0×10^{-29} m^3:

$$V = -\frac{3}{4} \times \frac{(1.0 \times 10^{-29} \ m^3)^2}{(3.0 \times 10^{-9} \ m)^6} \times I = -1.0 \times 10^{-7} \times I$$

If we suppose that the ionization energy of the phenyl group is about 5 eV (about 500 kJ mol^{-1}), this energy is approximately -23 mJ mol^{-1}.

11.9 Hydrogen bonding

Strong interactions of the type X–H⋯Y (with X, Y = N or O) are responsible for the formation of well-defined three-dimensional structures in proteins and nucleic acids. We need to understand the origin of the strength of these very important interactions.

The strongest intermolecular interaction arises from the formation of a **hydrogen bond**, in which a hydrogen atom lies between two strongly electronegative atoms and binds them together. The bond is normally denoted X–H⋯Y, with X and Y being N, O, or F. Unlike the other interactions we have considered, hydrogen bonding is not universal but is restricted to molecules that contain these atoms.

The most elementary description of the formation of a hydrogen bond is that it is the result of a Coulombic interaction between the partly exposed positive charge of a proton bound to an electron-withdrawing X atom (in the fragment X–H) and the negative charge of a lone pair on the second atom Y, as in $^{\delta-}$X–H$^{\delta+}$:Y$^{\delta-}$. A slightly more sophisticated version of the electrostatic description is to regard hydrogen bond formation as the formation of a Lewis acid–base complex in which the partly exposed proton of the X–H group is the Lewis acid and :Y, with its lone pair, is the Lewis base, as in X–H + :Y → X–H:Y.

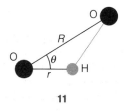

11

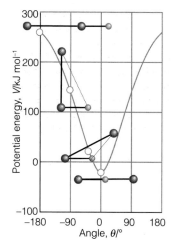

Fig. 11.28 The variation of the energy of interaction (on the electrostatic model) of a hydrogen bond as the angle between the O–H and :O groups is changed.

A brief illustration

A common hydrogen bond is that formed between O–H groups and O atoms, as in liquid water and ice. In *Exercise* 11.42, you are invited to use the electrostatic model to calculate the dependence of the potential energy of interaction on the OOH angle, denoted θ in (**11**), and the results are plotted in Fig. 11.28. We see that at $\theta = 0$ when the OHO atoms lie in a straight line; the molar potential energy is −19 kJ mol^{-1}. Note how sharply the energy depends on angle: it is negative only with ±12° of linearity.

Molecular orbital theory provides an alternative description that is more in line with the concept of delocalized bonding and the ability of an electron pair to bind more than one pair of atoms (Section 10.6). Thus, if the X–H bond is regarded as formed from the overlap of an orbital on X, ψ_X, and a hydrogen 1s orbital, ψ_H, and the lone pair on Y occupies an orbital on Y, ψ_Y, then when the two molecules are close together, we can build three molecular orbitals from the three basis orbitals:

$$\psi = c_1\psi_X + c_2\psi_H + c_3\psi_Y$$

One of the molecular orbitals is bonding, one almost nonbonding, and the third antibonding (Fig. 11.29). These three orbitals need to accommodate four electrons (two from the original X–H bond and two from the lone pair of Y), so two enter the bonding orbital and two enter the nonbonding orbital. Because the antibonding orbital remains empty, the net effect—depending on the precise location of the almost nonbonding orbital—may be a lowering of energy.

Experimental evidence and theoretical arguments have been presented in favor of both the electrostatic and molecular orbital view of hydrogen bonding. For example, recent experiments suggest that the hydrogen bonds in ice have significant covalent character, providing support for the molecular orbital treatment. However, the matter has not yet been resolved.

Hydrogen bond formation dominates all other interactions between electrically neutral molecules when it can occur (Table 11.4). It has a typical strength of the order of 20 kJ mol^{-1}, as can be inferred from the enthalpy of vaporization of water, 40.7 kJ mol^{-1}, for vaporization involves the breaking of two hydrogen bonds to each water molecule. Hydrogen bonding accounts for the rigidity of molecular solids such as sucrose and ice; the low vapor pressure, high viscosity, and surface tension of liquids such as water; the secondary structure of proteins (the formation of helices and sheets of polypeptide chains); the structure of DNA and hence

Table 11.4 Interaction potential energies

Interaction type	Distance dependence of potential energy	Typical energy (kJ mol^{-1})	Comment
Ion–ion	$1/r$	250	Only between ions
Hydrogen bond		20	Occurs in X–H···Y, where X, Y = N, O, or F
Ion–dipole	$1/r^2$	15	
Dipole–dipole	$1/r^3$	2	Between stationary polar molecules
	$1/r^6$	0.6	Between rotating polar molecules
London (dispersion)	$1/r^6$	2	Between all types of molecules and ions

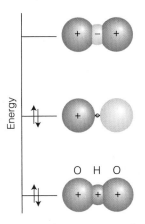

Fig. 11.29 A schematic portrayal of the molecular orbitals that can be formed from an X, H, and Y orbital and that give rise to an X–H···Y hydrogen bond. The lowest-energy combination is fully bonding, the next nonbonding, and the uppermost is antibonding. The antibonding orbital is not occupied by the electrons provided by the X–H bond and the :Y lone pair, so the configuration shown may result in a net lowering of energy in certain cases (namely, when the X and Y atoms are N, O, or F).

the transmission of genetic information; and the attachment of drugs to receptors sites in proteins (*Case study* 11.2). Hydrogen bonding also contributes to the solubility in water of species such as ammonia and compounds containing hydroxyl groups and to the hydration of anions. In this last case, even ions such as Cl^- and HS^- can participate in hydrogen bond formation with water, for their charge enables them to interact with the hydroxylic protons of H_2O.

11.10 The total interaction

To treat the myriad interactions in biological assemblies quantitatively, we need simple formulas that express the strengths of the attractions and repulsions.

Table 11.4 summarizes the strengths and distance dependence of the attractive interactions that we have considered so far. The total attractive interaction energy between rotating molecules that cannot participate in hydrogen bonding is the sum of the contributions from the dipole–dipole, dipole–induced-dipole, and dispersion interactions. Only the dispersion interaction contributes if both molecules are nonpolar. All three interactions vary as the inverse sixth power of the separation, so we may write

$$V = -\frac{C}{r^6} \tag{11.23}$$

where C is a coefficient that depends on the identity of the molecules and the type of interaction between them. As we have remarked, the energy of a hydrogen bond X–H···Y is typically 20 kJ mol^{-1} and occurs on contact for X, Y = N, O, or F.

Repulsive terms become important and begin to dominate the attractive forces when molecules are squeezed together (Fig. 11.30), for instance, during the impact of a collision, under the force exerted by a weight pressing on a substance, or simply as a result of the attractive forces drawing the molecules together. These repulsive interactions arise in large measure from the Pauli exclusion principle, which forbids pairs of electrons being in the same region of space. The repulsions increase steeply with decreasing separation in a way that can be deduced only by very extensive, complicated molecular structure calculations. In many cases, however, progress can be made by using a greatly simplified representation of

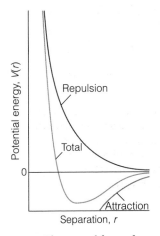

Fig. 11.30 The general form of an intermolecular potential energy curve (the graph of the potential energy of two closed shell species as the distance between them is changed). The attractive (negative) contribution has a long range, but the repulsive (positive) interaction increases more sharply once the molecules come into contact.

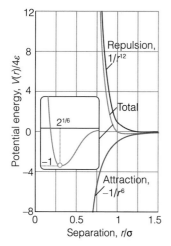

Fig. 11.31 The Lennard-Jones potential is another approximation to the true intermolecular potential energy curves. It models the attractive component by a contribution that is proportional to $1/r^6$ and the repulsive component by a contribution that is proportional to $1/r^{12}$. Specifically, these choices result in the Lennard-Jones (12,6) potential. Although there are good theoretical reasons for the former, there is plenty of evidence to show that $1/r^{12}$ is only a very poor approximation to the repulsive part of the curve.

the potential energy, where the details are ignored and the general features are expressed by a few adjustable parameters.

One such approximation is to express the short-range repulsive potential energy as inversely proportional to a high power of r:

$$V = +\frac{C^*}{r^n} \qquad (11.24)$$

where C^* is another constant (the asterisk signifies repulsion). Typically, n is set equal to 12, in which case the repulsion dominates the $1/r^6$ attractions strongly at short separations because then $C^*/r^{12} \gg C/r^6$. The sum of the repulsive interaction with $n = 12$ and the attractive interaction given by eqn 11.23 is called the **Lennard-Jones (12,6) potential**. It is normally written in the form

$$V = 4\varepsilon \left\{ \left(\frac{\sigma}{r}\right)^{12} - \left(\frac{\sigma}{r}\right)^{6} \right\} \qquad \text{Lennard-Jones (12,6) potential} \qquad (11.25)$$

and is drawn in Fig. 11.31. The two parameters are ε (epsilon), the depth of the well, and σ, the separation at which $V = 0$; some typical values are listed in Table 11.5. The well minimum occurs at $r = 2^{1/6}\sigma$. Although the (12,6) potential has been used in many calculations, there is plenty of evidence to show that $1/r^{12}$ is a very poor representation of the repulsive potential and that the exponential form $e^{-r/\sigma}$ is superior. An exponential function is more faithful to the exponential decay of atomic wavefunctions at large distances and hence to the distance dependence of the overlap that is responsible for repulsion. However, a disadvantage of the exponential form is that it is slower to compute, which is important when considering the interactions between the large numbers of atoms in liquids and macromolecules. A further computational advantage of the (12,6) potential is that once r^6 has been calculated, r^{12} is obtained simply by taking the square.

Self-test 11.12 At what separation does the minimum of the potential energy curve occur for a Lennard-Jones potential? *Hint:* Solve for r after setting the first derivative of the potential energy function to zero.

Answer: $r = 2^{1/6}\sigma$

Table 11.5 Lennard-Jones parameters for the (12,6) potential

	$\varepsilon/(\text{kJ mol}^{-1})$	σ/pm
Ar	128	342
Br₂	536	427
C₆H₆	454	527
Cl₂	368	412
H₂	34	297
He	11	258
Xe	236	406

With the advent of atomic force microscopy (AFM), in which the force between a molecular sized probe and a surface is monitored (see *In the laboratory* 9.2), it has become possible to measure directly the forces acting between molecules. The force, F, is the negative slope of the potential energy ($F = -dV/dr$), so for a Lennard-Jones potential between individual molecules we write

$$F = -\frac{dV}{dr} = \frac{24\varepsilon}{\sigma} \left\{ 2\left(\frac{\sigma}{r}\right)^{13} - \left(\frac{\sigma}{r}\right)^{7} \right\} \qquad (11.26)$$

The net attractive force is greatest (from $dF/dr = 0$) at $r = \left(\frac{26}{7}\right)^{1/6}\sigma$, or 1.244σ, and at that distance is equal to $-144\left(\frac{7}{26}\right)^{7/6}\varepsilon/13\sigma$, or $-2.397\varepsilon/\sigma$. For typical parameters, the magnitude of this force is about 10 pN.

Case study 11.2   *Molecular recognition in biology and pharmacology*

Molecular interactions are responsible for the assembly of many biological structures. Hydrogen bonding and hydrophobic interactions are primarily responsible for the three-dimensional structures of biopolymers, such as proteins, nucleic acids, and cell membranes. The binding of a ligand, or *guest*, to a biopolymer, or *host*, is also governed by molecular interactions. Examples of biological *host–guest complexes* include enzyme–substrate complexes, antigen–antibody complexes, and drug–receptor complexes. In all these cases, a site on the guest contains functional groups that can interact with complementary functional groups of the host. For example, a hydrogen bond donor group of the guest must be positioned near a hydrogen bond acceptor group of the host for tight binding to occur. It is generally true that many specific intermolecular contacts must be made in a biological host–guest complex and, as a result, a guest binds only hosts that are chemically similar. The strict rules governing molecular recognition of a guest by a host control every biological process, from metabolism to immunological response, and provide important clues for the design of effective drugs for the treatment of disease.

Interactions between nonpolar groups can be important in the binding of a guest to a host. For example, many enzyme active sites have hydrophobic pockets that bind nonpolar groups of a substrate. Coulombic interactions can be important in the interior of a biopolymer host, where the relative permittivity can be much lower than that of the aqueous exterior. For example, at physiological pH, amino acid side chains containing carboxylic acid or amine groups are negatively and positively charged, respectively, and can attract each other. Dipole–dipole interactions are also possible because many of the building blocks of biopolymers are polar, including the peptide link, –CONH– (see *Example* 11.6). However, hydrogen bonding interactions are by far the most prevalent in biological host–guest complexes. Many effective drugs bind tightly and inhibit the action of enzymes that are associated with the progress of a disease. In many cases, a successful inhibitor will be able to form the same hydrogen bonds with the binding site that the normal substrate of the enzyme can form, except that the drug is chemically inert toward the enzyme. This strategy has been used in the design of drugs for the treatment of HIV-AIDS. Here we describe the properties of a drug that fights HIV infection, highlighting the importance of molecular interactions.

For mature HIV particles to form in cells of the host organism, several large proteins encoded by the viral genetic material must be cleaved by a protease enzyme. The drug Crixivan (**12**) is a competitive inhibitor of HIV protease and has several molecular features that optimize binding to the enzyme's active site. First, the hydroxyl group highlighted in (**12**) displaces an H_2O molecule that acts as the nucleophile in the hydrolysis of the substrate. Second, the carbon atom to which the key –OH group is bound has a tetrahedral geometry that mimics the structure of the transition state of the peptide hydrolysis reaction. However, the tetrahedral moiety in the drug is not cleaved by the enzyme. Third, the inhibitor is anchored firmly to the active site by a network of hydrogen bonds involving the carbonyl groups of the drug, a water molecule, and peptide NH groups from the enzyme, as shown in (**12**).

12

Levels of structure

The concept of the 'structure' of a macromolecule takes on different meanings at the different levels at which we think about the arrangement of the chain or network of monomers. The term **configuration** refers to the structural features that can be changed only by breaking chemical bonds and forming new ones. Thus, the chains –A–B–C– and –A–C–B– have different configurations. The term **conformation** refers to the spatial arrangement of the different parts of a chain, and one conformation can be changed into another by rotating one part of a chain around a bond.

In the following sections we explore the molecular interactions responsible for the different levels of structure of biological macromolecules (primary, secondary, etc., as explained in *Fundamentals* F.1) and assemblies (such as biological membranes; see *Fundamentals* F.1). We draw from the concepts developed in Sections 11.5–11.10 and describe computational techniques that can help with the prediction of the three-dimensional structure of polypeptides and polynucleotides.

11.11 Minimal order: gases and liquids

Many biochemical processes take place in the aqueous intracellular space, so we need to understand the structure of liquids in general and of water in particular.

The form of matter with the least order is a gas. In a perfect gas there are no inter-molecular interactions and the distribution of molecules is completely random. In a real gas there are weak attractions and repulsions that have minimal effect on the relative locations of the molecules but that cause deviations from the perfect gas law for the dependence of pressure on the volume, temperature, and amount. Normally there is no need to consider such deviations in biological applications.

The attractions between molecules are responsible for the condensation of gases into liquids at low temperatures. First, at low enough temperatures the molecules of a gas have insufficient kinetic energy to escape from each other's attraction and they stick together. Second, although molecules attract each other when they are a few diameters apart, as soon as they come into contact, they repel

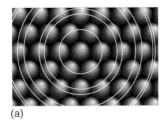
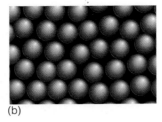

(a)   (b)

Fig. 11.32 (a) In a perfect crystal at $T = 0$, the distribution of molecules (or ions) is highly regular, and the radial distribution function has a series of sharp peaks that show the regular organization of rings of neighbors around any selected central molecule or ion. (b) In a liquid, there remain some elements of structure close to each molecule, but the greater the distance, the less the correlation. The radial distribution function now shows a pronounced (but broadened) peak corresponding to the nearest neighbors of the molecule of interest (which are only slightly more disordered than in the solid) and a suggestion of a peak for the next ring of molecules, but little structure at greater distances.

each other. This repulsion is responsible for the fact that liquids and solids have a definite bulk and do not collapse to an infinitesimal point. The molecules are held together by molecular interactions, but their kinetic energies are comparable to their potential energies. As a result, although the molecules of a liquid are not free to escape completely from the bulk, the whole structure is very mobile and we can speak only of the *average* relative locations of molecules.

The average locations of the molecules in a liquid are described in terms of the **radial distribution function**, $g(r)$. This function is defined so that $g(r)\mathrm{d}r$ is the probability that a molecule will be found at a distance between r and $r + \mathrm{d}r$ from another molecule.[5] It follows that if $g(r)$ passes through a maximum at a radius of, for instance, 0.5 nm, then the most probable distance (regardless of direction) at which a second molecule will be found will be at 0.5 nm from the first molecule.

In a crystal, $g(r)$ is an array of sharp spikes, representing the certainty (in the absence of defects and thermal motion) that particles lie at definite locations. This regularity continues out to large distances (to the edge of the crystal, billions of molecules away), so we say that crystals have **long-range order**. When the crystal melts, the long-range order is lost and wherever we look at long distances from a given particle there is equal probability of finding a second particle. Close to the first particle, however, there may be a remnant of order (Fig. 11.32). Its nearest neighbors might still adopt approximately their original positions, and even if they are displaced by newcomers, the new particles might adopt their vacated positions. It may still be possible to detect, on average, a sphere of nearest neighbors at a distance r_1 and perhaps beyond them a sphere of next-nearest neighbors at r_2. The existence of this **short-range order** means that $g(r)$ can be expected to have a broad but pronounced peak at r_1, a smaller and broader peak at r_2, and perhaps some more structure beyond that. As an illustration, Fig. 11.33 shows the radial distribution function for water at a series of temperatures. The shells of local structure shown are unmistakable. Closer analysis shows that any given H_2O molecule is surrounded by other molecules at the corners of a tetrahedron, similar to the arrangement in ice (Fig. 11.34). The form of $g(r)$ at 100°C shows that the intermolecular forces (in this case, largely hydrogen bonds) are strong enough to affect the local structure right up to the boiling point.

[5] Recall the analogous quantity used to describe the distance of an electron from an atom, Section 9.8.

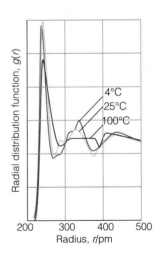

Fig. 11.33 The experimentally determined radial distribution function of the oxygen atoms in liquid water at three temperatures. Note the expansion as the temperature is raised.

Fig. 11.34 A fragment of the crystal structure of ice. Each O atom is at the center of a tetrahedron of four O atoms at a distance of 276 pm. The central O atom is attached by two short O–H bonds to two H atoms and by two relatively long O···H bonds to two neighboring H_2O molecules. Overall, the structure consists of planes of hexagonal puckered rings of H_2O molecules (like the chair form of cyclohexane).

11.12 Random coils

The next stage for understanding the link between the structure and properties of a biological macromolecule is to consider the least organized structure of a chain of atoms, a dynamically active random coil.

Unlike the molecules of a liquid, the atoms and subunits of a macromolecule are tied together by chemical bonds. However, the atoms may still have considerable freedom of location on account of the ability of the units to rotate relative to their neighbors. A **random coil** is a disorganized conformation of a flexible macromolecule. The simplest model of a random coil is a **freely jointed chain**, in which any bond is free to make any angle with respect to the preceding one (Fig. 11.35). We assume that the residues occupy zero volume, so different parts of the chain can occupy the same region of space. The model is obviously an oversimplification because a bond is actually constrained to a cone of angles around a direction defined by its neighbor. In a hypothetical one-dimensional freely jointed chain all the residues lie in a straight line, and the angle between neighbors is either 0° or 180°. The residues in a three-dimensional freely jointed chain are not restricted to lie in a line or a plane.

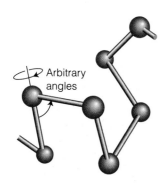

Arbitrary angles

Fig. 11.35 A freely jointed chain is like a three-dimensional random walk, each step being in an arbitrary direction but of the same length.

The probability, $f(r)dr$, that the distance between the ends of a three-dimensional freely jointed chain of N units of length l lies in the range r to $r + dr$ is[6]

$$f(r)dr = 4\pi \left(\frac{a}{\pi^{1/2}} \right)^3 r^2 e^{-a^2 r^2} dr, \quad a = \left(\frac{3}{2Nl^2} \right)^{1/2}$$

| Distribution of the separation of the ends of a three-dimensional chain | (11.27) |

In some coils, the ends may be far apart, whereas in others their separation is small. Note that it is very unlikely that the two ends will be found either very close together ($r = 0$), because the factor r^2 vanishes, or stretched out in an almost straight line, because the exponential factor then vanishes. An alternative interpretation of $f(r)$ is to regard each coil in a sample as ceaselessly writhing from one conformation to another; then $f(r)dr$ is the probability that at any instant the chain will be found with the separation of its ends between r and $r + dr$.

(a) Measures of size

There are several measures of the geometrical size of a three-dimensional random coil. The **root mean square separation**, R_{rms}, is a measure of the average separation of the ends of the coil:

$$R_{rms} = N^{1/2} l$$

| Root mean square separation | (11.28) |

We see that as the number of residues N (each of length l) increases, the root mean square separation of its ends increases as $N^{1/2}$, and consequently the volume

---

[6] See our *Physical chemistry* (2010) for full derivations of eqns 11.27, 11.28, and 11.31.

of the coil increases as $N^{3/2}$. The **contour length**, R_c, is the length of the macromolecule measured along its backbone from atom to atom:

$$R_c = Nl \qquad \boxed{\text{Contour length}} \quad (11.29)$$

Another convenient measure of size is the **radius of gyration** of the macromolecule, the radius of a thin hollow spherical shell of the same mass and moment of inertia as the molecule (Fig. 11.36). For example, a solid sphere of radius R has $R_g = (\frac{3}{5})^{1/2}R$ and a long thin rod of length l has $R_g = l/12^{1/2}$ for rotation about an axis perpendicular to the long axis. For a random coil,

$$R_g = \left(\frac{N}{6}\right)^{1/2} l \qquad \boxed{\text{Radius of gyration}} \quad (11.30)$$

and we see that, for specified values of N and l, $R_{rms} > R_g$ (Fig. 11.37). Table 11.6 lists some experimental values of R_g.

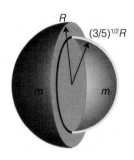

Fig. 11.36 A spherical molecule of radius R and the smaller hollow spherical shell that has the same rotational characteristics. The radius of the hollow shell is the radius of gyration R_g of the molecule.

| A brief illustration |

With a powerful microscope it is possible to see that a long piece of double-stranded DNA is flexible and writhes as if it were a random coil. However, small segments of the macromolecule resist bending, so it is more appropriate to visualize DNA as a freely jointed chain with N and l as the number and length, respectively, of these rigid units. The length l, the *persistence length*, is approximately 45 nm, corresponding to approximately 130 base pairs. It follows that for a piece of DNA with $N = 200$, we estimate (by using 10^3 nm = 1 μm)

from eqn 11.29: $R_c = 200 \times 45$ nm $= 9.0$ μm

from eqn 11.28: $R_{rms} = (200)^{1/2} \times 45$ nm $= 0.64$ μm

from eqn 11.30: $R_g = \left(\frac{200}{6}\right)^{1/2} \times 45$ nm $= 0.26$ μm

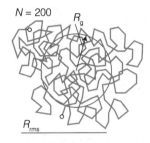

Fig. 11.37 A random coil in three dimensions. This one contains about 200 units. The root mean square distance between the ends (R_{rms}) and the radius of gyration (R_g) are indicated.

The random coil model ignores the role of the solvent: a 'poor' solvent will tend to cause the coil to tighten so that solute–solvent contacts are minimized; a 'good' solvent does the opposite. Therefore, calculations based on this model are better regarded as lower bounds to the dimensions for a coil in a good solvent and as an upper bound for a coil in a poor solvent.

(b) Conformational entropy

Because a random coil is the least structured conformation of an idealized polymer chain, it corresponds to the state of greatest entropy. Any stretching of the coil introduces order and reduces the entropy. Conversely, the formation of a random coil from a more extended form is a spontaneous process (provided enthalpy contributions do not interfere). The change in **conformational entropy**, the entropy arising from the arrangement of bonds, when a coil containing N bonds of length l is stretched or compressed by nl is

$$\Delta S = -\tfrac{1}{2}kN\ln\{(1+v)^{1+v}(1-v)^{1-v}\} \quad v = n/N \qquad \boxed{\text{Conformational entropy}} \quad (11.31)$$

where k is Boltzmann's constant and the maximum value of n is N, corresponding to maximum extension. This function is plotted in Fig. 11.38, and we see that minimum extension—fully coiled—corresponds to maximum entropy.

Table 11.6 Radii of gyration of biological macromolecules and assemblies

	$M/(\text{kg mol}^{-1})$	R_g/nm
DNA	4×10^3	117.0
Myosin	493	46.8
Serum albumin	66	2.98
Tobacco mosaic virus	3.9×10^4	92.4

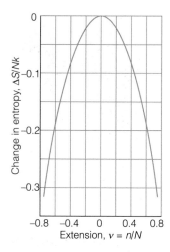

Fig. 11.38 The change in molar entropy of a freely jointed chain as its extension changes; $v = 1$ corresponds to complete extension; $v = 0$, the conformation of highest entropy, corresponds to the random coil.

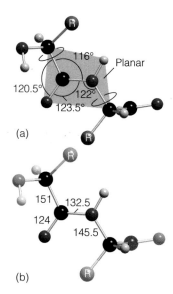

Fig. 11.39 The (a) angles and (b) bond lengths (pm) that characterize the peptide link. The C–NH–CO–C atoms define a plane (the C–N bond has partial double-bond character), but there is rotational freedom around the C–CO and N–C bonds.

A brief illustration

Suppose that $N = 1000$ and $l = 150$ pm. The change in entropy when the (one-dimensional) random coil is stretched through 1500 pm (corresponding to $n = 1500$ pm/150 pm = 10 and $v = 1/100$) is $\Delta S = -0.050k$. The change in molar entropy is therefore $\Delta S_m = -0.050R$ or -0.42 J K^{-1} mol^{-1} (we have used $R = N_A k$).

11.13 Proteins

We now need to understand how proteins attain complex structures.

For a protein to function correctly, it needs to have a well-defined conformation. For example, an enzyme has its greatest catalytic efficiency only when it is in a specific conformation. In this section we explore the covalent and noncovalent interactions that cause polypeptides to fold into complex assemblies.

(a) The secondary structure of a protein

The origin of the secondary structure of a protein is found in the rules formulated by Linus Pauling and Robert Corey in 1951. The essential feature is the stabilization of structures by hydrogen bonds involving the peptide link. The latter can act both as a donor of the H atom (the NH part of the link) and as an acceptor (the CO part). The **Corey–Pauling rules** are as follows (Fig. 11.39):

1. The four atoms of the peptide link lie in a relatively rigid plane. The planarity of the link is due to delocalization of π electrons over the O, C, and N atoms and the maintenance of maximum overlap of their p orbitals (see *Exercise* 10.41).

2. The N, H, and O atoms of a hydrogen bond lie in a straight line (with displacements of H tolerated up to not more than 30° from the N–O vector).

3. All NH and CO groups are engaged in hydrogen bonding.

The rules are satisfied by two structures. One, in which hydrogen bonding between peptide links leads to a helical structure, is the **α helix**. The other, in which hydrogen bonding between peptide links leads to a planar structure, is the **β sheet**;[7] this form is the secondary structure of the protein fibroin, the constituent of silk.

The α-helix is illustrated in Fig. 11.40. Each turn of the helix contains 3.6 amino acid residues, so the period of the helix corresponds to five turns (18 residues). The pitch of a single turn (the distance between points separated by 360°) is 544 pm. The N–H···O bonds lie parallel to the axis and link every fourth group (so residue i is linked to residues $i - 4$ and $i + 4$). All the R groups point away from the major axis of the helix.

There is freedom for the helix to be arranged as either a right- or a left-handed screw, but the overwhelming majority of natural polypeptides are right-handed on account of the preponderance of the L-configuration of the naturally occurring amino acids, as we explain below. The reason for their preponderance is not known.

A polypeptide chain adopts a conformation corresponding to a minimum Gibbs energy, which depends on the **conformational energy**, the energy of

---

[7] The sheet is often called the pleated sheet.

interaction between different parts of the chain, and the energy of interaction between the chain and surrounding solvent molecules. In the aqueous environment of biological cells, the outer surface of a protein molecule is covered by a mobile sheath of water molecules, and its interior contains pockets of water molecules. These water molecules play an important role in determining the conformation that the chain adopts through hydrophobic interactions and hydrogen bonding to amino acids in the chain.

The simplest calculations of the conformational energy of a polypeptide chain ignore entropy and solvent effects and concentrate on the total potential energy of all the interactions between nonbonded atoms. For example, these calculations predict that a right-handed α-helix of L-amino acids is marginally more stable than a left-handed helix of the same amino acids.

To calculate the energy of a conformation, we need to make use of many of the molecular interactions described earlier in the chapter and also of some additional interactions:

1. *Bond stretching.* Bonds are not rigid, and it may be advantageous for some bonds to stretch and others to be compressed slightly as parts of the chain press against one another.

If we liken the bond to a spring, then the potential energy takes the form corresponding to a Hooke's law of force (restoring force proportional to the displacement; eqn 9.28) and is

$$V_{stretch} = \tfrac{1}{2}k_{f,stretch}(R - R_e)^2$$

Contribution of bond stretching to the conformational energy (11.32)

where R_e is the equilibrium bond length and $k_{f,stretch}$ is the stretching force constant, a measure of the stiffness of the bond in question.

Self-test 11.13 The equilibrium bond length of a carbon–carbon single bond is 152 pm. Given a C–C force constant of 400 N m^{-1}, how much energy, in kilojoules per mole, would it take to stretch the bond to 165 pm?

Answer: 3.38×10^{-20} J, equivalent to 20.3 kJ mol^{-1}

---

2. *Bond bending.* An O–C–H bond angle (or some other angle) may open out or close in slightly to enable the molecule as a whole to fit together better.

If the equilibrium bond angle is θ_e, we write

$$V_{bend} = \tfrac{1}{2}k_{f,bend}(\theta - \theta_e)^2$$

Contribution of bond bending to the conformational energy (11.33)

where $k_{f,bend}$ is the bending force constant, a measure of how difficult it is to change the bond angle.

Self-test 11.14 Theoretical studies have estimated that the lumiflavin isoalloxazine ring system (**13**) has an energy minimum at the bending angle of 15°, but that it requires only 8.5 kJ mol^{-1} to increase the angle to 30°. If there are no other compensating interactions, what is the force constant for lumiflavin bending?

Answer: 1.26×10^{-22} J deg^{-2}, equivalent to 75.6 J mol^{-1} deg^{-2}

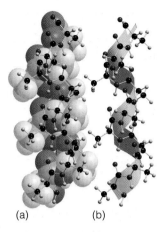

Fig. 11.40 The polypeptide α-helix, with poly-L-alanine as an example. There are 3.6 residues per turn and a translation along the helix of 150 pm per residue, giving a pitch of 544 pm. The diameter (ignoring side chains) is about 600 pm.

13 Lumiflavin

Fig. 11.41 The definition of the torsional angles ψ and ϕ between two peptide units.

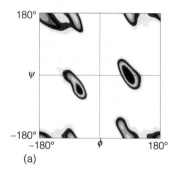

(a)

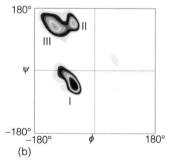

(b)

Fig. 11.42 Contour plots of potential energy against the torsional angles ψ and ϕ, also known as Ramachandran plots, for (a) a glycyl residue of a polypeptide chain and (b) an alanyl residue. The glycyl diagram is symmetrical, but that for alanyl is unsymmetrical, with the area shaded in red corresponding to an α-helix. (After T. Hovmoller et al., *Acta Cryst.* **D58**, 768 (2002).)

3. *Bond torsion*. There is a barrier to internal rotation of one bond relative to another (just like the barrier to internal rotation in ethane).

Because the planar peptide link is relatively rigid, the geometry of a polypeptide chain can be specified by the two angles that two neighboring planar peptide links make to each other. Figure 11.41 shows the two angles ϕ and ψ commonly used to specify this relative orientation. The sign convention is that a positive angle means that the front atom must be rotated clockwise to bring it into an eclipsed position relative to the rear atom. For an all-*trans* form of the chain, all ϕ and ψ are 180°. A helix is obtained when all the ϕ are equal and when all the ψ are equal. For a right-handed α-helix, all $\phi = 57°$ and all $\psi = 47°$. For a left-handed α-helix, both angles are positive. The torsional contribution to the total potential energy is

$$V_{\text{torsion}} = A(1 + \cos 3\phi) + B(1 + \cos 3\psi)$$
Contribution of bond torsion to the conformational energy   (11.34)

in which A and B are constants of the order of 1 kJ mol⁻¹. Because only two angles are needed to specify the conformation of a helix, and they range from −180° to +180°, the torsional potential energy of the entire molecule can be represented on a **Ramachandran plot**, a contour diagram in which one axis represents ϕ and the other represents ψ.

4. *Interaction between partial charges*. If the partial charges Q_i and Q_j on the atoms i and j are known, a Coulombic contribution of the form given in eqn 11.13 can be included, using the partial charges quoted in Table 11.2.

The interaction between partial charges does away with the need to take dipole–dipole interactions into account, for they are taken care of by dealing with each partial charge explicitly.

5. *Dispersive and repulsive interactions*. The interaction energy of two atoms separated by a distance r (which we know once ϕ and ψ are specified) can be given by the Lennard-Jones (12,6) form, eqn 11.25.

6. *Hydrogen bonding*. In some models of structure, the interaction between partial charges is judged to take into account the effect of hydrogen bonding.

In other models, hydrogen bonding is added as another interaction of the form

$$V_{\text{H–bonding}} = \frac{E}{r^{12}} - \frac{F}{r^{10}}$$
Contribution of hydrogen bonding to the conformational energy   (11.35)

The total potential energy of a given conformation (ϕ,ψ) can be calculated by summing the contributions given by eqns 11.32 through 11.35 and the contributions from Coulombic and dispersion interactions for all bond angles (including torsional angles) and pairs of atoms in the molecule. Figure 11.42 shows the potential energy contours for the helical form of polypeptide chains formed from the nonchiral amino acid glycine (R = H) and the chiral amino acid L-alanine (R=CH₃). The contours were computed by summing all the contributions described above for each choice of angles and then plotting contours of equal potential energy. The glycine map is symmetrical, with minima corresponding to the formation of right- and left-handed helices. In contrast, the map for L-alanine is unsymmetrical, with the lowest being consistent with the formation of an α-helix.

A β sheet is formed by hydrogen bonding between two extended polypeptide chains (large absolute values of the torsion angles ϕ and ψ). Some of the R groups point above and some point below the sheet. Two types of structures can be distinguished from the pattern of hydrogen bonding between the constituent chains.

In an **antiparallel β sheet** (Fig. 11.43a), $\phi = -139°$, $\psi = +113°$, and the N–H–O atoms of the hydrogen bonds form a straight line. This arrangement is a consequence of the antiparallel arrangement of the chains: every N–H bond on one chain is aligned with a C–O bond from another chain. Antiparallel β sheets are very common in proteins. In a **parallel β sheet** (Fig. 11.43b), $\phi = -119°$ and $\psi = +113°$, and the N–H–O atoms of the hydrogen bonds are not perfectly aligned. This arrangement is a result of the parallel arrangement of the chains: each N–H bond on one chain is aligned with an N–H bond of another chain and, as a result, each C–O bond of one chain is aligned with a C–O bond of another chain. These structures are not common in proteins.

Although we do not know all the rules that govern protein folding, X-ray diffraction studies of water-soluble natural proteins and synthetic polypeptides show that some amino acid residues appear in helical segments more frequently than in sheets, whereas others exhibit the opposite behavior. Table 11.7 summarizes the available data.

(b) Higher-order structures of proteins

In an aqueous environment, chains fold in such a way as to place nonpolar R groups in the interior (which is often not very accessible to solvent) and charged R groups on the surface (in direct contact with the polar solvent). A wide variety of structures can result from these broad rules. Among them, a **four-helix bundle** (Fig. 11.44), which is found in proteins such as cytochrome b_{562} (an electron-transport protein, Atlas P5), forms when each helix has a nonpolar region along its length. The four nonpolar regions pack together to form a nonpolar interior. Similarly, interconnected β sheets may interact to form a **β barrel** (Fig. 11.45), the interior of which is populated by nonpolar R groups and which has an exterior rich in charged residues. The retinol-binding protein of blood plasma (Atlas P11), which is responsible for transporting vitamin A, is an example of a β barrel structure.

Factors that promote the folding of proteins include covalent –S–S– **disulfide links** between cysteine residues (**14**), Coulombic interactions between ions (which depend on the degree of protonation of groups and therefore on the pH), hydrogen bonding (such as O–H···O), van der Waals interactions, and hydrophobic interactions. The clustering of nonpolar, hydrophobic amino acids into the interior of a protein is driven primarily by hydrophobic interactions (Section 2.7).

Proteins with $M > 50$ kg mol^{-1} are often found to be aggregates of two or more polypeptide chains. Hemoglobin, which consists of four myoglobin-like chains (Fig. 11.46 and Atlas P7), is an example of a quaternary structure. Myoglobin (Atlas P10) is an oxygen-storage protein. The subtle differences that arise when four such molecules coalesce to form hemoglobin result in the latter being an oxygen transport protein, able to load O_2 cooperatively and to unload it cooperatively too (see *Case studies* 4.1 and 10.4).

Proteins can also self-assemble into rather large aggregates. Collagen (Atlas P4), the most abundant protein in mammals and responsible for imparting mechanical strength to tissues and organs, consists of three long helices wound around each other. The protein actin forms thin, rodlike filaments that, when associated with several copies of the protein myosin, play an important role in the mechanism of muscle contraction. The microtubules that participate in the

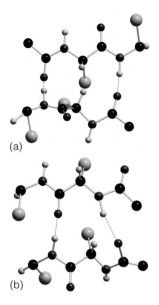

(a)

(b)

Fig. 11.43 (a) An antiparallel β sheet ($\phi = -139°$, $\psi = +113°$), in which the N–H–O atoms of the hydrogen bonds form a straight line. (b) A parallel β sheet ($\phi = -119°$ and $\psi = +113°$), in which the N–H–O atoms of the hydrogen bonds are not perfectly aligned.

14

Table 11.7 Relative frequencies of amino acid residues in helices and sheets

Amino acid	α helix	β sheet
Alanine	1.29	0.90
Arginine	0.96	0.99
Asparagine	0.90	0.76
Aspartic acid	1.04	0.72
Cysteine	1.11	0.74
Glutamic acid	1.44	0.75
Glutamine	1.27	0.80
Glycine	0.56	0.92
Histidine	1.22	1.08
Isoleucine	0.97	1.45
Leucine	1.30	1.02
Lysine	1.23	0.77
Methionine	1.47	0.97
Phenylalanine	1.07	1.32
Proline	0.52	0.64
Serine	0.82	0.95
Threonine	0.82	1.21
Tryptophan	0.99	1.14
Tyrosine	0.72	1.25
Valine	0.91	1.49

Data from T.E. Creighton, *Proteins: structures and molecular properties*, W. H. Freeman and Co., New York (1992).

Fig. 11.44 A four-helix bundle forms from the interactions between nonpolar amino acids on the surfaces of each helix, with the polar amino acids exposed to the aqueous environment of the solvent.

Fig. 11.45 Eight antiparallel β sheets, each represented by an arrow and linked by short random coils fold together as a barrel. Nonpolar amino acids are in the interior of the barrel.

separation of chromosomes during cell division, provide structural rigidity in cells, and participate in the motile function of flagella are hollow cylinders formed by aggregation of the protein tubulin.

Not all protein aggregates are beneficial. In patients afflicted with sickle-cell anemia, hemoglobin molecules aggregate into rods, rendering the red blood cell unable to transport O_2 efficiently. Also, the presence of aggregates of proteins in the brain appears to be associated with several serious conditions. For example, the *amyloid plaques* found in postmortem analysis of the brains of patients with Alzheimer's disease are a mixture of damaged neurons and aggregates of the β amyloid protein, which is an extended antiparallel β sheet.

11.14 Nucleic acids

Of crucial biological importance are the conformations adopted by nucleic acids, the key components of the mechanism of storage and transfer of genetic information in biological cells.

We saw in *Fundamentals* F.1 that DNA and RNA are polynucleotides, polymers of base–sugar–phosphate units linked by phosphodiester bonds, that self-assemble

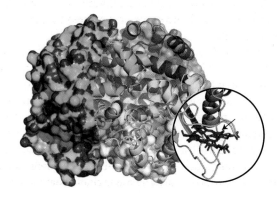

Fig. 11.46 A hemoglobin molecule consists of four myoglobin-like units. An O_2 molecule attaches to the ion atom in the heme group indicated by the arrow.

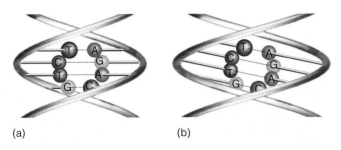

(a)   (b)

Fig. 11.47 The structural features of the most abundant forms of DNA in the cell: (a) B-DNA, (b) A-DNA.

into complex three-dimensional structures. An example of secondary structure in nucleic acids is the winding of two polynucleotide chains around each other to form a DNA double helix, as shown in Fig. 11.47.

Figure 11.47 also shows that different forms of the double helix are possible. In B-DNA, the most abundant form of DNA in the cell (Fig. 11.47a), the rodlike double helix is right-handed with a diameter of 2.37 nm and a pitch of 3.54 nm. The base pairs are approximately parallel to each other and perpendicular to the long axis of the rod. In A-DNA (Fig. 11.47b), the double helix is right-handed but slightly wider, with a diameter of approximately 2.55 nm and a pitch of 2.53 nm. The base pairs are parallel to each other but not perpendicular to the long axis of the helix. Double-stranded RNA and hybrid RNA-DNA, the assembly of one strand of ribonucleic acid strand with a DNA strand, assume the A form. A third form of DNA, called Z-DNA, is a left-handed helix with a diameter of 1.84 nm, a pitch of 4.56 nm, and a slightly tilted arrangement of the base pairs relative to the long axis of the helix. The physiological role of Z-DNA is not certain.

We saw in Section 3.5 that base pairing by hydrogen bonding is largely responsible for the thermal stability of DNA. A more subtle interaction that confers stability to DNA is **base stacking**, in which dispersion interactions bring together the planar systems of bases. Experiments show that stacking interactions are stronger between C–G base pairs than between A–T base pairs. It follows that two factors render DNA sequences rich in C–G base pairs more stable than sequences rich in A–T base pairs: more hydrogen bonds between the bases (Section 3.5) and stronger stacking interactions between base pairs. Some drugs with planar π systems, shown as a gray rectangle in the illustration, are effective because they intercalate between base pairs through stacking interactions, causing the helix to unwind slightly and altering the function of DNA (Fig. 11.48).

Because a long stretch of DNA is flexible, it can undergo further folding into a variety of tertiary structures. Two examples are shown in Fig. 11.49. Supercoiled DNA is found in the chromosome and can be visualized as the twisting of closed circular DNA (ccDNA), much like the twisting of a rubber band. Before it can participate in the transmission of genetic information, supercoiled DNA must be uncoiled. Both coiling and uncoiling are catalyzed by enzymes belonging to the topoisomerase family.

There are important differences in the chemical compositions of RNA and DNA that translate into different secondary and tertiary structures. In RNA the sugar is β-D-ribose (Atlas S1), whereas in DNA it is β-D-2-deoxyribose (Atlas S2). Although adenine, cytosine, and guanine are found in both DNA and RNA, in RNA uracil (Atlas B5) replaces thymine. As in DNA, the secondary and tertiary structures of RNA arise primarily from the pattern of hydrogen bonding between bases of one or more chains. The extra –OH group in β-D-ribose imparts enough steric strain to a polynucleotide chain that stable double helices cannot form in RNA. Therefore, RNA exists primarily as single chains that can fold into complex

Fig. 11.48 Some drugs with planar π systems, shown as a rectangle, intercalate between base pairs of DNA.

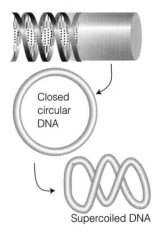

Closed circular DNA

Supercoiled DNA

Fig. 11.49 A long section of DNA may form closed circular DNA (ccDNA) by covalent linkage of the two ends of the chain. Twisting of ccDNA leads to the formation of supercoiled DNA.

Fig. 11.50 The structure of a transfer RNA (tRNA).

structures by formation of A–U and G–C base pairs. One example of this effect is the structure of transfer RNA (tRNA), shown schematically in Fig. 11.50, in which base-paired regions are connected by loops and coils. Transfer RNAs help assemble polypeptide chains during protein synthesis in the cell.

11.15 Polysaccharides

To understand the connection between structure and biological function of carbohydrates, we need to examine the conformations adopted by their polymers.

We saw in *Fundamentals* F.1 that polysaccharides are polymers of simple carbohydrates. Carbohydrate units are linked together in polysaccharides by **glycosidic bonds** that form between hydroxyl groups and result in C–O–C ether moieties. The orientation of one linked ring relative to another depends on which hydroxyl groups are linked and on their stereochemistry. Consider the α and β isomers of glucose (**15a** and **15b**, respectively), which differ in the configuration of the C1 carbon. Linking the C1 and C4 carbons by glycosidic bonds, so-called *1,4-glycosidic bonds*, results in either a bent (**16**) or a linear (**17**) chain, depending on whether the monomer is α- or β-glucose, respectively. Branched structures are also possible when a monomer makes three glycosidic bonds, as shown in (**18**).

15a α-D-Glucose

15b β-D-Glucose

16 A bent glycosidic chain

17 A linear glycosidic chain

18 A branched glycosidic chain

Like polypeptides and polynucleotides, polysaccharides also possess different levels of structure. In cellulose, linear chains of glucose, such as those shown in (**17**), interact through hydrogen bonds involving hydroxyl groups and ring oxygen atoms. The resulting structure is a thin but strong fiber that is used to construct the wall of a plant cell. In amylase, which stores glucose molecules for future use by the plant cells, a bent chain, such as that in (**16**), coils into a helical structure held together by hydrogen bonds. Glycogen, which stores glucose in animals and microbes, and amylopectin, which—like amylase—performs the same function in plants, also feature α-1,4-linkages, but because of branching points (as in **18**), these polymers do not adopt regular secondary structures.

11.16 Micelles and biological membranes

We need to understand the factors that optimize the self-assembly of cell membranes.

We saw in *Fundamentals* F.1 that phospholipids are amphipathic molecules that can group together through hydrophobic interactions to form bilayer structures and cell membranes (Fig. F.1). Here we explore details of the self-assembly of amphipathic molecules into a variety of structures with significance to biology and medicine.

(a) Micelles

In aqueous environments amphipathic molecules can group together as **micelles**, in which hydrophobic tails congregate, leaving hydrophilic heads exposed to the solvent (Fig. 11.51). Micelles are important in industry and biology on account of their solubilizing function: matter can be transported by water after it has been dissolved in their hydrocarbon interiors.

Micelles form only above a certain concentration of amphiphiles called the **critical micelle concentration** (CMC) and above the **Krafft temperature**. Nonionic amphipathic molecules may cluster together in clumps of 1000 or more, but ionic species tend to be disrupted by the electrostatic repulsions between head groups and are normally limited to groups of fewer than about 100. The interior of a micelle is like a droplet of oil, and experiments show that the hydrophobic tails are mobile, but slightly more restricted than in the bulk.

Different molecules tend to form micelles of different shapes. For example, ionic species such as sodium dodecyl sulfate (SDS) and cetyl trimethylammonium bromide (CTAB) form rods at moderate concentrations, whereas sugar molecules form small, approximately spherical micelles. Broadly speaking, the shapes of micelles vary with the shape of the constituent molecules, their concentration, and the temperature. A useful predictor of the shape of the micelle, the **surfactant parameter**, N_s, is defined as

$$N_s = \frac{V}{Al} \qquad \boxed{\text{Surfactant parameter}} \qquad (11.36)$$

where V is the volume of the hydrophobic tail, A is the area of the hydrophilic head group, and l is the maximum length of the tail. Table 11.8 summarizes the dependence of micelle shape on the surfactant parameter.

Under certain experimental conditions, a **liposome** may form, with an inward pointing inner surface of molecules surrounded by an outward pointing outer layer (Fig. 11.52). Liposomes may be used to carry nonpolar drug molecules in blood. **Reverse micelles** form in nonpolar solvents, with small polar head groups

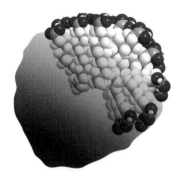

Fig. 11.51 A representation of a spherical micelle. The hydrophilic groups are represented by the red spheres and the hydrophobic hydrocarbon chains are represented by the stalks. The latter are mobile.

Table 11.8 Variation of micelle shape with the surfactant parameter

Value or range of the surfactant parameter, N_s	Micelle shape
< 0.33	Spherical
0.33–0.50	Cylindrical rods
0.50–1.00	Vesicles
1.00	Planar bilayers
> 1.00	Reverse micelles and other shapes

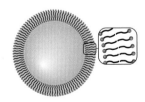

Fig. 11.52 The cross-sectional structure of a spherical liposome.

in a micellar core and more voluminous hydrophobic tails extending into the organic bulk phase. These spherical aggregates can solubilize water in organic solvents by creating a pool of trapped water molecules in the micellar core.

(b) Bilayers, vesicles, and membranes

Some micelles at concentrations well above the CMC form extended parallel sheets two molecules thick, called **planar bilayers**. The individual molecules lie perpendicular to the sheets, with hydrophilic groups on the outside in aqueous solution and on the inside in nonpolar media. When segments of planar bilayers fold back on themselves, **unilamellar vesicles** may form where the spherical hydrophobic bilayer shell separates an inner aqueous compartment from the external aqueous environment.

Bilayers show a close resemblance to biological membranes and are often a useful model on which to base investigations of biological structure. However, actual membranes are highly sophisticated structures, in which phospholipid molecules form layers instead of micelles because the hydrocarbon chains are too bulky to allow packing into nearly spherical clusters.

The bilayer is a highly mobile structure. Not only are the hydrocarbon chains ceaselessly twisting and turning in the region between the polar groups, but the phospholipid and cholesterol molecules migrate over the surface. It is better to think of the membrane as a viscous fluid rather than a permanent structure, with a viscosity about 100 times that of water. In common with diffusional behavior in general (Section 8.5), the average distance a phospholipid molecule diffuses is proportional to the square root of the time; more precisely, for a molecule confined to a two-dimensional plane, the average distance traveled in a time t is equal to $(4Dt)^{1/2}$, where D is the diffusion constant. Typically, a phospholipid molecule migrates through about 1 μm in about 1 min.

(c) Interactions between proteins and biological membranes

Peripheral proteins are proteins attached to the bilayer. **Integral proteins** are proteins embedded in the mobile but viscous bilayer. Examples include complexes I–IV of oxidative phosphorylation (Section 5.10), ion channels, and ion pumps (Section 5.3). Integral proteins may span the depth of the bilayer and consist of tightly packed α helices or, in some cases, β sheets containing hydrophobic residues that sit comfortably within the hydrocarbon region of the bilayer. The hydrophobicity of a residue can be assessed by measuring the Gibbs energy of transfer of the corresponding amino acid from an aqueous solution to the interior of a membrane (Table 11.9). Amino acids with negative values of the Gibbs energy of transfer are likely to be found in the membrane-spanning regions of integral proteins.

There are two views of the motion of integral proteins in the bilayer. In the **fluid mosaic model** shown in Fig. 11.53, the proteins are mobile, but their diffusion coefficients are much smaller than those of the lipids. In the **lipid raft model**, a number of lipid and cholesterol molecules form ordered structures, or 'rafts,' that envelop proteins and help carry them to specific parts of the cell.

The mobility of the bilayer enables it to flow around a molecule close to the outer surface, to engulf it, and to incorporate it into the cell by the process of *endocytosis*. Alternatively, material from the cell interior wrapped in cell membrane may coalesce with the cell membrane itself, which then withdraws and ejects the material in the process of *exocytosis*. An important function of the proteins embedded in the bilayer, however, is to act as devices for transporting matter into and out of the cell in a more subtle manner, as discussed in Section 8.6.

Table 11.9 Gibbs energies of transfer of amino acid residues in a helix from the interior of a membrane to water

Amino acid	$\Delta_{transfer}G/$ $(kJ\ mol^{-1})$
Phenylalanine	15.5
Methionine	14.3
Isoleucine	13.0
Leucine	11.8
Valine	10.9
Cysteine	8.4
Tryptophan	8.0
Alanine	6.7
Threonine	5.0
Glycine	4.2
Serine	2.5
Proline	−0.8
Tyrosine	−2.9
Histidine	−12.6
Glutamine	−17.2
Asparagine	−20.2
Glutamic acid	−34.4
Lysine	−37.0
Aspartic acid	−38.6
Arginine	−51.7

Data from D.M. Engelman, T.A. Steitz, and A. Goldman, *Ann. Rev. Biophys. Biophys. Chem.* **15**, 330 (1986).

11.17 Computer-aided simulation

To understand the various approaches to the prediction of structure, we need to see how to take into account a balance of interactions that give a biological macromolecule its native conformation or hold a drug and receptor together.

We saw in Chapter 10 that ideas derived from quantum mechanics can be used to predict the structures and the physical and chemical properties of molecules. Semi-empirical, *ab initio*, and density functional methods work very well for molecules of modest size but require too much computational power and time to be suitable for predicting the structures of macromolecules. The problem is particularly acute when the surrounding water plays an important role in governing structure. For this reason, biochemists often rely on other techniques to generate three-dimensional models of proteins, nucleic acids, lipid bilayers, and drug–receptor complexes. Computational methods based on the principles of classical physics lead to the visual representation of atomic motions in biopolymers, thereby opening a window onto the molecular factors that are responsible for such dynamic processes as protein folding and enzyme catalysis. Yet other strategies can give insight into the structural features of a drug that optimize its docking to a receptor site.

(a) Molecular mechanics calculations

We saw in Section 11.13 that the conformational energy, V_C, of a biopolymer can be calculated by adding the contributions from steric interactions (bond stretching, bending, and torsion and dispersive interactions), electrostatic interactions, and hydrogen bonding:

$$V_C = V_{stretch} + V_{bend} + V_{torsion} + V_{Coulomb}$$
$$+ V_{LJ} + V_{H\text{-bonding}}$$

Conformational energy (11.37)

In a **molecular mechanics** simulation, the locations of the atoms are changed until the conformation with the lowest value of V_C is found. For a macromolecule, a plot of the conformational energy against bond distance or bond angle often shows several local minima and a global minimum, which is associated with the preferred conformation (Fig. 11.54). Commercially available molecular modeling software packages include schemes for modifying and searching for these minima systematically.

Molecular mechanics calculations are fast and do not require a great deal of computing power. However, they are of limited utility because the structure corresponding to the global minimum is a snapshot of the molecule at $T = 0$. That is, only the potential energy is included in the calculation; contributions to the total energy from kinetic energy are excluded. Also, the method does not handle interactions with a solvent.

(b) Molecular dynamics and Monte Carlo simulations

Biological macromolecules (like all except the smallest molecules) are flexible and move ceaselessly. Atomic fluctuations and side-chain motions have amplitudes of 1–500 pm and characteristic times ranging from 1 fs to 0.1 s. Rigid body motions, such as the motions of helices and subunits, have amplitudes of 0.1–1.0 pm and characteristic times of 1 ns to 1 s. Folding transitions and the formation of quaternary structure from large structures have amplitudes greater than 0.5 nm and occur over a time span of from 100 ns to several hours.

In a **molecular dynamics** simulation, the molecule is set in motion by treating it as though it has been heated to a specified temperature and the possible

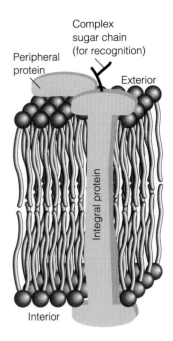

Fig. 11.53 In the fluid mosaic model of a biological cell membrane, integral proteins diffuse through the lipid bilayer. In the alternative lipid raft model, a number of lipid and cholesterol molecules envelop and transport the protein around the membrane.

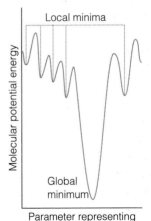

Fig. 11.54 For large molecules, a plot of potential energy against the molecular geometry often shows several local minima and a global minimum.

trajectories of all atoms under the influence of the intermolecular potentials are calculated. To appreciate what is involved, we consider the motion of an atom in one dimension. We show in the following *Justification* that after a time interval Δt, the position of an atom changes from x_{i-1} to a new value x_i given by

$$x_i = x_{i-1} + v_{i-1}\Delta t \tag{11.38}$$

where v_{i-1} is the velocity of the atom when it was at x_{i-1}, its location at the start of the interval. The velocity at x_i is related to v_{i-1}, the velocity at the start of the interval, by

$$v_i = v_{i-1} - m^{-1}\left.\frac{dV_C(x)}{dx}\right|_{x_{i-1}}\Delta t \tag{11.39}$$

where the derivative of the conformational energy $V_C(x)$ is evaluated at x_{i-1}. The time interval Δt is approximately 1 fs (10^{-15} s), which is shorter than the average time for the fastest atomic motions in a macromolecule. The calculation of x_i and v_i is then repeated for tens of thousands of such steps.

Justification 11.5 *The atomic trajectories according to molecular dynamics*

Consider an atom of mass m moving along the x direction with an initial velocity v_1 given by $v_i = \Delta x / \Delta t$. If the initial and new positions of the atom are x_1 and x_2, then $\Delta x = x_2 - x_1$ and $x_2 = x_1 + v_1\Delta t$. This expression generalizes to eqn 11.38 for the calculation of a position x_i from a previous position x_{i-1} and velocity v_{i-1}.

The atom moves under the influence of a force arising from interactions with other atoms in the molecule. From Newton's second law of motion, we write the force F_1 at x_1 as $F_1 = ma_1$, where the acceleration a_1 at x_1 is given by $a_1 = \Delta v / \Delta t$. If the initial and new velocities are v_1 and v_2, then $\Delta v = v_2 - v_1$ and

$$v_2 = v_1 + a_1\Delta t = v_1 + \frac{F_1}{m}\Delta_t$$

Because $F = -dV/dx$, the force acting on the atom is related to the potential energy of interaction with other nearby atoms, the conformational energy $V_C(x)$, by

$$F_1 = -\left.\frac{dV_C(x)}{dx}\right|_{x_1}$$

where the derivative is evaluated at x_1. It follows that

$$v_2 = v_1 - m^{-1}\left.\frac{dV_C(x)}{dx}\right|_{x_1}\Delta t$$

This expression generalizes to eqn 11.39 for the calculation of a velocity v_i from a previous velocity v_{i-1}.

Self-test 11.15 Consider a particle of mass m connected to a stationary wall with a spring of force constant k_f. Write an expression for the velocity of this particle once it is set into motion in the x direction from an equilibrium position x_0.

Answer: $v_i = v_{i-1} + (k_f/m)(x_{i-1} - x_0)\Delta t$

Commercially available software packages use versions of eqns 11.38 and 11.39 to calculate the trajectories of a large number of atoms in three dimensions. The trajectories correspond to the conformations that the molecule can sample at the temperature selected for the simulation. At very low temperatures, the molecule cannot overcome some of the potential energy barrier given by eqn 11.37, atomic motion is restricted, and only a few conformations are possible. At high temperatures, more potential energy barriers can be overcome and more conformations are accessible. Computational methods also allow for the simulation of a solvent cage around the macromolecule.

In the **Monte Carlo method**, the atoms of a macromolecule are moved through small but otherwise random distances, and the change in conformational energy, ΔV_C, is calculated. If the conformational energy is not greater than before the change, then the conformation is accepted. However, if the conformational energy is greater than before the change, it is necessary to check if the new conformation is reasonable and can exist in equilibrium with structures of lower conformational energy at the temperature of the simulation. To make progress, we use the Boltzmann distribution (*Fundamentals* F.3) to write that at equilibrium, the ratio of populations of two states with energy separation ΔV_C is $e^{-\Delta V_C/kT}$, where k is Boltzmann's constant. Because we are testing the viability of a structure with a higher conformational energy than the previous structure in the calculation, $\Delta V_C > 0$ and the exponential factor varies between 0 and 1. In the Monte Carlo method, the exponential factor is compared with a random number between 0 and 1; if the factor is larger than the random number, the conformation is accepted; if the factor is not larger, the conformation is rejected.

Molecular dynamics and Monte Carlo simulations are much faster than quantum chemical calculations and can handle with relative ease the effect of solvent on the structure of a biopolymer. However, neither method is likely to yield the native structure of a large biopolymer from its sequence because of the very large number of states that must be sampled during the calculation. Nevertheless, the methods can be used to predict the effect of a minor change in the sequence of a nucleic acid or protein of known structure. Because in such a case the chemical substitution is not expected to result in a large deviation from the native structure, the calculation needs to sample only a manageable (but still large) number of conformations. This approach allows for the systematic investigation of a very large number of biopolymers, potentially leading to the determination of the chemical rules for stabilization of biomolecular structure. In much the same vein, the combination of molecular dynamics and Monte Carlo simulations can be used to investigate the thermodynamics of interaction between a drug and a biopolymer.

(c) Quantitative structure–activity relationships

Computational approaches are having a considerable impact on the processes of drug discovery. To devise efficient therapies, it is necessary to know how to characterize and optimize both the three-dimensional structure of the drug and the molecular interactions between the drug and its target. Computational studies of the types described in Chapter 10 and this chapter can identify regions of a molecule that have high or low electron densities and result in specific interactions between the host protein and the guest agent. The graphical representation of numerical results brings these interactions vividly to life and allows modifications to be analyzed in the hope of improving specificity.

In *structure-based design*, new drugs are developed on the basis of the known structure of the receptor site of a known target. However, in many cases a number of so-called *lead compounds* are known to have some biological activity but little information is available about the target. To design a molecule with improved pharmacological efficacy, **quantitative structure–activity relationships** (QSAR) are often established by correlating data on activity of lead compounds with molecular properties, also called *molecular descriptors*, which can be determined either experimentally or computationally.

The first stage of the QSAR method consists of compiling molecular descriptors for a very large number of lead compounds. Descriptors such as molar mass, molecular dimensions and volume, and relative solubility in water and nonpolar solvents are available from routine experimental procedures. Quantum mechanical descriptors determined by calculations of the type described in Chapter 10 include bond orders and HOMO and LUMO energies.

In the second stage of the process, biological activity is expressed as a function of the molecular descriptors. An example of a QSAR equation is:

$$\text{activity} = c_0 + c_1 d_1 + c_2 d_1^2 + c_3 d_2 + c_4 d_2^2 + \cdots \qquad \boxed{\text{A QSAR correlation equation}} \qquad (11.40)$$

where d_i is the value of the descriptor and c_i is a coefficient calculated by fitting the data by regression analysis. The quadratic terms account for the fact that biological activity can have a maximum or minimum value at a specific descriptor value. For example, a molecule might not cross a biological membrane and become available for binding to targets in the interior of the cell if it is too hydrophilic, in which case it will not partition into the hydrophobic layer of the cell membrane, or too hydrophobic, for then it may bind too tightly to the membrane. It follows that the activity will peak at some intermediate value of a parameter that measures the relative solubility of the drug in water and organic solvents.

In the final stage of the QSAR process, the activity of a drug candidate can be estimated from its molecular descriptors and the QSAR equation either by interpolation or extrapolation of the data. The predictions are more reliable when a large number of lead compounds and molecular descriptors are used to generate the QSAR equation.

The traditional QSAR technique has been refined into **3D QSAR**, in which sophisticated computational methods are used to gain further insight into the three-dimensional features of drug candidates that lead to tight binding to the receptor site of a target. The process begins by using a computer to superimpose three-dimensional structural models of lead compounds and looking for common features, such as similarities in shape, location of functional groups, and electrostatic potential plots. The key assumption of the method is that common structural features are indicative of molecular properties that enhance binding of the drug to the receptor. The collection of superimposed molecules is then placed inside a three-dimensional grid of points. An atomic probe, typically an sp^3-hybridized carbon atom, visits each grid point and two energies of interaction are calculated: E_{steric}, the steric energy reflecting interactions between the probe and electrons in uncharged regions of the drug, and E_{elec}, the electrostatic energy arising from interactions between the probe and a region of the molecule carrying a partial charge. The measured equilibrium constant for binding of the drug to the target, K_{bind}, is then assumed to be related to the interaction energies at each point r by the 3D QSAR equation

$$\log K_{\text{bind}} = c_0 + \sum_r \{c_{\text{steric}}(r) E_{\text{steric}}(r) + c_{\text{elec}}(r) E_{\text{elec}}(r)\} \qquad \boxed{\text{A 3D-QSAR equation}} \qquad (11.41)$$

where the $c(r)$ are coefficients calculated by regression analysis, with the coefficients c_{steric} and c_{elec} reflecting the relative importance of steric and electrostatic

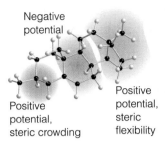

Negative potential

Positive potential, steric crowding

Positive potential, steric flexibility

Fig. 11.55 A 3D QSAR analysis of the steroids to human corticosteroid-binding globulin (CBG). The tinted regions indicate areas in the protein's binding site with positive and negative electrostatic potentials and with little or much steric crowding. [Adapted from P. Krogsgaard-Larsen, T. Liljefors, and U Marsden (cd.) *A textbook of drug design and discovery*, Taylor & Francis, London (2002).]

interactions, respectively, at the grid point *r*. Visualization of the regression analysis is facilitated by coloring each grid point according to the magnitude of the coefficients. Figure 11.55 shows results of a 3D QSAR analysis of the binding of steroids, molecules with the carbon skeleton shown, to human corticosteroid-binding globulin (CBG). Indeed, we see that the technique lives up to the promise of opening a window into the chemical nature of the binding site even when its structure is not known.

Checklist of key concepts

1. In ultracentrifugation, a sample is exposed to a strong centrifugal field generated by rotation at high speeds and the molar mass of a biopolymer is calculated from the sedimentation constant.

2. MALDI-TOF mass spectrometry is a technique for the determination of molar masses in which a sample is ionized in the gas phase and the mass-to-charge number ratios of all ions are measured.

3. In laser light scattering the size and shape of a macromolecule are obtained from analysis of the intensity of light scattered by the sample.

4. X-ray crystallography is a collection of X-ray diffraction techniques based on applications of Bragg's law to the determination of the three-dimensional structures of small and large molecules, including biopolymers.

5. Unit cells are classified into seven crystal systems according to their rotational symmetries.

6. Crystal planes are specified by a set of Miller indices (*hkl*).

7. In X-ray crystallography the electron density is calculated from the intensities I_h of scattered X-rays and the structure factors.

8. Crystals of proteins amenable to analysis by X-ray diffraction techniques can be made by adding a large amount of a salt, such as $(NH_4)_2SO_4$, to a solution containing a charged protein. Detergents are often used to crystallize hydrophobic proteins.

9. A van der Waals interaction between closed-shell molecules is inversely proportional to the sixth power of their separation.

10. A polar molecule is a molecule with a permanent electric dipole moment; the magnitude of a dipole moment is the product of the partial charge and the separation.

11. The following interactions are important in biological self-assembly: charge–charge, charge–dipole, dipole–dipole, dipole–induced-dipole, dispersion (London), hydrogen bonding.

12. A hydrogen bond is an interaction of the form X–H···Y, where X and Y are N, O, or F.

13. The Lennard-Jones (12,6) potential is a model of the total intermolecular potential energy.

14. The relative locations of molecules in a liquid are reported in terms of the radial distribution function, $g(r)$.

15. The least structured model of a macromolecule, such as a long stretch of DNA or a denatured protein, is as a random coil.

16. The conformational entropy of a random coil is the entropy arising from the arrangement of bonds.

17. The secondary structure of a polypeptide chain can be specified by the two angles, ϕ and ψ, that two neighboring planar peptide links make to each other.

18. The different forms of double-helical DNA (B-, A-, and Z-) differ in diameter, pitch, and tilt of the base pairs relative to the long axis of the helix.

19. Supercoiled DNA is formed by twisting of closed circular DNA (ccDNA).

20. RNA exists primarily as single chains that can fold into complex structures by formation of base pairs.

☐ 21. Carbohydrate units are linked together in polysaccharides by glycosidic bonds between hydroxyl groups; bent, linear, or branched chains can result depending on which hydroxyl groups are linked.

☐ 22. Micelles form in aqueous environments when the hydrophobic tails of amphipathic molecules congregate and their hydrophilic heads are exposed to the surrounding water molecules.

☐ 23. In the fluid mosaic model of the cell membrane, integral proteins are mobile. In the lipid raft model, a number of lipid and cholesterol molecules form ordered structures, or 'rafts', that envelop proteins and help carry them to specific parts of the cell.

☐ 24. A biopolymer adopts a conformation corresponding to a minimum Gibbs energy, which depends on the conformational energy, the energy of interaction between different parts of the polymer, and the energy of interaction between the polymer and surrounding solvent molecules.

☐ 25. In a molecular mechanics simulation, the locations of the atoms are changed until the conformation with the lowest value of the total potential energy is found.

☐ 26. In a molecular dynamics simulation, the molecule is set in motion by supposing that it has been heated to a specified temperature and the possible trajectories of all atoms under the influence of the intermolecular potentials are calculated.

☐ 27. In a Monte Carlo simulation, the atoms of a macromolecule are moved through small but otherwise random distances, and the change in conformational energy is calculated.

☐ 28. In the QSAR technique, pharmacological activity is correlated with a variety of molecular characteristics.

Checklist of key equations

Property	Equation	Comment
Molar mass obtained from:		
sedimentation rate	$M = SRT/bD$, $S = s/r\omega^2$	
equilibrium sedimentation	$M = [2RT/\{(r_2^2 - r_1^2)b\omega^2\}]\ln(c_2/c_1)$	
Mass-to-charge ratio in a TOF spectrometer	$m/z = 2e\mathcal{E}d(t/l)^2$	
Molar mass from the Rayleigh ratio	$R(\theta) = KP(\theta)c_M M$	
Separation between crystal planes	$1/d^2 = h^2/a^2 + k^2/b^2 + l^2/c^2$	Orthogonal lattice
Bragg's law	$\lambda = 2d\sin\theta$	
Fourier synthesis	$\rho(x) = (1/V)\left\{ F_0 + 2\sum_{h=1}^{\infty} F_h \cos(2h\pi x) \right\}$	
Coulomb's law in any medium	$V = Q_1 Q_2/4\pi\varepsilon r$	
Dipole moment in terms of electronegativity difference	$\mu/D \approx \Delta\chi$	
Charge–dipole interaction energy	$V = -Q_2\mu_1/(4\pi\varepsilon_0 r^2)$	See 7
	$V = -Q_2\mu_1\cos\theta/(4\pi\varepsilon_0 r^2)$	See 8
Dipole–dipole interaction energy	$V = \mu_1\mu_2(1 - 3\cos^2\theta)/(4\pi\varepsilon_0 r^3)$	See 9
	$V = -2\mu_1^2\mu_2^2/\{3(4\pi\varepsilon_0)^2 kTr^6\}$	Freely rotating dipoles
Polarizability volume	$\alpha' = \alpha/4\pi\varepsilon_0$	
Dipole–induced-dipole interaction energy	$V = -\mu_1^2\alpha_2/(4\pi\varepsilon_0 r^6)$	
London formula	$V = -\frac{3}{2} \times (\alpha_1'\alpha_2'/r^6) \times \{I_1 I_2/(I_1 + I_2)\}$	
Lennard-Jones (12,6) potential energy	$V = 4\varepsilon\{(\sigma/r)^{12} - (\sigma/r)^6\}$	
Measures of size:		Random coil
root mean square separation	$R_{rms} = N^{1/2}l$	
contour length	$R_c = Nl$	
radius of gyration	$R_g = (N/6)^{1/2}l$	
Conformational entropy	$\Delta S = \frac{1}{2}kN\ln\{(1 + v)^{1+v}(1 - v)^{1-v}\}$	Random coil
Conformational energy	$V_C = V_{stretch} + V_{bend} + V_{torsion} + V_{Coulomb} + V_{LJ} + V_{H\text{-}bonding}$	
bond stretching	$V_{stretch} = \frac{1}{2}k_{f,stretch}(R - R_e)^2$	
bond bending	$V_{bend} = \frac{1}{2}k_{f,bend}(\theta - \theta_e)^2$	
bond torsion	$V_{torsion} = A(1 + \cos 3\phi) + B(1 + \cos 3\psi)$	
hydrogen bonding	$V_{H\text{-}bonding} = E/r^{12} - F/r^{10}$	
Surfactant parameter	$N_s = V/Al$	Definition
A QSAR equation	$\text{Activity} = c_0 + c_1 d_1 + c_2 d_1^2 + c_3 d_2 + c_4 d_2^2 + \ldots$	

Discussion questions

11.1 What features in an X-ray diffraction pattern suggest a helical conformation for a biological macromolecule?

11.2 Describe the phase problem in X-ray diffraction and explain how it may be overcome.

11.3 Explain how the permanent dipole moment and the polarizability of a molecule arise.

11.4 Describe the formation of a hydrogen bond in terms of (a) an electrostatic interaction and (b) molecular orbital theory.

11.5 Distinguish between contour length, root-mean-square separation, and radius of gyration of a random coil.

11.6 Identify the terms in and limit the generality of the following expressions: (a) $V = -Q_2\mu_1/4\pi\varepsilon_0 r^2$, (b) $V = Q_2\mu_1\cos\theta/4\pi\varepsilon_0 r^2$, (c) $V = \mu_2\mu_1(1 - 3\cos^2\theta)/4\pi\varepsilon_0 r^3$, (d) $R_{rms} = N^{1/2}l$, and (e) $R_g = (N/6)^{1/2}l$.

11.7 Distinguish between an α helix, an anti-parallel β sheet, and a parallel β sheet.

11.8 Which amino acids have side chains that can interact with molecules (such as other amino acids or enzyme substrates) at pH 7 through (a) Coulombic interactions, (b) hydrogen bonding,

or (c) hydrophobic interactions (Section 2.7)? *Hint*: Consult data from Tables 4.6 and 11.9.

11.9 Why are DNA sequences rich in C–G base pairs more stable than sequences rich in A–T base pairs?

11.10 Discuss the factors that lead to bent, linear, and branched structures in polysaccharides.

11.11 (a) Distinguish between micelles, liposomes, bilayers, vesicles, and membranes. (b) Discuss the role of the surfactant parameter as a predictor of the shape of a micelle.

11.12 It is observed that the critical micelle concentration of sodium dodecyl sulfate in aqueous solution decreases as the concentration of added sodium chloride increases. Explain this effect.

11.13 Distinguish between the fluid mosaic and lipid raft models for motion of integral proteins in a biological membrane.

11.14 Distinguish between molecular mechanics, molecular dynamics, and Monte Carlo calculations. Why are these methods generally more popular in biochemical research than the quantum mechanical procedures discussed in Chapter 10?

Exercises

11.15 The data from a sedimentation equilibrium experiment performed at 300 K on a macromolecular solute in aqueous solution show that a graph of $\ln c$ against r^2 is a straight line with slope 729 cm^{-2}. The rotational rate of the centrifuge was 50 000 r.p.m. The specific volume of the solute is $v_s = 0.61$ cm^3 g^{-1}. Calculate the molar mass of the solute. *Hint*: Use eqn 11.3 and take $\rho = 1.00$ g cm^{-3}.

11.16 Find the drift speed of a particle of radius 20 m and density 1750 kg m^{-3} that is settling from suspension in water (density = 1000 kg m^{-3}) under the influence of gravity alone. The viscosity of water is 8.9×10^4 kg m^{-1} s^{-1}.

11.17 At 20°C the diffusion coefficient of a macromolecule is found to be 8.3×10^{-11} m^2 s^{-1}. Its sedimentation constant is 3.2 Sv in a solution of density 1.06 g cm^{-3}. The specific volume of the macromolecule is 0.656 cm^3 g^{-1}. Determine the molar mass of the macromolecule.

11.18 Calculate the speed of operation (in r.p.m.) of an ultracentrifuge needed to obtain a readily measurable concentration gradient in a sedimentation equilibrium experiment. Take that gradient to be a concentration at the bottom of the cell about five times greater than that at the top. Use $r_{top} = 5.0$ cm, $r_{bottom} = 7.0$ cm, $M \approx 10^5$ g mol^{-1}, $\rho v_s \approx 0.75$, $T = 298$ K.

11.19 Mass spectrometry can be used for sizing DNA molecules. To appreciate the power of the technique, consider the analysis by MALDI-TOF of a mixture of fragments of pBR 322 DNA. It was observed that the time of flight, t, varied with n_{bp}, the number of base pairs, as follows:

$t/\mu s$	39.03	66.43	96.28	121.25	154.01
n_{bp}	9	34	76	123	201

$t/\mu s$	189.67	217.23	247.81	269.05
n_{bp}	307	404	527	622

(a) Plot n_{bp} against t and then against t^2. Which plot is linear? Explain the physical origin of the linear relationship. (b) What time of flight would be observed for a fragment with 238 base pairs?

11.20 Draw a set of points as a rectangular array based on unit cells of side a and b, and mark the planes with Miller indices (10), (01), (11), (12), (23), (41), and (41).

11.21 Repeat *Exercise* 11.20 for an array of points in which the a and b axes make 60° to each other.

11.22 In a certain unit cell, planes cut through the crystal axes at $(2a,3b,c)$, (a,b,c), $(6a,3b,3c)$, and $(2a,-3b,-3c)$. Identify the Miller indices of the planes.

11.23 Draw an orthorhombic unit cell and mark on it the (100), (010), (001), (011), (101), and (101) planes.

11.24 (a) Calculate the separations of the planes (111), (211), and (100) in a crystal in which the cubic unit cell has sides of length 532 pm. (b) Calculate the separations of the planes (123) and (236) in an orthorhombic crystal in which the unit cell has sides of lengths 0.754, 0.623, and 0.433 nm.

11.25 The glancing angle of a Bragg reflection from a set of crystal planes separated by 97.3 pm is 19.85°. Calculate the wavelength of the X-rays.

11.26 Construct the electron density along the x-axis of a crystal given the following structure factors:

h	0	1	2	3	4	
F_h	+30.0	+8.2	+6.5	+4.1	+5.5	

h	5	6	7	8	9	
F_h	−2.4	+5.4	+3.2	−1.0	+1.1	

h	10	11	12	13	14	15
F_h	+6.5	+5.2	−4.3	−1.2	+0.1	+2.1

11.27 Consider the electrostatic model of the hydrogen bond. The N–C distance of the hydrogen bonded groups in proteins, such as occur in an α helix, is 0.29 nm. How much energy (in kJ mol^{-1}) is required to break the hydrogen bond (a) in a vacuum ($\varepsilon_r = 1$), (b) in a membrane (essentially a liquid hydrocarbon with $\varepsilon_r = 2.0$), and (c) in water ($\varepsilon_r = 80.0$)?

11.28 Estimate the dipole moment of an HCl molecule from the electronegativities of the elements and express the answer in debye and coulomb meters (C m).

11.29 The technique of vector addition can be used to predict the dipole moment of a molecule. The resultant μ_{res} of two dipole moments μ_1 and μ_2 that make an angle θ to each other is approximately

$$\mu_{res} \approx (\mu_1^2 + \mu_2^2 + 2\mu_1\mu_2 \cos\theta)^{1/2}$$

(a) Calculate the resultant of two dipoles of magnitude 1.50 D and 0.80 D that make an angle 109.5° to each other. (b) Estimate the ratio of the electric dipole moments of *ortho* (1,2-) and *meta* (1,3-) disubstituted benzenes.

11.30 Calculate the electric dipole moment of a glycine molecule using the partial charges in Table 11.2 and the locations of the atoms shown in (**19**).

(34,146,−98)
(−199,−1,−100)
(−101,−11,−126)
(−195,70,−38)
(−86,118,37)
(34,146,−98)

(49,−107,88)
(129,−146,126)
(82,−15,34)
(199,16,−38)

19 Glycine

11.31 (a) Plot the magnitude of the electric dipole moment of hydrogen peroxide as the H–O–O–H (azimuthal) angle ϕ changes. Use the dimensions shown in (**20**). (b) Devise a way for depicting how the angle as well as the magnitude changes.

90°   97
149
H ϕ H

20 Hydrogen peroxide

11.32 Calculate the molar energy required to reverse the direction of a water molecule located (a) 100 pm and (b) 300 pm from a Li$^+$ ion initially with the O atom closest to the ion. Take the dipole moment of water as 1.85 D.

11.33 Show, by following the procedure in *Justification* 11.4, that eqn 11.17 describes the potential energy of two electric dipole moments in the orientation shown in structure (**9**) of the text.

11.34 (a) What are the units of the polarizability α? (b) Show that the units of polarizability volume are cubic meters (m^3).

11.35 The magnitude of the electric field at a distance r from a point charge Q is equal to $Q/4\pi\varepsilon_0 r^2$. How close to a water molecule (of polarizability volume 1.48×10^{-30} m^3) must a proton approach before the dipole moment it induces is equal to the permanent dipole moment of the molecule (1.85 D)?

11.36 Phenylanine (**21** and Atlas A14) is a naturally occurring amino acid with a benzene ring. What is the energy of interaction between its benzene ring and the electric dipole moment of a neighboring peptide group? Take the distance between the groups as 4.0 nm and treat the benzene ring as benzene itself and the phenyl group as benzene molecules. The dipole moment of the peptide group is $\mu = 2.7$ D and the polarizability volume of benzene is $\alpha' = 1.04 \times 10^{-29}$ m^3.

H$_2$N   OH

21 Phenylalanine

11.37 Now consider the London interaction between the benzene rings of two Phe residues (see *Exercise* 11.36). Estimate the potential energy of attraction between two such rings (treated as benzene molecules) separated by 4.0 nm. For the ionization energy, use $I = 5.0$ eV.

11.38 In a region of the oxygen-storage protein myoglobin, the OH group of a tyrosine residue is hydrogen bonded to the N atom of a histidine residue in the geometry shown in (**22**). Use the partial charges in Table 11.2 to estimate the potential energy of this interaction.

HO   NH$_2$ O
                              OH
O
Tyr   97.5
              H
              104.3   His   NH$_2$
              N
              NH

22 Tyr–His

11.39 Given that force is the negative slope of the potential energy, calculate the distance dependence of the force acting between two nonbonded groups of atoms in a polypeptide chain that have a London dispersion interaction with each other. What is the separation at which the force is zero? *Hint*: Calculate the slope by considering the potential energy at R and $R + \delta R$, with $\delta R \ll R$, and evaluating $\{V(R + \delta R) - V(R)\}/\delta R$. You should use the expansion in *Justification* 11.4 together with

$$(1 \pm x + \cdots)^6 = 1 \pm 6x + \cdots$$
$$(1 \pm x + \cdots)^{12} = 1 \pm 12x + \cdots$$

At the end of the calculation, let δR become vanishingly small.

11.40 Repeat *Exercise* 11.39 by noting that $F = -dV/dr$ and differentiating the expression for V.

11.41 Acetic acid vapor contains a proportion of planar, hydrogen-bonded dimers (**23**). The apparent dipole moment of molecules in pure gaseous acetic acid increases with increasing temperature. Suggest an interpretation of the latter observation.

23 Acetic acid dimer

11.42 Consider the arrangement shown in Fig. 11.28 for a system consisting of an O–H group and an O atom, and then use the electrostatic model of the hydrogen bond to calculate the dependence of the molar potential energy of interaction on the angle θ. Set the partial charges on H and O to $+0.45e$ and $-0.83e$, respectively, and take $R = 200$ pm and $r = 95.7$ pm.

11.43 Considering the pattern of hydrogen bonding in β sheets and your answer to *Exercise* 11.42, explain why parallel β sheets are not common in proteins.

11.44 We can explore bond torsion in ethane to understand the barrier to internal rotation of one bond relative to another in saturated carbon chains, such as those found in lipids. The potential energy of a CH_3 group in ethane as it is rotated around the C–C bond can be written $V = \frac{1}{2}V_0(1 + \cos 3\phi)$, where ϕ is the azimuthal angle (**24**) and $V_0 = 11.6$ kJ mol^{-1}. (a) What is the change in potential energy between the *trans* and fully eclipsed conformations? (b) Show that for small variations in angle, the torsional (twisting) motion around the C–C bond can be expected to be that of a harmonic oscillator. (c) Estimate the vibrational frequency of this torsional oscillation.

24 Ethane

11.45 A certain macromolecule consists of 700 segments, each 0.90 nm long. If the chain were ideally flexible, what would be the r.m.s. separation of the ends of the chain?

11.46 Calculate the contour length (the length of the extended chain) and the root mean square separation (the end-to-end distance) for a macromolecule consisting of C–C links and with a molar mass of 280 kg mol^{-1}.

11.47 The radius of gyration of a macromolecule is found to be 7.3 nm. The chain consists of C–C links. Assume the chain is randomly coiled and estimate the number of links in the chain.

11.48 The radius of gyration of a solid sphere of radius R is $R_g = (\frac{3}{5})^{1/2}R$. (a) Write an expression for the molar volume of a spherical macromolecule in terms of its radius and then show that

$$R_g/\text{nm} = 0.0566\,902 \times \{(v_s/\text{cm}^3\,\text{g}^{-1})(M/\text{g mol}^{-1})\}^{1/3}$$

where v_s is the specific volume (the reciprocal of the density) and M the molar mass. (b) Use the information below and the expression for the radius of gyration of a solid sphere from part (a) to classify the species below as globular or rod-like.

	$M/(\text{g mol}^{-1})$	$v_s/(\text{cm}^3\,\text{g}^{-1})$	R_g/nm
Serum albumin	66×10^3	0.752	2.98
Bushy stunt virus	10.6×10^6	0.741	12.0
DNA	4×10^6	0.556	117.0

11.49 Suppose that a rodlike DNA molecule of length 250 nm undergoes a conformational change to a closed-circular (cc) form. (a) Use the information in *Exercise* 11.48 and an incident wavelength $\lambda = 488$ nm to calculate the ratio of scattering intensities by each of these conformations, $I_{\text{rod}}/I_{\text{cc}}$, when $\theta = 20°$, $45°$, and $90°$. (b) Suppose that you wish to use light scattering as a technique for the study of conformational changes in DNA molecules. Based on your answer to part (a), at which angle would you conduct the experiments? Justify your choice.

11.50 What is the change in conformational entropy when a random coil is stretched from fully coiled by 10 per cent?

11.51 The success of a molecular mechanics or molecular dynamics simulation depends on the proper choice of expressions for the calculation of the conformational energy. Suppose you distrusted the Lennard-Jones (12,6) potential for assessing a particular polypeptide conformation and replaced the repulsive term by an exponential function of the form $e^{-r/\sigma}$. (a) Sketch the form of the potential energy and locate the distance at which it is a minimum. (b) Identify the distance at which the exponential-6 potential energy is a minimum.

11.52 Derivatives of the compound TIBO (**25**) inhibit the enzyme reverse transcriptase, which catalyzes the conversion of retroviral RNA to DNA. A QSAR analysis of the activity A of a number of TIBO derivatives suggests the following equation:

$$\log A = b_0 + b_1 S + b_2 W$$

25 TIBO derivatives

where S is a parameter related to the drug's solubility in water and W is a parameter related to the width of the first atom in a substituent X shown in (**25**). (a) Use the following data to determine the values of b_0, b_1, and b_2. *Hint*: The QSAR equation relates one dependent variable, $\log A$, to two independent variables, S and W. To fit the data, you must use the mathematical procedure of *multiple regression*, which can be performed with mathematical software or an electronic spreadsheet.

X	H	Cl	SCH_3	OCH_3	CN
$\log A$	7.36	8.37	8.3	7.47	7.25
S	3.53	4.24	4.09	3.45	2.96
W	1.00	1.80	1.70	1.35	1.60

X	CHO	Br	CH_3	CCH
$\log A$	6.73	8.52	7.87	7.53
S	2.89	4.39	4.03	3.80
W	1.60	1.95	1.60	1.60

(b) What should be the value of W for a drug with $S = 4.84$ and $\log A = 7.60$?

Projects

11.53 Molecular orbital calculations may be used to predict the dipole moments of molecules.

(a) Using molecular modeling software and the computational method recommended by your instructor (extended Hückel, semi-empirical, *ab initio*, or DFT methods), calculate the dipole moment of the peptide link, modeled as a *trans-N*-methylacetamide (**26**).

26 *trans-N*-methylacetamide

(b) Plot the energy of interaction between two dipoles with dipole moments calculated in part (a) against the angle θ for $r = 3.0$ nm (see eqn 11.17).

(c) Compare the maximum value of the dipole–dipole interaction energy from part (b) to 20 kJ mol⁻¹, a typical value for the energy of a hydrogen-bonding interaction in biological systems. Comment on the similarity or disparity between the two values.

11.54 Molecular orbital calculations can be used to predict structures of intermolecular complexes. Hydrogen bonds between purine and pyrimidine bases are responsible for the double helix structure of DNA. Consider methyl adenine (**27**, with R = CH₃) and methyl thymine (**28**, with R = CH₃) as models of two bases that can form hydrogen bonds in DNA (where R would be replaced by deoxyribose).

27　　　　　**28**

(a) Using molecular modeling software and the computational method recommended by your instructor (extended Hückel,

semi-empirical, *ab initio*, or DFT methods), calculate the atomic charges of all atoms in methyl adenine and methyl thymine.

(b) Based on your tabulation of atomic charges, identify the atoms in methyl adenine and methyl thymine that are likely to participate in hydrogen bonds.

(c) Draw all possible adenine–thymine pairs that can be linked by hydrogen bonds, keeping in mind that linear arrangements of the A–H⋯B fragments are preferred in DNA. For this step, you may want to use your molecular modeling software to align the molecules properly.

(d) Which of the pairs that you drew in part (c) occur naturally in DNA molecules?

(e) Repeat parts (a)–(d) for cytosine and guanine, which also form base pairs in DNA.

11.55 Now you will use molecular mechanics software of your instructor's choice to gain some appreciation for the complexity of the calculations that lead to plots such as those in Fig. 11.42. Our model for the protein is the dipeptide (**29**) in which the terminal methyl groups replace the rest of the polypeptide chain.

29

(a) Draw three initial conformers of (**29**) with R = H: one with $\phi = 75°$, $\psi = -65°$, a second with $\phi = \psi = 180°$, and a third with $\phi = 65°$, $\psi = 35°$. Use a molecular mechanics routine to optimize the geometry of each conformer and measure the total potential energy and the final ϕ and ψ angles in each case. Did all of the initial conformers converge to the same final conformation? If not, what do these final conformers represent? Rationalize any observed differences in total potential energy of the final conformers.

(b) Use the approach in part (a) to investigate the case R = CH₃, with the same three initial conformers as starting points for the calculations. Rationalize any similarities and differences between the final conformers of the dipeptides with R = H and R = CH₃.

PART 4 Biochemical spectroscopy

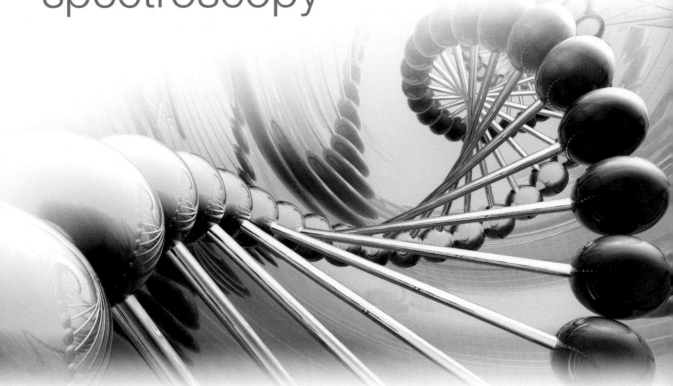

We now begin our study of molecular spectroscopy, the analysis of the electromagnetic radiation emitted, absorbed, or scattered by molecules. The starting point for the discussion in the next two chapters is the observation summarized in Chapter 9 that photons of radiation ranging from the infrared to the ultraviolet bring information to us about molecules as a result of electronic and vibrational transitions. In Chapter 12 we describe techniques used to study these transitions in biological systems and see how electronic transitions prepare molecules for such important light-induced processes as vision and photosynthesis. In Chapter 13 we see that the combined effect of an external magnetic field and molecular excitation with photons in the radiofrequency or microwave ranges leads to important spectroscopic techniques, collectively known as magnetic resonance spectroscopy, that are widely used in biochemical studies and diagnostic procedures. In short, molecular spectra are complicated but contain a great deal of information, including bond lengths, bond angles, and bond strengths, that can be used to analyze biological systems ranging in size from small co-factors to biopolymers and to whole biological cells. Along the way, we also see how molecular spectra complement information on biomolecular structure obtained from the diffraction techniques discussed in Chapter 11.

Optical spectroscopy and photobiology

12

In this chapter we consider electromagnetic radiation as a probe that provides information on molecular structure that is complementary to that provided by X-ray diffraction (Chapter 11). Indeed, there are several reasons why **spectroscopy**, the study of absorption, emission, and scattering of electromagnetic radiation, is sometimes the only suitable technique at the disposal of a biochemist. In the first place, the sample might be a mixture of molecules, in which case sharp X-ray diffraction images are not obtained. Even if all the molecules in the sample are identical, it might prove impossible to obtain a single crystal of sufficient quality. Furthermore, although work on proteins and nucleic acids has shown how immensely interesting and motivating X-ray diffraction data can be, the information is incomplete. For instance, what can be said about the shape of the molecule in its natural environment, a biological cell? What can be said about the response of its shape to changes in its environment? To answer these questions, we begin the chapter with a discussion of the general principles of molecular spectroscopy with radiation of frequencies that span over eight orders of magnitude, from radiofrequencies (10^8 Hz) up to the ultraviolet (10^{16} Hz). We focus on *vibrational spectra*, which report on molecular vibrations excited by the absorption or scattering of electromagnetic radiation, and *ultraviolet and visible spectra*, which probe the electronic distribution in a molecule and result from the absorption or emission of ultraviolet and visible radiation.

An understanding of the ability of molecules to absorb light is essential for understanding how light can induce physical and chemical change, and we end the chapter with a description of light as an initiator of many biochemical reactions. As remarked in the *Prolog*, essentially all the energy required for the sustenance of life on Earth is absorbed during photosynthesis in plants, algae, and some bacteria. Here we see how these organisms optimize the rates of the reactions that capture and make initial use of solar energy. But light also plays additional roles in biology and medicine, so we describe vision, damage of DNA by ultraviolet radiation, and one of many laser-based therapies now available.

General features of spectroscopy

There are three varieties of spectroscopy:

- **emission spectroscopy**, in which a molecule undergoes a transition from a state of high energy, E_2, to a state of lower energy, E_1, and emits the excess energy as a photon (Fig. 12.1)
- **absorption spectroscopy**, in which the absorption of radiation is monitored as the frequency of the radiation is swept over a range

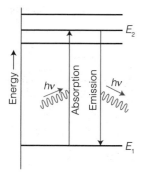

Fig. 12.1 In emission spectroscopy, a molecule returns to a lower state (typically the ground state) from an excited state and emits the excess energy as a photon. The same transition can be observed in absorption, when the incident radiation supplies a photon that can excite the molecule from its ground state to an excited state.

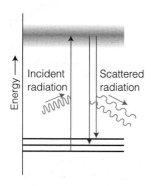

Fig. 12.2 In Raman spectroscopy, an incident photon is scattered from a molecule with either an increase in frequency (if the radiation collects energy from the molecule) or with a lower frequency if it loses energy to the molecule. The process can be regarded as taking place by an excitation of the molecule to a wide range of states (represented by the shaded band) and the subsequent return of the molecule to a lower state; the net energy change is then carried away by the photon.

- **Raman spectroscopy**, in which an intense, monochromatic (single-frequency) incident beam is passed through the sample and the frequencies present in the radiation scattered by the sample are recorded (Fig. 12.2).

The energy of a photon emitted or absorbed, and therefore the frequency, v (nu), of the radiation emitted or absorbed, is given by the Bohr frequency condition (Section 9.1):

$$hv = |E_1 - E_2| \qquad \boxed{\text{Bohr frequency condition}} \quad (12.1)$$

Here E_1 and E_2 are the energies of the two states between which the transition occurs and h is Planck's constant.[1] This relation is often expressed in terms of the wavelength, λ (lambda), of the radiation by using the relation

$$\lambda = \frac{c}{v} \qquad \boxed{\begin{array}{l}\text{Relation between}\\ \text{wavelength and frequency}\end{array}} \quad (12.2a)$$

where c is the speed of light or, in terms of the wavenumber, $\tilde{v}$ (nu tilde):

$$\tilde{v} = \frac{1}{\lambda} = \frac{v}{c} \qquad \boxed{\text{Definition of wavenumber}} \quad (12.2b)$$

The units of wavenumber are almost always chosen as reciprocal centimeters (cm^{-1}), so we can picture the wavenumber of radiation as the number of complete wavelengths per centimeter. The frequencies, wavelengths, and wavenumbers of the various regions of the electromagnetic spectrum were summarized in Fig. F.7. In this chapter we concentrate on vibrational and electronic transitions, which can be excited by the absorption of infrared and ultraviolet–visible radiation, respectively.

In Raman spectroscopy, molecular energy levels are explored by examining the frequencies present in the radiation scattered by molecules. About 1 in 10^7 of the incident photons collide with the molecules, give up some of their energy, and emerge with a lower energy. These scattered photons constitute the lower-frequency **Stokes radiation** from the sample. Other incident photons may collect energy from the molecules (if they are already vibrationally excited) and emerge as higher-frequency **anti-Stokes radiation**. The component of radiation scattered into the forward direction without change of frequency is called **Rayleigh radiation**. Raman spectra may be examined using visible and ultraviolet lasers, in which case a diffraction grating is used to distinguish between Rayleigh, Stokes, and anti-Stokes radiation. In Fourier-transform Raman spectrometers, radiation scattered by the sample passes through a Michelson interferometer.

In the laboratory 12.1   *Experimental techniques*

To design and interpret spectroscopic measurements on biological systems, we need to become acquainted with the instruments that generate and detect electromagnetic radiation in the infrared, visible, and ultraviolet regions. A **spectrometer** is an instrument that detects the characteristics of light scattered, emitted, or absorbed by atoms and molecules. Figure 12.3 shows the general layout of an absorption spectrometer operating in the ultraviolet and visible regions of the spectrum. Radiation from an appropriate source is directed toward a sample. In most spectrometers, light transmitted, emitted,

---

[1] Raman scattering is a special case, and we deal with it later.

or scattered by the sample is collected by mirrors or lenses and strikes a dispersing element that separates radiation into different frequencies. The intensity of light at each frequency is then analyzed by a suitable detector.

The source in a spectrometer typically produces radiation spanning a range of frequencies, but in a few cases (including lasers) it generates nearly monochromatic radiation. For the far-infrared ($35 \text{ cm}^{-1} < \tilde{v} < 200 \text{ cm}^{-1}$), the source is commonly a mercury arc inside a quartz envelope, most of the radiation being generated by the hot quartz. A *Nernst filament* or *globar* is used to generate radiation in the mid-infrared ($200 \text{ cm}^{-1} < \tilde{v} < 4000 \text{ cm}^{-1}$) and consists of a heated ceramic filament containing lanthanoid oxides. For the visible region of the spectrum, a *tungsten–iodine lamp* is used, which gives out intense white light. A discharge through deuterium gas or xenon in quartz is still widely used for the near-ultraviolet.

The dispersing element of choice in modern instruments operating in the ultraviolet and visible ranges uses a *diffraction grating*, a glass or ceramic plate into which fine grooves have been cut about 1000 nm apart (a spacing comparable to the wavelength of visible light) and covered with a reflective aluminum coating. The grating causes interference between waves reflected from its surface, and constructive interference occurs at specific angles that depend on the wavelength of the radiation being used. Thus, each wavelength of light is directed into a specific direction (Fig. 12.4). In a *monochromator*, a narrow exit slit allows only a narrow range of wavelengths to reach the detector and rotating the grating on an axis perpendicular to the incident and diffracted beams allows different wavelengths to be analyzed; in this way, the absorption or emission spectrum is built up one narrow wavelength range at a time. In a *polychromator* there is no slit and a broad range of wavelengths can be analyzed simultaneously by *array detectors*, such as those discussed below.

Modern spectrometers operating in the infrared and near-infrared almost always use Fourier transform techniques of spectral detection and analysis. The heart of a Fourier-transform (FT) spectrometer is a *Michelson interferometer*, a device for analyzing the frequencies present in a composite signal. The total signal from a sample is like a chord played on a piano, and the Fourier transform of the signal is equivalent to the separation of the chord into its individual notes, its spectrum. A major advantage of the Fourier transform procedure is that all the radiation emitted by the source is monitored continuously. This is in contrast to a conventional spectrometer, in which a monochromator discards most of the generated radiation. As a result, Fourier transform spectrometers have a higher sensitivity than conventional spectrometers.

The detector is a device that converts radiation into an electric current or potential difference for appropriate signal processing and display. Detectors may consist of a single radiation sensing element or of several small elements arranged in one- or two-dimensional arrays. A common detector is the *photodiode*, a solid-state device that conducts electricity when struck by photons because light-induced electron transfer reactions in the detector material create mobile charge carriers (negatively charged electrons and positively charged 'holes'). With appropriate choice of material, photodiodes can be used to detect light spanning a wide range of wavelengths. For example, silicon is sensitive in the visible region and germanium is used in most spectrometers operating in the near-infrared region of the spectrum.

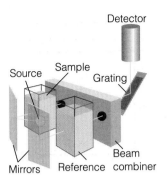

Fig. 12.3 The layout of a typical absorption spectrometer, in which the exciting beams of radiation pass alternately through a sample and a reference cell, and the detector is synchronized with them so that the relative absorption can be determined.

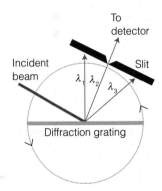

Fig. 12.4 A beam of light is dispersed by a diffraction grating into three component wavelengths λ_1, λ_2, and λ_3. In the configuration shown, only radiation with λ_2 passes through a narrow slit and reaches the detector. Rotating the diffraction grating in the direction shown by the double arrows allows other wavelengths to reach the detector.

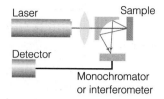

Fig. 12.5 A common arrangement adopted in Raman spectroscopy. A laser beam first passes through a lens and then through a small hole in a mirror with a curved reflecting surface. The focused beam strikes the sample and scattered light is both deflected and focused by the mirror. The spectrum is analyzed by a monochromator or an interferometer.

A *charge-coupled device* (CCD) is a two-dimensional array of millions of small photodiode detectors. With a CCD, a wide range of wavelengths that emerge from a polychromator are detected simultaneously, thus eliminating the need to measure light intensity one narrow wavelength range at a time. CCD detectors are used widely to measure absorption, emission, and Raman scattering.

The most common detectors found in commercial infrared spectrometers are sensitive in the mid-infrared region. An example is the mercury–cadmium–telluride (MCT) detector, a *photovoltaic device* for which the potential difference changes on exposure to infrared radiation.

In a typical Raman experiment, a laser beam is passed through the sample and the radiation scattered from the front face of the sample is monitored (Fig. 12.5). This detection geometry allows for the study of gases, pure liquids, solutions, suspensions, and solids.

12.1 The intensities of spectroscopic transitions: empirical aspects

To put spectrometers to good use in biochemical studies, we need to understand the factors that control the intensity of a spectroscopic transition.

We now focus on absorption spectroscopy.

(a) The Beer–Lambert law

The intensity of absorption of radiation at a particular wavelength passing through a uniform sample is related to the concentration [J] of the absorbing species J by the empirical **Beer–Lambert law** (Fig. 12.6, and commonly simply 'Beer's law'; we first encountered the law in *In the laboratory* 6.1 as a way of monitoring the concentrations of species in reactions. The law is commonly written

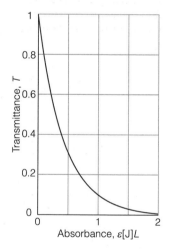

Fig. 12.6 The intensity of light transmitted by a uniform absorbing sample decreases exponentially with the path length through the sample (the length is proportional to the absorbance when $\varepsilon[J]$ is a constant).

$$I = I_0 10^{-\varepsilon[J]L} \qquad \text{Beer–Lambert law} \qquad (12.3)$$

where I_0 and I are the incident and transmitted intensities, respectively, L is the length of the sample, and ε (epsilon) is the **molar absorption coefficient**, which depends on the wavelength of the incident radiation. The dimensions of ε are l/(concentration × length), and it is normally convenient to express it in cubic decimeters (liters) per mole per centimeter ($dm^3 \ mol^{-1} \ cm^{-1}$), which are sensible when [J] is expressed in moles per cubic decimeter and L is in centimeters. The Beer–Lambert law is an empirical result. However, as we show in the following *Justification*, it is simple to account for its form. The law may be expressed in terms of the **absorbance**, A, of the sample or the **transmittance**, T:

$$A = \log \frac{I_0}{I} \qquad \text{Definition of absorbance} \qquad (12.4a)$$

$$T = \frac{I}{I_0} \qquad \text{Definition of transmittance} \qquad (12.4b)$$

It then follows from eqn 12.3 that

$$A = \varepsilon[J]L \qquad A = -\log T \qquad T = 10^{-\varepsilon[J]L} \qquad (12.5)$$

Justification 12.1 *The Beer–Lambert law*

We think of the sample as consisting of a stack of infinitesimal slices, like sliced bread (Fig. 12.7). The thickness of each slice is dx. The change in intensity, dI, that occurs when electromagnetic radiation passes through one particular slice is proportional to the thickness of the slice, the concentration of the absorber J, and the intensity of the incident radiation at that slice of the sample, so $dI \propto [J]I\,dx$. Because dI is negative (the intensity is reduced by absorption), we can write

$$dI = -\kappa[J]I\,dx$$

where κ (kappa) is the proportionality coefficient. Division by I gives

$$\frac{dI}{I} = -\kappa[J]\,dx$$

This expression applies to each successive slice. To obtain the intensity that emerges from a sample of thickness L when the intensity incident on one face of the sample is I_0, we sum all the successive changes. Because a sum over infinitesimally small increments is an integral, we write

$$\int_{I_0}^{I} \frac{dI}{I} = -\kappa \int_0^L [J]\,dx$$

If the concentration is uniform, [J] is independent of location and can be taken outside the integral, and we obtain

$$\ln \frac{I}{I_0} = -\kappa[J]L$$

Because the relation between natural and common logarithms is $\ln x = \ln 10 \times \log x$, we can write $\varepsilon = \kappa/\ln 10$ and obtain

$$\log \frac{I}{I_0} = -\varepsilon[J]L$$

which, on taking antilogarithms, is the Beer–Lambert law (eqn 12.3).

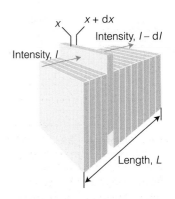

Fig. 12.7 To establish the theoretical basis of the empirical Beer–Lambert law, the sample is supposed to be sliced into a large number of planes. The reduction in intensity caused by one plane is proportional to the intensity incident on it (after passing through the preceding planes), the thickness of the plane, and the concentration of absorbing species.

Example 12.1 *The molar absorption coefficient of tryptophan*

Radiation of wavelength 280 nm passed through 1.0 mm of a solution that contained an aqueous solution of the amino acid tryptophan at a concentration of 0.50 mmol dm^{-3}. The light intensity is reduced to 54 per cent of its initial value (so $T = 0.54$). Calculate the absorbance and the molar absorption coefficient of tryptophan at 280 nm. What would be the transmittance through a cell of thickness 2.0 mm?

Strategy From $A = -\log T = \varepsilon[J]L$, it follows that

$$\varepsilon = -\frac{\log T}{[J]L}$$

For the transmittance through the thicker cell, we use $T = 10^{-A}$ and the value of ε calculated here.

Solution The molar absorption coefficient is

$$\varepsilon = -\frac{\log 0.54}{(5.0 \times 10^{-4}\ \mathrm{mol\ dm^{-3}}) \times (1.0\ \mathrm{mm})} = 5.4 \times 10^2\ \mathrm{dm^3\ mol^{-1}\ mm^{-1}}$$

These units are convenient for the rest of the calculation (but the outcome could be reported as $5.4 \times 10^3\ \mathrm{dm^3\ mol^{-1}\ cm^{-1}}$ if desired). The absorbance is

$$A = -\log 0.54 = 0.27$$

The absorbance of a sample of length 2.0 mm is

$$A = (5.4 \times 10^2\ \mathrm{dm^3\ mol^{-1}\ mm^{-1}}) \times (5.0 \times 10^{-4}\ \mathrm{mol\ dm^{-3}}) \times (2.0\ \mathrm{mm}) = 0.54$$

It follows that the transmittance is now

$$T = 10^{-A} = 10^{-0.54} = 0.29$$

That is, the emergent light is reduced to 29 per cent of its incident intensity.

Self-test 12.1 The transmittance of an aqueous solution that contained the amino acid tyrosine at a molar concentration of 0.10 mmol dm^{-3} was measured as 0.14 at 240 nm in a cell of length 5.0 mm. Calculate the molar absorption coefficient of tyrosine at that wavelength and the absorbance of the solution. What would be the transmittance through a cell of length 1.0 mm?

Answer: $1.7 \times 10^4\ \mathrm{dm^3\ mol^{-1}\ cm^{-1}}$, $A = 0.17$, $T = 0.68$

One measure of the intensity of a transition is the maximum value of the molar absorption coefficient, ε_{max}. However, because absorption bands generally spread over a range of wavenumbers, the absorption at a single wavenumber might not give a true indication of the intensity. The latter is best reported as the *integrated absorption coefficient*, $\mathcal{A}$, the area under the plot of the molar absorption coefficient against wavenumber (Fig. 12.8).

(b) The determination of concentration

Beer's law is used to determine the concentrations of species of known molar absorption coefficients. To do so, we measure the absorbance of a sample and rearrange the first relation in eqn 12.5 into

$$[J] = \frac{A}{\varepsilon L}$$

The determination of concentration   (12.6)

It follows from this equation that we can observe the appearance or depletion of a species during a reaction by monitoring changes in the absorbance of the reaction mixture. This was the expression used in Chapter 6 to monitor concentrations to establish a rate law.

In biological applications, it is common to make measurements of absorbance at two wavelengths and use them to find the individual concentrations of two components A and B in a mixture. For this analysis, we write the total absorbance at a given wavelength as

$$A = A_A + A_B = \varepsilon_A[A]L + \varepsilon_B[B]L = (\varepsilon_A[A] + \varepsilon_B[B])L$$

Then, for two measurements of the total absorbance at wavelengths 1 and 2 at which the molar absorption coefficients are 1 and 2 (Fig. 12.9), we have

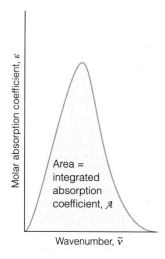

Fig. 12.8 The integrated absorption coefficient of a transition is the area under a plot of the molar absorption coefficient against the wavenumber of the incident radiation.

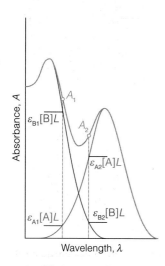

Fig. 12.9 The concentrations of two absorbing species in a mixture can be determined from their molar absorption coefficients and the measurement of their absorbances at two different wavelengths lying within their joint absorption region.

$$A_1 = (\varepsilon_{A1}[A] + \varepsilon_{B1}[B])L \qquad A_2 = (\varepsilon_{A2}[A] + \varepsilon_{B2}[B])L$$

We can solve these two simultaneous equations for the two unknowns (the molar concentrations of A and B) and find

$$[A] = \frac{\varepsilon_{B2}A_1 - \varepsilon_{B1}A_2}{(\varepsilon_{A1}\varepsilon_{B2} - \varepsilon_{A2}\varepsilon_{B1})L}$$

$$[B] = \frac{\varepsilon_{A1}A_2 - \varepsilon_{A2}A_1}{(\varepsilon_{A1}\varepsilon_{B2} - \varepsilon_{A2}\varepsilon_{B1})L}$$

Determination of two concentrations   (12.7)

There may be a wavelength, λ°, called the **isosbestic wavelength**,[2] at which the molar extinction coefficients of the two species are equal; we write this common value as ε°. The total absorbance of the mixture at the isosbestic wavelength is

$$A^\circ = \varepsilon^\circ([A] + [B])L$$

Absorbance at the isosbestic point   (12.8)

Even if A and B are interconverted in a reaction of the form $A \rightarrow B$ or its reverse, then because their total concentration remains constant, so does A°. As a result, one or more **isosbestic points**, which are invariant points in the absorption spectrum, may be observed (Fig. 12.10). It is very unlikely that three or more species would have the same molar extinction coefficients at a single wavelength. Therefore, the observation of an isosbestic point, or at least not more than one such point, is compelling evidence that a solution consists of only two solutes in equilibrium with each other with no intermediates.

12.2 The intensities of transitions: theoretical aspects

The intensity of a spectroscopic transition depends on a variety of factors, including the form of the wavefunctions of the initial and final states of the molecule and the population of the initial energy levels.

(a) The transition dipole moment

Whether or not an absorption band has a large integrated absorption coefficient (and, consequently, can be driven by the surrounding electromagnetic field) depends on a quantity called the **transition dipole moment**, μ_{fi}. The underlying classical idea is that, for the molecule to be able to interact with the electromagnetic field and absorb or create a photon of frequency ν, it must possess, at least transiently, a dipole oscillating at that frequency. This transient dipole is expressed quantum mechanically as

$$\mu_{fi} = \int \psi_f^\star \mu \psi_i \, d\tau$$

Transition dipole moment   (12.9)

where μ is the electric dipole moment operator, and ψ_i and ψ_f are the wavefunctions for the initial and final states, respectively. The size of the transition dipole can be regarded as a measure of the charge redistribution that accompanies a transition: a transition will be active (and generate or absorb photons) only if the accompanying charge redistribution is dipolar (Fig. 12.11). The intensity of the transition is proportional to the square of the transition dipole moment.

[2] The name isosbestic comes from the Greek words for 'the same' and 'extinguished'.

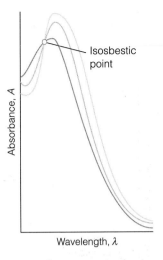

Fig. 12.10 One or more isosbestic points are formed when there are two interrelated absorbing species in solution. The three curves correspond to three different stages of the reaction $A \rightarrow B$.

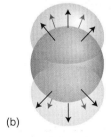

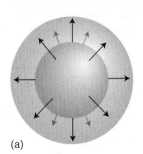

Fig. 12.11 The transition moment is a measure of the magnitude of the shift in charge during a transition. (a) A spherical redistribution of charge as in this transition has no associated dipole moment and does not give rise to electromagnetic radiation. (b) This redistribution of charge has an associated dipole moment.

A brief illustration

For a one-electron, one-dimensional system, like (to a good approximation) a carotene molecule, $\mu_x = -ex$, so

$$\mu_{x,\text{fi}} = -e \int \psi_f^* x \psi_i \, d\tau$$

For a conjugated hydrocarbon of N carbon atoms and length $L = (N-1)l_{CC}$, the first excitation energy is from the π orbital with $n = \frac{1}{2}N$ to the one above, so

$$\mu_{x,\text{fi}} = -e \left(\frac{2}{L}\right) \int_0^L \sin\left(\frac{(\frac{1}{2}N + 1)\pi x}{L}\right) x \sin\left(\frac{\frac{1}{2}N\pi x}{L}\right) dx$$

For N an odd number, this expression evaluates (using mathematical software) to $2eL/\pi^2 = 0.2eL$; virtually the same numerical value is obtained when N is an even number. This value suggests that the electron migrates through a distance of about 20 per cent of the length of the molecule when the transition takes place.

A **selection rule** is a statement about when the transition dipole can be non-zero. A **gross selection rule** specifies the general features a molecule must have if it is to have a spectrum of a given kind. For instance, we shall see that a molecule gives a vibrational absorption spectrum only if its electric dipole moment changes as the molecule vibrates. Once the gross selection rule has been recognized, we consider the **specific selection rule**, a statement about which changes in quantum number may occur in a transition. A transition that is permitted by a specific selection rule is classified as **allowed**. Transitions that are disallowed by a specific selection rule are called **forbidden**. Forbidden transitions sometimes occur weakly because the selection rule is based on an approximation that turns out to be slightly invalid.

(b) Stimulated and spontaneous transitions

The intensity of an absorption line is related to the rate at which energy from electromagnetic radiation at a specified frequency is absorbed by a molecule. Albert Einstein identified three contributions to the rates of transitions between states. **Stimulated absorption** is a transition from a low energy state to one of higher energy that is driven by the electromagnetic field oscillating at the transition frequency. Einstein reasoned that the more intense the electromagnetic field (the more intense the incident radiation), the greater the rate at which transitions are induced and hence the stronger the absorption by the sample, so he wrote the rate of stimulated absorption as

rate of stimulated absorption $= NBI$

where N is the number of molecules in the lower state, the constant B is the **Einstein coefficient of stimulated absorption**, and I is the intensity of radiation at the frequency of the transition. If B is large, then a given intensity of incident radiation will induce transitions strongly and the sample will be strongly absorbing.

Einstein considered that the radiation was also able to induce the molecule in the upper state to undergo a transition to the lower state and hence to generate a photon of frequency ν. Thus, he wrote the rate of this **stimulated emission** as

rate of stimulated emission $= N'B'I$

where N' is the number of molecules in the excited state and B' is the **Einstein coefficient of stimulated emission**. Note that only radiation of the same frequency as the transition can stimulate an excited state to fall to a lower state. However, Einstein realized that stimulated emission was not the only means by which the excited state could generate radiation and return to the lower state, and suggested that an excited state could undergo **spontaneous emission** at a rate that was independent of the intensity of the radiation (of any frequency) that is already present. He therefore wrote the total rate of transition from the upper to the lower state as

overall rate of emission $= N'(A + B'I)$

The constant A is the **Einstein coefficient of spontaneous emission**. It can be shown that the coefficients of stimulated absorption and emission are equal and that the coefficient of spontaneous emission is related to them by

$$A = \left(\frac{8\pi h v^3}{c^3} \right) B \qquad \boxed{\text{Relation between coefficients}} \quad (12.10)$$

The presence of v^3 in this relation implies that spontaneous emission can be largely ignored at the relatively low frequencies of vibrational transitions but may be important for transitions in the visible and ultraviolet regions. We shall see later (Sections 12.9 and 12.10) that spontaneous emission accounts for the phenomena of fluorescence and phosphorescence. Stimulated emission underlies the functioning of lasers ('laser' is an acronym formed from 'light amplification by the stimulated emission of radiation').

(c) Populations and intensities

The intensity of a spectroscopic transition depends on the number of molecules that are in the initial state. If we confine our attention to vibrational and electronic spectroscopy of molecules close to room temperature, then the situation is very simple: *almost all vibrational absorptions and all electronic absorptions occur from the ground state of a molecule*, because according to the Boltzmann distribution that is the only state significantly populated at room temperature. However, molecules can be prepared in short-lived excited states as a result of chemical reaction, electric discharge, or irradiation with an intense light source, including sunlight. In these cases the populations may be quite different from those at thermal equilibrium, and absorption and emission spectra—if they can be recorded quickly enough—then arise from transitions from all the populated levels.

A brief illustration

In *Fundamentals* F.3(b), we saw that the ratio of populations of states of energies E and E' is given by

$$\frac{N'}{N} = e^{-\Delta E/kT} \qquad \Delta E = E' - E$$

where k is Boltzmann's constant. Hence, for two vibrational states of a molecule separated by 45 zJ, which corresponds to 2300 cm^{-1}, we calculate $N'/N = 1.9 \times 10^{-5}$ at $T = 300$ K. We conclude that, at normal temperatures, almost all the molecules are in the ground vibrational state. It follows that the great majority of molecules are also in the ground electronic state because electronic states are typically more widely separated than vibrational states.

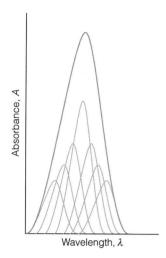

Fig. 12.12 An electronic absorption band consists of many superimposed bands that merge together to give a single broad band with unresolved vibrational structure.

(d) Linewidths

In condensed media, the 'width' of an electronic transition (that is, the range of wavenumbers, wavelengths, or frequencies over which there is a substantial absorption) results from the simultaneous excitation of molecular vibrations. When a vibration is stimulated by the migration of the electron from one orbital to another, the transition occurs at higher wavenumber than when no vibrational excitation occurs, so the absorption is slightly displaced. If several vibrational modes are stimulated, each with a different frequency, and if several vibrational levels of each one are stimulated, then the absorption occurs over a range of frequencies and the electronic transition appears as a broad feature in the spectrum (Fig. 12.12). We treat the excitation of vibration during an electronic transition in more detail in Section 12.6.

Even if only a single vibrational mode and a single level of that mode is stimulated during the electronic transition, the absorption would not be infinitely narrow. An important source of the broadening of the individual lines is the finite lifetime of the states involved in the transition. When the Schrödinger equation is solved for a system that is changing with time, it is found that the states of the system do not have precisely defined energies. If a state decays exponentially as $e^{-t/\tau}$ with a time constant τ (tau), which is called the **lifetime** of the state, then its energy levels are blurred by δE, where

$$\delta E \approx \frac{\hbar}{\tau} \qquad \boxed{\text{Lifetime broadening}} \qquad (12.11a)$$

We see that the shorter the lifetime of a state, the less well defined its energy. The energy spread inherent to the states of systems that have finite lifetimes is called **lifetime broadening**.[3] When we express the energy spread as a wavenumber by writing $\delta E = hc\delta\tilde{\nu}$ and use the values of the fundamental constants, the practical form of this relation becomes

$$\delta\tilde{\nu} \approx \frac{5.3 \text{ cm}^{-1}}{\tau/\text{ps}} \qquad (12.11b)$$

Only if τ is infinite can the energy of a state be specified exactly (with $\delta E = 0$). However, no excited state has an infinite lifetime; therefore, all states are subject to some lifetime broadening, and the shorter the lifetimes of the states involved in a transition, the broader the spectral lines.

Self-test 12.2 What is the width (expressed as a wavenumber) of a transition from a state with a lifetime of 5.0 ps?

Answer: 1.1 cm^{-1}

Two processes are principally responsible for the finite lifetimes of excited states and hence for the widths of transitions to or from them. One is **collisional deactivation**, which arises from collisions between molecules. The second is spontaneous emission. Because the rate of spontaneous emission cannot be changed (without changing the molecule), it is a natural limit to the lifetime of an excited state. The resulting lifetime broadening is the **natural linewidth** of

---

[3] Lifetime broadening is also called *uncertainty broadening*.

the transition. We have seen (eqn 12.10) that the rate of spontaneous emission increases as ν^3, so the natural lifetimes of electronic states (which are excited with ultraviolet to visible radiation) are very much shorter than for vibrational transitions (which are excited with infrared radiation). It follows that the natural linewidths of electronic transitions are much greater than those of vibrational transitions. For example, a typical electronic excited state natural lifetime is about 10^{-8} s (10 ns), corresponding to a natural width of about 5×10^{-4} cm^{-1} (equivalent to 15 MHz).

In the laboratory 12.2) *Biosensor analysis*

Biosensor analysis is a very sensitive and sophisticated optical technique that is now used routinely to measure the kinetics and thermodynamics of interactions between biopolymers. A biosensor detects changes in the optical properties of a surface in contact with a biopolymer.

The mobility of delocalized valence electrons accounts for the electrical conductivity of metals, and these mobile electrons form a **plasma**, a dense gas of charged particles. Bombardment of the plasma by light or an electron beam can cause transient changes in the distribution of electrons, with some regions becoming slightly more dense than others. Coulomb repulsion in the regions of high density causes electrons to move away from each other, so lowering their density. The resulting oscillations in electron density, called **plasmons**, can be excited both in the bulk and on the surface of a metal. Plasmons in the bulk may be visualized as waves that propagate through the solid. A surface plasmon also propagates away from the surface, but the amplitude of the wave, also called an **evanescent wave**, decreases sharply with distance from the surface. The decay constant for the decrease in amplitude is approximately the wavelength of the light being used.

Biosensor analysis is based on the phenomenon of **surface plasmon resonance** (SPR), the absorption of energy from an incident beam of electromagnetic radiation by surface plasmons. Absorption, or 'resonance', can be observed with appropriate choice of the wavelength and angle of incidence of the excitation beam. It is common practice to use a monochromatic beam and to vary the angle of incidence θ (Fig. 12.13). The beam passes through a prism that strikes one side of a thin film of gold or silver. The angle corresponding to light absorption depends on the refractive index of the medium in direct contact with the opposing side of the metallic film. This variation of the resonance angle with the state of the surface arises from the ability of the evanescent wave to interact with material a short distance away from the surface. For example, changing the identity and quantity of material on the surface changes the resonance angle. Hence, biosensor analysis can be used in the study of the binding of molecules to a surface or binding of ligands to a biopolymer attached to the surface; this interaction mimics the biological recognition processes that occur in cells. Examples of complexes amenable to analysis include antibody–antigen and protein–DNA interactions. The most important advantage of biosensor analysis is its sensitivity: it is possible to measure the deposition of nanograms of material on to a surface. The main disadvantage of the technique is its requirement for immobilization of at least one of the components of the system under study.

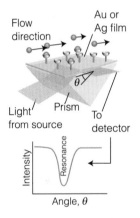

Fig. 12.13 The experimental arrangement for the observation of surface plasmon resonance, as explained in the text.

A brief comment
The refractive index, n_r, of the medium, the ratio of the speed of light in a vacuum, c, to its speed c' in the medium is $n_r = c/c'$. A beam of light changes direction ('bends') when it passes from a region of one refractive index to a region with a different refractive index.

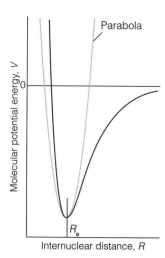

Fig. 12.14 A molecular potential energy curve can be approximated by a parabola near the bottom of the well. A parabolic potential results in harmonic oscillation. At high vibrational excitation energies the parabolic approximation is poor.

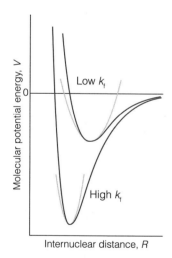

Fig. 12.15 A low value of the force constant k_f indicates a loose bond; a high value indicates a stiff bond. Although the value of k_f is not directly related to the strength of the bond, this illustration indicates that it is likely that a strong bond (one with a deep minimum) has a large force constant.

Vibrational spectra

All molecules are capable of vibrating, and complicated molecules may do so in a large number of different ways. Even a benzene molecule, with 12 atoms, can vibrate in 30 different modes, some of which involve the periodic swelling and shrinking of the ring and others its buckling into various distorted shapes. A molecule as big as a protein can vibrate in tens of thousands of different ways, twisting, stretching, and buckling in different regions and in different manners. Vibrations can be excited by the absorption of electromagnetic radiation. The observation of the frequencies at which this absorption occurs gives very valuable information about the identity of the molecule and provides quantitative information about the flexibility of its bonds.

12.3 The vibrations of diatomic molecules

We need to know how to treat the vibrations of diatomic molecules quantitatively because the vibrational characteristics of even the largest biological molecules can be understood in terms of the harmonic motion of each atom relative to its neighbors.

We base our discussion on Fig. 12.14, which shows a typical potential energy curve of a diatomic molecule as its bond is lengthened by pulling one atom away from the other or pressing it into the other. In regions close to the equilibrium bond length R_e (at the minimum of the curve) we can approximate the potential energy by a parabola (a curve of the form $y = x^2$) and write

$$V = \tfrac{1}{2}k_f(R - R_e)^2 \qquad \boxed{\text{Harmonic oscillator potential}} \qquad (12.12)$$

where k_f is the **force constant** of the bond (units: newtons per metre, N m^{-1}), as in the discussion of vibrations in Section 9.6. The steeper the walls of the potential (the stiffer the bond), the greater is the force constant (Fig. 12.15).

The potential energy in eqn 12.12 has the same form as that for the harmonic oscillator, so we can use the solutions of the Schrödinger equation given in Section 9.6. The only complication is that both atoms joined by the bond move, so the 'mass' of the oscillator has to be interpreted carefully. Detailed calculation shows that for two atoms of masses m_A and m_B joined by a bond of force constant k_f, the energy levels are[4]

$$E_v = (v + \tfrac{1}{2})h\nu \qquad v = 0, 1, 2, \ldots \qquad \boxed{\begin{array}{c}\text{Harmonic oscillator} \\ \text{energy levels}\end{array}} \qquad (12.13a)$$

where

$$\nu = \frac{1}{2\pi}\left(\frac{k_f}{\mu}\right)^{1/2} \qquad \mu = \frac{m_A m_B}{m_A + m_B} \qquad (12.13b)$$

and μ is called the **effective mass** of the molecule (some call it the *reduced mass*). Figure 12.16 (a repeat of Fig. 9.38) illustrates these energy levels: we see that they form a uniform ladder of separation $h\nu$ between neighbors.

At first sight it might be puzzling that the effective mass appears rather than the total mass of the two atoms. However, the presence of μ is physically plausible. If atom A were as heavy as a brick wall, it would not move at all during the

[4] We have previously warned about the importance of distinguishing between the quantum number v (vee) and the frequency ν (nu).

vibration and the vibrational frequency would be determined by the lighter, mobile atom. Indeed, if A were a brick wall, we could neglect m_B compared with m_A in the denominator of μ and find $\mu \approx m_B$, the mass of the lighter atom. This is approximately the case in HI, for example, where the I atom barely moves and $\mu \approx m_H$. In the case of a homonuclear diatomic molecule, for which $m_A = m_B = m$, the effective mass is half the mass of one atom: $\mu = \frac{1}{2}m$.

Self-test 12.3 Carbon monoxide is a poisonous gas because it binds strongly to hemoglobin, preventing the transport of oxygen by blood. The bond in a $^{12}C^{16}O$ molecule has a force constant of 1860 N m^{-1}. Calculate the vibrational frequency, v, of the molecule and the energy separation between any two neighboring vibrational energy levels.

Answer: 64.32 THz; 42.62 zJ; 1 zJ = 10^{-21} J

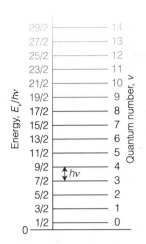

Fig. 12.16 The energy levels of a harmonic oscillator. The quantum number v ranges from 0 to infinity, and the permitted energy levels form a uniform ladder with spacing hv.

Equation 12.13 suggests that substitution of one or more of the atoms in a bond with different isotopes to give species known as **isotopologs** (such as CH_3OH and CH_2DOH) changes the vibrational frequency. The effect arises primarily from a change in the effective mass. The force constant is not affected by isotopic substitution because the key factors that determine the strength of a bond—the electronic structure and nuclear charges of the bonded atoms—do not change as neutrons are added to or removed from the nuclei.

Example 12.2 *The effect of isotopic substitution on the vibrational frequency of O_2*

Predict the vibrational frequency of $^{18}O_2$, given that the vibrational frequency of $^{16}O_2$ is 47.37 THz.

Strategy Use eqn 12.13b, with the same value of k_f for both molecules, to express the ratio $v(^{18}O_2)/v(^{16}O_2)$ in terms of the ratio $m(^{16}O)/m(^{18}O)$. Then calculate $v(^{18}O_2)$ from the known value of $v(^{16}O_2)$.

Solution From eqn 12.13b, the vibrational frequencies of $^{16}O_2$ and $^{18}O_2$ are given by

$$v(^{16}O_2) = \frac{1}{2\pi}\left(\frac{k_f}{\mu(^{16}O_2)}\right)^{1/2} \qquad \mu(^{16}O_2) = \frac{1}{2}m(^{16}O)$$

$$v(^{18}O_2) = \frac{1}{2\pi}\left(\frac{k_f}{\mu(^{18}O_2)}\right)^{1/2} \qquad \mu(^{18}O_2) = \frac{1}{2}m(^{18}O)$$

where we have used the fact that the force constant k_f is the same for both molecules. It follows that

$$\frac{v(^{18}O_2)}{v(^{16}O_2)} = \left(\frac{\mu(^{16}O_2)}{\mu(^{18}O_2)}\right)^{1/2} = \left(\frac{m(^{16}O)}{m(^{18}O)}\right)^{1/2} = \left(\frac{16.00 m_u}{18.00 m_u}\right)^{1/2} = \left(\frac{16.00}{18.00}\right)^{1/2}$$

(This ratio evaluates to 0.9428.) Therefore,

$$v(^{18}O_2) = \left(\frac{16.00}{18.00}\right)^{1/2} \times v(^{16}O_2) = \left(\frac{16.00}{18.00}\right)^{1/2} \times 47.37 \text{ THz} = 44.66 \text{ THz}$$

That is, substitution with heavier isotopes leads to a decrease in the vibrational frequency of the O=O bond.

A note on good practice To calculate the vibrational frequency precisely, we need to specify the nuclide. Also, the mass to use is the actual atomic mass, not the element's molar mass. In this *Example*, the units canceled.

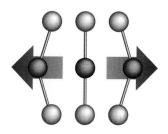

Fig. 12.17 The oscillation of a molecule, even if it is nonpolar, may result in an oscillating dipole that can interact with the electromagnetic field. Here we see a representation of a bending mode of CO_2.

Self-test 12.4 From your answer to *Self-test* 12.3, predict the vibrational frequency of the $^{13}C^{16}O$ molecule.

Answer: 62.89 THz

12.4 Vibrational transitions

To prepare for a discussion of the spectra of biological macromolecules, we need to describe the selection rules that govern vibrational transitions.

Because a typical vibrational excitation energy is of the order of 10^{-20}–10^{-19} J, the frequency of the radiation should be of the order of 10^{13}–10^{14} Hz (from $\Delta E = h\nu$). This frequency range corresponds to infrared radiation, so vibrational transitions are observed by **infrared spectroscopy**. In infrared spectroscopy, transitions are normally expressed in terms of their wavenumbers and lie typically in the range 300–3000 cm^{-1}.

(a) Infrared transitions

The gross selection rule for infrared absorption spectra is that *the electric dipole moment of the molecule must change during the vibration*. The basis of this rule is that the molecule can shake the electromagnetic field into oscillation only if it has an electric dipole moment that oscillates as the molecule vibrates (Fig. 12.17). The molecule need not have a permanent dipole: the rule requires only a *change* in dipole moment, possibly from zero. The stretching motion of a homonuclear diatomic molecule does not change its electric dipole moment from zero, so the vibrations of such molecules neither absorb nor generate radiation. We say that a homonuclear diatomic molecule is **infrared inactive** because its dipole moment remains zero however long the bond. A heteronuclear diatomic molecule, which has a dipole moment that changes as the bond lengthens and contracts, is **infrared active**.

The gross selection rule for infrared absorption plays an important role in discussions of climate change. The Earth's average temperature is maintained by an energy balance between solar radiation it absorbs and the infrared radiation it emits back into space, with most of the intensity of the latter in the range 200–2500 cm^{-1}. The trapping of infrared radiation by certain gases in the atmosphere warms the Earth, raises the average surface temperature well above the freezing point of water, and creates an environment in which life is possible. As you will be aware, there is currently great concern that human activity has led to significant increases in the concentrations of certain gases in the atmosphere, such as CO_2 and CH_4, which absorb infrared radiation and so result in the further warming of the planet with the potential of serious damage to the biosphere.

Example 12.3 *Identifying species that contribute to climate change*

Identify which of the following constituents of the atmosphere absorb infrared radiation: O_2, N_2, H_2O, CO_2, and CH_4. Is there a basis for the concern that increased levels of atmospheric CO_2 and CH_4 lead to climate change?

Strategy Molecules that are infrared active (that is, have vibrational spectra) have dipole moments that change during the course of a vibration. Therefore, judge whether a distortion of the molecule can change its dipole moment (including changing it from zero).

Solution Only N_2 and O_2 do not possess vibrational modes that result in a change of dipole moment, so CO_2, H_2O, and CH_4 are infrared active. Not all the modes of complicated molecules are infrared active. For example, a vibration of CO_2 in which the O–C–O bonds stretch and contract symmetrically is inactive because it leaves the dipole moment unchanged (at zero). A bending motion of the molecule, however, is active and can absorb radiation. It follows that the continued release of CO_2 and CH_4 into the atmosphere can contribute to climate change. Water also contributes, but it is already present in large amounts.

Self-test 12.5 Ethene, $CH_2=CH_2$, is a hormone responsible for the ripening of fruit, and nitric oxide, NO, is a neurotransmitter. Are these molecules infrared active?

Answer: Both are infrared active.

The specific selection rule for infrared absorption spectra is

$$\Delta v = \pm 1 \qquad \text{Vibrational selection rule} \qquad (12.14)$$

The change in energy for the transition from a state with quantum number v to one with quantum number $v + 1$ is

$$\Delta E = (v + \tfrac{3}{2})hv - (v + \tfrac{1}{2})hv = hv \qquad \text{Vibrational transition energies} \qquad (12.15)$$

It follows that absorption occurs when the incident radiation provides photons with this energy and therefore when the incident radiation has a frequency v given by eqn 12.13b (and wavenumber $\tilde{v} = v/c$). Molecules with stiff bonds (large k_f) joining atoms with low masses (small μ) have high vibrational frequencies. Bending modes are usually less stiff than stretching modes, so bends tend to occur at lower frequencies in the spectrum than stretches. At room temperature, almost all the molecules are in their vibrational ground states initially (the state with $v = 0$). Therefore, the most important spectral transition is from $v = 0$ to $v = 1$.

Self-test 12.6 The force constant of the bond in the CO group of a peptide link is approximately 1.2 kN m^{-1}. At what wavenumber would you expect it to absorb? *Hint:* For the effective mass, treat the group as a $^{12}C^{16}O$ molecule; see *Self-test* 12.3.

Answer: 1.7×10^3 cm^{-1}

The vibrational energies in eqn 12.13 are only approximate because they are based on a parabolic approximation to the actual potential energy curve. A parabola cannot be correct at all extensions because it does not allow a molecule to dissociate. At high vibrational excitations the swing of the atoms allows the molecule to explore regions of the potential energy curve where the parabolic approximation is poor. The motion then becomes **anharmonic**, in the sense that the restoring force is no longer proportional to the displacement. Because the actual curve is less confining than a parabola, we can anticipate that the energy levels become less widely spaced at high excitation (Fig. 12.18). The anharmonic nature of the motion accounts for the appearance of additional weak absorption lines called **overtones** corresponding to the transitions with $\Delta v = +2, +3, \ldots$

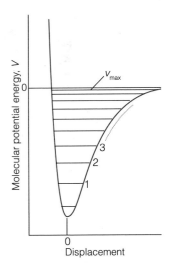

Fig. 12.18 The vibrational energy levels associated with the general shape of a molecular potential energy curve are less widely spaced at high excitation. The number of levels is finite, terminating at v_{max}.

These overtones appear because the usual selection rule is derived from the properties of harmonic oscillator wavefunctions, which are only approximately valid in the presence of anharmonicity.

(b) Raman transitions

Now we turn to **vibrational Raman spectroscopy**, in which the incident photon leaves some of its energy in the vibrational modes of the molecule it strikes or collects additional energy from a vibration that has already been excited. The gross selection rule for vibrational Raman transitions is that *the molecular polarizability must change as the molecule vibrates*. The polarizability plays a role in vibrational Raman spectroscopy because the molecule must be squeezed and stretched by the incident radiation in order that a vibrational excitation may occur during the photon–molecule collision. Both homonuclear and heteronuclear diatomic molecules swell and contract during a vibration, and the control of the nuclei over the electrons, and hence the molecular polarizability, changes too. Both types of diatomic molecule are therefore vibrationally Raman active. It follows that the information available from vibrational Raman spectra adds to that from infrared spectroscopy.

The specific selection rule for vibrational Raman transitions is the same as for infrared transitions ($\Delta v = \pm 1$). The photons that are scattered with a lower wavenumber than that of the incident light, the Stokes lines, are those for which $\Delta v = +1$. The anti-Stokes lines (for which $\Delta v = -1$) are less intense than the Stokes lines because very few molecules are in an excited vibrational state initially.

12.5 The vibrations of polyatomic molecules

We need to see how the concepts developed in previous sections can be used to interpret the information contained in the infrared and Raman spectra of biological macromolecules.

How many modes of vibration, N_{vib}, does a polyatomic molecule have? We can answer this question by thinking about how each atom may change its location, and we show in the following *Justification* that

for nonlinear molecules: $N_{vib} = 3N - 6$

for linear molecules: $N_{vib} = 3N - 5$

Justification 12.2   *The number of vibrational modes*

Each atom may move relative to any of three perpendicular axes. Therefore, the total number of such displacements in a molecule consisting of N atoms is $3N$. Three of these displacements correspond to the translational motion of the molecule as a whole. The remaining $3N - 3$ displacements are 'internal' modes of the molecule that leave its center of mass unchanged. Three angles are needed to specify the orientation of a nonlinear molecule in space (Fig. 12.19). Therefore three of the $3N - 3$ internal displacements leave all bond angles and bond lengths unchanged but change the orientation of the molecule as a whole. These three displacements are therefore rotations. That leaves $3N - 6$ displacements that can be identified as vibrational modes. A similar calculation for a linear molecule, which requires only two angles to specify its orientation in space, gives $3N - 5$ as the number of vibrational modes.

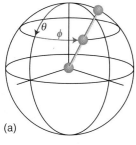

(a)

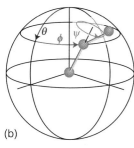

(b)

Fig. 12.19 (a) The orientation of a linear molecule requires the specification of two angles (the latitude and longitude of its axis). (b) The orientation of a nonlinear molecule requires the specification of three angles (the latitude and longitude of its axis and the angle of twist—the azimuthal angle—around that axis).

A brief illustration

A water molecule, H_2O, is triatomic ($N = 3$) and nonlinear and has three modes of vibration. Naphthalene, $C_{10}H_8$ ($N = 18$), has 48 distinct modes of vibration (some are degenerate in the sense of having the same frequency). Any diatomic molecule ($N = 2$) has one vibrational mode; carbon dioxide ($N = 3$) has four vibrational modes.

(a) Normal modes

The description of the vibrational motion of a polyatomic molecule is much simpler if we consider combinations of the stretching and bending motions of individual bonds. For example, although we could describe two of the four vibrations of a CO_2 molecule as individual carbon–oxygen bond stretches, v_L and v_R in Fig. 12.20, the description of the motion is much simpler if we use two combinations of these vibrations. One combination is v_1 in Fig. 12.21: this combination is the **symmetric stretch**. The other combination is v_3, the **antisymmetric stretch**, in which the two O atoms always move in the same directions and opposite to the C atom. The two modes are independent in the sense that if one is excited, then its motion does not excite the other. They are two of the four 'normal modes' of the molecule, its independent, collective vibrational displacements. The two other normal modes are the **bending modes**, v_2. In general, a **normal mode** is an independent, synchronous motion of atoms or groups of atoms that may be excited without leading to the excitation of any other normal mode. The number

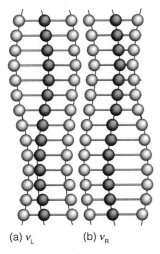

(a) v_L (b) v_R

Fig. 12.20 The stretching vibrations of a CO_2 molecule can be represented in a number of ways. In this representation (a) one O=C bond (on the left) vibrates and the remaining O atom is stationary, and (b) the C=O bond (on the right) vibrates while the other O atom is stationary. Because the stationary atom is linked to the C atom, it does not remain stationary for long. That is, if one vibration begins, it rapidly stimulates the other to occur.

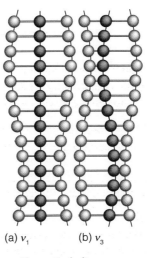

(a) v_1 (b) v_3

Fig. 12.21 Alternatively, linear combinations of the two modes can be taken to give these two normal modes of the molecule. The mode in (a) is the symmetric stretch and that in (b) is the antisymmetric stretch. The two modes are independent, and if either of them is stimulated, the other remains unexcited. Normal modes greatly simplify the description of the vibrations of the molecule.

of normal modes of vibration is the same as the number of vibrational modes calculated above, for normal modes are linear combinations of vibrational displacements of atoms.

The four normal modes of CO_2, and the $3N - 6$ (or $3N - 5$) normal modes of polyatomic molecules in general, are the key to the description of molecular vibrations. Each normal mode behaves like an independent harmonic oscillator and the energies of the vibrational levels are given by the same expression as in eqn 12.13, but with an effective mass that depends on the extent to which each of the atoms contributes to the vibration. Atoms that do not move, such as the C atom in the symmetric stretch of CO_2, do not contribute to the effective mass. The force constant also depends in a complicated way on the extent to which bonds bend and stretch during a vibration. Typically, a normal mode that is largely a bending motion has a lower force constant (and hence a lower frequency) than a normal mode that is largely a stretching motion.

(Self-test 12.7) How many normal modes of vibration are there in (a) ethyne (HC≡CH) and (b) a protein molecule of 4000 atoms?

Answer: (a) 7, (b) 11 994

(b) Infrared transitions

The gross selection rule for the infrared activity of a normal mode is that *the motion corresponding to a normal mode must give rise to a changing dipole moment*. Deciding whether this is so can sometimes be done by inspection. For example, the symmetric stretch of CO_2 leaves the dipole moment unchanged (at zero), so this mode is infrared inactive and makes no contribution to the molecule's infrared spectrum. The antisymmetric stretch, however, changes the dipole moment because the molecule becomes unsymmetrical as it vibrates, so this mode is infrared active. Both bending modes are also infrared active: they are accompanied by a changing dipole moment as the molecule oscillates between a linear (nonpolar) and bent (polar) geometry (as in Fig. 12.17). The fact that these modes do absorb infrared radiation enables carbon dioxide to absorb infrared radiation emitted from the surface of the Earth (see *Example* 12.3).

(Self-test 12.8) Dinitrogen monoxide (nitrous oxide, N_2O) is another minor constituent of the atmosphere that can contribute to global warming; it has also been used as an anesthetic. State the ways in which the infrared spectrum of dinitrogen monoxide will differ from that of carbon dioxide.

Answer: Different frequencies on account of different atomic masses and force constants; all four modes infrared active

---

(Self-test 12.9) Consider the normal modes of methane, CH_4, some of which are shown in Fig. 12.22. Which of the modes are infrared active?

Answer: The modes denoted (c) and (d) are infrared active.

To a good approximation, some of the normal modes of organic molecules can be regarded as motions of individual functional groups. Others are better

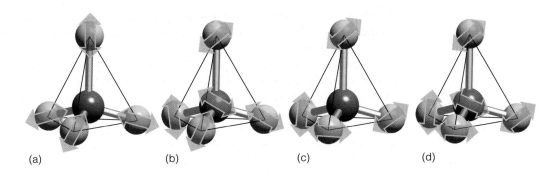

(a)   (b)   (c)   (d)

regarded as collective motions of the molecule as a whole. The latter are generally of relatively low frequency and occur below about 1500 cm^{-1} in the spectrum. The resulting whole-molecule region of the absorption spectrum is called the **fingerprint region** of the spectrum, for it is characteristic of the molecule. The matching of the fingerprint region with a spectrum of a known compound in a library of infrared spectra is a very powerful way of confirming the presence of a particular substance.

The characteristic vibrations of functional groups that occur outside the fingerprint region are very useful for the identification of an unknown compound. Most of these vibrations can be regarded as stretching modes, for the lower frequency bending modes usually occur in the fingerprint region and so are less readily identified. The characteristic wavenumbers of some functional groups are listed in Table 12.1.

(c) Raman transitions

The gross selection rule for the vibrational Raman spectrum of a polyatomic molecule is that the *normal mode of vibration is accompanied by a changing polarizability*. However, it is often quite difficult to judge by inspection when this is so. The symmetric stretch of CO_2, for example, alternately swells and contracts the molecule: this motion changes its polarizability, so the mode is Raman active. The other modes of CO_2 leave the polarizability unchanged (although that is hard to justify pictorially), so they are Raman inactive.

In some cases it is possible to make use of a very general rule about the infrared and Raman activity of vibrational modes:

The **exclusion rule** states that if the molecule has a centre of inversion, then no mode can be both infrared and Raman active.

(A mode may be inactive in both.) A molecule has a center of inversion if it looks unchanged when each atom is projected through a single point and out an equal distance on the other side (Fig. 12.23). Because we can often judge intuitively when a mode changes the molecular dipole moment, we can use this rule to identify modes that are not Raman active. The rule applies to CO_2 but to neither H_2O nor CH_4 because they have no center of inversion.

Fig. 12.22 Four representative normal modes of a tetrahedral molecule.

Table 12.1 Typical vibrational wavenumbers

Vibration type	$\tilde{\nu}/cm^{-1}$
C–H	2850–2960
C–H	1340–1465
C–C stretch, bend	700–1250
C=C stretch	1620–1680
C≡C stretch	2100–2260
O–H stretch	3590–3650
C=O stretch	1640–1780
C≡N stretch	2215–2275
N–H stretch	3200–3500
Hydrogen bonds	3200–3570

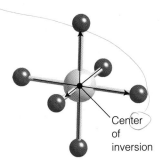

Center of inversion

Fig. 12.23 In an inversion operation, we consider every point in a molecule, and project them all through the center of the molecule out to an equal distance on the other side.

Self-test 12.10 One vibrational mode of benzene is a 'breathing mode' in which the ring alternately expands and contracts. Can it be vibrationally Raman active?

Answer: Yes

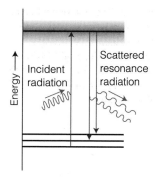

Fig. 12.24 In the resonance Raman effect, the incident radiation has a frequency corresponding to an actual electronic excitation of the molecule. A photon is emitted when the molecule returns to a state close to the ground state.

A modification of the basic Raman effect involves using incident radiation that nearly coincides with the frequency of an electronic transition of the sample (Fig. 12.24; compare with Fig. 12.2, where the incident radiation does not coincide with an electronic transition). The technique is then called **resonance Raman spectroscopy**. It is characterized by a much greater intensity in the scattered radiation. Furthermore, because it is often the case that only a few vibrational modes contribute to the more intense scattering, the spectrum is greatly simplified.

Case study 12.1 *Vibrational spectroscopy of proteins*

Insight into the vibrational spectrum of the peptide link, –CONH–, can be obtained by accounting for the major features in the infrared spectrum of N-methylacetamide, $CH_3CONHCH_3$ (Fig. 12.25). Above 2800 cm^{-1} we find a cluster of three bands, labeled (a), that correspond, in order of increasing wavenumber, to the symmetric and antisymmetric methyl C–H stretches from the C-methyl group, the symmetric and antisymmetric methyl C–H stretches from the N-methyl group, and the broad N–H stretch. In the fingerprint region we find two bands associated with the amide group and, more generally, with peptide groups in proteins. The *amide I band*, labeled (b), consists mostly of a CO stretch and occurs in the range 1640–1670 cm^{-1}. The *amide II band*, labeled (c), is a combination of a CO stretch and an NH bend and occurs in the range 1620–1650 cm^{-1}.

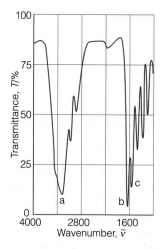

Fig. 12.25 The infrared spectrum of a thin liquid film of N-methylacetamide.

The vibrational spectra of proteins are rich in information because of the large number of absorption bands that can be associated not only with the peptide link but also the amino acid side chains (Fig. 12.26). However, biochemists focus primarily on the amide I and II bands of the peptide link because their wavenumbers are sensitive to hydrogen bonding and thus indicative of secondary structure. Hydrogen bonding between the CO group of one peptide link with the NH group of another leads to a shift of the amide I band to lower wavenumber because the delocalized N–H⋯O=C bond lowers the force constant of the C=O bond. On the other hand, hydrogen bonding constrains the bending motion of the N–H group, effectively increasing the C–N–H bending force constant and shifting the wavenumber of the amide II band to higher values. Furthermore, experiments have shown that the wavenumbers of the amide I and II bands are slightly different in α helices, β sheets, and random coils (Table 12.2). It follows that vibrational spectroscopy can be used to monitor conformational changes in proteins.

Table 12.2 Typical vibrational wavenumbers for the amide I and II bands in polypeptides

Vibration type	Vibrational wavenumber ($\tilde{v}$/cm^{-1})		
	α helix	β sheet	Random coil
Amide I	1653	1640	1656
Amide II	1545	1525	1535

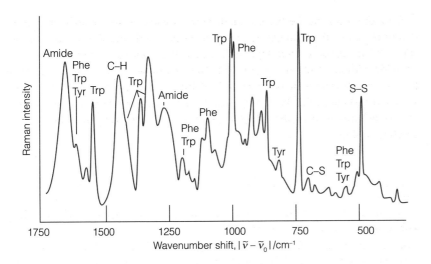

Fig. 12.26 The vibrational Raman spectrum of lysozyme in water. (From *Raman spectroscopy*, D.A. Long. Copyright 1977, McGraw-Hill, Inc. Used with the permission of the McGraw-Hill Book Company.)

It is difficult to find features in the complex infrared and conventional Raman spectra of proteins that can be assigned to co-factors. Biochemists often turn to resonance Raman spectroscopy to study co-factors that absorb strongly in the ultraviolet and visible regions of the spectrum. Examples include the heme groups in hemoglobin (Atlas P7) and the cytochromes (Atlas P5) and the pigments β-carotene (Atlas E1) and chlorophyll (Atlas R3), which capture solar energy during plant photosynthesis.

The resonance Raman spectra of Fig. 12.27 show vibrational transitions from only the few pigment molecules that are bound to very large proteins dissolved in an aqueous buffer solution. This selectivity arises from the fact that water (the solvent), amino acid residues, and the peptide group do not have electronic transitions at the laser wavelengths used in the experiment, so their conventional Raman spectra are weak compared to the enhanced spectra of the pigments. Comparison of the top and bottom spectra also shows that, with proper choice of excitation wavelength, it is possible to examine individual classes of pigments bound to the same protein: excitation at 488 nm, where β-carotene absorbs strongly, shows vibrational bands from β-carotene only, whereas excitation at 407 nm, where chlorophyll *a* and β-carotene absorb, reveals features from both types of pigments.

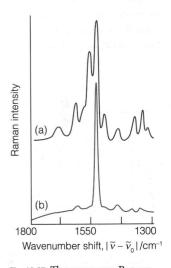

Fig. 12.27 The resonance Raman spectra of a protein complex that is responsible for some of the initial electron transfer events in plant photosynthesis. The Raman shift is the difference between the wavenumber of the scattered light and the wavenumber of the exciting laser radiation. (a) Laser excitation of the sample at 407 nm shows Raman bands due to both chlorophyll *a* and β-carotene molecules bound to the protein because both pigments absorb light at this wavelength. (b) Laser excitation at 488 nm shows Raman bands from β-carotene only because chlorophyll *a* does not absorb light very strongly at this wavelength. (Adapted from D.F. Ghanotakis et al., *Biochim. Biophys. Acta* **974**, 44 [1989].)

In the laboratory 12.3   *Vibrational microscopy*

Whereas scanning probe microscopy (*In the laboratory* 9.2) is a good probe of atoms and molecules on surfaces, conventional optical microscopy, which uses light to carry information about the specimen, can be used to study a wider variety of samples, from solids to flowing liquids. Hence, there is great interest in new modes of optical microscopy that can probe specimens as small as single molecules.

In conventional optical microscopy a beam of light is focused on to a specimen by a condenser lens and light transmitted or reflected by the sample is collected by the objective lens (Fig. 12.28). The magnified image of the specimen

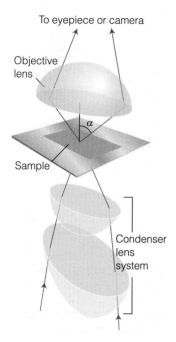

Fig. 12.28 The light path in a typical microscope. Light is focused by the condenser lens (typically a system of two lenses), scattered by the sample, and refocused by an objective lens. The ability of the objective lens to resolve two objects into distinct images depends on the numerical aperture, which is related to the refractive index of the lens material and the angle α, as discussed in the text.

is either viewed directly with the help of an eyepiece or captured by a video camera and displayed on a monitor. The image is constructed from a pattern of diffracted light waves that emanate from the specimen and reach the objective lens. As a result, some information about the specimen is lost by destructive interference of scattered light waves. Ultimately, this 'diffraction limit' prevents the study of samples that are much smaller than the wavelength of light used as a probe. In practice, two objects will appear as distinct images under a microscope if the distance between their centers is greater than the 'Airy radius', $\gamma_{Airy} = 0.61\lambda/a$, where λ is the wavelength of the incident beam of radiation and a is the aperture of the objective lens, which is defined as $a = n_r \sin\alpha$ where n_r is the refractive index of the material lying between the object and the objective lens, and α is the half-angle of the widest cone of scattered light that can collected by the lens (so the lens collects light beams within a cone of angle 2α; see Fig. 12.28).

It is now possible to combine optical microscopes with infrared and Raman spectrometers to obtain vibrational spectra of very small specimens. The techniques of **vibrational microscopy** provide details of cellular events that cannot be observed with electron microscopy.

In infrared and Raman microscopes the sample is moved by very small increments along a plane perpendicular to the direction of illumination and the process is repeated until vibrational spectra for all sections of the sample are obtained. The size of a sample that can be studied by vibrational microscopy depends on a number of factors, such as the area of illumination, the power of the radiation delivered to the illuminated area, and the wavelength of the incident radiation. Up until the diffraction limit is reached, the smaller the area that is illuminated, the smaller the area from which a spectrum can be obtained. High radiant power is required to increase the rate of arrival of photons at the detector from small illuminated areas. For this reason, lasers and synchrotron radiation are the preferred radiation sources. Use of the best equipment makes it possible to examine areas as small as 9 μm^2 by vibrational microscopy.

For Raman microscopy, the most common spectrometer system consists of a visible laser coupled to a polychromator and a CCD detector, although near-infrared Fourier transform spectrometers are also used. The CCD detector can be used in a variation of Raman microscopy known as *Raman imaging*: a special optical filter allows only one Stokes line to reach the two-dimensional detector, which then contains a map of the distribution of the intensity of that line in the illuminated area.

Fourier transform spectrometers are common in infrared microscopy. Figure 12.29 shows the infrared spectra of a single mouse cell, living and dying. Both spectra have features at 1545 cm^{-1} and 1650 cm^{-1} that are due to the peptide carbonyl groups of proteins and a feature at 1240 cm^{-1} that is due to the phosphodiester ($-PO_2^-$) groups of lipids. The dying cell shows an additional absorption at 1730 cm^{-1}, which is due to the ester carbonyl group from an unidentified compound. From a plot of the intensities of individual absorption features as a function of position in the cell it has been possible to map the distribution of proteins and lipids during cell division and cell death.

Vibrational microscopy has also been used in biomedical and pharmaceutical laboratories. Examples include the determination of the size and distribution of a drug in a tablet, the observation of conformational changes in proteins of cancerous cells on administration of antitumor drugs, and the measurement of differences between diseased and normal tissue, such as diseased arteries and the white matter from brains of patients suffering from multiple sclerosis.

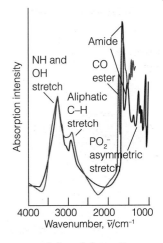

Fig. 12.29 Infrared absorption spectra of a single mouse cell: (red line) living cell, (blue line) dying cell. (Adapted from N. Jamin et al., *Proc. Natl. Acad. Sci. USA* **95**, 4837 (1998).)

Ultraviolet and visible spectra

The energy needed to change the distribution of an electron in a molecule is of the order of several electronvolts. Consequently, the photons emitted or absorbed when such changes occur lie in the visible and ultraviolet regions of the spectrum, which spread from about 14 000 cm⁻¹ for red light to 21 000 cm⁻¹ for blue, and on to 50 000 cm⁻¹ for ultraviolet radiation (Table 12.3). Indeed, many of the colors of the objects in the world around us, including the green of vegetation, the colors of flowers and of synthetic dyes, and the colors of pigments and minerals, stem from transitions in which an electron makes a transition from one orbital of a molecule or ion into another orbital. The change in location of an electron that takes place when chlorophyll absorbs red and blue light (leaving green to be reflected) is the primary energy-harvesting step by which our planet captures energy from the Sun and uses it to drive the nonspontaneous reactions of photosynthesis (*Case study* 12.3). In some cases the relocation of an electron may be so extensive that it results in the breaking of a bond and the dissociation of the molecule: such processes give rise to the numerous reactions of photochemistry, including the reactions that sustain or damage the atmosphere.

White light is a mixture of light of all different colors. The removal, by absorption, of any one of these colors from white light results in the 'complementary color' being observed. For instance, the absorption of red light from white light by an object results in that object appearing green, the complementary color of red. Conversely, the absorption of green results in the object appearing red. The pairs of complementary colors are neatly summarized by the artist's color wheel shown in Fig. 12.30, where complementary colors lie opposite each other along a diameter.

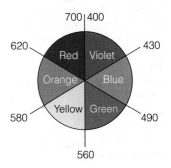

Fig. 12.30 An artist's color wheel: complementary colors are opposite each other on a diameter. The numbers correspond to wavelengths of light in nanometers.

Table 12.3 Color, frequency, and energy of light

Color	λ/nm	ν/(10¹⁴ Hz)	$\tilde{\nu}$/(10⁴ cm⁻¹)	E/eV	E/(kJ mol⁻¹)
Infrared	1000	3.00	1.00	1.24	120
Red	700	4.28	1.43	1.77	171
Orange	620	4.84	1.61	2.00	193
Yellow	580	5.17	1.72	2.14	206
Green	530	5.66	1.89	2.34	226
Blue	470	6.38	2.13	2.64	254
Violet	420	7.14	2.38	2.95	285
Near-ultraviolet	300	10.0	3.33	4.13	399
Far-ultraviolet	200	15.0	5.00	6.20	598

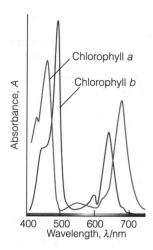

Fig. 12.31 The absorption spectra of chlorophylls *a* and *b* (Atlas R3), the main pigments in plants, in the visible region. Note that the chlorophylls absorb in the orange–red and blue regions and that green light is not absorbed significantly.

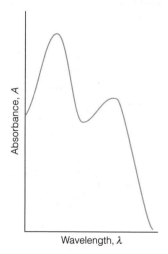

Fig. 12.32 An electronic absorption band of a species in solution is typically very broad and consists of several broad bands.

It should be stressed, however, that the perception of color is a very subtle phenomenon. Although an object may appear green because it absorbs red light, it may also appear green because it absorbs all colors from the incident light *except* green. This is the origin of the color of vegetation, because chlorophyll absorbs in two regions of the spectrum, leaving green to be reflected (Fig. 12.31). Moreover, an absorption band may be very broad, and although it may be a maximum at one particular wavelength, it may have a long tail that spreads into other regions (Fig. 12.32). In such cases, it is very difficult to predict the perceived color from the location of the absorption maximum.

12.6 The Franck–Condon principle

To understand how Nature makes use of colored materials in such important processes as photosynthesis and vision, we need to know the factors that control the intensity of electronic transitions and the shapes of absorption bands.

Whenever an electronic transition takes place, it is accompanied by the excitation of vibrations of the molecule. In the electronic ground state of a molecule, the nuclei take up locations in response to the Coulombic forces acting on them. These forces arise from the electrons and the other nuclei. After an electronic transition, when an electron has migrated to a different part of the molecule, the nuclei are subjected to different Coulombic forces from the surrounding electrons. The molecule may respond to the sudden change in forces by bursting into vibration. As a result, some of the energy used to redistribute an electron is in fact used to stimulate the vibrations of the absorbing molecules. Therefore, instead of a single, sharp, and purely electronic absorption line being observed, the absorption spectrum consists of many lines. This **vibrational structure** of an electronic transition can be resolved if the sample is gaseous, but as we remarked in Section 12.2d, in a liquid or solid the lines usually merge together and result in a broad, almost featureless band.

The vibrational structure of a band is explained by the **Franck–Condon principle**:

Because nuclei are so much more massive than electrons, an electronic transition takes place faster than the nuclei can respond.

In an electronic transition, electron density is lost rapidly from some regions of the molecule and is built up rapidly in others. As a result, the initially stationary nuclei suddenly experience a new force field. They respond by beginning to vibrate, and (in classical terms) swing backwards and forwards from their original separation, which they maintained during the rapid electronic excitation. The equilibrium separation of the nuclei in the initial electronic state therefore becomes a **turning point**, one of the end points of a nuclear swing, in the final electronic state (Fig. 12.33).

To predict the most likely final vibrational state we draw a vertical line from the minimum of the lower curve (the starting point for the transition) up to the point at which the line intersects the curve representing the upper electronic state (the turning point of the newly stimulated vibration). This procedure gives rise to the name **vertical transition** for a transition that takes place in accord with the Franck–Condon principle. In practice, the electronically excited molecule may be formed in one of several excited vibrational states, so the absorption occurs at several different frequencies. As remarked above, in a condensed medium, the individual transitions merge together to give a broad, largely featureless band of absorption.

12.7 Chromophores

Biological systems contain organic compounds and complexes of metal ions with characteristic electronic transitions. We need to see how to investigate these transitions and use ultraviolet and visible spectroscopy to elucidate biochemical processes.

The absorption of a photon can often be traced to the excitation of an electron that is localized on a small group of atoms. For example, an absorption at about 290 nm is normally observed when a carbonyl group is present, as in the peptide link. Groups with characteristic optical absorptions are called **chromophores** (from the Greek for 'color bringer'), and their presence often accounts for the colors of many substances.

A d-metal complex may absorb light as a result of transfer of an electron between d orbitals split by a ligand field (see Section 10.8). The energy separation between d orbitals in a complex is not very large, so **d–d transitions** between sets of orbitals typically occur in the visible region of the spectrum. Also possible is the transfer of an electron from the ligands into the d orbitals of the central atom, or vice versa. In such **charge-transfer transitions** the electron moves through a considerable distance, which means that the redistribution of charge as measured by the transition dipole moment may be large and the absorption correspondingly intense. This mode of chromophore activity is shown by the copper-containing site of the bacterial protein azurin: the charge redistribution that accompanies the migration of an electron from a sulfur atom of a cysteine ligand to the Cu^{2+} ion accounts for its intense blue color (resulting from absorption in the range 500–700 nm).

The transition responsible for absorption in carbonyl compounds can be traced to the lone pairs of electrons on the O atom. One of these electrons may be excited into an empty π^* orbital of the carbonyl group (Fig. 12.34), which gives rise to an **n-to-π^* transition**, where n denotes a nonbonding orbital (an orbital that is neither bonding nor antibonding, such as that occupied by a lone pair). Typical absorption energies are about 4 eV.

A C=C double bond acts as a chromophore because the absorption of a photon excites a π electron into an antibonding π^* orbital (Fig. 12.35). The chromophore activity is therefore due to a **π-to-π^* transition**. Its energy is around 7 eV for an unconjugated double bond, which corresponds to an absorption at 180 nm (in the ultraviolet). When the double bond is part of a conjugated chain, the energies of the molecular orbitals lie closer together and the transition shifts into the visible region of the spectrum (see Section 10.11). Many of the reds and yellows of vegetation are due to transitions of this kind. For example, the carotenes, long polyenes present in green leaves (but concealed by the intense absorption of the chlorophyll until the latter decays in the fall), collect some of the solar radiation incident on the leaf by a π-to-π^* transition in their long conjugated hydrocarbon chains. A similar type of absorption is responsible for the primary process of vision (*Case study* 12.2).

Table 12.4 lists values of ε_{max} and λ_{max} (the wavelength at which $\varepsilon = \varepsilon_{max}$) for a number of biological molecules. The band positions and intensities are both sensitive to molecular interactions. For example, the ultraviolet spectrum of an α helix has two π-to-π^* transitions instead of one. The effect is due to **exciton coupling**, which can be traced to interactions between transition dipoles and leads to excited states with lower and higher energies with respect to the energy of the monomer excited state.

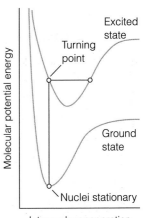

Fig. 12.33 According to the Franck–Condon principle, the most intense electronic transition is from the ground vibrational state to the vibrational state that lies vertically above it in the upper electronic state. Transitions to other vibrational levels also occur, but with lower intensity.

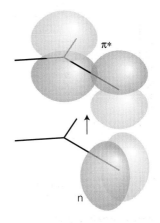

Fig. 12.34 A carbonyl group acts as a chromophore primarily on account of the excitation of a nonbonding O lone-pair electron to an antibonding CO π^* orbital.

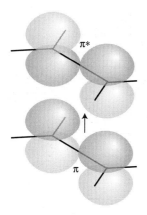

Fig. 12.35 A carbon–carbon double bond acts as a chromophore. One of its important transitions is the π-to-π* transition illustrated here, in which an electron is promoted from a π orbital to the corresponding antibonding orbital.

Table 12.4 Electronic absorption properties of amino acids, purine, and pyrimidine bases in water at pH = 7

Compound	λ_{max}/nm	ε_{max}/(10^3 dm^3 mol^{-1} cm^{-1})
Tryptophan	280	5.6
Tyrosine	274	1.4
Phenylalanine	257	0.2
Adenine	260	13.4
Guanine	275	8.1
Cytosine	267	6.1
Uracil	260	9.5

12.8 Optical activity and circular dichroism

The electronic spectra of biopolymers can reveal additional structural details when experiments are conducted with **polarized light**, electromagnetic radiation with electric and magnetic fields that oscillate only in certain directions. Light is **plane polarized** when the electric and magnetic fields each oscillate in a single plane (Fig. 12.36). The plane of polarization may be oriented in any direction around the direction of propagation (the x-direction in Fig. 12.36), with the electric and magnetic fields perpendicular to that direction (and perpendicular to each other). An alternative mode of polarization is **circular polarization**, in which the electric and magnetic fields rotate around the direction of propagation in either a clockwise or a counterclockwise sense but remain perpendicular to it and each other.

When plane-polarized radiation passes through samples of certain kinds of matter, the plane of polarization is rotated around the direction of propagation. This rotation is the phenomenon of **optical activity**. Optical activity is observed when the molecules in the sample are **chiral**, which means they are distinguishable from their mirror image (Fig. 12.37). In many cases, organic chiral compounds are easy to identify because they contain a carbon atom to which are bonded four different groups. The amino acid alanine, $NH_2CH(CH_3)COOH$, is an example. Mirror image pairs of chiral molecules, which are called **enantiomers** (from the Greek words for 'both parts'), rotate light of a given frequency through exactly the same angle but in opposite directions.

Chiral molecules have a second characteristic: they absorb left and right circularly polarized light to different extents. In a circularly polarized ray of light, the electric field describes a helical path as the wave travels through space (Fig. 12.38), and the rotation may be either clockwise or counterclockwise. The differential absorption of left- and right-circularly polarized light is called **circular dichroism**. In terms of the absorbances for the two components, A_L and A_R, the circular dichroism of a sample of molar concentration [J] is reported as

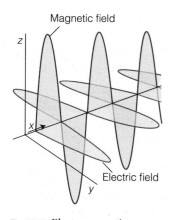

Fig. 12.36 Electromagnetic radiation consists of a wave of electric and magnetic fields perpendicular to the direction of propagation (in this case the x-direction) and mutually perpendicular to each other. This illustration shows a plane-polarized wave, with the electric and magnetic fields oscillating in the xy and xz planes, respectively.

$$\Delta\varepsilon = \varepsilon_L - \varepsilon_R = \frac{A_L - A_R}{[J]L}$$

Circular dichroism   (12.16)

where L is the path length through the sample.

Circular dichroism is a useful adjunct to visible and ultraviolet spectroscopy. For example, CD spectra give information about secondary structure of

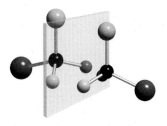

Fig. 12.37 A chiral molecule is one that is not superimposable on its mirror image. A carbon atom attached to four different groups is an example of a chiral center in a molecule. Such molecules are optically active.

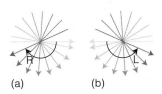

Fig. 12.38 In circularly polarized light, the electric field at different points along the direction of propagation rotates. The arrays of arrows in these illustrations show the view of the electric field when looking toward the oncoming ray: (a) right circularly polarized, (b) left circularly polarized light.

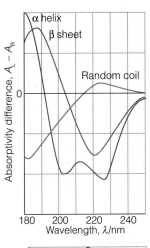

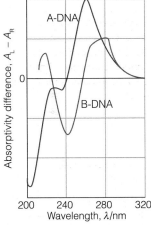

Fig. 12.39 Representative CD spectra of polypeptides and polynucleotides: (a) random coils, α helices, and β sheets have different CD features in the spectral region where the peptide link absorbs; (b) B- and A-DNA can be distinguished on the basis of CD spectroscopy in the spectral region where the bases absorb.

polypeptides and nucleic acids. Consider a helical polypeptide. Not only are the individual monomer units chiral, but so is the helix. Therefore, we expect the α helix to have a unique CD spectrum related to the secondary structure of the polypeptide. Because β sheets and random coils also have distinguishable spectral features (Fig. 12.39a), circular dichroism is a very important technique for the study of protein conformation. Circular dichroism is also a powerful tool for the study of nucleic acids (Fig. 12.39b).

The **Raman optical activity** (ROA) technique depends on the detection of a small difference in the intensity of Raman scattering when left and right circularly polarized incident radiation is used (the *incident circularly polarized*, ICP, technique). The amide III region (*Case study* 12.1) at 1230–1310 cm^{-1} is important for ROA studies because the coupling between N–H and C$_\alpha$–H deformations is very sensitive to the local molecular geometry. The signal depends largely on the nature of the polypeptide skeleton rather than the details of the amino acid side chains, so it reveals information about the secondary structure. Thus, amide I contributions to ROA close to 1650 cm^{-1} are a good indicator of the presence of an α-helix regions in polypeptides; the ROA technique can also provide a signature at close to 1240 cm^{-1} of β-sheet regions (Fig. 12.40).

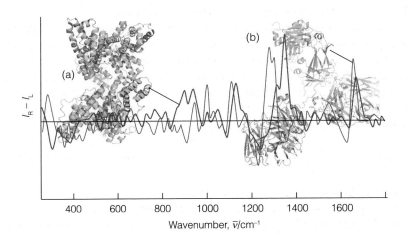

Fig. 12.40 The ROA spectrum of (a) human serum globulin, a protein with many α-helical regions, and (b) human immunoglobulin, a protein with many β-sheet regions. From L.D. Barron et al., *J. Mol. Structure*, **7**, 834 (2007).

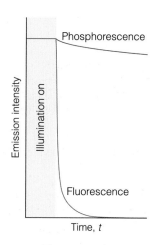

Fig. 12.41 The empirical (observation-based) distinction between fluorescence and phosphorescence is that the former is extinguished very quickly after the exciting source is removed, whereas the latter continues with relatively slowly diminishing intensity.

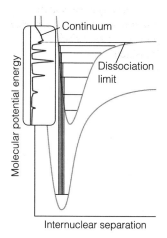

Fig. 12.42 When absorption occurs to unbound states of the upper electronic state, the molecule dissociates and the absorption is a continuum. Below the dissociation limit the electronic spectrum has a normal vibrational structure.

Radiative and nonradiative decay

In most cases, the excitation energy of a molecule that has absorbed a photon is degraded into the disordered thermal motion of its surroundings. However, one process by which an electronically excited molecule can discard its excess energy is by **radiative decay**, in which an electron makes a transition into a lower-energy orbital and in the process generates a photon. As a result, an observer sees the sample glowing (if the emitted radiation is in the visible region of the spectrum).

There are two principal modes of radiative decay, fluorescence and phosphorescence (Fig. 12.41). In **fluorescence**, the spontaneously emitted radiation ceases very soon after the exciting radiation is extinguished. In **phosphorescence**, the spontaneous emission may persist for long periods (even hours, but characteristically seconds or fractions of seconds). The difference suggests that fluorescence is an immediate conversion of absorbed light into re-emitted radiant energy and that phosphorescence involves the storage of energy in a reservoir from which it slowly leaks.

Other than thermal degradation, a nonradiative fate for an electronically excited molecule is **dissociation**, or fragmentation (Fig. 12.42). The onset of dissociation can be detected in an absorption spectrum by seeing that the vibrational structure of a band terminates at a certain energy. Absorption occurs in a continuous band above this **dissociation limit**, the highest frequency before the onset of continuous absorption, because the final state is unquantized translational motion of the fragments. Locating the dissociation limit is a valuable way of determining the bond dissociation energy.

12.9 Fluorescence

Figure 12.43 is a simple example of a **Jablonski diagram**, a schematic portrayal of molecular electronic and vibrational energy levels, which shows the sequence of steps involved in fluorescence. The initial absorption takes the molecule to an excited electronic state, and if the absorption spectrum were monitored, it would look like the one shown in Fig. 12.44a. The excited molecule is subjected to collisions with the surrounding molecules, and as it gives up energy it steps down the ladder of vibrational levels. The surrounding molecules, however, might be unable to accept the larger energy needed to lower the molecule to the ground electronic state. The excited state might therefore survive long enough to generate a photon and emit the remaining excess energy as radiation. The downward electronic transition is **vertical**, which means it is in accord with the Franck–Condon principle, and the fluorescence spectrum has a vibrational structure characteristic of the lower electronic state (Fig. 12.44b).

Fluorescence occurs at a lower frequency than that of the incident radiation for two reasons. First, fluorescence radiation is emitted after some vibrational energy has been discarded into the surroundings. The vivid oranges and greens of fluorescent dyes are an everyday manifestation of this effect: they absorb in the ultraviolet and blue and fluoresce in the visible. The mechanism also suggests that the intensity of the fluorescence ought to depend on the ability of the solvent molecules to accept the electronic and vibrational quanta. It is indeed found that a solvent composed of molecules with widely spaced vibrational levels (such as water) may be able to accept the large quantum of electronic energy and so

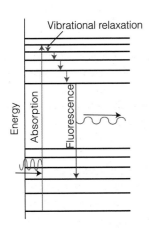

Fig. 12.43 A Jablonski diagram showing the sequence of steps leading to fluorescence. After the initial absorption the upper vibrational states undergo radiationless decay—the process of vibrational relaxation—by giving up energy to the surroundings. A radiative transition then occurs from the ground state of the upper electronic state. In practice, the separation of the ground states of the electronic states (the lowest horizontal line in each set) is 10 to 100 times greater than the separation of the vibrational levels.

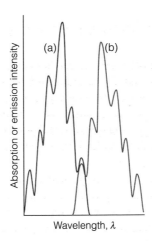

Fig. 12.44 The absorption spectrum (a) shows a vibrational structure characteristic of the upper state. The fluorescence spectrum (b) shows a structure characteristic of the lower state; it is also displaced to lower frequencies and resembles a mirror image of the absorption.

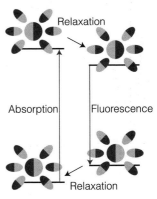

Fig. 12.45 The solvent can shift the fluorescence spectrum relative to the absorption spectrum. On the left we see that the absorption occurs with the solvent (the ellipses) in the arrangement characteristic of the ground electronic state of the molecule (the sphere). However, before fluorescence occurs, the solvent molecules relax into a new arrangement, and that arrangement is preserved during the subsequent radiative transition.

decrease the intensity of the solute's fluorescence. The second reason for the shift in frequency between absorption and fluorescence peaks is the possibility that the solvent interacts differently with the solute in the ground and excited states (for instance, the hydrogen bonding pattern might differ). Because the solvent molecules do not have time to rearrange during the fast electronic transition, the absorption occurs in an environment characteristic of the solvated ground state; however, the fluorescence occurs in an environment characteristic of the solvated excited state (Fig. 12.45).

12.10 Phosphorescence

Figure 12.46 is a Jablonski diagram showing the events leading to phosphorescence. The first steps are the same as in fluorescence, but the presence of a triplet state plays a decisive role. A **triplet state** is a state in which two electrons in different orbitals have parallel spins: the ground state of O_2, which was discussed in *Case study* 10.1, is an example. The name 'triplet' reflects the (quantum mechanical) fact that the total spin of two parallel electron spins ($\uparrow\uparrow$) can adopt only three orientations with respect to an axis. An ordinary spin-paired state ($\uparrow\downarrow$) is called a **singlet state** because the pair has zero net spin angular momentum and such a resultant cannot adopt different orientations in space.

The ground state of a typical phosphorescent molecule is a singlet because its electrons are all paired; the excited state to which the absorption excites the molecule is also a singlet. The peculiar feature of a phosphorescent molecule, however, is that it possesses an excited triplet state with an energy similar to that of the excited singlet state and into which the excited singlet state may convert. Hence, if there is a mechanism for unpairing two electron spins (and so converting ↑↓ into ↑↑), then the molecule may undergo **intersystem crossing** and become a triplet state. The unpairing of electron spins is possible because the angular momentum needed to convert a singlet state into a triplet state may be acquired from the orbital motion of the electrons. The mixing of spin and orbital angular momentum is called **spin–orbit coupling** and is enhanced by the presence of heavy atoms such as sulfur and phosphorus. We can understand this increase by thinking about the source of the orbital magnetic field. To do so, imagine that we are riding on the electron as it orbits the nucleus. From our viewpoint, the nucleus appears to orbit around us (rather as the pre-Copernicans thought the Sun revolved around the Earth). If the nucleus has a high atomic number, it will have a high charge, we shall be at the center of a strong electric current, and we experience a strong magnetic field. If the nucleus has a low atomic number, we experience a feeble magnetic field arising from the low current that encircles us.

After an excited singlet molecule crosses into a triplet state, it continues to discard energy into the surroundings and to step down the ladder of vibrational states. However, it is now stepping down the triplet's ladder, and at the lowest vibrational energy level it is trapped. The solvent cannot extract the final, large quantum of electronic excitation energy. Moreover, the molecule cannot radiate its energy because return to the ground state is forbidden: detailed analysis shows that a triplet state cannot convert radiatively into a singlet state. This rule stems from the fact that light does not affect the spin directly, so the spin of one electron cannot reverse in direction relative to the other electron during the absorption or emission of a photon. The radiative transition, however, is not totally forbidden because the spin–orbit coupling responsible for the intersystem crossing also breaks this rule. The molecules are therefore able to emit weakly and the emission may continue long after the original excited state was formed.

The mechanism of phosphorescence summarized in Fig. 12.46 accounts for the observation that the excitation energy seems to become trapped in a slowly leaking reservoir. It also suggests (as is confirmed experimentally) that phosphorescence should be most intense from solid samples: energy transfer is then less efficient and the intersystem crossing has time to occur as the singlet excited state loses vibrational energy. The mechanism also suggests that the phosphorescence efficiency should depend on the extent of spin–orbit coupling in the molecule: both the yield of the triplet state and its decay rate are increased by the presence of a moderately heavy atom (with its ability to flip electron spins).

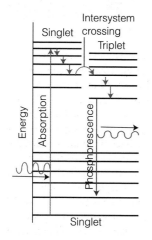

Fig. 12.46 The sequence of steps leading to phosphorescence. The important step is the intersystem crossing from an excited singlet to an excited triplet state. The triplet state acts as a slowly radiating reservoir because the return to the ground state is very slow.

In the laboratory 12.4 *Fluorescence microscopy*

Apart from a small number of cofactors, such as the chlorophylls and flavins, the majority of the building blocks of proteins and nucleic acids do not fluoresce strongly. Four notable exceptions are the amino acids tryptophan ($\lambda_{abs} \approx 280$ nm and $\lambda_{fluor} \approx 348$ nm in water), tyrosine ($\lambda_{abs} \approx 274$ nm and $\lambda_{fluor} \approx 303$ nm in water), and phenylalanine ($\lambda_{abs} \approx 257$ nm and $\lambda_{fluor} \approx 282$ nm in water) and the oxidized form of the sequence serine–tyrosine–glycine (**1**) found in the green fluorescent protein (GFP) of certain jellyfish. The wild type of GFP from *Aequora Victoria* absorbs strongly at 395 nm and emits maximally at 509 nm.

In **fluorescence microscopy**, images of biological cells at work are obtained by attaching a large number of fluorescent molecules to proteins, nucleic acids, and membranes and then measuring the distribution of fluorescence intensity within the illuminated area. A common fluorescent label is GFP. With proper filtering to remove light due to Rayleigh scattering of the incident beam, it is possible to collect light from the sample that contains only fluorescence from the label. However, great care is required to eliminate fluorescent impurities from the sample.

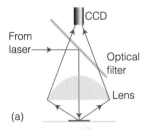

1 The chromophore of GFP

In the laboratory 12.5 *Single-molecule spectroscopy*

Fluorescence and vibrational microscopy with conventional spectrometers and microscopes can provide only as much molecular detail as allowed by the diffraction limit. Most molecules—including biopolymers—have dimensions that are much smaller than visible wavelengths, so special techniques had to be developed to visualize single molecules with optical microscopes. Here we outline the most popular strategies comprising a collection of tools known as **single-molecule spectroscopy**.

The bulk of the work done in single-molecule spectroscopy is based on fluorescence microscopy done with laser excitation of the specimen. The laser is the radiation source of choice because it provides the high intensity required to increase the rate of arrival of photons at the detector from small illuminated areas. Two techniques are commonly used to circumvent the diffraction limit. First, the concentration of the sample is kept so low that, on average, only one fluorescent molecule is in the illuminated area. Second, special strategies are used to illuminate very small volumes. In **near-field scanning optical microscopy** (NSOM), a very thin metal-coated fiber is used to deliver light to a small area. It is possible to construct fibers with tip diameters in the range of 50 to 100 nm, which are indeed smaller than visible wavelengths. The fiber tip is placed very close to the sample, in a region known as the *near field*, where, according to classical physics, photons do not diffract.

In **far-field confocal microscopy**, laser light focused by an objective lens is used to illuminate about 1 μm³ of a very dilute sample placed beyond the near field. This illumination scheme is limited by diffraction and, as a result, data from far-field microscopy have less structural detail than data from NSOM. However, far-field microscopes are very easy to construct and the technique can be used to probe single molecules as long as there is one molecule, on average, in the illuminated area.

In the **wide-field epifluorescence method**, a CCD detects fluorescence excited by a laser and scattered back from the sample (Fig. 12.47a). If the fluorescing molecules are well separated in the specimen, then it is possible to obtain a map of the distribution of fluorescent molecules in the illuminated area. For example, Fig. 12.47b shows how epifluorescence microscopy can be used

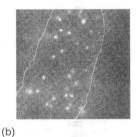

Fig. 12.47 (a) Layout of an epifluorescence microscope. Laser radiation is diverted to a sample by a special optical filter that reflects radiation with a specified wavelength (in this case, the laser excitation wavelength) but transmits radiation with other wavelengths (in this case, wavelengths at which the fluorescent label emits). A CCD detector analyzes the spatial distribution of the fluorescence signal from the illuminated area. (b) Observation of fluorescence from single MHC proteins that have been labeled with a fluorescent marker and are bound to the surface of a cell (the area shown has dimensions of 12 μm × 12 μm). (Image provided by Professor W.E. Moerner, Stanford University.)

to observe single molecules of the major histocompatibility (MHC) protein on the surface of a cell.

Single-molecule spectroscopy has been used to address important problems in biology. One notable example is the visualization of some of the steps involved in the synthesis of ATP by the enzyme ATPase, which we discussed in Chapter 5.

Photobiology

So far, we have considered the decay of excited electronic states of molecules by the emission of light or degradation into thermal motion ('heat'). However, in photochemical reactions the energy in excited states can also be used to drive chemical reactions. The most important of all are the photochemical processes that capture the Sun's radiant energy. Some of these reactions lead to the heating of the atmosphere during the daytime by absorption in the ultraviolet region as a result of reactions like those depicted in Fig. 12.48. Others include the absorption of red and blue light by chlorophyll and the subsequent use of the energy to bring about the photosynthesis of carbohydrates from carbon dioxide and water. Indeed, without light-initiated chemical processes the world would be simply a warm, sterile rock. **Photobiology** is the study of biochemical reactions that are initiated by the absorption of light. In the following sections we explore the mechanisms of some important photobiological processes: photosynthesis, vision, light-induced DNA damage, and light-based therapies.

12.11 The kinetics of decay of excited states

To treat photobiology quantitatively we often invoke concepts of chemical kinetics, so we need to see how the mathematical techniques discussed in Chapters 6–8 can be used to describe the fates of excited electronic states as they participate in such processes as vision and photosynthesis.

A molecule acquires enough energy to react by absorbing a photon. However, not every excited molecule may form a specific primary product (atoms, radicals, or

Fig. 12.48 The temperature profile through the atmosphere and some of the reactions that take place in each region.

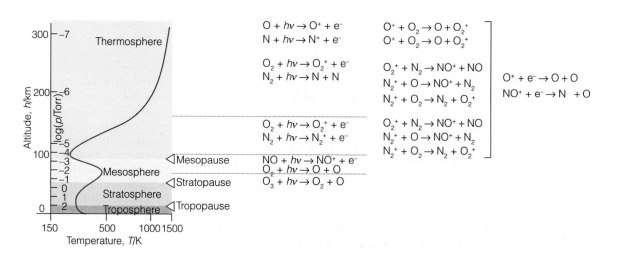

ions, for instance) because we have seen that there are many ways in which the excitation may be lost other than by dissociation or ionization. We therefore speak of the **primary quantum yield**, ϕ (phi), which is the number of events (physical changes or chemical reactions) that lead to primary products (photons, atoms, or ions, for instance) divided by the number of photons absorbed by the molecule in the same time interval:

$$\phi = \frac{\text{number of events}}{\text{number of photons absorbed}} \qquad \boxed{\text{Primary quantum yield}} \quad (12.17)$$

If each molecule that absorbs a photon undergoes dissociation (for instance), then $\phi = 1$. If none does, because the excitation energy is lost before the molecule has time to dissociate, then $\phi = 0$.

If we divide the numerator and denominator of eqn 12.17 by the time interval during which the photochemical event occurs, we see that primary quantum yield is also the rate of radiation-induced primary events divided by the rate of photon absorption. Furthermore, if we equate the rate of photon absorption with the intensity, I_{abs}, of light absorbed by the molecule, we may write

$$\phi = \frac{\text{rate}}{I_{abs}} \qquad (12.18)$$

A molecule in an excited state must either decay to the ground state or form a photochemical product. Therefore, the total number of molecules deactivated by radiative processes, nonradiative processes, and photochemical reactions must be equal to the number of excited species produced by absorption of light. We conclude that the sum of primary quantum yield ϕ_i for *all* physical changes and photochemical reactions i must be equal to 1, regardless of the number of reactions involving the excited state. It follows that

$$\sum_i \phi_i = \sum_i \frac{\text{rate}_i}{I_{abs}} = 1 \qquad (12.19)$$

One successfully excited molecule might initiate the consumption of more than one reactant molecule. We therefore need to introduce the **overall quantum yield**, Φ (uppercase phi), which is the number of reactant molecules that react for each photon absorbed. In the photochemical dissociation of HI, for example, the processes are

$$HI + h\nu \rightarrow H + I$$
$$H + HI \rightarrow H_2 + I$$
$$I + I + M \rightarrow I_2 + M$$

(where M is a 'third body', an inert species that removes excess energy). The overall quantum yield is 2 because the absorption of one photon leads to the destruction of two HI molecules.

In many cases, the proper description of the rates and mechanisms of photochemical reactions also requires knowledge of processes such as fluorescence and phosphorescence that can deactivate an excited state before the reaction has a chance to occur. Electronic absorption takes place in about 10^{-16}–10^{-15} s, and because fluorescence lifetimes are typically 10^{-12}–10^{-6} s, an excited singlet state can initiate very fast photochemical reactions in the range from femtoseconds (10^{-15} s, the time it takes to excite a molecule) to picoseconds (10^{-12} s, the lifetime of the excited state). Examples of such ultrafast reactions are the initial events of vision and photosynthesis (*Case studies* 12.2 and 12.3). Typical phosphorescence lifetimes for large organic molecules are 10^{-6}–10^{-1} s, respectively. As a consequence, excited triplet states can be photochemically important. Indeed, because the phosphorescence lifetime is several orders of magnitude longer than the time required for most typical reactions, species in excited triplet states can undergo a very large number of collisions with other reactants before they lose their energy by radiation or are deactivated nonradiatively.

We begin our exploration of the interplay between reaction rates and excited state decay rates by considering the mechanism of deactivation of an excited singlet state in the absence of a chemical reaction. The following steps are involved:

Absorption: $S + h v_i \rightarrow S^*$ $\qquad v_{abs} = I_{abs}$

Fluorescence: $S^* \rightarrow S + h v_f$ $\qquad v_f = k_f[S^*]$

Intersystem crossing: $S^* \rightarrow T^*$ $\qquad v_{ISC} = k_{ISC}[S^*]$

Internal conversion: $S^* \rightarrow S$ $\qquad v_{IC} = k_{IC}[S^*]$

in which S is an absorbing species, S^* is an excited singlet state, T^* is an excited triplet state, and $h v_i$ and $h v_f$ are the energies of the incident and fluorescent photons, respectively. From the methods developed in Chapter 7 and the rates of the steps that form and destroy the excited singlet state S^*, we write the rate of formation and decay of S^* as

Rate of formation of $[S^*] = I_{abs}$

Rate of decay of $[S^*] = -k_F[S^*] - k_{ISC}[S^*] - k_{IC}[S^*] = -(k_F + k_{ISC} + k_{IC})[S^*]$

It follows that the excited state decays by a first-order process, so when the light is turned off, the concentration of S^* varies with time t as

$$[S^*]_t = [S^*]_0 e^{-t/\tau_0} \tag{12.20}$$

where the **observed fluorescence lifetime**, τ_0, is defined as

$$\tau_0 = \frac{1}{k_F + k_{ISC} + k_{IC}} \qquad \boxed{\text{Observed fluorescence lifetime}} \tag{12.21}$$

We show in the following *Justification* that the quantum yield of fluorescence is

$$\phi_F = \frac{k_F}{k_F + k_{ISC} + k_{IC}} \qquad \boxed{\text{Quantum yield of fluorescence}} \tag{12.22}$$

Justification 12.3   *The quantum yield of fluorescence*

Most fluorescence measurements are conducted by illuminating a relatively dilute sample with a continuous and intense beam of light. It follows that $[S^*]$ is small and constant, so we may invoke the steady-state approximation (Section 7.3(c)) and write

$$\frac{d[S^*]}{dt} = I_{abs} - k_F[S^*] - k_{ISC}[S^*] - k_{IC}[S^*] = I_{abs} - (k_F + k_{ISC} + k_{IC})[S^*] = 0$$

Consequently,

$$I_{abs} = (k_F + k_{ISC} + k_{IC})[S^*]$$

By using this expression and eqn 12.18, the quantum yield of fluorescence is written as

$$\phi_F = \frac{\text{rate of fluorescence}}{I_{abs}} = \frac{k_F[S^*]}{(k_F + k_{ISC} + k_{IC})[S^*]}$$

which, by canceling the $[S^*]$, simplifies to eqn 12.22.

The observed fluorescence lifetime can be measured with a pulsed laser technique. First, the sample is excited with a short light pulse from a laser using a wavelength at which S absorbs strongly. Then, the exponential decay of the fluorescence intensity after the pulse is monitored. From eqns 12.21 and 12.22, it follows that

$$\tau_0 = \frac{1}{k_F + k_{ISC} + k_{IC}} = \left(\frac{k_F}{k_F + k_{ISC} + k_{IC}}\right) \times \frac{1}{k_F} = \frac{\phi_F}{k_F} \qquad (12.23)$$

A brief illustration

In water, the fluorescence quantum yield and observed fluorescence lifetime of tryptophan are $\phi_F = 0.20$ and $\tau_0 = 2.6$ ns, respectively. It follows from eqn 12.23 that the fluorescence rate constant k_F is

$$k_F = \frac{\phi_F}{\tau_0} = \frac{0.20}{2.6 \times 10^{-9}\,\text{s}} = 7.7 \times 10^7\,\text{s}^{-1}$$

12.12 Fluorescence quenching

The dependence of the fluorescence intensity on the presence of other species gives valuable information about photobiological processes and can also be used to measure molecular distances in biological systems.

Now we consider the kinetic information about photochemical processes that can be obtained by 'quenching' studies. **Fluorescence quenching** is the nonradiative removal of the excitation energy from a fluorescent molecule and the elimination of its fluorescence. Quenching may be either a desired process, such as in energy or electron transfer, or an undesired side reaction that can decrease the quantum yield of a desired photochemical process. Quenching effects may be studied by monitoring the fluorescence of a species involved in the photochemical reaction.

(a) The experimental analysis

The **Stern–Volmer equation**, which is derived in the following *Justification*, relates the fluorescence quantum yields $\phi_{F,0}$ and ϕ_F measured in the absence and presence, respectively, of a quencher Q at a molar concentration $[Q]$:

$$\frac{\phi_{F,0}}{\phi_F} = 1 + \tau_0 k_Q[Q] \qquad \boxed{\text{Stern–Volmer equation}} \qquad (12.24)$$

This equation tells us that a plot of $\phi_{F,0}/\phi_F$ against $[Q]$ should be a straight line with slope $\tau_0 k_Q$. Such a plot is called a **Stern–Volmer plot** (Fig. 12.49). The method may also be applied to the quenching of phosphorescence.

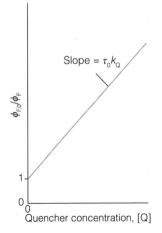

Fig. 12.49 The format of a Stern–Volmer plot and the interpretation of the slope in terms of the rate constant for quenching and the observed fluorescence lifetime in the absence of quenching.

Justification 12.4 *The Stern–Volmer equation*

The addition of a quencher, Q, opens an additional channel for deactivation of S*:

Quenching: $S^* + Q \rightarrow S + Q$ Rate of quenching $= k_Q[Q][S^*]$

The steady-state approximation for [S*] now gives

$$\frac{d[S^*]}{dt} = I_{abs} - (k_F + k_{ISC} + k_{IC} + k_Q[Q])[S^*] = 0$$

and the fluorescence quantum yield in the presence of the quencher is

$$\phi_F = \frac{k_F}{k_F + k_{ISC} + k_{IC} + k_Q[Q]}$$

We can identify the fluorescence lifetime in the presence of quencher as $\tau = 1/(k_F + k_{ISC} + k_{IC} + k_Q[Q])$. When [Q] = 0, the quantum yield is

$$\phi_{F,0} = \frac{k_F}{k_F + k_{ISC} + k_{IC}}$$

It follows that

$$\frac{\phi_{F,0}}{\phi_F} = \left(\frac{k_F}{k_F + k_{ISC} + k_{IC}}\right) \times \left(\frac{k_F + k_{ISC} + k_{IC} + k_Q[Q]}{k_F}\right)$$

$$= \frac{k_F + k_{ISC} + k_{IC} + k_Q[Q]}{k_F + k_{ISC} + k_{IC}}$$

$$= 1 + \frac{k_Q}{k_F + k_{ISC} + k_{IC}}[Q]$$

By using eqn 12.23, this expression simplifies to eqn 12.24.

Because the fluorescence intensity and lifetime are both proportional to the fluorescence quantum yield (specifically, from eqn 12.23, $\tau = \phi_F/k_F$), plots of $I_{F,0}/I_F$ and τ_0/τ (where the subscript 0 indicates a measurement in the absence of quencher) against [Q] should also be linear with the same slope and intercept as those shown for eqn 12.24.

Example 12.4 *Determining the quenching rate constant*

The quenching of tryptophan fluorescence by dissolved O_2 gas was monitored by measuring emission lifetimes at 348 nm in aqueous solutions. Determine the quenching rate constant for this process from the following data:

$[O_2]/(10^{-2}\ mol\ dm^{-3})$	0	2.3	5.5	8	10.8
$\tau/(10^{-9}\ s)$	2.6	1.5	0.92	0.71	0.57

Strategy We rewrite the Stern–Volmer equation (eqn 12.24) for use with life-time data and then fit the data to a straight line.

Solution On substitution of τ_0/τ for $\phi_{F,0}/\phi_F$ in eqn 12.24 and after rearrangement, we obtain

$$\frac{1}{\tau} = \frac{1}{\tau_0} + k_Q[Q] \qquad (12.25)$$

Figure 12.50 shows a plot of $1/\tau$ against $[O_2]$ and the results of a fit to eqn 12.25. The slope of the line is 1.3×10^{10}, so $k_Q = 1.3 \times 10^{10} \ dm^3 \ mol^{-1} \ s^{-1}$.

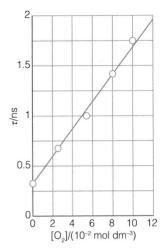

(Self-test 12.11) From the data above, predict the value of $[O_2]$ required to decrease the intensity of tryptophan emission to 50 per cent of the unquenched value.

Answer: $3.0 \times 10^{-2} \ mol \ dm^{-3}$

Fig. 12.50 The Stern–Volmer plot of the data for *Example* 12.5.

(b) Mechanisms of quenching

Three common mechanisms for quenching of an excited singlet (or triplet) state are

Collisional deactivation: $S^* + Q \rightarrow S + Q$

Electron transfer: $S^* + Q \rightarrow S^+ + Q^-$ or $S^- + Q^+$

Resonance energy transfer: $S^* + Q \rightarrow S + Q^*$

The quenching rate constant itself does not give much insight into the mechanism of quenching. However, there are some criteria that govern the relative efficiencies of collisional deactivation, energy transfer, and electron transfer. Energy transfer is a special case and we treat it in detail shortly. For now, we consider collisional deactivation and light-induced electron transfer.

Collisional quenching is particularly efficient when the quencher is a heavy species, such as iodide ion, that receives energy from the fluorescing species and then decays nonradiatively to the ground state. This fact may be used to determine the accessibility of amino acid residues of a folded protein to solvent. For example, fluorescence from a tryptophan residue is quenched by iodide ion when the residue is on the surface of the protein and hence accessible to the solvent. Conversely, residues in the hydrophobic interior of the protein are not quenched effectively by I^-.

According to the Marcus theory of electron transfer discussed in Sections 8.9–8.12, the rate of electron transfer from ground or excited states, and therefore in this context the rate of fluorescence quenching, depends on

1. The distance between the donor and acceptor, with electron transfer becoming more efficient as the distance between donor and acceptor decreases.

2. The reaction Gibbs energy, $\Delta_r G$, with electron transfer becoming more efficient as the reaction becomes more exergonic. For example, efficient photo-oxidation of S requires that the reduction potential of S^* be lower than the reduction potential of Q.

3. The reorganization energy, the energy cost incurred by molecular rearrangements of donor, acceptor, and medium during electron transfer. The electron transfer rate is predicted to increase if this reorganization energy is matched closely by the reaction Gibbs energy.

Electron transfer can be studied by time-resolved spectroscopy (*In the laboratory* 7.2) because the oxidized and reduced products often have electronic absorption spectra distinct from those of their neutral parent compounds. Therefore, the rapid appearance of such known features in the absorption spectrum after excitation by a laser pulse may be taken as an indication of quenching by electron transfer.

12.13 **Fluorescence resonance energy transfer**

Now we turn to resonance energy transfer. We visualize the process $S^* + Q \rightarrow S + Q^*$ as follows. The oscillating electric field of the incoming electromagnetic radiation induces an oscillating electric dipole moment in S. Energy is absorbed by S if the frequency of the incident radiation, v, is such that $v = \Delta E_S/h$, where ΔE_S is the energy separation between the ground and excited electronic states of S and h is Planck's constant. This is the 'resonance condition' for absorption of radiation. The oscillating dipole on S now can affect electrons bound to a nearby Q molecule by inducing an oscillating dipole moment in the latter. If the frequency of oscillation of it is such that $v = E_Q/h$ (where ΔE_Q is the energy separation between the ground and excited electronic states of Q), then Q will absorb energy from S.

The efficiency, η_T, of resonance energy transfer is defined as

$$\eta_T = 1 - \frac{\phi_F}{\phi_{F,0}}$$ Efficiency of energy transfer (12.26)

According to the **Förster theory** of resonance energy transfer, which was proposed by T. Förster in 1959, energy transfer is efficient when

1. the energy donor and acceptor are separated by a short distance (of the order of nanometers);

2. photons emitted by the excited state of the donor can be absorbed directly by the acceptor.

For donor–acceptor systems that are held rigidly either by covalent bonds or by a protein 'scaffold', η_T increases with decreasing distance, R, according to

$$\eta_T = \frac{R_0^6}{R_0^6 + R^6}$$ Förster efficiency (12.27)

where R_0 is a parameter (with units of distance) that is characteristic of each donor–acceptor pair. Equation 12.27 has been verified experimentally, and values of R_0 are available for a number of donor–acceptor pairs (Table 12.5).

The emission and absorption spectra of molecules span a range of wavelengths, so the second requirement of the Förster theory is met when the emission spectrum of the donor molecule overlaps significantly with the absorption spectrum

Table 12.5 Values of R_0 for some donor–acceptor pairs*

Donor	Acceptor	R_0/nm
Naphthalene	Dansyl	2.2
Dansyl	ODR	4.3
Pyrene	Coumarin	3.9
1.5-I-AEDANS	FITC	4.9
Tryptophan	1.5-I-AEDANS	2.2
Tryptophan	Haem	2.9

*Abbreviations: dansyl, 5-dimethylamino-l-naphthalenesulfonic acid; FITC, fluorescein,5-isothiocyanate; 1.5-I-AEDANS: 5-((((2-iodoacetyl)amino)ethyl)amino)naphthalene-1-sulfonic acid; ODR, octadecyl-rhodamine.

of the acceptor. In the overlap region, photons emitted by the donor have the proper energy to be absorbed by the acceptor (Fig. 12.51).

If the donor and acceptor molecules diffuse in solution or in the gas phase, Förster theory predicts that the efficiency of quenching by energy transfer increases as the average distance traveled between collisions of donor and acceptor decreases. That is, the quenching efficiency increases with concentration of quencher, as predicted by the Stern–Volmer equation.

In many cases, it is possible to prove that energy transfer is the predominant mechanism of quenching if the excited state of the acceptor fluoresces or phosphoresces at a characteristic wavelength. In a pulsed laser experiment, the rise in fluorescence intensity from Q* with a time constant that is the same as that for the decay of the fluorescence of S* is often taken as indication of energy transfer from S to Q.

Equation 12.26 forms the basis for **fluorescence resonance energy transfer** (FRET), in which the dependence of the energy transfer efficiency, η_T, on the distance, R, between energy donor and acceptor can be used to measure distances in biological systems. In a typical FRET experiment, a site on a biopolymer or membrane is labeled covalently with an energy donor and another site is labeled covalently with an energy acceptor. In certain cases, the donor or acceptor may be natural constituents of the system, such as amino acid groups, co-factors, or enzyme substrates. The distance between the labels is then calculated from the known value of R_0 and eqn 12.27. Several tests have shown that the FRET technique is useful for measuring distances ranging from 1 to 9 nm.

Fig. 12.51 According to the Förster theory, the rate of energy transfer from a molecule S* in an excited state to a quencher molecule Q is optimized at radiation frequencies in which the emission spectrum of S* overlaps with the absorption spectrum of Q, as shown in the shaded region.

A brief illustration

As an illustration of the FRET technique, consider a study of the protein rhodopsin (*Case study* 12.2). When an amino acid on the surface of rhodopsin was labeled covalently with the energy donor 1.5-I-AEDANS (2), the fluorescence quantum yield of the label decreased from 0.75 to 0.68 due to quenching by the visual pigment 11-*cis*-retinal (Atlas E3 and 3). From eqn 12.26, we calculate $\eta_T = 1 - (0.68/0.75) = 0.093$, and from eqn 12.27 and the known value of $R_0 = 5.4$ nm for the 1.5-I-AEDANS/11-*cis*-retinal pair we calculate $R = 7.9$ nm. Therefore, we take 7.9 nm to be the distance between the surface of the protein and 11-*cis*-retinal.

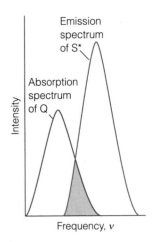

2 1.5-I-AEDANS

3 11-*cis*-Retinal

Case study 12.2   *Vision*

The eye is an exquisite photochemical organ that acts as a transducer, converting radiant energy into electrical signals that travel along neurons. Here we concentrate on the events taking place in the human eye, but similar processes occur in all animals. Indeed, a single type of protein, rhodopsin, is the primary receptor for light throughout the animal kingdom, which indicates that vision emerged very early in evolutionary history, no doubt because of its enormous value for survival.

Photons enter the eye through the cornea, pass through the ocular fluid that fills the eye, and fall on the retina. The ocular fluid is principally water, and passage of light through this medium is largely responsible for the *chromatic*

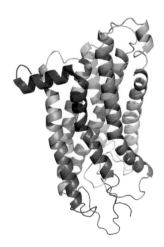

4 A xanthophyll

Fig. 12.52 The structure of rhodopsin, showing the α helices that anchor retinal, the visual pigment.

aberration of the eye, the blurring of the image as a result of different frequencies being brought to slightly different focuses. The chromatic aberration is reduced to some extent by the tinted region called the *macular pigment*, which covers part of the retina. The pigments in this region are the carotene-like xanthophylls (**4**), which absorb some of the blue light and hence help to sharpen the image. They also protect the photoreceptor molecules from too great a flux of potentially dangerous high-energy photons. The xanthophylls have delocalized electrons that spread along the chain of conjugated double bonds, and the π-to-π* transition lies in the visible.

About 57 per cent of the photons that enter the eye reach the retina; the rest are scattered or absorbed by the ocular fluid. Here the primary act of vision takes place, in which the chromophore of a rhodopsin molecule absorbs a photon in another π-to-π* transition. A rhodopsin molecule consists of an opsin protein molecule to which is attached an 11-*cis*-retinal molecule (Atlas E3 and **3**). The latter resembles half a carotene molecule, showing Nature's economy in its use of available materials. The attachment is by the formation of a protonated Schiff's base, utilizing the CHO group of the chromophore and the terminal NH_2 group of the side chain of a lysine residue from opsin (**5**). The free 11-*cis*-retinal molecule absorbs in the ultraviolet, but attachment to the opsin protein molecule shifts the absorption into the visible region. The rhodopsin molecules are situated in the membranes of special cells (the 'rods' and the 'cones') that cover the retina. The opsin molecule is anchored into the cell membrane by two hydrophobic groups and largely surrounds the chromophore (Fig. 12.52).

Immediately after the absorption of a photon, the 11-*cis*-retinal molecule undergoes photoisomerization into all-*trans*-retinal (Atlas E2 and **6**). Photoisomerization takes about 200 fs, and about 67 pigment molecules isomerize for every 100 photons that are absorbed. The process occurs because the π-to-π* excitation of an electron loosens one of the π-bonds (the one indicated by the arrow in **5**), its torsional rigidity is lost, and one part of the molecule swings around into its new position. At that point, the molecule returns to its ground state but is now trapped in its new conformation. The straightened tail of all-*trans*-retinal results in the molecule taking up more space than 11-*cis*-retinal did, so the molecule presses against the coils of the opsin molecule that surrounds it. In about 0.25–0.50 ms from the initial absorption event, the rhodopsin molecule is activated both by the isomerization of retinal

5

hv

6

and deprotonation of its Schiff's base tether to opsin, forming an intermediate known as *metarhodopsin II*.

In a sequence of biochemical events known as the *biochemical cascade*, metarhodopsin II activates the protein transducin (Atlas P13), which in turn activates a phosphodiesterase enzyme that hydrolyzes cyclic guanine monophosphate (cGMP) to GMP. The reduction in the concentration of cGMP causes cGMP-gated ion channels to close, and the result is a sizable change in the transmembrane potential. The pulse of electric potential travels through the optical nerve and into the optical cortex, where it is interpreted as a signal and incorporated into the web of events we call 'vision'.

The resting state of the rhodopsin molecule is restored by a series of non-radiative chemical events powered by ATP. The process involves the escape of all-*trans*-retinal as all-*trans*-retinol (in which –CHO has been reduced to –CH₂OH) from the opsin molecule by a process catalyzed by the enzyme rhodopsin kinase and the attachment of another protein molecule, arrestin. The free all-*trans*-retinol molecule now undergoes enzyme-catalyzed isomerization into 11-*cis*-retinol followed by dehydrogenation to form 11-*cis*-retinal, which is then delivered back into an opsin molecule. At this point, the cycle of excitation, photoisomerization, and regeneration is ready to begin again.

Case study 12.3) *Photosynthesis*

Up to about 1 kW m⁻² of solar radiation reaches the Earth's surface, with the exact intensity depending on latitude, time of day, and weather. A significant amount of this energy is harnessed during photosynthesis. Other photochemical processes also occur in both photosynthetic and nonphotosynthetic organisms. Among the beneficial processes in humans are vision and the biosynthesis of vitamin D_3 from 7-dehydrocholesterol in skin. Other processes, such as DNA damage caused by prolonged exposure to ultraviolet radiation, are deleterious to both higher and lower organisms. When controlled carefully, however, these potentially harmful photochemical processes may be turned into beneficial forms of therapy.

A large proportion of solar radiation with wavelengths below 400 nm and above 1000 nm is absorbed by atmospheric gases such as ozone and O_2, which absorb ultraviolet radiation, and CO_2 and H_2O, which absorb infrared radiation (see *Example 12.3*). As a result, plants, algae, and some species of bacteria have evolved photosynthetic apparatus that capture visible and near-infrared radiation. Plants use radiation in the wavelength range 400–700 nm to drive the endergonic reduction of CO_2 with concomitant oxidation of water to O_2 ($\Delta_r G^{\ominus} = +2880$ kJ mol⁻¹). We have already examined the thermodynamics of plant photosynthesis (Section 5.11); here we shall describe the kinetics of the capture and utilization of solar energy.

In the chloroplast, chlorophylls *a* and *b* (Atlas R3) and carotenoids (of which β-carotene, Atlas E1, is an example) bind to integral proteins called *light-harvesting complexes*, which absorb solar energy and transfer it to protein complexes known as *reaction centers*, where light-induced electron transfer reactions occur. The combination of a light harvesting complex and a reaction center complex is called a **photosystem**. Plants have two photosystems, photosystems I and II, that drive the reduction of NADP$^+$ by water (Section 5.11):

$$2\,NADP^+ + 2\,H_2O \xrightarrow{\text{light}} O_2 + 2\,NADPH + 2H^+$$

Light-harvesting complexes bind large numbers of pigments in order to provide a sufficiently large area for capture of radiation. In photosystems I and II, absorption of a photon raises a chlorophyll or carotenoid molecule to an excited singlet state and within 0.1–5 ps the energy hops to a nearby pigment by the Förster mechanism (Section 12.13). About 100–200 ps later, which corresponds to thousands of hops within the light-harvesting complex, more than 90 per cent of the absorbed energy reaches the reaction center. There, a chlorophyll *a* dimer becomes electronically excited and initiates ultrafast electron transfer reactions. For example, the transfer of an electron from the excited singlet state of P680, the chlorophyll dimer of the photosystem II reaction center, to its immediate electron acceptor, a pheophytin *a* molecule,[5] occurs within 3 ps. Once the excited state of P680 has been quenched efficiently by this first reaction, subsequent steps that lead to the oxidation of water and reduction of plastoquinone occur more slowly, with reaction times varying from 200 ps to 1 ms. The electrochemical reactions within the photosystem I reaction center also occur in this time regime.

In summary, the initial energy and electron transfer events of photosynthesis are under tight kinetic control. Photosynthesis captures solar energy efficiently because the excited singlet state of chlorophyll is quenched rapidly by processes that occur with time constants that are much shorter than the fluorescence lifetime, which is about 5 ns in diethyl ether at room temperature.

Case study 12.4   *Damage of DNA by ultraviolet radiation*

Ozone trapped in the Earth's *stratosphere*, a region spanning from 15 km to 50 km above the surface of the Earth, partially shields the biosphere from harmful ultraviolet radiation in the 'UVB range', 290–320 nm. The depletion of stratospheric ozone by reactions with atmospheric pollutants (most notably the chlorofluorocarbons) has increased the amount of UVB radiation at the Earth's surface. Because the physiological consequences of prolonged exposure to UVB radiation include DNA damage, genetic mutations, cell destruction, sunburn, and skin cancers, there is concern that the depletion of the protective ozone layer may lead to an increase in mortality not only of animals but also the plants and lower organisms that form the base of the food chain.

The principal mechanism of DNA damage involves the photodimerization of adjacent thymine bases to yield either a cyclobutane–thymine dimer or a 6,4 photoproduct (Fig. 12.53). The former has been linked directly to cell death, and the latter may lead to DNA mutations and, consequently, to the formation of tumors.

There are several natural mechanisms for protection from and repair of photochemical damage. For example, the enzyme DNA photolyase, present in organisms from all kingdoms but not in humans, catalyzes the destruction

---

[5] Pheophytin *a* is a chlorophyll *a* molecule where the central Mg^{2+} ion is replaced by two protons, which are bound to two of the pyrrole nitrogens in the ring.

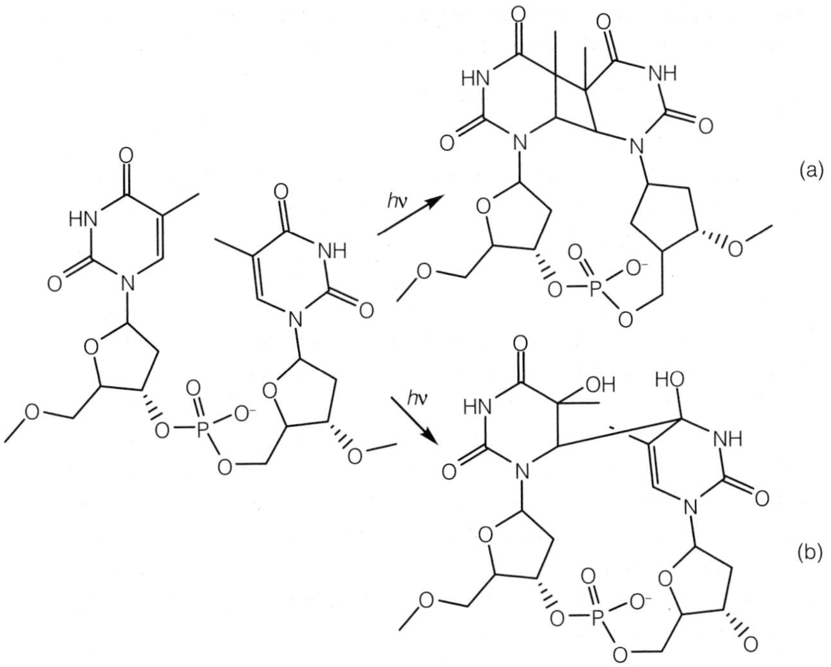

Fig. 12.53 The photodimerization of thymine bases to form either (a) a cyclobutane–thymine dimer or (b) a 6,4 photoproduct.

of cyclobutane thymine dimers. Also, ultraviolet radiation can induce the production of the pigment melanin (in a process more commonly known as 'tanning'), which shields the skin from damage. However, repair and protective mechanisms become increasingly less effective with persistent and prolonged exposure to solar radiation.

Case study 12.5 *Photodynamic therapy*

The reactions of a molecule that does not absorb light directly can be made to occur if another absorbing molecule is present because the latter may be able to transfer its energy to the former during a collision. An example of this *photosensitization* is the reaction used to generate excited state O_2 in a type of treatment known as **photodynamic therapy** (PDT). In PDT, laser radiation is absorbed by a drug that, in its first excited triplet state ^3P, photosensitizes the formation of an excited singlet state of O_2, $^1O_2$, from its triplet ground state, $^3O_2$. The $^1O_2$ molecules are very reactive and destroy cellular components, and it is thought that cell membranes are the primary cellular targets. Hence, the photochemical cycle below leads to the shrinkage (and sometimes total destruction) of diseased tissue.

Absorption: $P + h\nu \rightarrow P^*$

Intersystem crossing: $P^* \rightarrow {}^3P$

Photosensitization: $^3P + {}^3O_2 \rightarrow P + {}^1O_2$

Oxidation reactions: $^1O_2 + \text{reactants} \rightarrow \text{products}$

The photosensitizer is hence a 'photocatalyst' for the production of $^1O_2$. It is common practice to use a porphyrin photosensitizer, such as compounds derived from hematoporphyrin (7). However, much effort is being expended to develop better drugs with enhanced photochemical properties.

7 Hematoporphyrin

A potential PDT drug must meet many criteria. From the point of view of pharmacological effectiveness, the drug must be soluble in tissue fluids so it can be transported to the diseased organ through blood and secreted from the body through urine. The therapy should also result in very few side effects. The drug must also have unique photochemical properties. It must be activated photochemically at wavelengths that are not absorbed by blood and skin. In practice, this means that the drug should have a strong absorption band at $\lambda > 650$ nm. Drugs based on hematoporphyrin do not meet this criterion very well, so novel porphyrin and related macrocycles with more desirable electronic properties are being synthesized and tested. At the same time, the quantum yield of triplet formation and of $^1O_2$ formation must be high so many drug molecules can be activated and many oxidation reactions can occur during a short period of laser irradiation. Photodynamic therapy has been used successfully in the treatment of macular degeneration, a disease of the retina that leads to blindness, and in a number of cancers, including those of the lung, bladder, skin, and esophagus.

Checklist of key concepts

☐ 1. Spectroscopy is the analysis of the electromagnetic radiation emitted, absorbed, or scattered by atoms and molecules.

☐ 2. A spectrometer consists of a source of radiation, a dispersing element (or an interferometer), and a detector.

☐ 3. In a Raman spectrum lines shifted to lower frequency than the incident radiation are called Stokes lines and lines shifted to higher frequency are called anti-Stokes lines.

☐ 4. The intensity of a transition is proportional to the square of the transition dipole moment.

☐ 5. A selection rule is a statement about when the transition dipole can be nonzero.

☐ 6. A gross selection rule specifies the general features a molecule must have if it is to have a spectrum of a given kind.

☐ 7. A specific selection rule is a statement about which changes in quantum number may occur in a transition.

☐ 8. The gross selection rule for infrared absorption spectra is that the electric dipole moment of the molecule must change during the vibration.

☐ 9. The specific selection rule for vibrational transitions is $\Delta v = \pm 1$.

☐ 10. The gross selection rule for the vibrational Raman spectrum of a polyatomic molecule is that the normal mode of vibration is accompanied by a changing polarizability.

☐ 11. The exclusion rule states that if the molecule has a center of inversion, then no modes can be both infrared and Raman active.

☐ 12. In resonance Raman spectroscopy, radiation that nearly coincides with the frequency of an electronic transition is used to excite the sample and the result is a much greater intensity in the scattered radiation.

☐ 13. In conventional microscopy, the diffraction limit prevents the study of specimens that are much smaller than the wavelength of light used as a probe.

☐ 14. In vibrational microscopy, an infrared or Raman spectrometer is combined with a microscope to yield the vibrational spectrum of molecules in small specimens, such as single cells.

☐ 15. The Franck–Condon principle states that because nuclei are so much more massive than electrons, an electronic transition takes place faster than the nuclei can respond.

☐ 16. A chromophore is a group with characteristic optical absorption: chromophores include d-metal complexes, the carbonyl group, and the carbon–carbon double bond.

☐ 17. Chiral molecules may show optical activity and circular dichroism, the differential absorption of left- and right-circularly polarized light.

☐ 18. In fluorescence, the spontaneously emitted radiation ceases quickly after the exciting radiation is extinguished.

☐ 19. In phosphorescence, the spontaneous emission may persist for long periods; the process involves intersystem crossing into a triplet state.

☐ 20. In fluorescence microscopy, images of biological cells at work are obtained by attaching a large number of fluorescent molecules to proteins, nucleic acids, and membranes, and then measuring the distribution of fluorescence intensity within the illuminated area. Special techniques permit the observation of fluorescence from single molecules in cells.

☐ 21. The primary quantum yield of a photochemical reaction is the number of events producing specified primary products for each photon absorbed; the overall quantum yield is the number of reactant molecules that react for each photon absorbed.

☐ 22. Collisional deactivation, electron transfer, and resonance energy transfer are common fluorescence quenching processes. The rate constants of electron and resonance energy transfer decrease with increasing separation between donor and acceptor molecules.

☐ 23. Fluorescence resonance energy transfer (FRET) forms the basis of a technique for measuring distances between molecules in biological systems.

Checklist of key equations

Property	Equation	Comment
Beer–Lambert law	$I = I_0 e^{-\varepsilon[J]L}$	Uniform solution
Absorbance	$A = \varepsilon[J]L$	Definition
Transmittance	$T = I/I_0$	Definition
Transition dipole moment	$\boldsymbol{\mu}_{fi} = \int \psi_f^* \boldsymbol{\mu} \psi_i \, d\tau$	Definition
Lifetime broadening	$\delta E \approx \hbar/\tau$	In practice: $\delta\tilde{v} \approx (5.3\ \mathrm{cm}^{-1})/(\tau/\mathrm{ps})$
Vibrational selection rule	$\Delta v = \pm 1$	Harmonic oscillator model
Number of vibrational modes	(a) $3N-6$, (b) $3N-5$	(a) Nonlinear molecules, (b) linear molecules
Primary quantum yield	$\phi = \mathrm{rate}/I_{abs}$	
Observed fluorescence lifetime	$\tau_0 = \phi_F/k_F$	
Stern–Volmer equation	$\phi_{F,0}/\phi_F = 1 + \tau_0 k_Q[Q]$	
Energy transfer efficiency	$\eta_T = 1 - \phi_F/\phi_{F,0}$	Definition
Förster theory	$\eta_T = R_0^6/(R_0^6 + R^6)$	

Discussion questions

12.1 Describe the physical origins of linewidths in absorption and emission spectra.

12.2 (a) Discuss the physical origins of the gross selection rules for infrared spectroscopy and Raman spectroscopy. (b) Suppose that you wish to characterize the normal modes of benzene in the gas phase. Why is it important to obtain both infrared absorption and Raman spectra of your sample?

12.3 Explain how color can arise from molecules.

12.4 Explain the origin of the Franck–Condon principle and how it leads to the appearance of vibrational structure in an electronic transition.

12.5 Provide examples of common chromophores.

12.6 Describe the mechanisms of photon emission by fluorescence and phosphorescence.

12.7 (a) Summarize the main features of the Förster theory of resonance energy transfer. (b) Discuss FRET and photosynthetic light harvesting in terms of Förster theory.

Exercises

12.8 Express a wavelength of 670 nm as (a) a frequency and (b) a wavenumber.

12.9 What is (a) the wavenumber and (b) the wavelength of the radiation used by an FM radio transmitter broadcasting at 92.0 MHz?

12.10 When light of wavelength 410 nm passes through 2.5 mm of a solution of the dye responsible for the yellow of daffodils at a concentration 0.433 mmol dm^{-3}, the transmission is 71.5 per cent.

Calculate the molar absorption coefficient of the coloring matter at this wavelength and express the answer in centimeters squared per mole (cm^2 mol^{-1}).

12.11 An aqueous solution of a triphosphate derivative of molar mass 602 g mol^{-1} was prepared by dissolving 30.2 mg in 500 cm^3 of water and a sample was transferred to a cell of length 1.00 cm. The absorbance was measured as 1.011. (a) Calculate the molar absorption coefficient. (b) Calculate the transmittance, expressed as a percentage, for a solution of twice the concentration.

12.12 A swimmer enters a gloomier world (in one sense) on diving to greater depths. Given that the mean molar absorption coefficient of seawater in the visible region is 6.2×10^{-5} dm^3 mol^{-1} cm^{-1}, calculate the depth at which a diver will experience **(a)** half the surface intensity of light and **(b)** one-tenth that intensity.

12.13 Consider a solution of two unrelated substances A and B. Let their molar absorption coefficients be equal at a certain wavelength and write their total absorbance A. Show that we can infer the concentration of A and B from the total absorbance at some other wavelength provided we know the molar absorption coefficients at that different wavelength. (See eqn 12.7.)

12.14 The molar absorption coefficients of tryptophan and tyrosine at 240 nm are 2.00×10^3 dm^3 mol^{-1} cm^{-1} and 1.12×10^4 dm^3 mol^{-1} cm^{-1}, respectively, and at 280 nm they are 5.40×10^3 dm^3 mol^{-1} cm^{-1} and 1.50×10^3 dm^3 mol^{-1} cm^{-1}. The absorbance of a sample obtained by hydrolysis of a protein was measured in a cell of thickness 1.00 cm and was found to be 0.660 at 240 nm and 0.221 at 280 nm. What are the concentrations of the two amino acids?

12.15 A solution was prepared by dissolving tryptophan and tyrosine in 0.15 M NaOH(aq) and a sample was transferred to a cell of length 1.00 cm. The two amino acids share the same molar absorption coefficient at 294 nm (2.38×10^3 dm^3 mol^{-1} cm^{-1}), and the absorbance of the solution at that wavelength is 0.468. At 280 nm the molar absorption coefficients are 5.23×10^3 and 1.58×10^3 dm^3 mol^{-1} cm^{-1}, respectively and the total absorbance of the solution is 0.676. What are the concentrations of the two amino acids? *Hint*: It would be sensible to use the result derived in *Exercise* 12.13, but this specific example could be worked through without using that general case.

12.16 In many cases it is possible to assume that an absorption band has a Gaussian line shape (one proportional to e^{-x^2}) centered on the band maximum. **(a)** Assume such a line shape and show that

$$A = \int \varepsilon(\tilde{\nu})\mathrm{d}\tilde{\nu} \approx 1.0645 \varepsilon_{max} \Delta \tilde{\nu}_{1/2}$$

where $\Delta \tilde{\nu}_{1/2}$ is the width at half-height. **(b)** The electronic absorption bands of many molecules in solution have half-widths at half-height of about 5000 cm^{-1}. Estimate the integrated absorption coefficients of bands for which **(i)** $\varepsilon_{max} \approx 1 \times 10^4$ dm^3 mol^{-1} cm^{-1} and **(ii)** $\varepsilon_{max} \approx 5 \times 10^2$ dm^3 mol^{-1} cm^{-1}.

12.17 *Ozone absorbs ultraviolet radiation in a part of the electromagnetic spectrum energetic enough to disrupt DNA in biological organisms and absorbed by no other abundant atmospheric constituent. This spectral range, denoted UVB, spans wavelengths from about 290 nm to 320 nm. **(a)** The abundance of ozone is typically inferred from measurements of UV absorption and is often expressed in terms of *Dobson units* (DU): 1 DU is equivalent to a layer of pure ozone 10 μm thick at 1 atm and 0°C. Compute the absorbance of UV radiation at 300 nm expected for an ozone abundance of 300 DU (a typical value) and 100 DU (a value reached during seasonal Antarctic ozone depletions) given a molar absorption coefficient of 476 dm^3 mol^{-1} cm^{-1}. **(b)** The molar extinction coefficient of ozone over the UVB range is given in the table below. Compute the integrated absorption coefficient of ozone over the wavelength range 290–320 nm. *Hint*: $\varepsilon(\tilde{\nu})$ can be fitted to an exponential function quite well.

λ/nm	292.0	296.3	300.8	305.4
ε/(dm^3 mol^{-1} cm^{-1})	1512	865	477	257
λ/nm		310.1	315.0	320.0
ε/(dm^3 mol^{-1} cm^{-1})		135.9	69.5	34.5

12.18 The Beer–Lambert law is derived on the basis that the concentration of absorbing species is uniform (see *Justification* 12.1). Suppose instead that the concentration falls exponentially as $[J] = [J]_0 e^{-x/\lambda}$. Derive an expression for the variation of I with sample length: suppose that $l \gg \lambda$. *Hint*: Work through *Justification* 12.1, but use this expression for the concentration.

12.19 Assume that the electronic states of the π electrons of a conjugated molecule can be approximated by the wavefunctions of a particle in a one-dimensional box and that the dipole moment can be related to the displacement along this length by $\mu = -ex$. Show that the transition probability for the transition $n = 1 \rightarrow n = 2$ is nonzero, whereas that for $n = 1 \rightarrow n = 3$ is zero. *Hint*: The following relations will be useful:

$$\sin x \sin y = \tfrac{1}{2}\cos(x - y) - \tfrac{1}{2}\cos(x + y)$$

$$\int x \cos ax\,\mathrm{d}x = \frac{1}{a^2}\cos ax + \frac{x}{a}\sin ax$$

12.20 Estimate the lifetime of a state that gives rise to a line of width **(a)** 0.1 cm^{-1}, **(b)** 1 cm^{-1}, and **(c)** 1.0 GHz.

12.21 A molecule in a liquid undergoes about 1×10^{13} collisions in each second. Suppose that **(a)** every collision is effective in deactivating the molecule vibrationally and **(b)** that one collision in 200 is effective. Calculate the width (in cm^{-1}) of vibrational transitions in the molecule.

12.22 Suppose that the C=O group in a peptide bond can be regarded as isolated from the rest of the molecule. Given that the force constant of the bond in a carbonyl group is 908 N m^{-1}, calculate the vibrational frequency of **(a)** $^{12}C=^{16}O$ and **(b)** $^{13}C=^{16}O$.

12.23 The hydrogen halides have the following fundamental vibrational wavenumbers:

	HF	HCl	HBr	HI
$\tilde{\nu}$/cm^{-1}	4141.3	2988.9	2649.7	2309.5

(a) Calculate the force constants of the hydrogen–halogen bonds. **(b)** From the data in part (a), predict the fundamental vibrational wavenumbers of the deuterium halides.

12.24 Which of the following molecules may show infrared absorption spectra: **(a)** H_2, **(b)** HCl, **(c)** CO_2, **(d)** H_2O, **(e)** CH_3CH_3, **(f)** CH_4, **(g)** CH_3Cl, and **(h)** N_2?

12.25 How many normal modes of vibration are there for **(a)** NO_2, **(b)** N_2O, **(c)** cyclohexane, and **(d)** hexane?

12.26 Consider the vibrational mode that corresponds to the uniform expansion of the benzene ring. Is it **(a)** Raman or **(b)** infrared active?

12.27 Suppose that three conformations are proposed for the nonlinear molecule H_2O_2 (**8, 9**, and **10**). The infrared absorption spectrum of gaseous H_2O_2 has bands at 870, 1370, 2869, and

8   **9**   **10**

---

* Adapted from a problem supplied by Charles Trapp and Carmen Giunta.

3417 cm^{-1}. The Raman spectrum of the same sample has bands at 877, 1408, 1435, and 3407 cm^{-1}. All bands correspond to fundamental vibrational wavenumbers, and you may assume that (i) the 870 and 877 cm^{-1} bands arise from the same normal mode and (ii) the 3417 and 3407 cm^{-1} bands arise from the same normal mode. (a) If H_2O_2 were linear, how many normal modes of vibration would it have? (b) Determine which of the proposed conformations is inconsistent with the spectroscopic data. Explain your reasoning.

12.28 The compound $CH_3CH=CHCHO$ has a strong absorption in the ultraviolet at 46 950 cm^{-1} and a weak absorption at 30 000 cm^{-1}. Justify these features in terms of the structure of the compound.

12.29 Figure 12.54 shows the UV–visible absorption spectra of a selection of amino acids. Suggest reasons for their different appearances in terms of the structures of the molecules.

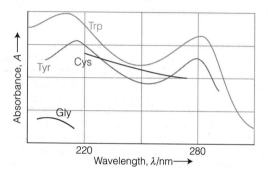

Fig. 12.54

12.30 Suppose that you are a color chemist and have been asked to intensify the color of a dye without changing the type of compound and that the dye in question is a polyene. (a) Would you choose to lengthen or to shorten the chain? (b) Would the modification to the length shift the apparent color of the dye toward the red or the blue?

12.31 Dansyl chloride (**11**), which absorbs maximally at 330 nm and fluoresces maximally at 510 nm, can be used to label amino acids in fluorescence microscopy and FRET studies. Tabulated below is the variation of the fluorescence intensity of an aqueous solution of dansyl chloride with time after excitation by a short laser pulse (with I_0 the initial fluorescence intensity):

t/ns	5.0	10.0	15.0	20.0
I_f/I_0	0.45	0.21	0.11	0.05

11

(a) Calculate the observed fluorescence lifetime of dansyl chloride in water. (b) The fluorescence quantum yield of dansyl chloride in water is 0.70. What is the fluorescence rate constant?

12.32 Consider some of the precautions that must be taken when conducting single-molecule spectroscopy experiments. (a) What is the molar concentration of a solution in which there is, on average, one solute molecule in 1.0 μm^3 (1.0 fL) of solution? (b) It is important to use pure solvents in single-molecule spectroscopy because optical signals from fluorescent impurities in the solvent may mask optical signals from the solute. Suppose that water containing a fluorescent impurity of molar mass 100 g mol^{-1} is used as solvent and that analysis indicates the presence of 0.10 mg of impurity per 1.0 kg of solvent. On average, how many impurity molecules will be present in 1.0 μm^3 of solution? You may take the density of water as 1.0 g cm^{-3}. Comment on the suitability of this solvent for single-molecule spectroscopy experiments.

12.33 Light-induced degradation of molecules, also called *photobleaching*, is a serious problem in single-molecule spectroscopy. A molecule of a fluorescent dye commonly used to label biopolymers can withstand about 106 excitations by photons before light-induced reactions destroy its π system and the molecule no longer fluoresces. For how long will a single dye molecule fluoresce while being excited by 1.0 mW of 488 nm radiation from a laser? You may assume that the dye has an absorption spectrum that peaks at 488 nm and that every photon delivered by the laser is absorbed by the molecule.

12.34 Consider a unimolecular photochemical reaction with rate constant $k_r = 1.7 \times 10^4$ s^{-1} that involves a reactant with an observed fluorescence lifetime of 1.0 ns and an observed phosphorescence lifetime of 1.0 ms. Is the excited singlet state or the excited triplet state the most likely precursor of the photochemical reaction?

12.35 In a photochemical reaction A → 2 B + C, the quantum yield with 500 nm light is 2.1×10^2 mol einstein^{-1} (1 einstein = 1 mol photons). After exposure of 300 mmol of A to the light, 2.28 mmol of B is formed. How many photons were absorbed by A?

12.36 In an experiment to measure the quantum yield of a photochemical reaction, the absorbing substance was exposed to 490 nm light from a 100 W source for 45 min. The intensity of the transmitted light was 40 per cent of the intensity of the incident light. As a result of irradiation, 0.344 mol of the absorbing substance decomposed. Determine the quantum yield.

12.37 When benzophenone is illuminated with ultraviolet radiation, it is excited into a singlet state. This singlet changes rapidly into a triplet, which phosphoresces. Triethylamine acts as a quencher for the triplet. In an experiment in methanol as solvent, the phosphorescence intensity I_{phos} varied with amine concentration as shown below. A time-resolved laser spectroscopy experiment had also shown that the half-life of the fluorescence in the absence of quencher is 29 μs. What is the value of k_Q?

[Q]/(mol dm^{-3})	0.0010	0.0050	0.0100
I_{phos}/(arbitrary units)	0.41	0.25	0.16

12.38 The fluorescence intensity I_f of a solution of a plant pigment illuminated by 330 nm radiation was studied in the presence of a quenching agent, with the following results

[Q]/(mmol dm^{-3})	1.0	2.0	3.0	4.0	5.0
I_f/I_{abs}	0.31	0.18	0.13	0.10	0.081

In a second series of experiments, the fluorescence lifetimes of the pigment were determined by time-resolved spectroscopy:

[Q]/(mmol dm^{-3})	1.0	2.0	3.0	4.0	5.0
τ/ns	76	45	32	25	20

Determine the quenching rate constant and the half-life of the fluorescence.

12.39 The Förster theory of resonance energy transfer and the basis for the FRET technique can be tested by performing fluorescence measurements on a series of compounds in which an energy donor and an energy acceptor are covalently linked by a rigid molecular linker of variable and known length. L. Stryer and R.P. Haugland, *Proc. Natl. Acad. Sci.* USA **58**, 719 (1967), collected the following data on a family of compounds with the general composition dansyl-(L-prolyl)$_n$-naphthyl, in which the distance R between the naphthyl donor and the dansyl acceptor was varied by increasing the number of prolyl units in the linker:

R/nm	1.2	1.5	1.8	2.8	3.1	3.4	3.7	4.0	4.3	4.6
η_T	0.99	0.94	0.97	0.82	0.74	0.65	0.40	0.28	0.24	0.16

Are the data described adequately by the Förster theory (eqns 12.26 and 12.27)? If so, what is the value of R_0 for the naphthyl–dansyl pair?

12.40 An amino acid on the surface of a protein was labeled covalently with 1.5-I-AEDANS and another was labeled covalently with FITC. The fluorescence quantum yield of 1.5-I-AEDANS decreased by 10 per cent due to quenching by FITC. What is the distance between the amino acids? *Hint*: see Table 21.6.

12.41 The flux of visible photons reaching Earth from the North Star is about 4×10^3 mm^{-2} s^{-1}. Of these photons, 30 per cent are absorbed or scattered by the atmosphere and 25 per cent of the surviving photons are scattered by the surface of the cornea of the eye. A further 9 per cent are absorbed inside the cornea. The area of the pupil at night is about 40 mm^2 and the response time of the eye is about 0.1 s. Of the photons passing through the pupil, about 43 per cent are absorbed in the ocular medium. How many photons from the North Star are focused onto the retina in 0.1 s? For a continuation of this story, see R.W. Rodieck, *The first steps in seeing*, Sinauer, Sunderland (1998).

12.42 In light-harvesting complexes, the fluorescence of a chlorophyll molecule is quenched by nearby chlorophyll molecules. Given that for a pair of chlorophyll *a* molecules $R_0 = 5.6$ nm, by what distance should two chlorophyll *a* molecules be separated to shorten the fluorescence lifetime from 1 ns (a typical value for monomeric chlorophyll *a* in organic solvents) to 10 ps?

12.43 The light-induced electron transfer reactions in photosynthesis occur because chlorophyll molecules (whether in monomeric or dimeric forms) are better reducing agents in their electronic excited states. Justify this observation with the help of molecular orbital theory.

12.44 The emission spectrum of a porphyrin dissolved in O_2-saturated water shows a strong band at 650 nm and a weak band at 1270 nm. In separate experiments, it was observed that the electronic absorption spectrum of the porphyrin sample showed bands at 420 nm and 550 nm and the electronic absorption spectrum of O_2-saturated water showed no bands in the visible range of the spectrum (and therefore no emission spectrum when excited in the same range). Based on these data alone, make a preliminary assignment of the emission band at 1270 nm. Propose additional experiments that test your hypothesis.

Projects

12.45 At the current stage of your study, you have enough knowledge of physical chemistry and biochemistry to begin reading the current literature with a critical eye. Consult monographs, journal articles, and reliable internet resources, such as those listed in the web site for this text, and write a brief report (similar in length and depth of coverage to one of the many *Case studies* in this text) on each of the following topics.

(a) In *confocal Raman microscopy*, light must pass through several holes of very small diameter before reaching the detector. In this way light that is out of focus does not interfere with an image that is in focus. Prepare a brief report on the advantages and disadvantages of confocal Raman microscopy over conventional Raman microscopy in the study of biological systems. *Hint*: A good place to start is P. Colarusso, L.H. Lidder, I.W. Levin, E.N. Lewis, Raman and IR microspectroscopy. In *Encyclopedia of spectroscopy and spectrometry* (ed. J.C. Lindon, G.E. Tranter, and J.L. Holmes), 3, 1945. Academic Press, San Diego (2000).

(b) We have seen throughout the text that it is possible to observe the cooperativity of biopolymer denaturation by determining the extent of denaturation as a function of some parameter that affects its stability, such as temperature or denaturant concentration. Prepare a report summarizing the use of a spectroscopic technique in the study of protein denaturation. Your report should include (i) a description of experimental methods, (ii) a discussion of the information that can be obtained from the measurements, (iii) an example from the literature of the use of the technique in protein stability work, and (iv) a brief discussion of the advantages and disadvantages of the technique of your choice over differential scanning calorimetry (*In the laboratory* 1.1), a very popular technique for the study of biopolymer stability.

12.46 The protein hemerythrin (Her) is responsible for binding and carrying O_2 in some invertebrates. Each protein molecule has two Fe^{2+} ions that are in very close proximity and work together to bind one molecule of O_2. The Fe_2O_2 group of oxygenated hemerythrin is colored and has an electronic absorption band at 500 nm.

(a) Figure 12.55 shows the UV–visible absorption spectrum of a derivative of hemerythrin in the presence of different concentrations of CNS^- ions. What may be inferred from the spectrum?

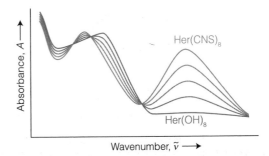

Fig. 12.55

(b) The resonance Raman spectrum of oxygenated hemerythrin obtained with laser excitation at 500 nm has a band at 844 cm^{-1} that has been attributed to the O—O stretching mode of bound $^{16}O_2$. Why is resonance Raman spectroscopy and not infrared spectroscopy the method of choice for the study of the binding of O_2 to hemerythrin?

(c) Proof that the 844 cm^{-1} band in the resonance Raman spectrum of oxygenated hemerythrin arises from a bound O_2 species may be obtained by conducting experiments on samples of hemerythrin that have been mixed with $^{18}O_2$ instead of $^{16}O_2$. Predict the fundamental vibrational wavenumber of the $^{18}O-^{18}O$ stretching mode in a sample of hemerythrin that has been treated with $^{18}O_2$.

(d) The fundamental vibrational wavenumbers for the O–O stretching modes of O_2, O_2^- (superoxide anion), and O_2^{2-} (peroxide anion) are 1555, 1107, and 878 cm^{-1}, respectively. **(i)** Explain this trend in terms of the electronic structures of O_2, O_2^-, and O_2^{2-}. *Hint*: Review *Case study 10.1*. **(ii)** What are the bond orders of O_2, O_2^-, and O_2^{2-}?

(e) Based on the data given in part (d), which of the following species best describes the Fe_2O_2 group of hemerythrin: $Fe_2^{2+}O_2$, $Fe^{2+}Fe^{3+}O_2^-$, or $Fe_2^{3+}O_2^{2-}$? Explain your reasoning.

(f) The resonance Raman spectrum of hemerythrin mixed with $^{16}O^{18}O$ has two bands that can be attributed to the O–O stretching mode of bound oxygen. Discuss how this observation may be used to exclude one or more of the four proposed schemes (**12–15**) for binding of O_2 to the Fe_2 site of hemerythrin.

Fe
 \
  O — O
       \
        Fe

12

Fe ⟨ O / O ⟩ Fe

13

Fe_O_/Fe
      |
      O

14

Fe- -Fe—O—O

15

12.47 As an example of the steps taken in biosensor analysis, consider the association of two proteins, A and B. In a typical experiment, a stream of solution containing a known concentration of A flows above the sensor's surface to which B is attached covalently. Figure 12.56 shows that the kinetics of binding of A to B may be followed by monitoring the time dependence of the surface plasmon resonance (SPR) signal, denoted by R, which is typically the shift in resonance angle. Typically, the system is first allowed to reach equilibrium, which is denoted by the plateau in Fig. 12.56. Then a solution containing no A is flowed above the surface and the AB complex dissociates. Now we see that analysis of the decay of the SPR signal reveals the kinetics of dissociation of the AB complex.

(a) First, show that the equilibrium constant for formation of the AB complex can be measured directly from data of the type displayed in Fig. 12.56. Consider the equilibrium

$$A + B \rightleftharpoons AB \qquad K = k_{on}/k_{off}$$

where k_{on} and k_{off} are, respectively, the rate constants for formation and dissociation of the AB complex and K is the equilibrium constant for

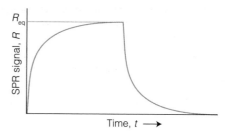

Fig. 12.56

formation of the AB complex. Write an expression for dR/dt and then show that

$$R_{eq} = R_{max}\left(\frac{a_0 K}{a_0 K + 1}\right)$$

where R_{eq} is the value or R at equilibrium, R_{max} is the maximum value that R can have, and a_0 is the total concentration of A. To make progress with the derivation, consider that (i) in a typical SPR experiment, the flow rate of A is sufficiently high that $[A] = a_0$ is essentially constant, (ii) we can write $[B] = b_0 - [AB]$, where b_0 is the total concentration of B, (iii) the SPR signal is often observed to be proportional to $[AB]$, and (iv) the maximum value that R can have is $R_{max} \propto b_0$, which would be measured if all B molecules were ligated to A.

(b) Discuss how a plot of a_0/R_{eq} against a_0 can be used to evaluate R_{max} and K.

(c) Show that, for the association part of the experiment in Fig. 12.56, $R(t) = R_{eq}(1 - e^{-k_{obs}t})$.

(d) Derive an expression for $R(t)$ that applies to the dissociation part of the experiment in Fig. 12.56.

12.48 The Beer–Lambert law states that the absorbance of a sample at a wavenumber is proportional to the molar concentration $[J]$ of the absorbing species J and to the length L of the sample (eqn 12.5). In this problem you will show that the intensity of fluorescence emission from a sample of J is also proportional to $[J]$ and L. Consider a sample of J that is illuminated with a beam of intensity $I_0(\tilde{v})$ at the wavenumber $\tilde{v}$. Before fluorescence can occur, a fraction of $I_0(\tilde{v})$ must be absorbed and an intensity $I(\tilde{v})$ will be transmitted. However, not all the absorbed intensity is emitted, and the intensity of fluorescence depends on the fluorescence quantum yield, ϕ_F, the efficiency of photon emission. The fluorescence quantum yield ranges from 0 to 1 and is proportional to the ratio of the integral of the fluorescence spectrum over the integrated absorption coefficient. Because of a shift of magnitude $\Delta\tilde{v}$, fluorescence occurs at a wavenumber $\tilde{v}_F$, with $\tilde{v}_F + \Delta\tilde{v} = \tilde{v}$. It follows that the fluorescence intensity at v_F, $I_F(\tilde{v}_F)$, is proportional to ϕ_F and to the intensity of exciting radiation that is absorbed by J, $I_{abs}(\tilde{v}) = I_0(\tilde{v}) - I(\tilde{v})$.

(a) Use the Beer–Lambert law to express $I_{abs}(\tilde{v})$ in terms of $I_0(\tilde{v})$, $[J]$, L, and $\varepsilon(\tilde{v})$, the molar absorption coefficient of J at $\tilde{v}$.

(b) Use your result from part (a) to show that $I_F(\tilde{v}) \propto I_0(\tilde{v})\varepsilon(\tilde{v})\phi_F[J]L$.

(c) In fluorescence excitation spectroscopy, the intensity of emitted radiation at a constant emission wavelength (typically the wavelength at which emission is maximal) is monitored while the excitation wavelength is scanned. Use your results from parts (a) and (b) to justify the statement that for a system consisting of a single species, the resulting excitation spectrum is identical to the absorption spectrum of the emitting species.

(d) Discuss how fluorescence excitation spectroscopy may be used to provide evidence for resonance energy transfer between a donor and acceptor molecule.

The following projects require the use of molecular modeling software.

12.49 We saw in *Example* 12.3 that water, carbon dioxide, and methane are able to absorb some of the Earth's infrared emissions, whereas nitrogen and oxygen cannot. The semiempirical, *ab initio*, and DFT methods discussed in Chapter 10 can also be used to simulate vibrational spectra, and from the results of the calculation it is possible to determine the correspondence between a vibrational frequency and the atomic displacements that give rise to a normal mode.

(a) Using molecular modeling software and the computational method of your instructor's choice, visualize the vibrational normal modes of CH_4, CO_2, and H_2O in the gas phase.

(b) Which vibrational modes of CH_4, CO_2, and H_2O are responsible for absorption of infrared radiation?

12.50 Use molecule (**16**) as a model of the *trans* conformation of the chromophore found in rhodopsin. In this model, the methyl group bound to the nitrogen atom of the protonated Schiff's base replaces the protein.

(a) Using molecular modeling software and the computational method of your instructor's choice, calculate the energy separation between the HOMO and LUMO of (**16**).

(b) Repeat the calculation for the 11-*cis* form of (**16**).

(c) Based on your results from parts (a) and (b), do you expect the experimental frequency for the π-to-π* visible absorption of the *trans* form of (**16**) to be higher or lower than that for the 11-*cis* form of (**16**)?

16

13

Magnetic resonance

One of the most widely used and helpful forms of spectroscopy, and a technique that has transformed the practice of chemistry, biochemistry, and medicine, makes use of an effect that is familiar from classical physics. When two pendulums are joined by the same slightly flexible support and one is set in motion, the other is forced into oscillation by the motion of the common axle, and energy flows between the two. The energy transfer occurs most efficiently when the frequencies of the two oscillators are identical. The condition of strong effective coupling when the frequencies are identical is called **resonance**, and the excitation energy is said to 'resonate' between the coupled oscillators.

Resonance is the basis of a number of everyday phenomena, including the response of radios to the weak oscillations of the electromagnetic field generated by a distant transmitter. Historically, spectroscopic techniques that measure transitions between nuclear and electron spin states have carried the term 'resonance' in their names because they have depended on matching a set of energy levels to a source of monochromatic radiation and observing the strong absorption that occurs at resonance.

In this chapter we explore **magnetic resonance**, a form of spectroscopy that when originally developed (and in some cases still) depends on matching a set of energy levels to a source of monochromatic radiation in the radiofrequency and microwave ranges and observing the strong absorption by magnetic nuclei in **nuclear magnetic resonance** (NMR) or by unpaired electrons in **electron paramagnetic resonance** (EPR) that occurs at resonance. Nuclear magnetic resonance is a radiofrequency technique; EPR is a microwave technique.

A growing number of structures of biopolymers are now determined by NMR. So powerful is the technique that a clever variation, known as **magnetic resonance imaging** (MRI), makes possible the spectroscopic characterization of living tissue and has become a major diagnostic tool in medicine.

Principles of magnetic resonance

The application of resonance that we describe here depends on the fact that electrons and many nuclei possess spin angular momentum (Table 13.1). An electron in a magnetic field can take two orientations, corresponding to $m_s = +\frac{1}{2}$ (denoted α or $\uparrow$) and $m_s = -\frac{1}{2}$ (denoted β or $\downarrow$). A nucleus with **nuclear spin quantum number** I (the analog of s for electrons and that can be an integer or a half-integer) may take $2I + 1$ different orientations relative to an arbitrary axis. These orientations are distinguished by the quantum number m_I, which can take on the values $m_I = I, I-1, \ldots, -I$. A proton has $I = \frac{1}{2}$ (the same spin as an electron) and can adopt either of two orientations ($m_I = +\frac{1}{2}$ and $-\frac{1}{2}$). A ^{14}N nucleus has $I = 1$ and can adopt any of

Table 13.1 Nuclear constitution and the nuclear spin quantum number

Number of protons	Number of neutrons	I
Even	Even	0
Odd	Odd	Integer (1, 2, 3, ...)
Even	Odd	Half-integer ($\frac{1}{2}, \frac{3}{2}, \frac{5}{2}, ...$)
Odd	Even	Half-integer ($\frac{1}{2}, \frac{3}{2}, \frac{5}{2}, ...$)

Table 13.2 Nuclear spin properties

Nucleus	Natural abundance/percent	Spin, I	g_I	$\gamma_N/(10^7\ \mathrm{T^{-1}\ s^{-1}})$
^{1}H	99.98	$\frac{1}{2}$	5.5857	26.752
^{2}H (D)	0.0156	1	0.857 44	4.1067
^{12}C	98.99	0		—
^{13}C	1.11	$\frac{1}{2}$	1.4046	6.7272
^{14}N	99.64	1	0.403 56	1.9328
^{16}O	99.96	0		—
^{17}O	0.037	$\frac{5}{2}$	−0.7572	3.627
^{19}F	100	$\frac{1}{2}$	5.2567	25.177
^{31}P	100	$\frac{1}{2}$	2.2634	10.840
^{35}Cl	75.4	$\frac{3}{2}$	0.5479	2.624
^{37}Cl	24.6	$\frac{3}{2}$	0.4561	2.184

three orientations ($m_I = +1, 0, -1$). Spin-$\frac{1}{2}$ nuclei include protons (^{1}H) and ^{13}C, ^{19}F, and ^{31}P nuclei (Table 13.2). As for electrons, the state with $m_I = +\frac{1}{2}$ (↑) is denoted α and that with $m_I = -\frac{1}{2}$ (↓) is denoted β.

13.1 Electrons and nuclei in magnetic fields

To understand the principles of EPR and NMR we need to understand the magnetic properties of electrons and nuclei.

An electron possesses a magnetic moment due to its spin, and this moment interacts with an external magnetic field. That is, an electron behaves like a tiny bar magnet. The orientation of this magnet is determined by the value of m_s, and in a magnetic field $\mathcal{B}_0$ the two orientations have different energies. These energies are given by

$$E_{m_s} = -g_e \gamma \hbar \mathcal{B}_0 m_s \qquad \boxed{\text{Energy of an electron in a magnetic field}} \qquad (13.1)$$

where γ is the **magnetogyric ratio** of the electron,

$$\gamma = -\frac{e}{2m_e} \qquad \boxed{\text{Magnetogyric ratio}} \qquad (13.2)$$

and g_e is a factor, the **g-value of the electron**, which is close to 2.0023 for a free electron.[1] The energies are sometimes expressed in terms of the **Bohr magneton**

$$\mu_B = \frac{e\hbar}{2m_e} \qquad \mu_B = 9.274 \times 10^{-24}\,\text{J T}^{-1}$$
Bohr magneton   (13.3)

a fundamental unit of magnetism. The symbol T, for tesla, is the unit for reporting the intensity of a magnetic field ($1\,\text{T} = 1\,\text{kg s}^{-2}\text{A}^{-1}$). It follows from eqns 13.1 and 13.3 that

$$E_{m_s} = g_e\mu_B\mathcal{B}_0 m_s$$
Alternative expression for the energy
of an electron in a magnetic field   (13.4)

For an electron, the β state lies below the α state.

A nucleus with nonzero spin also has a magnetic moment and behaves like a tiny magnet. The orientation of this magnet is determined by the value of m_I, and in a magnetic field $\mathcal{B}_0$ the $2I + 1$ orientations of the nucleus have different energies. These energies are given by

$$E_{m_I} = -\gamma_N\hbar\mathcal{B}_0 m_I$$
Energy of a nucleus
in a magnetic field   (13.5)

where γ_N is the **nuclear magnetogyric ratio**. For spin-$\frac{1}{2}$ nuclei with positive magnetogyric ratios (such as ^{1}H), the α state lies below the β state. The energy is sometimes written in terms of the **nuclear magneton**, μ_N,

$$\mu_N = \frac{e\hbar}{2m_p} \qquad \mu_N = 5.051 \times 10^{-27}\,\text{J T}^{-1}$$
Nuclear magneton   (13.6)

and an empirical constant called the **nuclear g-factor**, g_I, when it becomes

$$E_{m_I} = -g_I\mu_N\mathcal{B}_0 m_I$$
Alternative expression for the energy
of a nucleus in a magnetic field   (13.7)

Nuclear g-factors are experimentally determined dimensionless quantities that vary between −6 and +6 (see Table 13.2). Positive values of γ_N (and g_I) indicate that the nuclear magnet lies in the same direction as the nuclear spin (this is the case for protons). Negative values indicate that the magnet points in the opposite direction. A nuclear magnet is about 2000 times weaker than the magnet associated with electron spin. Two very common nuclei, ^{12}C and ^{16}O, have zero spin and hence are not affected by external magnetic fields.

The energy separation of the two spin states of an electron (Fig. 13.1) is

$$\Delta E = E_\alpha - E_\beta = \tfrac{1}{2}g_e\mu_B\mathcal{B}_0 - (-\tfrac{1}{2}g_e\mu_B\mathcal{B}_0) = g_e\mu_B\mathcal{B}_0$$
Energy difference
between the spin
states of an electron
in a magnetic field   (13.8)

We infer from the Boltzmann distribution (*Fundamentals* F.3) that the populations of the α and β states, N_α and N_β, are proportional to $e^{-E_\alpha/kT}$ and $e^{-E_\beta/kT}$, respectively, so the ratio of populations at equilibrium is

$$\frac{N_\alpha}{N_\beta} = e^{-(E_\alpha - E_\beta)/kT} \tag{13.9}$$

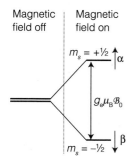

Fig. 13.1 The energy levels of an electron in a magnetic field. Resonance occurs when the energy separation of the levels matches the energy of the photons in the electromagnetic field.

---

[1] The 2 comes from Dirac's relativistic theory of the electron; the 0.0023 comes from additional correction terms.

Because $E_\alpha - E_\beta > 0$ (the β state lies below the α state), $N_\alpha/N_\beta < 1$ and there are slightly more β spins than α spins. If the sample is exposed to radiation of frequency ν, the energy separations come into resonance with the radiation when the frequency satisfies the **resonance condition**:

$$h\nu = g_e \mu_B \mathcal{B}_0 \quad \text{or} \quad \nu = \frac{g_e \mu_B \mathcal{B}_0}{h}$$

Resonance condition for an electron   (13.10)

At resonance there is strong coupling between the electron spin and the radiation, and strong absorption occurs as the spins flip from β (low energy) to α (high energy). We refer to these transitions as electron paramagnetic resonance (EPR), or electron spin resonance (ESR), transitions.

The behavior of nuclei is very similar. The energy separation of the two states of a spin-$\frac{1}{2}$ nucleus (Fig. 13.2) is

$$\Delta E = E_\beta - E_\alpha = \tfrac{1}{2}\gamma_N \hbar \mathcal{B}_0 - (-\tfrac{1}{2}\gamma_N \hbar \mathcal{B}_0) = \gamma_N \hbar \mathcal{B}_0$$

Energy difference between the spin states of a nucleus in a magnetic field   (13.11)

Because for nuclei with positive γ_N the α state lies below the β state, $E_\beta - E_\alpha > 0$ and it follows from eqn 13.9 that $N_\beta/N_\alpha < 1$: there are slightly more α spins than β spins (the opposite of an electron). If the sample is exposed to radiation of frequency ν, the energy separations come into resonance with the radiation when the frequency satisfies the resonance condition:

$$h\nu = \gamma_N \hbar \mathcal{B}_0 \quad \text{or} \quad \nu = \frac{\gamma_N \mathcal{B}_0}{2\pi}$$

Resonance condition for a nucleus   (13.12)

At resonance there is strong coupling between the nuclear spins and the radiation, and strong absorption occurs as the spins flip from α (low energy) to β (high energy). We refer to these transitions as nuclear magnetic resonance (NMR) transitions.

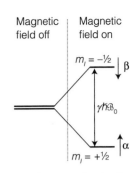

Magnetic field off | Magnetic field on

$m_l = -\frac{1}{2}$ β

$\gamma \hbar \mathcal{B}_0$

$m_l = +\frac{1}{2}$ α

Fig. 13.2 The energy levels of a spin-$\frac{1}{2}$ nucleus (for example, ^{1}H or ^{13}C) in a magnetic field. Resonance occurs when the energy separation of the levels matches the energy of the photons in the electromagnetic field.

Self-test 13.1 Calculate the frequency at which radiation comes into resonance with proton spins in a 12 T magnetic field.

Answer: 510 MHz

13.2 The intensities of NMR and EPR transitions

To appreciate the power of NMR and EPR for investigating biochemical structures and reactions, we need to understand the factors that control the intensities of spin-flipping transitions.

The intensity of an NMR transition depends on a number of factors. We show in the following *Justification* that

$$\text{intensity} \propto (N_\alpha - N_\beta)\mathcal{B}_0 \tag{13.13}$$

where

$$N_\alpha - N_\beta \approx \frac{N\gamma_N \hbar \mathcal{B}_0}{2kT} \tag{13.14}$$

with N the total number of spins ($N = N_\alpha + N_\beta$). It follows that decreasing the temperature increases the intensity by increasing the population difference. By combining eqns 13.13 and 13.14, we see that the intensity is proportional to $\mathcal{B}_0^2$ so NMR transitions can be enhanced significantly by increasing the strength of the applied magnetic field. Similar arguments apply to EPR transitions. We also conclude that absorptions of nuclei with large magnetogyric ratios (^{1}H, for instance) are more intense than those with small magnetogyric ratios (^{13}C, for instance).

Justification 13.1   *Intensities in NMR spectra*

From the general considerations of transition intensities in Section 12.2, we know that the rate of absorption of electromagnetic radiation is proportional to the population of the lower energy state (N_α in the case of a proton NMR transition) and the rate of stimulated emission is proportional to the population of the upper state (N_β). At the low frequencies typical of magnetic resonance, we can neglect spontaneous emission as it is very slow. Therefore, the net rate of absorption is proportional to the difference in populations, and we can write

rate of absorption $\propto N_\alpha - N_\beta$

The intensity of absorption, the rate at which energy is absorbed, is proportional to the product of the rate of absorption (the rate at which photons are absorbed) and the energy of each photon, and the latter is proportional to the frequency v of the incident radiation (through $E = hv$). At resonance, this frequency is proportional to the applied magnetic field (through $v = \gamma_N \mathcal{B}_0/2\pi$), so we can write

intensity of absorption $\propto (N_\alpha - N_\beta)\mathcal{B}_0$

To write an expression for the population difference, we begin with eqn 13.9, written as

$$\frac{N_\beta}{N_\alpha} = e^{-\Delta E/kT} \approx 1 - \frac{\Delta E}{kT} = 1 - \frac{\gamma_N \mathcal{B}_0}{kT}$$

A brief comment
The Taylor expansion
(*Mathematical toolkit* 3.2)
of an exponential function
used in *Justification* 13.1 is
$e^{-x} = 1 - x + \frac{1}{2}x^2 - \cdots$. If $x \ll 1$,
then $e^{-x} \approx 1 - x$.

where $\Delta E = E_\beta - E_\alpha$. The expansion of the exponential term is appropriate for $\Delta E \ll kT$, a condition usually met for electron and nuclear spins. It follows after rearrangement that

$$\frac{N_\alpha - N_\beta}{N_\alpha + N_\beta} = \frac{N_\alpha(1 - N_\beta/N_\alpha)}{N_\alpha(1 + N_\beta/N_\alpha)} = \frac{1 - N_\beta/N_\alpha}{1 + N_\beta/N_\alpha}$$

$$\approx \frac{1 - (1 - \gamma_N \hbar \mathcal{B}_0/kT)}{1 + (1 - \gamma_N \hbar \mathcal{B}_0/kT)} \approx \frac{\gamma_N \hbar \mathcal{B}_0/kT}{2}$$

Then, with $N_\alpha + N_\beta = N$, the total number of spins, we have

$$N_\alpha - N_\beta \approx \frac{N \gamma_N \hbar \mathcal{B}_0}{2kT}$$

The essence of this result is that the population difference is proportional to the applied field. Consequently, the intensity of absorption at resonance is proportional to $\mathcal{B}_0^2$, as stated in the text.

The information in NMR spectra

In its simplest form NMR is the observation of the frequency at which magnetic nuclei in molecules come into resonance with an electromagnetic field when the molecule is exposed to a strong magnetic field. When applied to proton spins, the technique is occasionally called **proton magnetic resonance** (^{1}H-NMR). In the early days of the technique the only nuclei that could be studied were protons (which behave like relatively strong magnets because γ_N is large), but now a wide variety of nuclei, especially ^{13}C and ^{31}P, are investigated routinely.

An NMR spectrometer consists of a magnet that can produce a uniform, intense field and the appropriate sources of radiofrequency radiation (Fig. 13.3). In simple instruments the magnetic field is provided by an electromagnet; for serious work, a superconducting magnet capable of producing fields of the order of 10 T and more is used. The use of high magnetic fields has two advantages. One is that the field increases the intensities of transitions (eqn 13.13). Second, a high field simplifies the appearance of certain spectra. Proton resonance occurs at about 400 MHz in fields of 9.4 T, so NMR is a radiofrequency technique (400 MHz corresponds to a wavelength of 75 cm).

In the following sections we describe the chemical factors that control the appearance of NMR spectra. The discussion will set the stage for the exploration of powerful techniques that make use of radiofrequency pulses and form the basis for all modern applications of NMR in biochemistry.

13.3 The chemical shift

We need to understand the molecular origins of the local magnetic field experienced by nuclei to see how careful analysis of the NMR spectrum reveals details of the structure of a biological molecule and its environment.

We need to know that an applied magnetic field induces the circulation of electronic currents. These currents give rise to a magnetic field that, in diamagnetic substances, opposes the applied field and, in paramagnetic substances, augments the applied field. It follows that in an NMR experiment, the applied magnetic field can induce a circulating motion of the electrons in the molecule, and that motion gives rise to a small additional magnetic field, $\delta \mathcal{B}$. This additional field is proportional to the applied field, and it is conventional to express it as

$$\delta \mathcal{B} = -\sigma \mathcal{B}_0 \tag{13.15}$$

where the dimensionless quantity σ (sigma) is the **shielding constant**. The shielding constant may be positive or negative according to whether the induced field adds to or subtracts from the applied field. The ability of the applied field to induce the circulation of electrons through the nuclear framework of the molecule depends on the details of the electronic structure near the magnetic nucleus of interest, so nuclei in different chemical groups have different shielding constants.

Because the total local field is

$$\mathcal{B}_{\mathrm{loc}} = \mathcal{B}_0 + \delta \mathcal{B} = (1 - \sigma)\mathcal{B}_0$$

the resonance condition is

$$\nu = \frac{\gamma_N \mathcal{B}_{\mathrm{loc}}}{2\pi} = \frac{\gamma_N}{2\pi}(1 - \sigma)\mathcal{B}_0 \tag{13.16}$$

Resonance condition in terms of the shielding constant

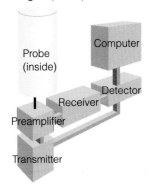

Fig. 13.3 The layout of a typical NMR spectrometer. The link from the transmitter to the detector indicates that the high frequency of the transmitter is subtracted from the high-frequency received signal to give a low-frequency signal for processing.

A brief comment
A superconductor is a material that conducts electricity with zero resistance and can sustain large currents, an important requirement for a strong magnet. A magnetic field of 10 T is indeed very strong: a small magnet, for example, gives a magnetic field of only a few millitesla.

Because σ varies with the environment, different nuclei (even of the same element in different parts of a molecule) come into resonance at different frequencies.

(a) The δ scale

The **chemical shift** of a nucleus is the difference between its resonance frequency and that of a reference standard. The standard for protons is the proton resonance in tetramethylsilane, $Si(CH_3)_4$, commonly referred to as TMS, which bristles with protons and dissolves without reaction in many solutions. Other references are used for other nuclei. For ^{13}C, the reference frequency is the ^{13}C resonance in TMS, and for ^{31}P it is the ^{31}P resonance in 85 per cent $H_3PO_4(aq)$. The separation of the resonance of a particular group of nuclei from the standard increases with the strength of the applied magnetic field because the induced field is proportional to the applied field, and the stronger the latter the greater the shift.

Chemical shifts are reported on the δ **scale**, which is defined as

$$\delta = \frac{v - v^{\circ}}{v^{\circ}} \times 10^6 \qquad \text{The } \delta \text{ scale} \quad (13.17)$$

where v° is the resonance frequency of the standard. The advantage of the δ scale is that shifts reported on it are independent of the applied field (because both numerator and denominator are proportional to the applied field). The resonance frequencies themselves, however, do depend on the applied field through

$$v = v^{\circ} + (v^{\circ}/10^6)\delta \qquad \text{The resonance frequency in terms of the } \delta \text{ scale} \quad (13.18)$$

A brief illustration

The protons belonging to the methyl group ($-CH_3$) of the amino acid alanine have a resonance at $\delta = 1.39$. In a spectrometer operating at 500 MHz (1 MHz $= 10^6$ Hz) the shift relative to the reference is

$$v - v^{\circ} = \frac{500 \text{ MHz}}{10^6} \times 1.39 = 500 \text{ Hz} \times 1.39 = 695 \text{ Hz}$$

In a spectrometer operating at 100 MHz, the shift relative to the reference would be only 139 Hz.

A note on good practice
In much of the literature that uses NMR, chemical shifts are reported in parts per million, ppm, in recognition of the factor of 10^6 in the definition. This practice is unnecessary.

Self-test 13.2 The protons belonging to the $-CH_2$ group of the amino acid glycine have a resonance at $\delta = 3.97$. What is the shift of the resonance from TMS at an operating frequency of 350 MHz?

Answer: 1.39 kHz

If $\delta > 0$, we say that the nucleus is **deshielded**; if $\delta < 0$, then it is **shielded**. A positive δ indicates that the resonance frequency of the group of nuclei in question is higher than that of the standard. Hence $\delta > 0$ indicates that the local magnetic field is stronger than that experienced by the nuclei in the standard under the same conditions. Figure 13.4 shows some typical chemical shifts.

Nuclear magnetic resonance spectra are plotted with δ increasing from right to left. Consequently, in a given applied magnetic field the resonance frequency

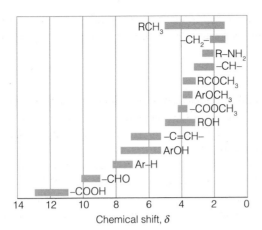

Fig. 13.4 The range of typical chemical shifts for ^{1}H resonances.

also increases from right to left. In a continuous wave (CW) spectrometer, in which the radiofrequency is held constant and the magnetic field is varied (a 'field sweep experiment'), the spectrum is displayed with the applied magnetic field increasing from left to right: a nucleus with a small chemical shift experiences a relatively low local magnetic field, so it needs a higher applied magnetic field to bring it into resonance with the radiofrequency field. Consequently, the right-hand (low chemical shift) end of the spectrum was previously known as the 'high-field end' of the spectrum.

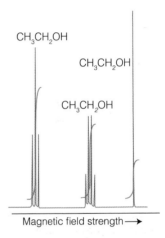

Fig. 13.5 The NMR spectrum of ethanol. The red letters denote the protons giving rise to the resonance peak and the step-like curves are the integrated signals for each group of lines.

A brief illustration

The existence of a chemical shift explains the general features of the NMR spectrum of ethanol shown in Fig. 13.5. The CH$_3$ protons form one group of nuclei with $\delta = 1$. The two CH$_2$ protons are in a different part of the molecule, experience a different local magnetic field, and hence resonate at $\delta = 3$. Finally, the OH proton is in another environment and has a chemical shift of $\delta = 4$.

We can use the relative intensities of the signal (the areas under the absorption lines) to help distinguish which group of lines corresponds to which chemical group, and spectrometers can **integrate** the absorption—that is, determine the areas under the absorption signal—automatically (as is shown in Fig. 13.5). In ethanol the group intensities are in the ratio 3:2:1 because there are three CH$_3$ protons, two CH$_2$ protons, and one OH proton in each molecule. Counting the number of magnetic nuclei as well as noting their chemical shifts is valuable analytically because it helps us identify the compound present in a sample and to identify substances in different environments.

(b) Contributions to the shift

The observed shielding constant is the sum of three contributions:

$$\sigma = \sigma(\text{local}) + \sigma(\text{neighbor}) + \sigma(\text{solvent}) \tag{13.19}$$

The **local contribution**, $\sigma(\text{local})$, is essentially the contribution of the electrons of the atom that contains the nucleus in question. The **neighboring group contribution**, $\sigma(\text{neighbor})$, is the contribution from the groups of atoms that form the

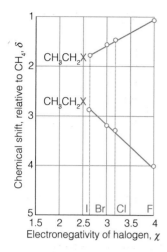

Fig. 13.6 The variation of chemical shift with the electronegativity of the halogen in the haloalkanes. Note that although the chemical shift of the immediately adjacent protons becomes more positive (the protons are deshielded) as the electronegativity increases, that of the next nearest protons decreases.

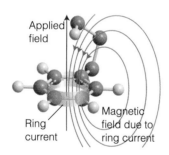

Fig. 13.7 The shielding and deshielding effects of the ring current induced in the benzene ring by the applied field. Protons attached to the ring are deshielded, but a proton attached to a substituent that projects above the ring is shielded.

rest of the molecule. The **solvent contribution**, σ(solvent), is the contribution from the solvent molecules.

The local contribution is broadly proportional to the electron density of the atom containing the nucleus of interest. It follows that the shielding is decreased if the electron density on the atom is reduced by the influence of an electronegative atom nearby. That reduction in shielding translates into an increase in deshielding and hence to an increase in the chemical shift δ as the electronegativity of a neighboring atom increases (Fig. 13.6). That is, as the electronegativity increases, δ increases for protons adjacent to the electronegative atom. Another contribution to σ(local) arises from the ability of the applied field to force the electrons to circulate through the molecule by making use of orbitals that are unoccupied in the ground state and is large in molecules with low-lying excited states and is dominant for atoms other than hydrogen. This contribution is zero in free atoms and around the axes of linear molecules (such as ethyne, HC≡CH), where the electrons can circulate freely and a field applied along the internuclear axis is unable to force them into other orbitals.

The neighboring group contribution arises from the currents induced in nearby groups of atoms. The strength of the additional magnetic field the proton experiences is inversely proportional to the cube of the distance r between H and the neighboring group. A special case of a neighboring group effect is found in aromatic compounds. The field induces a **ring current**, a circulation of electrons around the ring, when it is applied perpendicular to the molecular plane. Protons in the plane are deshielded (Fig. 13.7), but any that happen to lie above or below the plane (as members of substituents of the ring) are shielded.

A solvent can influence the local magnetic field experienced by a nucleus in a variety of ways. Some of these effects arise from specific interactions between the solute and the solvent (such as hydrogen-bond formation and other forms of Lewis acid–base complex formation). Moreover, if there are steric interactions that result in a loose but specific interaction between a solute molecule and a solvent molecule, then protons in the solute molecule may experience shielding or deshielding effects according to their location relative to the solvent molecule (Fig. 13.8). We shall see that the NMR spectra of species that contain protons with widely different chemical shifts are easier to interpret than those in which the shifts are similar, so the appropriate choice of solvent may help to simplify the appearance and interpretation of a spectrum.

13.4 The fine structure

We need to know how to interpret the features of an NMR spectrum so that we can translate the data into the three-dimensional structure of a biological molecule.

The splitting of the groups of resonances into individual lines in Fig. 13.5 is called the **fine structure** of the spectrum. It arises because each magnetic nucleus contributes to the local field experienced by the other nuclei and modifies their resonance frequencies. The strength of the interaction is expressed in terms of the **spin–spin coupling constant**, J, and reported in hertz (Hz). Spin coupling constants are an intrinsic property of the molecule and independent of the strength of the applied field.

(a) The appearance of fine structure

Consider first a molecule that contains two spin-$\frac{1}{2}$ nuclei A and X. Suppose the spin of X is α, then A will resonate at a certain frequency as a result of the

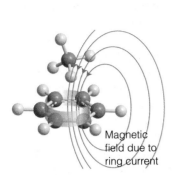

Fig. 13.8 An aromatic solvent (benzene here) can give rise to local currents that shield or deshield a proton in a solute molecule. In this relative orientation of the solvent and solute, the proton on the solute molecule is shielded.

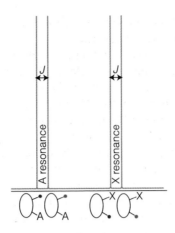

Fig. 13.9 The effect of spin–spin coupling on an NMR spectrum of two spin-$\frac{1}{2}$ nuclei with widely different chemical shifts. Each resonance is split into two lines separated by J. Red circles indicate α spins, green circles indicate β spins.

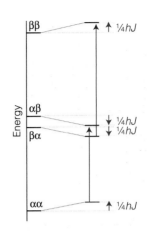

Fig. 13.10 The energy levels of a two-proton system in the presence of a magnetic field. The levels on the left apply in the absence of spin–spin coupling. Those on the right are the result of allowing for spin–spin coupling. The only allowed transitions differ in frequency by J.

combined effect of the external field, the shielding constant, and the spin–spin interaction of nucleus A with nucleus X. As we show in the following *Justification*, instead of a single line from A, the spectrum consists of a doublet of lines separated by a frequency J (Fig. 13.9). The same splitting occurs in the X resonance: instead of a single line it is a doublet with splitting J (the same value as for the splitting of A).

Justification 13.2   *The structure of an AX spectrum*

First, neglect spin–spin coupling. The total energy of two protons in a magnetic field $\mathcal{B}$ is the sum of two terms like eqn 13.11 but with $\mathcal{B}_0$ modified to $(1 - \sigma)\mathcal{B}_0$:

$$E = -\gamma_N \hbar (1 - \sigma_A)\mathcal{B}_0 m_A - \gamma_N \hbar (1 - \sigma_X)\mathcal{B}_0 m_X$$

Here σ_A and σ_X are the shielding constants of A and X, respectively. The four energy levels predicted by this formula are shown on the left of Fig. 13.10. The spin–spin coupling energy is normally written

$$E_{\text{spin–spin}} = hJm_A m_X$$

There are four possibilities, depending on the values of the quantum numbers m_A and m_X:

	$E_{\text{spin–spin}}$
$\alpha_A \alpha_X$	$+\frac{1}{4}hJ$
$\alpha_A \beta_X$	$-\frac{1}{4}hJ$
$\beta_A \alpha_X$	$-\frac{1}{4}hJ$
$\beta_A \beta_X$	$+\frac{1}{4}hJ$

The resulting energy levels are shown on the right in Fig. 13.10.

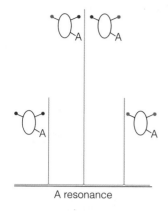

A resonance

Fig. 13.11 The origin of the 1:2:1 triplet in the A resonance of an AX_2 species. The two X nuclei may have the $2^2 = 4$ spin arrangements: ($\uparrow\uparrow$), ($\uparrow\downarrow$), ($\downarrow\uparrow$), and ($\downarrow\downarrow$). The middle two arrangements are responsible for the coincident resonances of A.

1 A fragment of Pascal's triangle

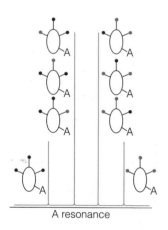

A resonance

Fig. 13.12 The origin of the 1:3:3:1 quartet in the A resonance of an AX_3 species, where A and X are spin-$\frac{1}{2}$ nuclei with widely different chemical shifts. There are $2^3 = 8$ arrangements of the spins of the three X nuclei and their effects on the A nucleus give rise to four groups of resonances.

Now consider the transitions. When an A nucleus changes its spin from α to β, the X nucleus remains in its same spin state, which may be either α or β. The two transitions are shown in the illustration, and we see that they differ in frequency by *J*. Alternatively, the X nucleus can undergo a transition from α to β; now the A nucleus remains in its same spin state, which may be either α or β, and we again get two transitions that differ in frequency by *J*.

If there is another X nucleus in the molecule with the same chemical shift as the first X (corresponding to an AX_2 species), the resonance of A is split into a doublet by one X, and each line of the doublet is split again by the same amount (Fig. 13.11) by the second X. This splitting results in three lines in the intensity ratio 1:2:1 (because the central frequency can be obtained in two ways). As in the AX case discussed above, the X resonance of the AX_2 species is split into a doublet by A.

Three equivalent X nuclei (an AX_3 species) split the resonance of A into four lines of intensity ratio 1:3:3:1 (Fig. 13.12). The X resonance remains a doublet as a result of the splitting caused by A. In general, *N* equivalent spin-$\frac{1}{2}$ nuclei split the resonance of a nearby spin or group of equivalent spins into *N* + 1 lines with an intensity distribution given by Pascal's triangle (**1**). Subsequent rows of this triangle are formed by adding together the two adjacent numbers in the line above.

Self-test 13.3 Complete the next line of the triangle, the pattern arising from five equivalent protons.

Answer: 1:5:10:10:5:1

Example 13.1 *Accounting for the fine structure in a spectrum*

Account for the fine structure in the ^{1}H-NMR spectrum of the C–H protons of ethanol.

Strategy Refer to Pascal's triangle to determine the effect of a group of *N* equivalent protons on a proton, or (equivalently) a group of protons, of interest.

Solution The three protons of the CH_3 group split the single resonance of the CH_2 protons into a 1:3:3:1 quartet with a splitting *J*. Likewise, the two protons of the CH_2 group split the single resonance of the CH_3 protons into a 1:2:1 triplet. Each of these lines is split into a doublet to a small extent by the OH proton.

Self-test 13.4 What fine structure can be expected for the C–H protons in alanine?

Answer: A 1:3:3:1 quartet for the –CH group and a doublet for the –CH$_3$ group

The spin–spin coupling constant of two nuclei joined by *N* bonds is normally denoted ^NJ, with subscripts for the types of nuclei involved. Thus, $^1J_{CH}$ is the coupling constant for a proton joined directly to a ^{13}C atom, and $^2J_{CH}$ is the coupling constant when the same two nuclei are separated by two bonds (as in ^{13}C–C–H). A typical value of $^1J_{CH}$ is between 10^2 and 10^3 Hz; the value of $^2J_{CH}$ is about 10 times less, between about 10 and 10^2 Hz. Both ^3J and ^4J give detectable effects in a spectrum, but couplings over larger numbers of bonds can generally be ignored.

A brief illustration

Figure 13.13 shows the ^{1}H-NMR spectrum of diethyl ether, $(CH_3CH_2)_2O$. The resonance at $\delta = 3.4$ corresponds to CH_2 in an ether; that at $\delta = 1.2$ corresponds to CH_3 in CH_3CH_2. As we saw in *Example* 13.1, the fine structure of the CH_2 group (a 1:3:3:1 quartet) is characteristic of splitting caused by CH_3; the fine structure of the CH_3 resonance is characteristic of splitting caused by CH_2. The spin–spin coupling constant is $J = -60$ Hz (the same for each group). If the spectrum had been recorded with a spectrometer operating at five times the magnetic field strength, the groups of lines would have been observed to be five times farther apart in frequency (but the same δ values). No change in spin–spin splitting would be observed.

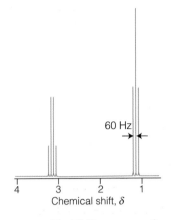

Fig. 13.13 The NMR spectrum of diethyl ether considered in the *brief illustration*.

The magnitude of $^3J_{HH}$ depends on the dihedral angle, ϕ, between the two C–H bonds (**2**). The variation is expressed quite well by the **Karplus equation**:

$$^3J_{HH} = A + B\cos\phi + C\cos 2\phi \qquad \boxed{\text{Karplus equation}} \qquad (13.20)$$

Typical values of A, B, and C are +7 Hz, −1 Hz, and +5 Hz, respectively, for an HCCH fragment. Figure 13.14 shows the angular variation the equation predicts. It follows that the measurement of $^3J_{HH}$ in a series of related compounds can be used to determine their conformations.

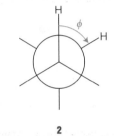

2

A brief illustration

Many three-dimensional structures of biological macromolecules are determined by NMR spectroscopy. As a first illustration of the power of NMR in structural biology consider how an analysis of H–N–C–H couplings in polypeptides can help to reveal their conformation. For $^3J_{HH}$ coupling in such a group, $A = +5.1$ Hz, $B = −1.4$ Hz, and $C = +3.2$ Hz. For an α helix, ϕ is close to 120°, which would give $^3J_{HH} \approx 4$ Hz. For a β sheet, ϕ is close to 180°, which would give $^3J_{HH} \approx 10$ Hz. Consequently, small coupling constants indicate an α helix, whereas large couplings indicate a β sheet.

The coupling constant $^1J_{CH}$ also depends on the hybridization of the C atom:

	sp	sp^2	sp^3
$^1J_{CH}$/Hz:	250	160	125

(b) The origin of fine structure

Spin–spin coupling in molecules in solution can be explained in terms of the **polarization mechanism**, in which the interaction is transmitted through the bonds.

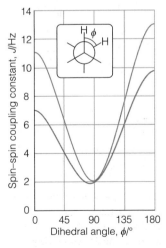

Fig. 13.14 The variation of $^3J_{HH}$ with angle, according to the Karplus equation. The orange line is for H–C–C–H and the green line is for H–N–C–H.

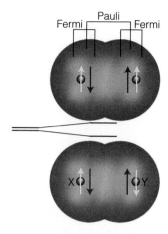

Fig. 13.15 The polarization mechanism for spin–spin coupling ($^1J_{HH}$). The two arrangements have slightly different energies. In this case, J is positive, corresponding to a lower energy when the nuclear spins are antiparallel.

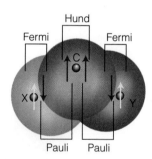

Fig. 13.16 The polarization mechanism for $^2J_{HH}$ spin–spin coupling. The spin information is transmitted from one bond to the next by a version of the mechanism that accounts for the lower energy of electrons with parallel spins in different atomic orbitals (Hund's rule of maximum multiplicity). In this case, $J < 0$, corresponding to a lower energy when the nuclear spins are parallel.

The simplest case to consider is that of $^1J_{XY}$ where X and Y are spin-$\frac{1}{2}$ nuclei joined by an electron-pair bond (Fig. 13.15). The coupling mechanism depends on the fact that in some atoms it is favorable for the nucleus and a nearby electron spin to be parallel (both α or both β), but in others it is favorable for them to be antiparallel (one α and the other β). The electron–nucleus coupling is magnetic in origin and may be either a dipolar interaction (Section 11.6) between the magnetic moments of the electron and nuclear spins or a **Fermi contact interaction**, an interaction that depends on the very close approach of an electron to the nucleus and hence can occur only if the electron occupies an s orbital. We shall suppose that it is energetically favorable for an electron spin and a nuclear spin to be antiparallel (as is the case for a proton and an electron in a hydrogen atom), either $\alpha_e\beta_N$ or $\beta_e\alpha_N$, where we are using the labels e and N to distinguish the electron and nucleus spins.

If the X nucleus is α_X, a β electron of the bonding pair will tend to be found nearby (because that is energetically favorable for it). The second electron in the bond, which must have α spin if the other is β, will be found mainly at the far end of the bond (because electrons tend to stay apart to reduce their mutual repulsion). Because it is energetically favorable for the spin of Y to be antiparallel to an electron spin, a Y nucleus with β spin has a lower energy than a Y nucleus with α spin:

$$\text{low energy: } \alpha_X\beta_e\ldots\alpha_e\beta_Y \qquad \text{high energy: } \alpha_X\beta_e\ldots\alpha_e\alpha_Y$$

The opposite is true when X is β, for now the α spin of Y has the lower energy:

$$\text{low energy: } \beta_X\alpha_e\ldots\beta_e\alpha_Y \qquad \text{high energy: } \beta_X\alpha_e\ldots\beta_e\beta_Y$$

In other words, antiparallel arrangements of nuclear spins ($\alpha_X\beta_Y$ and $\beta_X\alpha_Y$) lie lower in energy than parallel arrangements ($\alpha_X\alpha_Y$ and $\beta_X\beta_Y$) as a result of their magnetic coupling with the bond electrons. That is, $^1J_{HH}$ is positive, for then hJm_Xm_Y is negative when m_X and m_Y have opposite signs.

To account for the value of $^2J_{XY}$, as in H–C–H, we need a mechanism that can transmit the spin alignments through the central C atom (which may be ^{12}C, with no nuclear spin of its own). In this case (Fig. 13.16), an X nucleus with α spin polarizes the electrons in its bond, and the α electron is likely to be found closer to the C nucleus. The more favorable arrangement of two electrons on the same atom is with their spins parallel (Hund's rule, Section 9.11), so the more favorable arrangement is for the α electron of the neighboring bond to be close to the C nucleus. Consequently, the β electron of that bond is more likely to be found close to the Y nucleus and therefore that nucleus will have a lower energy if it is α:

$$\text{low energy: } \alpha_X\beta_e\ldots\alpha_e[C]\alpha_Y\ldots\beta_e\alpha_Y \qquad \text{high energy: } \alpha_X\beta_e\ldots\alpha_e[C]\alpha_e\ldots\beta_e\beta_Y$$
$$\text{low energy: } \beta_X\alpha_e\ldots\beta_e[C]\beta_e\ldots\alpha_e\beta_Y \qquad \text{high energy: } \beta_X\alpha_e\ldots\beta_e[C]\beta_e\ldots\alpha_e\alpha_Y$$

Hence, according to this mechanism, the lower energy of Y will be obtained if its spin is parallel ($\alpha_X\alpha_Y$ and $\beta_X\beta_Y$) to that of X. That is, $^2J_{HH}$ is negative, for then hJm_Xm_Y is negative when m_X and m_Y have the same sign.

The coupling of nuclear spin to electron spin by the Fermi contact interaction is most important for proton spins, but it is not necessarily the most important mechanism for other nuclei. These nuclei may also interact by a dipolar mechanism with the electron magnetic moments and with their orbital motion, and there is no simple way of specifying whether J will be positive or negative.

13.5 Conformational conversion and chemical exchange

> We need to understand how to analyze spectra to determine rates of dynamical events of biological importance, such as conformational changes and proton exchange between molecules.

The appearance of an NMR spectrum is changed if magnetic nuclei can jump rapidly between different environments. Consider a molecule, such as N,N-dimethylformamide, that can jump between conformations; in its case, the methyl shifts depend on whether they are *cis* or *trans* to the carbonyl group (Fig. 13.17). When the jumping rate is low, the spectrum shows two sets of lines, one each from molecules in each conformation. When the interconversion is fast, the spectrum shows a single line at the mean of the two chemical shifts. At intermediate inversion rates, the line is very broad. This maximum broadening occurs when the lifetime, τ (tau), of a conformation gives rise to a linewidth that is comparable to the difference of resonance frequencies, δv, and both broadened lines blend together into a very broad line. Coalescence of the two lines occurs when

$$\tau = \frac{2^{1/2}}{\pi \delta v}$$

Condition for coalescence of two NMR lines   (13.21)

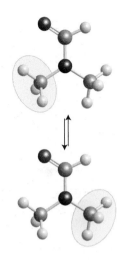

Fig. 13.17 When a molecule changes from one conformation to another, the positions of its protons are interchanged and jump between magnetically distinct environments.

Example 13.2   *Interpreting line broadening*

The NO group in N,N-dimethylnitrosamine, $(CH_3)_2N-NO$, rotates about the N–N bond and, as a result, the magnetic environments of the two CH_3 groups are interchanged. The two CH_3 resonances are separated by 390 Hz in a 600 MHz spectrometer. At what rate of interconversion will the resonance collapse to a single line?

Strategy Use eqn 13.21 for the average lifetimes of the conformations. The rate of interconversion is the inverse of their lifetime.

Solution With $\delta v = 390$ Hz,

$$\tau = \frac{2^{1/2}}{\pi \times (390 \text{ s}^{-1})} = 1.2 \text{ ms}$$

It follows that the signal will collapse to a single line when the interconversion rate exceeds about 830 s^{-1}.

Self-test 13.5   What would you deduce from the observation of a single line from the same molecule in a 300 MHz spectrometer?

Answer: Conformation lifetime less than 2.3 ms

A similar explanation accounts for the loss of fine structure in solvents able to exchange protons with the sample. For example, amino and hydroxyl protons are able to exchange with water protons. When this **chemical exchange** occurs, a molecule ROH, such as serine or tyrosine, with an α-spin proton (we write this ROH$_\alpha$) rapidly converts to ROH$_\beta$ and then perhaps to ROH$_\alpha$ again because the protons provided by the solvent molecules in successive exchanges have random spin orientations. Therefore, instead of seeing a spectrum composed of contributions from both ROH$_\alpha$ and ROH$_\beta$ molecules (that is, a spectrum showing a doublet structure due to the OH proton), we see a spectrum that shows no splitting caused by coupling of the OH proton (as in Fig. 13.5). The effect is observed when the lifetime of a molecule due to this chemical exchange is so short that the lifetime broadening is greater than the doublet splitting. Because this splitting is often very small (a few hertz), a proton must remain attached to the same molecule for longer than about 0.1 s for the splitting to be observable. In water, the exchange rate is much faster than that, so alcohols show no splitting from the OH protons. In dry dimethylsulfoxide (DMSO), the exchange rate may be slow enough for the splitting to be detected.

Pulse techniques in NMR

Modern methods of detecting the energy separation between nuclear spin states are more sophisticated than simply looking for the frequency at which resonance occurs. One of the best analogies that has been suggested to illustrate the difference between the old and new ways of observing an NMR spectrum is that of detecting the spectrum of vibrations of a bell. If we hit a bell with a hammer, we obtain a clang composed of all the frequencies that the bell can produce. The equivalent in NMR is to monitor the radiation nuclear spins emit as they return to equilibrium after the appropriate stimulation. The resulting **Fourier-transform NMR** (FT-NMR) gives greatly increased sensitivity, so opening up the entire periodic table to the technique.

13.6 **Time- and frequency-domain signals**

Multiple-pulse FT-NMR gives biochemists unparalleled control over the information content and display of spectra, and to take full advantage of the technique, we need to understand how radiofrequency pulses work to excite a spin system and how the signal is monitored and interpreted.

It is sometimes useful to compare the quantum mechanical and classical pictures of magnetic nuclei pictured as tiny bar magnets. A bar magnet in an externally applied magnetic field undergoes the motion called **precession** as it twists around the direction of the field (Fig. 13.18). The rate of precession is proportional to the strength of the applied field and is in fact equal to $(\gamma_N/2\pi)\mathcal{B}_0$, which in this context is called the **Larmor precession frequency**, ν_L.

The quantum mechanical description in Section 13.1 indicates that a spin-$\frac{1}{2}$ nucleus is like a bar magnet with two possible orientations with respect to the direction of the field, one with low energy (the α state) and the other with high energy (the β state). We can merge the classical and quantum mechanical pictures by visualizing an α or β spin as precessing around its cone of possible orientations at the Larmor frequency (Fig. 13.19): the stronger the field, the more rapid is the rate of precession. If we were to imagine stepping onto a platform, a so-called **rotating frame**, that rotates around the direction of the applied field at the Larmor frequency, then all the spins would appear to be stationary on their respective cones.

Now suppose that somehow we have arranged all the spins in a sample to have exactly the same angle around the field direction at an instant. We saw in Section 13.1 that there are more α spins than β spins. The imbalance means that there is a net nuclear magnetic moment, the **magnetization**, M, that we represent by a vector pointing in the same direction as the vector representing the applied field and with a length proportional to the population difference (Fig. 13.20).

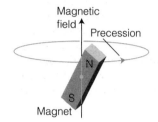

Fig. 13.18 A bar magnet in a magnetic field undergoes the motion called *precession*. A nuclear spin (and an electron spin) has an associated magnetic moment and behaves in the same way. The frequency of precession is called the Larmor precession frequency and is proportional to the applied field and the magnitude of the magnetic moment.

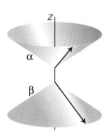

Fig. 13.19 The interactions between the α and β states of a proton and an external magnetic field may be visualized as the precession of the vectors representing the angular momentum.

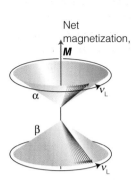

Fig. 13.20 The magnetization of a sample of protons is the resultant of all their magnetic moments. In the presence of a field, the spins precess around their cones (that is, there is an energy difference between the α and β states) and there are slightly more α spins than β spins. As a result, there is a net magnetization M along the z-axis.

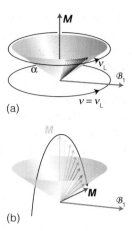

Fig. 13.21 (a) In a resonance experiment, a circularly polarized radiofrequency magnetic field $\mathcal{B}_1$ is applied in the xy-plane (the magnetization vector lies along the z-axis). (b) If we step into a frame rotating at the Larmor frequency, the radiofrequency field appears to be stationary if its frequency is the same as the Larmor frequency. When the two frequencies coincide, the magnetization vector of the sample begins to rotate around the direction of the $\mathcal{B}_1$ field.

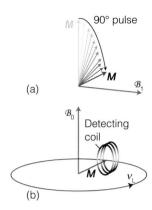

Fig. 13.22 (a) If the radiofrequency field is applied for a certain time, the magnetization vector is rotated into the xy-plane. (b) To an external stationary observer (the coil), the magnetization vector is rotating at the Larmor frequency and can induce a signal in the coil.

We now consider the effect of a radiofrequency field circularly polarized in the plane perpendicular to the direction of the applied field in the sense that the magnetic component of the electromagnetic field (the only component we need to consider) is rotating around the direction of the applied field $\mathcal{B}_0$. The strength of the rotating magnetic field is $\mathcal{B}_1$. Suppose we choose the frequency of this field to be equal to the Larmor frequency of the spins, $\nu_L = (\gamma_N/2\pi)\mathcal{B}_0$. It follows from eqn 13.12 that this choice is equivalent to selecting the resonance condition in the conventional experiment. The nuclei now experience a steady $\mathcal{B}_1$ field because the rotating magnetic field is in step with the precessing spins (Fig. 13.21). Just as the spins precess about the strong static field $\mathcal{B}_0$ at a frequency $\gamma_N\mathcal{B}_0/2\pi$, so in the rotating frame they precess about the direction of $\mathcal{B}_1$ at a frequency $\gamma_N\mathcal{B}_1/2\pi$. If the $\mathcal{B}_1$ field is applied in a pulse of duration $\pi/2\gamma_N\mathcal{B}_1$, the magnetization tips through 90° in the rotating frame and we say that we have applied a **90° pulse** (or a 'π/2 pulse'). The duration of the pulse depends on the strength of the $\mathcal{B}_1$ field but is typically of the order of microseconds. Now imagine stepping out of the rotating frame. To a stationary external observer (the role played by a radiofrequency coil, Fig. 13.22), the magnetization vector is now rotating at the Larmor frequency in the plane perpendicular to the direction of the applied magnetic field. The rotating magnetization induces in the coil a signal that oscillates at the Larmor frequency.

As time passes, the individual spins move out of step (partly because they are precessing at slightly different rates, as we explain later), so the magnetization vector shrinks exponentially with a time constant T_2 and induces an ever weaker signal in the detector coil. The form of the signal that we can expect is therefore the oscillating–decaying **free-induction decay** (FID) shown in Fig. 13.23.

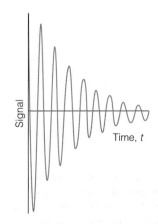

Fig. 13.23 A simple free-induction decay of a sample of spins with a single resonance frequency.

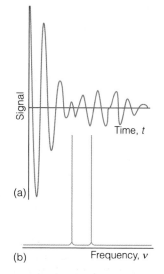

Fig. 13.24 (a) A free-induction decay signal of a sample of an AX species and (b) its analysis into its frequency components.

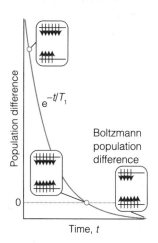

Fig. 13.25 The spin–lattice relaxation time is the time constant for the exponential return of the population of the spin states to their equilibrium (Boltzmann) distribution.

Now consider a two-spin system. We can think of the magnetization vector of an AX spin system with $J = 0$ as consisting of two parts, one formed by the A spins and the other by the X spins. When the 90° pulse is applied, both magnetization vectors are tipped into the perpendicular plane. However, because the A and X nuclei precess at different frequencies, they induce two signals in the detector coils, and the overall FID curve may resemble that in Fig. 13.24a. The composite FID curve is the analog of the struck bell emitting a rich tone composed of all the frequencies at which it can vibrate.

The problem we must address is how to recover the resonance frequencies present in a free-induction decay. We know that the FID curve is a sum of oscillating functions, so the problem is to analyze it into its component frequencies by carrying out a Fourier transformation. When the signal in Fig. 13.24a is transformed in this way, we get the frequency-domain spectrum shown in Fig. 13.24b. One line represents the Larmor frequency of the A nuclei and the other that of the X nuclei.

13.7 **Spin relaxation**

Because careful analysis of the decay process reveals details of molecular structure and of interactions between molecules, we need to understand how a spin system returns to equilibrium after the application of a radiofrequency pulse.

As resonant absorption continues, the population of the upper state rises to match that of the lower state. From eqn 13.13, we can expect the intensity of the absorption signal to decrease with time as the populations of the spin states equalize. This decrease due to the progressive equalization of populations is called **saturation**.

The fact that saturation is often not observed must mean that there are non-radiative processes by which β nuclear spins can become α spins again and hence help to maintain the population difference between the two sites. The nonradiative return to an equilibrium distribution of populations in a system (eqn 13.9) is an aspect of the process called **relaxation**. If we were to imagine forming a system of spins in which all the nuclei were in their β state, then the system returns exponentially to the equilibrium distribution (a small excess of α spins over β spins) with a time constant called the **spin–lattice relaxation time**, T_1 (Fig. 13.25).

However, there is another, more subtle aspect of relaxation. Let us go back to the classical picture of magnetic nuclei with the spins in the artificial arrangement shown in Fig. 13.20, all lying at the same azimuthal angle on their respective cones. If each spin has a slightly different Larmor frequency (because they experience slightly different local magnetic fields), then they will gradually fan out, and at thermal equilibrium all the bar magnets will lie at *random* angles around the direction of the applied field. The time constant for the exponential return of the system into this random arrangement is called the **spin–spin relaxation time**, T_2 (Fig. 13.26). For spins to be truly at thermal equilibrium, therefore, not only is the ratio of populations of the spin states given by eqn 13.9, but the spin orientations must be random around the field direction.

What causes each type of relaxation? In each case the spins are responding to local magnetic fields that act to twist them into different orientations. However, there is a crucial difference between the two processes.

The best kind of local magnetic field for inducing a transition from β to α (as in spin–lattice relaxation) is one that fluctuates at a frequency close to the resonance frequency. Such a field can arise from the tumbling motion of the molecule in the fluid sample. If the tumbling motion of the molecule is slow compared to the resonance frequency, it will give rise to a fluctuating magnetic field that oscillates too slowly to induce transitions, so T_1 will be long. If the molecule tumbles much faster than the resonance frequency, then it will give rise to a fluctuating magnetic field that oscillates too rapidly to induce transitions, so T_1 will again be long. Only if the molecule tumbles at about the resonance frequency will the fluctuating magnetic field be able to induce transitions effectively, and only then will T_1 be short. The rate of molecular tumbling increases with temperature and with reducing viscosity of the solvent, so we can expect a dependence like that shown in Fig. 13.27.

The best kind of local magnetic field for causing spin–spin relaxation is one that does not change very rapidly. Then each molecule in the sample lingers in its particular local magnetic environment for a long time, and the orientations of the spins have time to become randomized around the applied field direction. If the molecules move rapidly from one magnetic environment to another, the effects of different magnetic fields average out and the randomization does not take place as quickly. In other words, slow molecular motion corresponds to short T_2 and fast motion corresponds to long T_2 (as shown in Fig. 13.27). Detailed calculation shows that when the motion is fast, the two relaxation times are equal, as has been drawn in the illustration.

Spin relaxation studies—using advanced techniques that utilize complicated sequences of pulses of radiofrequency energy to drive spins into special orientations and then monitoring their return to equilibrium—have two main applications. First, they reveal information about the mobility of molecules or parts of molecules. For example, by studying spin relaxation times of protons in the hydrocarbon chains of lipid bilayers, it is possible to build up a detailed picture of the motion of these chains and hence come to an understanding of the dynamics of cell membranes. Second, relaxation times depend on the separation of the nucleus from the source of the magnetic field that is causing its relaxation: that source may be another magnetic nucleus in the same molecule. By studying the relaxation times, we can determine the internuclear distances within the molecule and use them to build up a model of its shape.

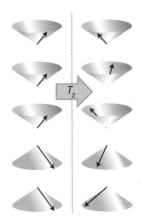

Fig. 13.26 The spin–spin relaxation time is the time constant for the exponential return of the spins to a random distribution around the direction of the magnetic field. No change in populations of the two spin states is involved in this type of relaxation, so no energy is transferred from the spins to the surroundings.

In the laboratory 13.1   *Magnetic resonance imaging*

One of the most striking applications of nuclear magnetic resonance is in physiology and medicine, where special radiofrequency pulse sequences are used to identify the distribution of protons in an organism. To understand this technique, we need to see how NMR techniques are modified to allow the study of three-dimensional objects, such as a human body. **Magnetic resonance imaging** (MRI) is a portrayal of the distribution of protons in a three-dimensional object. The technique relies on the application of specific pulse sequences to an object in a spatially varying magnetic field. If an object containing hydrogen nuclei (a tube of water or a human body) is placed in an NMR spectrometer and exposed to a *homogeneous* magnetic field (a field that has the same value throughout the sample), then a single resonance signal will be detected. Now consider a flask of water in a magnetic field that varies linearly in the z-direction according to $\mathcal{B}_0 + \mathcal{G}_z z$, where $\mathcal{G}_z$ is the field gradient along the z-direction (Fig. 13.28). Then the water protons will be resonant at the frequencies

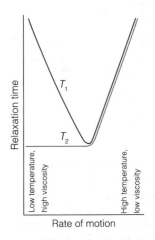

Fig. 13.27 The variation of the two relaxation times with the rate at which the molecules move (either by tumbling or migrating through the solution). The horizontal axis can be interpreted as representing temperature or viscosity. Note that the two relaxation times coincide when the motion is rapid.

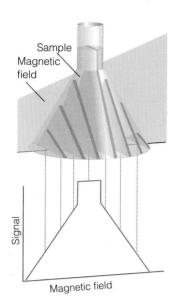

Fig. 13.28 In a magnetic field that varies linearly over a sample, all the protons within a given slice (that is, at a given field value) come into resonance and give a signal of the corresponding intensity. The resulting intensity pattern is a map of the number of protons in all the slices and portrays the shape of the sample. Changing the orientation of the field shows the shape along the corresponding direction, and computer manipulation can be used to build up the three-dimensional shape of the sample.

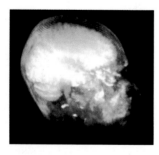

Fig. 13.29 The great advantage of MRI is that it can display soft tissue, such as in this cross-section through a patient's head (http://www.sv.vt.edu/mri/mri.html).

$$v(z) = \frac{\gamma_{\mathrm{N}}}{2\pi}(\mathcal{B}_0 + \mathcal{G}_z z)$$

Similar equations may be written for gradients along the x- and y-directions. Exposure of the sample to radiation of frequency $v(z)$ results in a signal with an intensity that is proportional to the number of protons at the position z. This procedure is an example of **slice selection**, the use of radiofrequency radiation that excites nuclei in a specific region, or slice, of the sample. It follows that the intensity of the NMR signal will be a projection of the number of protons on a line parallel to the field gradient. The image of a three-dimensional object such as a flask of water can be obtained if the slice selection technique is applied at different orientations (Fig. 13.28). In **projection reconstruction**, the projections can be analyzed on a computer to reconstruct the three-dimensional distribution of protons in the object.

A common problem with these techniques is image contrast, which must be optimized in order to show spatial variations in water content in the sample. One strategy for solving this problem takes advantage of the fact that the relaxation times of water protons are shorter for water in biological tissues than for the pure liquid. Furthermore, relaxation times from water protons are also different in healthy and diseased tissues. A T_1-**weighted image** is obtained by obtaining data before spin–lattice relaxation can return the spins in the sample to equilibrium. Under these conditions, differences in signal intensities are directly related to differences in T_1. A T_2-**weighted image** is obtained by collecting data after the system has relaxed extensively but not completely. In this way, signal intensities are strongly dependent on variations in T_2. However, allowing so much of the decay to occur leads to weak signals even for those protons with long spin–spin relaxation times. Another strategy involves the use of **contrast agents**, which are paramagnetic compounds that shorten the relaxation times of nearby protons. The technique is particularly useful for enhancing image contrast and for diagnosing disease if the contrast agent is distributed differently in healthy and diseased tissues.

The MRI technique is used widely to detect physiological abnormalities and to observe metabolic processes. With **functional MRI** (fMRI), blood flow in different regions of the brain can be studied and related to the mental activities of the subject. The technique is based on differences in the magnetic properties of deoxygenated and oxygenated hemoglobin. In *Example* 10.3 we saw that when the Fe(II) atom of hemoglobin is oxygenated and its coordination number changes from 5 to 6, it is converted from a high-spin d^6 ($d_{xy}^2 d_{yz}^1 d_{zx}^1 d_{x^2-y^2}^1 d_{z^2}^1$) configuration, in which the maximum number of electrons have parallel spins, to a low-spin d^6 ($d_{xy}^2 d_{yz}^2 d_{zx}^2$) configuration. The more paramagnetic deoxygenated hemoglobin affects the proton resonances of tissue differently from the oxygenated protein. Because there is enhanced blood flow in active regions of the brain compared to inactive regions, changes in the intensities of proton resonances due to changes in levels of oxygenated hemoglobin can be related to brain activity.

A special advantage of MRI is that it can image *soft* tissues (Fig. 13.29), whereas X-rays are largely used for imaging hard, bony structures and abnormally dense regions, such as tumors. In fact, the invisibility of hard structures in MRI is an advantage as it allows the imaging of structures encased by bone, such as the brain and the spinal cord. X-rays are known to be dangerous on account of the ionization they cause; the high magnetic fields used in MRI may also be dangerous, but apart from anecdotes about the extraction of loose fillings from teeth, there is no convincing evidence of their harmfulness and the technique is considered safe.

13.8 Proton decoupling

Because biological macromolecules contain a large number of proton spins, we need to see how special pulse sequences can simplify the appearance of a carbon-13 spectrum and reveal such important information as the three-dimensional arrangement of the carbon backbones of proteins, nucleic acids, and lipids.

Carbon-13 is a **dilute-spin species** in the sense that it is unlikely that more than one ^{13}C nucleus will be found in any given small molecule (provided the sample has not been enriched with that isotope; the natural abundance of ^{13}C is only 1.1 per cent). Even in large molecules, although more than one ^{13}C nucleus may be present, it is unlikely that they will be close enough to give an observable splitting. Hence, it is not normally necessary to take into account $^{13}C-^{13}C$ spin–spin coupling within a molecule.

Protons are **abundant-spin species** in the sense that a molecule is likely to contain many of them. If we were observing a ^{13}C-NMR spectrum, we would obtain a very complex spectrum on account of the coupling of the one ^{13}C nucleus with many of the protons that are present. To avoid this difficulty, ^{13}C-NMR spectra are normally observed using the technique of **proton decoupling**. Thus, if the CH_3 protons of ethanol are irradiated with a second, strong, resonant radiofrequency pulse, they undergo rapid spin reorientations and the ^{13}C nucleus senses an average orientation. As a result, its resonance is a single line and not a 1:3:3:1 quartet. Proton decoupling has the additional advantage of enhancing sensitivity because the intensity is concentrated into a single transition frequency instead of being spread over several transition frequencies. If care is taken to ensure that the other parameters on which the strength of the signal depends are kept constant, the intensities of proton-decoupled spectra are proportional to the number of ^{13}C nuclei present.

13.9 The nuclear Overhauser effect

The technique described here is of considerable usefulness for the determination of the conformations of proteins and other biological macromolecules in their natural aqueous environments.

Consider a very simple AX system in which the two spins interact by a magnetic dipole–dipole interaction. We expect two lines in the spectrum, one from A and the other from X. However, when we irradiate the system with radiofrequency radiation at the resonance frequency of X using such a high intensity that we *saturate* the transition (that is, we equalize the populations of the X levels), we find that the A resonance is modified. It may be enhanced, diminished, or even converted into an emission rather than an absorption. That modification of one resonance by saturation of another is called the **nuclear Overhauser effect** (NOE).

To understand the effect, we need to think about the populations of the four levels of an AX system (Fig. 13.30). At thermal equilibrium, the population of the $\alpha_A\alpha_X$ level is the greatest, and that of the $\beta_A\beta_X$ level is the least; the other two levels

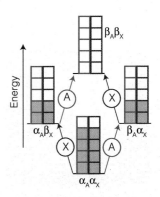

Fig. 13.30 The energy levels of an AX system and an indication of their relative populations. The squares denote notional populations. The transitions of A and X are marked.

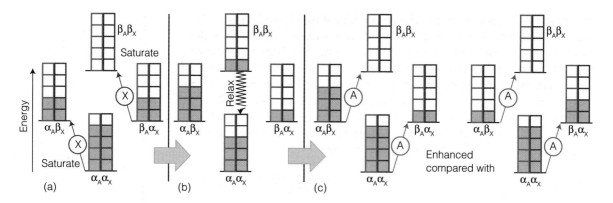

Fig. 13.31 (a) When the X transition is saturated, the populations of its two states are equalized and the population excess and deficit become as shown (using the same symbols as in Fig. 13.30). (b) Dipole–dipole relaxation relaxes the populations of the highest and lowest states, and they regain their original populations. (c) The A transitions reflect the difference in populations resulting from the preceding changes and are enhanced compared with those shown in Fig. 13.30.

have the same energy and an intermediate population. The thermal equilibrium absorption intensities reflect these populations, as the illustration shows. Now consider the combined effect of saturating the X transition and spin relaxation. When we saturate the X transition, the populations of the X levels are equalized, but at this stage there is no change in the populations of the A levels. If that were all that happened, all we would see would be the loss of the X resonance and no effect on the A resonance.

Now consider the effect of spin relaxation. Relaxation can occur in a variety of ways if there is a dipolar interaction between the A and X spins. One possibility is for the magnetic field acting between the two spins to cause them both to flop from α to β, so the $\alpha_A\alpha_X$ and $\beta_A\beta_X$ states regain their thermal equilibrium populations. However, the populations of the $\alpha_A\beta_X$ and $\beta_A\alpha_X$ levels remain unchanged at the values characteristic of saturation. As we see from Fig. 13.31, the population difference between the states joined by transitions of A is now greater than at equilibrium, so the resonance absorption is enhanced. Another possibility is for the dipolar interaction between the two spins to cause α to flip to β and β to flop to α. This transition equilibrates the populations of $\alpha_A\beta_X$ and $\beta_A\alpha_X$ but leaves the $\alpha_A\alpha_X$ and $\beta_A\beta_X$ populations unchanged (Fig. 13.32). Now we see from the illustration that the population differences in the states involved in the A transitions are decreased, so the resonance absorption is diminished.

Which effect wins? Does NOE enhance the A absorption or does it diminish it? As in the discussion of relaxation times in Section 13.7, the efficiency of the

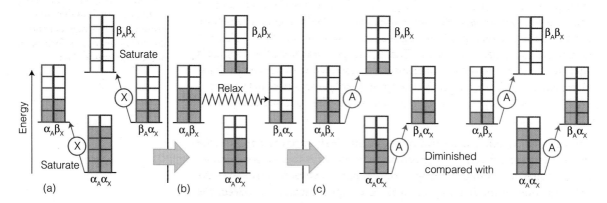

Fig. 13.32 (a) When the X transition is saturated, just as in Fig. 13.31 the populations of its two states are equalized and the population excess and deficit become as shown. (b) Dipole–dipole relaxation relaxes the populations of the two intermediate states, and they regain their original populations. (c) The A transitions reflect the difference in populations resulting from the preceding changes and are diminished compared with those shown in Fig. 13.30.

intensity-enhancing $\beta_A\beta_X \leftrightarrow \alpha_A\alpha_X$ relaxation is high if the dipole field is modulated at the transition frequency, which in this case is close to 2ω; likewise, the efficiency of the intensity-diminishing $\alpha_A\beta_X \leftrightarrow \beta_A\alpha_X$ relaxation is high if the dipole field is stationary (as there is no frequency difference between the initial and final states). A large molecule rotates so slowly that there is very little motion at 2ω, so we expect intensity decrease (Fig. 13.33). A small molecule rotating rapidly can be expected to have substantial motion at 2ω and a consequent enhancement of the signal. In practice, the enhancement lies somewhere between the two extremes and is reported in terms of the parameter η (eta), where

$$\eta = \frac{I - I_0}{I_0} \qquad \boxed{\text{NOE enhancement parameter}} \qquad (13.22)$$

Here I_0 is the normal intensity and I is the NOE intensity of a particular transition; theoretically, η lies between -1 (diminution) and $+\frac{1}{2}$ (enhancement).

The value of η depends strongly on the separation of the two spins involved in the NOE, for the strength of the dipolar interaction between two spins separated by a distance r is proportional to $1/r^3$ and its effect depends on the square of that strength, and therefore on $1/r^6$. This sharp dependence on separation is used to build up a picture of the conformation of a protein by using NOE to identify which nuclei can be regarded as neighbors (Fig. 13.34). The enormous importance of this procedure is that we can determine the conformation of polypeptides in an aqueous environment and do not need to try to make the single crystals that are essential for an X-ray diffraction investigation.

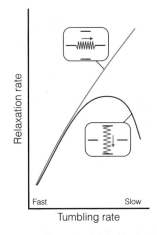

Fig. 13.33 The relaxation rates of the two types of relaxation (as indicated by the small diagrams) as a function of the tumbling rate of the molecule.

| **In the laboratory 13.2** | *Two-dimensional NMR* |

An NMR spectrum contains a great deal of information and, if many protons are present, is very complex. The complexity would be reduced if we could use two axes to display the data, with resonances belonging to different groups lying at different locations on the second axis. This separation is essentially what is achieved in **two-dimensional NMR**.

Much modern NMR work makes use of **correlation spectroscopy** (COSY) in which a clever choice of pulses and Fourier transformation techniques makes it possible to determine all spin–spin couplings in a molecule. A typical outcome for an AX system is shown in Fig. 13.35. The diagram shows contours of equal signal intensity on a plot of intensity against the frequency coordinates ν_1 and ν_2. The **diagonal peaks** are signals centered on (δ_A, δ_A) and (δ_X, δ_X), and lie along the diagonal where $\nu_1 = \nu_2$. That is, the spectrum along the diagonal is equivalent to the one-dimensional spectrum obtained with the conventional NMR technique. The **cross peaks** (or *off-diagonal peaks*) are signals centered on (δ_A, δ_X) and (δ_X, δ_A), and owe their existence to the coupling between the A and X nuclei.

Although information from two-dimensional NMR spectroscopy is trivial in an AX system, it can be of enormous help in the interpretation of more complex spectra, leading to a map of the couplings between spins and to the determination of the bonding network in complex molecules. Indeed, the spectrum of a biological macromolecule that would be impossible to interpret in one-dimensional NMR can often be interpreted reasonably rapidly by two-dimensional NMR. In *Case study* 13.1 we illustrate the procedure by assigning the resonances in the COSY spectrum of an amino acid.

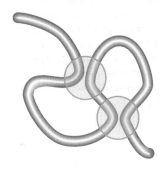

Fig. 13.34 If an NOE experiment shows that the protons within each of the two circles are coupled by a dipolar interaction, we can be confident that those protons are close together and therefore infer the conformation of the polypeptide chain.

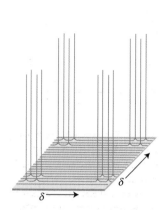

Fig. 13.35 A representation of the two-dimensional NMR spectrum obtained by application of the COSY pulse sequence to an AX spin system.

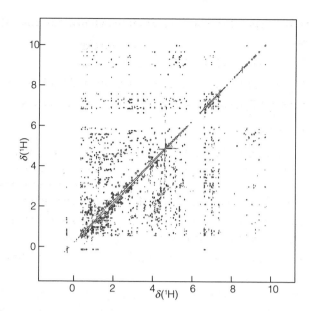

Fig. 13.36 Proton NOESY spectrum of the SH3 domain of the Fyn protein. Adapted from P.J. Hore, *Nuclear magnetic resonance*, Oxford Chemistry Primers 32, Oxford University Press, Oxford (1995).

We have seen that the nuclear Overhauser effect can provide information about internuclear distances through analysis of enhancement patterns in the NMR spectrum before and after saturation of selected resonances. In **nuclear Overhauser effect spectroscopy** (NOESY) a map of all possible NOE interactions is obtained by again using a proper choice of radiofrequency pulses and Fourier transformation techniques. Like a COSY spectrum, a NOESY spectrum consists of a series of diagonal peaks that correspond to the one-dimensional NMR spectrum of the sample. The off-diagonal peaks indicate which nuclei are close enough to each other to give rise to a nuclear Overhauser effect. NOESY data reveal internuclear distances up to about 0.5 nm.

Figure 13.36 shows an example of a two-dimensional proton NOESY spectrum of a protein. Although working in more than one dimension provides more information, even for a relatively small protein the number of off-diagonal peaks is very large. To simplify the spectrum further it is now common to use genetic engineering protocols to express proteins under conditions where specific amino acids are enriched in ^{13}C or ^{15}N, which have $I = \frac{1}{2}$ and can be investigated by NMR spectroscopy. As a result of this **isotope labeling** technique, spectral features—such as off-diagonal peaks in the ^{15}N NOESY spectrum of the labeled protein—become more prominent and easier to interpret. The technique can also be used to simplify COSY spectra.

Case study 13.1 *The COSY spectrum of isoleucine*

Figure 13.37 is a portion of the COSY spectrum of the amino acid isoleucine (**3**), showing the resonances associated with the protons bound to the carbon atoms. We begin the assignment process by considering which protons should

3 Isoleucine

be interacting by spin–spin coupling. From the known molecular structure, we conclude

- the C_a–H proton is coupled only to the C_b–H proton
- the C_b–H protons are coupled to the C_a–H, C_c–H, and C_d–H protons
- the inequivalent C_d–H protons are coupled to the C_b–H and C_e–H protons.

We now note that:

- the resonance with $\delta = 3.6$ shares a cross-peak with only one other resonance at $\delta = 1.9$, which in turn shares cross-peaks with resonances at $\delta = 1.4, 1.2$, and 0.9; this identification is consistent with the resonances at $\delta = 3.6$ and 1.9 corresponding to the C_a–H and C_b–H protons, respectively
- the proton with resonance at $\delta = 0.8$ is not coupled to the C_b–H protons, so we assign the resonance at $\delta = 0.8$ to the C_e–H protons
- the resonances at $\delta = 1.4$ and 1.2 do not share cross-peaks with the resonance at $\delta = 0.9$
- in the light of the expected couplings, we assign the resonance at $\delta = 0.9$ to the C_c–H protons and the resonances at $\delta = 1.4$ and 1.2 to the inequivalent C_d–H protons

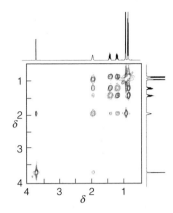

Fig. 13.37 Proton COSY spectrum of isoleucine. (*Case study* 13.1 and this spectrum were adapted from K.E. van Holde, W.C. Johnson, and P.S. Ho, *Principles of physical biochemistry*, p. 508, Prentice Hall, Upper Saddle River (1998).)

The information in EPR spectra

The magnetic moment of an electron is much bigger than that of any nucleus, so even quite modest fields can require high frequencies to induce EPR transitions. Much work is done using fields of about 0.3 T, when resonance occurs at about 9 GHz, corresponding to microwave radiation with a wavelength of 3 cm. Electron paramagnetic resonance is much more limited than NMR because it is applicable only to species with unpaired electrons, which include radicals (perhaps resulting from electron transfer reactions or prepared by radiation damage) and d-metal complexes, including such biologically active species as hemoglobin. But the limitations of EPR can also represent a great advantage of the technique over other spectroscopic methods, for with EPR it is possible to focus attention on a single species, such as a tyrosine radical, in a large biopolymer, such as cytochrome c oxidase. By contrast, it is very difficult (and sometimes impossible) to identify features due to a single amino acid or co-factor in the NMR or IR spectrum of a large biological macromolecule.

Both Fourier-transform (FT) and continuous-wave (CW) EPR spectrometers are available. The FT-EPR instrument is like an FT-NMR spectrometer except that pulses of microwaves are used to excite electron spins in the sample. The layout of the more common CW-EPR spectrometer is shown in Fig. 13.38. It consists of a microwave source (a klystron or a Gunn oscillator), a cavity in which the sample is inserted in a glass or quartz container, a microwave detector, and an electromagnet with a field that can be varied in the region of 0.3 T. The EPR spectrum is obtained by monitoring the microwave absorption as the field is changed, and a typical spectrum (of the benzene radical anion, $C_6H_6^-$) is shown in Fig. 13.39. The peculiar appearance of the spectrum, which is in fact the first

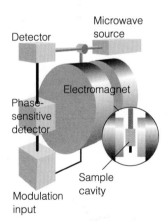

Fig. 13.38 The layout of a continuous-wave EPR spectrometer. A typical magnetic field is 0.3 T, which requires microwaves of frequency 9 GHz (wavelength 3 cm) for resonance.

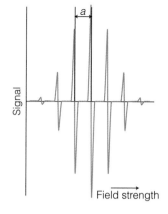

Fig. 13.39 The EPR spectrum of the benzene radical anion, $C_6H_6^-$, in fluid solution. The quantity a is the hyperfine splitting of the spectrum; the center of the spectrum is determined by the g-value of the radical.

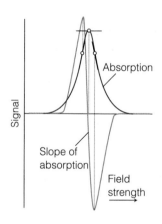

Fig. 13.40 When phase-sensitive detection is used, the signal is the first derivative of the absorption intensity. Note that the peak of the absorption corresponds to the point where the derivative passes through zero.

4 A tyrosine radical

derivative of the absorption, arises from the detection technique, which is sensitive to the slope of the absorption curve (Fig. 13.40).

13.10 The g-value

To begin to interpret the EPR spectra of organic radicals that can form during biological processes we need to compare the spectrum of the sample with that of a free electron.

Equation 13.10 gives the resonance frequency for a transition between the $m_s = -\frac{1}{2}$ and $m_s = +\frac{1}{2}$ levels of a 'free' electron in terms of the g-value $g_e \approx 2.0023$. The magnetic moment of an unpaired electron in a radical also interacts with an external field, but the g-value is different from that of a free electron on account of local magnetic fields induced in the molecular framework of the radical. Consequently, the resonance condition is normally written as

$$h\nu = g\mu_B \mathcal{B}_0 \qquad \boxed{\text{Resonance condition in EPR spectroscopy}} \quad (13.23)$$

where g is the **g-value** of the radical. Many organic radicals have g-values close to 2.0027; inorganic radicals have g-values typically in the range 1.9–2.1; paramagnetic d-metal complexes have g-values in a wider range (for example, 0 to 6).

The deviation of g from $g_e = 2.0023$ depends on the ability of the applied field to induce local electron currents in the radical, and therefore its value gives some information about electronic structure. In that sense, the g-value plays a similar role in EPR as the shielding constant plays in NMR. Because g-values differ very little from g_e in many radicals (for example, 2.003 for H, 1.999 for NO_2, 2.01 for ClO_2), their main use in biochemical applications is to aid the identification of the species present in a sample.

> **A brief illustration**
>
> Recent EPR studies have shown that the amino acid tyrosine participates in a number of biological electron transfer reactions, including the oxidation of water to O_2 in plant photosystem II, the reduction of O_2 to water in cytochrome c oxidase, and the reduction of ribonucleotides to deoxyribonucleotides catalyzed by the enzyme ribonucleotide reductase. During the course of these electron transfer reactions, a tyrosine radical forms (**4**). The center of the EPR spectrum of the tyrosine radical in cytochrome c oxidase of the bacterium *P. denitrificans* occurs at 344.50 mT in a spectrometer operating at 9.6699 GHz (radiation belonging to the X band of the microwave region). Its g-value is therefore
>
> $$g = \frac{h\nu}{\mu_B \mathcal{B}_0} = \frac{(6.626\,08 \times 10^{-34}\,\text{J s}) \times (9.6699 \times 10^9\,\text{s}^{-1})}{(9.2740 \times 10^{-24}\,\text{J T}^{-1}) \times (0.344\,50\,\text{T})} = 2.0055$$

> **Self-test 13.6** At what magnetic field would the tyrosine radical come into resonance in a spectrometer operating at 34.000 GHz (radiation belonging to the Q band of the microwave region)?
>
> **Answer:** 1.2113 T

13.11 Hyperfine structure

The second step in the interpretation of the EPR spectra of organic radicals is to take into account the effect that magnetic nuclei have on the energy levels of unpaired electrons.

The most important features of EPR spectra are their **hyperfine structure**, the splitting of individual resonance lines into components. In general in spectroscopy, the term 'hyperfine structure' means the structure of a spectrum that can be traced to interactions of the electrons with nuclei other than as a result of the latter's point electric charge. The source of the hyperfine structure in EPR is the magnetic interaction between the electron spin and the magnetic dipole moments of the nuclei present in the radical.

Consider the effect on the EPR spectrum of a single H nucleus located somewhere in a radical. The proton spin is a source of magnetic field, and depending on the orientation of the nuclear spin, the field it generates adds to or subtracts from the applied field. The total local field is therefore

$$\mathcal{B}_{loc} = \mathcal{B} + am_I \qquad m_I = \pm\tfrac{1}{2}$$

> The role of the hyperfine coupling constant (13.24)

where a is the **hyperfine coupling constant**. Half the radicals in a sample have $m_I = +\tfrac{1}{2}$, so half resonate when the applied field satisfies the condition

$$h\nu = g\mu_B(\mathcal{B} + \tfrac{1}{2}a) \quad \text{or} \quad \mathcal{B} = \frac{h\nu}{g\mu_B} - \tfrac{1}{2}a$$

> Resonance condition in the presence of hyperfine structure (13.25a)

The other half (which have $m_I = -\tfrac{1}{2}$) resonate when

$$h\nu = g\mu_B(\mathcal{B} - \tfrac{1}{2}a) \quad \text{or} \quad \mathcal{B} = \frac{h\nu}{g\mu_B} + \tfrac{1}{2}a$$

> Resonance condition in the presence of hyperfine structure (13.25b)

Therefore, instead of a single line, the spectrum shows two lines of half the original intensity separated by a and centered on the field determined by g (Fig. 13.41).

If the radical contains a ^{14}N atom ($I = 1$), its EPR spectrum consists of three lines of equal intensity because the ^{14}N nucleus has three possible spin orientations and each spin orientation is possessed by one third of all the radicals in the sample. In general, a spin-I nucleus splits the spectrum into $2I + 1$ hyperfine lines of equal intensity.

When there are several magnetic nuclei present in the radical, each one contributes to the hyperfine structure. In the case of equivalent protons (for example, the two CH_2 protons in the radical $CH_3CH_2\cdot$) some of the hyperfine lines are coincident. It is not hard to show that if the radical contains N equivalent protons, then there are $N = 1$ hyperfine lines with an intensity distribution given by Pascal's triangle (1). The spectrum of the benzene radical anion in Fig. 13.39, which has seven lines with intensity ratio 1:6:15:20:15:6:1, is consistent with a radical containing six equivalent protons. More generally, if the radical contains N equivalent nuclei with spin quantum number I, then there are $2NI + 1$ hyperfine lines with an intensity distribution given by a modified version of Pascal's triangle. For instance, the hyperfine interaction with two equivalent ^{14}N ($I = 1$) nuclei gives rise to five lines with intensities in the ratio 1:2:3:2:1.

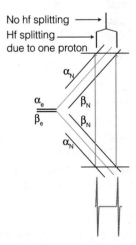

Fig. 13.41 The hyperfine interaction between an electron and a spin-$\tfrac{1}{2}$ nucleus results in four energy levels in place of the original two. As a result, the spectrum consists of two lines (of equal intensity) instead of one. The intensity distribution can be summarized by a simple stick diagram. The diagonal lines show the energies of the states as the applied field is increased, and resonance occurs when the separation of states matches the fixed energy of the microwave photon.

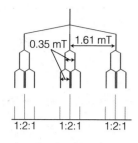

Fig. 13.42 The analysis of the hyperfine structure of radicals containing one ^{14}N nucleus ($I = 1$) and two equivalent protons with the hyperfine splitting shown.

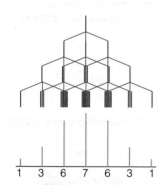

Fig. 13.43 The analysis of the hyperfine structure of radicals containing three equivalent ^{14}N nuclei.

Example 13.3 *Predicting the hyperfine structure of an EPR spectrum*

We shall see (*In the laboratory* 13.3) that radicals containing the ^{14}N nucleus can be used to investigate biological macromolecules and aggregates. A radical has one ^{14}N nucleus ($I = 1$) with hyperfine constant 1.61 mT and two equivalent protons ($I = \frac{1}{2}$) with hyperfine constant 0.35 mT. Predict the form of the EPR spectrum.

Strategy We consider the hyperfine structure that arises from each type of nucleus or group of equivalent nuclei in succession. So, split a line with one nucleus, then split each of those lines by a second nucleus (or group of nuclei), and so on. It is best to start with the nucleus with the largest hyperfine splitting; however, any choice could be made, and the order in which nuclei are considered does not affect the conclusion.

Solution The ^{14}N nucleus gives three hyperfine lines of equal intensity separated by 1.61 mT. Each line is split into doublets of spacing 0.35 mT by the first proton, and each line of these doublets is split into doublets with the same 0.35 mT splitting (Fig. 13.42). The central lines of each split doublet coincide, so the proton splitting gives 1:2:1 triplets of internal splitting 0.35 mT. Therefore, the spectrum consists of three equivalent 1:2:1 triplets.

Self-test 13.7 Predict the form of the EPR spectrum of a radical containing three equivalent ^{14}N nuclei and no other magnetic nuclei.

Answer: Fig. 13.43

The hyperfine structure of an EPR spectrum is a kind of fingerprint that helps to identify the radicals present in a sample. The interaction between the unpaired electron and the hydrogen nucleus responsible for hyperfine structure is either a dipolar interaction or the Fermi contact interaction described in Section 13.4. In the case of the contact interaction, the magnitude of the splitting depends on the distribution of the unpaired electron near the magnetic nuclei present, so the spectrum can be used to map the molecular orbital occupied by the unpaired electron. For example, because the hyperfine splitting in $C_6H_6^-$ is 0.375 mT and one proton is close to a C atom with one sixth the unpaired electron density (because the electron is spread uniformly around the ring), the hyperfine splitting caused by a proton in the electron spin entirely confined to a single adjacent C atom should be 6×0.375 mT $= 2.25$ mT. If in another aromatic radical we find a hyperfine splitting constant a, then the **spin density**, ρ (rho), the probability that an unpaired electron is on the atom, can be calculated from the **McConnell equation**:

$$a = Q\rho$$

McConnell equation   (13.26)

with $Q = 2.25$ mT. In this equation, ρ is the spin density on a C atom and a is the hyperfine splitting observed for the H atom to which it is attached.

In the laboratory 13.3 *Spin probes*

The appearance of the EPR spectrum of a radical changes as its motion is restricted, and we need to see how to take advantage of this effect in biochemical investigations. Figure 13.44 shows the variation of the line shape of the EPR spectrum of the di-*tert*-butyl nitroxide radical (**5**) with temperature. At 292 K the spectrum consists of three sharp peaks arising from hyperfine coupling to the neighboring ^{14}N nucleus. However, the spectral lines broaden when the temperature is lowered to 77 K. At high temperatures, the radical tumbles freely and the motion becomes restricted as the temperature decreases. It follows that we can use the line shape of the EPR spectrum as a probe of the mobility of the radical.

A **spin probe** (or *spin label*) is a radical with an EPR spectrum that reports on the dynamical properties of the biopolymer. The ideal spin probe is one with an EPR spectrum that broadens significantly as its motion is restricted to a relatively small extent. Nitroxide spin probes have been used to show that the hydrophobic interiors of biological membranes, once thought to be rigid, are in fact very fluid and individual lipid molecules move laterally through the sheet-like structure of the membrane. The EPR spectrum also can reveal whether a nitroxide spin probe is free in solution, positioned as a guest within a macromolecular host, or intercalated within micelles (Section 11.16). For example, hyperfine coupling constants to the ^{14}N nucleus can change if the N–O group is exposed to the solvent or buried in the assembly.

Just as chemical exchange can broaden proton NMR spectra (Section 13.5), electron exchange between two radicals can broaden EPR spectra, therefore the distance between two spin probe molecules may be measured from the line widths of their EPR spectra. The effect can be used in a number of biochemical studies. For example, the kinetics of association of two polypeptides labeled with the synthetic amino acid 2,2,6,6,-tetramethylpiperidine-1-oxyl-4-amino-4-carboxylic acid (**6**) can be studied by measuring the line width of the EPR spectrum of the label as a function of time. Alternatively, the thermodynamics of association may be studied by examining the temperature dependence of the EPR line width.

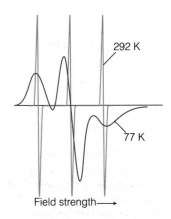

Fig. 13.44 EPR spectra of the di-*tert*-butyl nitroxide radical at 292 K and 77 K. (Based on information from J.R. Bolton.)

5 di-*tert*-butyl nitroxide radical

6 A spin-labeled amino acid

Checklist of key concepts

1. Resonance is the condition of strong effective coupling when the frequencies of two oscillators are identical.

2. Nuclear magnetic resonance (NMR) is the observation of the frequency at which magnetic nuclei in molecules come into resonance with an electromagnetic field when the molecule is exposed to a strong magnetic field; NMR is a radiofrequency technique.

3. Electron paramagnetic resonance (EPR) is the observation of the frequency at which an electron spin comes into resonance with an electromagnetic

field when the molecule is exposed to a strong magnetic field; EPR is a microwave technique.

4. The intensity of an NMR or EPR transition increases with the difference in population of α and β states and the strength of the applied magnetic field.

5. The chemical shift of a nucleus is the difference between its resonance frequency and that of a reference standard.

6. The observed shielding constant is the sum of a local contribution, a neighboring group contribution, and a solvent contribution.

7. The fine structure of an NMR spectrum is the splitting of the groups of resonances into individual lines; the strength of the interaction is expressed in terms of the spin–spin coupling constant, J.

8. N equivalent spin-$\frac{1}{2}$ nuclei split the resonance of a nearby spin or group of equivalent spins into $N + 1$ lines with an intensity distribution given by Pascal's triangle.

9. Spin–spin coupling in molecules in solution can be explained in terms of the polarization mechanism, in which the interaction is transmitted through the bonds.

10. The Fermi contact interaction is a magnetic interaction that depends on the very close approach of an electron to the nucleus and can occur only if the electron occupies an s orbital.

11. Coalescence of the two lines occurs in conformational interchange or chemical exchange when the lifetime of the states is related to their resonance frequency difference.

12. In Fourier-transform NMR, the spectrum is obtained by mathematical analysis of the free-induction decay of magnetization, the response of nuclear spins in a sample to the application of one or more pulses of radiofrequency radiation.

13. Relaxation is the nonradiative return to an equilibrium distribution of populations in a system with random relative spin orientations; the system returns exponentially to the equilibrium distribution with a time constant called the spin–lattice relaxation time, T_1.

14. The spin–spin relaxation time, T_2, is the time constant for the exponential return of the system into random relative orientations.

15. Magnetic resonance imaging (MRI) is a portrayal of the concentrations of protons in an object. The technique relies on the application of specific pulse sequences to an object in an inhomogeneous magnetic field (a field with values that vary inside the sample).

16. With functional MRI, blood flow in different regions of the brain can be studied and related to the mental activities of the subject. The technique is based on differences in the magnetic properties of deoxygenated and oxygenated hemoglobin and their effects on proton resonances.

17. In proton decoupling of ^{13}C-NMR spectra, protons are made to undergo rapid spin reorientations and the ^{13}C nucleus senses an average orientation. As a result, its resonance is a single line and not a group of lines.

18. The nuclear Overhauser effect (NOE) is the modification of one resonance by the saturation of another.

19. In two-dimensional NMR, spectra are displayed in two axes, with resonances belonging to different groups lying at different locations on the second axis.

20. In correlation spectroscopy (COSY) all spin–spin couplings in a molecule are determined.

21. In nuclear Overhauser effect spectroscopy (NOESY) internuclear distances up to about 0.5 nm are determined.

22. The hyperfine structure of an EPR spectrum is its splitting of individual resonance lines into components by the magnetic interaction of the electron and nuclei with spin.

23. If a radical contains N equivalent nuclei with spin quantum number I, then there are $2NI + 1$ hyperfine lines with an intensity distribution given by a modified version of Pascal's triangle.

24. A spin probe is a radical with an EPR spectrum that reports on the dynamical properties of the biopolymer.

Checklist of key equations

Property	Equation	Comment
Energy in a magnetic field	$E_{m_s} = g_e \mu_B \mathcal{B}_0 m_s$	Free electron
	$E_{m_I} = -g_I \mu_N \mathcal{B}_0 m_I$	Nucleus
Resonance condition	$h\nu = g_e \mu_B \mathcal{B}_0$	Free electron
	$h\nu = \gamma_N \hbar \mathcal{B}_0$	Nucleus
δ Scale for chemical shifts	$\delta = \{(\nu - \nu^o)/\nu^o\} \times 10^6$	Definition
Karplus equation	$^3 J_{HH} = A + B \cos\phi + C \cos 2\phi$	
Condition for coalescence of two NMR lines	$\tau = 2^{1/2}/\pi \delta \nu$	
Nuclear Overhauser enhancement parameter	$\eta = (I - I_0)/I_0$	Definition
EPR resonance condition	$h\nu = g \mu_B \mathcal{B}_0$	
McConnell equation	$a = Q\rho$	

Discussion questions

13.1 To what extent are all spectroscopic techniques resonance techniques, and why are magnetic resonance techniques best so-called?

13.2 Discuss the origins of the local, neighboring group, and solvent contributions to the shielding constant.

13.3 Describe the significance of the chemical shift in relation to the terms 'high-field' and 'low-field'.

13.4 Discuss how the Fermi contact interaction and the polarization mechanism contribute to spin–spin couplings in NMR.

13.5 Suggest a reason why the relaxation times of ^{13}C nuclei are typically much longer than those of ^{1}H nuclei.

13.6 Suggest a reason why the spin–lattice relaxation time of benzene (a small molecule) in a mobile, deuterated hydrocarbon solvent increases whereas that of an oligopeptide (a large molecule) decreases.

13.7 Discuss the origin of the nuclear Overhauser effect and how it can be used to measure distances between protons in a biopolymer.

13.8 Discuss the origins of diagonal and cross-peaks in the COSY spectrum of an AX system.

13.9 Explain how the EPR spectrum of an organic radical can be used to pinpoint the molecular orbital occupied by the unpaired electron.

13.10 Suggest how spin probes could be used to estimate the depth of a crevice in a biopolymer, such as the active site of an enzyme.

Exercises

13.11 Calculate the energy separation between the spin states of an electron in a magnetic field of 0.300 T.

13.12 The nucleus ^{32}S has a spin of $\frac{3}{2}$ and a nuclear g-factor of 0.4289. Calculate the energies of the nuclear spin states in a magnetic field of 7.500 T.

13.13 Equations 13.5–13.7 define the magnetogyric ratio and the g-factor of a nucleus. Given that g is a dimensionless number, what are the units of γ_N expressed in (a) tesla and hertz, and (b) SI base units?

13.14 The magnetogyric ratio of ^{31}P is 1.0840×10^8 T^{-1} s^{-1}. What is the g-factor of the nucleus?

13.15 Calculate the value of $(N_\beta - N_\alpha)/N$ for electrons in a field of (a) 0.30 T and (b) 1.1 T.

13.16 Calculate the resonance frequency and the corresponding wavelength for an electron in a magnetic field of 0.330 T, the magnetic field commonly used in EPR.

13.17 Calculate the value of $(N_\alpha - N_\beta)/N$ for (a) protons and (b) carbon-13 nuclei in a field of 10 T.

13.18 The first generally available NMR spectrometers operated at a frequency of 60 MHz; today it is not uncommon to use a spectrometer that operates at 800 MHz. What are the relative population differences ($\delta N/N$) of ^{13}C spin states in these two spectrometers at 25°C?

13.19 The magnetogyric ratio of ^{19}F is 2.5177×10^8 T^{-1} s^{-1}. Calculate the frequency of the nuclear transition in a field of 8.200 T.

13.20 Calculate the resonance frequency of an ^{14}N nucleus ($I = 1$, $g = 0.4036$) in a 15.00 T magnetic field.

13.21 Calculate the magnetic field needed to satisfy the resonance condition for unshielded protons in a 500.0 MHz radiofrequency field.

13.22 What is the shift of the resonance from TMS of a group of protons with $\delta = 6.33$ in a polypeptide in a spectrometer operating at 420 MHz?

13.23 What are the relative values of the chemical shifts observed for nuclei in the spectrometers mentioned in *Exercise* 13.18 in terms of (a) δ values and (b) frequencies?

13.24 To determine the structures of biopolymers by NMR spectroscopy, biochemists use spectrometers that operate at the highest available frequencies. Use your results from *Exercises* 13.18 and 13.23 to justify this choice.

13.25 The chemical shift of the CH$_3$ protons in acetaldehyde (ethanal) is $\delta = 2.20$ and that of the CHO proton is 9.80. What is the difference in local magnetic field between the two regions of the molecule when the applied field is (a) 1.5 T and (b) 6.0 T?

13.26 Using the information in Fig. 13.4, state the splitting (in hertz, Hz) between the methyl and aldehydic proton resonances in a spectrometer operating at (a) 300 MHz and (b) 500 MHz.

13.27 What would be the nuclear magnetic resonance spectrum for a proton resonance line that was split by interaction with seven identical protons?

13.28 What would be the nuclear magnetic resonance spectrum for a proton resonance line that was split by interaction with (a) two and (b) three equivalent nitrogen nuclei (the spin of a nitrogen nucleus is 1)?

13.29 Repeat *Justification* 13.2 for an AX$_2$ spin-$\frac{1}{2}$ system and deduce the pattern of lines expected in the spectrum.

13.30 Sketch the appearance of the ^{1}H-NMR spectrum of acetaldehyde (ethanal) using $J = 2.90$ Hz and the data in Fig. 13.4 in a spectrometer operating at (a) 300 MHz and (b) 500 MHz.

13.31 Sketch the form of an A$_3$M$_2$X$_4$ spectrum, where A, M, and X are protons with distinctly different chemical shifts and $J_{AM} > J_{AX} < J_{MX}$.

13.32 Formulate the version of Pascal's triangle that you would expect in an NMR spectrum for a collection of N spin-1 nuclei, with N up to 5.

13.33 Show that the coupling constant as expressed by the Karplus equation passes through a minimum when $\cos \phi = B/4C$. *Hint:* Evaluate the first derivative with respect to ϕ and set the result equal to 0. To confirm that the extremum is a minimum, go on to evaluate the second derivative and show that it is positive.

13.34 A proton jumps between two sites with $\delta = 2.7$ and $\delta = 4.8$. At what rate of interconversion will the two signals collapse to a single line in a spectrometer operating at 500 MHz?

13.35 NMR spectroscopy may be used to determine the equilibrium constant for dissociation of a complex between a small molecule, such as an enzyme inhibitor I, and a protein, such as an enzyme E:

$$EI \rightleftharpoons E + I \qquad K_I = [E][I]/[EI]$$

In the limit of slow chemical exchange, the NMR spectrum of a proton in I would consist of two resonances: one at ν_I for free I and another at ν_{EI} for bound I. When chemical exchange is fast, the NMR spectrum of the same proton in I consists of a single peak with a resonance frequency ν given by

$$\nu = f_I \nu_I + f_{EI} \nu_{EI}$$

where $f_I = [I]/([I] + [EI])$ and $f_{EI} = [EI]/([I] + [EI])$ are, respectively, the fractions of free I and bound I. For the purposes of analyzing the data, it is also useful to define the frequency differences $\delta\nu = \nu - \nu_I$ and $\Delta\nu = \nu_{EI} - \nu_I$. Show that when the initial concentration of I, $[I]_0$, is much greater than the initial concentration of E, $[E]_0$, a plot of $[I]_0$ versus $1/\delta\nu$ is a straight line with slope $[E]_0\Delta\nu$ and y-intercept $-K_I$.

13.36 The duration of a 90° pulse depends on the strength of the $\mathcal{B}_1$ field. If a 90° pulse requires 10 μs, what is the strength of the $\mathcal{B}_1$ field?

13.37 Interpret the following features of the NMR spectra of hen lysozyme: (a) saturation of a proton resonance assigned to the side chain of methionine-105 changes the intensities of proton resonances assigned to the side chains of tryptophan-28 and tyrosine-23; (b) saturation of proton resonances assigned to tryptophan-28 did not affect the spectrum of tyrosine-23.

13.38 You are designing an MRI spectrometer. What field gradient (in microtesla per meter, μT m^{-1}) is required to produce a separation of 100 Hz between two protons separated by the long diameter of a human kidney (taken as 8 cm) given that they are in environments with $\delta = 3.4$? The radiofrequency field of the spectrometer is at 400 MHz and the applied field is 9.4 T.

13.39 Suppose that a uniform disk-shaped organ is in a linear field gradient and that the MRI signal is proportional to the number of protons in a slice of width δx at each horizontal distance x from the center of the disk. Sketch the shape of the absorption intensity for the MRI image of the disk before any computer manipulation has been carried out.

13.40 Figure 13.45 shows the proton COSY spectrum of 1-nitropropane. Account for the appearance of off-diagonal peaks in the spectrum.

13.41 The proton chemical shifts for the NH, $C_\alpha H$, and $C_\beta H$ groups of alanine are 8.25, 4.35, and 1.39, respectively. Sketch the COSY spectrum of alanine between $\delta = 1.00$ and 8.50.

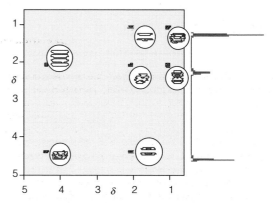

Fig. 13.45 Proton COSY spectrum of 1-nitropropane. The circles show enhanced views of the spectral features. (Spectrum provided by Prof. G. Morris.)

13.42 The center of the EPR spectrum of atomic hydrogen lies at 329.12 mT in a spectrometer operating at 9.2231 GHz. What is the g-value of the electron in the atom?

13.43 A radical containing two equivalent protons shows a three-line spectrum with an intensity distribution 1:2:1. The lines occur at 330.2 mT, 332.5 mT, and 334.8 mT. What is the hyperfine coupling constant for each proton? What is the g-value of the radical given that the spectrometer is operating at 9.319 GHz?

13.44 Predict the intensity distribution in the hyperfine lines of the EPR spectra of (a) ·CH$_3$ and (b) ·CD$_3$.

13.45 The benzene radical anion has $g = 2.0025$. At what field should you search for resonance in a spectrometer operating at (a) 9.302 GHz and (b) 33.67 GHz?

13.46 The EPR spectrum of a radical with two equivalent nuclei of a particular kind is split into five lines of intensity ratio 1:2:3:2:1. What is the spin of the nuclei?

13.47 Formulate the version of Pascal's triangle that you would expect to represent the hyperfine structure in an EPR spectrum for a collection of N spin-$\frac{3}{2}$ nuclei, with N up to 5.

13.48 (a) Sketch the EPR spectra of the di-*tert*-butyl nitroxide radical (5) at 292 K in the limits of very low concentration (at which electron exchange is negligible), moderate concentration (at which electron exchange effects begin to be observed), and high concentration (at which electron exchange effects predominate). (b) Discuss how the observation of electron exchange between nitroxide spin probes can inform the study of lateral mobility of lipids in a biological membrane.

Projects

13.49 Consult library and reliable internet resources, such as those listed on the website for this text, and write a brief report (similar in length and depth of coverage to one of the many *Case studies* in this text) summarizing the use of NMR or EPR spectroscopy in the study of protein denaturation. Your report should include (a) a description of experimental methods, (b) a discussion of the information that can be obtained from the measurements, (c) an example from the chemical or biological literature of the use of the technique in protein stability work, and (d) a brief discussion of the advantages and disadvantages of the technique of your choice over differential scanning calorimetry (*In the laboratory* 1.1) and the techniques you described in *Exercise* 12.45b.

13.50 The following pulse sequence is used in the *inversion recovery technique*: a 180° pulse is followed by a time interval τ, then a 90° pulse, acquisition of a FID curve, and Fourier transformation. A 180° pulse is achieved by applying a $\mathcal{B}_1$ field for twice as long as for a 90° pulse, so the magnetization vector precesses through 180° and points in the $-z$-direction.

(a) If a 180° pulse requires 12.5 μs, what is the strength of the $\mathcal{B}_1$ field?

(b) Draw a series of diagrams showing the effect of the pulse sequence described in part (a) on a sample of equivalent nuclei. The first diagram can be drawn with ease because we already know that the 180° pulse tips the magnetization vector toward the $-z$-direction. The second diagram should show the effect of spin–lattice relaxation on the magnitude of the magnetization vector after a time interval $0 < \tau < T_1$ has elapsed. The third diagram should show the effect of the 90° pulse on the magnetization vector.

(c) Why is an FID signal generated after application of the 90° pulse?

(d) How does the intensity of the spectrum (obtained by Fourier transformation of the FID curve) vary with the time interval τ, with $0 < \tau < T_1$?

(e) Use your results from parts (a)–(d) to show that the inversion recovery technique can be used to measure spin–lattice relaxation times.

The following project requires the use of molecular modeling software. The website for this text contains links to freeware and to other sites where you may perform molecular orbital calculations directly from your web browser.

13.51 The molecular electronic structure methods described in Chapter 10 may be used to predict the spin density distribution in a radical. Recent EPR studies have shown that the amino acid tyrosine participates in a number of biological electron transfer reactions, including the processes of water oxidation to O_2 in plant photosystem II and of O_2 reduction to water in cytochrome *c* oxidase. During the course of these electron transfer reactions a tyrosine radical forms, with spin density delocalized over the side chain of the amino acid.

(a) The phenoxy radical shown in (7) is a suitable model of the tyrosine radical. Using molecular modeling software and the computational method of your instructor's choice, calculate the spin densities at the O atom and at all of the C atoms in (7).

(b) Predict the form of the EPR spectrum of (7).

7 Phenoxy radical

Resource section 1:
Atlas of structures

In this section are displayed structures of biologically significant molecules that occur throughout the text. They are arranged as follows:

Section A   Amino acids
Section B   Bases
Section C   Carboxylic acids
Section E   Polyenes
Section L   Lipids
Section M   Miscellaneous
Section N   Nucleotides
Section P   Proteins
Section R   Porphyrin-based ring complexes
Section S   Saccharides
Section T   Nucleic acids

For proteins, we give the appropriate Protein Data Bank reference.

A Amino acids

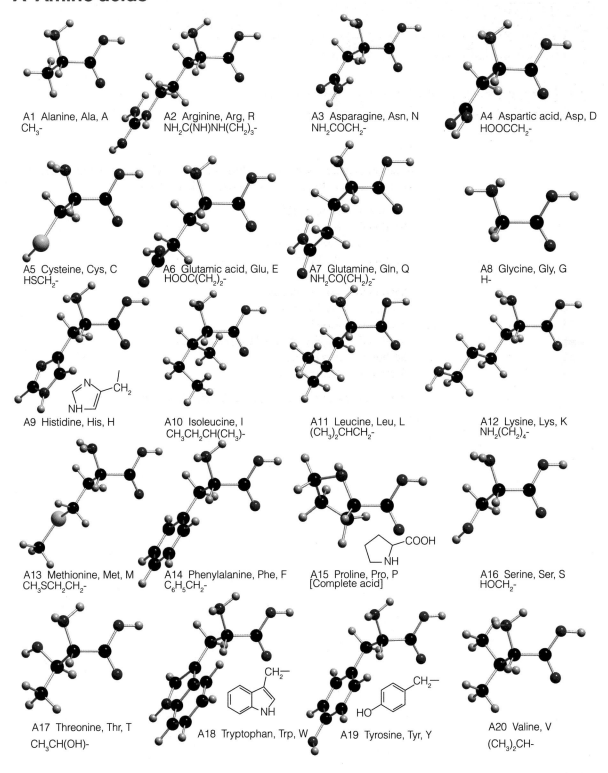

A1 Alanine, Ala, A
CH_3-

A2 Arginine, Arg, R
$NH_2C(NH)NH(CH_2)_3$-

A3 Asparagine, Asn, N
NH_2COCH_2-

A4 Aspartic acid, Asp, D
$HOOCCH_2$-

A5 Cysteine, Cys, C
$HSCH_2$-

A6 Glutamic acid, Glu, E
$HOOC(CH_2)_2$-

A7 Glutamine, Gln, Q
$NH_2CO(CH_2)_2$-

A8 Glycine, Gly, G
H-

A9 Histidine, His, H

A10 Isoleucine, I
$CH_3CH_2CH(CH_3)$-

A11 Leucine, Leu, L
$(CH_3)_2CHCH_2$-

A12 Lysine, Lys, K
$NH_2(CH_2)_4$-

A13 Methionine, Met, M
$CH_3SCH_2CH_2$-

A14 Phenylalanine, Phe, F
$C_6H_5CH_2$-

A15 Proline, Pro, P
[Complete acid]

A16 Serine, Ser, S
$HOCH_2$-

A17 Threonine, Thr, T
$CH_3CH(OH)$-

A18 Tryptophan, Trp, W

A19 Tyrosine, Tyr, Y

A20 Valine, V
$(CH_3)_2CH$-

B Bases

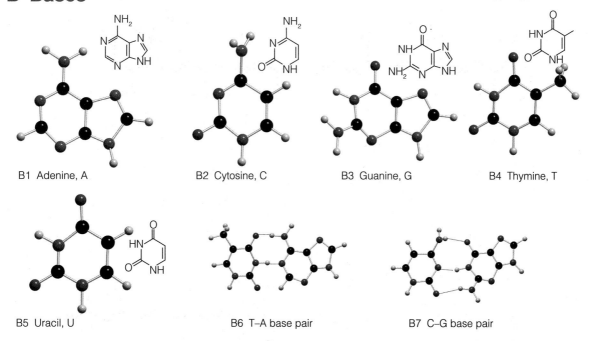

B1 Adenine, A

B2 Cytosine, C

B3 Guanine, G

B4 Thymine, T

B5 Uracil, U

B6 T–A base pair

B7 C–G base pair

C Carboxylic acids

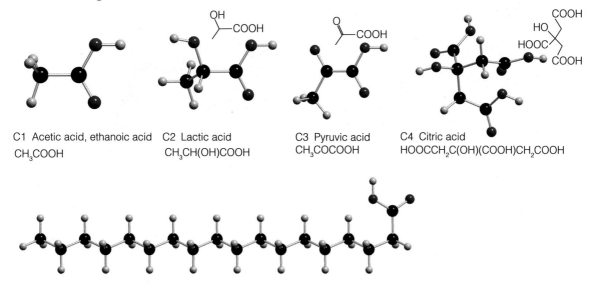

C1 Acetic acid, ethanoic acid
CH_3COOH

C2 Lactic acid
$CH_3CH(OH)COOH$

C3 Pyruvic acid
$CH_3COCOOH$

C4 Citric acid
$HOOCCH_2C(OH)(COOH)CH_2COOH$

C5 Stearic acid
$CH_3(CH_2)_{16}COOH$

E Polyenes

E1 β-Carotene
$C_{40}H_{56}$

E2 all-*trans*-Retinal
$C_{20}H_{28}O$

E3 11-*cis*-Retinal
$C_{20}H_{28}O$

L Lipids

L1 Cholesterol

R = $(CH_2)_{14}CH_3$
R' = $(CH_2)_7CH=CH(CH_2)_7CH_3$

L2 Phosphatidylcholine (lecithin)

M Miscellaneous

M1 Ascorbic acid, Vitamin C
$C_6H_8O_6$

M2 Uric acid
$C_5H_4N_4O_3$

M3 α-Tocopherol, Vitamin E
$C_{29}H_{50}O_2$

M4 Glutathione
$C_{10}H_{17}N_3O_6S$

M5 Ubiquinone, ubiquinone 50, coenzyme Q
$C_{59}H_{90}O_4$

N Nucleotides

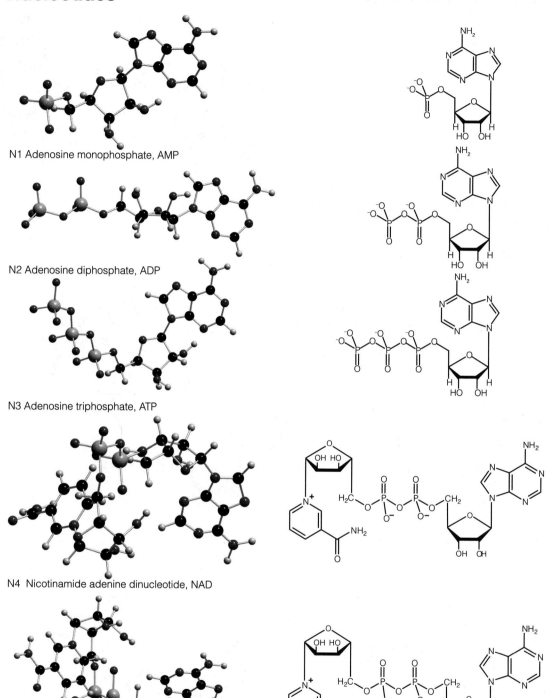

N1 Adenosine monophosphate, AMP

N2 Adenosine diphosphate, ADP

N3 Adenosine triphosphate, ATP

N4 Nicotinamide adenine dinucleotide, NAD

N5 Nicotinamide adenine dinucleotide phosphate, NADP

N Nucleotides continued

N6 Coenzyme A

N7 Flavin adenosine dinucleotide, FAD

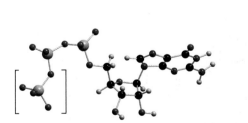

N8 Guanosine di- and triphosphate, GDP [and GTP]

P Proteins

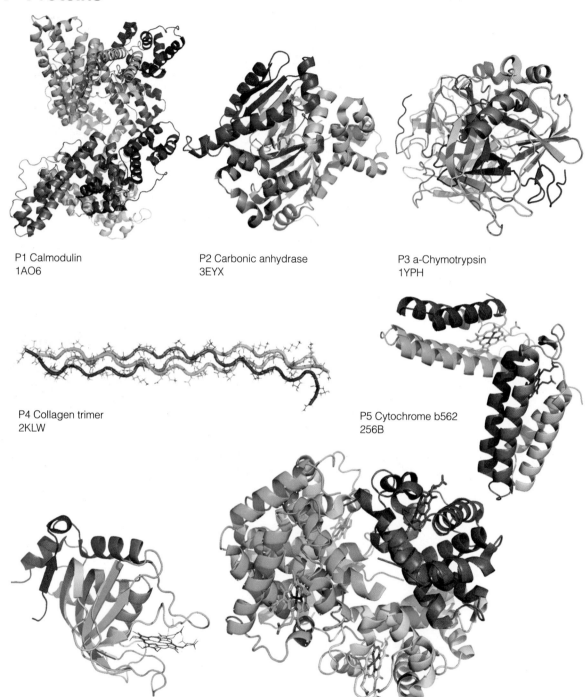

P1 Calmodulin
1AO6

P2 Carbonic anhydrase
3EYX

P3 a-Chymotrypsin
1YPH

P4 Collagen trimer
2KLW

P5 Cytochrome b562
256B

P6 Haem, heme in situ
2UYD

P7 Haemoglobin, hemoglobin, Hb
3HRW

P Proteins continued

P8 Retinol binding protein
1CRB

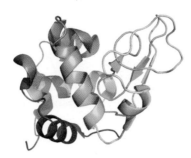

P9 Lysozyme
3A3Q

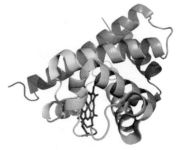

P10 Myoglobin, Mb
3HRW

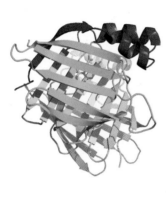

P11 Retinol binding protein
3FA6

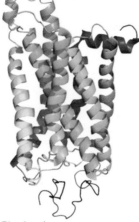

P12 Rhodopsin
3C9M

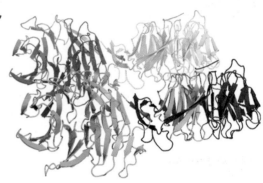

P13 Transducin
2CE9

R Porphyrin-based ring complexes

R1 Porphine
$C_{20}H_{14}N_4$

R2 Haem, heme

Chlorophyll b

X:

R3 Chlorophyll a [inset: b]

S Saccharides

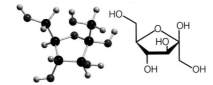 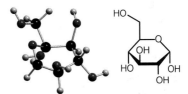

S1 Ribose, β-D-ribofuranose
$C_5H_{10}O_5$

S2 Deoxyribose, β-D-2-deoxyribose
$C_5H_{10}O_4$

S3 Fructose (fructofuranose)
$C_6H_{12}O_6$

S4 Glucose, α-D-glucose
$C_6H_{12}O_6$

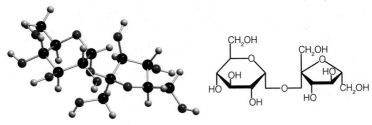

S6 Sucrose, β-D-fructofuranosyl-α-D-glucopyranoside
$C_{12}H_{22}O_{11}$

T Nucleic acids

T1 t-RNA
2OUE

Resource section 2: Units

Table 1 The SI base units

Physical quantity	Symbol for quantity	Base unit
Length	l	meter, m
Mass	m	kilogram, kg
Time	t	second, s
Electric current	I	ampere, A
Thermodynamic temperature	T	kelvin, K
Amount of substance	n	mole, mol
Luminous intensity	I_v	candela, cd

Table 2 A selection of derived units

Physical quantity	Derived unit*	Name of derived unit
Force	$1 \, kg \, m \, s^{-2}$	newton, N
Pressure	$1 \, kg \, m^{-1} \, s^{-2}$	pascal, Pa
	$1 \, N \, m^{-2}$	
Energy	$1 \, kg \, m^2 \, s^{-2}$	joule, J
	$1 \, N \, m$	
	$1 \, Pa \, m^3$	
Power	$1 \, kg \, m^2 \, s^{-3}$	watt, W
	$1 \, J \, s^{-1}$	

*Equivalent definitions in terms of derived units are given following the definition in terms of base units.

Table 3 Common SI prefixes

Prefix	z	a	f	p	n	μ	m	c	d	k	M	G	T
Name	zepto	atto	femto	pico	nano	micro	milli	centi	deci	kilo	mega	giga	tera
Factor	10^{-21}	10^{-18}	10^{-15}	10^{-12}	10^{-9}	10^{-6}	10^{-3}	10^{-2}	10^{-1}	10^3	10^6	10^9	10^{12}

Table 4 Some common units

Physical quantity	Name of unit	Symbol for unit	Value*
Time	minute	min	60 s
	hour	h	3600 s
Length	ångström	Å	10^{-10} m
Volume	liter	L, l	1 dm^3
Mass	tonne	t	10^3 kg
Pressure	bar	bar	10^5 Pa
	atmosphere	atm	101.325 kPa
Energy	electronvolt	eV	$1.602\ 176 \times 10^{-19}$ J
			96.485 31 kJ mol^{-1}

*All values in the final column are exact, except for the definition of 1 eV.

Resource section 3: Data

Table 1 Thermodynamic data for organic compounds (all values relate to 298.15 K)

	$M/\mathrm{g\,mol^{-1}}$	$\Delta_f H^\circ/\ \mathrm{kJ\,mol^{-1}}$	$\Delta_f G^\circ/\ \mathrm{kJ\,mol^{-1}}$	$S_m^\circ/\ \mathrm{J\,K^{-1}\,mol^{-1}}$	$C_{p,m}^\circ/\ \mathrm{J\,K^{-1}\,mol^{-1}}$	$\Delta_c H^\circ/\ \mathrm{kJ\,mol^{-1}}$
C(s) (graphite)	12.011	0	0	5.740	8.527	−393.51
C(s) (diamond)	12.011	+1.895	+2.900	+2.377	6.113	−395.40
$CO_2(g)$	44.010	−393.51	−394.36	213.74	37.11	
Hydrocarbons						
$CH_4(g)$, methane	16.04	−74.81	−50.72	186.26	35.31	−890
$CH_3(g)$, methyl	15.04	+145.69	+147.92	194.2	38.70	
$C_2H_2(g)$, ethyne	26.04	+226.73	+209.20	200.94	43.93	−1300
$C_2H_4(g)$, ethene	28.05	+52.26	+68.15	219.56	43.56	−1411
$C_2H_6(g)$, ethane	30.07	−84.68	−32.82	229.60	52.63	−1560
$C_3H_6(g)$, propene	42.08	+20.42	+62.78	267.05	63.89	−2058
$C_3H_6(g)$, cyclopropane	42.08	−103.85	−23.49	269.91	73.5	−2220
$C_4H_8(g)$, 1-butene	56.11	−0.13	+71.39	305.71	85.65	−2717
$C_4H_8(g)$, *cis*-2-butene	56.11	−6.99	+65.95	300.94	78.91	−2710
$C_4H_8(g)$, *trans*-2-butene	56.11	−11.17	+63.06	296.59	87.82	−2707
$C_4H_{10}(g)$, butane	58.13	−126.15	−17.03	310.23	97.45	−2878
$C_5H_{12}(g)$, pentane	72.15	−146.44	−8.20	348.40	120.2	−3537
$C_5H_{12}(l)$	72.15	−173.1				
$C_6H_6(l)$, benzene	78.12	+49.0	+124.3	173.3	136.1	−3268
$C_6H_6(g)$	78.12	+82.93	+129.72	269.31	81.67	−3320
$C_6H_{12}(l)$, cyclohexane	84.16	−156	26.8		156.5	−3902
$C_6H_{14}(l)$, hexane	86.18	−198.7		204.3		−4163
$C_6H_5CH_3(g)$, methylbenzene (toluene)	92.14	+50.0	+122.0	320.7	103.6	−3953
$C_7H_{16}(l)$, heptane	100.21	−224.4	+1.0	328.6	224.3	
$C_8H_{18}(l)$, octane	114.23	−249.9	+6.4	361.1		−5471
$C_8H_{18}(l)$, iso-octane	114.23	−255.1				−5461
$C_{10}H_8(s)$, naphthalene	128.18	+78.53				−5157
Alcohols and phenols						
$CH_3OH(l)$, methanol	32.04	−238.86	−166.27	126.8	81.6	−726
$CH_3OH(g)$	32.04	−200.66	−166.27	239.81	43.89	−764
$C_2H_5OH(l)$, ethanol	46.07	−277.69	−174.78	160.7	111.46	−1368
$C_2H_5OH(g)$	46.07	−235.10	−168.49	282.70	65.44	−1409
$C_6H_5OH(s)$, phenol	94.12	−165.0	−50.9	146.0		−3054

	M/g mol^{-1}	$\Delta_f H^\circ$/ kJ mol^{-1}	$\Delta_f G^\circ$/ kJ mol^{-1}	S_m°/ J K^{-1} mol^{-1}	$C_{p,m}^\circ$/ J K^{-1} mol^{-1}	$\Delta_c H^\circ$/ kJ mol^{-1}
Carboxylic acids, hydroxy acids, and esters						
HCOOH(l), formic	46.03	−424.72	−361.35	128.95	99.04	−255
CH$_3$COOH(l), ethanoic	60.05	−484.3	−389.9	159.8	124.3	−875
CH$_3$COOH(aq)	60.05	−485.76	−396.46	178.7		
CH$_3$CO$_2$(aq)	59.05	−486.01	−369.31	86.6	−6.3	
CH$_3$(CO)COOH(l), pyruvic	88.06					−950
CH$_3$(CH$_2$)$_2$COOH(l), butanoic	88.10	−533.8				
CH$_3$COOC$_2$H$_5$(l), ethyl acetate	88.10	−479.0	−332.7	259.4	170.1	−2231
(COOH)$_2$(s), oxalic	90.04	−827.2			117	−254
CH$_3$CH(OH)COOH(s), lactic	90.08	−694.0	−522.9			−1344
HOOCCH$_2$CH$_2$COOH(s), succinic	118.09	−940.5	−747.4	153.1	167.3	
C$_6$H$_5$COOH(s), benzoic	122.13	−385.1	−245.3	167.6	146.8	−3227
CH$_3$(CH$_2$)$_8$COOH(s), decanoic	172.27	−713.7				
C$_6$H$_8$O$_6$(s), ascorbic	176.12	−1164.6				
HOOCCH$_2$C(OH)(COOH) CH$_2$COOH(s), citric	192.12	−1543.8	−1236.4			−1985
CH$_3$(CH$_2$)$_{10}$COOH(s), dodecanoic	200.32	−774.6			404.3	
CH$_3$(CH$_2$)$_{14}$COOH(s), hexadecanoic	256.41	−891.5				
C$_{18}$H$_{36}$O$_2$(s), stearic	284.48	−947.7			501.5	
Alkanals and alkanones						
HCHO(g), methanal	30.03	−108.57	−102.53	218.77	35.40	−571
CH$_3$CHO(l), ethanal	44.05	−192.30	−128.12	160.2		−1166
CH$_3$CHO(g)	44.05	−166.19	−128.86	250.3	57.3	−1192
CH$_3$COCH$_3$(l), propanone	58.08	−248.1	−155.4	200.4	124.7	−1790
Sugars						
C$_5$H$_{10}$O$_5$(s), D-ribose	150.1	−1051.1				
C$_5$H$_{10}$O$_5$(s), D-xylose	150.1	−1057.8				
C$_6$H$_{12}$O$_6$(s), α-D-glucose	180.16	−1273.3	−917.2	212.1		−2808
C$_6$H$_{12}$O$_6$(s), β-D-glucose	180.16	−1268				
C$_6$H$_{12}$O$_6$(s), β-D-fructose	180.16	−1265.6				−2810
C$_6$H$_{12}$O$_6$(s), α-D-galactose	180.16	−1286.3	−918.8	205.4		
C$_{12}$H$_{22}$O$_{11}$(s), sucrose	342.30	−2226.1	−1543	360.2		−5645
C$_{12}$H$_{22}$O$_{11}$(s), lactose	342.30	−2236.7	−1567	386.2		
Amino acids[1]						
L-Glycine						
solid	75.07	−528.5	−373.4	103.5	99.2	−969
aqueous solution	75.07	−469.8	−315.0	111.0		
L-Alanine	89.09	−604.0	−369.9	129.2	122.2	−1618
L-Serine	105.09	−732.7	−508.8	149.2	135.6	−1455
L-Proline	115.13	−515.2		164.0	151.2	
L-Valine	117.15	−617.9	−359.0	178.9	168.8	−2922
L-Threonine	119.12	−807.2	−550.2	152.7	147.3	−2053
L-Cysteine	121.16	−534.1	−340.1	169.9	162.3	−1651

	$M/\text{g mol}^{-1}$	$\Delta_f H^{\ominus}/$ kJ mol^{-1}	$\Delta_f G^{\ominus}/$ kJ mol^{-1}	$S_m^{\ominus}/$ $\text{J K}^{-1}\text{mol}^{-1}$	$C_{p,m}^{\ominus}/$ $\text{J K}^{-1}\text{mol}^{-1}$	$\Delta_c H^{\ominus}/$ kJ mol^{-1}
L-Leucine	131.17	−637.4	−347.7	211.8	200.1	−3582
L-Isoleucine	131.17	−637.8	−347.3	208.0	188.3	−3581
L-Asparagine	132.12	−789.4	−530.1	174.5	160.2	−530
L-Aspartic acid	133.10	−973.3	−730.1	170.1	155.2	−1601
L-Glutamine	146.15	−826.4	−532.6	195.0	184.2	−2570
L-Glutamic acid	147.13	−1009.7	−731.4	188.2	175.0	−2244
L-Methionine	149.21	−577.5	−505.8	231.5	290.0	−2782
L-Histidine	155.16	−466.7				
L-Phenylalanine	165.19	−466.9	−211.7	213.6	203.0	−4647
L-Tyrosine	181.19	−685.1	−385.8	214.0	216.4	−4442
L-Tryptophan	204.23	−415.3	−119.2	251.0	238.1	−5628
L-Cystine	240.32	−1032.7	−685.8	280.6	261.9	−3032
Peptides						
$NH_2CH_2CONHCH_2COOH(s)$, glycylglycine	132.12	747.7	487.9	180.3	164.0	1972
$NH_2CH(CH_3)CONHCH_2COOH$, alanylglycine	146.15		489.9	213.4	182.4	2619
Other nitrogen compounds						
$CH_3NH_2(g)$, methylamine	31.06	−22.97	+32.16	243.41	53.1	−1085
$(NH_2)_2CO(s)$, urea	60.06	−333.1	−197.33	104.60	93.14	−632
$C_6H_5NH_2(l)$, aniline	93.13	+31.1				−3393
$C_4H_5N_3O(s)$, cytosine	111.10	−221.3			132.6	
$C_4H_4N_2O_2(s)$, uracil	112.09	−429.4				
$C_5H_6N_2O_2(s)$, thymine	126.11	−462.8			150.8	
$C_5H_5N_5(s)$, adenine	135.14	+96.9	+299.6	151.1	147.0	
$C_5H_5N_5O(s)$, guanine	151.13	−183.9	+47.4	160.3		

[1]See the *Atlas of structures*, Section A, for the molecular structures of the amino acids. Unless otherwise noted, data relate to the substance in the solid state.

Table 2 Thermodynamic data (all values relate to 298.15 K)*

	$M/(\text{g mol}^{-1})$	$\Delta_f H^{\ominus}/(\text{kJ mol}^{-1})$	$\Delta_f G^{\ominus}/(\text{kJ mol}^{-1})$	$S_m^{\ominus}/(\text{J K}^{-1}\text{mol}^{-1})$	$C_{p,m}^{\ominus}/(\text{J K}^{-1}\text{mol}^{-1})$
Aluminum					
Al(s)	26.98	0	0	28.33	24.35
Al(l)	26.98	+10.56	+7.20	39.55	24.21
Al(g)	26.98	+326.4	+285.7	164.54	21.38
$Al^{3+}(g)$	26.98	+5483.17			
$Al^{3+}(aq)$	26.98	−531	−485	−321.7	
$Al_2O_3(s,)$	101.96	−1675.7	−1582.3	50.92	79.04
$AlCl_3(s)$	133.24	−704.2	−628.8	110.67	91.84
Argon					
Ar(g)	39.95	0	0	154.84	20.786

	$M/(\text{g mol}^{-1})$	$\Delta_f H^\circ/(\text{kJ mol}^{-1})$	$\Delta_f G^\circ/(\text{kJ mol}^{-1})$	$S_m^\circ/(\text{J K}^{-1}\,\text{mol}^{-1})$	$C_{p,m}^\circ/(\text{J K}^{-1}\,\text{mol}^{-1})$
Antimony					
Sb(s)	121.75	0	0	45.69	25.23
SbH$_3$(g)	153.24	+145.11	+147.75	232.78	41.05
Arsenic					
As(s,α)	74.92	0	0	35.1	24.64
As(g)	74.92	+302.5	+261.0	174.21	20.79
As$_4$(g)	299.69	+143.9	+92.4	314	
AsH$_3$(g)	77.95	+66.44	+68.93	222.78	38.07
Barium					
Ba(s)	137.34	0	0	62.8	28.07
Ba(g)	137.34	+180	+146	170.24	20.79
Ba^{2+}(aq)	137.34	−537.64	−560.77	+9.6	
BaO(s)	153.34	−553.5	−525.1	70.43	47.78
BaCl$_2$(s)	208.25	−858.6	−810.4	123.68	75.14
Beryllium					
Be(s)	9.01	0	0	9.50	16.44
Be(g)	9.01	+324.3	+286.6	136.27	20.79
Bismuth					
Bi(s)	208.98	0	0	56.74	25.52
Bi(g)	208.98	+207.1	+168.2	187.00	20.79
Bromine					
Br$_2$(l)	159.82	0	0	152.23	75.689
Br$_2$(g)	159.82	+30.907	+3.110	245.46	36.02
Br(g)	79.91	+111.88	+82.396	175.02	20.786
Br$^-$(g)	79.91	−219.07			
Br$^-$(aq)	79.91	−121.55	−103.96	+82.4	−141.8
HBr(g)	90.92	−36.40	−53.45	198.70	29.142
Cadmium					
Cd(s,γ)	112.40	0	0	51.76	25.98
Cd(g)	112.40	+112.01	+77.41	167.75	20.79
Cd^{2+}(aq)	112.40	−75.90	−77.612	−73.2	
CdO(s)	128.40	−258.2	−228.4	54.8	43.43
CdCO$_3$(s)	172.41	−750.6	−669.4	92.5	
Caesium: see cesium					
Calcium					
Ca(s)	40.08	0	0	41.42	25.31
Ca(g)	40.08	+178.2	+144.3	154.88	20.786
Ca^{2+}(aq)	40.08	−542.83	−553.58	−53.1	
CaO(s)	56.08	−635.09	−604.03	39.75	42.80
CaCO$_3$(s) (calcite)	100.09	−1206.9	−1128.8	92.9	81.88

	$M/(\text{g mol}^{-1})$	$\Delta_f H^\circ/(\text{kJ mol}^{-1})$	$\Delta_f G^\circ/(\text{kJ mol}^{-1})$	$S_m^\circ/(\text{J K}^{-1}\text{ mol}^{-1})$	$C_{p,m}^\circ/(\text{J K}^{-1}\text{ mol}^{-1})$
$CaCO_3(s)$ (aragonite)	100.09	−1207.1	−1127.8	88.7	81.25
$CaF_2(s)$	78.08	1219.6	−1167.3	68.87	67.03
$CaCl_2(s)$	110.99	−795.8	−748.1	104.6	72.59
$CaBr_2(s)$	199.90	−682.8	−663.6	130	
Carbon (for 'organic' compounds, see Table 1)					
$C(s)$ (graphite)	12.011	0	0	5.740	8.527
$C(s)$ (diamond)	12.011	+1.895	+2.900	2.377	6.133
$C(g)$	12.011	+716.68	+671.26	158.10	20.838
$C_2(g)$	24.022	+831.90	+775.89	199.42	43.21
$CO(g)$	28.011	−110.53	−137.17	197.67	29.14
$CO_2(g)$	44.010	−393.51	−394.36	213.74	37.11
$CO_2(aq)$	44.010	−413.80	−385.98	117.6	
$H_2CO_3(aq)$	62.03	−699.65	−623.08	187.4	
$HCO_3^-(aq)$	61.02	−691.99	−586.77	+91.2	
$CO_3^{2-}(aq)$	60.01	−677.14	−527.81	−56.9	
$CCl_4(l)$	153.82	−135.44	−65.21	216.40	131.75
$CS_2(l)$	76.14	+89.70	+65.27	151.34	75.7
$HCN(g)$	27.03	+135.1	+124.7	201.78	35.86
$HCN(l)$	27.03	+108.87	+124.97	112.84	70.63
$CN^-(aq)$	26.02	+150.6	+172.4	+94.1	
Cesium					
$Cs(s)$	132.91	0	0	85.23	32.17
$Cs(g)$	132.91	+76.06	+49.12	175.60	20.79
$Cs^+(aq)$	132.91	−258.28	−292.02	+133.05	−10.5
Chlorine					
$Cl_2(g)$	70.91	0	0	223.07	33.91
$Cl(g)$	35.45	+121.68	+105.68	165.20	21.840
$Cl^-(g)$	35.45	−233.13			
$Cl^-(aq)$	35.45	−167.16	−131.23	+56.5	−136.4
$HCl(g)$	36.46	−92.31	−95.30	186.91	29.12
$HCl(aq)$	36.46	−167.16	−131.23	56.5	−136.4
Chromium					
$Cr(s)$	52.00	0	0	23.77	23.35
$Cr(g)$	52.00	+396.6	+351.8	174.50	20.79
$CrO_4^{2-}(aq)$	115.99	−881.15	−727.75	+50.21	
$Cr_2O_7^{2-}(aq)$	215.99	−1490.3	−1301.1	+261.9	
Copper					
$Cu(s)$	63.54	0	0	33.150	24.44
$Cu(g)$	63.54	+338.32	+298.58	166.38	20.79
$Cu^+(aq)$	63.54	+71.67	+49.98	+40.6	
$Cu^{2+}(aq)$	63.54	+64.77	+65.49	−99.6	
$Cu_2O(s)$	143.08	−168.6	−146.0	93.14	63.64

	$M/(\text{g mol}^{-1})$	$\Delta_f H^\circ/(\text{kJ mol}^{-1})$	$\Delta_f G^\circ/(\text{kJ mol}^{-1})$	$S_m^\circ/(\text{J K}^{-1}\text{ mol}^{-1})$	$C_{p,m}^\circ/(\text{J K}^{-1}\text{ mol}^{-1})$
CuO(s)	79.54	−157.3	−129.7	42.63	42.30
CuSO$_4$(s)	159.60	−771.36	−661.8	109	100.0
CuSO$_4$·H$_2$O(s)	177.62	−1085.8	−918.11	146.0	134
CuSO$_4$·5H$_2$O(s)	249.68	−2279.7	−1879.7	300.4	280
Deuterium					
D$_2$(g)	4.028	0	0	144.96	29.20
HD(g)	3.022	+0.318	−1.464	143.80	29.196
D$_2$O(g)	20.028	−249.20	−234.54	198.34	34.27
D$_2$O(l)	20.028	−294.60	−243.44	75.94	84.35
HDO(g)	19.022	−245.30	−233.11	199.51	33.81
HDO(l)	19.022	−289.89	−241.86	79.29	
Fluorine					
F$_2$(g)	38.00	0	0	202.78	31.30
F(g)	19.00	+78.99	+61.91	158.75	22.74
F$^-$(aq)	19.00	−332.63	−278.79	−13.8	−106.7
HF(g)	20.01	−271.1	−273.2	173.78	29.13
Gold					
Au(s)	196.97	0	0	47.40	25.42
Au(g)	196.97	+366.1	+326.3	180.50	20.79
Helium					
He(g)	4.003	0	0	126.15	20.786
Hydrogen (see also deuterium)					
H$_2$(g)	2.016	0	0	130.684	28.824
H(g)	1.008	+217.97	+203.25	114.71	20.784
H$^+$(aq)	1.008	0	0	0	0
H$_2$O(l)	18.015	−285.83	−237.13	69.91	75.291
H$_2$O(g)	18.015	−241.82	−228.57	188.83	33.58
H$_2$O$_2$(l)	34.015	−187.78	−120.35	109.6	89.1
Iodine					
I$_2$(s)	253.81	0	0	116.135	54.44
I$_2$(g)	253.81	+62.44	+19.33	260.69	36.90
I(g)	126.90	+106.84	+70.25	180.79	20.786
I$^-$(aq)	126.90	−55.19	−51.57	+111.3	−142.3
HI(g)	127.91	+26.48	+1.70	206.59	29.158
Iron					
Fe(s)	55.85	0	0	27.28	25.10
Fe(g)	55.85	+416.3	+370.7	180.49	25.68
Fe^{2+}(aq)	55.85	−89.1	−78.90	−137.7	
Fe^{3+}(aq)	55.85	−48.5	−4.7	−315.9	
Fe$_3$O$_4$(s) (magnetite)	231.54	−1184.4	−1015.4	146.4	143.43
Fe$_2$O$_3$(s) (hematite)	159.69	−824.2	−742.2	87.40	103.85

	$M/(\text{g mol}^{-1})$	$\Delta_f H^{\circ}/(\text{kJ mol}^{-1})$	$\Delta_f G^{\circ}/(\text{kJ mol}^{-1})$	$S_m^{\circ}/(\text{J K}^{-1}\,\text{mol}^{-1})$	$C_{p,m}^{\circ}/(\text{J K}^{-1}\,\text{mol}^{-1})$
FeS(s,α)	87.91	−100.0	−100.4	60.29	50.54
FeS_2(s)	119.98	−178.2	−166.9	52.93	62.17
Krypton					
Kr(g)	83.80	0	0	164.08	20.786
Lead					
Pb(s)	207.19	0	0	64.81	26.44
Pb(g)	207.19	+195.0	+161.9	175.37	20.79
Pb^{2+}(aq)	207.19	−1.7	−24.43	+10.5	
PbO(s, yellow)	223.19	−217.32	−187.89	68.70	45.77
PbO(s, red)	223.19	−218.99	−188.93	66.5	45.81
PbO_2(s)	239.19	−277.4	−217.33	68.6	64.64
Lithium					
Li(s)	6.94	0	0	29.12	24.77
Li(g)	6.94	+159.37	+126.66	138.77	20.79
Li^{+}(aq)	6.94	−278.49	−293.31	+13.4	+68.6
Magnesium					
Mg(s)	24.31	0	0	32.68	24.89
Mg(g)	24.31	+147.70	+113.10	148.65	20.786
Mg^{2+}(aq)	24.31	−466.85	−454.8	−138.1	
MgO(s)	40.31	−601.70	−569.43	26.94	37.15
$MgCO_3$(s)	84.32	−1095.8	−1012.1	65.7	75.52
$MgCl_2$(s)	95.22	−641.32	−591.79	89.62	71.38
$MgBr_2$(s)	184.13	−524.3	−503.8	117.2	
Mercury					
Hg(l)	200.59	0	0	76.02	27.983
Hg(g)	200.59	+61.32	+31.82	174.96	20.786
Hg^{2+}(aq)	200.59	+171.1	+164.40	−32.2	
Hg_2^{2+}(aq)	401.18	+172.4	+153.52	+84.5	
HgO(s)	216.59	−90.83	−58.54	70.29	44.06
Hg_2Cl_2(s)	472.09	−265.22	−210.75	192.5	102
$HgCl_2$(s)	271.50	−224.3	−178.6	146.0	
HgS(s, black)	232.65	−53.6	−47.7	88.3	
Neon					
Ne(g)	20.18	0	0	146.33	20.786
Nitrogen					
N_2(g)	28.013	0	0	191.61	29.125
N(g)	14.007	+472.70	+455.56	153.30	20.786
NO(g)	30.01	+90.25	+86.55	210.76	29.844
N_2O(g)	44.01	+82.05	+104.20	219.85	38.45
NO_2(g)	46.01	+33.18	+51.31	240.06	37.20
N_2O_4(g)	92.01	+9.16	+97.89	304.29	77.28

	$M/(\text{g mol}^{-1})$	$\Delta_f H^\ominus/(\text{kJ mol}^{-1})$	$\Delta_f G^\ominus/(\text{kJ mol}^{-1})$	$S_m^\ominus/(\text{J K}^{-1}\,\text{mol}^{-1})$	$C_{p,m}^\ominus/(\text{J K}^{-1}\,\text{mol}^{-1})$
$N_2O_5(s)$	108.01	−43.1	+113.9	178.2	143.1
$N_2O_5(g)$	108.01	+11.3	+115.1	355.7	84.5
$HNO_3(l)$	63.01	−174.10	−80.71	155.60	109.87
$HNO_3(aq)$	63.01	−207.36	−111.25	146.4	−86.6
$NO_3^-(aq)$	62.01	−205.0	−108.74	+146.4	−86.6
$NH_3(g)$	17.03	−46.11	−16.45	192.45	35.06
$NH_3(aq)$	17.03	−80.29	−26.50	113.3	
$NH_4^+(aq)$	18.04	−132.51	−79.31	+113.4	+79.9
$NH_2OH(s)$	33.03	−114.2			
$HN_3(l)$	43.03	+264.0	+327.3	140.6	
$NH_3(g)$	43.03	+294.1	+328.1	238.97	43.68
$N_2H_4(l)$	32.05	+50.63	+149.43	121.21	98.87
$NH_4NO_3(s)$	80.04	−365.56	−183.87	151.08	139.3
$NH_4Cl(s)$	53.49	−314.43	−202.87	94.6	84.1
Oxygen					
$O_2(g)$	31.999	0	0	205.138	29.355
$O(g)$	15.999	+249.17	+231.73	161.06	21.912
$O_3(g)$	47.998	+142.7	+163.2	238.93	39.20
$OH^-(aq)$	17.007	−229.99	−157.24	−10.75	−148.5
Phosphorus					
$P(s, wh)$	30.97	0	0	41.09	23.840
$P(g)$	30.97	+314.64	+278.25	163.19	20.786
$P_2(g)$	61.95	+144.3	+103.7	218.13	32.05
$P_4(g)$	123.90	+58.91	+24.44	279.98	67.15
$PH_3(g)$	34.00	+5.4	+13.4	210.23	37.11
$PCl_3(g)$	137.33	−287.0	−267.8	311.78	71.84
$PCl_3(l)$	137.33	−319.7	−272.3	217.1	
$PCl_5(g)$	208.24	−374.9	−305.0	364.6	112.8
$PCl_5(s)$	208.24	−443.5			
$H_3PO_3(s)$	82.00	−964.4			
$H_3PO_3(aq)$	82.00	−964.8			
$H_3PO_4(s)$	94.97	−1279.0	−1119.1	110.50	106.06
$H_3PO_4(l)$	94.97	−1266.9			
$H_3PO_4(aq)$	94.97	−1277.4	−1018.7	−222	
$PO_4^{3-}(aq)$	94.97	−1277.4	−1018.7	−222	
$P_4O_{10}(s)$	283.89	−2984.0	−2697.0	228.86	211.71
$P_4O_6(s)$	219.89	−1640.1			
Potassium					
$K(s)$	39.10	0	0	64.18	29.58
$K(g)$	39.10	+89.24	+60.59	160.336	20.786
$K^+(g)$	39.10	+514.26			
$K^+(aq)$	39.10	−252.38	−283.27	+102.5	+21.8

	$M/(\text{g mol}^{-1})$	$\Delta_f H^{\ominus}/(\text{kJ mol}^{-1})$	$\Delta_f G^{\ominus}/(\text{kJ mol}^{-1})$	$S_m^{\ominus}/(\text{J K}^{-1}\text{ mol}^{-1})$	$C_{p,m}^{\ominus}/(\text{J K}^{-1}\text{ mol}^{-1})$
KOH(s)	56.11	−424.76	−379.08	78.9	64.9
KF(s)	58.10	−576.27	−537.75	66.57	49.04
KCl(s)	74.56	−436.75	−409.14	82.59	51.30
KBr(s)	119.01	−393.80	−380.66	95.90	52.30
KI(s)	166.01	−327.90	−324.89	106.32	52.93
Silicon					
Si(s)	28.09	0	0	18.83	20.00
Si(g)	28.09	+455.6	+411.3	167.97	22.25
SiO$_2$(s,α)	60.09	−910.93	−856.64	41.84	44.43
Silver					
Ag(s)	107.87	0	0	42.55	25.351
Ag(g)	107.87	+284.55	+245.65	173.00	20.79
Ag$^+$(aq)	107.87	+105.58	+77.11	+72.68	+21.8
AgBr(s)	187.78	−100.37	−96.90	107.1	52.38
AgCl(s)	143.32	−127.07	−109.79	96.2	50.79
Ag$_2$O(s)	231.74	−31.05	−11.20	121.3	65.86
AgNO$_3$(s)	169.88	−124.39	−33.41	140.92	93.05
Sodium					
Na(s)	22.99	0	0	51.21	28.24
Na(g)	22.99	+107.32	+76.76	153.71	20.79
Na$^+$(aq)	22.99	−240.12	−261.91	+59.0	+46.4
NaOH(s)	40.00	−425.61	−379.49	64.46	59.54
NaCl(s)	58.44	−411.15	−384.14	72.13	50.50
NaBr(s)	102.90	−361.06	−348.98	86.82	51.38
NaI(s)	149.89	−287.78	−286.06	98.53	52.09
Sulfur					
S(s,α) (rhombic)	32.06	0	0	31.80	22.64
S(s,β) (monoclinic)	32.06	+0.33	+0.1	32.6	23.6
S(g)	32.06	+278.81	+238.25	167.82	23.673
S$_2$(g)	64.13	+128.37	+79.30	228.18	32.47
S^{2-}(aq)	32.06	+33.1	+85.8	−14.6	
SO$_2$(g)	64.06	−296.83	−300.19	248.22	39.87
SO$_3$(g)	80.06	−395.72	−371.06	256.76	50.67
H$_2$SO$_4$(l)	98.08	−813.99	−690.00	156.90	138.9
H$_2$SO$_4$(aq)	98.08	−909.27	−744.53	20.1	−293
SO$_4^{2-}$(aq)	96.06	−909.27	−744.53	+20.1	−293
HSO$_4^-$(aq)	97.07	−887.34	−755.91	+131.8	−84
H$_2$S(g)	34.08	−20.63	−33.56	205.79	34.23
H$_2$S(aq)	34.08	−39.7	−27.83	121	
HS$^-$(aq)	33.072	−17.6	+12.08	+62.08	
SF$_6$(g)	146.05	−1209	−1105.3	291.82	97.28

	$M/(\text{g mol}^{-1})$	$\Delta_f H^{\ominus}/(\text{kJ mol}^{-1})$	$\Delta_f G^{\ominus}/(\text{kJ mol}^{-1})$	$S_m^{\ominus}/(\text{J K}^{-1}\,\text{mol}^{-1})$	$C_{p,m}^{\ominus}/(\text{J K}^{-1}\,\text{mol}^{-1})$
Tin					
$Sn(s,\beta)$	118.69	0	0	51.55	26.99
$Sn(g)$	118.69	+302.1	+267.3	168.49	20.26
$Sn^{2+}(aq)$	118.69	−8.8	−27.2	−17	
$SnO(s)$	134.69	−285.8	−256.8	56.5	44.31
$SnO_2(s)$	150.69	−580.7	+519.6	52.3	52.59
Xenon					
$Xe(g)$	131.30	0	0	169.68	20.786
Zinc					
$Zn(s)$	65.37	0	0	41.63	25.40
$Zn(g)$	65.37	+130.73	+95.14	160.98	20.79
$Zn^{2+}(aq)$	65.37	−153.89	−147.06	−112.1	+46
$ZnO(s)$	81.37	−348.28	−318.30	43.64	40.25

*Entropies and heat capacities of ions are relative to $H^+(aq)$ and are given with a sign.

Table 3a Standard potentials at 298.15 K in electrochemical order

Reduction half-reaction	$E^{\ominus}/V$	Reduction half-reaction	$E^{\ominus}/V$
Strongly oxidizing		$Cu^{2+} + e^- \rightarrow Cu^+$	+0.16
$H_4XeO_6 + 2\,H^+ + 2\,e^- \rightarrow XeO_3 + 3\,H_2O$	+3.0	$Sn^{4+} + 2\,e^- \rightarrow Sn^{2+}$	+0.15
$F_2 + 2\,e^- \rightarrow 2\,F^-$	+2.87	$AgBr + e^- \rightarrow Ag + Br^-$	+0.07
$O_3 + 2\,H^+ + 2\,e^- \rightarrow O_2 + H_2O$	+2.07	$Ti^{4+} + e^- \rightarrow Ti^{3+}$	0.00
$S_2O_8^{2-} + 2\,e^- \rightarrow 2\,SO_4^{2-}$	+2.05	$2\,H^+ + 2\,e^- \rightarrow H$	0, by definition
$Ag^{2+} + e^- \rightarrow Ag^+$	+1.98	$Fe^{3+} + 3\,e^- \rightarrow Fe$	−0.04
$Co^{3+} + e^- \rightarrow Co^{2+}$	+1.81	$O_2 + H_2O + 2\,e^- \rightarrow HO_2^- + OH^-$	−0.08
$HO_2 + 2\,H^+ + 2\,e^- \rightarrow 2\,H_2O$	+1.78	$Pb^{2+} + 2\,e^- \rightarrow Pb$	−0.13
$Au^+ + e^- \rightarrow Au$	+1.69	$In^+ + e^- \rightarrow In$	−0.14
$Pb^{4+} + 2\,e^- \rightarrow Pb^{2+}$	+1.67	$Sn^{2+} + 2\,e^- \rightarrow Sn$	−0.14
$2\,HClO + 2\,H^+ + 2\,e^- \rightarrow Cl_2 + 2\,H_2O$	+1.63	$AgI + e^- \rightarrow Ag + I$	−0.15
$Ce^{4+} + e^- \rightarrow Ce^{3+}$	+1.61	$Ni^{2+} + 2\,e^- \rightarrow Ni$	−0.23
$2\,HBrO + 2\,H^+ + 2\,e^- \rightarrow Br_2 + 2\,H$	+1.60	$Co^{2+} + 2\,e^- \rightarrow Co$	−0.28
$MnO_4^- + 8\,H^+ + 5\,e^- \rightarrow Mn^{2+} + 4\,H_2O$	+1.51	$In^{3+} + 3\,e^- \rightarrow In$	−0.34
$Mn^{3+} + e^- \rightarrow Mn^{2+}$	+1.51	$Tl^+ + e^- \rightarrow Tl$	−0.34
$Au^{3+} + 3\,e^- \rightarrow Au$	+1.40	$PbSO_4 + 2\,e^- \rightarrow Pb + SO_4^{2-}$	−0.36
$Cl_2 + 2\,e^- \rightarrow 2\,Cl^-$	+1.36	$Ti^{3+} + e^- \rightarrow Ti^{2+}$	−0.37
$Cr_2O_7^{2-} + 14\,H^+ + 6\,e^- \rightarrow 2\,Cr^{3+} + 7\,H_2O$	+1.33	$Cd^{2+} + 2\,e^- \rightarrow Cd$	−0.40
$O_3 + H_2O + 2\,e^- \rightarrow O_2 + 2\,OH^-$	+1.24	$In^{2+} + e^- \rightarrow In^+$	−0.40
$O_2 + 4\,H^+ + 4\,e^- \rightarrow 2\,H_2O$	+1.23	$Cr^{3+} + e^- \rightarrow Cr^{2+}$	−0.41
$ClO_4^- + 2\,H^+ + 2\,e^- \rightarrow ClO_3^- + H_2O$	+1.23	$Fe^{2+} + 2\,e^- \rightarrow Fe$	−0.44
$MnO_2 + 4\,H^+ + 2\,e^- \rightarrow Mn^{2+} + 2\,H_2O$	+1.23	$In^{3+} + 2\,e^- \rightarrow In^+$	−0.44

Reduction half-reaction	E°/V	Reduction half-reaction	E°/V
$Br_2 + 2\,e^- \rightarrow 2\,Br^-$	+1.09	$S + 2\,e^- \rightarrow S^{2-}$	−0.48
$Pu^{4+} + e^- \rightarrow Pu^{3+}$	+0.97	$In^{3+} + e^- \rightarrow In^{2+}$	−0.49
$NO_3^- + 4\,H^+ + 3\,e^- \rightarrow NO + 2\,H_2O$	+0.96	$U^{4+} + e^- \rightarrow U^{3+}$	−0.61
$2\,Hg^{2+} + 2\,e^- \rightarrow Hg_2^{2+}$	+0.92	$Cr^{3+} + 3\,e^- \rightarrow Cr$	−0.74
$ClO^- + H_2O + 2\,e^- \rightarrow Cl^- + 2\,OH^-$	+0.89	$Zn^{2+} + 2\,e^- \rightarrow Zn$	−0.76
$Hg^{2+} + 2\,e^- \rightarrow Hg$	+0.86	$Cd(OH)_2 + 2\,e^- \rightarrow Cd + 2\,OH^-$	−0.81
$NO_3^- + 2\,H^+ + e^- \rightarrow NO_2 + H_2O$	+0.80	$2\,H_2O + 2\,e^- \rightarrow H_2 + 2\,OH^-$	−0.83
$Ag^+ + e^- \rightarrow Ag$	+0.80	$Cr^{2+} + 2\,e^- \rightarrow Cr$	−0.91
$Hg_2^{2+} + 2\,e^- \rightarrow 2\,Hg$	+0.79	$Mn^{2+} + 2\,e^- \rightarrow Mn$	−1.18
$Fe^{3+} + e^- \rightarrow Fe^{2+}$	+0.77	$V^{2+} + 2\,e^- \rightarrow V$	−1.19
$BrO^- + H_2O + 2\,e^- \rightarrow Br^- + 2\,OH^-$	+0.76	$Ti^{2+} + 2\,e^- \rightarrow Ti$	−1.63
$Hg_2SO_4 + 2\,e^- \rightarrow 2\,Hg + SO_4^{2-}$	+0.62	$Al^{3+} + 3\,e^- \rightarrow Al$	−1.66
$MnO_4^{2-} + 2\,H_2O + 2\,e^- \rightarrow MnO_2 + 4\,OH^-$	+0.60	$U^{3+} + 3\,e^- \rightarrow U$	−1.79
$MnO_4^- + e^- \rightarrow MnO_4^{2-}$	+0.56	$Mg^{2+} + 2\,e^- \rightarrow Mg$	−2.36
$I_2 + 2\,e^- \rightarrow 2\,I^-$	+0.54	$Ce^{3+} + 3\,e^- \rightarrow Ce$	−2.48
$Cu^+ + e^- \rightarrow Cu$	+0.52	$La^{3+} + 3\,e^- \rightarrow La$	−2.52
$I_3^- + 2\,e^- \rightarrow 3\,I^-$	+0.53	$Na^+ + e^- \rightarrow Na$	−2.71
$NiOOH + H_2O + e^- \rightarrow Ni(OH)_2OH^-$	+0.49	$Ca^{2+} + 2\,e^- \rightarrow Ca$	−2.87
$IAg_2CrO_4 + 2\,e^- \rightarrow 2\,Ag + CrO_4^{2-}$	+0.45	$Sr^{2+} + 2\,e^- \rightarrow Sr$	−2.89
$O_2 + 2\,H_2O + 4\,e^- \rightarrow 4\,OH^-$	+0.40	$Ba^{2+} + 2\,e^- \rightarrow Ba$	−2.91
$ClO_4^- + H_2O + 2\,e^- \rightarrow ClO_3^- + 2\,OH^-$	+0.36	$Ra^{2+} + 2\,e^- \rightarrow Ra$	−2.92
$[Fe(CN)_6]^{3-} + e^- \rightarrow [Fe(CN)_6]^{4-}$	+0.36	$Cs^+ + e^- \rightarrow Cs$	−2.92
$Cu^{2+} + 2\,e^- \rightarrow Cu$	+0.34	$Rb^+ + e^- \rightarrow Rb$	−2.93
$Hg_2Cl_2 + 2\,e^- \rightarrow 2\,Hg + 2\,Cl^-$	+0.27	$K^+ + e^- \rightarrow K$	−2.93
$AgCl + e^- \rightarrow Ag + Cl^-$	+0.22	$Li^+ + e^- \rightarrow Li$	−3.05
$Bi^{3+} + 3\,e^- \rightarrow Bi$	+0.20	*Strongly reducing*	

Table 3b Standard potentials at 298.15 K in alphabetical order

Reduction half-reaction	E°/V	Reduction half-reaction	E°/V
$Ag^+ + e^- \rightarrow Ag$	+0.80	$I_2 + 2\,e^- \rightarrow 2\,I^-$	+0.54
$Ag^{2+} + e^- \rightarrow Ag^+$	+1.98	$I_3^- + 2\,e^- \rightarrow 3\,I^-$	+0.53
$AgBr + e^- \rightarrow Ag + Br^-$	+0.0713	$In^+ + e^- \rightarrow In$	−0.14
$AgCl + e^- \rightarrow Ag + Cl^-$	+0.22	$In^{2+} + e^- \rightarrow In^+$	−0.40
$Ag_2CrO_4 + 2\,e^- \rightarrow 2\,Ag + CrO_4^{2-}$	+0.45	$In^{3+} + 2\,e^- \rightarrow In^+$	−0.44
$AgF + e^- \rightarrow Ag + F^-$	+0.78	$In^{3+} + 3\,e^- \rightarrow In$	−0.34
$AgI + e^- \rightarrow Ag + I^-$	−0.15	$In^{3+} + e^- \rightarrow In^{2+}$	−0.49
$Al^{3+} + 3\,e^- \rightarrow Al$	−1.66	$K^+ + e^- \rightarrow K$	−2.93
$Au^+ + e^- \rightarrow Au$	+1.69	$La^{3+} + 3\,e^- \rightarrow La$	−2.52
$Au^{3+} + 3\,e^- \rightarrow Au$	+1.40	$Li^+ + e^- \rightarrow Li$	−3.05
$Ba^{2+} + 2\,e^- \rightarrow Ba$	−2.91	$Mg^{2+} + 2\,e^- \rightarrow Mg$	−2.36

Reduction half-reaction	$E^{\ominus}/V$	Reduction half-reaction	$E^{\ominus}/V$
$Be^{2+} + 2\,e^- \rightarrow Be$	-1.85	$Mn^{2+} + 2\,e^- \rightarrow Mn$	-1.18
$Bi^{3+} + 3\,e^- \rightarrow Bi$	$+0.20$	$Mn^{3+} + e^- \rightarrow Mn^{2+}$	$+1.51$
$Br_2 + 2\,e^- \rightarrow 2\,Br^-$	$+1.09$	$MnO_2 + 4\,H^+ + 2\,e^- \rightarrow Mn^{2+} + 2\,H_2O$	$+1.23$
$BrO^- + H_2O + 2\,e^- \rightarrow Br^- + 2\,OH^-$	$+0.76$	$MnO_4^- + 8\,H^+ + 5\,e^- \rightarrow Mn^{2+} + 4\,H_2O$	$+1.51$
$Ca^{2+} + 2\,e^- \rightarrow Ca$	-2.87	$MnO_4^- + e^- \rightarrow MnO_4^{2-}$	$+0.56$
$Cd(OH)_2 + 2\,e^- \rightarrow Cd + 2\,OH^-$	-0.81	$MnO_4^{2-} + 2\,H_2O + 2\,e^- \rightarrow MnO_2 + 4\,OH^-$	$+0.60$
$Cd^{2+} + 2\,e^- \rightarrow Cd$	-0.40	$Na^+ + e^- \rightarrow Na$	-2.71
$Ce^{3+} + 3\,e^- \rightarrow Ce$	-2.48	$Ni^{2+} + 2\,e^- \rightarrow Ni$	-0.23
$Ce^{4+} + e^- \rightarrow Ce^{3+}$	$+1.61$	$NiOOH + H_2O + e^- \rightarrow Ni(OH)_2 + OH^-$	$+0.49$
$Cl_2 + 2\,e^- \rightarrow 2\,Cl^-$	$+1.36$	$NO_3^- + 2\,H^+ + e^- \rightarrow NO_2 + H_2O$	$+0.80$
$ClO^- + H_2O + 2\,e^- \rightarrow Cl^- + 2\,OH^-$	$+0.89$	$NO_3^- + 3\,H^+ + 3\,e^- \rightarrow NO + 2\,H_2O$	$+0.96$
$ClO_4^- + 2\,H^+ + 2\,e^- \rightarrow ClO_3^- + H_2O$	$+1.23$	$NO_3^- + H_2O + 2\,e^- \rightarrow NO_2^- + 2\,OH^-$	$+0.10$
$ClO_4^- + H_2O + 2\,e^- \rightarrow ClO_3^- + 2\,OH^-$	$+0.36$	$O_2 + 2\,H_2O + 4\,e^- \rightarrow 4\,OH^-$	$+0.40$
$Co^{2+} + 2\,e^- \rightarrow Co$	-0.28	$O_2 + 4\,H^+ + 4\,e^- \rightarrow 2\,H_2O$	$+1.23$
$Co^{3+} + e^- \rightarrow Co^{2+}$	$+1.81$	$O_2 + e^- \rightarrow O_2^-$	-0.56
$Cr^{2+} + 2\,e^- \rightarrow Cr$	-0.91	$O_2 + H_2O + 2\,e^- \rightarrow HO_2^- + OH^-$	-0.08
$Cr_2O_7^{2-} + 14\,H^+ + 6\,e^- \rightarrow 2\,Cr^{3+} + 7\,H_2O$	$+1.33$	$O_3 + 2\,H^+ + 2\,e^- \rightarrow O_2 + H_2O$	$+2.07$
$Cr^{3+} + 3\,e^- \rightarrow Cr$	-0.74	$O_3 + H_2O + 2\,e^- \rightarrow O_2 + 2\,OH^-$	$+1.24$
$Cr^{3+} + e^- \rightarrow Cr^{2+}$	-0.41	$Pb^{2+} + 2\,e^- \rightarrow Pb$	-0.13
$Cs^+ + e^- \rightarrow Cs$	-2.92	$Pb^{4+} + 2\,e^- \rightarrow Pb^{2+}$	$+1.67$
$Cu^+ + e^- \rightarrow Cu$	$+0.52$	$PbSO_4 + 2\,e^- \rightarrow Pb + SO_4^{2-}$	-0.36
$Cu^{2+} + 2\,e^- \rightarrow Cu$	$+0.34$	$Pt^{2+} + 2\,e^- \rightarrow Pt$	$+1.20$
$Cu^{2+} + e^- \rightarrow Cu^+$	$+0.16$	$Pu^{4+} + e^- \rightarrow Pu^{3+}$	$+0.97$
$F_2 + 2\,e^- \rightarrow 2\,F^-$	$+2.87$	$Ra^{2+} + 2\,e^- \rightarrow Ra$	-2.92
$Fe^{2+} + 2\,e^- \rightarrow Fe$	-0.44	$Rb^+ + e^- \rightarrow Rb$	-2.93
$Fe^{3+} + 3\,e^- \rightarrow Fe$	-0.04	$S + 2\,e^- \rightarrow S^{2-}$	-0.48
$Fe^{3+} + e^- \rightarrow Fe^{2+}$	$+0.77$	$S_2O_8^{2-} + 2\,e^- \rightarrow SO_4^{2-}$	$+2.05$
$[Fe(CN)_6]^{3-} + e^- \rightarrow [Fe(CN)_6]^{4-}$	$+0.36$	$Sn^{2+} + 2\,e^- \rightarrow Sn$	-0.14
$2\,H^+ + 2\,e^- \rightarrow H_2$	0, by definition	$Sn^{4+} + 2\,e^- \rightarrow Sn^{2+}$	$+0.15$
$2\,H_2O + 2\,e^- \rightarrow H_2 + 2\,OH^-$	-0.83	$Sr^{2+} + 2\,e^- \rightarrow Sr$	-2.89
$2\,HBrO + 2\,H^+ + 2\,e^- \rightarrow Br_2 + 2\,H_2O$	$+1.60$	$Ti^{2+} + 2\,e^- \rightarrow Ti$	-1.63
$2\,HClO + 2\,H^+ + 2\,e^- \rightarrow Cl_2 + 2\,H_2O$	$+1.63$	$Ti^{3+} + e^- \rightarrow Ti^{2+}$	-0.37
$H_2O_2 + 2\,H^+ + 2\,e^- \rightarrow 2\,H_2O$	$+1.78$	$Ti^{4+} + e^- \rightarrow Ti^{3+}$	0.00
$H_4XeO_6 + 2\,H^+ + 2\,e^- \rightarrow XeO_3 + 3\,H_2O$	$+3.0$	$Tl^+ + e^- \rightarrow Tl$	-0.34
$Hg_2^{2+} + 2\,e^- \rightarrow 2\,Hg$	$+0.79$	$U^{3+} + 3\,e^- \rightarrow U$	-1.79
$Hg_2Cl_2 + 2\,e^- \rightarrow 2\,Hg + 2\,Cl^-$	$+0.27$	$U^{4+} + e^- \rightarrow U^{3+}$	-0.61
$Hg^{2+} + 2\,e^- \rightarrow Hg$	$+0.86$	$V^{2+} + 2\,e^- \rightarrow V$	-1.19
$2\,Hg^{2+} + 2\,e^- \rightarrow Hg_2^{2+}$	$+0.92$	$V^{3+} + e^- \rightarrow V^{2+}$	-0.26
$Hg_2SO_4 + 2\,e^- \rightarrow 2\,Hg + SO_4^{2-}$	$+0.62$	$Zn^{2+} + 2\,e^- \rightarrow Zn$	-0.76

Table 3c Biological standard potentials at 298.15 K in electrochemical order

Reduction half-reaction	E^*/V
$O_2 + 4\,H^+ + 4\,e^- \rightarrow 2\,H_2O$	+0.81
$NO_3^- + 2\,H^+ + 2\,e^- \rightarrow NO_2^- + H_2O$	+0.42
$Fe^{3+}(\text{cyt}\,f) + e^- \rightarrow Fe^{2+}(\text{cyt}\,f)$	+0.36
$Cu^{2+}(\text{plastocyanin}) + e^- \rightarrow Cu^+(\text{plastocyanin})$	+0.35
$Cu^{2+}(\text{azurin}) + e^- \rightarrow Cu^+(\text{azurin})$	+0.30
$O_2 + 2\,H^+ + 2\,e^- \rightarrow H_2O_2$	+0.30
$Fe^{3+}(\text{cyt}\,c_{551}) + e^- \rightarrow Fe^{2+}(\text{cyt}\,c_{551})$	+0.29
$Fe^{3+}(\text{cyt}\,c) + \rightarrow Fe^{2+}(\text{cyt}\,c)$	+0.25
$Fe^{3+}(\text{cyt}\,b) + e^- \rightarrow Fe^{2+}(\text{cyt}\,b)$	+0.08
Dehydroascorbic acid $+ 2\,H^+ + 2\,e^- \rightarrow$ ascorbic acid	+0.08
Coenzyme Q $+ 2\,H^+ + 2\,e^- \rightarrow$ coenzyme QH_2	+0.04
$Fumarate^{2-} + 2\,H^+ + 2\,e^- \rightarrow succinate^{2-}$	+0.03
Vitamin K_1(ox) $+ 2\,H^+ + 2\,e^- \rightarrow$ vitamin K_1(red)	−0.05
$Oxaloacetate^{2-} + 2\,H^+ + 2\,e^- \rightarrow malate^{2-}$	−0.17
$Pyruvate^- + 2\,H^+ + 2\,e^- \rightarrow lactate^-$	−0.18
Ethanal $+ 2\,H^+ + 2\,e^- \rightarrow$ ethanol	−0.20
Riboflavin(ox) $+ 2\,H^+ + 2\,e^- \rightarrow$ riboflavin (red)	−0.21
FAD $+ 2\,H^+ + 2\,e^- \rightarrow FADH_2$	−0.22
Glutathione (ox) $+ 2\,H^+ + 2\,e^- \rightarrow$ glutathione (red)	−0.23
Lipoic acid (ox) $+ 2\,H^+ + 2\,e^- \rightarrow$ lipoic acid (red)	−0.29
$NAD^+ + H^+ + 2\,e^- \rightarrow NADH$	−0.32
Cystine $+ 2\,H^+ + 2\,e^- \rightarrow 2$ cysteine	−0.34
Acetyl $-$ CoA $+ 2\,H^+ + 2\,e^- \rightarrow$ ethanal $+$ CoA	−0.41
$2\,H_2O + 2\,e^- \rightarrow H_2 + 2\,OH^-$	−0.42
Ferredoxin (ox) $+ e^- \rightarrow$ ferredoxin (red)	−0.43
$O_2 + e^- \rightarrow O_2^-$	−0.4

Answers to odd-numbered exercises

Fundamentals

F.13   −459.67°F

F.15   2.52 mmol

F.17   4.18 bar

F.19   388 K

F.21   0.50 m³

F.23   26 J

F.25   2.3 kJ

F.27   31.2 V

F.29   3.26 m

F.31   0.37

F.33   (a) 638 m s⁻¹, 1.26 km s⁻¹, 2.30 km s⁻¹
         (b) 319 m s⁻¹, 627 m s⁻¹, 1.15 km s⁻¹

Chapter 1

E1.11   39 J

E1.13 -1.0×10^2 J

E1.15 $9.2\bar{0} \times 10^2$ kJ, 6.1×10^2 s

E1.17   −13.0 J

E1.19   (b) 37.1 J K⁻¹ mol⁻¹, 28.8 J K⁻¹ mol⁻¹

E1.21 $b + 2cT$

E1.23   (a) $\Delta H_m(T) = aT + \dfrac{bT^2}{2} - \dfrac{c}{T} - 16.1$ kJ mol⁻¹

E1.25   (a) +1.9 kJ mol⁻¹
         (b) +30.6 kJ mol⁻¹

E1.27   +301 kJ

E1.29 $40.\bar{6}$ kJ mol⁻¹, $37.\bar{5}$ kJ mol⁻¹

E1.31 $-234\bar{6}$ kJ mol⁻¹, $-105\bar{1}$ kJ mol⁻¹

E1.33   15.2 kJ g⁻¹, 34.0 kJ g⁻¹

E1.35   (a) $-24.\bar{7}$ kJ
         (b) 7.9 m
         (c) $+39.\bar{0}$ kJ
         (d) $12.\bar{4}$ m

E1.37   (a) −1560 kJ mol⁻¹
         (b) slightly less efficient

E1.39   (a) −23.47 kJ mol⁻¹
         (b) −93.9 kJ mol⁻¹
         (c) −2810.44 kJ mol⁻¹
         (d) +306.94 kJ mol⁻¹

E1.41   (a) (i) $+55\bar{2}$ kJ mol⁻¹   (ii) −2.9 kJ mol⁻¹
         (b) $263.\bar{5}$ K

E1.43   279 J K⁻¹ mol⁻¹, −2805 kJ mol⁻¹, less exothermic

E1.45 $\Delta_r U^{\ominus}(298\ \text{K}) + \Delta_r C_V^{\ominus} \times (T - 298\ \text{K})$

Chapter 2

E2.9   5.03 kJ K⁻¹

E2.11   122 J K⁻¹, 130 J K⁻¹, 606 J K⁻¹, 858 J K⁻¹

E2.13 4.0×10^{-4} J K⁻¹ mol⁻¹

E2.15   5.11 J K⁻¹

E2.17   0.95 J K⁻¹ mol⁻¹

E2.19   (b) +34 kJ K⁻¹ mol⁻¹

E2.23   537 J K⁻¹ mol⁻¹

E2.25   −198.72 J K⁻¹, −32.99 kJ

E2.27   0.41 g

E2.29 8.1×10^{23} molecules of ATP

Chapter 3

E3.9   (a) +2.03 kJ mol⁻¹
         (b) +1.50 J mol⁻¹

E3.11   (a) +1.7 kJ mol⁻¹
         (b) −20 kJ mol⁻¹

E3.13   (a) 2.4 kg
         (b) 32 kg
         (c) 2.5 g
         (d) 135.6 bar

E3.15   (b) 0.758 Pa

E3.21   (b) −20.5 kJ mol⁻¹, −126 kJ mol⁻¹, −372 J K⁻¹ mol⁻¹
         (c) 348 K

E3.23   (a) 1.32 dm³
         (b) 61.2 kPa

E3.25 2.41×10^{-3}

E3.27   (a) −1.31 kJ mol⁻¹, spontaneous
         (b) +4.38 J K⁻¹ mol⁻¹

E3.29   2.30 kPa

E3.31   (a) 0.056 mg N_2, 0.014 mg N_2
         (b) 0.17 mg N_2

E3.33   (a) 1.36 mmol dm⁻³
         (b) 33.9 mmol dm⁻³

0.27°C

−0.09°C

3.39 4.9×10^3 mol dm^{-3}

Chapter 4

E4.9   **(a)** 2.9×10^{-5}
    (b) 1.2×10^9
    (c) 1.8×10^2

E4.11 -294 kJ mol^{-1}

E4.13 3.5×10^3, 2.3×10^2, 36

E4.15   6.8 kJ mol^{-1}

E4.17 -25.1 kJ mol^{-1}

E4.19   **(a)** 0
    (b) -61 kJ mol^{-1}
    (c) $+18$ kJ mol^{-1}

E4.21   **(a)** 41%
    (b) 75%

P4.23   **(a)** $K_{\text{Mb}} = 2.33$ torr $= 0.311$ kPa, $K_{\text{Hb}} = 34.7$ torr $= 4.62$ kPa

E4.25   **(a)** -5798 kJ mol^{-1}
    (b) (i) -16.5 MJ
        (ii) -16.9 MJ

E4.27   **(a)** -13 kJ mol^{-1}
    (b) more exergonic

E4.29 -15.0 kJ mol^{-1}, -38.9 J K^{-1} mol^{-1}

E4.33   **(b)** 14.870
    (c) 3.67×10^{-8} mol dm^{-3}, 3.67×10^{-8} mol dm^{-3}
    (d) 7.45
    (e) 14.870

E4.35   **(a)** 9.60×10^{-3} mol dm^{-3}, 2.02
    (b) 0.025 mol dm^{-3}, 12.40
    (c) 1.26

E4.37   **(a)** 9.1
    (b) 4.83
    (c) none of the Br$^-$ is protonated

E4.39   **(a)** 2.0, 12.0, 0.083
    (b) 3.9, 10.1, 0.87
    (c) 5.0, 9.0, 6.1×10^{-5}
    (d) 5.0, 9.0, 6.1×10^{-5}
    (e) 2.6, 11.4, 0.024

E4.41   **(a)** 6.5
    (b) 2.1
    (c) 1.5

E4.43   **(a)** 6.9×10^{-2} mol dm^{-3}, 1.4×10^{-13} mol dm^{-3}, 8.1×10^{-2} mol dm^{-3}, 6.5×10^{-5} mol dm^{-3}, 0.069 mol dm^{-3}
    (b) 1.65×10^{-3} mol dm^{-3}, 2.78

E4.45   **(a)** $\text{pI} = \text{pH} = \frac{1}{2}(\text{p}K_{\text{a1}} + \text{p}K_{\text{a2}})$
    (b) $\text{pI} = \text{pH} = \frac{1}{2}(\text{p}K_{\text{a1}} + \text{p}K_{\text{a2}})$
    (c) $\text{pI} = \text{pH} = \frac{1}{2}(\text{p}K_{\text{a2}} + \text{p}K_{\text{a3}})$

E4.49   **(a)** $\dfrac{K_{\text{d}}}{1 + K_{\text{d}}}$
    (b) $K = \exp\{-(\Delta_{\text{d}}H^{\circ} - T\Delta_{\text{d}}S^{\circ})/RT\}$
    (c) 317 K
    (d) 9 kJ mol^{-1}

E4.51   **(a)** (i) 0.142   (ii) 0.858
    (b) (i) 0.142   (ii) 0.858   (iii) 0.716   (iv) 0.68

Chapter 5

For notational simplicity, we have used both the molality concentration expression $a_{\text{J}} = \gamma_{\text{J}} b_{\text{J}}/b^{\circ}$, where $b^{\circ} = 1$ mol kg^{-1} [5.1a], and $a_{\text{J}} = \gamma_{\text{J}} b_{\text{J}}$ [5.1b] where b_{J} is the unitless magnitude of molality. The convention of eqn 5.1b is most often used in calculations of ionic strength, while the convention of eqn 5.1a appears in Nernst equation computations.

E5.9   0.90

E5.11 $\gamma_{\pm} = (\gamma_+ \gamma_-^2)^{1/3}$

E5.13 $B = 2.01$

E5.15   hydrolysis of 1 mol of ATP does supply sufficient Gibbs energy to transport 3 mol of sodium cation and 2 mol of potassium cation

E5.17   Yes

E5.21 $v = 2$

E5.23 $v = 4$, $+0.56$ V, $-21\bar{6}$ kJ mol^{-1}, 7.3×10^{37}

E5.25   41 mV

E5.29   **(a)** $v = 2$
    (b) $v = 1$
    (c) $v = 2$

E5.31   **(a)** $+0.94$ V
    (b) $E_{\text{cell}}/\text{V} = 1.51 - 0.094656 \times \text{pH}$

E5.33   **(a)** decreases
    (b) increases
    (c) decreases

E5.35   **(a)** -440 kJ mol^{-1}
    (b) $+29.7$ kJ mol^{-1}
    (c) -313 kJ mol^{-1}

E5.37 $+1.15$ V
    (a) -444 kJ mol^{-1}, -505 kJ mol^{-1}
    (b) -442 kJ mol^{-1}, -442 kJ mol^{-1}

E5.39 $+1.15$ V, $+0.08$ V, $+0.82$ V, -0.33 V

E5.41   reduced lipoic acid

E5.43   **(a)** $+0.34$ V
    (b) -0.09 V

E5.47 -131.25 kJ mol^{-1}, -167.10 kJ mol^{-1}, $+56.7$ J K^{-1} mol^{-1}

E5.49   1 mol

E5.51   **(a)** -27 kJ mol^{-1}
    (b) eight

Chapter 6

E6.7   1.6 per cent

E6.9   **(a)** 1.5 mol dm^{-3} s^{-1}, 0.73 mol dm^{-3} s^{-1}, 1.5 mol dm^{-3} s^{-1}
    (b) mol^{-2} dm^6 s^{-1}

E6.13   0.92 g dm^{-3} h^{-1}

E6.17 3.19×10^{-6} Pa^{-1} s^{-1}

E6.19 3.67×10^{-3} min^{-1}

E6.21   **(a)** first
    (b) 30.27 dm^3 mol^{-1} s^{-1}

E6.23 $\dfrac{1}{[A]^2} = \dfrac{1}{[A]_0^2} + 2k_{\text{r}}t$

E6.25   **(a)** $k_{\text{r}}t = \dfrac{2x(A_0 - x)}{A_0^2(A_0 - 2x)^2}$
    (b) $\left(\dfrac{2x}{A_0^2(A_0 - 2x)}\right) + \left(\dfrac{1}{A_0^2}\right)\ln\left(\dfrac{A_0 - 2x}{A_0 - x}\right)$

E6.27   3067 a

E6.29 $1.44 \times 10^{-9} t_{1/2}$, 1.64 min

E6.31   **(a)** 0.043 mol dm^{-3}, 0.138 mol dm^{-3}
           (b) 0.0001 mol dm^{-3}, 0.0951 mol dm^{-3}

E6.33 $\dfrac{2^{n-1} - 1}{(\frac{4}{3})^{n-1} - 1}$

E6.35   85.6 kJ mol^{-1}, 3.65×10^{11} mol dm^{-3} s^{-1}

E6.37 $E_a = 52$ kJ mol^{-1}

E6.39   30.1 kJ mol^{-1}

E6.41   47.8 kJ mol^{-1}

Chapter 7

E7.11 $[A] = \dfrac{k'([A]_0 + [B]_0) + \{k_r[A]_0 - k'[B]_0\}e^{-(k_r+k')t}}{k_r + k'}$,

           $[B] = \dfrac{(k_r[A]_0 - k'[B]_0)(1 + e^{-(k_r+k')t})}{k_r + k'}$

E7.13 $\tau^{-1} = 4k_r[A]_{eq} + k'_r$

E7.15 $1.\overline{7} \times 10^7$ s^{-1}, 2.8×10^9 mol^{-1} dm^3 s^{-1}, $1.\overline{6} \times 10^2$

E7.17 $1.\overline{6} \times 10^2$

E7.19   first-order in H_2O_2 and in Br$^-$, second-order overall

E7.21   (ii) Both the pre-equilibrium approximation and the steady-state approximation predict that the reaction is first-order in A, first-order in B, and second-order overall

E7.23 $[A^-] = \dfrac{k_a[HA][B]}{k'_a[BH^+] + k_b[HA]}$, $\dfrac{k_a k_b[HA]^2[B]}{k'_a[BH^+] + k_b[HA]}$

E7.25 $N = N_0 e^{(b-d) \times (t-1750\,y)}$, the Malthus model does seem to describe the data as an exponential growth, $k_r = b - d = 0.0095$ y^{-1}

E7.27   **(a)** (i) 0.18   (ii) 0.30
           (b) (i) 3.9×10^{-18} (ii) 6.0×10^{-6}

E7.29 1.5×10^{15}

E7.31 -3 kJ mol^{-1}

E7.33 $12\overline{6}$ kJ mol^{-1}

E7.35 -33.8 J K^{-1} mol^{-1}, $+27.6$ kJ mol^{-1}, 37.7 kJ mol^{-1}

E7.37   **(b)** $+61.4$ kJ mol^{-1}

E7.39   1.08 dm^6 mol^{-2} min^{-1}

Chapter 8

E8.9   rate of formation of P $= k_b[ES] = \dfrac{k_b[E]_0[S]}{[S] + \dfrac{1}{K}}$, $k'_a \gg k_b$

E8.13 $[S] = K_M$.

E8.15   2.31 μmol dm^{-3} s^{-1}, 1.11 μmol dm^{-3}, $1.1\overline{6} \times 10^2$ s^{-1}, 1.0×10^2 μmol^{-1} dm^3 s^{-1}

E8.17   **(a)** $\dfrac{[S]}{v} = \dfrac{[S]}{v_{max}} + \dfrac{K_M}{v_{max}}$
           (c) 279 pmol dm^{-3} s^{-1}, 86.4 μmol dm^{-3}

E8.19   **(b)** 3.0, 0.91 fmol dm^{-3}

E8.21   Sequential mechanism, 5.10 mol s^{-1} (kg protein)$^{-1}$, 0.259 mmol dm^{-3}, 0.0189 mol dm^{-3}, 0.0173 mol dm^{-3}

E8.23   Phenylbutyrate ion is a competitive inhibitor of carboxypeptidase. Benzoate ion is an uncompetitive inhibitor of carboxypeptidase.

E8.27 $t = x^2/2D$ or $x = (2Dt)^{1/2}$
           (a) 27 h
           (b) 2.7×10^3 h
           (c) 3.0×10^3 a

E8.29 $r^2 = 6Dt$ or $t = r^2/6D$, 1.7×10^{-2} s

E8.31   0.234

E8.33 1×10^6 steps

E8.35   62.3 μm s^{-1}

E8.37   **(a)** 1.33×10^{-9} m^2 s^{-1}, 184 pm
           (b) 1

E8.39   **(a)** $16.\overline{5}$ nm^{-1}

Chapter 9

E9.11   **(a)** 6.6×10^{-19} J, 4.0×10^2 kJ mol^{-1}
           (b) 3.3×10^{-20} J, 20 kJ mol^{-1}
           (c) 1.3×10^{-33} J, 7.8×10^{-13} kJ mol^{-1}

E9.13   **(a)** 6.6×10^{-31} m
           (b) 6.6×10^{-39} m
           (c) 99.7 pm
           (d) 3.5×10^{-36} m

E9.15   **(a)** 5.35 pm
           (b) 2.2×10^{-24} m s^{-1}

E9.17   0.90 nm

E9.19   **(a)** $1.\overline{77} \times 10^{-4}$
           (b) $5.\overline{92} \times 10^{-5}$

E9.21   **(a)** 0.196
           (b) 0.609
           (c) 0.196

E9.23   **(a)** 3.30×10^{-19} J
           (b) 4.95×10^{-14} s^{-1}

E9.25   **(a)** $1.\overline{1} \times 10^{10}$ s^{-1}
           (b) 0.24, 4

E9.27   **(a)** 5.275×10^{-34} J s, 7.89×10^{-19} J
           (b) 5.2×10^{14} Hz

E9.31   **(a)** 6.89×10^{13} s^{-1}
           (b) 4.35 μm

E9.33   **(a)** 6.432×10^{13} s^{-1}
           (b) 2146 cm^{-1}
           (c) 2099 cm^{-1}, $\tilde{v}_{^{12}C^{18}O} = 2094$ cm^{-1}, $\tilde{v}_{^{13}C^{18}O} = 2046$ cm^{-1}

E9.35   16 orbitals

E9.37 $\frac{1}{2}$

E9.39   **(a)** 1.9 a_0 and 7.1 a_0
           (b) 1.87 a_0, 6.61 a_0, and 15.5 a_0

E9.41   **(a)** ang. mom. $= 0$
           (b) ang. mom. $= 0$
           (c) ang. mom. $= \sqrt{6}\,\hbar$
           (d) ang. mom. $= \sqrt{2}\,\hbar$
           (e) ang. mom. $= \sqrt{2}\,\hbar$

E9.43   **(b)** Mg, Ti

Chapter 10

E10.19 $N(0.644A - 0.245B)$

E10.25 C_2 and CN are stabilized by anion formation. NO, O_2, and F_2 are stabilized by cation formation

E10.27   **(a)** g, u, g

E10.29 N_2

E10.35 $2\alpha + 2\beta, 4\alpha + 4.48\beta$, lower

E10.37 **(b)** $1.518\beta, 8.913$ eV

E10.39 **(b)** 5, 1

E10.41 square planar arrangement

Chapter 11

E11.15 3.40×10^3 kg mol^{-1}

E11.17 31 kg mol^{-1}

E11.19 **(a)** plot of n_{bp} against t^2 is linear
(b) 167 μs

E11.25 66.1 pm

E11.27 **(a)** $47.\overline{9}$ kJ mol^{-1}
(b) 24 kJ mol^{-1}
(c) 0.60 kJ mol^{-1}

E11.29 **(a)** 1.45 D
(b) $\mu_{ortho}/\mu_{meta} = \sqrt{3}$

E11.31 $2\mu_{O-H} \cos(\phi/2)$
(a) 2.13 D
(b) $2 \arccos(\mu_{H-O-O-H}/3.02$ D)

E11.35 196 pm

E11.37 -4.2×10^{-3} J mol^{-1}

P11.39 $R = 2^{1/6} \sigma$

E11.45 24 nm

E11.47 1.3×10^4

E11.49 serum albumin and bushy stunt virus resemble solid spheres, but DNA does not

E11.51 -0.042 J K^{-1} mol^{-1}

E11.53 **(a)** $b = \begin{pmatrix} 0.957 \\ 0.362 \\ 3.59 \end{pmatrix} \begin{matrix} b_0 \\ b_1 \\ b_2 \end{matrix}$
(b) $W = 1.362$

Chapter 12

E12.9 **(a)** 0.307 m^{-1}
(b) 3.26 m

E12.11 **(a)** 1.01×10^4 dm^3 mol^{-1} cm^{-1}
(b) 0.951%

E12.13 $[B] = \dfrac{A_2 - r_A A_1}{(\Delta \varepsilon_2)L}$, $[A] = \dfrac{r_B A_1 - A_2}{(\Delta \varepsilon_2)L}$

E12.15 99.5 μmol dm^{-3}, 96.3 μmol dm^{-3}

E12.17 **(a)** 6.37, 2.12
(b) 1.74×10^6 dm^3 mol^{-1} cm^{-2}

E12.21 **(a)** $\delta\tilde{\nu} = 53$ cm^{-1}.
(b) $\delta\tilde{\nu} = 0.27$ cm^{-1}.

E12.23 **(a)** 967.0, 515.6, 411.8, 314.2
(b) 3002.2, 2143.7, 1885.8, 1640.2

E12.25 **(a)** 3
(b) 4

(c) 48
(d) 54

E12.27 **(a)** 7
(b) structure **5** is inconsistent with these absorptions.

E12.31 **(a)** $6.5\overline{4}$ ns
(b) 0.11 ns^{-1}

E12.33 0.4 ns

E12.35 3.3×10^{18}

E12.37 $\dfrac{1}{I_{phos}} = \dfrac{1}{I_{abs}} + \dfrac{k_Q[Q]}{k_{phos}I_{abs}}$, 5.2×10^6 dm^3 mol^{-1} s^{-1}

E12.39 3.5 nm

E12.41 3×10^3

E12.47 **(a)** $\dfrac{R_{eq}}{R_{max}} = \dfrac{a_0 K}{a_0 K + 1}$
(b) $R_{max} = 1/slope$ and $K = slope/intercept$
(c) $R(t) = R_{eq}(1 - e^{-k_{obs}t})$, where $k_{obs} = k_{on}a_0 + k_{off}$
(d) $R(t) = R_{max} e^{-k_{obs}t}$, where $k_{obs} = k_{off}$

E12.49 **(b)** All modes

Chapter 13

E13.11 5.57×10^{-24} J

E13.13 **(a)** T^{-1}Hz
(b) A s kg^{-1}

E13.15 **(a)** 6.72×10^{-4}
(b) 2.47×10^{-3}

E13.17 **(a)** 3.4×10^{-5}
(b) 8.6×10^{-6}

E13.19 328.5 MHz

E13.21 11.74 T

E13.23 **(a)** independent
(b) 13

E13.25 **(a)** 9.5 μT
(b) 46 μT

E13.27 1:7:21:35:35:21:7:1

E13.33 $\cos\phi = B/4C$

E13.35 $[I]_0 = \dfrac{[E]_0 \Delta\nu}{\delta\nu} - K_I$

E13.43 $\left. \begin{matrix} \mathcal{B}_3 - \mathcal{B}_2 = (334.8 - 332.5)\text{mT} = 2.3 \text{ mT} \\ \mathcal{B}_2 - \mathcal{B}_1 = (332.5 - 330.2)\text{mT} = 2.3 \text{ mT} \end{matrix} \right\} a = 2.3 \text{ mT}$

$g = \dfrac{h\nu}{\mu_B \mathcal{B}_0}$ [13.23] $= (7.14478 \times 10^{-11} \text{ T Hz}^{-1}) \times \dfrac{9.319 \times 10^9 \text{ Hz}}{332.5 \times 10^{-3} \text{ T}}$

$= 2.002\overline{5}$

E13.45 **(a)** 331.9 mT
(b) 1.201 T

E13.51 **(b)** seven lines separated by $\frac{1}{2} \times (0.675$ mT), 1:2:3:4:3:2:1

Index of Tables

Index

General data and fundamental constants

Quantity	Symbol	Value	Power of 10	Units
Speed of light	c	2.997 924 58*	10^8	$m\,s^{-1}$
Elementary charge	e	1.602 176	10^{-19}	C
Faraday's constant	$F = N_A e$	9.648 53	10^4	$C\,mol^{-1}$
Boltzmann constant	k	1.380 65	10^{-23}	$J\,K^{-1}$
Gas constant	$R = N_A k$	8.314 47		$J\,K^{-1}\,mol^{-1}$
		8.314 47	10^{-2}	$dm^3\,bar\,K^{-1}\,mol^{-1}$
		8.205 74	10^{-2}	$dm^3\,atm\,K^{-1}\,mol^{-1}$
		6.236 37	10^1	$dm^3\,Torr\,K^{-1}\,mol^{-1}$
Planck's constant	h	6.626 08	10^{-34}	J s
	$\hbar = h/2\pi$	1.054 57	10^{-34}	J s
Avogadro's constant	N_A	6.022 14	10^{23}	mol^{-1}
Atomic mass constant	m_u	1.660 54	10^{-27}	kg
Mass				
electron	m_e	9.109 38	10^{-31}	kg
proton	m_p	1.672 62	10^{-27}	kg
neutron	m_n	1.674 93	10^{-27}	kg
Vacuum permittivity	$\varepsilon_0 = 1/c^2\mu_0$	8.854 19	10^{-12}	$J^{-1}\,C^2\,m^{-1}$
	$4\pi\varepsilon_0$	1.112 65	10^{-10}	$J^{-1}\,C^2\,m^{-1}$
Vacuum permeability	μ_0	4π	10^{-7}	$J\,s^2\,C^{-2}\,m^{-1}\,(= T^2\,J^{-1}\,m^3)$
Magneton				
Bohr	$\mu_B = e\hbar/2m_e$	9.274 01	10^{-24}	$J\,T^{-1}$
nuclear	$\mu_N = e\hbar/2m_p$	5.050 78	10^{-27}	$J\,T^{-1}$
g value	g_e	2.002 32		
Proton magnetic moment	μ_p	1.410 61	10^{-26}	$J\,T^{-1}$
Magnetogyric ratio				
electron	$\gamma_e = -e/2m_e$	−8.7941	10^{10}	$C\,kg^{-1}$
proton	$\gamma_p = 2\mu_p/\hbar$	2.675 222	10^8	$C\,kg^{-1}$
Bohr radius	$a_0 = 4\pi\varepsilon_0\hbar^2/m_e e^2$	5.291 77	10^{-11}	m
Rydberg constant	$\mathcal{R} = m_e e^4/8h^3 c\varepsilon_0^2$	1.097 37	10^5	cm^{-1}
Standard acceleration of free fall	g	9.806 65*		$m\,s^{-2}$
Gravitational constant	G	6.673	10^{-11}	$N\,m^2\,kg^{-2}$

*Exact value